SETON HALL UNIVERSITY
QH465.A1 D52 1988
DNA replication and mutagenesis   MAIN

3 3073 00255342 6

DNA Replication and Mutagenesis

DNA Replication and Mutagenesis

Edited by

Robb E. Moses
Department of Cell Biology
Baylor College of Medicine
Texas Medical Center
Houston, Texas

William C. Summers
Department of Therapeutic Radiobiology
Yale University School of Medicine
New Haven, Connecticut

American Society for Microbiology
Washington, D.C.

Copyright © 1988 American Society for Microbiology
1913 I Street, N.W.
Washington, DC 20006

Library of Congress Cataloging-in-Publication Data

DNA replication and mutagenesis / edited by Robb E. Moses, William C. Summers.
    p. cm.
  Papers originating from a meeting sponsored by the American Society for Microbiology in the fall of 1987.
    Includes index.
    ISBN 1-55581-003-9
  1. Mutagenesis—Congresses. 2. DNA—Synthesis—Congresses. 3. Microbial genetics—Congresses.   I. Moses, Robb E. II. Summers, William C. III. American Society for Microbiology.
QH465.A1D52 1988
575.2'92—dc19   88-17053
                                                                                                CIP

ISBN 1-55581-003-9

All Rights Reserved
Printed in the United States of America

Contents

Contributors ix
Acknowledgments xvii
Introduction xix

I. ENZYMOLOGY OF DNA REPLICATION

Enzymology of DNA Replication: Introduction 3

1. Purified Protein System for Initiation of Replication from the Origin of the *Escherichia coli* Chromosome • *Arthur Kornberg, Tania A. Baker, LeRoy L. Bertsch, David Bramhill, Kazuhisa Sekimizu, Elmar Wahle, and Benjamin Yung* 6
2. DNA Polymerase III Holoenzyme. Mechanism and Regulation of a True Replicative Complex • *Charles S. McHenry, Henry Tomasiewicz, Mark A. Griep, Jens P. Fürste, and Ann M. Flower* ... 14
3. τ and γ Subunits of DNA Polymerase III Holoenzyme. Their Relationship to Each Other and the *dnaX* Gene • *Suk-Hee Lee, Michael P. Spector, and James R. Walker* 27
4. Collection of Informative Bacteriophage T4 DNA Polymerase Mutants • *Linda J. Reha-Krantz* 34
5. Multienzyme Complex for DNA Replication in HeLa Cells • *Robert J. Hickey, Linda H. Malkas, Nina Pedersen, Congjun Li, and Earl F. Baril* 41
6. Human DNA Polymerase β. Expression in *Escherichia coli* and Characterization of the Recombinant Enzyme • *John Abbotts, Dibyendu N. SenGupta, Barbara Z. Zmudzka, Steven G. Widen, and Samuel H. Wilson* 55
7. Human Placental DNA Polymerases δ and α • *Marietta Y. W. T. Lee* 68

II. DNA REPLICATION SYSTEMS

DNA Replication Systems: Introduction 83

8. Formation and Propagation of the Bacteriophage T7 Replication Fork • *Hiroshi Nakai, Benjamin B. Beauchamp, Julie Bernstein, Hans E. Huber, Stanley Tabor, and Charles C. Richardson* 85
9. Reverse Genetics and *Saccharomyces cerevisiae* DNA Replication and Repair • *Judith L. Campbell, Martin Budd, Dave Burbee, Karen Sitney, Kevin Sweder, and Fred Heffron* 98
10. Origin Methylation in the Replication of P1 Plasmids • *Ann L. Abeles and Stuart J. Austin* 103
11. R1 Plasmid Replication In Vitro. *repA*- and *dnaA*-Dependent Initiation at *oriR* • *Hisao Masai and Ken-ichi Arai* 113

12. Processive DNA Synthesis In Vitro by the φ29 DNA Polymerase. Structural and Functional Comparison with Other DNA Polymerases • *Luis Blanco, Antonio Bernad, and Margarita Salas* 122
13. N4 DNA Replication In Vivo and In Vitro • *Gordon Lindberg, Margaret Sue Pearle, and Lucia B. Rothman-Denes* 130
14. Repressorlike Effect of a Membrane Protein on the Initiation of *Bacillus subtilis* DNA Replication • *J. Laffan and W. Firshein* 140
15. Control of Plasmid Mini-F *oriS* and Chromosomal *oriC* Replication in *Escherichia coli* • *Bruce C. Kline, Malcolm S. Shields, Ross A. Aleff, and Jeffrey E. Tam* 146
16. Protein-Protein Interactions of the *Escherichia coli* Single-Stranded DNA-Binding Protein • *Ralph R. Meyer, Steven M. Ruben, Suzanne E. VanDenBrink-Webb, Phyllis S. Laine, Frederick W. Perrino, and Diane C. Rein* 154
17. DNA Replication Mutants from Hamster V79 Cells Exhibiting Hypersensitivity to Aphidicolin • *Philip K. Liu and Thomas Norwood* 163
18. Viral DNA Synthesis and Recombination in the Transformation Pathway of Polyomavirus • *David L. Hacker, Karen H. Friderici, Claudette Priehs, Susan Kalvonjian, and Michele M. Fluck* 173
19. Fidelity of DNA Synthesis by Human Cell Extracts during In Vitro Replication from the Simian Virus 40 Origin • *John D. Roberts and Thomas A. Kunkel* 182

III. MECHANISMS OF MISINCORPORATION
Mechanisms of Misincorporation: Introduction 193
20. Fidelity of Base Selection by DNA Polymerases. Site-Specific Incorporation of Base Analogs • *Bradley D. Preston, Richard A. Zakour, B. Singer, and Lawrence A. Loeb* 196
21. Oligonucleotides with Site-Specific Structural Anomalies as Probes of Mutagenesis Mechanisms and DNA Polymerase Function • *G. Peter Beardsley, Thomas Mikita, Alton B. Kremer, and James M. Clark* 208
22. Structural Model for the Editing Decision of *Escherichia coli* DNA Polymerase I • *Catherine M. Joyce, Jonathan M. Friedman, Lorena Beese, Paul S. Freemont, and Thomas A. Steitz* 220
23. DNA Replication Errors. Frameshift Errors Produced by *Escherichia coli* Polymerase I • *Lynn S. Ripley and Catherine Papanicolaou* 227
24. Effect of Accessory Replication Proteins on the Misincorporation of a Carcinogen-Modified Nucleotide by Bacteriophage T4 DNA Polymerase during DNA Replication In Vitro • *MaryClaire Shiber and Navin K. Sinha* 237
25. Essential Function of the Editing Subunit of DNA Polymerase III in *Salmonella typhimurium* • *Russell Maurer, Edward Lancy, Miriam Lifsics, and Patricia Munson* 247
26. Very-High-Frequency Mutagenesis Induced by 5-Bromo-2'-Deoxyuridine and Deoxynucleotide Pool Imbalance in Syrian Hamster Cells • *Elliot R. Kaufman* 254

27. DNA Structure and Mutation Hot Spots • *Robert P. P. Fuchs, Anne-Marie Freund, Marc Bichara, and Nicole Koffel-Schwartz* ... 263

IV. BYPASS SYNTHESIS

Bypass Synthesis: Introduction 275

28. Possible Roles of RecA Protein and DNA Polymerase III Holoenzyme in UV Mutagenesis in *Escherichia coli* • *B. A. Bridges, C. Kelly, U. Hübscher, and S. G. Sedgwick* 277

29. Molecular Mechanisms of DNA Synthesis Fidelity and Isolation of a Possible SOS-Induced Polymerase • *Myron F. Goodman, John Petruska, Michael S. Boosalis, Cynthia Bonner, Sandra K. Randall, Lawrence C. Sowers, and Lynn Mendelman* 284

30. Bypass and Termination at Lesions during In Vitro DNA Replication. Implication for SOS Mutagenesis • *Zvi Livneh, Hasia Shwartz, Dana Hevroni, Orna Shavitt, Yaakov Tadmor, and Orna Cohen* 296

31. DNA Polymerase III Is Required for Mutagenesis • *Sharon Bryan, Michael E. Hagensee, and Robb E. Moses* 305

32. In Vitro Model of Mutagenesis • *Janet Sahm, Edith Turkington, Diane LaPointe, and Bernard Strauss* 314

33. *Escherichia coli mutT* Mutator. Increased dGTP Misincorporation Opposite Template Adenines during DNA Replication • *Roel M. Schaaper and Ronnie L. Dunn* 325

34. Genetic Characterization of the SOS Mutator Effect in *Escherichia coli* • *G. Maenhaut-Michel* 332

35. Blocking and Recovery of DNA Polymerase Readthrough on Templates Bearing Chemically Modified Cytosines • *Tadayoshi Bessho, Kazuo Negishi, and Hikoya Hayatsu* 339

V. GENETIC CONTROL OF MUTAGENESIS

Genetic Control of Mutagenesis: Introduction 347

36. RecA-Mediated Cleavage Activates UmuD for UV and Chemical Mutagenesis • *Takehiko Nohmi, John R. Battista, and Graham C. Walker* 349

37. Is Mismatch Repair Involved in UV-Induced Mutagenesis in *Saccharomyces cerevisiae*? • *Friederike Eckardt-Schupp, Fred Ahne, Eva-Maria Geigl, and Wolfram Siede* 355

38. Is Exonuclease I Involved in the Regulation of the SOS Response in *Escherichia coli*? • *Sidney R. Kushner and Gregory J. Phillips* 362

39. Mutations Suppressing Loss of Replication Control. Genetic Analysis of Bacteriophage λ-Dependent Replicative Killing, Replication Initiation, and Mechanism of Mutagenesis • *Sidney Hayes* 367

40. Use of Predesigned Plasmids To Study Deletions. Strategies for Dealing with a Complex Problem • *Elias Balbinder* 378

VI. DAMAGE-DIRECTED MUTAGENESIS

Damage-Directed Mutagenesis: Introduction 395

41. UV Mutagenesis in *Escherichia coli*. UmuC-Independent Targeted Mutations, Altered Spectrum in Exconjugants, and Mutagenesis Resulting from a Single T-T Cyclobutane Dimer • *J. E. LeClerc, J. R. Christensen, R. B. Christensen, P. V. Tata, S. K. Banerjee, and C. W. Lawrence* 397

42. SOS Mutagenesis in *Escherichia coli* Occurs Primarily, Perhaps Exclusively, at Sites of DNA Damage • *Eric Eisenstadt* 403

43. Experimental System for the Study of Site-Specific Mutagenesis in Mammalian Cells and Bacteria • *M. Moriya, M. Takeshita, K. Peden, and A. P. Grollman* 410

44. Shuttle Plasmids Derived from Epstein-Barr Virus for the Molecular Analysis of Mutagenesis in Human Cells • *Norman R. Drinkwater, Kristin A. Eckert, Caroline A. Ingle, and Donna K. Klinedinst* 416

45. DNA Damage Mediated by Reducing Sugars In Vitro and In Vivo • *Annette T. Lee and Anthony Cerami* 423

46. Mutagenic Activity of Smokeless-Tobacco Extracts • *Charles W. Berry* 428

47. Comparisons between the Cellular and Herpesvirus Forms of dUTP Nucleotidohydrolase and Uracil-DNA Glycosylase • *Ronald Lirette and Sal Caradonna* 434

48. Uracil-DNA Glycosylase Activity. Relationship to Proposed Biased Mutation Pressure in the Class *Mollicutes* • *Marshall V. Williams and J. Dennis Pollack* 440

49. Expression of *Escherichia coli* DNA Alkylation Repair Genes in Human Cells • *G. William Rebeck, Bruce Derfler, Patrick Carroll, and Leona Samson* 445

VII. INDUCED MUTAGENESIS

Induced Mutagenesis: Introduction 455

50. Endogenous Cellular Contributions to Mammalian Mutagenesis • *Peter Herrlich* 457

51. Evidence of Inducible Error-Prone Mechanisms in Diploid Human Fibroblasts • *Veronica M. Maher, Kenji Sato, Suzanne Kateley-Kohler, Harvey Thomas, Sonya Michaud, J. Justin McCormick, Marcus Kraemer, Hans J. Rahmsdorf, and Peter Herrlich* 465

52. Analysis of Induced Mutagenesis in Mammalian Cells, Using a Simian Virus 40-Based Shuttle Vector • *Kathleen Dixon, Emmanuel Roilides, Ruth Miskin, and Arthur S. Levine* 472

53. Damage-Induced Genes in Mammalian Cells and Their Role in Mutagenesis • *Nella A. Greggio, Peter M. Glazer, and William C. Summers* 479

Index 485

Contributors

John Abbotts • Laboratory of Biochemistry, National Cancer Institute, Bethesda, MD 20892

Ann L. Abeles • Laboratory of Chromosome Biology, BRI-Basic Research Program, NCI-Frederick Cancer Research Facility, Frederick, MD 21701

Fred Ahne • Gesellschaft für Strahlen- und Umweltforschung, Institut für Strahlenbiologie, D-8042 Neuherberg, Federal Republic of Germany

Ross A. Aleff • Department of Biochemistry and Molecular Biology, Mayo Clinic/Foundation, Rochester, MN 55905

Ken-ichi Arai • Department of Molecular Biology, DNAX Research Institute of Molecular and Cellular Biology, Palo Alto, CA 94304

Stuart J. Austin • Laboratory of Chromosome Biology, BRI-Basic Research Program, NCI-Frederick Cancer Research Facility, Frederick, MD 21701

Tania A. Baker • Department of Biochemistry, Stanford University, Stanford, CA 94305

Elias Balbinder • Department of Biochemistry, Biophysics and Genetics, University of Colorado Health Sciences Center, Denver, CO 80262

S. K. Banerjee • Department of Biophysics, University of Rochester School of Medicine and Dentistry, Rochester, NY 14642

Earl F. Baril • Cell Biology Group, Worcester Foundation for Experimental Biology, Shrewsbury, MA 01545

John R. Battista • Biology Department, Massachusetts Institute of Technology, Cambridge, MA 02139

G. Peter Beardsley • Departments of Pediatrics and Pharmacology, Yale University School of Medicine, New Haven, CT 06510

Benjamin B. Beauchamp • Department of Biological Chemistry and Molecular Pharmacology, Harvard Medical School, Boston, MA 02115

Lorena Beese • Department of Molecular Biophysics and Biochemistry and Howard Hughes Medical Institute, Yale University, New Haven, CT 06511

Antonio Bernad • Centro de Biología Molecular (CSIC-UAM), Universidad Autónoma, Canto Blanco, 28049 Madrid, Spain

Julie Bernstein • Department of Biological Chemistry and Molecular Pharmacology, Harvard Medical School, Boston, MA 02115

Charles W. Berry • Department of Microbiology, Baylor College of Dentistry, Dallas, TX 75246

LeRoy L. Bertsch • Department of Biochemistry, Stanford University, Stanford, CA 94305

Tadayoshi Bessho • Faculty of Pharmaceutical Sciences, Okayama University, Tsushima, Okayama 700, Japan

Marc Bichara • Groupe de Cancérogénèse et de Mutagénèse Moléculaire et Structurale, Institut de Biologie Moleculaire et Cellulaire, Centre National de la Recherche Scientifique, 67084 Strasbourg, France

Luis Blanco • Centro de Biología Molecular (CSIC-UAM), Universidad Autónoma, Canto Blanco, 28049 Madrid, Spain

Cynthia Bonner • Department of Biological Sciences, Molecular Biology Section, University of Southern California, Los Angeles, CA 90089-1340

Michael S. Boosalis • Department of Biological Sciences, Molecular Biology Section, University of Southern California, Los Angeles, CA 90089-1340

David Bramhill • Department of Biochemistry, Stanford University, Stanford, CA 94305

B. A. Bridges • MRC Cell Mutation Unit, University of Sussex, Falmer, Brighton BN1 9RR, United Kingdom

Sharon Bryan • Department of Cell Biology, Baylor College of Medicine, Houston, TX 77030

Martin Budd • Braun Laboratories, California Institute of Technology, Pasadena, CA 91125

Dave Burbee • Department of Molecular Biology, Scripps Clinic and Research Foundation, La Jolla, CA 92037

Judith L. Campbell • Braun Laboratories, California Institute of Technology, Pasadena, CA 91125

Sal Caradonna • Department of Biochemistry, University of Medicine and Dentistry of New Jersey, Piscataway, NJ 08854

Patrick Carroll • Laboratory of Toxicology, Harvard School of Public Health, Boston, MA 02115

Anthony Cerami • Laboratory of Medical Biochemistry, Rockefeller University, New York, NY 10021

J. R. Christensen • Department of Microbiology and Immunology, University of Rochester School of Medicine and Dentistry, Rochester, NY 14642

R. B. Christensen • Department of Biophysics, University of Rochester School of Medicine and Dentistry, Rochester, NY 14642

James M. Clark • National Institute of Environmental Health Sciences, Research Triangle Park, NC 27709

Orna Cohen • Department of Biochemistry, The Weizmann Institute of Science, Rehovot 76100, Israel

Bruce Derfler • Laboratory of Toxicology, Harvard School of Public Health, Boston, MA 02115

Kathleen Dixon • Section on Viruses and Cellular Biology, National Institute of Child Health and Human Development, Bethesda, MD 20892

Norman R. Drinkwater • McArdle Laboratory for Cancer Research, University of Wisconsin, Madison, WI 53706

Ronnie L. Dunn • Laboratory of Molecular Genetics, National Institute of Environmental Health Sciences, Research Triangle Park, NC 27709

Friederike Eckardt-Schupp • Gesellschaft für Strahlen- und Umweltforschung, Institut für Strahlenbiologie, D-8042 Neuherberg, Federal Republic of Germany

Kristin A. Eckert • McArdle Laboratory for Cancer Research, University of Wisconsin, Madison, WI 53706

Eric Eisenstadt • Office of Naval Research, Arlington, VA 22217

W. Firshein • Molecular Biology and Biochemistry Department, Wesleyan University, Middletown, CT 06457

Contributors

Ann M. Flower • Department of Biochemistry, Biophysics and Genetics, University of Colorado Health Sciences Center, Denver, CO 80262

Michele M. Fluck • Department of Microbiology and Public Health, Michigan State University, East Lansing, MI 48824-1101

Paul S. Freemont • Imperial Cancer Research Fund Laboratories, London, United Kingdom

Anne-Marie Freund • Groupe de Cancérogénèse et de Mutagénèse Moléculaire et Structurale, Institut de Biologie Moleculaire et Cellulaire, Centre National de la Recherche Scientifique, 67084 Strasbourg, France

Karen H. Friderici • Department of Microbiology and Public Health, Michigan State University, East Lansing, MI 48824-1101

Jonathan M. Friedman • Department of Molecular Biophysics and Biochemistry and Howard Hughes Medical Institute, Yale University, New Haven, CT 06511

Robert P. P. Fuchs • Groupe de Cancérogénèse et de Mutagénèse Moléculaire et Structurale, Institut de Biologie Moleculaire et Cellulaire, Centre National de la Recherche Scientifique, 67084 Strasbourg, France

Jens P. Fürste • Department of Biochemistry, Biophysics and Genetics, University of Colorado Health Sciences Center, Denver CO 80262

Eva-Maria Geigl • Gesellschaft für Strahlen- und Umweltforschung, Institut für Strahlenbiologie, D-8042 Neuherberg, Federal Republic of Germany

Peter M. Glazer • Departments of Therapeutic Radiology, Human Genetics, and Molecular Biophysics and Biochemistry, Yale University School of Medicine, New Haven, CT 06510

Myron F. Goodman • Department of Biological Sciences, Molecular Biology Section, University of Southern California, Los Angeles, CA 90089-1340

Nella A. Greggio • Department of Pediatrics, School of Medicine, University of Padua, Padua, Italy

Mark A. Griep • Department of Biochemistry, Biophysics and Genetics, University of Colorado Health Sciences Center, Denver, CO 80262

A. P. Grollman • Department of Pharmacological Sciences, State University of New York at Stony Brook, Stony Brook, NY 11794

David L. Hacker • Department of Microbiology and Public Health, Michigan State University, East Lansing, MI 48824-1101

Michael E. Hagensee • Department of Cell Biology, Baylor College of Medicine, Houston, TX 77030

Hikoya Hayatsu • Faculty of Pharmaceutical Sciences, Okayama University, Tsushima, Okayama 700, Japan

Sidney Hayes • Department of Microbiology, College of Medicine, University of Saskatchewan, Saskatoon, Saskatchewan S7N 0W0, Canada

Fred Heffron • Department of Molecular Biology, Scripps Clinic and Research Foundation, La Jolla, CA 92037

Peter Herrlich • Kernforschungszentrum Karlsruhe, Institut für Genetik und Toxikologie, and Institut für Genetik, Universität Karlsruhe, D-7500 Karlsruhe, Federal Republic of Germany

Dana Hevroni • Department of Biochemistry, The Weizmann Institute of Science, Rehovot 76100, Israel

Robert J. Hickey • Cell Biology Group, Worcester Foundation for Experimental Biology, Shrewsbury, MA 01545

Hans E. Huber • Department of Biological Chemistry and Molecular Pharmacology, Harvard Medical School, Boston, MA 02115

U. Hübscher • Institute for Pharmacology and Biochemistry, University of Zurich-Irchel, Zurich, Switzerland

Caroline A. Ingle • McArdle Laboratory for Cancer Research, University of Wisconsin, Madison, WI 53706

Catherine M. Joyce • Department of Molecular Biophysics and Biochemistry, Yale University Medical School, New Haven, CT 06510

Susan Kalvonjian • Department of Microbiology and Public Health, Michigan State University, East Lansing, MI 48824-1101

Suzanne Kateley-Kohler • Carcinogenesis Laboratory, Department of Microbiology, and Department of Biochemistry, Michigan State University, East Lansing, MI 48824

Elliot R. Kaufman • Department of Genetics, University of Illinois College of Medicine at Chicago, Chicago, IL 60612

C. Kelly • MRC Cell Mutation Unit, University of Sussex, Falmer, Brighton BN1 9RR, United Kingdom

Bruce C. Kline • Department of Biochemistry and Molecular Biology, Mayo Clinic/Foundation, Rochester, MN 55905

Donna K. Klinedinst • McArdle Laboratory for Cancer Research, University of Wisconsin, Madison, WI 53706

Nicole Koffel-Schwartz • Group de Cancérogénèse et de Mutagénèse Moléculaire et Structurale, Institut de Biologie Moleculaire et Cellulaire, Centre National de la Recherche Scientifique, 67084 Strasbourg, France

Arthur Kornberg • Department of Biochemistry, Stanford University, Stanford, CA 94305

Marcus Kraemer • Kernforschungszentrum Karlsruhe, Institut für Genetik und für Toxikologie, and Institut für Genetik der Universität Karlsruhe, D-7500 Karlsruhe, Federal Republic of Germany

Alton B. Kremer • Ortho Pharmaceuticals, Raritan, NJ 08869

Thomas A. Kunkel • Laboratory of Molecular Genetics, National Institute of Environmental Health Sciences, Research Triangle Park, NC 27709

Sidney R. Kushner • Department of Genetics, University of Georgia, Athens, GA 30602

J. Laffan • Molecular Biology and Biochemistry Department, Wesleyan University, Middletown, CT 06457

Phyllis S. Laine • Department of Biological Sciences, University of Cincinnati, Cincinnati, OH 45221

Edward Lancy • Department of Molecular Biology and Microbiology, Case Western Reserve University, Cleveland, OH 44106

Diane LaPointe • Department of Molecular Genetics and Cell Biology, University of Chicago, Chicago, IL 60637

C. W. Lawrence • Department of Biophysics, University of Rochester School of Medicine and Dentistry, Rochester, NY 14642

J. E. LeClerc • Department of Biochemistry, University of Rochester School of Medicine and Dentistry, Rochester, NY 14642

Annette T. Lee • Laboratory of Medical Biochemistry, Rockefeller University, New York, NY 10021

Marietta Y. W. T. Lee • Department of Medicine, University of Miami School of Medicine, Miami, FL 33101

Suk-Hee Lee • Memorial Sloan-Kettering Cancer Center, New York, NY 10021

Arthur S. Levine • Section on Viruses and Cellular Biology, National Institute of Child Health and Human Development, Bethesda, MD 20892

Congjun Li • Cell Biology Group, Worcester Foundation for Experimental Biology, Shrewsbury, MA 01545

Miriam Lifsics • Department of Molecular Biology and Microbiology, Case Western Reserve University, Cleveland, OH 44106

Gordon Lindberg • Department of Biochemistry and Molecular Biology, University of Chicago, Chicago, IL 60637

Ronald Lirette • Department of Biochemistry, University of Medicine and Dentistry of New Jersey, Piscataway, NJ 08854

Philip K. Liu • Department of Environmental Health Sciences, Case Western Reserve University School of Medicine, Cleveland, OH 44106

Zvi Livneh • Department of Biochemistry, The Weizmann Institute of Science, Rehovot 76100, Israel

Lawrence A. Loeb • Joseph Gottstein Memorial Cancer Research Laboratory, Department of Pathology, University of Washington, Seattle, WA 98195

G. Maenhaut-Michel • Laboratoire de Biophysique et Radiobiologie, Département de Biologie Moléculaire, Université libre de Bruxelles, B-1640 Rhode St. Genèse, Belgium

Veronica M. Maher • Carcinogenesis Laboratory, Department of Microbiology, and Department of Biochemistry, Michigan State University, East Lansing, MI 48824

Linda H. Malkas • Cell Biology Group, Worcester Foundation for Experimental Biology, Shrewsbury, MA 01545

Hisao Masai • Department of Molecular Biology, DNAX Research Institute of Molecular and Cellular Biology, Palo Alto, CA 94304

Russell Maurer • Department of Molecular Biology and Microbiology, Case Western Reserve University, Cleveland, OH 44106

J. Justin McCormick • Carcinogenesis Laboratory, Department of Microbiology, and Department of Biochemistry, Michigan State University, East Lansing, MI 48824

Charles S. McHenry • Department of Biochemistry, Biophysics and Genetics, University of Colorado Health Sciences Center, Denver, CO 80262

Lynn Mendelman • Department of Biological Sciences, Molecular Biology Section, University of Southern California, Los Angeles, CA 90089-1340

Ralph R. Meyer • Department of Biological Sciences, University of Cincinnati, Cincinnati, OH 45221

Sonya Michaud • Carcinogenesis Laboratory, Department of Microbiology, and Department of Biochemistry, Michigan State University, East Lansing, MI 48824

Thomas Mikita • Departments of Pediatrics and Pharmacology, Yale University School of Medicine, New Haven, CT 06510

Ruth Miskin • Biochemistry Department, Weizmann Institute of Science, Rehovot 76100, Israel

M. Moriya • Department of Pharmacological Sciences, State University of New York at Stony Brook, Stony Brook, NY 11794

Robb E. Moses • Department of Cell Biology, Baylor College of Medicine, Houston, TX 77030

Patricia Munson • Department of Molecular Biology and Microbiology, Case Western Reserve University, Cleveland, OH 44106

Hiroshi Nakai • Department of Biological Chemistry and Molecular Pharmacology, Harvard Medical School, Boston, MA 02115

Kazuo Negishi • Faculty of Pharmaceutical Sciences, Okayama University, Tsushima, Okayama 700, Japan

Takehiko Nohmi • Biology Department, Massachusetts Institute of Technology, Cambridge, MA 02139

Thomas Norwood • Department of Pathology, University of Washington, Seattle, WA 98195

Catherine Papanicolaou • Department of Microbiology and Molecular Genetics, New Jersey Medical School, University of Medicine and Dentistry of New Jersey, Newark, NJ 07103

Margaret Sue Pearle • Department of Biochemistry and Molecular Biology, University of Chicago, Chicago, IL 60637, and Northwestern University School of Medicine, Chicago, IL 60611

K. Peden • Howard Hughes Medical Institute, Department of Molecular Biology and Genetics, Johns Hopkins University School of Medicine, Baltimore, MD 21205

Nina Pedersen • Cell Biology Group, Worcester Foundation for Experimental Biology, Shrewsbury, MA 01545

Frederick W. Perrino • Department of Pathology, University of Washington School of Medicine, Seattle, WA 98195

John Petruska • Department of Biological Sciences, Molecular Biology Section, University of Southern California, Los Angeles, CA 90089-1340

Gregory J. Phillips • Department of Molecular Biology, Princeton University, Princeton, NJ 08544

J. Dennis Pollack • Department of Medical Microbiology and Immunology and Comprehensive Cancer Center, Ohio State University, Columbus, OH 43210

Bradley D. Preston • Joseph Gottstein Memorial Cancer Research Laboratory, Department of Pathology, University of Washington, Seattle, WA 98195

Claudette Priehs • Department of Microbiology and Public Health, Michigan State University, East Lansing, MI 48824-1101

Hans J. Rahmsdorf • Kernforschungszentrum Karlsruhe, Institut für Genetik und für Toxikologie, and Institut für Genetik der Universität Karlsruhe, D-7500 Karlsruhe, Federal Republic of Germany

Sandra K. Randall • Department of Biological Sciences, Molecular Biology Section, University of Southern California, Los Angeles, CA 90089-1340

G. William Rebeck • Laboratory of Toxicology, Harvard School of Public Health, Boston, MA 02115

Linda J. Reha-Krantz • Department of Genetics, University of Alberta, Edmonton, Alberta T6G 2E9, Canada

Diane C. Rein • Department of Biological Sciences, University of Cincinnati, Cincinnati, OH 45221

Charles C. Richardson • Department of Biological Chemistry and Molecular Pharmacology, Harvard Medical School, Boston, MA 02115

CONTRIBUTORS

Lynn S. Ripley • Department of Microbiology and Molecular Genetics, New Jersey Medical School, University of Medicine and Dentistry of New Jersey, Newark, NJ 07103

John D. Roberts • Laboratory of Molecular Genetics, National Institute of Environmental Health Sciences, Research Triangle Park, NC 27709

Emmanuel Roilides • Section on Viruses and Cellular Biology, National Institute of Child Health and Human Development, Bethesda, MD 20892

Lucia B. Rothman-Denes • Departments of Biochemistry and Molecular Biology and of Molecular Genetics and Cell Biology, University of Chicago, Chicago, IL 60637

Steven M. Ruben • Department of Molecular Oncology, Roche Institute of Molecular Biology, Nutley, NJ 07110

Janet Sahm • Department of Molecular Genetics and Cell Biology, University of Chicago, Chicago, IL 60637

Margarita Salas • Centro de Biología Molecular (CSIC-UAM), Universidad Autónoma, Canto Blanco, 28049 Madrid, Spain

Leona Samson • Laboratory of Toxicology, Harvard School of Public Health, Boston, MA 02115

Kenji Sato • Carcinogenesis Laboratory, Department of Microbiology, and Department of Biochemistry, Michigan State University, East Lansing, MI 48824

Roel M. Schaaper • Laboratory of Molecular Genetics, National Institute of Environmental Health Sciences, Research Triangle Park, NC 27709

S. G. Sedgwick • National Institute for Medical Research, The Ridgeway, Mill Hill, London, United Kingdom

Kazuhisa Sekimizu • Department of Biochemistry, Stanford University, Stanford, CA 94305

Dibyendu N. SenGupta • Laboratory of Biochemistry, National Cancer Institute, Bethesda, MD 20892

Orna Shavitt • Department of Biochemistry, The Weizmann Institute of Science, Rehovot 76100, Israel

MaryClaire Shiber • Waksman Institute of Microbiology, Rutgers University, Piscataway, NJ 08854

Malcolm S. Shields • Department of Biochemistry and Molecular Biology, Mayo Clinic/Foundation, Rochester, MN 55905

Hasia Shwartz • Department of Biochemistry, The Weizmann Institute of Science, Rehovot 76100, Israel

Wolfram Siede • Department of Pathology, Stanford University, Stanford, CA 94305

B. Singer • Donner Laboratory, University of California, Berkeley, CA 94720

Navin K. Sinha • Waksman Institute of Microbiology, Rutgers University, Piscataway, NJ 08854

Karen Sitney • Braun Laboratories, California Institute of Technology, Pasadena, CA 91125

Lawrence C. Sowers • Department of Biological Sciences, Molecular Biology Section, University of Southern California, Los Angeles, CA 90089-1340

Michael P. Spector • Department of Microbiology, University of Texas at Austin, Austin, TX 78712

Thomas A. Steitz • Department of Molecular Biophysics and Biochemistry and Howard Hughes Medical Institute, Yale University, New Haven, CT 06511

Bernard Strauss • Department of Molecular Genetics and Cell Biology, University of Chicago, Chicago, IL 60637

William C. Summers • Departments of Therapeutic Radiology, Human Genetics, and Molecular Biophysics and Biochemistry, Yale University School of Medicine, New Haven, CT 06510

Kevin Sweder • Braun Laboratories, California Institute of Technology, Pasadena, CA 91125

Stanley Tabor • Department of Biological Chemistry and Molecular Pharmacology, Harvard Medical School, Boston, MA 02115

Yaakov Tadmor • Department of Biochemistry, The Weizmann Institute of Science, Rehovot 76100, Israel

M. Takeshita • Department of Pharmacological Sciences, State University of New York at Stony Brook, Stony Brook, NY 11794

Jeffrey E. Tam • Department of Biochemistry and Molecular Biology, Mayo Clinic/Foundation, Rochester, MN 55905

P. V. Tata • Department of Biophysics, University of Rochester School of Medicine and Dentistry, Rochester, NY 14642

Harvey Thomas • Carcinogenesis Laboratory, Department of Microbiology, and Department of Biochemistry, Michigan State University, East Lansing, MI 48824

Henry Tomasiewicz • Department of Biochemistry, Biophysics and Genetics, University of Colorado Health Sciences Center, Denver, CO 80262

Edith Turkington • Department of Molecular Genetics and Cell Biology, University of Chicago, Chicago, IL 60637

Suzanne E. VanDenBrink-Webb • Department of Rheumatology, University of Connecticut Health Sciences Center, VA Medical Center, South Newington, CT 06111

Elmar Wahle • Department of Biochemistry, Stanford University, Stanford, CA 94305

Graham C. Walker • Biology Department, Massachusetts Institute of Technology, Cambridge, MA 02139

James R. Walker • Department of Microbiology, University of Texas at Austin, Austin, TX 78712

Steven G. Widen • Laboratory of Biochemistry, National Cancer Institute, Bethesda, MD 20892

Marshall V. Williams • Department of Medical Microbiology and Immunology and Comprehensive Cancer Center, Ohio State University, Columbus, OH 43210

Samuel H. Wilson • Laboratory of Biochemistry, National Cancer Institute, Bethesda, MD 20892

Benjamin Yung • Department of Biochemistry, Stanford University, Stanford, CA 94305

Richard A. Zakour • Molecular and Applied Genetics Laboratory, Allied Corp., Morristown, NJ 07960

Barbara Z. Zmudzka • Laboratory of Biochemistry, National Cancer Institute, Bethesda, MD 20892

Acknowledgments

The editors gratefully acknowledge the support of the following organizations: Beckman Instruments, Inc., Fullerton, Calif.; Bellco Glass, Inc., Vineland, N.J.; Bethesda Research Laboratories, Gaithersburg, Md.; Boehringer-Mannheim Corp., Indianapolis, Ind.; Merck Sharp & Dohme Research Laboratories, Rahway, N.J.; Smith Kline and French Laboratories, King of Prussia, Pa.; and The Upjohn Co., Kalamazoo, Mich. Portions of this work were supported by Public Health Service grant GM38507 from the National Institutes of Health and by grant DNB8704342 from the National Science Foundation.

Introduction

Mutagenesis is the end result of changes occurring and remaining in the genome of organisms. These may occur in DNA unrelated to the replication process or they may occur in processes related to DNA replication. Most mutagenesis appears to be the result of functions during DNA replication. It is our purpose in this book to investigate these functions and the contribution of each of them to mutagenesis of the genome. This is a good time to review what is known and to anticipate what will be found in the near future because of the development of several pertinent techniques: (i) in vitro DNA replication systems have advanced to the point where the major proteins involved in the act of replication in procaryotes and in some phage systems can be identified and where definitions are beginning in eucaryotic systems; (ii) the products of the in vitro systems of DNA replication can be shown to be biologically active; and (iii) recombinant DNA techniques allow analysis of the results of mutagenesis in a definitive manner. Therefore, investigators can now study the molecular biology of mutagenesis in environments which appear to duplicate those in the cell in terms of both enzymology and genetics. By use of these techniques, coupled with the development of rapid screening methods dependent on reporter functions in the DNA itself, it is now possible to screen at the molecular level for mutagenesis and for the cellular functions affecting mutagenesis.

This book originated from a meeting on DNA replication and mutagenesis sponsored by the American Society for Microbiology in the fall of 1987. The promise of analyzing mutagenesis in the in vitro DNA replication systems which catalyze authentic reactions offers fulfillment of a goal which has tantalized investigators. On the other hand, the definition of the components in the in vitro systems for DNA replication is not complete, nor is our understanding of the action of the individual proteins. Nevertheless, the roles of the major components have been assigned, in some cases multiple roles to an individual protein. It is clear from the information contained in this book that the results form a complex array of DNA-altering or -damaging agents. However, the tools are available to analyze and correlate cause and effect for a bewildering array of agents. It has been possible in some cases to identify discrete effects of the DNA environment on mutagenesis. This analysis can be extended to the in vitro reactions as well as the cellular environment.

A concept which has gained support in the past few years is that cellular functions not only participate in but also may be *required* for mutagenesis in the cell. This is clear for procaryotes and is likely to turn out to be the case for eucaryotes as well. This observation might seem somewhat surprising at first, but it is simplistic to have assumed that mutagenesis is a passive process in which the cell is battered from without by noxious agents. In responding to this stress as well as going about its day-to-day business of replicating DNA, the cell calls on numerous activities. These activities must have a finite error rate and therefore should participate in allowing alterations in the genome to become established or fixed.

At no point is the interplay of cellular functions and mutagenesis better illustrated than in the phenomenon of induction of cell functions by DNA damage. Cells which have been exposed to certain DNA-damaging agents (perhaps most or nearly all) show an altered response on subsequent exposure. This is manifest in a variety of findings, notably in improved survival, activation of several functions such as induction of prophage, and increased rate of mutagenesis. This process is clearly established and best studied at present only in procaryotic systems, but there is sufficient circumstantial evidence to warrant careful investigation in eucaryotic cells. Cellular functions are required for the induction process and therefore are at least indirectly required for the increased mutagenesis seen in this situation. More recently, it has been shown that some of the cellular functions may contribute directly to the mutagenesis.

From the standpoint of genetic diversity, a given "background" mutagenesis during the DNA replication may give selective advantage. The increased mutagenesis noted following induction can be rationalized as a response to accept high levels of mistakes in replicating the damaged genome in order to allow a few cells to escape and survive, even at the cost of mutations. The mechanisms by which induction contributes to mutagenesis will remain an open question.

DNA replication in procaryotes is complex and incompletely understood, but some generalizations are possible. A key control point in DNA replication is initiation of replication at origins. With the successful development of the *oriC* replication system in *Escherichia coli* as well as several phage systems, common features are beginning to emerge. The origins are regions of complex quaternary structure. They frequently have direct or inverted DNA sequence repeats; sites for binding of specific proteins which are required for initiation have been defined. A second generalization is that the specificity of DNA replication resides in the origin. The requirements peculiar to *E. coli* are not reflected in the phage system, which may be completely or partially dependent on host functions. A third generalization is that replication is catalyzed by multiprotein complexes. The synthesis activity of DNA polymerase alone is not sufficient to catalyze DNA replication. More than one DNA polymerase may participate in some steps of DNA replication. In addition to priming functions, other proteins associated with the DNA polymerase activity are required for unwinding of the parental duplex DNA. Various organisms have solved these problems in several ways: *E. coli* apparently has separate proteins encoding primase and helicase functions, whereas T7 bacteriophage combines these functions in a single protein (the gene 4 product). In the more complex replisomes, there is a subdivision of protein activities into a "core" synthesis unit, which may contain several proteins in addition to the DNA polymerase activity, and additional "accessory" proteins, which perform helicase or priming functions or supply DNA-binding proteins.

In vitro DNA replication systems developed for procaryotes and the adenovirus and simian virus 40 DNA replication systems promise to allow definition of required components of DNA replication in eucaryotes. The elucidation of these functions and the manner in which the activities are distributed within protein domains will be important for models of mutagenesis. For example, DNA polymerases in eucaryotic systems do not appear to contain editorial exonuclease activities in some instances. Such activities may be found in other subunits which are associated with the DNA polymerase. Alternatively, the activity may be "cryptic" in the DNA polymerase activity assayed in vitro and yet may be functional in the cell.

Mutagenesis may be divided into three general phases related to DNA replication: a misincorporation phase, a repair phase, and a fixation phase. An initial "error" must be made by the replication system, and there must be failure to remove or correct this error with subsequent replication continuing downstream from the altered base. If the change in nucleotide sequence is not bypassed, the misincorporation is an abortive event from the viewpoint of mutagenesis.

The misincorporation phase of mutagenesis can be subdivided into at least two steps: an incorporation step and an editing step. The incorporation step may involve simple mispairing, utilization of a base analog, or incorporation of a base in a region of defective template function (for example, an apurinic site). Once this event has occurred, the replication apparatus must recognize the error and remove it in a proofreading function. The proofreading function resides at least partly in exonuclease activities, which are either intrinsic to DNA polymerase activity or associated with it in the replisome, but other functions may be involved in editing incorporation. Editing functions are commonly modeled in terms of mismatched termini. In fact, we know very little about the recognition basis for removal of a mismatched base. We do not understand the distortions which are recognized or the free energies involved in the recognition. We cannot conclude that the mismatched base is removed prior to *any* additional incorporation or recognize the spatial relationships of editing function to the active site of synthesis.

Mutagenesis also involves a prefixation phase in which there is a failure of mismatch correction or repair. The biological impact of this process has been supported in recent studies which show that the gene products involved in mismatch repair are encoded by genes recognized to have an effect on the general mutation level in the cell. In *E. coli* mutations at some genes increase the overall rate of mutagenesis in the cell. In some instances these genes, known as mutator genes, code for products involved in mismatch correction. The models of mismatch correction and the available biochemical data indicate that the enzymatic functions discriminate "correct" and "incorrect" strands of DNA on the basis of methylation pattern. Thus, there is a bias in the system for preserving parental information. Little is known about the discrete act of recognition of the mismatch. However, the enzymatic entities involved in the repair are at least partially defined in procaryotic systems. There are suggestive data to support mismatch repair in eucaryotic systems. Once again, the genetics of eucaryotic systems have not permitted the definition of mutator genes.

A final requirement for establishing the mutation is a subsequent round of DNA replication. This may be viewed as the last step of fixation, a result requiring each of the steps subsequent to misincorporation.

Taken together, the steps of mutagenesis, misincorporation, mismatch repair, and fixation can account approximately for the levels of mutagenesis observed in the cell. In procaryotic systems, stable changes in the nucleotide sequence seem to be in the range of 10^{-8} to 10^{-9} base changes per base per replication cycle. It appears that mismatch repair lowers the frequency of mutation by 10^1 to 10^2 and that the editorial function associated with the DNA polymerases may lower the incorporation by a factor of 10^2. That leaves the needed misinsertion rate of 10^{-5} to 10^{-6} which is observed in synthesis systems lacking editing or mismatch repair.

While the development of the in vitro DNA replication system has been a major advance for the study of mutagenesis and has yielded results in procaryotes, other developments have allowed analysis of mutagenesis in eucaryotic systems. Shuttle vectors can be rescued from the eucaryotic cell and analyzed for mutagenesis in a procaryotic system. A second advance has been the use of reporter functions or small stretches of nucleotides which can easily be analyzed. These systems have allowed the analysis of mutagenesis at the nucleotide level in eucaryotic cell environments. Sensitivity is such that events occurring at the frequency of cellular mutagenesis can be scored.

From the comments above, it is clear that a number of genes are involved in directly effecting mutagenesis in the cell. This conclusion is best supported in *E. coli*, where there is now evidence that cellular functions are required for mutagenesis. Thus, cellular activities not only act against mutagenesis but also act for mutagenesis. Some of the functions involved in mutagenesis may be constitutively present, such as the synthesis activity of a DNA polymerase; other functions involved in mutagenesis are known to be induced. The functions that are induced are dependent on the *recA-lexA* genetic axis. The probable inducing signal for this system is single-stranded DNA, and it is known that the LexA protein is cleaved by proteolytic action of the RecA protein in the early stages of induction. Twenty or more *E. coli* genes are derepressed, including the *recA* gene and the *umuCD* gene. There is evidence that the *recA* gene product may participate directly in mutagenesis by affecting the ability of the DNA polymerase to synthesize past the DNA damage with a higher rate of mispairing. In addition, the UmuC and UmuD proteins may be directly involved in favoring bypass synthesis. In *E. coli*, there appears to be a specific requirement for DNA polymerase III at this stage, although DNA polymerase I may play other roles in mutagenesis.

It is not clear whether a parallel system exists in eucaryotic cells. A number of eucaryotic genes can be induced, and there may be evidence of "global" regulation of some eucaryotic gene families. It is also not clear whether there is a response on a unified basis to DNA damage to the same extent as in procaryotes, nor are there indications that such a response directly favors mutagenesis.

On the basis of procaryotic models, it seems safe to conclude that cellular functions in eucaryotes are required for misincorporation to occur, both at the misinsertion step and in a failure of proofreading function. It also seems likely that cellular functions are directly required for bypass synthesis in mutation fixation. It is at this step that the possibility of altered or induced cellular functions seems greatest. During mutagenesis, following induction, a cellular function which is ordinarily present may be modified to have an altered role in DNA replication.

What is the outlook for the next several years? An increasingly defined analysis of mutagenesis should utilize the power of in vitro DNA replication systems. The systems will become ever more faithful copies of the system existing in the cell, and the role of additional proteins will be defined. Altered protein-protein interactions may prove to be critical for mutagenesis. The effect of the local environment on mutagenesis at a given site in the genome will be analyzed. Some of the principles indicating what determines a "hot spot" for mutagenesis and what favors a mutagenic response following some types of damage should be defined. An improved correlation of the spectrum of mutagenesis in response to specific damage should develop. Since there are numerous DNA replication systems, any significant differences between these systems will become apparent and should reflect differences between the biological systems from which they are derived. The cellular proteins involved directly in favoring mutagenesis and the modifications to these proteins favoring mutagenesis can now be studied. Finally, the meaning of induction by DNA damage may at last be explained at the molecular level.

Robb E. Moses and William C. Summers

I. Enzymology of DNA Replication

Enzymology of DNA Replication: Introduction

It is the goal of this part to summarize recent observations in the enzymology of the replication of DNA. Many proteins required for DNA replication have been identified and purified. In procaryotic systems these proteins are the products of numerous genes. The end goal of purification of the enzymes involved in DNA replication has been achieved by two major routes: purification of proteins from in vitro systems which are able to reproduce certain aspects of the biological system and identification of the products of genes which are required for DNA replication as defined by temperature-sensitive alleles. The analysis of the enzymes of DNA replication has been based on simple circular single-stranded phage systems, more complex linear and circular double-stranded DNA phage systems, and both procaryotic and eucaryotic cellular systems.

In general, these systems have established partial reactions representing an incomplete picture of the overall replication process in cells. For example, initiation of DNA replication in the single-stranded DNA phages mimics the cellular microinitiation process for the discontinuous replication process (Okazaki piece initiation) but may be quite different, as with M13 replication, where host RNA polymerase is required for the microinitiation process. More recently, replication systems reflecting "macroinitiation" of cellular DNA replication have been developed and demonstrate the requirements for initiation of new rounds of synthesis in the cell: a requirement for the DnaA protein and a requirement for *oriC* of the *E. coli* chromosome. Thus, it has been possible to fractionate steps in the overall DNA replication process of the procaryotic cells.

The major conclusion from these combined studies is that the process of DNA replication is a sequential, highly controlled process involving multienzyme complexes (replisomes). DNA polymerases are the central functions in the role of DNA synthesis, but their activity is modulated by interactions with numerous other proteins. In the case of *E. coli* there are at least six subunits in the holoenzyme in addition to the synthesis subunit of the complex (α). These are τ, γ, β, δ, ϵ, and θ. The α subunit is the product of the *dnaE* (*polC*) gene of *E. coli* and is responsible for the synthesis activity of the holoenzyme. The roles of the additional proteins have been clarified to varying degrees. It appears that some of these proteins may play more than one role. In some instances, different degrees of "stringency" for the gene product may be displayed, as in the case of ε, which codes for an apparent $3' \rightarrow 5'$ exonuclease activity responsible for editing. A deficiency in this subunit may result in a mutator phenotype. On the other hand, other mutations in this gene product are lethal, raising the question of additional physiological action. In all cases but one (θ), the gene assignment for recognized subunits has been made.

In eucaryotic cells, the enzymology is somewhat less defined. The apparent replication enzyme, DNA polymerase α, has been demonstrated to exist in a "holoenzyme" complex with other associated proteins. This finding and additional biochemical and genetic observations mark α as the probable replicase of eucaryotic cells. In addition to DNA polymerases β and γ, it now appears that there is a fourth DNA polymerase species, δ, in higher eucaryotic cells. This DNA polymerase, which closely resembles α in many respects, seems to be distinguishable on the basis of inhibition studies and the presence of an intrinsic $3' \rightarrow 5'$ exonuclease activity.

For both procaryotic and eucaryotic systems, a general picture is emerging of a synthesis subunit interacting with multiple other proteins which control the synthesis activity, reducing mutations through editorial exonuclease activity and increasing processivity of the enzyme. The latter is the hallmark of the replicating complex and distinguishes the holoenzyme from the "simple" enzymatic activity of DNA polymerase I or the α subunit of DNA polymerase III in *E. coli*.

Recently, evidence has accumulated to suggest that the replicating complex is constructed as a dimer. This model, originally put forth by Alberts, Kornberg, and McHenry with their co-workers, has been substantiated by isolation of apparent dimeric complexes and by genetic evidence suggesting interaction of the α subunits of DNA polymerase III in *E. coli*.

While the above specific and general conclusions can be reached, numerous questions remain.

(i) In a general sense we do not yet understand what enzymatic functions in the holoenzyme contribute to the special features of the replication complex that allow the synthesis subunit to be so effective. This is quite likely the result of numerous protein-protein and protein-DNA interactions.

(ii) The "priority" of the reactions in DNA replication is only now beginning to be established through partial reactions. It appears that we have the tools to analyze the sequence of processes in both macro- and microinitiation, but the specific requirements of each of the stages are far from defined.

(iii) It is not clear whether all of the proteins involved in DNA replication have been defined. Conversely, there may be some false leads, or proteins apparently required by phage systems may not be required for host DNA replication.

(iv) At present, we have a limited understanding of the regulation of expression of the gene products involved in DNA replication. Whether these are constitutively expressed, induced by DNA damage, or induced by other growth conditions remains to be explored.

(v) We have little understanding of the regulation of interaction of the proteins involved in forming the replication complex (replisome). The highly ordered reactions and the apparent sequence of partial reactions which have already been defined suggest that this is not a random mass interaction but rather might be regulated. The basis for the interaction of the subunits or for possible dimerization of the synthesis subunits of the replicating complex has not been established.

(vi) It is not clear what mechanisms the cell uses to exclude like enzymatic activities from the replication process, nor do we understand whether this is an active or passive control. For example, *E. coli* possesses three well-defined DNA polymerases, yet only one of these, the product of the *dnaE* gene, ordinarily can serve as the synthetic subunit of the replication complex.

(vii) Lastly, the hierarchy of "backup" roles among the enzymes of DNA replication is not clear. Evidence is accumulating that in some situations a secondary enzyme may substitute for a primary one in the face of a deficiency of the primary enzyme. These alternative pathways of DNA replication offer intriguing routes of exploration.

The Editors

Chapter 1

Purified Protein System for Initiation of Replication from the Origin of the *Escherichia coli* Chromosome

Arthur Kornberg, Tania A. Baker, LeRoy L. Bertsch, David Bramhill, Kazuhisa Sekimizu, Elmar Wahle, and Benjamin Yung

Understanding the regulation of cell growth and proliferation requires detailed knowledge of the various motifs used by biological switches of chromosome replication. The switches that function to regulate individual genes and groups of genes are revealing recurring themes of how proteins and nucleic acids respond to environmental stimuli to change gene expression. However, an understanding of the molecular mechanism of a switch that regulates the firing of an origin of replication is not yet in hand. This is due partially to lack of knowledge of the essential steps that must occur at origins, for it is these very steps that will be the targets of regulation.

Toward this end, we have been studying the enzymology of replication from the *Escherichia coli* chromosomal origin, *oriC*. *oriC* has been defined as a 245-base-pair (bp) sequence that is necessary and sufficient to impart autonomous replication to plasmids lacking another functional origin. *oriC* must therefore contain all the information required for the assembly of functional replication forks. Furthermore, since replication in *E. coli* is regulated at the point of initiation and timing of replication of *oriC* plasmids within the cell cycle is like that of the chromosome, the target of this timing mechanism must also be within *oriC*. (For reviews on replication and its regulation see references 9 and 12.)

Comparison of the nucleotide sequence of *oriC* from *E. coli* with that from related bacteria revealed several important features (Fig. 1) (13). It is composed of two types of sequence: highly conserved portions, which contain very few changes, and intervening regions, regarded as "spacers," in which the nucleotide sequence is randomized but deletions and insertions are not tolerated. The conserved regions contain two classes of elements whose functions are being clarified at the molecular level: (i) four repeats of a

---

Arthur Kornberg, Tania A. Baker, LeRoy L. Bertsch, David Bramhill, Kazuhisa Sekimizu, Elmar Wahle, and Benjamin Yung • Department of Biochemistry, Stanford University, Stanford, California 94305.

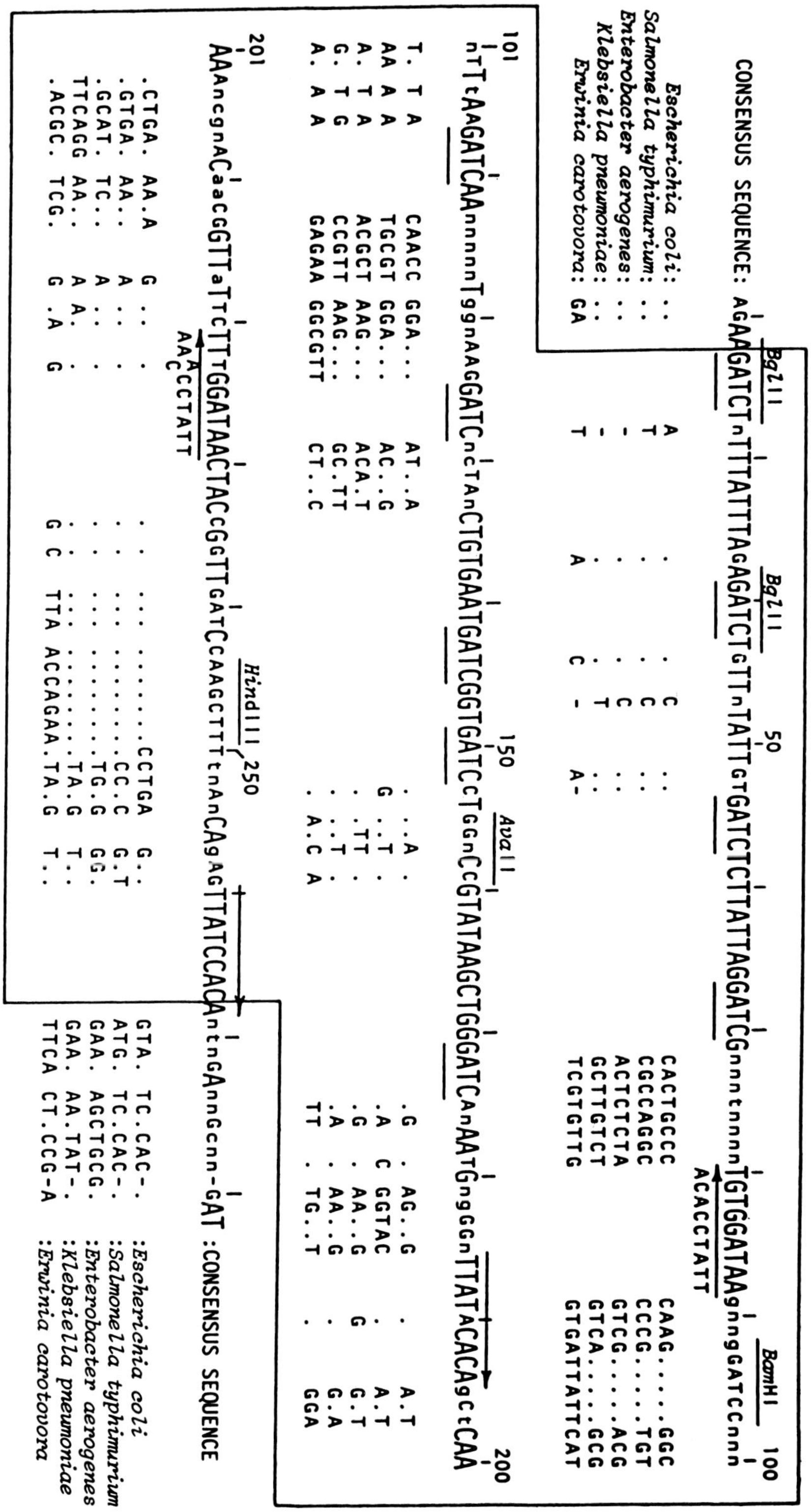

FIGURE 1. Consensus sequence of the minimal origin of enteric bacterial chromosomes (courtesy of J. Zyskind).

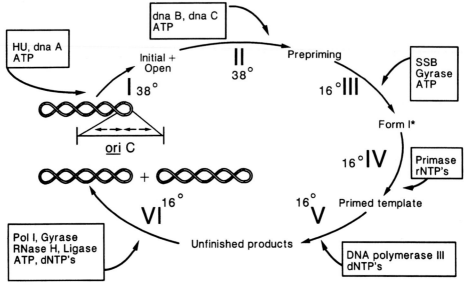

FIGURE 2. Stages of *oriC* plasmid replication.

9-mer recognition sequence serve for the tight binding of the essential initiator DnaA protein, and (ii) three tandem repeats of a 13-mer sequence, also recognized by DnaA protein, are the first regions of the duplex to be opened for insertion of the replication enzymes to form forks. Finally, the sequence GATC, which statistically should occur only once, occurs 11 times (three of them introducing each of the three 13-mers); GATC is the recognition sequence of the DNA adenine methylase. Methylation of these sequences seems to affect origin function in vivo and in vitro, but the basis of this effect is not yet understood.

How is the *oriC* sequence recognized and used by the dozen replication proteins?

STAGES IN THE CYCLE OF *oriC* REPLICATION

By purifying the factors from the crude soluble enzyme system capable of specifically replicating plasmids containing *oriC*, we have identified 12 proteins that are intimately involved in *oriC*-directed replication (6–8a). Their actions and the overall replication reaction have been divided into six major stages (Fig. 2).

Initiation requires that *oriC* be part of a supercoiled template.

(I) The DnaA protein binds its 9-mer recognition sites within *oriC*; 20 to 40 monomers bind cooperatively to form a large complex encompassing about 200 bp of DNA (3). DNase I footprinting established that the 9-mers (DnaA boxes) are especially recognized. The 10-bp periodicity of protection and the electron microscopic analysis of the complex suggest that the DNA is wrapped around the complex as in a nucleosome. Although binding studies show that ATP is not required for 9-mer binding, formation of an open complex required that DnaA protein contain a tightly bound ATP (see Fig. 2 for more detailed description of the early stages). The histonelike protein HU stimulates the early initiation stages. Stability of this DnaA-*oriC* complex requires that it be isolated at a high temperature (30 to 38°C); it is cold sensitive.

(II) Once bound to *oriC*, DnaA allows the entry of the DnaB helicase into the complex in a reaction that requires DnaC protein (5). This stage depends on ATP and involves an opening of the DNA duplex, which will be discussed in detail below (11). The resulting prepriming complex is now stable to low temperature (e.g., 16°C). Upon incubation of this prepriming complex with ATP and single-strand DNA-binding protein (SSB), DnaB helicase can leave the origin, unwinding the template in both directions (1).

(III) With gyrase present to relieve the topological strain associated with unzipping the double helix, unwinding by DnaB protein can proceed unchecked almost entirely around the circle. When this unwinding is allowed to proceed in the absence of replication, a highly underwound, single-stranded DNA is the product, Form I* (2).

(IV) Once unwinding by DnaB is initiated and long before it is as extensive as in Form I*, the template becomes competent for priming by primase (1). These first primers are restricted to the origin region if primase is present at the start of DnaB unwinding; when unwinding is extensive, primers are distributed all over the template.

(V) The primers are then elongated by the multisubunit DNA polymerase III holoenzyme (2,4).

(VI) Finally, primers are excised by DNA polymerase I with the help of RNase H. The gaps are then filled by polymerase I or polymerase III holoenzyme and the nicks sealed by ligase. Gyrase decatenates and supercoils the daughter molecules, generating two plasmids identical to the starting template and in good yield (4).

THE INITIATOR DnaA PROTEIN OPENS THE ORIGIN FOR DnaB HELICASE ACTION AND BIDIRECTIONAL REPLICATION

Electron microscopic studies of the prepriming complex and early unwinding bubbles indicated that DnaB protein enters the *oriC*-DnaA protein complex at the left extremity of the origin (Fig. 1) not previously bound by DnaA protein (5). Examination of this region revealed that it contained three repeats of the 13-bp sequence GATCTNTTNTTTT (10/11, 10/11, and 9/11 matches with two unspecified bases) (D. Bramhill and A. Kornberg, submitted for publication). Using this consensus sequence, similar repeats were found near the DnaA boxes in the origin of pSC101 and the promoter regions of the genes for DnaA and the 16-kilodalton protein (the gene immediately to the right of *oriC*). Attempts to observe binding of DnaB or DnaC proteins to this sequence in the absence of DnaA-dependent prepriming complex formation were unsuccessful (D. Bramhill, unpublished data). Interactions between DnaA protein and this region were then examined in more detail.

Under the conditions of the replication reaction, DnaA protein can open the duplex at the 13-mer sequences. This opening can be observed by incubating DnaA protein, a supercoiled template, and ATP (under appropriate buffer conditions) followed by a 5-s digestion with a carefully titrated amount of endonuclease P1 that evinces a single-strand specificity. P1 resembles S1 nuclease but is far more active at the neutral pH required for the replication reaction. About 40 to 60% of the template is linearized by this treatment. (Low levels of HU protein are included to reduce DnaA-independent cleavage by P1; the DnaA- or ATP-independent cleavage is determined separately as a background value and subtracted.) Like replication, this duplex-opening reaction required that DnaA be in the ATP-bound form; in a typical experiment, ATP · DnaA gave 46% linearization, while the ADP · DnaA form gave only 6%, the background level. Mapping of the location of the cleavages showed that they are clustered within the 13-mers (Bramhill and Kornberg, submitted). The observation that the predominant products after P1 digestion are linear

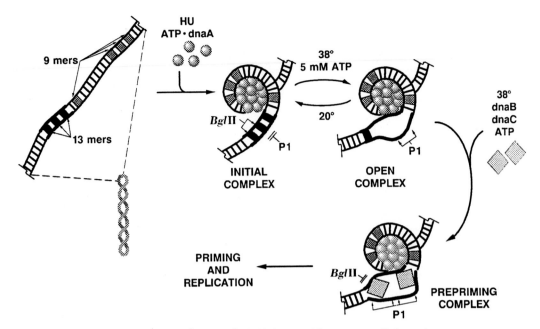

FIGURE 3. Proposed scheme for initiation at *oriC* on a supercoiled template.

suggests that the structure created by DnaA protein is a single-stranded bubble, unprotected by protein on either strand. We call it the *open complex* (Fig. 3). (The first stage [Fig. 2] of DnaA interacting with the *oriC* has been divided into two stages in Fig. 3, the initial complex and the open complex.)

This duplex opening by DnaA protein allows the entrance of DnaB protein into the 13-mer region to form the *prepriming complex*. In addition to the electron microscopic evidence described above, localization of DnaB to the 13-mers can be shown in two ways. One is by P1 sensitivity. After addition of DnaB and DnaC proteins, the P1 sensitivity of the 13-mers can be observed at low temperature (i.e., 16°C), whereas with DnaA protein alone, P1 sensitivity can be observed only above 25°C. The pattern of cleavages is also slightly changed. The other way is by restriction at the two *Bgl*II sites within the 13-mers. These sequences become resistant to digestion after addition of DnaB and DnaC proteins.

Deletion analysis further established the interaction between DnaA and the 13-mers and suggested the detailed functions of these three tandem repeats, all of which are essential for replication in vivo. The three 13-mers (L, left; M, middle; R, right) were sequentially deleted from left to right and the resulting constructions were assayed for their capacities to (i) form an open complex (P1^s at 38°C), (ii) form a prepriming complex (P1^s at 16°C), and (iii) sustain replication (Table 1). A dramatic difference was seen in the behavior of these deletions in the three assays. Deletions lacking the left or left and middle repeats show a level of duplex opening near that of the wild type. Only upon deleting the final 13-mer does the reaction fall to the level of the vector lacking *oriC*. Thus, one 13-mer is sufficient for duplex opening by DnaA protein. When assaying for prepriming complex formation or replication, however, all three 13-mers are needed. Deletion of the left 13-mer reduces the level of prepriming complex 5-fold and of replication more than 10-fold. Deletion of the left and middle 13-mers reduces the

TABLE 1
Effect of Deletions of the 13-mers on *oriC* Function

13-mers present in *oriC*	Open complex (%)[a]	Prepriming complex (%)[a]	Replication (pmol)
L M R	100	100	178
M R	97	19	16
R	92	9	4
None (L, M, and R deleted)	16	5	4
Vector	15	4	4

[a] Percentage values are expressed relative to the maximum value (100%) representing linearization of about half of the added template molecules.

activity of the plasmids to levels indistinguishable from that of the vector lacking *oriC*. While one 13-mer can be opened by DnaA protein, indicating that a 13-mer is the unit of recognition, three 13-mers in a row are necessary for introduction of DnaB protein. A similar conclusion was drawn from the fact that a higher transition temperature (by 3 to 4°C) is required for forming the prepriming complex than for the open complex. Thus, melting of the entire 13-mer region, about 40 bp, is required to start the assembly of replication forks at *oriC*.

FORMATION OF AN RNA-DNA HYBRID BY TRANSCRIPTION ACTIVATES *oriC* FOR DUPLEX OPENING

Studies of the pathway for *oriC* initiation and replication were generally made in the absence of RNA polymerase; primase fulfilled the requirements for initiation in vitro. However, in vivo, the rifampin sensitivity of initiation after a period when protein synthesis is no longer required implies the existence of a transcription essential to initiation. Furthermore, *rpoB* mutants, encoding altered β subunits of RNA polymerase, suppress temperature-sensitive mutants in the *dnaA* gene, again implying RNA polymerase action in initiation (see reference 12).

In the crude in vitro system for replication of *oriC* plasmids, rifampin sensitivity of the reaction also implied a requirement for RNA polymerase. Initial attempts at staging the reaction indicated that its activity was required very early. However, on purification and reconstitution of the replication system, RNA polymerase was not always essential; its contribution depended on the levels of the other proteins and on the reaction conditions (10). Three of the factors that modulate the dependence on RNA polymerase are the level of the histonelike protein HU, the reaction temperature, and the superhelix density (the balance between topoisomerase I and gyrase). HU protein at a level that coats one-third or more of the template, or a reaction temperature of less than 25°C, or a superhelix density more relaxed than -0.03 can each block the replication reaction in the absence of transcription by RNA polymerase. Replication in the presence of HU or at 20°C is completely dependent on transcription.

All three of these conditions specifically inhibit the DnaA-dependent duplex-opening reaction, parallel to their inhibition of replication (T. Baker and A. Kornberg, unpublished data). Since these conditions appeared not to inhibit the preceding reaction stage (i.e., DnaA protein binding to the template), we conclude that they inhibit replication by preventing the opening of the duplex. Thus, the role of RNA polymerase transcription is at the point of assisting duplex opening under conditions where DnaA protein, acting alone, is insufficient. Apparently, the large region of the duplex (about 40 bp) that must be melted by DnaA protein for the entrance of DnaB helicase makes duplex opening energetically very costly and highly sensitive to agents that stabilize the helix (e.g., low temperature, relaxation, and HU protein binding).

How does RNA polymerase counteract the inhibition by these "freezing" agents? The activation absolutely depends on transcription. Titration of RNA polymerase for

TABLE 2
Transcriptional Activation of an *oriC* Template Coated with HU Protein

RNA polymerase	DNA synthesis (pmol)[a]	RNA synthesis (pmol)[b]
None	17	23
Holoenzyme	188	232
Core	20	264

[a] DNA synthesis was assayed on the *oriC* plasmid pCM959 in the presence of inhibitory levels of HU.
[b] RNA synthesis was assayed with activated calf thymus DNA.

TABLE 3
Effect of RNase Treatments on an Activated Template

Treatment	Replication activity (pmol)[a]	RNA remaining (pmol)[a]
No addition	120	648
RNase A	83	69
RNase A + RNase H	9	53

[a] Replication activity and RNA remaining were determined at 2 min after addition of the RNase.

activation (most commonly replication on a template inhibited by high levels of HU) revealed that about five RNA polymerase molecules per template was optimal; furthermore, the holoenzyme form was required (Table 2). When amounts of holoenzyme and core RNA polymerase that gave equal RNA synthesis on activated calf thymus DNA (a template that does not require the sigma subunit for initiation) were compared for the ability to activate replication, the core form was virtually inert. Apparently, the interaction of RNA polymerase with promoters is required. All four ribonucleotide triphosphates were also required, as were time and temperature (T. Baker and A. Kornberg, unpublished data). The phage RNA polymerases of T3 and T7 substituted readily for *E. coli* RNA polymerase on plasmids containing promoters for these polymerases. Thus, what is required for activation is transcription rather than a special interaction between RNA polymerase and DnaA protein or some DNA structure (e.g., promoter or terminator) within the origin.

RNA polymerase can act first, before addition of any replication proteins. The transcription and replication stages could be separated by the use of rifampin. When transcription was allowed for a period of time (less than a minute) before addition of rifampin (added just before the replication proteins), there was no inhibition. However, when rifampin was added at the onset of the transcription phase, the reaction was completely inhibited. Clearly, transcription acts first and the template retains a "memory" of the transcription event.

This memory of transcription strongly suggested that the RNA product was involved in the activation reaction. Other experiments showed that activation worked only in *cis*, and attempts to use free RNA as a substitute for transcription have not been successful. It seemed likely, therefore, that an RNA-DNA hybrid was responsible for activation. The behavior of the activated templates after treatment with RNases confirmed this conclusion (Table 3). Transcription was allowed to proceed in the absence of replication proteins for 5 min. RNase A or RNase A together with RNase H was added and samples were removed after 2 min to determine the amount of activation and RNA remaining. Treatment of the activated template with RNase A reduced replication activity less than twofold, while removing more than 90% of the RNA. Gel electrophoresis of the RNA confirmed that it was converted into short pieces by this treatment, leaving only a small amount hybridized to the template. Incubation with RNase H (specific for RNA in RNA-DNA hybrids), in addition to RNase A, completely inactivated the template. Gel analysis confirmed that the hybrids were destroyed at the same time, showing that an RNA-DNA hybrid is necessary for the activation. The hybrid, isolated from the small RNA pieces and from RNA polymerase by gel filtration, retained activity, confirming that it is sufficient.

The molecular mechanism by which an RNA-DNA hybrid counteracts the inhibition of duplex opening is still unclear. Further experiments are needed to establish the precise length and location of the hybrid required. It seems likely that alterations of the template structure by the hybrid must be involved. The results should be of broad interest considering that transcriptional activation is appearing as a recurring theme in the initiation of replication in procaryotes (λ, ColE1) and in eucaryotes (adenovirus, simian virus 40), as well as in recombinational processes.

SUMMARY

DnaA protein locates *oriC* and directs the assembly of replication forks. To achieve this, the initiator protein first binds in a sequence-dependent manner to the four "DnaA boxes" located there. Then, in a step highly sensitive to configuration of the duplex, it separates the strands to allow the entry of DnaB helicase. Entry of DnaB helicase can be thought of as the "committed step" to replication; priming and replication follow easily once DnaB is established. In this scheme, DnaA protein appears to play a role similar to that of the sigma subunit in RNA polymerase. DnaA directs the "sequence-blind" elongation enzymes to the "replication promoter" or origin as sigma does for the RNA polymerase core. Plasmids and phage of *E. coli* encode their own initator proteins, specific to their own origins as "alternate sigma factors," to attract the replication machinery away from *oriC*.

LITERATURE CITED

1. Baker, T. A., B. E. Funnell, and A. Kornberg. 1987. Helicase action of dnaB protein during replication from the *Escherichia coli* chromosomal origin *in vitro*. *J. Biol. Chem.* 262:6877–6885.
2. Baker, T. A., K. Sekimizu, B. E. Funnell, and A. Kornberg. 1986. Extensive unwinding of the plasmid template during staged enzymatic initiation of DNA replication from the origin of the *E. coli* chromosome. *Cell* 45:53–64.
3. Fuller, R. S., B. E. Funnell, and A. Kornberg. 1984. The dnaA protein complex with the *E. coli* chromosomal replication origin (*oriC*) and other DNA sites. *Cell* 38:889–900.
4. Funnell, B. E., T. A. Baker, and A. Kornberg. 1986. Complete enzymatic replication of plasmids containing the origin of the *Escherichia coli* chromosome. *J. Biol. Chem.* 261:5616–5624.
5. Funnell, B. E., T. A. Baker, and A. Kornberg. 1987. *In vitro* assembly of a prepriming complex at the origin of the *Escherichia coli* chromosome. *J. Biol. Chem.* 262:10327–10334.
6. Kaguni, J. M., and A. Kornberg. 1984. Replication initiated at the origin (*oriC*) of the *E. coli* chromosome reconstituted with purified enzymes. *Cell* 38:183–190.
7. Kornberg, A. 1980. *DNA Replication*. W. H. Freeman & Co., San Francisco.
8. Kornberg A. 1982. *DNA Replication, 1982 Supplement*. W. H. Freeman & Co., San Francisco.
8a. Kornberg, A. 1988. DNA replication. *J. Biol. Chem.* 263:1–4.
9. McMacken, R., L. Silver, and C. Georgopoulos. 1987. DNA replication, p. 564–612. *In* F. C. Neidhardt, J. L. Ingraham, K. B. Low, B. Magasanik, M. Schaechter, and H. E. Umbarger (ed.), *Escherichia coli and Salmonella typhimurium: Cellular and Molecular Biology*, vol. 1. American Society for Microbiology, Washington, D.C.
10. Ogawa, T., T. A. Baker, A. van der Ende, and A. Kornberg. 1985. Initiation of enzymatic replication at the origin of the *Escherichia coli* chromosome: contributions of RNA polymerase and primase. *Proc. Natl. Acad. Sci. USA* 82:3562–3566.
11. Sekimizu, K., D. Bramhill, and A. Kornberg. 1987. ATP activates dnaA protein in initiating replication of plasmids bearing the origin of the *E. coli* chromosome. *Cell* 50:259–265.
12. von Meyenburg, K., and F. G. Hansen. 1987. Regulation of chromosome replication, p. 1555–1577. *In* F. C. Neidhardt, J. L. Ingraham, K. B. Low, B. Magasanik, M. Schaechter, and H. E. Umbarger (ed.), *Escherichia coli and Salmonella typhimurium: Cellular and Molecular Biology*, vol. 2. American Society for Microbiology, Washington, D.C.
13. Zyskind, J. W., J. M. Cleary, W. S. Brusilow, N. E. Harding, and D. W. Smith. 1983. Chromosomal replication origin from the marine bacterium *Vibrio harveyi* functions in *Escherichia coli*: *oriC* consensus sequence. *Proc. Natl. Acad. Sci. USA* 80:1164–1168.

Chapter 2

DNA Polymerase III Holoenzyme
Mechanism and Regulation of a True Replicative Complex

Charles S. McHenry, Henry Tomasiewicz, Mark A. Griep, Jens P. Fürste, and Ann M. Flower

The DNA polymerase III holoenzyme is the complex multisubunit DNA polymerase that is responsible for the synthesis of the majority of the *Escherichia coli* chromosome. It contains at least seven different subunits, α, τ, γ, β, δ, ε, and θ, ranging from 130,000 to 10,000 daltons (for a review, see references 27 and C. S. McHenry, *Annu. Rev. Biochem.*, in press). This enzyme contains a catalytic core, termed DNA polymerase III, that is capable of efficiently filling gaps in activated DNA (19, 21). The core is composed of α, the catalytic subunit (*dnaE* gene product); ε, which contains the proofreading exonuclease (*mutD* gene product); and θ, a subunit without an identified structural gene or function (28, 36, 38).

The polymerase III core is inert, without the remaining auxiliary subunits, on natural chromosomal templates such as that provided by the bacteriophage G4 (1, 10).

Much attention has been paid to these auxiliary subunits since they confer on the core polymerase the special properties that distinguish true replicative complexes from simpler DNA polymerases. The auxiliary subunits that have been established to date include β (*dnaN*), τ and γ (both products of the *dnaX* gene), and δ (3, 12, 17, 26, 29, 32, 41).

The addition of τ to DNA polymerase III forms DNA polymerase III', causes the polymerase to dimerize, and increases its processivity sixfold (8, 26). DNA polymerase III*, which also contains the γ and δ subunits plus additional components that are either subunits or tightly associated proteins, exhibits a processivity at least 20-fold greater than that of the core polymerase and gains the ability to interact with the *E. coli* single-stranded DNA-binding protein (8).

The most striking difference between the DNA polymerase III holoenzyme and simpler DNA polymerases is its ability to form a stable initiation complex with primed single-stranded DNA at the expense of ATP hydrolysis (2, 13, 14, 42). Upon formation of initiation complexes, the β subunit be-

---

Charles S. McHenry, Henry Tomasiewicz, Mark A. Griep, Jens P. Fürste, and Ann M. Flower • Department of Biochemistry, Biophysics and Genetics, University of Colorado Health Sciences Center, Denver, Colorado 80262.

comes immersed such that elongation becomes resistant to the inhibitory action of anti-β immunoglobulin G (IgG) (15). Upon addition of the required deoxynucleoside triphosphates, the complex can processively replicate an entire 5,000-nucleotide G4 circle without dissociating (7). These complexes show remarkable stability in comparison to the labile DNA polymerase III holoenzyme. They can be isolated by gel filtration at room temperature, a process that takes 40 min. The stability of these complexes creates kinetic barriers with respect to the expected rate (1/s) for cycling between Okazaki fragments during synthesis of the lagging strand of the E. coli replication fork. It is the purpose of this chapter to report recent findings that influence our understanding of how the holoenzyme overcomes these difficulties at natural replication forks and how it may coordinate leading- with lagging-strand synthesis. We will also pursue other recent studies that pertain to the physical properties of the replicative complex and its genetic regulation.

KINETIC BARRIER FOR HOLOENZYME RECYCLING

Starting from a preformed initiation complex, the DNA polymerase III holoenzyme can replicate an entire 5,000-nucleotide G4 molecule in under 15 s without dissociating, yet it remains stably bound to the completed RFII product for over 40 min (15). The stability of these complexes varies somewhat with reaction conditions, the presence of excess subunits, and the enzyme preparation used, but it remains a severe problem relative to the cellular requirements for polymerase cycling during Okazaki fragment synthesis. In vivo, one Okazaki fragment must be synthesized approximately every second at the E. coli replication fork. Our in vitro elongation rate approximates the required elongation velocity, but the slow recycling from a completed strand to the next primed template presents a severe kinetic barrier.

It occurred to us that modification of the theories regarding dimeric replication complexes (18, 26, 37) could provide a solution to the problem of recycling of the DNA polymerase III holoenzyme. We had previously demonstrated that the more complex forms of DNA polymerase III were dimeric (26). We reasoned that the release of the lagging-strand polymerase upon completion of the synthesis of an Okazaki fragment might be facilitated by the leading-strand polymerase being in a productive elongation conformation. Such information could be communicated by allosteric interactions between polymerase halves. Support for this model came from an investigation of the effect of an ATP analog on initiation complex formation.

THE DNA POLYMERASE III HOLOENZYME FUNCTIONS AS AN ASYMMETRIC DIMER

We have gained further insight into the functioning of the dimeric holoenzyme by the use of ATPγS (16). This analog substitutes for ATP in the formation of initiation complexes between the DNA polymerase III holoenzyme and primed DNA. Like ATP, it is hydrolyzed to ADP during and subsequent to initiation complex formation. Initiation complexes formed in the presence of ATPγS are indistinguishable from those formed with ATP by their resistance to antibody directed against the β subunit of the DNA polymerase III holoenzyme or by their ability to elongate without any further nucleotide-mediated activation (16).

However, the most striking result of these studies was the twofold difference in the maximal extent to which ATP and ATPγS can support initiation complex formation. ATPγS will support only half as much complex formation as ATP (Fig. 1A). This effect is also seen upon reversal of the

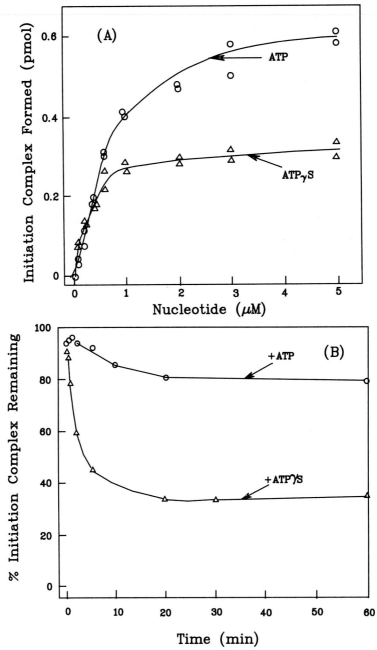

FIGURE 1. (A) ATP supports twice as much initiation complex formation between the DNA polymerase III holoenzyme and primed DNA as ATPγS. (B) ATPγS induces the dissociation of half of the initiation complex formed in the presence of ATP. Details of this experiment have been reported previously (16).

reaction. If maximal initiation complex is formed in the presence of ATP and isolated free of nucleotide by gel filtration, the resulting complex can be 50% dissociated by ATPγS (Fig. 1B). ATP causes no more dissociation than the addition of a buffer control. The extent of dissociation does not vary with ATPγS concentration once sufficient ATPγS has been added to saturate its binding site on holoenzyme. Thus, this is not an equilibrium effect that can be pushed to complete dissociation by ATPγS. This argues for two distinct populations of holoenzyme in solution: one that can form initiation complexes in the presence of ATPγS and one that forms complexes only via ATP and is dissociated by ATPγS.

Keeping in mind our notions concerning the role of a dimeric holoenzyme in the natural replicative reaction, we proposed that these results were indicative not of two distinct holoenzyme populations in solution but of a difference between the two halves of a dimeric holoenzyme (16, 27). This asymmetry could have a structural basis. One half of the holoenzyme could contain a different subunit composition or a covalent modification. Alternatively, the asymmetry could be induced by allosteric interactions between the two halves of the enzyme when bound to the replication fork. In any case, this asymmetry could be used to solve the problems imposed by the difference in functional requirements of polymerases at the replication fork. The leading-strand polymerase, once bound to its template, need not dissociate until the chromosome is completely replicated, a process that takes 40 min in *E. coli*. The lagging-strand polymerase, however, must synthesize an Okazaki fragment each second and recycle to the next primer synthesized at the replication fork. An asymmetric DNA polymerase dimer might have a higher affinity for the template in the leading-strand half and a diminished affinity in the lagging half that would permit efficient recycling. We are also aware that this geometry might permit the dimeric holoenzyme to associate with the priming apparatus, helicases, and other enzymes and cellular structures associated with the replication fork in one large replicative complex.

IMPLICATIONS OF THE ASYMMETRIC DIMERIC REPLICATION HYPOTHESIS

Communication between two halves of a dimeric replicative complex has implications not only for the mechanism of cycling on the lagging strand but also for the coordination of leading- with lagging-strand synthesis. Perhaps this communication is useful in causing the opposing polymerase to stall if it encounters a barrier such as a chemically damaged region, an annealed oligonucleotide, or a tightly associated protein. Causing the opposing polymerase to pause for a few seconds until a barrier is removed would prevent it from extending thousands of nucleotides beyond the blocked polymerase. Extensive single-stranded structures created by noncoordinated polymerases would be disadvantageous to the cell for several reasons. They would create packaging problems for the cell, they would exceed the cell's capacity to provide adequate single-stranded DNA-binding protein, and they would be exposed to cellular nucleases. A coordinated dimeric polymerase could minimize these potential difficulties.

Another implication of a dimeric polymerase pertains to topological considerations that have not, to our knowledge, been addressed previously. With a monomeric polymerase, the only topological problem is in advance of the polymerase on leading-strand DNA. This induction of positively supercoiled DNA and the relief of this tension by DNA gyrase have received considerable attention (6, 31). In the nascent DNA product, there are no serious topological problems because the polymerase can revolve around the DNA template as dictated by the helical nature of the product (Fig. 2A).

With a dimeric, *highly processive* DNA polymerase that maintains *continuous* molecular contacts with the product strand, such rotation is not possible (Fig. 2B). The attachment of one complex to the other blocks the required rotation. (Any fixation of the polymerase, e.g., membrane attachment, would create the same topological problem.) On the lagging strand this presumably is not a problem, because the template is free to rotate in single-stranded regions behind the duplex (Fig. 2B). On the leading strand, the enormous length of the product precludes this solution. Here, negative supercoils should accumulate to the same extent that positive supercoils accumulate in advance of the replication fork (Fig. 2B). This superhelical tension in the nascent product could be relieved by *E. coli* topoisomerase I. This enzyme (ω protein) can relieve negative but not positive superhelical turns (22). Other roles for topoisomerase I have been demonstrated in the Marians laboratory for plasmid decatenation during replication (30).

COOPERATIVE INTERACTIONS BETWEEN THE TWO HOLOENZYME HALVES

The asymmetric dimer model makes several testable predictions. One is that the two halves are not merely stuck to one another but are functionally interacting. This was tested by examination of the binding of ATP and ATPγS, the two nucleotides that support initiation complex formation. Both bind with strong positive cooperativity. Instead of the straight line obtained for a Scatchard plot in a noncooperative system, the plot exhibits a maximum, a feature diagnostic of positive cooperativity. We calculate a Hill coefficient of approximately 1.5, indicating strong allosteric interactions between the ATP binding sites. In addition to the allosteric basis for holoenzyme asymmetry, recent data suggest a structural basis as well.

RELATIONSHIP BETWEEN τ AND γ

The *dnaX* gene encodes two proteins of approximately 71,000 and 52,000 daltons (17, 32). Both of these proteins contain common sequences. The smaller protein was known from previous work to be the γ subunit of the DNA polymerase III holoenzyme (12, 41). It is encoded by the amino-terminal portion of the *dnaX* gene. The larger protein was proposed to be τ on the basis of its electrophoretic mobility (17). We confirmed this conclusion by using a τ-specific monoclonal antibody to demonstrate that τ is overproduced by strains carrying the *dnaX* gene on a plasmid and to specifically immunoprecipitate the 71,000-dalton product of *dnaX* (11). The specificity of the anti-τ monoclonal antibody is demonstrated in Fig. 3. Even though most of the γ sequence is contained within τ, this antibody reacts exclusively with τ. The antibody must be directed toward a sequence within the unique carboxyl-terminal domain of τ that is lacking from γ.

Sequencing of the *dnaX* gene has indicated that it contains one continuous open reading frame encoding a protein of approximately 71,000 daltons (9, 44). The mechanism by which γ arises from τ is uncertain. Proteolysis of τ to yield γ has been proposed (17, 32). If true, this cleavage must be tightly controlled to yield adequate γ for cellular needs without excessively diminishing the quantity of the essential τ protein. Alternatively, the generation of γ could be regulated translationally through a frameshifting mechanism (A. M. Flower and C. S. McHenry, unpublished results; McHenry, in press).

From the sequence of τ, several potentially significant similarities with other proteins have been determined. Previously,

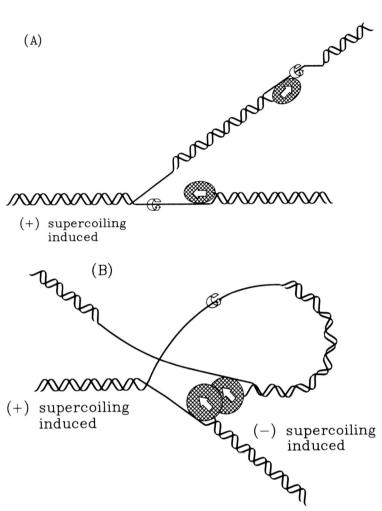

FIGURE 2. Topological constraints induced during leading-strand synthesis by a dimeric polymerase. (A) With a monomeric polymerase, the enzyme is free to rotate around the template strand approximately once per 10 bases inserted, avoiding topological constraints in the nascent product. (B) When a highly processive polymerase becomes fixed in a dimeric complex it cannot rotate around the product strand, creating an underwinding of the leading strand. However, the shorter lagging strand is free to relieve negative superhelical tension by rotating around the single-stranded region preceding the duplex.

a similarity between τ and a conserved ATPase sequence was found (44). We have found more extensive similarity in adjacent regions of the amino-terminal segment of τ with the major membrane ATPase of *E. coli*, underscoring the relationship between these proteins (Fig. 4). The holoenzyme contains an ATPase that is essential for tightly clamping the holoenzyme on the template in processive form. Recently, a τ–β-galactosidase fusion protein has been shown to have feeble ATPase activity (20). The putative ATPase site of τ overlaps a region of similarity with the bacteriophage Mu gene *B* protein (25, 39). The gene *B* protein, like γ, is involved in the ATP-dependent formation

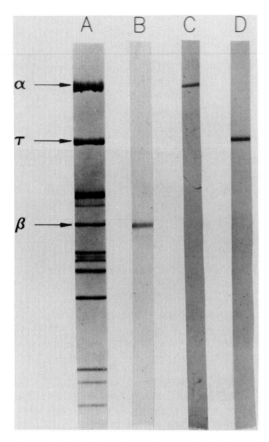

FIGURE 3. Specificity of monoclonal antibody directed against τ. All four lanes contain the resolved holoenzyme subunits. Lane A, Stained holoenzyme permitting visualization of all bands. Lanes B, C, and D, Immunostaining of the β, α, and τ subunits, respectively, with specific antibodies. DNA polymerase III holoenzyme was subjected to sodium dodecyl sulfate-polyacrylamide gel electrophoresis and detected by the immunoblot method. Electrophoretic transfer of the resolved proteins from the gel to a nitrocellulose sheet was conducted as described (15). After transfer, individual lanes were cut out for processing. Lanes to be treated with antibody were blocked with bovine serum albumin saline (BSA-saline). Strips were treated with antibody, washed in 0.05% Tween 20 in Tris-saline, and then treated with peroxidase-conjugated goat anti-mouse IgG in BSA-saline. The strips were washed and visualized by incubation with o-dianisidine. β and α (lanes B and C, respectively) were detected with specific antibodies as previously described (15, 43). Holoenzyme (10 μg) in lane A was stained to provide markers for subunit positions as indicated in the margin. In lane D, holoenzyme (7 μg) was treated with anti-τ IgG. The antibody solutions were used at concentrations of about 1,000 U/ml.

of a multiprotein complex with DNA (39). Additional similarities have also been found. Toward a more central segment, similarity with the 32,000-dalton cowpea mosaic virus has been detected. This raises the possibility of an autoproteolytic mechanism being involved in the generation of γ from τ. Perhaps a τ-τ pair forms and one τ cleaves the other and inactivates its proteolytic activity guaranteeing a 1:1 ratio of the two proteins. Toward the carboxyl-terminus end that is unique to τ, we observe similarity with a number of proteins that interact with nucleic acids. These include ribosomal proteins, RNA polymerases, and reverse transcriptases. The strongest similarity in this zone is with histone H1 from sea urchin. This similarity arises primarily from the content of lysines and alanines for the two proteins, but it may suggest a tight DNA-binding domain within τ that is disrupted in γ. This could have functional significance, as discussed below.

BOTH τ AND γ ARE CONTAINED WITHIN THE SAME HOLOENZYME ASSEMBLIES

Treatment of DNA polymerase III holoenzyme with anti-τ IgG and secondary precipitating reagents permits τ and all holoenzyme activity to be quantitatively removed from solution. Thus, we know that all active holoenzyme molecules contain τ. Examination of the immunoprecipitated subunits from radiolabeled holoenzyme indicated that roughly equivalent amounts of both γ and τ were precipitated (Fig. 5). Thus, γ and τ reside in the same holoenzyme assemblies. If proteolysis had been artifactual, one would expect some subassemblies to contain only γ and others only τ.

The existence of these two related proteins within the same holoenzyme assembly may provide a basis for a structural difference between the two halves of the holoenzyme, in addition to the allosteric interac-

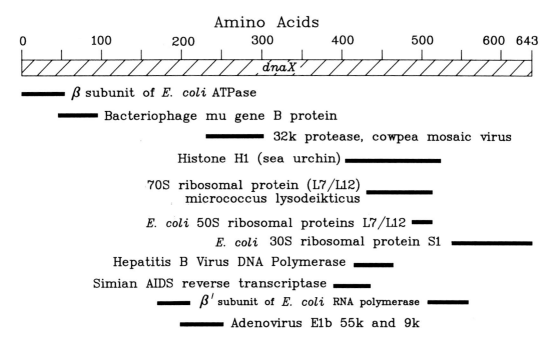

FIGURE 4. Similarities between the τ subunit of the DNA polymerase III holoenzyme and other proteins. The protein homology analysis was performed using the Intelligenetics IFind program to select proteins with regions of similarity to the *dnaX* protein products. The search was performed using a 20-amino-acid window and a gap penalty of 2. Selected proteins with the highest scores were examined using the Align program of Intelligenetics. This program allows two proteins to be aligned in regions of maximum similarity. Alteration of the parameters in this analysis (window, gap penalty) allowed the alignments to be made looking for either regions of high similarity or longer regions of less similarity. The adenovirus E1b 55K and 9K proteins had the best scores of 10 from the IFind search (completely random proteins usually score 2 to 4). The alignment showed an 18-amino-acid stretch with 61% similarity or a 25-amino-acid stretch with 64%. The histone H1 from sea urchin had the second best IFind score, 9, and had a 20-amino-acid stretch of 55% similarity. The area of similarity could be lengthened to 123 amino acids, but the similarity dropped to 28%. The gene *B* protein from bacteriophage Mu had an IFind score of 8 and an alignment of 25 amino acids with 48% similarity. The ribosomal proteins all scored a 7 with IFind, and the alignment ranged from 7 amino acids with 100% similarity (50S L7/L12) to 120 amino acids with 23% similarity (30S S1). The three polymerases also scored 7 on IFind, and they range from 11 amino acids with 73% similarity (simian AIDS virus) to 54 amino acids with 30% similarity (also simian AIDS). The hepatitis B virus polymerase and *E. coli* RNA polymerase were intermediate between these two values. The β subunit of the *E. coli* ATPase and the 32K protease from cowpea mosaic virus also both scored 7. The 32K protease was 67% homologous over a 12-amino-acid stretch or 38% over 40 amino acids, while the ATPase was 80% homologous over 10 amino acids or 36% over 36 amino acids.

tions we have observed, and may predetermine which half is the leading-strand polymerase and which is the lagging-strand half. The leading-strand half may use the basic domain of τ to form a tight clamp on the DNA, and the lagging-strand half might contain the analogous γ protein bound by similar molecular interactions with the other holoenzyme components. Perhaps the γ-driven complex is capable of more efficient cycling, mediated in part by looser interactions with the DNA.

PHYSICAL PROPERTIES OF THE REPLICATIVE POLYMERASE

Hydrodynamic studies of DNA polymerase III and its higher-order assemblies indicate that it is very irregular in shape. This

FIGURE 5. Immunoprecipitation of τ and associated proteins by monoclonal anti-τ IgG. ^{125}I-holoenzyme was added to 20 μl of E. coli cell lysate before the addition of anti-τ IgG. The resulting solutions were treated with goat anti-mouse affinity gel (20 μl of 50% suspension, gentle vortexing, 1 h, 4°C) and then centrifuged, and the pellets were washed. Pellets were resuspended in sodium dodecyl sulfate gel buffer, boiled, clarified by centrifugation, and subjected to sodium dodecyl sulfate-polyacrylamide gel electrophoresis. Bands were detected by autoradiography. Holoenzyme was radioactively labeled by the Iodobead method (24) with modifications. The enzyme was dialyzed to remove dithiothreitol, which interferes with the iodination reaction. Iodination was permitted to proceed for only 6 min, and dithiothreitol was added back immediately to preserve holoenzyme structure. The labeled enzyme was purified by gel filtration.

becomes apparent by examining the migration relative to protein standards on gel filtration or during sedimentation (26). For example, DNA polymerase III (167,000 daltons) behaves as though it is larger than catalase (244,000 daltons) on gel filtration but migrates considerably more slowly than this standard on sedimentation. One can also observe a marked shift in relative mobility of polymerase III' relative to urease (a roughly spherical enzyme of the same molecular weight). This is due to irregularity in shape causing the polymerase to be excessively excluded from the pores of a gel filtration matrix and excessively retarded because of friction during sedimentation.

From these data, Perrin shape factors of 1.31 and 1.55 can be calculated for polymerase III and polymerase III', respectively. In contrast, a spherical protein would have a value of 1. From these values, these polymerases can be modeled as elongated prolate ellipsoids with axial ratios of 6:1 (for polymerase III) and 10:1 for polymerase III' (5). It is interesting that upon dimerizing the polymerase becomes even more irregular in shape. This elongated shape could permit extensive contacts with the linear template-primer.

DNA polymerase III holoenzyme, in the form in which we isolate it (34), migrates during gel filtration with thyroglobulin (670,000 daltons). However, care must be taken in interpreting a molecular mass for the reasons described above.

Several years ago we demonstrated that β is a dimer in solution by itself (14). Recently, however, we have gained evidence that β may function within each holoenzyme half as a monomer, or at least may enter holoenzyme complexes as a monomer. We have synthesized a series of fully active β derivatives with fluorophores attached to a uniquely reactive sulfhydryl group. These have been made in anticipation of fluorescence energy transfer studies in which we intend to map out the various intersubunit distances and subunit-primer distances within the holoenzyme complex. Since β typically exists as a dimer in solution, the attached fluorophores on adjacent subunits interact. This permitted the demonstration that the β dimer dissociated at about 10 nM. However, the presence of assay concentrations of Mg^{2+} shifted the dissociation constant to 100 nM, thus favoring monomers at experimentally and physiologically significant β concentrations. Gel filtration experiments confirmed this result with unsubstituted β. Our fluorescent β derivatives will also permit sorting out whether β dimerizes in the holoenzyme and other interactions with holoenzyme subunits.

GENETIC REGULATION OF THE DNA POLYMERASE III HOLOENZYME

Recently, we initiated a parallel program in which we plan to examine the genetic regulation of the elongation apparatus. We began by determining the sequence of two holoenzyme structural genes, *dnaX* (9) and *dnaE*. The sequence of *dnaE* indicated that it encodes a protein of 130,000 daltons, consistent with the denatured molecular weight of the isolated α subunit (40). The location of the *dnaE* coding sequences relative to adjacent open reading frames suggested that it may reside in an operon, translationally coupled with at least two upstream genes (Fig. 6). The termination codons of *lpxB* and orf$_{23}$ overlap with orf$_{23}$ and *dnaE*, respectively.

lpxB has been demonstrated by Raetz and co-workers (33) to be the structural gene for lipid A disaccharide synthase. The enzyme catalyzes the synthesis of the lipid A core for the *E. coli* lipopolysaccharide, an essential component of the outer membrane.

Since the products of orf$_{23}$ and orf$_{37}$ can be visualized on gels, we know these are real genes. Thus, *dnaE* may reside on a complex operon with *lpxB*, apparently related only in that both genes encode enzymes required for cell division. The operon that contains the genes coding for the σ subunit of RNA polymerase, the DNA primase, and the S21 ribosomal protein has been referred to as the macromolecular synthesis operon because of their role in the synthesis of macromolecules (4, 23). Since both lipid A disaccharide synthase and α are involved in macromolecular synthesis, we have termed this operon "macromolecular synthesis II." This may be a point for global control of replication and cell division.

Even though we have evidence that *dnaE* resides on an operon with upstream genes, we have identified a promoter directly in front of *dnaE* that can direct its transcription. It is common to find internal promoters in complex operons. This permits discoordinate regulation under specific conditions. Deletion of all sequences upstream of this putative promoter permits expression of *dnaE*. Removal of 8 bases from the −35 region abolishes *dnaE* expression. Potential *dnaA* binding sites are found upstream of this internal promoter. Tandem sequences with similarity to *dnaA* binding sites are also

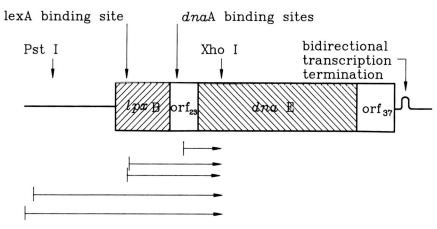

FIGURE 6. Preliminary structure of the macromolecular synthesis II operon.

observed upstream of a *dnaX* promoter (9). The *dnaN* gene, which encodes the β holoenzyme subunit, resides in an operon with *dnaA* (35). Thus, *dnaA* protein levels may play a role in coordinating the level of holoenzyme subunit synthesis.

Transcript maps indicate that some transcription originates from the region of the internal promoter during balanced growth, but most *dnaE* transcripts extend upstream to a site preceding *lpxB*. This provides strong support for our suggestion, initially based only on sequence data, that *dnaE* resides in a complex operon.

Yet another internal initiation site is observed within the *lpxB* gene. We do not know yet whether this represents transcriptional initiation or an RNA processing site, but we see a sequence with similarity to the consensus *lexA* binding site. This reveals a possible mechanism for SOS induction of the *dnaE* gene product. It might be advantageous during periods of stress (during SOS induction) when cell division is inhibited to synthesize *dnaE* without the coordinate expression of *lpxB*.

We plan to further test the regulatory coordination within this potential operon, particularly addressing the regulatory signals that control the biosynthesis of these otherwise very different enzymes, both of which are essential for cell growth and division.

ACKNOWLEDGMENTS. This work was supported by grant MV348 from the American Cancer Society and Public Health Service grants GM36255 and GM35695 from the National Institutes of Health. J.P.F. and M.A.G. were supported by postdoctoral fellowships from the Deutsche Forschungsgemeinschaft and the American Cancer Society, respectively.

LITERATURE CITED

1. Bouché J.-P., K. Zechel, and A. Kornberg. 1975. *dna*G gene product, a rifampicin-resistant RNA polymerase initiates the conversion of a single-stranded coliphage DNA to its duplex replicative form. *J. Biol. Chem.* **250**:5995–6001.
2. Burgers, P. M. J., and A. Kornberg. 1982. ATP activation of DNA polymerase III holoenzyme of *Escherichia coli*: ATP-dependent formation of an initiation complex with a primed template. *J. Biol. Chem.* **257**:11468–11473.
3. Burgers, P. M. J., A. Kornberg, and Y. Sakakibara. 1981. The *dna*N gene codes for the β subunit of DNA polymerase III holoenzyme of Escherichia coli. *Proc. Natl. Acad. Sci. USA* **78**:5391–5395.
4. Burton, Z. F., C. A. Gross, K. K. Watanabe, and R. R. Burgess. 1983. The operon that encodes the sigma subunit of RNA polymerase also encodes ribosome protein S21 and DNA primase in *E. coli* K12. *Cell* **32**:335–349.
5. Cantor, C. R., and P. R. Shimmel. 1982. *Biophysical Chemistry*, Part II. *Techniques for the Study of Biological Structure and Function.* W. H. Freeman & Co., San Francisco.
6. Cozzarelli, N. R. 1980. DNA topoisomerases. *Cell* **22**:327–328.
7. Fay, P. J., K. O. Johanson, and C. S. McHenry. 1981. Size classes of products synthesized processively by DNA polymerase III and DNA polymerase III holoenzyme of Escherichia coli. *J. Biol. Chem.* **256**:976–983.
8. Fay, P. J., K. O. Johanson, C. S. McHenry, and R. Bambara. 1982. Size classes of products synthesized processively by two subassemblies of Escherichia coli DNA polymerase III holoenzyme. *J. Biol. Chem.* **257**:5692–5699.
9. Flower, A. M., and C. S. McHenry. 1986. The adjacent *dna*Z and *dna*X genes of Escherichia coli are contained within one continuous open reading frame. *Nucleic Acids Res.* **14**:8091–8101.
10. Godson, G. N. 1974. Isolation of four new φX-like phages and comparison with φX174. Evolution of φX174. *Virology* **58**:272–289.
11. Hawker, J. R., and C. S. McHenry. 1987. Monoclonal antibodies specific for the τ subunit of the DNA polymerase III holoenzyme of Escherichia coli. *J. Biol. Chem.* **262**:12722–12727.
12. Hübscher, U., and A. Kornberg. 1980. The *dna*Z protein: the γ subunit of DNA polymerase III holoenzyme of Escherichia coli. *J. Biol. Chem.* **255**:11698–11703.
13. Hurwitz, J., and S. Wickner. 1974. Involvement of 2 protein factors and ATP in in-vitro DNA synthesis catalyzed by DNA polymerase III of Escherichia coli. *Proc. Natl. Acad. Sci. USA* **71**:6–10.
14. Johanson, K. O., and C. S. McHenry. 1980. Purification and characterization of the β subunit of the DNA polymerase III holoenzyme of Escherichia coli. *J. Biol. Chem.* **255**:10984–10990.

15. Johanson, K. O., and C. S. McHenry. 1982. The β subunit of the DNA polymerase III holoenzyme becomes inaccessible to antibody after formation of an initiation complex with primed DNA. *J. Biol. Chem.* 257:12310–12315.
16. Johanson, K. O., and C. S. McHenry. 1984. Adenosine 5'-O-3-thio-triphosphate can support the formation of an initiation complex between the DNA polymerase III holoenzyme and primed DNA. *J. Biol. Chem.* 259:4589–4595.
17. Kodaira, M., S. B. Biswas, and A. Kornberg. 1983. The *dna*X gene encodes the DNA polymerase III holoenzyme τ subunit, precursor of the γ subunit, the *dna*Z gene product. *Mol. Gen. Genet.* 192:80–86.
18. Kornberg, A. 1982. DNA Replication, 1982 Supplement. W. H. Freeman & Co., San Francisco.
19. Kornberg, T., and M. L. Gefter. 1972. Deoxyribonucleic acid synthesis in cell-free extracts. *J. Biol. Chem.* 247:5369–5375.
20. Lee, S.-H., and J. R. Walker. 1987. Escherichia coli *dna*X product, the τ subunit of DNA polymerase III, is a multifunctional protein with single-stranded DNA-dependent ATPase activity. *Proc. Natl. Acad. Sci. USA* 84:2713–2717.
21. Livingston, D. M., D. C. Hinkle, and C. C. Richardson. DNA polymerase III of Escherichia coli: purification and properties. *J. Biol. Chem.* 250:461–469.
22. Lui, L. F., and J. C. Wang. 1979. Interaction between DNA and Escherichia coli DtsNA topoisomerase formation of complexes between the protein and superhelical and nonsuperhelical duplex DNAs. *J. Biol. Chem.* 254:11082–11088.
23. Lupski, J. R., A. A. Ruiz, and G. N. Godson. 1983. Regulation of the *rps*U-*dna*G-*rpo*D macromolecular synthesis operon and the initiation of DNA replication in Escherichia coli K-12. *Mol. Gen. Genet.* 189:48–57.
24. Markwell, M. A. K. 1982. A new solid-state reagent to iodinate proteins. *Anal. Biochem.* 125:427–432.
25. Maxwell, A., R. Craigie, and K. Mizuuchi. 1987. Protein of bacteriophage Mu is an ATPase that preferentially stimulates intermolecular DNA strand transfer. *Proc. Natl. Acad. Sci. USA* 84:699–703.
26. McHenry, C. S. 1982. Purification and characterization of DNA polymerase III': identification of γ as a subunit of the DNA polymerase III holoenzyme. *J. Biol. Chem.* 257:2657–2663.
27. McHenry, C. S. 1985. DNA polymerase III holoenzyme of Escherichia coli: components and function of a true replicative complex. *Mol. Cell. Biochem.* 66:71–85.
28. McHenry, C. S., and W. Crow. 1979. DNA polymerase III of Escherichia coli: purification and identification of subunits. *J. Biol. Chem.* 254:1748–1753.
29. McHenry, C. S., and A. Kornberg. 1977. DNA polymerase III holoenzyme of Escherichia coli: purification and resolution into subunits. *J. Biol. Chem.* 252:6478–6484.
30. Minden, J. S., and K. J. Marians. 1986. Escherichia coli topoisomerase I can segretate replication PBR-322 daughter DNA molecules in-vitro. *J. Biol. Chem.* 261:11906–11917.
31. Mizuuchi, K., M. H. O'Dea, and M. Gellert. 1978. DNA gyrase: subunit structure and ATPase activity of the purified enzyme. *Proc. Natl. Acad. Sci. USA* 75:5960–5963.
32. Mullin, D. A., C. L. Woldringh, J. M. Henson, and J. Walker. 1983. Cloning of the Escherichia coli *dna*ZX region and identification of its products. *Mol. Gen. Genet.* 192:73–79.
33. Nishijima, M., C. E. Bulawa, and C. R. H. Raetz. 1981. Two interacting mutations causing temperature-sensitive phosphatidylglycerol synthesis in *Escherichia coli* membranes. *J. Bacteriol.* 145:113–121.
34. Oberfelder, R., and C. S. McHenry. 1987. Characterization of 2',3'-trinitrophenyl-ATP as an inhibitor of ATP-dependent initiation complex formation between the DNA polymerase III holoenzyme and primed DNA. *J. Biol. Chem.* 262:4190–4194.
35. Sakakibara, Y., H. Tsukano, and T. Sako. 1981. Organization and transcription of the *dna*A and *dna*N genes of *Escherichia coli*. *Gene* 13:47–55.
36. Scheuermann, R. H., and H. Echols. 1984. A separate editing exonuclease for DNA replication: the ε subunit of Escherichia coli DNA polymerase III holoenzyme. *Proc. Natl. Acad. Sci. USA* 81:7747–7751.
37. Sinha, N. K., C. F. Morris, and B. M. Alberts. 1980. Efficient in vitro replication of double-stranded DNA templates by a purified T4 bacteriophage replication system. *J. Biol. Chem.* 255:4290–4303.
38. Spanos, A., G. T. Yarranton, U. Hübscher, G. R. Banks, and S. Sedgwick. 1981. Detection of the catalytic activities of DNA polymerases and their associated exonucleases following SDS-polyacrylamide gel electrophoresis. *Nucleic Acids Res.* 9:1825–1839.
39. Surette, M. G., S. J. Buch, and G. Chaconas. 1987. Transpososomes: stable protein-DNA complexes involved in the in vitro transposition of bacteriophage Mu DNA. *Cell* 49:253–262.
40. Tomasiewicz, H. G., and C. S. McHenry. 1987. Sequence analysis of the *Escherichia coli dna*E gene. *J. Bacteriol.* 169:5735–5744.
41. Wickner, S., and J. Hurwitz. 1976. Involvement of Escherichia coli *dna*Z gene product in DNA

elongation in vitro. *Proc. Natl. Acad. Sci. USA* 73:1053–1057.
42. **Wickner, W., and A. Kornberg.** 1974. A holoenzyme form of DNA polymerase: isolation and properties. *J. Biol. Chem.* 249:6244–6249.
43. **Wu, Y. H., M. A. Franden, J. R. Hawker, and C. S. McHenry.** 1984. Monoclonal antibodies specific for the α subunit of the Escherichia coli DNA polymerase III holoenzyme. *J. Biol. Chem.* 259:12117–12122.
44. **Yin, K.-C., A. Blinkowa, and J. R. Walker.** 1986. Nucleotide sequence of the Escherichia coli replication gene *dna*X. *Nucleic Acids Res.* 14:6541–6549.

Chapter 3

τ and γ Subunits of DNA Polymerase III Holoenzyme
Their Relationship to Each Other and the *dnaX* Gene

Suk-Hee Lee, Michael P. Spector, and James R. Walker

DNA polymerase III holoenzyme of *Escherichia coli* is composed of a core, containing the α, ε, and θ subunits, and at least four auxiliary subunits (12, 13). The α subunit (the *dnaE* gene product) is the catalytic subunit possessing 5'→3' polymerizing activity (21). The ε subunit (*dnaQ* gene product) is a 3'→5' proofreading exonuclease (3, 19). The θ subunit function is unknown. The auxiliary proteins function to increase both the specificity of DNA polymerase III core for natural templates and the processivity of core enzyme (2, 8, 13, 16, 25), but their specific functions are still unclear.

The 71.1-kilodalton (kDa) τ and 56.5-kDa γ auxiliary subunits are products of the *dnaX* gene, which consists of one 643-codon reading frame (4, 9, 15, 26). (This gene originally was reported to be two separate, closely linked genes [6]; the τ product was referred to as DnaX protein and γ was called the DnaZ protein [9, 15].) τ (as a τ'-LacZ' fusion protein) is a single-strand DNA-stimulated ATPase (11). γ is required for DNA polymerase III-catalyzed replication in vitro (7, 23, 24). The relationship between the two proteins and their gene will be described.

THE *dnaX* GENE ENCODES BOTH THE τ AND γ SUBUNITS OF DNA POLYMERASE III HOLOENZYME

To help in the isolation of the *dnaX* products in quantities sufficient for characterization, the *dnaX* gene lacking 20 C-terminal codons was fused in phase to the *lacZ* gene lacking eight N-terminal codons. This fusion was then cloned under control of the *tac* promoter and *lac* Shine-Dalgarno sequence. This plasmid, pSL641, was transformed into a strain carrying *lacI*^{q} on a compatible plasmid. Induction by isopropyl-β-D-thiogalactopyranoside led to the overproduction of a 185-kDa fusion protein, designated τ'-LacZ', and a 56.5-kDa protein, which was found to be the γ protein (11). Antibody to a region which should have

---

Suk-Hee Lee • Memorial Sloan-Kettering Cancer Center, New York, New York 10021. *Michael P. Spector and James R. Walker* • Department of Microbiology, University of Texas at Austin, Austin, Texas 78712.

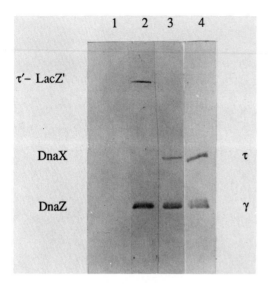

FIGURE 1. Antibody to a synthetic peptide present in both the DnaX and DnaZ proteins reacts with the τ and γ subunits of purified DNA polymerase III by Western blot analysis. Lanes: 1, extract of a control strain; 2, extract of a strain which overproduces τ'-LacZ' and γ; 3, extract of a strain which overproduces τ and γ; 4, purified DNA polymerase III (from C. McHenry). (Reprinted from reference 10 with permission from IRL Press.)

been contained in both τ and γ (amino acids 420 to 440 of the deduced sequence [26]) reacted with both the overproduced τ'-LacZ' fusion protein and γ protein as well as with τ and γ subunits of holoenzyme (11) (Fig. 1). Thus, this amino acid sequence not only is present in both τ and γ but also is found in each in a region capable of being recognized by the antibody, suggesting that τ and γ have regions of similar secondary structure.

PROCESSING OF THE τ'-LacZ FUSION PROTEIN IN VITRO

The 185-kDa fusion protein was isolated by affinity chromatography using its β-galactosidase activity (11). The τ'-LacZ' protein prepared by this method was stable during isolation and subsequent storage at −80°C. However, attempts to purify the fusion protein from soluble protein fractions using other methods, such as gel filtration, resulted in disappearance of the 185-kDa protein and concomitant appearance of 135- and 56.5-kDa proteins. More than 90% of the 185-kDa fusion protein was converted during gel filtration of an ammonium sulfate-precipitated extract on Sephacryl S200, presumably via proteolysis by an endogenous protease, to form a new 135-kDa protein which possessed β-galactosidase activity and increased amounts of the 56.5-kDa γ protein (Fig. 2). The β-galactosidase activity and size of the 135-kDa cleavage product indicated that it represented the C-terminal portion of τ' linked to β-galactosidase (11).

PURIFIED γ PROTEIN MIGRATED IDENTICALLY TO γ PROTEIN SYNTHESIZED IN VIVO DURING O'FARRELL TWO-DIMENSIONAL GEL ELECTROPHORESIS

The 135-kDa cleavage product and 56.5-kDa γ protein (probably containing both γ protein synthesized in vivo and γ formed in vitro by cleavage of τ'-LacZ') were further purified. For this, Sephacryl S200 column fractions containing both γ protein and 135-kDa protein were pooled and subjected to ion-exchange chromatography on DEAE-Sephacel (Fig. 2). The 135-kDa protein was further purified by affinity chromatography using its β-galactosidase activity, and the γ protein was further purified by phosphocellulose chromatography (Fig. 2) (10).

To determine whether the purified preparation of γ had characteristics similar to those of γ synthesized in vivo, cells with or without a plasmid which directs the overproduction of native τ and γ (11) were pulsed with [^{35}S]methionine. Extracts were then subjected to two-dimensional electrophoresis (2-DGE) (isoelectric focusing followed by sodium dodecyl sulfate-polyacrylamide gel electrophoresis [SDS-PAGE]) (17). Labeled γ protein was visible following autoradiogra-

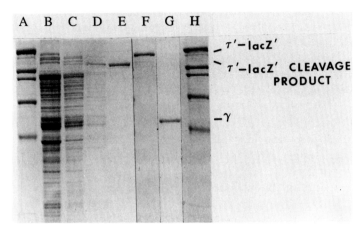

FIGURE 2. Purification of the cleavage products of the τ'-LacZ' protein. Proteins were separated by electrophoresis and stained by Coomassie blue. Lanes: A and H, molecular weight markers; B, extract containing overproduced τ'-LacZ' protein and γ; C, pooled fractions (from a gel filtration column) containing 135-kDa cleavage product and γ; D, 135-kDa cleavage product purified from C by DEAE-Sephacel chromatography; E, 135-kDa cleavage product purified from D by affinity chromatography; F, τ'-LacZ' purified from B by affinity chromatography; G, γ purified from C by phosphocellulose chromatography. (Reprinted from reference 11 with permission from the authors.)

phy as major and minor species in extracts from overproducing strains but was not detectable in extracts from nonoverproducing strains. Purified γ, mixed with nonoverproducing strain extract and subjected to 2-DGE and stained with Coomassie blue, migrated to the same position as the major γ species synthesized in vivo in overproducing strains. Thus, it is likely that the purified γ was identical to the major form of the in vivo protein and that cleavage of the τ'-LacZ' fusion protein in vitro hydrolyzes the same bond as in vivo cleavage (10).

RELATIONSHIP OF THE τ'-LacZ', 135-kDa AND γ PROTEINS TO EACH OTHER AND TO THE *dnaX* GENE

The N-terminal sequences of τ'-LacZ' and γ were identical to each other and to the deduced amino acid sequence (Fig. 3) (10, 26).

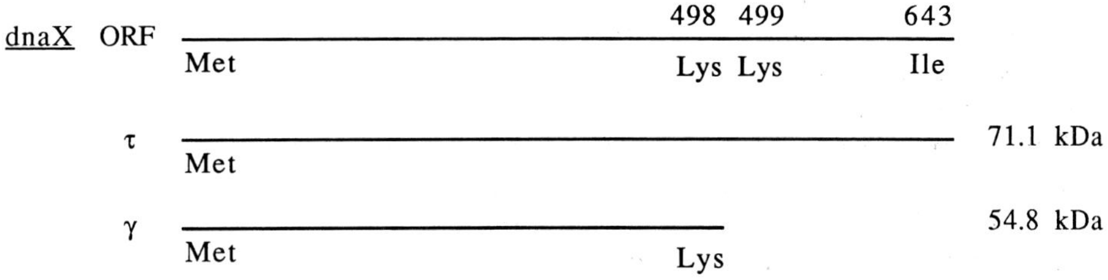

FIGURE 3. Diagram of the *dnaX* open reading frame, the τ product, and the 498-amino-acid γ cleavage product.

Translation of both initiates with the methionine codon at nucleotides 142 to 144 (26) and τ and γ are identical at their N-terminal ends.

The N-terminal sequence of the 135-kDa cleavage protein, representing a C-terminal portion of τ, matched the predicted amino acid sequence 499 through 508 (Fig. 3) (10, 26). This indicates that the in vitro cleavage of τ'-LacZ' occurred at a dibasic dipeptide (between lysine 498 and lysine 499) to generate γ protein. Since γ formed in vitro appears identical to the major form of γ synthesized in vivo, as seen on 2-DGE, it is likely that in vivo τ is cleaved between these two lysine residues, generating a 498-amino-acid γ protein.

PRECURSOR-PRODUCT RELATIONSHIP OF τ AND γ

To analyze the precursor-product relationship of τ and γ in vivo, a culture of *dnaX*-plasmid-containing cells was grown at 22°C, induced to overproduce τ and γ, and pulse-labeled with [^{35}S]methionine for 5, 10, 20, 40, or 80 s. The labeled τ and γ proteins were then precipitated with antibody and protein A-Sepharose. Precipitates were dissolved and separated by SDS-PAGE. Autoradiography indicated that intact γ is formed before intact τ (10). The γ band was detectable by 10 s of labeling, and only after an additional 10 s did both τ and γ become detectable. The earlier detection of γ might have resulted from the positions of the methionines relative to the C-terminal ends of τ and γ. One methionine is found only 13 residues upstream of the C-terminal end of γ, but the closest methionine to the τ C-terminal end is 65 residues away (26). Ryals et al. (18) estimated the rate of translation at 22°C to be about 4 to 6 amino acids polymerized per ribosome per second, and therefore a 10-s pulse would label only 40 to 60 amino acids. Thus, it is possible that the earlier detection of γ reflects the incorporation of label near the C-terminal portion of antibody-precipitable polypeptides. In any case, it is clear that the formation of γ does not require intact τ as a precursor (10). Perhaps cleavage occurs during translation of the τ protein; the protease could be ribosome associated or the cleavage could be autoproteolytic.

THE γ PROTEIN PROBABLY EXISTS IN TWO FORMS

As discussed above, the overproduction of γ yielded two forms, a major and a minor species. Radioimmune precipitation of τ and γ from extracts of an overproducing strain (after pulse-chase experiments and after steady-state labeling with [^{35}S]methionine), electrophoresis, and autoradiography revealed γ as a pair of bands, of which the minor one migrated more slowly. Two forms of γ are likely to occur in cells not overproducing τ or γ as well, since DNA polymerase III holoenzyme purified from wild-type cells contained two γ forms, as seen in Western blots (Fig. 1) (11). The significance of these two forms is still unknown. However, their existence suggests many possibilities, including: (i) γ might be modified prior to or following its formation from τ; (ii) this modification might play a role in regulating its formation from τ; and (iii) the two forms might have different functions in the holoenzyme complex.

THE τ SUBUNIT IS A SINGLE-STRANDED DNA-STIMULATED ATPASE

Yin et al. (26) have shown through sequence analysis that τ and γ possess a region between residues 39 and 59 that closely resembles the consensus ATP binding site A (20). In earlier studies, Biswas and Kornberg (1) reported that τ and γ subunits bind ATP and dATP. Lee and Walker (11) found that purified τ'-LacZ fusion protein possessed ATPase and dATPase activity which was stimulated over 10-fold by the presence of single-stranded DNA. The possible function of this ATPase activity will be discussed below.

POSSIBLE FUNCTION(S) OF τ AND γ IN VIVO

The τ protein is multifunctional, having at least five functions or activities. First, τ functions as a substrate for proteolysis to form γ protein (5, 11). This proteolysis occurs rapidly, perhaps even before synthesis of τ is complete, which suggests involvement of a ribosome-associated protease or autoproteolysis. Second, τ is an ATP- and dATP-binding protein (1) having ATPase and dATPase activity (11). Third, τ binds single-stranded DNA, since the ATPase and dATPase activity is stimulated severalfold in the presence of single-stranded DNA (11). However, the binding to single-stranded DNA has not been directly shown. Fourth, McHenry (14) found that τ is a subunit in the holoenzyme subassembly DNA polymerase III', composed of α, ε, θ, and τ. Furthermore, he found that τ causes DNA polymerase core to dimerize, as judged by molecular weight determination using a glycerol gradient (14). Fifth, τ association with core enzyme to form DNA polymerase III' increases the processivity of core enzyme, allowing the complex to replicate long stretches of single-stranded DNA in the presence of spermidine (14). Even though τ has been shown to be a component of DNA polymerase III holoenzyme and to increase the processivity of polymerase III' relative to core enzyme (5, 14), its exact function in holoenzyme is still unclear. McHenry (14) and M. E. O'Donnell (personal communication) have found that the presence of τ has little effect on in vitro replication of phage G4 single-stranded template when core, β, and γ-δ complex plus ATP are present. The importance of τ as a subunit of holoenzyme should become clear when more is known about leading- and lagging-strand DNA synthesis and the proposed asymmetry of the DNA polymerase III holoenzyme dimer (12). The function of the single-stranded DNA-stimulated ATPase (and dATPase) activity of τ is interesting. Possible functions for this activity include (i) stimulation of DNA polymerase III core or holoenzyme binding to single-stranded DNA; (ii) translocation of the holoenzyme along the single-stranded DNA template using the hydrolysis of ATP or dATP; and (iii) function as a helicase or in the stimulation of a helicase activity during replication.

The γ protein function is also unclear. The γ subunit also binds ATP or dATP (1), although no ATPase activity has been reported. However, γ has been found to be essential for in vitro replication of single-stranded DNA templates (2, 7, 8, 16, 23, 25). It complexes with the δ subunit to form a γ-δ complex which is required for formation of stable preinitiation complexes, in the presence of β subunit and ATP, on primed single-stranded DNA templates (15a). This supports the earlier findings of Wickner (22) that *dnaZ* protein (γ), elongation factor III (δ), and elongation factor I (β) in the presence of ATP formed a tight complex with oligo(dT)-primed poly(dA). Formation of the preinitiation complex is the only step requiring ATP, and since the γ and δ subunits of this complex have been shown to bind ATP (1), it is likely that one or both hydrolyze ATP (or dATP) in the formation of preinitiation complex. It is unclear whether β and γ-δ form a complex on primed single-strand template or γ-δ functions only to shuttle or transfer β to the primed template (15a). However, the finding of Hawker and McHenry (5) that τ and γ-δ are found within the same enzyme assemblies (antibody directed specifically against τ precipitated both γ and δ subunits as well as α subunit) suggests that the former is more likely to occur.

In summary, both τ and γ can be isolated as components of DNA polymerase III holoenzyme. The finding that both are the products of a single gene, *dnaX*, and that γ is formed by proteolytic cleavage of τ raises some important and interesting questions, such as that of the mechanism and regulation of τ cleavage both in vivo and in vitro.

ACKNOWLEDGMENTS. This work was supported by American Cancer Society grant NP169. Parts of the work were supported by Public Health Service grant GM34471 from the National Institutes of Health and Welch Foundation grant F949.

LITERATURE CITED

1. **Biswas, S. B., and A. Kornberg.** 1984. Nucleoside triphosphate binding to DNA polymerase III holoenzyme of *Escherichia coli*. A. Direct photoaffinity labeling study. *J. Biol. Chem.* **259**:7990–7993.
2. **Burgers, P. M. J., and A. Kornberg.** 1982. ATP activation of DNA polymerase III holoenzyme of *Escherichia coli*. Initiation complex stoichiometry and reactivity. *J. Biol. Chem.* **257**:11474–11478.
3. **Echols, H., C. Lu, and P. M. J. Burgers.** 1983. Mutator strains of *Escherichia coli* with defective exonucleolytic editing by DNA polymerase III holoenzyme. *Proc. Natl. Acad. Sci. USA* **80**:2189–2192.
4. **Fowler, A. M., and C. S. McHenry.** 1986. The adjacent *dnaZ* and *dnaX* genes of *E. coli* are contained within one continuous open reading frame. *Nucleic Acids Res.* **14**:8091–8101.
5. **Hawker, J. R., Jr., and C. S. McHenry.** 1987. Monoclonal antibodies specific for the τ subunit of the DNA polymerase III holoenzyme of *Escherichia coli*. *J. Biol. Chem.* **262**:12722–12727.
6. **Henson, J. M., H. Chu, C. A. Irwin, and J. R. Walker.** 1979. Isolation and characterization of *dnaX* and *dnaY* temperature sensitive mutants of *Escherichia coli*. *Genetics* **92**:1041–1060.
7. **Hübscher, U., and A. Kornberg.** 1980. The *dnaZ* protein is the γ subunit of DNA polymerase III holoenzyme of *Escherichia coli*. *J. Biol. Chem.* **255**:11698–11703.
8. **Hurwitz, J., and S. Wickner.** 1974. Involvement of 2 protein factors and ATP in in vitro DNA synthesis catalyzed by DNA polymerase III of *Escherichia coli*. *Proc. Natl. Acad. Sci. USA* **71**:6–10.
9. **Kodaira, M., S. B. Biswas, and A. Kornberg.** 1983. The *dnaX* gene encodes the DNA polymerase III holoenzyme τ subunit, precursor of the γ subunit, the *dnaZ* gene product. *Mol. Gen. Genet.* **192**:80–86.
10. **Lee, S.-H., P. Kanda, R. C. Kenney, and J. R. Walker.** 1987. Relation of the *Escherichia coli dnaX* gene to its two products—the τ and γ subunits of DNA polymerase III holoenzyme. *Nucleic Acids Res.* **15**:7663–7675.
11. **Lee, S.-H., and J. R. Walker.** 1987. *Escherichia coli* DnaX product, the τ subunit of DNA polymerase III, is a multifunctional protein with single-stranded DNA-dependent ATPase activity. *Proc. Natl. Acad. Sci. USA* **84**:2713–2717.
12. **McHenry, C.** 1985. DNA polymerase III holoenzyme of *Escherichia coli*: components and function of a true replicative complex. *Mol. Cell. Biochem.* **66**:71–85.
13. **McHenry, C., and A. Kornberg.** 1977. DNA polymerase III holoenzyme of *Escherichia coli*: purification and resolution into subunits. *J. Biol. Chem.* **252**:6478–6484.
14. **McHenry, C. S.** 1982. Purification and characterization of DNA polymerase III'. Identification of τ as a subunit of the DNA polymerase III holoenzyme. *J. Biol. Chem.* **257**:2657–2663.
15. **Mullin, D. A., C. L. Woldringh, J. M. Henson, and J. R. Walker.** 1983. Cloning of the *Escherichia coli dnaZX* region and identification of its products. *Mol. Gen. Genet.* **192**:73–79.
15a. **O'Donnell, M. E.** 1987. Accessory proteins bind a primed template and mediate rapid cycling of DNA polymerase III holoenzyme from *Escherichia coli*. *J. Biol. Chem.* **262**:16558–16565.
16. **O'Donnell, M. E., and A. Kornberg.** 1985. Complete replication of templates by *Escherichia coli* DNA polymerase III holoenzyme. *J. Biol. Chem.* **260**:12884–12889.
17. **O'Farrell, P. Z.** 1975. High resolution two-dimensional gel electrophoresis of proteins. *J. Biol. Chem.* **250**:4007–4021.
18. **Ryals, J., R. Little, and H. Bremer.** 1982. Temperature dependence of RNA synthesis parameters in *Escherichia coli*. *J. Bacteriol.* **151**:879–887.
19. **Scheurmann, R., S. Tam, P. M. J. Burgers, C. Lu, and E. Echols.** 1983. Identification of the ε subunit of *Escherichia coli* DNA polymerase III holoenzyme as the *dnaQ* gene product. A fidelity subunit for DNA replication. *Proc. Natl. Acad. Sci. USA* **80**:7085–7089.
20. **Walker, J. E., M. Saraste, M. J. Runswick, and N. J. Gay.** 1982. Distantly related sequence in the α- and β-subunits of ATP synthetase, myosin, kinases and other ATP-requiring enzymes and a common nucleotide binding fold. *EMBO J.* **1**:945–951.
21. **Welch, M. M., and C. S. McHenry.** 1982. Cloning and identification of the product of the *dnaE* gene of *Escherichia coli*. *J. Bacteriol.* **152**:351–356.

22. Wickner, S. 1976. Mechanism of DNA elongation catalyzed by *Escherichia coli* DNA polymerase III, *dnaZ* protein, and DNA elongation factors I and III. *Proc. Natl. Acad. Sci. USA* 73:3511–3515.
23. Wickner, S., and J. Hurwitz. 1976. Involvement of *Escherichia coli dnaZ* gene product in DNA elongation in vitro. *Proc. Natl. Acad. Sci. USA* 73:1053–1057.
24. Wickner, S. H., R. B. Wickner, and C. R. H. Raetz. 1976. Overproduction of *dna* gene products by *Escherichia coli* strains carrying hybrid ColE1 plasmids. *Biochem. Biophys. Res. Commun.* 70:389–396.
25. Wickner, W., and A. Kornberg. 1974. A holoenzyme form of DNA polymerase III: isolation and properties. *J. Biol. Chem.* 249:6244–6249.
26. Yin, K.-C., A. Blinkowa, and J. R. Walker. 1986. Nucleotide sequence of the *Escherichia coli* replication gene *dnaZX*. *Nucleic Acids Res.* 14:6541–6549.

Chapter 4

Collection of Informative Bacteriophage T4 DNA Polymerase Mutants

Linda J. Reha-Krantz

Here I describe our progress in studying bacteriophage T4 DNA polymerase structure-function relationships. The focus of our work has been the isolation of "informative" T4 DNA polymerase mutants (23, 25; L. J. Reha-Krantz, *J. Mol. Biol.*, in press). Because the locations of the active sites for the polymerization and 3'→5' exonuclease activities in the T4 DNA polymerase are unknown, mutations were not targeted to specific sites in the DNA polymerase gene; rather, the entire T4 genome was mutagenized and then mutants were selected that increased the frequency of base substitution replication errors (mutator phenotype). To date, 16 unique mutator strains have been isolated and all have mutations in the DNA polymerase gene.

DNA polymerase mutants isolated on the basis of the mutator phenotype could have alterations in polymerization or 3'→5' exonuclease activities, because both DNA polymerase functions are important in determining the fidelity of DNA replication (26). Although the DNA polymerase is not thought to actively select correct deoxyribonucleotides for insertion at the primer terminus, incorporation of correct deoxyribonucleotides is dependent on the "residence time" of the incoming deoxyribonucleotide such that phosphodiester bond formation is catalyzed only if the incoming deoxyribonucleotide is stabilized in the active site, presumably by hydrogen bonding between the template base and incoming base and other interactions with DNA and the DNA polymerase (1, 26). Changes in the DNA polymerase conformation that alter the protein environment at the primer terminus or at the deoxynucleoside triphosphate (dNTP) site could increase deoxyribonucleotide misinsertion by acting to increase the residence time of incorrect deoxyribonucleotides. An altered protein conformation could also result in an increase in water at the active site. It has been proposed that water may interact with the bases to diminish intrinsic hydrogen bonding between the template and incoming deoxyribonucleotide, and thus reduce free-energy differences between right and wrong base pairs which would affect both the insertion and 3'→5' exonuclease proofreading steps (20). Elegant studies have demonstrated the importance of the T4 DNA polymerase 3'→5' exonuclease activity in the fidelity of DNA replication; some T4 muta-

Linda J. Reha-Krantz • Department of Genetics, University of Alberta, Edmonton, Alberta T6G 2E9, Canada.

tor DNA polymerases have a reduction in the relative amount of 3'→5' exonuclease activity compared to polymerase activity, and some mutant DNA polymerases that replicate DNA more accurately than the wild-type DNA polymerase (antimutator phenotype) have increased 3'→5' activity (17). In addition to changes in the polymerase and/or 3'→5' exonuclease activities, the DNA polymerase mutator mutations could provide information on other DNA polymerase activities that function in the accurate replication of DNA that could not be predicted from our current information on DNA polymerase function. Thus, the mutator mutants could provide new insight into how a DNA polymerase works.

Before discussing the DNA polymerase mutator mutants further, it is necessary to summarize recent observations about the relatedness, at the protein sequence level, of DNA polymerases from several diverse organisms. When these studies were initiated, we assumed that the bacteriophage T4 DNA polymerase would be a good model for studying procaryotic DNA polymerases, but recently we (E. K. Spicer, J. Rush, C. Fung, L. J. Reha-Krantz, J. D. Karam, and W. H. Konigsberg, *J. Biol. Chem.*, in press; Reha-Krantz, in press) and others (J. Hall, *Trends Genet.*, in press) have observed that the T4 DNA polymerase shares several regions of colinear, similar protein sequence with the DNA polymerases of herpes, adeno, and pox viruses and with the human α DNA polymerase (T. Wang, personal communication). The degree of protein sequence similarity among the DNA polymerases is striking. For example, the DNA polymerases of T4 and herpes simplex virus type 1 (HSV-1) have seven consecutive amino acids in common in one region and have two other regions with five identical consecutive amino acids. Overall, for four colinear regions of the T4 and HSV-1 DNA polymerases that encompass 137 amino acids, or 15% of the total T4 DNA polymerase amino acids, 36% of the amino acids are identical. In contrast to the similarity between T4 and several other DNA polymerases, there are few shared protein sequences between the T4 DNA polymerase and the *Escherichia coli* DNA polymerase I, the α and ε subunits of the *E. coli* DNA polymerase III holoenzyme, or the bacteriophage T7 DNA polymerase; however, there is similarity between the *E. coli* DNA polymerase I and the T7 DNA polymerase (19) and between *E. coli* DNA polymerase I and the ε subunit of the DNA polymerase III holoenzyme (7). Thus, surprisingly, the bacteriophage T4 DNA polymerase appears to be a better model of eucaryotic and eucaryoticlike DNA polymerases than of procaryotic DNA polymerases.

We (E. Spicer, personal communication) believe that the regions of similar protein sequence shared by the various DNA polymerases are conserved amino acids that are critical for DNA polymerase function. This hypothesis is based on homologous protein sequences observed in the DNA polymerases from several diverse organisms and on the nucleotide sequences that encode the conserved sequences (Spicer et al., in press). It has been demonstrated for DNA sequences from diverse organisms that the rate of nucleotide substitution varies considerably, but the more important the function encoded by the DNA region, the lower the rate of nucleotide substitution (9, 13, 14, 16). We find, in comparing the nucleotide sequences of the conserved regions in T4 and HSV-1, that even though the DNA polymerase genes of T4 and HSV-1 reflect the A + T and G + C richness of these organisms, respectively, there have only been synonymous codon substitutions, usually nucleotide substitutions in the third position of codons. Thus, it appears that the regions of similar protein sequence are evolving under functional constraint.

At least three of the homologous regions are implicated in dNTP and pyrophosphate binding, because mutations in the herpes DNA polymerase that confer resistance

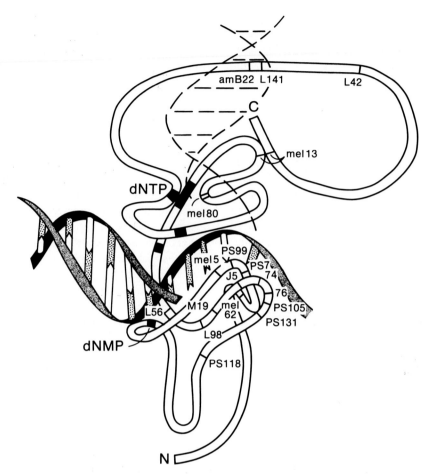

FIGURE 1. Model of T4 DNA polymerase. The T4 DNA polymerase is illustrated as a rope that has been folded randomly into two proposed domains: the N-terminal domain is predicted to contain the 3'→5' exonuclease activity, and the C-terminal domain is thought to contain the dNTP binding site and DNA polymerase activities. The DNA binding site is positioned between the two domains or alternatively, as illustrated by dashed lines, in the C-terminal domain. The scale is 25 amino acids per inch. Sites of amino acid substitutions that confer the mutator phenotype are indicated by lines on the rope, and lines also indicate the positions of three mutations that confer the antimutator phenotype (amB22, L141, L42). Similar amino acid sequences that have been detected in the DNA polymerases from several diverse organisms are illustrated as blackened sections on the rope.

to aphidicolin or nucleotide analogs or pyrophosphate analogs are located in these regions (11, 27). The mutations are in the central and C-terminal regions of the herpes DNA polymerase, and additional mutations implicated in dNTP and PP$_i$ binding have been mapped by recombination to the C-terminal region (5). Thus, the polymerase active site of the herpes DNA polymerase (5) and, by inference, of related DNA polymerases including the T4 DNA polymerase, is predicted to be in the C-terminal portion of the DNA polymerase (Fig. 1).

Even though there is little similarity between the T4 DNA polymerase and *E. coli* DNA polymerase I at the protein sequence level, I have proposed that there is at least one common protein sequence in the N-terminal region (Reha-Krantz, in press). This sequence contains five identical amino acids in a sequence of eight amino acids that, in *E. coli* DNA polymerase I, encodes part of the binding site for divalent metal ions and deoxynucleoside monophosphate (dNMP) that is associated with the $3' \rightarrow 5'$ exonuclease active site. In the Klenow fragment of *E. coli* DNA polymerase I, aspartate and glutamate residues in the proposed conserved sequence have been shown by X-ray crystallographic data to participate in binding Mg^{2+} (8, 18). If the aspartate and glutamate residues of the Klenow fragment are changed to alanines, the mutant Klenow fragment has polymerase activity but not exonuclease activity (8). Because T4 and the related DNA polymerases have this similar sequence of amino acids, including the aspartate and glutamate residues, the $3' \rightarrow 5'$ exonuclease activity in all the DNA polymerases is predicted to reside in the N-terminal region (Fig. 1).

WHAT ADDITIONAL INFORMATION IS PROVIDED BY THE T4 MUTATOR DNA POLYMERASE MUTATIONS?

The T4 DNA polymerase mutations that confer a mutator phenotype are located in three loose clusters in the N-terminal and central regions of the protein (25; Reha-Krantz, in press). It was predicted that amino acid substitutions that produce mutator DNA polymerases may affect the polymerase and/or $3' \rightarrow 5'$ exonuclease activities directly; however, none of the mutator mutations that have been isolated thus far are located in the conserved regions that are predicted to function in either nucleotide incorporation or $3' \rightarrow 5'$ exonuclease activity (Fig. 1). It may be that amino acid substitutions within the conserved sequences that could result in a mutator phenotype are forbidden because a nonfunctioning DNA polymerase would be produced. It is important to note that only mutant DNA polymerases that can still function in DNA replication, at least at a minimum level, can be isolated by the selection procedures used.

Some mutations that confer a mutator phenotype are located near regions of homology, and it can be proposed, as in Fig. 1, that sites far apart in the linear structure are actually near one another in the folded, three-dimensional structure. I emphasize that the structure in Fig. 1 is purely conjectural, but the drawing provides a way to illustrate the positions of the mutations that produce a decrease in the fidelity of DNA replication and the conserved amino acid sequences. The drawing also provides a model for future experiments.

Although the *E. coli* DNA polymerase I and the T4 DNA polymerase have few similarities at the protein sequence level, except for the predicted dNMP binding site, the two DNA polymerases may have similar structures. Thus, I have proposed that the T4 DNA polymerase, like the *E. coli* DNA polymerase I, has two separate protein domains: (i) $3' \rightarrow 5'$ exonuclease activity (dNMP site) is predicted to be located in the N-terminal domain, and (ii) the polymerase activity (dNTP site) is predicted to be located in a separate C-terminal domain (Fig. 1). This is the arrangement of activities that has been found for *E. coli* DNA polymerase I (8). In the model of the T4 DNA polymerase (Fig. 1), the dNMP and dNTP binding sites have been positioned on either side of the primer terminus; however, if the structure of the T4 DNA polymerase is found to resemble the structure of *E. coli* DNA polymerase I, then the DNA binding site would be in the C-terminal domain, perhaps in a position similar to the DNA helix illustrated with dashed lines (Fig. 1). Crystal structure analysis of the Klenow fragment of *E. coli* DNA polymerase I indicates that the C-terminal domain contains a deep cleft that may bind duplex DNA (8, 18). There is no evidence for the location of the DNA binding site in the T4 DNA polymerase yet, but two mutator strains, mel80 and mel13, may provide information. The mel80 strain apparently has two mutations that result in two amino acid substitutions. The mutant amino acids are separated by 10 amino acids in a protein region rich in both basic and aromatic amino acids which could interact with DNA. The mel13 strain is also a double mutant, but the two amino acid substitutions required to produce the mutator phenotype, one in the center of the DNA polymerase and the second near the C terminus, are separated by 425 amino acids (Reha-Krantz, in press). It is not known if the two mel13 mutations are located near one another in the folded protein and the mutator phenotype is due to a disruption of their interaction (Fig. 1) or if the amino acid substitutions exert their effects independently, but the requirement for two amino acid changes in order to produce a variant DNA polymerase with mutator activity may provide information on protein conformation and/or substrate interactions.

Most of the amino acid changes that produce mutant DNA polymerases with reduced fidelity are located in the proposed N-terminal domain (Fig. 1); however, biochemical characterization of the mutant DNA polymerases has indicated that not all of them have a decrease in $3' \rightarrow 5'$ exonuclease activity. The tsL56 and tsL98 DNA polymerases have reduced $3' \rightarrow 5'$ exonuclease activities (17), and they have been positioned near the proposed dNMP site in the model (Fig. 1). The tsM19 DNA polymerase, however, is apparently not defective in proofreading activity but has reduced discrimination for the base analog, 2-aminopurine deoxyribonucleoside triphosphate (23). The J5 and mel62 DNA polymerases have active $3' \rightarrow 5'$ exonuclease activities (22, 24), and the reduction in exonuclease activity for the mel74 DNA polymerase is not sufficient to account for the high mutator activity of the mel74 strain (22). Because the mutants from this region have defects in both exonuclease and insertion activities, I have suggested that the N-terminal region may also have a role in binding the primer terminus (Fig. 1); both exonuclease and insertion activities function at the primer terminus. Some of the region could also form a tether between the proposed exonuclease and polymerase domains (Reha-Krantz, in press).

The mutator mutations do not change amino acids within the conserved sequences, and, furthermore, many of the amino acid substitutions are dramatic. For example, four of the mutant sites involve proline residues. Pseudorevertants have been isolated at two of these sites, but only the proline substitutions produce the mutator phenotype (Reha-Krantz, in press). Thus, it appears that amino acid substitutions that are likely to produce protein conformational changes are required in order to produce variant DNA polymerases with reduced fidelity. How could a change in DNA polymerase conformation produce a mutant DNA polymerase? One possibility involves the hypothesis presented earlier that water in the active site decreases fidelity. An alteration in conformation or a significant difference in the size of the mutant amino acids compared to the wild-type amino acids may be compensated in the protein structure by the inclusion of water. If the amino acid substitutions result in an increase in water at the active site, then a decrease in fidelity due to misinsertion or proofreading would be predicted. Thus, some of the mutator mutations may prove to be useful in testing whether water is normally excluded at the primer terminus.

Enzymatic, genetic, and X-ray crystallographic analyses of the *E. coli* DNA polymerase I have produced a comprehensive "picture" of this important DNA polymerase (2, 4, 6–8, 10, 18); however, mutant selections have focused on isolating conditional lethal mutants (2, 6, 12) and on genetic engineering and targeted mutagenesis studies based on protein chemistry, DNA sequence, and crystallographic data (4, 7, 8). Conditional lethal bacteriophage T4 DNA polymerase mutants have also been isolated (3), but because little structural information about the T4 DNA polymerase is known, mutant selection has not been targeted to specific regions of the T4 DNA polymerase. Instead, mutant selection has been directed at the modification of T4 DNA polymerase activities without any preconceived ideas about protein structures that contribute to DNA polymerase function. The mutant selection strategies have been successful because T4 has just one DNA polymerase and it is essential for the replication of T4 DNA. Furthermore, the DNA polymerase is the major determinant of spontaneous mutation in T4 (25). Thus, because of the advantages in selecting T4 DNA polymerase mutants with alterations in DNA polymerase functions, studies of some of the mutants may provide new insights into DNA replication.

I will conclude this chapter by describing our recent observations on some of the mutator DNA polymerases that will likely provide new insights into DNA polymerase function. The observation is that single amino acid substitutions in the T4 DNA polymerase can produce DNA polymerase variants with the ability to replicate past noncoding DNA lesions (L. J. Reha-Krantz, manuscript in preparation). This ability is manifested in vivo, after DNA damage, by increased survival accompanied by increased mutagenesis. In vitro, there is direct evidence of stable incorporation at abasic sites. The wild-type T4 DNA polymerase, like many other DNA polymerases, does not synthesize past an abasic site, but "idles" at the site and preferentially attempts to insert dATP (21). In contrast, some of the mutator DNA polymerases can stably incorporate dAMP. We are now trying to determine the mechanism for stable incorporation across from abasic sites, but, presumably, the mutant DNA polymerases stabilize the dAMP inserted across from the abasic template and protect this unpaired primer terminus from degradation by the $3' \rightarrow 5'$ exonuclease activity.

SUMMARY

The picture of the T4 DNA polymerase is far from complete. X-ray crystallographic studies are needed and have been initiated (W. Konigsberg, personal communication). Biochemical characterization of more of the mutator DNA polymerases and additional genetic studies are required, particularly targeted mutagenesis of the proposed $3' \rightarrow 5'$ exonuclease and dNTP sites, of other homologous regions, and of regions defined by the mutator mutations. An expression vector carrying the T4 DNA polymerase gene has been constructed (15) and will be useful for all of the above studies. I am especially interested in the prospect that the T4 mutator strains will provide new insights into DNA polymerase function. It is also interesting that T4 DNA polymerase studies may be applicable to studies of eucaryotic and eucaryoticlike DNA polymerases. The three best experimental approaches for determining protein structure-function relationships are being used: (i) isolation of informative mutants, (ii) biochemical studies, and (iii) crystal structure; thus, I am optimistic that future experiments will be successful in developing a complete picture of the bacteriophage T4 DNA polymerase.

ACKNOWLEDGMENTS. I extend special thanks to E. Spicer for several helpful discussions and for sharing data prior to publication and to W. Konigsberg (Yale University).

Research was supported by operating grants from the Natural Sciences and Engineering Research Council, Canada, and from the National Cancer Institute of Canada. L.J.R.-K. is a Scholar of the Alberta Heritage Foundation for Medical Research.

LITERATURE CITED

1. **Clayton, L. K., M. F. Goodman, E. W. Branscomb, and D. J. Galas.** 1979. Error induction and

correction by mutant and wild type T4 DNA polymerases. *J. Biol. Chem.* 254:1902–1912.
2. De Lucia, P., and J. Cairns. 1969. Isolation of an *E. coli* strain with a mutation affecting DNA polymerase. *Nature* (London) 224:1164–1168.
3. Epstein, R. H., A. Bolle, C. M. Steinberg, E. Kellenberger, E. Boy de la Tour, R. Chevalley, R. S. Edgar, M. Susman, G. H. Denhardt, and A. Lielausis. 1963. Physiological studies of conditional lethal mutants of bacteriophage T4D. *Cold Spring Harbor Symp. Quant. Biol.* 28:375–392.
4. Freemont, P. S., D. L. Ollis, T. A. Steitz, and C. M. Joyce. 1986. A domain of the Klenow fragment of *Escherichia coli* DNA polymerase I has polymerase but no exonuclease activity. *Proteins* 1:66–73.
5. Gibbs, J. S., H. C. Chiou, J. D. Hall, D. W. Mount, M. J. Retondo, S. K. Weller, and D. M. Coen. 1985. Sequence and mapping analyses of the herpes simplex virus DNA polymerase gene predict a C-terminal substrate binding domain. *Proc. Natl. Acad. Sci. USA* 82:7969–7973.
6. Joyce, C. M., D. M. Fujii, H. S. Laks, C. M. Hughes, and N. D. F. Grindley. 1985. Genetic mapping and DNA sequence analysis of mutations in the *polA* gene of *Escherichia coli*. *J. Mol. Biol.* 186:283–292.
7. Joyce, C. M., D. L. Ollis, J. Rush, T. A. Steitz, W. H. Konigsberg, and N. D. F. Grindley. 1986. Relating structure to a function for DNA polymerase I of *Escherichia coli*. *UCLA Symp. Mol. Cell. Biol. New Ser.* 32:197–205.
8. Joyce, C. M., and T. A. Steitz. 1987. DNA polymerase I: from crystal structure to function via genetics. *Trends Biochem. Sci.* 12:288–292.
9. Kimura, M. 1977. Preponderance of synonymous changes as evidence for the neutral theory of molecular evolution. *Nature* (London) 267:275–276.
10. Kornberg, A. 1969. Active center of DNA polymerase. *Science* 163:1410–1418.
11. Larder, B. A., S. D. Kemp, and G. Darby. 1987. Related functional domains in virus DNA polymerases. *EMBO J.* 6:169–175.
12. Lehman, I. R., and D. G. Uyemura. 1976. DNA polymerase I: essential replication enzyme. *Science* 193:963–969.
13. Li, W.-H., C.-C. Luo, and C.-I. Wu. 1985. Evolution in DNA sequences, p. 1–92. *In* R. J. MacIntyre (ed.), *Molecular Evolutionary Genetics*. Plenum Publishing Corp., New York.
14. Li, W.-H., C.-I. Wu, and C.-C. Luo. 1985. A new method for estimating synonymous and nonsynonymous rates of nucleotide substitution considering the relative likelihood of nucleotide and codon changes. *Mol. Biol. Evol.* 2:150–174.
15. **Lin, T.-C., J. Rush, E. K. Spicer, and W. H. Konigsberg.** 1987. Cloning and expression of T4 DNA polymerase. *Proc. Natl. Acad. Sci. USA* 84:7000–7004.
16. **Miyata, T., T. Yasunaga, and T. Nishida.** 1980. Nucleotide sequence divergence and functional constraint in mRNA evolution. *Proc. Natl. Acad. Sci. USA* 77:7328–7332.
17. **Muzyczka, N., R. L. Poland, and M. J. Bessman.** 1972. Studies on the biochemical basis of mutation I. A comparison of the deoxyribonucleic acid polymerases of mutator, antimutator, and wild type strains of bacteriophage T4. *J. Biol. Chem.* 247:7116–7122.
18. **Ollis, D. L., P. Brick, R. Hamlin, N. G. Xuong, and T. A. Steitz.** 1985. Structure of large fragment of *Escherichia coli* DNA polymerase I complexed with dTMP. *Nature* (London) 313:762–766.
19. **Ollis, D. L., C. Kline, and T. A. Steitz.** 1985. Domain of *E. coli* DNA polymerase I showing sequence homology to T7 DNA polymerase. *Nature* (London) 313:818–819.
20. **Petruska, J., L. C. Sowers, and M. F. Goodman.** 1986. Comparison of nucleotide interactions in water, proteins, and vacuum: model for DNA polymerase fidelity. *Proc. Natl. Acad. Sci. USA* 83:1559–1562.
21. **Randall, S. K., R. Eritja, B. E. Kaplan, J. Petruska, and M. F Goodman.** 1987. Nucleotide insertion kinetics opposite abasic lesions in DNA. *J. Biol. Chem.* 262:6864–6870.
22. **Reha-Krantz, L. J.** 1987. Genetic and biochemical studies of the bacteriophage T4 DNA polymerase. *UCLA Symp. Mol. Cell. Biol. New Ser.* 47:501–509.
23. **Reha-Krantz, L. J., and M. J. Bessman.** 1981. Studies on the biochemical basis of mutation. VI. Selection and characterization of a new bacteriophage T4 mutator DNA polymerase. *J. Mol. Biol.* 145:677–695.
24. **Reha-Krantz, L. J., and J. K. J. Lambert.** 1985. Structure-function studies of the bacteriophage T4 DNA polymerase. Isolation of a novel suppressor mutant. *J. Mol. Biol.* 186:505–514.
25. **Reha-Krantz, L. J., E. M. Liesner, S. Parmaksizoglu, and S. Stocki.** 1986. Isolation of bacteriophage T4 DNA polymerase mutator mutants. *J. Mol. Biol.* 189:261–272.
26. **Sinha, N. K., and M. F. Goodman.** 1983. Fidelity of DNA replication, p. 131–137. *In* C. K. Mathews, E. M. Kutter, G. Mosig, and P. B. Berget (ed.), *Bacteriophage T4*. American Society for Microbiology, Washington, D.C.
27. **Tsurumi, T., K. Maeno, and Y. Nishiyama.** 1987. A single-base change within the DNA polymerase locus of herpes simplex virus type 2 can confer resistance to aphidicolin. *J. Virol.* 61:388–394.

Chapter 5

Multienzyme Complex for DNA Replication in HeLa Cells

Robert J. Hickey, Linda H. Malkas, Nina Pedersen, Congjun Li, and Earl F. Baril

DNA replication in mammalian cell nuclei most probably involves the concerted interactions of several enzymes and nonenzymatic proteins in addition to the DNA replicase. Through a series of elegant studies in procaryote systems it has been clearly shown that chromosome replication in *Escherichia coli* (15, 16) and for some of its bacteriophage (1, 15, 16; R. McMacken, C. Alfano, K. Mensa-Wilmot, K. Carroll, K. Stephans, M. Dodson, and H. Echols, unpublished data; H. Naki, B. Beauchamp, J. Bernstein, H. Huber, S. Tabor, and C. C. Richardson, unpublished data) requires the interaction of proteins in addition to the respective replicases. The number of proteins that are required varies with the replication system. For example, only three and seven proteins are required for the in vitro replication of T7 (Naki et al., unpublished data) and T4 (1) bacteriophage DNAs, respectively, while nine proteins are required for the initiation of in vitro replication of λ phage DNA (McMacken et al., unpublished data). Thirty

or more proteins are probably required for replication of the *E. coli* chromosome (15, 16). Also, the DNA replicase itself in *E. coli*, polymerase III, exists as a complex multisubunit holoenzyme structure rather than a single protein (15, 16; C. S. McHenry, A. Flower, and J. Hawher, unpublished data).

DNA polymerase α is the major DNA polymerase in proliferating animal cells and probably represents the primary replicase in these cells (4, 8). During the past few years, multiprotein forms of DNA polymerase α have been identified and purified to near homogeneity from *Drosophila melanogaster* embryos (13), HeLa cells (28), and calf thymus (19) by conventional procedures and from human KB cells (30) and calf thymus (5), using immunoaffinity chromatography. Although the structure-function relationships of the multiple subunits of the polymerase complexes have not yet been completely defined, some similarities to the *E. coli* polymerase III holoenzyme are beginning to emerge. There is now general agreement that the DNA polymerase α activity resides with a single polypeptide of about 180 kilodaltons (kDa) (8). Also, a single-strand specific $3' \rightarrow 5'$ exonuclease activity

---

Robert J. Hickey, Linda H. Malkas, Nina Pedersen, Congjun Li, and Earl F. Baril • Cell Biology Group, Worcester Foundation for Experimental Biology, Shrewsbury, Massachusetts 01545.

that has the potential for a proofreading activity has been shown to be associated with the highly purified multiprotein DNA polymerase α complexes from HeLa cells (24, 28) and *Drosophila* embryos (6). A 3'→5' exonuclease activity is also associated with a holoenzyme form of calf thymus DNA polymerase α, although no evidence has been presented in regard to its potential proofreading ability (19). A unique feature of the multiprotein DNA polymerase α complexes relative to DNA polymerase III holoenzyme from *E. coli* is the tight association of a DNA primase with the former. The primase has now been shown to be associated with DNA polymerase α isolated from a variety of eucaryotic cells, and the primase and polymerase active sites were shown to reside on separate polypeptides (4, 8). Other proteins in addition to the polymerase catalytic subunit, DNA primase, and 3'→5' exonuclease are also associated with the multiprotein forms of DNA polymerase α (13, 17, 28). Some of the proteins, such as the C1 and C2 primer recognition proteins from HeLa (17) and CV-1 (22) cells, may function as accessory proteins for the polymerase and/or the primase.

MULTIPROTEIN DNA POLYMERASE α COMPLEX FROM HeLa CELLS

DNA polymerase α activity in the combined nuclear extract-postmicrosomal supernatant (NE-PMS) solution from NE-PMS synchronized HeLa cells in the S phase of the cell cycle resides with a 640-kDa multiprotein complex (17, 28). The polymerase α activity from the extraction of nuclei at physiological salt concentrations (i.e., 0.15 M KCl) represents more than 90% of the extractable DNA polymerase α activity from isolated nuclei. In this regard it is of interest that most of the in vitro replication activity for simian virus 40 (SV40) DNA in the presence of purified large T antigen is also extractable from the nucleus of cells at physiological salt concentrations (18). We have also observed that the 0.15 M KCl nuclear extract from synchronized HeLa cells in the mid-S phase of the cell cycle has high in vitro replication activity for SV40 DNA in the presence of purified T antigen (E. Baril, L. Malkas, R. Hickey, C. Li, J. Vishwanatha, and S. Coughlin, *Cancer Cells*, in press).

The 640-kDa DNA polymerase α complex has been purified to near electrophoretic homogeneity by combined chromatography on coupled columns of DEAE-cellulose and native and denatured DNA-cellulose followed by chromatography on DEAE–Bio-Gel and polyacrylamide gel electrophoresis under nondenaturing conditions (28). More than 70% of the DNA polymerase activity from the NE-PMS fraction that is applied to a DEAE-cellulose column is eluted from DEAE–Bio-Gel by 0.15 M KCl (28). This fraction was operationally designated DNA polymerase α_2 based on its order of elution. In addition to DNA polymerase activity with activated DNA, this fraction contains high levels of DNA polymerase activity with primed single-stranded DNA templates, DNA primase, 3'→5' exonuclease, DNA ligase, and RNase H activities (17, 28; Baril et al., in press). There is also an A+T sequence recognition protein that is associated with the DNA polymerase α_2 chromatographic fraction. This protein has now been purified to electrophoretic homogeneity (L. Malkas and E. Baril, submitted for publication). It is a 62-kDa protein that specifically binds to the 17-base-pair A+T tract in the minimal origin for SV40 DNA replication that has been shown to be a bending locus (10) and essential for the initiation of SV40 replication (9, 10). Gel transfer studies under nondenaturing conditions showed that the DNA polymerase activity with activated and primed single-stranded DNA templates, primase, 3'→5' exonuclease, and the binding of the unique dinucleotide diadenosine 5',5'''-P1,P4-tetraphosphate (Ap$_4$A) activities migrate coincidentally with a single major protein band of 640 kDa (Fig. 1). Gel transfer experiments with fractions taken throughout the major protein band from nondenaturing gels and transferred to sodium dodecyl sulfate (SDS)-polyacrylamide gels for electrophoresis under denaturing conditions showed that the 640-kDa multiprotein polymerase complex is composed of at least 10 polypeptides in the range of 183 to 15 kDa (28). About 10% of the applied DNA polymerase is eluted from DEAE–Bio-Gel by 0.3 M KCl. This fraction contains only DNA polymerase activity with activated DNA and does not contain the other activities associated with the polymerase α_2 (17, 28). Polyacrylamide gel electrophoresis under nondenaturing and denaturing conditions showed that the polymerase activity in this fraction has a native molecular size of 270 kDa and contains the 183-kDa subunit associated with a 92-kDa protein.

Our finding that most of the DNA polymerase α in the NE-PMS fraction is recovered as a 640-kDa complex is in agreement with the finding of Holler et al. (11) that almost all of the polymerase activity in HeLa cell extracts migrated as a 640-kDa complex during activity gel analysis under nondenaturing conditions. The complex subunit structure of the 640-kDa DNA polymerase α complex consisting of polypeptides of 183, 92, 70, 69, 52, 47, 25, and 15 kDa is in agreement with the subunit structure recently observed for highly purified calf thymus polymerase α (19) and for immunoaffinity-purified human KB cell (30) and calf thymus polymerase (5). The subunit structure of the highly purified animal cell DNA polymerase α, in general, appears more complex than that observed for highly purified DNA polymerase α from *Drosophila* (13) and yeast DNA polymerase I, the yeast counterpart of polymerase α (20). Whether this has functional significance as in the case of the procaryotic systems (1, 15, 16; McMacken et al., unpublished data; Naki et al., unpublished data) is not clear at this time.

The subunits of the multiprotein DNA polymerase α complex have been resolved by using a combination of hydrophobic affinity chromatography and conventional enzyme purification procedures (2, 24, 27, 28). Six of the subunits have been highly purified and characterized and their function(s) have been partially defined (Table 1). The 183-kDa polypeptide is the polymerase α catalytic subunit (28). Polyclonal antibody prepared against the 183-kDa polypeptide specifically neutralizes the polymerase activity of the complex and reacted only with the 183-kDa polypeptide in immunoblots with HeLa cell extracts or with polymerase α_2 fractions under denaturing conditions (28). The 94-kDa subunit of the polymerase complex binds the unique dinucleotide Ap_4A

TABLE 1
Subunits of the 640-kDa DNA Polymerase α Complex from HeLa Cells[a]

Native molecular size (kDa)	Subunits	Function	Reference(s)
183	1	DNA polymerase α catalytic subunit	28
92	1	Unknown	
70	1	Primase	27
69	1	$3' \rightarrow 5'$ Exonuclease	24
52	1	Accessory protein (C2)	17, 22, 23
94	2	Unknown (binds Ap_4A)	2
96	4	Accessory protein (C1)	17, 22, 23

[a] Data are from Baril et al., in press.

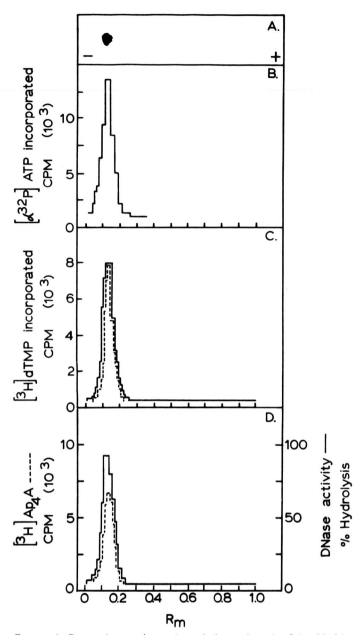

FIGURE 1. Preparative nondenaturing gel electrophoresis of the 640-kDa DNA polymerase α multiprotein complex. The peak activity of DNA polymerase α in the 150 mM KCl eluate from DEAE–Bio-Gel under nondenaturing conditions as described previously (28). Following electrophoresis, the gels were sliced into 1-mm sections, pulverized, and extracted overnight at 4°C with gentle agitation. Samples (20 μl) were removed from each extract to assay for primase (B), DNA polymerase α activity with activated (——) and heat-denatured (– – –) DNA templates (C), and [^{3}H]Ap$_4$A binding (– – –) and DNase activity (——) (D) according to assay procedures described previously (17, 28). The remainder of extract from the gel slice corresponding to the peak of the activities was reelectrophoresed on a 4% polyacrylamide gel under identical nondenaturing conditions (A). (Reprinted with permission from *J. Biol. Chem.*, copyright 1986, The American Society of Biological Chemists.)

with a high degree of specificity and was shown by photoaffinity labeling to be a dimer of identical 47-kDa subunits (2). Although the 94-kDa subunit binds Ap_4A and DNA polymerase α_2 utilizes the dinucleotide as primer with complementary single-stranded DNA templates, we have been unable to show a direct affect of Ap_4A on the activities associated with the multiprotein polymerase complex. Thus, the actual function of the 94-kDa polypeptide is unknown at this time. The DNA primase associated with the DNA polymerase α complex has been purified to homogeneity by hydrophobic chromatography on phenyl-Sepharose and hexylagarose followed by chromatography on Cibacron blue agarose and glycerol gradient centrifugation. The peak fraction of primase activity from the glycerol gradient showed a single major 70-kDa band by SDS-polyacrylamide gel electrophoresis (27). Thus, we tentatively ascribe the primase subunit to the 70-kDa polypeptide. The 69-kDa polypeptide associated with the multiprotein polymerase α complex is a single-stranded DNA-specific $3'\rightarrow 5'$ exonuclease (24). This subunit has been purified to electrophoretic homogeneity and was shown to specifically hydrolyze single-stranded DNA substrates, predominately in the $3'\rightarrow 5'$ direction but with a low rate of hydrolysis in the $5'\rightarrow 3'$ direction. The exonuclease efficiently recognizes and excises single-base mismatches and has potential for a proofreading function in vivo. The 52- and 24-kDa polypeptides of the 640-kDa DNA polymerase α complex represent the C2 and C1 accessory proteins, respectively (Table 1). The C2 protein is a single polypeptide and the 96-kDa C1 protein is a tetramer of identical 24-kDa subunits (28). These proteins have been shown by reconstitution studies to function as primer recognition proteins and permit the polymerase subunit to function efficiently with RNA or DNA primed single-stranded DNA templates (17, 23).

INTERACTION OF THE 640-kDa DNA POLYMERASE α COMPLEX WITH OTHER PROTEINS INVOLVED IN DNA REPLICATION

Li and Kelly (18) first demonstrated the complete replication of SV40 DNA in vitro in the presence of T antigen with concentrated, low-salt HeLa cell extracts (18). This assay is now being performed in a number of laboratories by using recombinant plasmid DNAs containing the replication origin of SV40 DNA. As discussed earlier, we have observed complete replication of SV40 DNA in vitro, using the nuclear extract or the postmicrosomal supernatant solution from synchronized HeLa cells in the S phase of the cell cycle (Baril et al., in press). The in vitro SV40 DNA replication activity is abolished, however, in the initial steps of the purification of the DNA polymerase α multiprotein complex. Among the components that are removed from the NE-PMS at the initial purification step are topoisomerase I and a DNA-dependent ATPase that bind to native DNA-cellulose and are both eluted by 0.3 M KCl (J. Vishwanatha and E. Baril, submitted for publication).

In an effort to analyze the possible physical and functional interactions of the 640-kDa DNA polymerase α complex with other proteins involved in DNA replication, the NE-PMS was subfractionated according to the scheme outlined in Fig. 2. The polyethylene glycol precipitation step in the presence of 2 M KCl is a key step since it removes about 50% of the protein from the NE-PMS fraction but retains all of the in vitro SV40 DNA replication activity, as well as polymerase α activities in the supernatant solution (Malkas and Baril, submitted). The differential centrifugation step onto a 2 M sucrose cushion was used since we had found that this is an effective procedure for isolating complexes of enzymes involved in DNA synthesis. As shown in Table 2, most of the DNA polymerase activity that functions with primed single-stranded DNA tem-

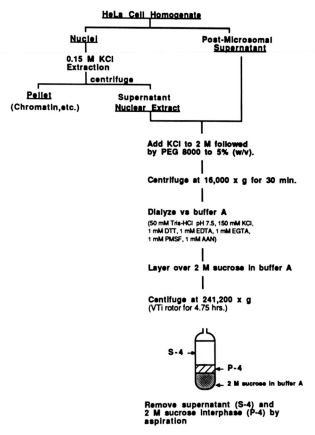

FIGURE 2. Flow diagram of the fractionation procedure used to isolate the postmicrosomal P-4 and S-4 fractions from HeLa cells. (Reprinted with permission from *Cancer Cells*, copyright 1988, Cold Spring Harbor Press.)

plates, DNA primase, DNA-dependent ATPase, and the T-antigen-dependent in vitro replication activity for SV40 DNA were all recovered at the interphase above the 2 M sucrose layer (P-4 fraction). However, not all of the activites are recovered in the P-4 fraction. DNA polymerase activity with activated DNA template, some RNase H, DNA-dependent ATPase, topoisomerase I (Fig. 3), and DNA ligase (Fig. 4) activities are recovered in the nonsedimentable S-4 fraction (Table 2). The P-4 fraction, however, contains most of the DNA polymerase activity that functions with primed single-stranded DNA templates and DNA primase activity that resides with the multiprotein polymerase α complex (28). In addition to topoisomerase I and DNA ligase activities, this fraction contains most of the DNA-dependent ATPase and the T-antigen-dependent in vitro replication activity for SV40 DNA. Metabolic labeling experiments indicated the absence of DNA and RNA in the P-4 fraction (Malkas and Baril, submitted).

Monoclonal antibodies to the purified multiprotein DNA polymerase complex were prepared in BALB/c mice. More than

TABLE 2
Enzyme Activities Recovered in the
P-4 and S-4 Fractions[a]

Activity	Subcellular fraction (total units)	
	P-4	S-4
DNA polymerase with:		
Activated DNA	5,648	3,365
Primed single-stranded DNA	2,916	1,320
DNA primase	1,224	29
ATPase		
Plus denatured DNA	16,000	14,380
Minus denatured DNA	1,468	18,725
Topoisomerase[b]		
Type I	+	+
Type II	−	−
DNA ligase[a]	+	+
RNase H[b]	1,530	346
A+T sequence recognition protein[c]	1,230	62
T-antigen-dependent in vitro synthesis of SV40 DNA[d]	75	<1

[a] See Table 1, footnote a.
[b] Topoisomerase and DNA ligase activities were not quantitated, but equal activities appear to be present in these subcellular fractions.
[c] One unit of RNase H activity equals 1 nmol of [³H]poly(A) converted to acid-soluble form per hour at 30°C. One unit of A-T sequence binding protein equals 10,000 cpm of [³H]poly(dA) or [³H]poly(dT) bound to nitrocellulose filters after a 30-min incubation period at 25°C.
[d] One unit is equal to the incorporation of 2 pmol of deoxynucleoside monophosphate into SV40 DNA per hour per 60 ng of pJLO DNA.

200 positive clones were isolated and are currently being subcloned and characterized. One of the characterized subclones (1A4) reacts specifically with a 640-kDa protein band in nondenaturing Western blot analysis of DNA polymerase α_2 and the P-4 fraction (Fig. 5), but no 640-kDa protein band was observed with the S-4 fraction (data not shown). Also, monoclonal antibody prepared against proteolytic fragments of human polymerase α (26) (i.e., SJK 132-20) showed a major protein band greater than 400 kDa in addition to the 180-kDa band for the polymerase α subunit in nondenaturing Western blot analysis of the P-4 but not the S-4 subfraction of NE-PMS (N. Pedersen, R. Hickey, and E. Baril, unpublished data). These results in conjunction with those reported by Holler et al. (11) are supportive evidence that the 640-kDa DNA polymerase α multiprotein complex exists as a distinct entity in cells. Moreover, they suggest that the 640-kDa polymerase complex exists with a sedimentable complex, possibly by loose associations with other proteins, that contains other enzymes for DNA synthesis and the in vitro replication activity for SV40 DNA.

It was previously found by glycerol gradient centrifugation that the 640-kDa DNA polymerase multiprotein complex and the polymerase α catalytic subunit were 10S and 6S, respectively (28). Sedimentation velocity analysis of the sedimentable P-4 complex in a 5 to 30% sucrose gradient in the presence of 0.5 M KCl, however, showed two incompletely resolved peaks of DNA polymerase activity (Fig. 6a). The DNA polymerase activity with activated and primed single-stranded DNA templates as well as the primase sedimented coincidentally as a 16S peak with a shoulder at 13S. In contrast, the polymerase α activity in the S-4 fraction sedimented at 8S, and there is little to no primase or DNA polymerase α activity with primed single-stranded DNA templates in this fraction. DNA ligase activity is present with both the 16S and 8S peaks from gradient centrifugation of the P-4 and S-4 fractions, respectively (Fig. 4). This is in agreement with the finding that DNA ligase activity is distributed between the P-4 and the nonsedimentable S-4 subfractions of the NE-PMS.

The P-4 and S-4 fractions showed different profiles during chromatography on Q Sepharose (Fig. 6b). About 30% of the protein in the P-4 fraction did not bind to Q-Sepharose and appeared in the column flowthrough. There was no DNA polymerase or primase activity in the flowthrough fraction, however. The flowthrough fraction did contain DNA ligase (Fig. 4) but not topoisomerase I or DNA-dependent ATPase activities (data not shown). The DNA polymerase activity with activated and

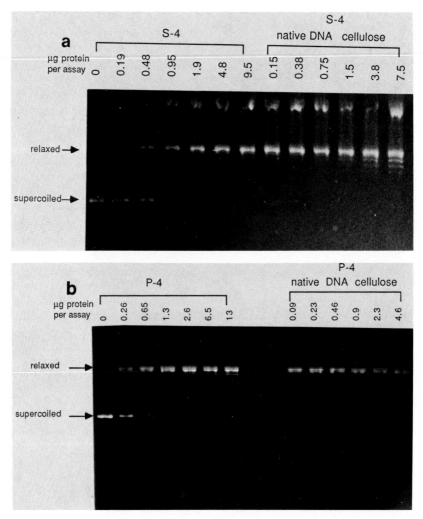

FIGURE 3. Agarose gel electrophoresis of the products from assay of topoisomerase I activity in the P-4 and S-4 fractions. The assay of topoisomerase I activity was performed according to a published procedure (18a), using pUC8-14 DNA and various amounts of protein from the respective cell fractions. Incubation was at 35°C for 10 min. (a) Assay of the S-4 fractions at 0, 0.19, 0.48, 0.95, 1.9, 4.8, and 9.5 μg of protein per assay and of the 0.3 M KCl eluate from chromatography of S-4 on native DNA-cellulose at 0.15, 0.38, 0.75, 1.5, 3.8, and 7.5 μg of protein per assay. (b) Assay of the P-4 fraction at 0, 0.26, 0.65, 1.3, 2.6, 6.5, and 13 μg of protein per assay and of the 0.3 M KCl eluate from chromatography of the P-4 fraction on a native DNA-cellulose column at 0.09, 0.23, 0.46, 0.9, 2.3, and 4.6 μg of protein per assay.

primed single-stranded DNA templates and primase were eluted as two incompletely resolved peaks by an increasing KCl gradient (Fig. 6b). These two peaks of activity also contained DNA ligase activity (Fig. 4). All of the activities coincided with the peaks of eluted protein, but the activities were higher in peak B. For the S-4 fraction about 50% of the protein and DNA polymerase activity appeared in the column flowthrough fraction (Fig. 6b). A single peak of DNA polymerase activity and a broad peak of protein were

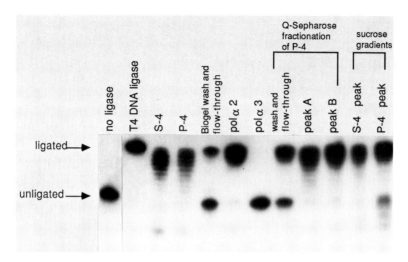

FIGURE 4. DNA ligase activity in the P-4, S-4, and related fractions in the isolation of the multienzyme complex. DNA ligase in the respective fractions was assayed by measuring the conversion of complementary ^{32}P-oligodecadeoxynucleotides to oligododecamerdeoxynucleotides as measured by sequencing gel electrophoresis (N. Pedersen, J. Goodchild, and E. Baril, unpublished data). Assays were performed in 50-μl volumes containing 50 mM Tris hydrochloride, pH 7.6, 10 mM dithiothreitol, 20 mM MgCl$_2$, 5 mM ATP, 65 pmol of oligododecadeoxynucleotide, 16 pmol of ^{32}P-oligodecadeoxynucleotide, and 2.2 to 25 μl of the respective fractions or 40 U of T4 DNA ligase as a positive control. Following incubation at 25°C for 4 h, the product was deproteinized in the presence of SDS (0.5%) and proteinase K (250 μg/ml), extracted with phenol, precipitated with ethanol, dissolved in 89 mM Tris base–89 mM boric acid–2 mM EDTA (pH 8.3)–80% formamide, heated at 65°C for 10 min, and electrophoresed on 15% polyacrylamide sequencing gel in the presence of 8 M urea.

eluted by the KCl gradient. The DNA primase and polymerase activities with primed single-stranded DNA templates associated with the eluted fractions were low.

Overall, these results indicate that the sedimentable P-4 fraction containing in vitro SV40 DNA replication activity and associated enzymes for DNA synthesis also contains higher molecular forms of polymerase-primase than does the nonsedimentable S-4 fraction. The immunological data also suggest that the P-4 complex contains the multiprotein polymerase α complex. In the P-4 fraction, however, the polymerase-primase exists as a 16S complex rather than the 10S form of 640-kDa polymerase α complex. Whether this is attributable to the formation of a multienzyme complex directly or to conformational changes that may occur in the interactions will require further analysis.

DNA POLYMERASE α FUNCTIONS IN SV40 DNA REPLICATION

The results of our studies strongly indicate that the 640-kDa multiprotein DNA polymerase α complex interacts with other enzymes and proteins to form a multienzyme complex for DNA synthesis. The functional evidence for this comes from the observation that this multienzyme complex contains in vitro replication activity for SV40 DNA in the presence of purified T

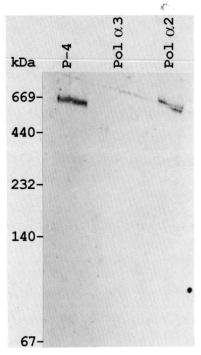

FIGURE 5. Nondenaturing Western blot analysis of the 640-kDa multiprotein DNA polymerase α complex. The P-4 (20 μg of protein), polymerase α_3 (35 μg of protein), and polymerase α_2 (25 μg of protein) fractions were electrophoresed on 3 to 10% nondenaturing gradient polyacrylamide gels under conditions previously described (17). Following electrophoresis the gels were electroblotted at 4°C for 12 h and 20 V/cm onto nitrocellulose. After incubation with blocking buffer for 30 min, the nitrocellulose strips were incubated with 500 μl of monoclonal antibody (IA4) produced to the purified DNA polymerase α_2 at room temperature for 1.5 h. Following a wash, the strips were incubated with alkaline phosphatase-conjugated sheep anti-mouse immunoglobulin G for 1 h, washed with 20 mM Tris hydrochloride (pH 7.5)-buffered 50 mM NaCl (saline), and developed with Nitro Blue Tetrazolium plus 5-bromo-4-chloro-3-indolylphosphate. The native molecular mass markers are thyroglobulin (669 kDa), ferritin (440 kDa), catalase 232 kDa), lactate dehydrogenase (140 kDa), and bovine serum albumin (67 kDa).

antigen. There is evidence, however, that DNA polymerase δ by itself or in association with other proteins functions in the replication of SV40 DNA (21). Thus, we investigated the nature of the DNA polymerase activity present in the P-4 and S-4 subfractions of the NE-PMS fraction from HeLa cells based on their sensitivities to N^2-(p-n-butylphenyl)-9-(2-deoxy-β-D-ribofuranosyl) guanine 5′-triphosphate) (BuPdGTP) and neutralizing monoclonal antibody (SJK 132-20) to human DNA polymerases α. A difference in sensitivity to these reagents has been used by others to distinguish between DNA polymerases α and δ (3, 7). The DNA polymerase activity in the P-4 and S-4 fractions and in the enzyme fractions that were purified from them was inhibited by the neutralizing antibody SJK 132-20 (Fig. 7a–c). Although the activity in the crude P-4 and S-4 fractions is neutralized to a lesser extent than that in the enzyme fractions partially purified from them, this is probably attributable to lack of availability of some of the determinants for interaction with the antibody. The DNA polymerase activity in the P-4 and S-4 fractions was also completely inhibited by BuPdGTP at concentrations 50 to 100 times lower than those reported to be required for inhibition of polymerase δ, i.e., at 0.5 to 10 μM (3, 7). These data suggest that all of the DNA polymerase activity present in the NE-PMS of HeLa cells and with the putative multienzyme complex of the P-4 fraction belongs to the α class of DNA polymerases. That this fraction functions in the in vitro replication of SV40 DNA in the presence of T antigen suggests, but does not prove, that only

FIGURE 6. Velocity centrifugation on a 10 to 30% sucrose gradient and Q-Sepharose chromatography of the P-4 and S-4 fractions from HeLa cells. (a) One-milliliter portions of the P-4 (6.6 mg of protein) and S-4 (6 mg of protein) fractions were loaded onto 5 to 30% sucrose gradients in buffer b (28) plus 0.5 M KCl prepared in polyallomer tubes for the SW41 rotor and layered over a 2 M sucrose cushion. After centrifugation at 36,000 rpm for 16 h, 1-ml fractions were collected by tube puncture and assayed as described in the text. Horse spleen apoferritin (17S) and yeast alcohol dehydrogenase (7S) were used as markers. (b) The P-4 (30 mg of protein) and S-4 (25 mg of protein) fractions were chromatographed on 1-ml columns of Q-Sepharose by the procedure described in the text. The units of enzyme activity are as previously described (28). (Reprinted with permission from *Cancer Cells*, copyright 1988, Cold Spring Harbor Press.)

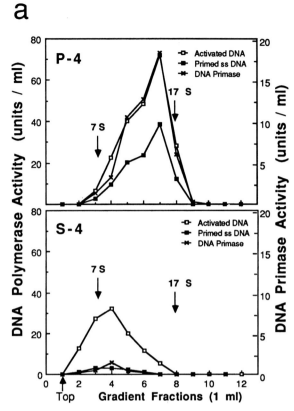
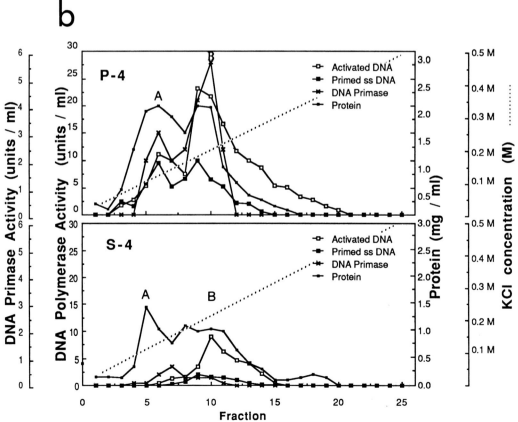

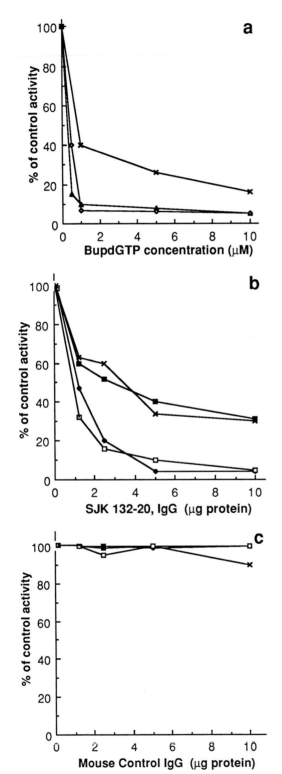

FIGURE 7. Inhibition of DNA polymerase activity in P-4 and S-4 fractions by BuPdGTP and monoclonal antibody (SJK 132-20) to human DNA polymerase α. (a) Inhibition by BuPdGTP was carried out according to a published procedure (3). The units per assay of DNA polymerase control activity for the P-4 fractions were 0.30 (▲) and 0.46 (◇) for Q-Sepharose peaks A and B and 0.30 for the S-4 fraction (X). (b) Assays of neutralization of DNA polymerase fractions by SJK 132-20 immunoglobulin G were performed according to a published procedure (27). The units per assay of DNA polymerase control activity for the fractions were: P-4, 0.10 (■); DEAE–Bio-Gel, 0.15 M KCl eluate of P-4, 0.06 (●); S-4, 0.04 (X); DEAE–Bio-Gel, 0.15 M KCl eluate of S-4, 0.07 (□). (c) Neutralization assays were performed as in panel b except that the designated amounts of purified mouse (control) immunoglobulin G were used. Symbols are the same as in panel b. (Reprinted with permission from *Cancer Cells*, copyright 1988, Cold Spring Harbor Press.)

DNA polymerase α is required for the replication of SV40 DNA in vitro. Our results are in agreement with those reported by Kelly et al. (unpublished data), who also observed that the major DNA polymerase that appears to be involved in the in vitro replication of SV40 DNA is polymerase α.

ACKNOWLEDGMENTS. We are grateful to Dr. George Wright for the generous gift of BuPdGTP. We thank Matthew Potter, Sandra Johnson, and Carol Savage for assistance in the preparation of this chapter.

This research was supported by Public Health Service grants P30-12708 and CA15187 from the National Institutes of Health. L.H.M. is the recipient of a postdoctoral fellowship from the National Institutes of Health.

LITERATURE CITED

1. **Alberts, B. M.** 1985. Protein machines mediate the basic genetic processes. *Trends Genet.* 1:26–31.
2. **Baril, E. F., P. Bonin, D. Burstein, K. Mara, and P. Zamecnik.** 1983. Resolution of the diadenosine

$5',5'''$-P^1,P^4-tetraphosphate binding subunit from a multiprotein form of HeLa cell DNA polymerase α. *Proc. Natl. Acad. Sci. USA* 80:4931–4935.
3. Byrnes, J. J. 1985. Differential inhibitors of DNA polymerases alpha and delta. *Biochem. Biophys. Res. Commun.* 132:628–634.
4. Campbell, J. L. 1986. Eukaryotic DNA replication. *Annu. Rev. Biochem.* 55:733–771.
5. Chang, L. M. S., E. Rafter, C. Augl, and F. J. Bollum. 1984. Purification of DNA polymerase-DNA primase complex from calf thymus glands. *J. Biol. Chem.* 259:14679–14687.
6. Cotterill, S. M., M. E. Reyland, L. A. Loeb, and I. R. Lehman. 1987. A cryptic proofreading $3'\rightarrow5'$ exonuclease associated with the polymerase subunit of the DNA polymerase-primase from *Drosophila melanogaster*. *Proc. Natl. Acad. Sci. USA* 84:5635–5639.
7. Crute, J. J., A. F. Wahl, and R. A. Bambara. 1986. Purification and characterization of two new high molecular weight forms of DNA polymerase α. *Biochemistry* 25:26–36.
8. Fry, M., and L. A. Loeb. 1986. *Animal Cell DNA Polymerases*. CRC Press, Inc., Boca Raton, Fla.
9. Gerard, R., and Y. Gluzman. 1986. Functional analysis of the role of the A+T-rich region and upstream flanking sequences in simian virus 40 DNA replication. *Mol. Cell. Biol.* 6:4570–4577.
10. Hertz, G. Z., M. R. Young, and J. E. Mertz. 1987. The A+T-rich sequence of the simian virus 40 origin is essential for replication and is involved in bending of the viral DNA. *J. Virol.* 61:2322–2325.
11. Holler, E., H. Fischer, and S. Helmut. 1985. Non-disruptive detection of DNA polymerases in nondenaturing polyacrylamide gels. *Eur. J. Biochem.* 151:311–317.
12. Kaguni, L. S., J. M. Rossignol, R. C. Conway, G. R. Banks, and I. R. Lehman. 1983. Association of DNA primase with the β/γ subunits of DNA polymerase α from *Drosophila melanogaster* embryos. *J. Biol. Chem.* 288:9037–9049.
13. Kaguni, L. S., J. M. Rossignol, R. C. Conway, and I. R. Lehman. 1983. Isolation of an intact DNA polymerase-primase from embryos of *Drosophila melanogaster*. *Proc. Natl. Acad. Sci. USA* 80:2221–2225.
14. Khan, N. N., G. E. Wright, L. W. Dudycz, and N. C. Brown. 1984. Butylphenyl dGTP: a selective and potent inhibitor of mammalian DNA polymerase alpha. *Nucleic Acids Res.* 12:3695–3706.
15. Kornberg, A. 1980. *DNA Replication*. W. H. Freeman & Co., San Francisco.
16. Kornberg, A. 1982. *DNA Replication, 1982 Supplement*. W. H. Freeman & Co., San Francisco.
17. Lamothe, P., B. Baril, A. Chi, L. Lee, and E. Baril. 1981. Accessory proteins for DNA polymerase α activity with single-strand DNA templates. *Proc. Natl. Acad. Sci. USA* 78:4723–4727.
18. Li, J. J., and T. J. Kelly, Jr. 1985. Simian virus 40 DNA replication in vitro: specificity of initiation and evidence for bidirectional replication. *Mol. Cell. Biol.* 5:1238–1246.
18a.Liu, L. F., and K. G. Miller. 1981. Eukaryotic DNA topoisomerases. Two forms of type I topoisomerases from HeLa cell nuclei. *Proc. Natl. Acad. Sci. USA* 78:3487–3491.
19. Ottiger, H., P. Frei, M. Hassig, and U. Hubscher. 1987. Mammalian DNA polymerase α: a replication competent holoenzyme form from calf thymus. *Nucleic Acids Res.* 15:4789–4806.
20. Plevani, P., G. Badaracco, C. Aool, and L. M. S. Chang. 1984. DNA polymerase I and DNA primase complex in yeast. *J. Biol. Chem.* 259:7532–7539.
21. Prelich, G., C.-K. Tan, M. Kostura, M. B. Matthews, A. G. So, K. M. Downey, and B. Stillman. 1967. Functional identity of proliferating cell nuclear antigen and a DNA polymerase-δ auxiliary protein. *Nature* (London) 326:518–519.
22. Pritchard, C. G., and M. L. DePamphilis. 1983. Preparation of DNA polymerase α C_1C_2 by reconstituting DNA polymerase α with its specific stimulatory cofactors C_1C_2. *J. Biol. Chem.* 258:9801–9809.
23. Pritchard, C. G., D. T. Weaver, E. F. Baril, and M. L. DePamphilis. 1983. DNA polymerase α cofactors C_1C_2 function as primer recognition proteins. *J. Biol. Chem.* 258:9810–9819.
24. Skarnes, W., P. Bonin, and E. Baril. 1986. Exonuclease activity associated with a multiprotein form of HeLa cell DNA polymerase α. Purification and properties of the exonuclease. *J. Biol. Chem.* 261:6629–6636.
25. Tan, C.-K., C. Castillo, A. G. So, and N. M. Downey. 1986. An auxiliary protein for DNA polymerase-δ from fetal calf thymus. *J. Biol. Chem.* 261:12310–12316.
26. Tanaka, S., S.-Z. Hu, T. S.-F. Wang, and D. Korn. 1982. Preparation and preliminary characterization of monoclonal antibodies against human polymerase α. *J. Biol. Chem.* 257:8386–8390.
27. Vishwanatha, J. K., and E. F. Baril. 1986. Resolution and purification of free primase activity from the DNA primase-polymerase α complex of HeLa cells. *Nucleic Acids Res.* 14:8467–8487.
28. Vishwanatha, J. K., S. A. Coughlin, M. Wesolowski-Owen, and E. F. Baril. 1986. A multiprotein form of DNA polymerase α from HeLa cells. Resolution of its associated catalytic activities. *J. Biol. Chem.* 261:6619–6628.
29. Vishwanatha, J. K., M. Yamaguchi, M. L. DePamphilis, and E. F. Baril. 1986. Selection of

template initiation sites and the lengths of RNA primers synthesized by DNA primase are strongly affected by its organization in a multiprotein DNA polymerase alpha complex. *Nucleic Acids Res.* 14: 7305–7323.

30. **Wang, T. S.-F., S.-Z. Hu, and D. Korn.** 1984. DNA primase from KB cells. Characterization of a primase tightly associated with immunoaffinity-purified DNA polymerase α. *J. Biol. Chem.* 259:1854–1865.

Chapter 6

Human DNA Polymerase β
Expression in *Escherichia coli* and Characterization of the Recombinant Enzyme

John Abbotts, Dibyendu N. SenGupta, Barbara Z. Zmudzka, Steven G. Widen, and Samuel H. Wilson

DNA polymerase β, a DNA repair polymerase of eucaryotic cells (for a review, see reference 6), is seen as a model enzyme for structure-function analysis of the nucleotidyltransferase reaction by DNA polymerases (21). This enzyme is the simplest DNA polymerase known in both size and catalytic repertoire. It is also the least accurate of eucaryotic polymerases, showing misinsertion error rates of 1/1,300 to 1/6,600 on natural DNA templates (9, 12). This suggests that DNA repair is a mutagenic process or that auxiliary cellular factors must enhance the fidelity of DNA repair synthesis. The human and rat β-polymerases are polypeptides of 335 amino acids, and secondary structure predictions suggest an ordinary globular structure with a high α-helix content (19, 27). The purified enzyme lacks exonuclease activities and detectable reverse reactions (5, 21), and the polymerase activity is fully distributive under most reaction conditions (2). Thus, the β-polymerase mechanism is a two-substrate, two-product reaction and follows ordered BiBi kinetics (21).

To examine physical biochemical properties and structure-function relationships of mammalian β-polymerase, we overexpressed the coding region of a human β-polymerase cDNA (19) in the λ p_L promoter-based bacterial expression system pRC23 (1) and purified the recombinant enzyme in milligram quantities. Studies revealed that the enzyme is a characteristic β-polymerase and is appropriate for structure-function studies of this enzyme. Here we describe the expression and characterization of the recombinant polymerase and discuss opportunities for further investigation of the enzyme and its role in DNA repair.

EXPRESSION AND PURIFICATION

Subcloning and Expression of Recombinant β-Polymerase

We reexamined the sequence of the 5' end of the human β-polymerase cDNA and found that the start of the coding region is 51

John Abbotts, Dibyendu N. SenGupta, Barbara Z. Zmudzka, Steven G. Widen, and Samuel H. Wilson • Laboratory of Biochemistry, National Cancer Institute, Bethesda, Maryland 20892.

nucleotides 5' of the start codon previously assigned (19). Sequence analysis of this region by the Sanger method had been complicated by a reproducible band compression such that adjacent G residues were scored as a single residue. Aberrant gel migration for this region also was observed with the Maxam-Gilbert method. Our corrected sequence of the first 205 nucleotide residues of the human cDNA is shown in Fig. 1, along with a comparison with the sequence of the rat β-polymerase cDNA (27). The 5' end of the rat cDNA corresponds to the C residue in the Arg-4 codon of the human cDNA. From this C residue to the ATG codon for Met-18, the start codon assigned previously, each nucleotide difference between human and rat is silent, consistent with the idea that coding properties of this sequence were conserved through protein function. The five nucleotides upstream of the codon for Met-1 have the canonical sequence, CCGCC, for translation initiation sites (8). An in-frame termination codon, TGA, occurs 30 nucleotides upstream from the Met-1 codon. The presence of this termination codon was confirmed in experiments described below.

The cDNA contains only two HaeIII sites, one 5 nucleotides 5' of the start codon (Fig. 1) and the other 2 nucleotides 3' of the termination codon. Hence, the coding region was excised from a plasmid subclone with HaeIII, and the resulting fragment was ligated into pRC23 (Fig. 2). The construction of the final plasmid, pEX17, was confirmed by sequencing ~50 residues of the 5' and 3' ends of the cDNA. With this plasmid, expression of β-polymerase is controlled by the lambda p_L promoter. The expression plasmid was used to transform *Escherichia coli* RRI containing the low-copy-number plasmid pRK248 cIts to ampicillin resistance. Transformants were grown at 30°C, heat induced, and cultured at 42°C.

Escherichia coli transformed with pEX17 and induced did not contain an abundant new dye-stained peptide by sodium dodecyl sulfate-polyacrylamide gel electrophoresis of a crude soluble extract. However, when Western blots of such gels were evaluated with a β-polymerase antiserum, an epitope peptide was detected at a molecular weight of ~39,000. This peptide was not observed in the absence of heat induction, when the cDNA was in reverse orientation in pRC23, or with pRC23 alone.

Demonstration of Upstream Stop Codon

The pRC23 expression system was used to evaluate the presence of a termination codon upstream of the Met-1 codon (Fig. 1). A *Kpn*I fragment containing the coding region and sequence 5' of the Met-1 codon (Fig. 1) was excised and subcloned into the variable-reading-frame derivative of pRC23 (1). This construction would lead to termination of translation at the upstream termination codon (Fig. 1), corresponding to residues 80 to 82 of the cDNA. As a control in the experiment, the HaeIII fragment of the cDNA described above also was cloned into the variable-reading-frame vector. Transformation of *E. coli* with these plasmids resulted in production of an epitope peptide with the HaeIII fragment, as expected, but not with the longer fragment. This indicates that the termination codon 30 nucleotides 5' of the Met-1 (Fig. 1) is indeed correct and strongly suggests that there is no further coding region upstream of Met-1. Finally, we found that the amount of epitope peptide in the induced cells was the same with our standard construct in pRC23, pEX17, and with the variable-reading-frame construct.

Purification and Characteristics of Recombinant β-Polymerase

Recombinant β-polymerase was purified (Table 1) from the soluble extract using step elutions from phosphocellulose and single-stranded DNA–cellulose columns and then Sephacryl-S200 gel filtration (J. Abbotts, D. N. SenGupta, B. Z. Zmudzka, S. G. Widen, V. Notario, and S. H. Wilson,

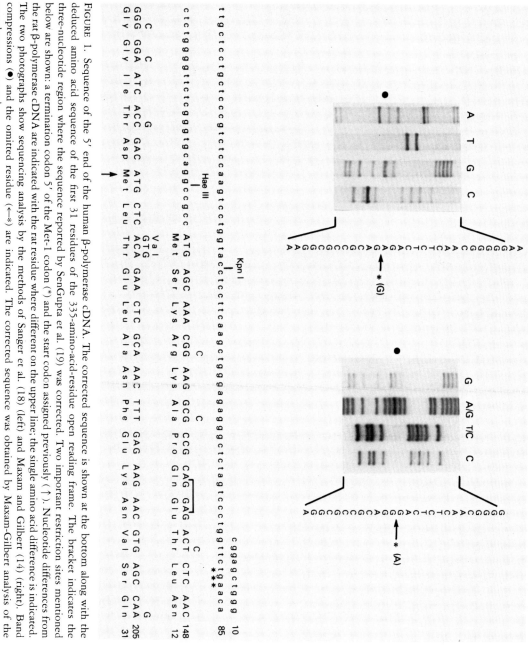

FIGURE 1. Sequence of the 5' end of the human β-polymerase cDNA. The corrected sequence is shown at the bottom along with the deduced amino acid sequence of the first 31 residues of the 335-amino-acid-residue open reading frame. The bracket indicates the three-nucleotide region where the sequence reported by SenGupta et al. (19) was corrected. Two important restriction sites mentioned below are shown: a termination codon 5' of the Met-1 codon (*) and the start codon assigned previously (↑). Nucleotide differences from the rat β-polymerase cDNA are indicated with the rat residue where different on the upper line; the single amino acid difference is indicated. The two photographs show sequencing analysis by the methods of Sanger et al. (18) (left) and Maxam and Gilbert (14) (right). Band compressions (●) and the omitted residue (←*) are indicated. The corrected sequence was obtained by Maxam-Gilbert analysis of the complementary strand.

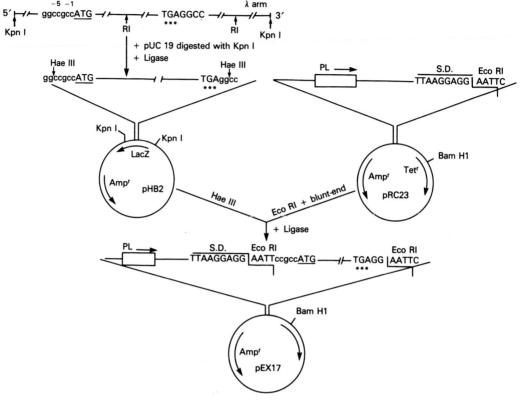

FIGURE 2. Construction of expression plasmid for human β-polymerase. The cDNA coding region contained an internal EcoRI site; therefore, the coding region was removed from the lambda recombinant clone (19) with KpnI. The coding region was excised from the KpnI fragment with HaeIII and subcloned in pUC19 to yield the intermediate plasmid pHB2. The insert junctions (about 50 nucleotides) in pHB2 were sequenced to confirm the construct, and the insert then was subcloned in pRC23 as illustrated.

Biochemistry, in press). The purification was monitored using DNA polymerase assays for β-polymerase activity, including poly (dA) and oligo(dT) as the template and primer, respectively, Mn^{2+}, and pH 8.8. The enzymatic activity formed a sharp symmetrical peak in each column profile. In the Sephacryl-S200 column fractions, the only DNA polymerase activity present formed a symmetrical peak corresponding to a globular protein of 39,000 M_r. When the purified enzyme was examined with the β-polymerase antibody in immunoblotting, the epitope peptide gave a strong signal corresponding in size to the epitope peptide in the crude extract. In the final fraction, the recovery of the 39,000-M_r protein from the soluble extract routinely was ~2 mg/40 g of cells.

The DNA polymerase specific activity of the final fraction (Table 2) was similar to that of the enzyme purified from mouse myeloma (21). The recombinant enzyme exhibited template-primer specificity characteristics of a β-polymerase, but unlike E. coli polymerase I. Hence, on the basis of its size, the presence of β-polymerase antibody epitope, and its template-primer specificity and specific activity, the recombinant enzyme was identified as β-polymerase. This suggests that β-polymerase does not require a eucaryotic cell posttranslational modification for activity. Similar observations on expression of another mammalian β-polymerase in E. coli have been made recently (T. Date, M. Yamaguchi, F. Hirose, Y. Nishimoto, K. Tanihara, and A. Matsukage, personal com-

TABLE 1
Purification of β-Polymerase from *E. coli* Carrying pEX17 and
Grown for 2.5 h at 42°C

Fraction	Volume (ml)	Total protein (mg)	β-Polymerase U
I. Crude lysate[a]	100	650	—[b]
II. Low-speed supernatant fraction	90	400	—
III. Phosphocellulose eluate	10	25	100
IV. Single-stranded DNA–cellulose eluate	10	2	100
V. Sephacryl-S200 peak (pool)	16	0.5	50

[a] The experiment was conducted with 10 g of frozen cell paste. One unit is 1 μmol of dNMP incorporated per h.
[b] —, Not determined.

munication). Compared with the work described here, those investigators used a different strategy for subcloning a rat cDNA and a different expression system, yet they were able to purify a fully active recombinant β-polymerase from an extract of transformed *E. coli*.

CHARACTERISTICS OF THE ENZYME

Processivity

We evaluated three questions about the DNA polymerase activity of recombinant human β-polymerase. Is the enzyme capable of processive synthesis? Does the enzyme have "pause sites" when replicating a long single-stranded template? Can the enzyme fill a short single-stranded gap in double-stranded DNA? Processivity was evaluated using a template-primer system composed of φX174 DNA (Fig. 3). A 16-residue oligonucleotide primer labeled with ^{32}P in the 5' end was annealed to single-stranded φX174 DNA; the single-stranded template region 95 nucleotides downstream of the primer is considered to be free of secondary structure on the basis of computer-derived predictions and previous enzymatic studies (24). The DNA polymerase incubations were conducted for 200 s at a 40-to-1 molar ratio of template-primer to enzyme. Products were then separated by gel electrophoresis, visualized by autoradiography, and quantified. Under these incubation conditions most of the product molecules represented one cycle of enzyme binding, synthesis, and termination. This system allows one to distinguish increased processivity from increased polymerase activity under different conditions, since the size distribution of products is distinct for the two modes of synthesis (3).

In the presence of magnesium, the enzyme was distributive, adding one nucleotide residue to a primer and then dissociating (Fig. 3, lanes 1 and 2). Even after a 20-min incubation allowing for many cycles of enzyme binding, synthesis, and termination, the large majority of products still represent

TABLE 2
Summary of General Properties of Purified Recombinant Human β-Polymerase

Property	Result
Specific activity (with activated DNA, Mg^{2+})	dNMP incorporation of 90 μmol/h per mg
5'→3' Exonuclease activity	None
3'→5' Exonuclease activity	None
Nicking activity against plasmid DNA	None
Native M_r by gel filtration (in 500 mM NaCl)	~39,000
M_r by sodium dodecyl sulfate-polyacrylamide gel electrophoresis	~39,000

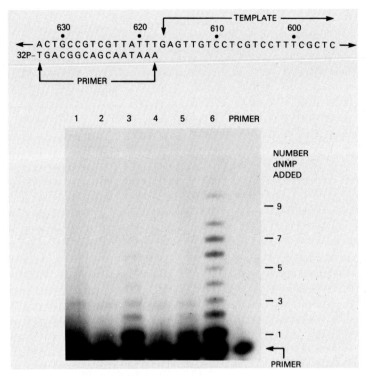

FIGURE 3. Processivity analysis of recombinant β-polymerase. Shown at the top is the synthetic primer labeled with ^{32}P on the 5′ end and hybridized to φX174 (+) strand DNA. The numbers correspond to the φX174 map. DNA synthesis reactions were carried out with a 1:40 molar ratio of polymerase β to template. Products were displayed by electrophoresis and visualized by autoradiography as shown. Reaction conditions were as follows: lane 1, 30 μM each dNTP, 5 mM MgCl$_2$, 200-s incubation; lane 2, 280 μM dNTPs, 6 mM MgCl$_2$, 200-s incubation; lane 3, 280 μM dNTPs, 6 mM MgCl$_2$, 20-min incubation; lane 4, 30 μM dNTPs, 0.5 mM MnCl$_2$, 200-s incubation; lane 5, 280 μM dNTPs, 2 mM MnCl$_2$, 200-s incubation; lane 6, 280 μM dNTPs, 2 mM MnCl$_2$, 20-min incubation.

just one nucleotide addition (Fig. 3, lane 3). In the presence of manganese and 30 μM deoxynucleoside triphosphate (dNTP), synthesis again was distributive (Fig. 3, lane 4). However, with manganese and 280 μM dNTP the enzyme exhibited a very low level of processive synthesis (Fig. 3, lane 5). In the reaction mixtures incubated for 20 min, approximately 5% of the extended primer molecules represented processive addition of 6 to 10 residues. These results indicate that the recombinant enzyme, like the natural enzymes, is generally distributive in its mode of synthesis, and they are partially consistent with results of Wang and Korn (23), who found by kinetic methods that β-polymerase from human KB cells was processive for incorporation of ~5 residues in reactions containing Mn^{2+}. The low level of processive synthesis found here could not have been detected by kinetic methods.

In addition, Kunkel and Loeb (10) found that β-polymerase conducts more in vitro replication in φX reversion assays in the presence of supersaturating levels of dNTP. Under these conditions, the mechanistic explanation for the effects of manganese compared with magnesium and of 280 compared with 30 μM dNTP are not clear. We note that high concentrations of dNTP are required to saturate the nucleotide binding site of *E. coli* polymerase I in equilibrium dialysis experiments (7), and the same could be true of β-polymerase. Saturation of a low-affinity binding site for dNTP could drive the enzyme toward forming a productive complex with the template-primer terminus. These ideas remain to be tested experimentally.

Gap-Filling Ability and Pause Sites

Synthesis was evaluated further using the same template-primer system as in Fig. 3. In contrast to the processivity assay, the ratio of enzyme to template-primer and the time of incubation were adjusted to obtain many cycles of enzyme binding, synthesis, and termination. The results are shown in Fig. 4. With the long single-stranded template, β-polymerase added as many as 50 nucleotides to the primer during a 20-min incubation and then extended these molecules into much longer products after a 60-min incubation. Nine prominent bands of product molecules corresponding to so-called pause sites were noted after the 20-min incubation, but these and the other shorter products were almost completely absent after the 60-min incubation. Most of the pause sites corresponded to an incoming G residue. Of the first seven cases where G was the incoming nucleotide, five were pause sites. This pattern was distinctly different from that seen with *E. coli* polymerase I (not shown) or reported for α-polymerase (24).

Gap-filling synthesis was examined by annealing an unlabeled oligonucleotide of 16 bases to the template, beginning 20 bases in front of the labeled primer. In the presence of this downstream oligonucleotide, the products of DNA synthesis from the labeled primer were not elongated beyond position 638, leaving a gap of 4 to 7 nucleotides before the 5′ terminus of the downstream oligomer. After incubation for a total of 60 min the pattern of labeled product molecules was unchanged, and only a very small amount of limited strand displacement of the downstream oligomer was observed (Fig. 4, lanes 4 and 5). Raising the sodium chloride concentration of the reaction mixture to 100 mM failed to alter the pattern of product molecules. When the experiment in Fig. 4, lanes 1 and 2, was repeated, but with primer 2 labeled instead of primer 1, elongation similar to that with labeled primer 1 was observed, as expected. Annealing unlabeled primer 1 to the template strand had no effect on elongation of primer 2. This confirms the presence of active enzyme in the reaction mixtures shown in Fig. 4, lanes 3 to 5, even though the gap of 4 to 7 nucleotides is not filled.

We were surprised by these results, since the literature contains reports of the ability of β-polymerase to both fill single-stranded gaps and conduct synthesis at nicks (16, 20, 22). We therefore tested natural β-polymerases purified from chick and calf, as well as the preparation from HeLa cells for which strand displacement synthesis had been reported (16). Using equal amounts (activities) of these enzymes, we observed that each showed behavior on the gapped φX174 template similar to that seen with the recombinant enzyme; annealing the second primer dramatically inhibited extension of the upstream primer. When autoradiograms were overexposed, all enzymes, including the recombinant, showed some very minor synthesis into the region of the downstream primer, but significant gap filling and displacement of this primer were not seen.

Figure 5 presents a hypothetical model consistent with these experimental results. As the single-stranded gap size was reduced by elongation of primer 1, the second primer and its elongation products were strongly preferred for productive enzyme binding and synthesis. There was little change in the size of the single-stranded gap after a 20-min incubation, as the polymerase preferred to

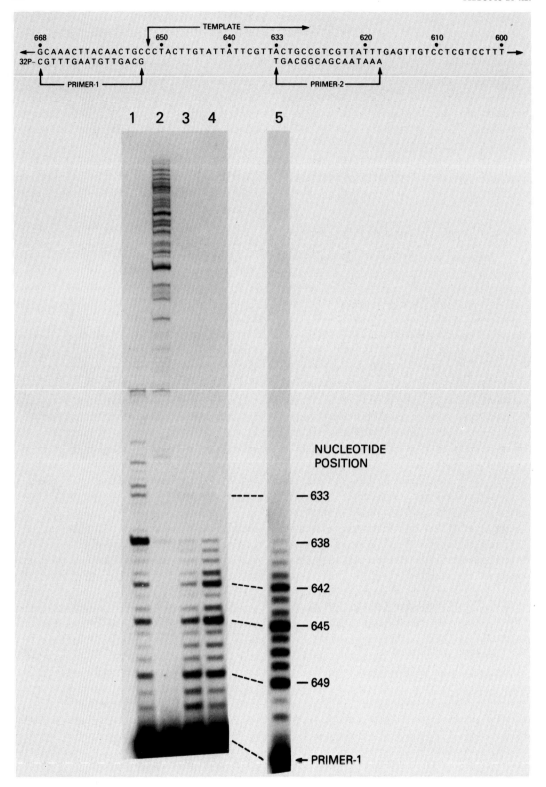

FIGURE 4. Pause site and gap-filling analysis of recombinant β-polymerase with φX174 DNA as template-primer. Photographs of autoradiograms are shown. At the top is shown synthetic primer 1,5' end labeled and hybridized to φX174 DNA. DNA synthesis reactions were carried out as described in the legend of Fig. 3, except that in lanes 3 to 5 both primer 1 and unlabeled primer 2 were hybridized to φX174 DNA. DNA synthesis reaction mixtures contained 20 mM Tris hydrochloride, 2 mM $MnCl_2$, 280 μM dNTPs, a 1:4 ratio of polymerase to template, and 10 mM NaCl for lanes 1 to 4. Additional conditions were as follows: lane 1, only primer 1 hybridized, 20-min incubation; lane 2, conditions identical to lane 1, 60-min incubation; lane 3, both primers hybridized, 20-min incubation; lane 4, both primers hybridized, 60-min incubation; lane 5, both primers hybridized, 100 mM NaCl, 60-min incubation. Nucleotide position refers to the template nucleotide opposite the last nucleotide of the product.

extend the downstream primers rather than fill the DNA gap. These results may have implications for DNA repair. Auxiliary proteins may be necessary to assist β-polymerase in completely filling single-stranded gaps or may direct this polymerase to gapped regions of DNA until they are filled completely.

Characterization of the recombinant enzyme shows that it is very similar to natural β-polymerase. Yet differences were observed in pause site analysis between the recombinant polymerase and the natural enzymes. The chick and calf enzymes showed identical patterns of pause sites on the φX174 template, and their patterns differed slightly from that of the recombinant enzyme (Fig. 6). The early strong pause sites for these enzymes are summarized in Table 3; some sites are common to all polymerases examined, but differences are also seen. The significance of these differences is unknown. The differences could be due to contaminating proteins in the natural enzyme preparations; all of the natural enzymes tested, calf thymus, HeLa, chick embryo, and Novikoff hepatoma, were less pure than the recombinant enzyme. An alternative explanation could be differential posttranslational modification between the polymerase expressed in bacteria and enzymes synthesized in eukaryotic cells.

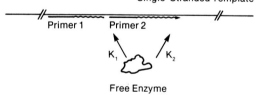

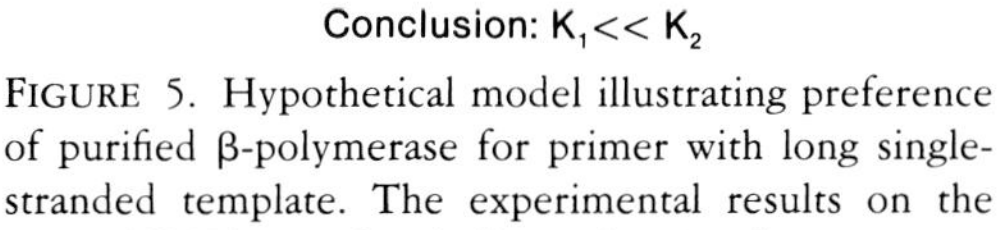

Conclusion: $K_1 \ll K_2$

FIGURE 5. Hypothetical model illustrating preference of purified β-polymerase for primer with long single-stranded template. The experimental results on the φX174 DNA template indicate that β-polymerase recognizes a region 4 to 7 nucleotides downstream of the primer and will utilize alternative primers if this downstream region is double-stranded. The overall association constants (K_1 and K_2) for productive binding of free enzyme are shown. Wavy lines represent synthesized DNA.

TABLE 3
Pause Sites for β-Polymerase in the First 40 Nucleotides Synthesized on the φX174 Template

Nucleotide position,[a] pause site for recombinant enzyme	Incoming dNTP	Nucleotide position, pause site for chick and calf enzymes
	T	650
649	G	649
645	A	
	T	644
642	A	
638	G	638
633	G	
632	A	
629	G	629
626	G	626
617	T	

[a] Nucleotide position refers to the template residue opposite the end of the product, as in Fig. 4. The autoradiogram in Fig. 6 was scanned by densitometry. Bands were compared with several neighboring bands, and those designated strong pause sites showed intensities at least twice as great.

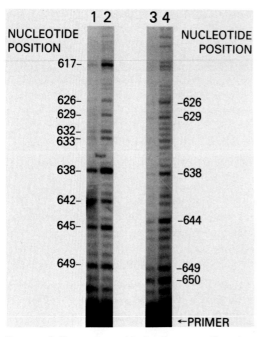

FIGURE 6. Pause sites with β-polymerase. Reaction conditions were as described in the legend of Fig. 4, with only primer 1 labeled and hybridized. Reaction mixtures contained 2 mM $MnCl_2$, 280 μM dNTPs, and a 1:4 molar ratio of polymerase to template. Products of synthesis by the recombinant β-polymerase are displayed in lane 1 (20-min incubation) and lane 2 (60-min incubation); products of synthesis by β-polymerase purified from chick embryo (26) and provided as a gift by Akio Matsukage are displayed in lane 3 (20 min) and lane 4 (60 min). Nucleotide position refers to the template residue opposite the end of the product, as shown in Fig. 4. β-Polymerase purified from calf showed a pause site pattern essentially identical to that of β-polymerase from chick (data not shown).

Modified Template Residues and β-Polymerase Elongation

We examined the effect on chain elongation of selected base modifications that changed hydrogen-bonding properties between the template and the incoming nucleotide. A synthetic template containing the sequence complementary to primer 1 used above plus 24 residues of single-stranded homopolymer template was used. Modified versions of the template included O^6-methyl-dG and N^3-methyl-dT. Recombinant β-polymerase was unable to synthesize past these two modified residues efficiently, even in the presence of all four dNTPs (Fig. 7). Product molecules did not accumulate upstream of the modified template residue, indicating that the enzyme did not detect the presence of the template modification until it attempted to incorporate an incoming nucleotide opposite the modification. At that point, the enzyme showed difficulty incorporating. Even when the polymerase did incorporate at position 30 opposite the modified base, it was not able to further elongate at a significant frequency. This indicates that the enzyme could not use a "loopout" mechanism to synthesize past the modified base and that the normal base pairing at the 3'-OH end of the primer is required for activity. Recombinant β-polymerase was able to synthesize to the end of the template with a normal T nucleotide replacing the methyl derivative at position 30 (data not shown).

Larson et al. (11) had previously shown that methylation produces pause sites for *E. coli* polymerase I but to our knowledge this is the first demonstration that methylated bases that alter hydrogen bonding cause pause sites with β-polymerase. Our results and those of Larson et al. suggest that it may be necessary to remove methyl groups from damaged DNA bases before DNA repair synthesis can occur. That DNA repair polymerases stall at sites opposite a methylated base may provide the cell with a defense against misincorporation, allowing alkyl removal before the extension of synthesis.

OPPORTUNITIES FOR FURTHER INVESTIGATIONS

The availability of the cDNA for human DNA polymerase β and the ability to produce milligram quantities of the protein significantly expand the scope of investigations that can be conducted with this enzyme. Opportunities should be available to further elucidate the role of this polymerase in

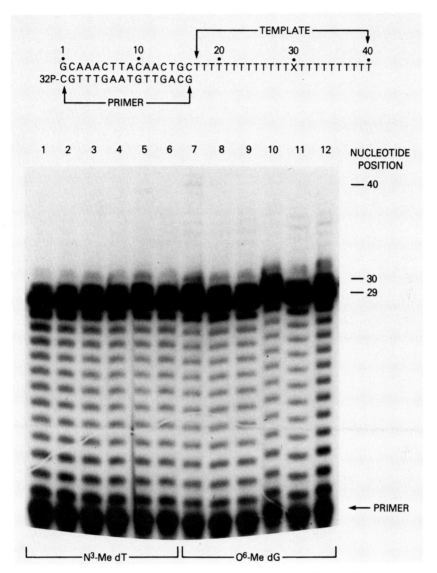

FIGURE 7. Pause site analysis of recombinant β-polymerase with a synthetic homopolymer template containing one methylated base residue. A photograph of an autoradiogram is shown. The templates-primers are illustrated at the top. DNA synthesis reaction mixtures contained a 1:8 molar ratio of enzyme to template and 2 mM MnCl$_2$. Incubation times were 60 min. For lanes 1 to 6, the template contained N^3-methyl-dT at position 30; for lanes 7 to 12, the template contained O^6-methyl-dG at position 30. All reaction mixtures contained 300 μM dATP. Other dNTP concentrations were as follows: lanes 1 and 7, no additional dNTPs; lanes 2 and 8, 600 μM total dATP; lanes 3 and 9, 300 μM dCTP; lanes 4 and 10, 300 μM dGTP; lanes 5 and 11, 300 μM dTTP; lanes 6 and 12, 100 μM each dCTP, dGTP, and dTTP.

DNA repair, with regard to studies of the protein as well as regulation of its expression in the cell.

Among the possibilities for future experiments on protein structure and function, the following investigations are under way. (i) Site-specific mutagenesis of recombinant proteins may allow identification of the polymerase active site. These studies would be facilitated if the protein could be crystallized. (ii) The same techniques might also allow the identification of protein regions in addition to the active site that might play a role in particular polymerase characteristics, such as replication fidelity or processivity. (iii) Considerable homology has been identified between DNA polymerase β and terminal deoxynucleotidyltransferase (13; R. S. Anderson, C. B. Lawrence, S. H. Wilson, and K. L. Beattie, *Gene*, in press). The availability of the cDNA may allow the production of hybrid enzymes with mixed protein domains and the identification of the protein regions responsible for these different activities.

In addition, the availability of milligram quantities of the enzyme might allow the use of DNA polymerase β affinity columns, similar to those used with T4 replication proteins (4), which could identify auxiliary proteins that enhance processivity, fidelity, or other features of β-polymerase. The use of defined template-primer systems, such as those reported here, might allow a more precise understanding of the activities of β-polymerase on damaged DNA. In that regard, we note that Goodman and colleagues (17) employed a defined template to examine the kinetics of dNTP insertion by *Drosophila* DNA polymerase α opposite a single depurinated site.

The availability of cDNAs for mammalian β-polymerases should allow further investigations of gene regulation, and the following projects are under way. (i) Message levels of the enzyme in cells can be examined to determine whether they change with the cell cycle or with exposure to DNA-damaging agents. (ii) Availability of the cDNA has allowed the identification of genomic DNA sequences 5' to the transcribed region in rat (25) and human (this laboratory, unpublished data). This could allow detection of sequences involved in regulatory control during normal conditions or conditions of DNA damage and detection of cellular proteins which control transcription of the β-polymerase message. It is important to note also that sequencing of this 5'-flanking genomic DNA segment and identification of the mRNA start site have provided confirmation that the open reading frame of the β-polymerase mRNA corresponds to the Met-1 residue designated in Fig. 1. (iii) DNA polymerase β has been mapped to human chromosome 8 (15). The availability of the cDNA and knowledge of the genomic 5' regulatory region might allow identification of the molecular bases for some diseases of DNA repair. If these investigations of structure-function relationships of human DNA polymerase β and of the cellular regulation of the enzyme can be carried out, they might provide a dramatic expansion in understanding of DNA repair phenomena.

ACKNOWLEDGMENTS. We thank Catherine Parrott for expert technical assistance, Vishram Kedar for assistance with experiments, and Joseph Shiloach for preparation of cultures of pEX17-transformed *E. coli*.

LITERATURE CITED

1. **Crowl, R., C. Seamans, P. Lomedico, and S. McAndrew.** 1985. Versatile expression vectors for high-level synthesis of cloned gene products in *Escherichia coli*. *Gene* **38**:31–38.
2. **Detera, S. D., S. P. Becerra, J. A. Swack, and S. H. Wilson.** 1981. Studies on the mechanism of DNA polymerase α: nascent chain elongation, steady state kinetics, and the initiation phase of DNA synthesis. *J. Biol. Chem.* **256**:6933–6943.
3. **Detera, S. D., and S. H. Wilson.** 1982. Studies on the mechanism of *Escherichia coli* DNA polymerase

I large fragment: chain termination and modulation by polynucleotides. *J. Biol. Chem.* 257:9770–9780.
4. Formosa, T., R. L. Burke, and B. M. Alberts. 1983. Affinity purification of bacteriophage T4 proteins essential for DNA replication and genetic recombination. *Proc. Natl. Acad. Sci. USA* 80:2442–2446.
5. Fry, M. 1983. Eukaryotic DNA polymerases, p. 39–92. *In* S. T. Jacob (ed.), *Enzymes of Nucleic Acid Synthesis and Modification*, vol. 1. CRC Press, Inc., Boca Raton, Fla.
6. Fry, M., and L. A. Loeb. 1986. *Animal Cell DNA Polymerases*. CRC Press, Inc., Boca Raton, Fla.
7. Kornberg, A. 1969. Active center of DNA polymerase. *Science* 163:1410–1418.
8. Kozak, M. 1984. Compilation and analysis of sequences upsteam from the translational start site in eukaryotic mRNAs. *Nucleic Acids Res.* 12:857–872.
9. Kunkel, T. A., and P. S. Alexander. 1986. The base substitution fidelity of eukaryotic DNA polymerases: mispairing frequencies, site preferences, insertion preferences, and base substitution by dislocation. *J. Biol. Chem.* 261:160–166.
10. Kunkel, T. A., and L. A. Loeb. 1981. Fidelity of mammalian DNA polymerases. *Science* 213:765–767.
11. Larson, K., J. Sahm, R. Shenkar, and B. Strauss. 1985. Methylation-induced blocks to in vitro replication. *Mutat. Res.* 150:77–84.
12. Loeb, L. A., and T. A. Kunkel. 1982. Fidelity of DNA synthesis. *Annu. Rev. Biochem.* 51:429–457.
13. Matsukage, A., K. Nishikawa, T. Ooi, Y. Seto, and M. Yamaguchi. 1987. Homology between mammalian DNA polymerase β and terminal deoxynucleotidyltransferase. *J. Biol. Chem.* 262:8960–8962.
14. Maxam, A. M., and W. Gilbert. 1980. Sequencing end-labeled DNA with base-specific chemical cleavages. *Methods Enzymol.* 65:499–560.
15. McBride, O. W., B. Z. Zmudzka, and S. H. Wilson. 1987. Chromosomal location of the human gene for DNA polymerase β. *Proc. Natl. Acad. Sci. USA* 84:503–507.
16. Mosbaugh, D. W., and S. Linn. 1983. Excision repair and DNA synthesis with a combination of HeLa DNA polymerase β and DNase V. *J. Biol. Chem.* 258:108–118.
17. Randall, S. K., R. Eritja, B. E. Kaplan, J. Petruska, and M. F. Goodman. 1987. Nucleotide insertion kinetics opposite abasic lesions in DNA. *J. Biol. Chem.* 262:6864–6870.
18. Sanger, F., S. Nicklen, and A. R. Coulson. 1977. DNA sequencing with chain-terminating inhibitors. *Proc. Natl. Acad. Sci. USA* 74:5463–5467.
19. SenGupta, D. N., B. Z. Zmudzka, P. Kumar, F. Cobianchi, J. Skowronski, and S. H. Wilson. 1986. Sequence of human DNA polymerase β mRNA obtained through cDNA cloning. *Biochem. Biophys. Res. Commun.* 136:341–347.
20. Siedlecki, J. A., R. Nowak, A. Soltyk, and B. Zmudzka. 1981. Net DNA synthesis catalyzed by calf thymus DNA polymerase β. *Acta Biochim. Pol.* 28:157–173.
21. Tanabe, K., E. W. Bohn, and S. H. Wilson. 1979. Steady-state kinetics of mouse DNA polymerase β. *Biochemistry* 18:3401–3406.
22. Wang, T. S.-F., and D. Korn. 1980. Reactivity of KB cell deoxyribonucleic acid polymerases α and β with nicked and gapped deoxyribonucleic acid. *Biochemistry* 19:1782–1790.
23. Wang, T. S.-F., and D. Korn. 1982. Specificity of the catalytic interaction of human DNA polymerase β with nucleic acid substrates. *Biochemistry* 21:1597–1608.
24. Weaver, D. T., and M. L. DePamphilis. 1982. Specific sequences in native DNA that arrest synthesis by DNA polymerase α. *J. Biol. Chem.* 257:2075–2086.
25. Yamaguchi, M., F. Hirose, Y. Hayashi, Y. Nishimoto, and A. Matsukage. 1987. Murine DNA polymerase β gene: mapping of transcription initiation sites and the nucleotide sequence of the putative promoter region. *Mol. Cell. Biol.* 7:2012–2018.
26. Yamaguchi, M., K. Tanabe, Y. N. Taguchi, M. Nishizawa, T. Takahashi, and A. Matsukage. 1980. Chick embryo DNA polymerase β: purified enzyme consists of a single $M_r = 40,000$ polypeptide. *J. Biol. Chem.* 255:9942–9948.
27. Zmudzka, B. Z., D. SenGupta, A. Matsukage, F. Cobianchi, P. Kumar, and S. H. Wilson. 1986. Structure of rat DNA polymerase β revealed by partial amino acid sequencing and cDNA cloning. *Proc. Natl. Acad. Sci. USA* 83:5106–5110.

Chapter 7

Human Placental DNA Polymerases δ and α

Marietta Y. W. T. Lee

Despite several decades of study, the mammalian polymerases remain a major challenge in terms of the elucidation of their enzymology and properties. Of the three types of DNA polymerase, α, β, and γ, it is generally accepted that polymerase α is involved in DNA replication (see references 8 and 18 for reviews). A general theme in the study of polymerase α is that it appears in multiple forms and is not readily obtained in pure form and in high yield. A consistent finding is that of multiple polypeptide compositions. General observations are of a group of large polypeptides of 120 to 180 kilodaltons (kDa) with which the polymerase activity is associated and a group of smaller polypeptides of 40 to 70 kDa; current evidence favors a "core" catalytic polypeptide of about 180 kDa with several smaller subunits (18). Besides having multiple subunits, DNA polymerase α may also be associated to varying degrees with other enzyme activities and protein factors required for the assembly of a functional replication complex.

It was long considered that mammalian

Marietta Y. W. T. Lee • Department of Medicine, University of Miami School of Medicine, Miami, Florida 33101.

DNA polymerase α characteristically does not possess an associated $3'\rightarrow 5'$ exonuclease activity. In 1976 an apparently new type of DNA polymerase was discovered in rabbit reticulocytes (6). This new DNA polymerase was unique among mammalian polymerases in that it possessed a tightly associated $3'\rightarrow 5'$ exonuclease activity, and it was named DNA polymerase δ to distinguish it from the currently known mammalian DNA polymerases (6, 7). This finding was contrary to prevailing views on mammalian DNA polymerases and was not immediately accepted, since numerous earlier studies had failed to reveal the existence of such an enzyme and, moreover, the apparent association of exonuclease activity could be explained by the copurification of a contaminating exonuclease activity. Perhaps for these reasons, the study of this enzyme has largely been localized to three laboratories at the University of Miami.

The existence of this enzyme became more firmly established when it was shown that it could be isolated from calf thymus (26–28) and also from human placenta (30, 31). The enzymatic properties of the enzyme, in relation to the $3'\rightarrow 5'$ exonuclease activity, were strikingly similar to the findings originally made with *Escherichia coli* polymerase I (pol I), including a preference for the hydrolysis of mismatched primer termini from double-stranded synthetic substrates (7, 25, 26, 30). Of particular note also is the selective inhibition of the 3'→5' exonuclease activity of polymerase δ by 5'-AMP (7, 25); this was shown also to be a property of *E. coli* pol I (39), which had previously been shown to have a 5'-AMP binding site (22). The possession of a 5'-AMP binding site by polymerase δ is a distinguishing feature which allows its separation from DNA polymerase α by ligand chromatography on 5'-AMP–agarose (13, 32). The 3'→5' exonuclease activity characteristically remained associated with DNA polymerase activity (6, 13, 26–28, 30) throughout a variety of chromatographic and electrophoretic separation procedures. These studies established the presence of an enzyme in eucaryotic cells which possessed the requisite properties to function as a proofreading DNA polymerase.

ISOLATION AND CHARACTERIZATION OF HUMAN PLACENTAL DNA POLYMERASE δ

Work in my laboratory has focused on the human placental form of DNA polymerase. The isolation of polymerase δ poses many of the problems previously encountered in the study of DNA polymerase α. These are the small amounts of enzyme protein available; the persistence of satellite polypeptides which could be contaminants, "accessory" proteins, or other subunits of a holoenzyme; the presence of multiple enzyme forms; and the susceptibility of the enzyme to proteolysis. The latter problem was solved by the use of a cocktail of protease inhibitors and by systematic monitoring for the presence of protease activity; most of the endogenous protease activity was found to be separated by hydrophobic chromatography, which was introduced at an early stage in the purification. Human placental DNA polymerase δ was routinely obtained as preparations which contained a single major polypeptide of about 170 kDa (30); nevertheless, it should be noted that our preparations often contained multiple lower-molecular-weight polypeptides, a phenomenon similar to that observed for a number of mammalian polymerase α preparations. However, we have not observed these accompanying polypeptides in amounts which could represent a stoichiometry expected of true subunits; this situation is similar to that reported for many preparations of DNA polymerase α (8, 18, 36, 41).

Our studies of the human enzyme have led to identification of the 170-kDa polypeptide as the native size of the polymerase catalytic subunit (30). This is based on activity staining following the renaturation of the enzyme in situ after sodium dodecyl sulfate-polyacrylamide gel electrophoresis (SDS-PAGE) (Fig. 1) as well as Western blotting of the 170-kDa polypeptide by murine antisera in both purified enzyme preparations and tissue extracts. This finding was of importance since it was the first direct evidence for the association of catalytic activity with a specific polypeptide for polymerase δ and also identified the enzyme as a high-molecular-weight polymerase similar in size to DNA polymerase α, thus bringing into focus the potential relationships of the two. Occasional labeling of smaller polypeptides was observed by activity staining (e.g., see Fig. 1), depending on the enzyme preparations; we have observed activity staining of immunoreactive polypeptides as small as 60 kDa. The susceptibility of the catalytic polypeptide of human DNA polymerase δ to proteolysis to smaller forms which are detectable by activity staining is interesting, as it suggests that the functional domain for catalytic activity is relatively small and capable of independent renaturation to a functional state.

A comparison of the polypeptide compositions of the different preparations of polymerase δ that have been reported is

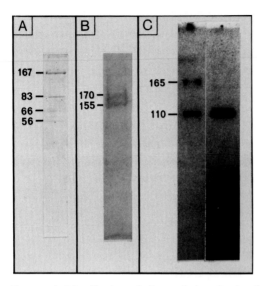

FIGURE 1. Identification of the catalytic subunit of DNA polymerase δ. (A) Purified DNA polymerase δ (step 8, reference 30) was run on a 10% acrylamide slab gel, which was then stained for protein by the silver staining procedure. (B) DNA polymerase δ (5 U), purified to step 5 (hydroxylapatite chromatography, reference 9), was run on a 7% acrylamide slab gel and then stained for activity. (C) Purified DNA polymerase δ (7 U, step 9, reference 30) was run on a 10% acrylamide slab gel and stained for activity, as shown in the left lane. The activity stain for purified *E. coli* pol I (110 kDa), run in parallel on the same gel, is shown in the right lane.

given in Table 1. Highly purified preparations of the reticulocyte enzyme have been obtained with a single polypeptide of 122 kDa (19) and of the calf thymus enzyme as a preparation with a native molecular weight of 173,000 and two polypeptides of 125 and 48 kDa (27). Two forms of calf thymus polymerases recently studied "δ1" and "δ2," are obviously cruder preparations with multiple polypeptides (13). We have observed the conversion of the human 170-kDa polypeptide to forms of 120 kDa and lower, and we have been able to isolate human polymerase δ as a 120-kDa polypeptide form. This is illustrated in Fig. 2, which shows the Western blotting of several DNA polymerase δ preparations with a murine polyclonal antiserum. Blotting of preparations stored for several months also led to the appearance of smaller immunoreactive polypeptides (Fig. 2). It seems likely that the differences in data for the human, calf thymus, and reticulocyte enzymes are not due to species differences but more probably reflect posthomogenization modification by proteolysis.

The enzymatic properties of the human form of DNA polymerase δ parallel those that have been described for the calf thymus and reticulocyte enzymes. All three enzymes exhibit proofreading ability in vitro (5, 26, 30), supporting the idea that the $3' \rightarrow 5'$ exonuclease activity may be one of the mechanisms by which eucaryotic cells maintain a high degree of fidelity during DNA replication. The fidelity of DNA synthesis by the calf thymus enzyme has also been examined using an M13 *lacZ*α nonsense codon reversion assay (24). Some differences have been reported for the template speci-

TABLE 1
Properties of DNA Polymerase δ

Source	Molecular size (kDa)	Polypeptides (kDa)	Activity stain[a] (kDa)	Reference
Rabbit bone marrow	122	122	ND	19
Fetal calf thymus	173	125, 48	ND	27
Calf thymus, "δ1"	240	245, 164, 120, 110, 60, 45	ND	13
Calf thymus, "δ2"	290	245, 135, 110, 60, 45	ND	13
Human placenta	170	167, 83, 66, 56	170	30
HeLa cell	—	220	ND	33

[a] ND, Not done.

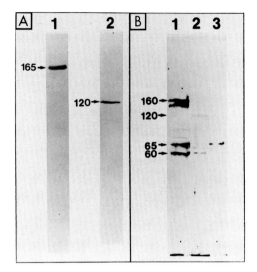

FIGURE 2. Immunoblots of several DNA polymerase δ preparations. Several DNA polymerase δ preparations were examined by immunoblotting with a murine polyclonal antiserum after electrophoresis on 5 to 15% gradient gels. (A) Lane 1, Immunoblot of a DNA polymerase δ preparation immediately after preparation (5 U of enzyme); lane 2, immunoblot of the same preparation (5 U) as in lane 1 after 2 weeks of storage at $-70°C$. (B) Lanes 1 to 3, Immunoblots of three separate DNA polymerase δ preparations which had been stored for periods of several months at $-70°C$ (10, 4, and 3 U of enzyme, respectively).

ficities of different DNA polymerase δ preparations. The fetal calf thymus (27) and human placenta enzymes (30) prefer activated alternating copolymers such as poly (dA-dT) and poly(dG-dC), while the bone marrow (19) and calf thymus δ1 and δ2 enzymes (13) show a preference for poly (dA)/oligo(dT) as the template/primer. The fetal calf thymus (27, 40) and human placenta (30) enzymes are not able to use poly(dA) and oligo(dT) unless an auxiliary protein is added. These differences may be due to differences in purification procedures and in the degree of purity of the preparations. The DNA polymerase δ preparations from fetal calf thymus (40) and human placenta (30) do not have associated primase activities, whereas the δ1 and δ2 enzyme preparations, which exhibit highly heterogeneous polypeptide compositions, do have primase activity (13), which potentially could originate from an incomplete separation from DNA polymerase α.

ASSOCIATION OF EXONUCLEASE WITH THE POLYMERASE ACTIVITY OF δ

A primary goal of our work was to attain the rigorous isolation of DNA polymerase δ in order to obtain definitive information regarding the nature of the association of the $3'\rightarrow5'$ exonuclease activity. As a working definition, we consider that the term "associated" in the context of DNA polymerase δ could mean an association of two proteins which could be present in the cell at different concentrations; the degree of the association under given conditions then depends on the association constant. We interpret the term "intrinsic" as meaning that the $3'\rightarrow5'$ exonuclease accompanies the DNA polymerase catalytic polypeptide as a subunit of a holoenzyme, in a strict conventional sense which implies a fixed molar ratio of the two polypeptides as well as a strong protein-protein interaction and a regulated expression of both subunits that precludes their appearance in a free state in the cell. This definition also includes the case in which the $3'\rightarrow5'$ exonuclease active site is located on the same polypeptide as the DNA polymerase activity.

The current evidence for the association of the $3'\rightarrow5'$ exonuclease with DNA polymerase δ favors the idea that this activity is intrinsic according to the definition given above. This evidence includes the failure to separate the two activities by a variety of chromatographic and electrophoretic procedures as well as ligand chromatography, the presence of both activities in preparations consisting a single major polypeptide, and the inhibition of both polymerase and $3'\rightarrow5'$ exonuclease activities by aphidicolin (19, 27, 30). We can suggest speculatively that $3'\rightarrow5'$ activity is likely to be located on the same polypeptide as the polymerase activity. Rigorous and direct identification of the location of the active site for the 3'→5' exonuclease activity remains to be accomplished and requires the development of active-site-labeling reagents that are specific for the exonuclease activity.

HUMAN PLACENTAL DNA POLYMERASE α

Despite the large variation in polypeptide sizes reported for DNA polymerase α from a variety of sources, there is a growing body of evidence that the catalytic polypeptide for this enzyme is of 180 kDa (8, 18, 21, 41, 42). The question of the relationships of polymerases δ and α became a concern as the similarities of polypeptide size became apparent, and we therefore undertook the isolation of DNA polymerase α from human placenta to make comparative studies of the two enzymes. Our studies of human placental DNA polymerase α, which is readily separated from polymerase δ activities by several chromatographic procedures, have led to the isolation to apparent homogeneity of a form which contains a 180-kDa polypeptide as the major component, accompanied by several smaller bands. This polypeptide is the catalytic polypeptide, as shown by activity staining following renaturation in situ after SDS-PAGE (M. Y. W. T. Lee, manuscript in preparation). This polymerase conforms in enzymatic properties to those previously described for DNA polymerase α from other mammalian sources.

PROTEIN ACTIVATING FACTOR OF HUMAN PLACENTAL DNA POLYMERASE δ

The original studies of calf thymus polymerase δ revealed the presence of a protein factor of 37 kDa which enhanced its activity (27). The calf thymus protein was further characterized as an auxiliary factor for polymerase δ (40) which markedly enhances its processivity (38), and it has recently been identified as being identical to proliferating cell nuclear antigen (PCNA) (4, 38), a cell-cycle-specific protein whose expression is tightly linked to the S phase of the cell cycle (9) and which is necessary for the replication of simian virus 40 in vitro (37). We have also isolated this factor from human placenta. It has an apparent molecular size of 36 kDa (30) and stimulates DNA polymerase δ but not α (Fig. 3). Polyclonal and monoclonal antibodies to the humn placenta factor were raised. These antibodies immunoblot the 36-kDa polypeptide (Fig. 3).

ANTIBODIES TO HUMAN PLACENTAL DNA POLYMERASE δ

We have prepared both polyclonal and monoclonal antibodies to human placental DNA polymerase δ. Polyclonal antisera to polymerase δ were raised in 11 BALB/c mice. Although the amounts of antisera obtained were small, the use of this animal was found to be effective in terms of both the smaller amounts of enzyme protein needed for immunization and the specificity and titer of the antibodies. Examination of these antisera showed that they invariably immunoblot a 170-kDa polypeptide in both crude tissue extracts and the purified enzyme preparations in a relatively specific manner. This is illustrated for three of our antisera in Fig. 4. Since these antisera are derived from different mice, their properties are not necessarily identical, and some of these also immunoblot smaller polypeptides; as shown in Fig. 4, two of the three antisera also blot a 36-kDa polypeptide. Immunoblotting of a two-dimensional SDS-PAGE gel of a purified polymerase δ preparation shows the blotting of a single polypeptide of the appropriate size (170 kDa) and isoelectric point (pH 5.8) for DNA polymerase δ (Fig. 5).

We have obtained a panel of monoclo-

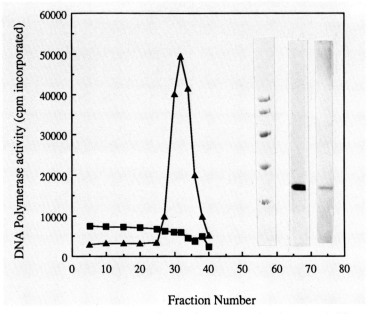

FIGURE 3. Protein activating factor of human DNA polymerase δ. The protein activating factor was purified and 20 μg of protein was run on a 5% nondenaturing gel. The gel was sliced into 40 fractions and the fractions extracted in buffer. The extracts were then assayed for ability to activate DNA polymerase δ (▲) or polymerase α (■), using poly(dA) and oligo(dT)$_{20:1}$ as the template and primer. (Insert) Extracts of fractions 29 to 35 were pooled and examined by SDS-PAGE, as shown in the center lane. The left lane shows the protein standards (β-galactosidase, 116 kDa; phosphorylase b, 97 kDa; bovine serum albumin, 68 kDa; ovalbumin, 45 kDa; carbonic anhydrase, 29 kDa). The right lane shows an immunoblot of the activating factor, using mouse monoclonal antibody 22.

nal antibodies to DNA polymerase δ (Lee, in preparation); the selection of these antibodies proved to be a complex task because of the small amounts of enzyme protein available and also because of the likely production of antibodies to other proteins which might be present in our polymerase δ preparations. The screening of these antibodies involved both identification of the polypeptides which are blotted and examination of the inhibition of DNA polymerase δ. A summary of the properties of these antibodies is given in Table 2. Included in the list are antibodies to the 170-kDa polypeptide of polymerase δ; these were identified by immunoblotting experiments. Also included in Table 2 are antibodies which do not blot the 170-kDa polypeptide but react with smaller polypeptides in highly purified DNA polymerase δ preparations. We have retained these for further study as they potentially represent antibodies to polypeptides associated with the catalytic polypeptide, rather than fortuitous contaminants. Of these, antibody 223 is notable as it inhibits polymerase δ and α activity although it does not immunoblot the catalytic polypeptide. In addition, we recently identified one of these monoclonal antibodies as one which inhibits both the DNA polymerase and exonuclease activities of DNA polymerase δ. It is of interest that two monoclonal antibodies which recognize a 36-kDa polypeptide were also obtained; one of them has been identified as being directed toward the activating factor for polymerase δ (see Fig. 3), and in

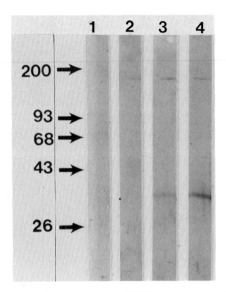

FIGURE 4. Immunoblotting of DNA polymerase δ by mouse polyclonal antisera. Partially purified DNA polymerase δ (step 5, reference 9) was immunoblotted using three separate mouse antisera at dilution of 1/1,000 (lanes 2 to 4). Lane 1, Control immunoblot with nonimmune mouse serum.

FIGURE 5. Immunoblotting of DNA polymerase δ after two-dimensional SDS-PAGE. Isoelectric focusing of DNA polymerase δ (10 U) was performed using LKB ampholines (pH 3.5 to 10). The gel was run in the second dimension on SDS-PAGE. The molecular size (vertical axis) of the immunoreactive polypeptide was estimated as 170 kDa, using prestained protein standards. The isoelectric point (horizontal axis) was estimated as pH 5.5, using trypsin inhibitor (pI, 4.55), carbonic anhydrase (pI, 6.57), and methyl red (pI, 3.75) as standards.

preliminary experiments it was found also to blot human proliferating cell nuclear antigen. These antibodies provide a set of potentially important reagents for future studies of polymerase δ.

COMPARISONS OF HUMAN PLACENTAL DNA POLYMERASES δ AND α

The availability of both purified DNA polymerases δ and α from human placenta and of antibodies to DNA polymerase δ has allowed us to examine the potential relationships of the two polymerases. A summary of our studies of the two proteins is given below.

They can be immunochemically distinguished. Individual polyclonal antisera and monoclonal antibodies to polymerase δ inhibit polymerase δ but not polymerase α and immunoblot the 170-kDa polymerase δ polypeptide but not the 180-kDa catalytic polypeptide of polymerase α. Polypeptides from polymerase δ alone could be immunoprecipitated from radioiodinated human polymerase α and δ preparations with polyclonal antisera to DNA polymerase δ (Fig. 6). The monoclonal antibodies to KB cell DNA polymerase α, SJK 287-38 and SJK 132-20, are potent inhibitors of human placental DNA polymerase α but do not inhibit DNA polymerase δ.

Their catalytic polypeptides are physically distinct. The 180-kDa DNA polymerase α catalytic polypeptide isolated from human placenta is distinctly larger than the 165- to 170-kDa DNA polymerase δ catalytic polypeptide when examined in the same SDS-PAGE gels.

Their enzymatic activities respond differently to various effectors. The two enzyme activities are differentially affected by dimethyl sulfoxide, which we have shown to activate polymerase δ and inhibit polymerase α (29). The nucleotide analog N^2-(p-n-butylphenyl)-9-(2-deoxy-β-D-ribofuranosyl)guanine 5'-triphosphate (BuPdGTP) and 2-(p-n-butylanilino)-9-(2-deoxy-β-D-ribofuranosyl)adenine 5'-triphosphate (BuAdATP) have

TABLE 2
Monoclonal Antibodies to DNA Polymerase δ

Antibody	Type[a]	Inhibition		Polypeptide immunoblotted (kDa)
		Polymerase δ	Polymerase α	
8	IgG1	+	−	170
236	IgG1	+	−	170
23	IgM	+	−	170
423	IgM	+	−	170
496	IgG1	−	−	170
548	IgG2b	−	−	125
69	IgG1	−	−	120
223	IgM	+	+	88
74	IgG1	−	−	62
61	IgM	−	−	36
22	IgM	−	−	36
51	IgM	−	−	21

[a] IgG1, Immunoglobulin G1.

differential inhibition patterns on DNA polymerases α and δ (31). DNA polymerase α is exquisitely sensitive to these analogs whereas polymerase δ is relatively insensitive. This differential sensitivity makes these analogs useful for the selective assay of the two enzyme activities in tissue extracts (31). Both enzymes are potently inhibited by aphidicolin; however, our data show that human polymerase δ is significantly more sensitive than polymerase α (31). Finally, human DNA polymerase δ (30), like the calf thymus enzyme (27, 40), is stimulated by the 36-kDa auxiliary protein, whereas human DNA polymerase α is not affected (Fig. 3).

The two enzymes exhibit different chromatographic properties and template primer specificities. In addition, purified human DNA polymerase δ does not have associated primase activities (30), in contrast to numerous findings for polymerase α, including findings for the human placental preparations that we have studied.

The summary given above indicates that human DNA polymerase δ can be distinguished operationally from DNA polymerase α by a number of different criteria. Our evidence is that human polymerase δ is a discrete protein; more precisely, the catalytic polypeptides of polymerases α and δ are different. This appears to eliminate the possibility that polymerases α and δ are derived from a common precursor originating from a single gene by proteolytic conversion or posttranslational modifications such as phosphorylation. That the polymerases are totally unrelated also seems unlikely. It should be noted that our immunochemical studies so far do not eliminate the possibility that polymerases α and δ may have some structural identity or share common epitopes. In fact, we have obtained preliminary evidence for monoclonal antibodies that are able to inhibit both DNA polymerases α and δ. Apart from teleological arguments of functional commonality, there are significant parallels in the properties of the two enzymes, for instance, their polypeptide sizes and compositions, similarities in terms of proteolytic modification to smaller active polypeptides, and common sensitivity to aphidicolin, all of which makes it a reasonable hypothesis that they do have some structural relationship. Thus, the key issue that remains is whether these two proteins are structurally related and, if so, the extent of this relationship. The definition of this relationship must come from structural information, and in this regard the cloning of the cDNAs for polymerase α (42a) and current efforts to clone

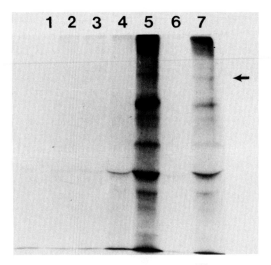

FIGURE. 6. Immunoprecipitation of ^{125}I-radioiodinated human DNA polymerases δ and α with two antisera to DNA polymerase δ. Partially purified human DNA polymerases δ and α (both purified to the stage of hydroxylapatite chromatography) were radioiodinated by the iodogen method. The ^{125}I-labeled enzymes (30 μl containing 120,000 dpm for polymerase α and 140,000 dpm for polymerase δ) were incubated with polyclonal mouse antisera to polymerase δ (10 μl) for 14 h at 4°C and then adsorbed to Pansorbin. The immunoprecipitates were washed, dissolved in SDS sample buffer, subjected to SDS-PAGE, and autoradiographed. Lanes: 1, polymerase α with nonimmune control mouse serum; 2, polymerase α with mouse antiserum A; 3, polymerase α with mouse antiserum B; 4, polymerase δ with nonimmune control mouse serum; 5, polymerase δ with mouse antiserum A; 6, polymerase δ with a purified nonspecific mouse monoclonal antibody; 7, polymerase δ with mouse antiserum B. The arrow marks the position of the 170-kDa band. The band immediately below it has a relative molecular mass of about 110 kDa.

polymerase δ should in the future provide a resolution.

FUNCTIONS OF DNA POLYMERASE δ

The sensitivity to aphidicolin has long been a major source of evidence that DNA polymerase α is involved in eucaryotic DNA replication, because of its apparent unique sensitivity and because aphidicolin also inhibits DNA synthesis in intact cells. The sensitivity of DNA polymerase δ to this inhibitor places a new light on this evidence, and it can now be argued that DNA polymerase δ is also a candidate for a role in DNA replication in eucaryotic cells. This raises the question of whether DNA polymerase δ represents a significant fraction of the cellular DNA polymerase activities. In studies of the ontogeny of DNA polymerases α and δ in rat heart, we have obtained data which indicate that polymerase δ is a significant fraction of the assayable DNA polymerase in heart extracts, from 30 to 40% by the use of BuAdATP and monoclonal antibodies as specific inhibitors of DNA polymerase α (43). Similar values have been found in other cell types (5, 13, 15, 20, 30). This by no means provides an estimate of the relative amounts of enzyme protein, but it does serve to establish that polymerase δ is unlikely to be an insignificant activity in terms of its catalytic potential in relation to DNA replication.

The question of the subcellular localization of DNA polymerase δ is important in the consideration of its potential role in cellular DNA replication. Using monoclonal antibodies to DNA polymerase δ, we have shown by optical immunofluorescence studies that it is indeed localized to the nucleus of both normal and neoplastic cells (Lee, in preparation). The existence of specific antibodies for both polymerases α and δ offers opportunities for future studies to delineate the specific functions of these two enzymes. In addition, the analogs BuAdATP and BuPdGTP, which are selective inhibitors of polymerase α (5, 31), are likely to be of continuing utility in these studies.

Recent studies have now provided significant evidence for a role of DNA polymerase δ in eucaryotic DNA replication. Evidence supporting a role of polymerase δ in DNA repair and replication (15, 16, 20) processes has also come from studies of the effects of BuPdGTP, aphidicolin, 2′,3′-dideoxythymidine 5′-triphosphate, and monoclonal antibodies to DNA polymerase α on these processes in permeabilized cell systems. The finding that PCNA is necessary for simian virus 40 replication in an in vitro cell-free system (37) and its identification as a polymerase δ activating factor which acts by stimulating its processivity (38, 40) provide provocative evidence for a replicative function for polymerase δ. It is striking also that a HeLa cell factor required for repair synthesis has been isolated and identified as DNA polymerase δ (33). These findings provide a strong impetus to further investigations to define the precise role of DNA polymerase δ at the replication fork.

WHITHER DNA POLYMERASE δ?

Current studies reflect a change in viewpoint regarding eucaryotic replicative DNA polymerases. The emergence of a clearer view of the complex nature of the procaryotic replication apparatus, in contrast with the simpler state of information on mammalian systems, which, until recently, was focused largely on the enzymology of polymerase α, now provides a perspective in terms of parallels in the properties of eucaryotic polymerases with those of the procaryotic systems. This is evidenced, for example, in the descriptions of accessory proteins involved in DNA replication and of polymerase α forms possessing associated primase activity (for reviews, see references 8 and 18). Thus, the discovery of polymerase δ, with its associated $3' \rightarrow 5'$ exonuclease activity, may be viewed retrospectively as part of a general direction in the study of eucaryotic DNA polymerases, in which activities with functions analogous to those described in procaryotic systems are emerging.

The challenge for the future is the rigorous elucidation of the replication apparatus of the eucaryotic cell and a definition of the role of these two polymerases. From both conceptual and speculative viewpoints, the advances in our understanding of the procaryotic and phage replication provide models for future explorations. The possibility that discrete DNA polymerases are involved in synthesis of the leading and lagging strands at the replication fork (23) leads to the suggestion that polymerases α and δ are logical candidates for these roles (3). The association of primase with DNA polymerase α points to a role in the discontinuous synthesis of Okazaki fragments in the lagging strand, while the enhancement of the processivity of DNA polymerase δ by PCNA (38) is consistent with a role in the continuous synthesis of DNA in the leading strand. In addition, it may be that there exists a functional and/or spatial coupling of the two enzymes within a replication complex which provides the coordinated synthesis of both leading and lagging strands necessary for the orderly progression of the replication fork. Future studies of the functions of polymerases δ and α in in vitro systems (14, 37) for replication of the simian virus 40 chromosome and in permeabilized cell systems, together with the availability of specific antibodies to the polymerases and their associated factors, should provide significant information on these possibilities.

The question of the relationship of polymerases α and δ is discussed above, but the definition of this relationship may be more complex, as shown by evidence that DNA polymerase α from a lower eucaryote, *Drosophila*, exhibits a cryptic $3' \rightarrow 5'$ exonuclease activity which is revealed when the 50-, 60-, and 73-kDa subunits are removed (12). In yeast, three forms of polymerase have been described, an α-like enzyme, pol I, and pol II and III, which, like polymerase δ, possess an associated $3' \rightarrow 5'$ exonuclease activity (2, 8, 10). There have been other reports of α-like polymerases with associated $3' \rightarrow 5'$ exonuclease activities from eucaryotic tissues, for example, mouse myeloma (11), cytomegalovirus-induced polymerase (34), herpesvirus-induced polymerase (35), adenovirus (17), and *Ustilago maydis* (1). Thus, is DNA polymerase δ a form of DNA polymerase α or vice versa? Considerations of nomenclature, at this point, are less important than the fact that polymerase δ can be regarded operationally as distinct in terms of its isolation and characterization as a discrete protein with a behavior different from that of DNA polymerase α, as well the growing realization that both polymerases are likely to serve crucial and distinct roles in DNA replication. Questions of the relationships of polymerases α and δ and indeed of the different forms of α will be resolved only by detailed structural comparisons of the two proteins. The molecular cloning of HeLa cell DNA polymerase α catalytic polypeptide has provided some provocative new findings (42a) as well as possibilities for more detailed understanding of structure-function relationships of the molecule. The deduced protein sequence has shown significant similarities with yeast pol I and the DNA polymerases of bacteriophages T4 and φ29, herpes virus, vaccinia virus, and adenovirus. Thus, there is every expectation that DNA polymerases δ and α will have strong structural similarities. Interestingly, the deduced amino acid sequence of DNA polymerase α shows a region which shares similarity with T4 gene 46 protein, which is an exonuclease (42a). Cloning of the cDNA for DNA polymerase δ is now a major necessity and is currently a major goal in this laboratory. Clearly, there will be a convergence of approaches and concepts in future studies of these two enzyme activities, both of which may be intimately associated with eucaryotic DNA replication, and the stage is set for a period of major progress in the definition of their respective functions at the level of the replication fork, as well as the functional identification of their accessory proteins.

ACKNOWLEDGMENTS. This work was supported by Public Health Service grant GM 31793 from the National Institutes of Health and was performed during the tenure of an Established Investigatorship of the American Heart Association to M. Y. W. T. Lee.

LITERATURE CITED

1. **Banks, G. R., W. K. Holloman, M. W. Kairis, G. T. Yarranton, and A. Spanos.** 1976. A DNA polymerase from *Ustilago maydis*. 1. Purification and properties of the polymerase activity. *Eur. J. Biochem.* **62**:131–142.
2. **Bauer, G. A., H. H. Heller, and P. M. J. Burgers.** 1988. DNA polymerase III from Saccharomyces cerevisiae. I. Purification and characterization. *J. Biol. Chem.* **263**:917–924.
3. **Blow, J.** 1987. Many strands converge. *Nature* (London) **326**:441–442.
4. **Bravo, R., R. Frank, P. A. Blundell, and H. MacDonald-Bravo.** 1987. Cyclin/PCNA is the auxiliary protein of DNA polymerase δ. *Nature* (London) **326**:515–517.
5. **Byrnes, J. J.** 1985. Differential inhibition of DNA polymerases α and δ. *Biochem. Biophys. Res. Commun.* **132**:628–634.
6. **Byrnes, J. J., K. M. Downey, V. L. Black, and A. G. So.** 1976. A new mammalian DNA polymerase with 3' to 5' exonuclease activity: DNA polymerase δ. *Biochemistry* **15**:2817–2823.
7. **Byrnes, J. J., K. M. Downey, B. G. Que, M. Y. W. Lee, and A. G. So.** 1977. Selective inhibition of the 3' to 5' exonuclease activity associated with DNA polymerases: a mechanism for mutagenesis. *Biochemistry* **16**:3740–3746.
8. **Campbell, J.** 1986. Eucaryotic DNA replication. *Annu. Rev. Biochem.* **55**:733–771.
9. **Celis, J. E., P. Madsen, A. Celis, H. V. Nielsen, and B. Gesser.** 1987. Cyclin (PCNA, auxiliary protein of DNA polymerase δ) is a central component of the pathways leading to DNA replication and cell division. *FEBS Lett.* **220**:1–7.
10. **Chang, L. M. S.** 1977. Polymerases from bakers' yeast. *J. Biol. Chem.* **252**:1873–1880.
11. **Chen, Y. C., E. W. Bohn, S. R. Planck, and S. H. Wilson.** 1979. Mouse DNA polymerase α. Subunit structure and identification of a species with associated exonuclease. *J. Biol. Chem.* **254**:11678–11687.
12. **Cotterill, S. M., M. E. Reyland, L. A. Loeb, and I. R. Lehman.** 1987. A cryptic proofreading 3' to 5' exonuclease associated with the polymerase subunit of the DNA polymerase-primase from Drosophila melanogaster. *Proc. Natl. Acad. Sci. USA* **84**:5635–5639.
13. **Crute, J. J., A. F. Wahl, and R. A. Bambara.** 1986. Purification and characterization of two new high molecular weight forms of DNA polymerase δ. *Biochemistry* **25**:26–36.

14. Decker, R. S., M. Yamaguchi, R. Possenti, M. K. Bradley, and M. L. DePamphilis. 1987. In vitro initiation of DNA replication in simian virus 40 chromosomes. *J. Biol. Chem.* 262:10863–10872.
15. Dresler, S. L., and M. G. Frattini. 1986. DNA replication and UV-induced DNA repair synthesis in human fibroblasts are much less sensitive than DNA polymerase α to inhibition by butylphenyldeoxyguanosine triphosphate. *Nucleic Acids Res.* 17:7093–7102.
16. Dresler, S. L., and K. S. Kimbro. 1987. 2'3' Dideoxythymidine 5' triphosphate inhibition of DNA replication and ultraviolet induced DNA repair synthesis in human cells: evidence for involvement of DNA polymerase δ. *Biochemistry* 26:2264–2268.
17. Field, J., R. M. Gronostajski, and J. Hurwitz. 1984. Properties of the adenovirus DNA polymerase. *J. Biol. Chem.* 259:9487–9495.
18. Fry, M., and L. A. Loeb. 1986. DNA polymerase α, p. 13–74. *In* M. Fry and L. A. Loeb (ed.), *Animal Cell DNA Polymerases*. CRC Press, Inc., Boca Raton, Fla.
19. Goscin, L. P., and J. J. Byrnes. 1982. DNA polymerase δ: one polypeptide, two activities. *Biochemistry* 21:2518–2524.
20. Hammond, R. A., J. J. Byrnes, and M. R. Miller. 1987. Identification of DNA polymerase δ in CV-1 cells: studies implicating both DNA polymerase δ and DNA polymerase α in DNA replication. *Biochemistry* 26:6817–6824.
21. Holmes, A. M., E. Cheriathundam, F. J. Bollum, and L. M. S. Chang. 1986. Immunological analysis of the polypeptide structure of calf thymus DNA polymerase-primase complex. *J. Biol. Chem.* 261:11924–11930.
22. Huberman, J. A., and A. Kornberg. 1970. Enzymatic synthesis of deoxyribonucleic acids. XXXV. A 3' hydroylribonucleotide binding site of Escherichia coli deoxyribonucleic acid polymerase. *J. Biol. Chem.* 245:5326–5334.
23. Kornberg, A. 1988. DNA replication. *J. Biol. Chem.* 263:1–4.
24. Kunkel, T. A., R. D. Sabatino, and R. A. Bambara. 1987. Exonucleolytic proofreading by calf thymus DNA polymerase δ. *Proc. Natl. Acad. Sci. USA* 84:4865–4869.
25. Lee, M. Y. W. T., J. J. Byrnes, K. M. Downey, and A. G. So. 1980. Mechanism of inhibition of deoxyribonucleic acid synthesis by 1-β-D-arabinofuranosyladenosine triphosphate and its potentiation by 6-mercaptopurine ribonucleoside 5'-monophosphate. *Biochemistry* 19:215–219.
26. Lee, M. Y. W. T., C. K. Tan, K. M. Downey, and A. G. So. 1981. Structural and functional properties of calf thymus DNA polymerase δ. *Prog. Nucleic Acid Res. Mol. Biol.* 26:83–96.
27. Lee, M. Y. W. T., C. K. Tan, K. M. Downey, and A. G. So. 1984. Further studies on calf thymus DNA polymerase δ purified to homogeneity by a new procedure. *Biochemistry* 23:1906–1913.
28. Lee, M. Y. W. T., C. K. Tan, A. G. So, and K. M. Downey. 1980. Purification of deoxyribonucleic acid polymerase δ from calf thymus: partial characterization of physical properties. *Biochemistry* 19:2096–2101.
29. Lee, M. Y. W. T., and N. L. Toomey. 1986. Differential effects of dimethylsulfoxide on the activities of human DNA polymerases α and δ. *Nucleic Acids Res.* 14:1719–1726.
30. Lee, M. Y. W. T., and N. L. Toomey. 1987. Human placental DNA polymerase δ: identification of a 170 kDa polypeptide by activity staining and immunoblotting. *Biochemistry* 26:1076–1085.
31. Lee, M. Y. W. T., N. L. Toomey, and G. E. Wright. 1985. Differential inhibition of human placental DNA polymerases δ and α by BuPdGTP and BuAdATP. *Nucleic Acids Res.* 13:8623–8630.
32. Lee, M. Y. W. T., and W. A. Whyte. 1984. Selective affinity chromatography of DNA polymerase with associated 3' to 5' exonuclease activities. *Anal. Biochem.* 138:291–297.
33. Nishida, C., P. Reinhardt, and S. Linn. 1988. DNA repair synthesis in human fibroblasts requires DNA polymerase δ. *J. Biol. Chem.* 263:501–510.
34. Nishiyama, Y., K. Maeno, and S. Yoshida. 1983. Characterization of human cytomegalovirus-induced DNA polymerase and the associated 3' to 5' exonuclease. *Virology* 124:221–231.
35. Ostrander, M., and K. C. Chang. 1981. Properties of herpes simplex virus type 1 and type 2 DNA polymerase. *Biochim. Biophys. Acta* 609:232–245.
36. Ottinger, H., P. Frei, M. Hassig, and U. Hubscher. 1987. Mammalian DNA polymerase α: a replication competent holoenzyme form from calf thymus. *Nucleic Acids Res.* 15:4789–4807.
37. Prelich, G., M. Kostura, D. R. Marshak, M. B. Mathews, and B. Stillman. 1987. The cell cycle regulated proliferating cell nuclear antigen is required for SV40 DNA replication in vitro. *Nature* (London) 326:471–475.
38. Prelich, G., C. K. Tan, M. Kostura, M. B. Mathews, A. G. So, K. M. Downey, and B. Stillman. 1987. Functional identity of proliferating cell nuclear antigen and a DNA polymerase δ auxiliary protein. *Nature* (London) 326:517–520.
39. Que, B. G., K. M. Downey, and A. G. So. 1976. Mechanism of selective inhibition of 3' to 5' exonuclease activity of Escherichia coli DNA polymerase I by nucleoside 5' monophosphates. *Biochemistry* 17:1603–1606.
40. Tan, C. K., C. Castillo, A. G. So, and K. M.

Downey. 1986. An auxiliary protein for DNA polymerse δ from fetal calf thymus. *J. Biol. Chem.* **261**:12310–12316.

41. Vishwanatha, J. K., S. A. Coughlin, M. Wesolowski-Owen, and E. F. Baril. 1986. A multiprotein form of DNA polymerase α from HeLa cells; resolution of its associated catalytic activities. *J. Biol. Chem.* **261**:6619–6628.

42. Wong, S. W., L. R. Paborsky, P. A. Fisher, T. S. F. Wang, and D. Korn. 1986. Structural and enzymological characterization of immunoaffinity-purified DNA polymerase α-DNA primase complex from KB cells. *J. Biol. Chem.* **261**:7958–7968.

42a. Wong, S. W., A. F. Wahl, P. M. Yuan, N. Arai, B. E. Pearson, K. Arai, D. Korn, M. W. Hunkapiller, and T. S. F. Wang. 1988. Human DNA polymerase α gene expression is cell proliferation dependent and its primary structure is similar to both prokaryotic and eukaryotic replicative DNA polymerases. *EMBO J.* **7**:37–47.

43. Zhang, S. J., and M. Y. W. T. Lee. 1987. Biochemical characterization and development of DNA polymerases α and δ in the neonatal rat heart. *Arch. Biochem. Biophys.* **252**:24–31.

II. DNA Replication Systems

DNA Replication Systems: Introduction

It is the goal of this part of the volume to present examples of DNA replication systems and to give an indication of the sample range and the current status of the studies in these systems.

As noted in Part I, biochemical studies of DNA replication have progressed rapidly since the establishment of in vitro replication systems mimicking biological processes. The establishment of these systems was dependent on several requirements: from the biochemical standpoint, a system had to demonstrate a specificity for origin and template of an identified DNA; from a genetic standpoint, it had to show a dependence on gene products which were identified as being required for DNA replication in the organism of study; lastly, the product of the in vitro systems had to demonstrate biological activity implying an accurate (low-mutagenesis) DNA replication process. These systems in turn have led to the development of in vitro models with which to study mutagenesis of DNA.

DNA replication in these systems is dependent on more than one protein. DNA polymerase alone is not sufficient to catalyze a specific DNA replication for each template. The specificity for DNA replication rests with initiation at the origin of DNA replication for each of the biological systems. The varieties of requirements beyond that are immense. In the better studied systems, it appears that some proteins enhance or define specificity in the in vitro system. For example, the Ssb protein enhances definition of the origin of replication in the small single-stranded DNA phage systems. These systems allow the probing of DNA replication or mutagenesis of DNA with controlled permutations.

A theme which is only now beginning to be exploited is that of extragenic suppression. There are examples in *Escherichia coli* of extragenic suppression for a deficiency in a gene product which is required for DNA replication. In a few instances these suppression pathways or alternative mechanisms of DNA synthesis are beginning to be elucidated.

The in vitro systems allow us to address the sequence of interactions during both the initiation and elongation phases of replication. In the elongation phase, there must be coupling of the leading- and lagging-strand synthesis. This has presented a conceptual problem, since it appears that the synthesis on the leading strand is such that facilitation of initiation and synthesis on the lagging strand might be required in order to keep pace. This control may be dependent on dimerization of the synthesis subunits of the DNA-replicating complex, and this dimerization may be dependent on accessory proteins of DNA replication.

Proteins in the in vitro systems may be demonstrated to have more than one role. An example of this is the gene 4 protein of T7, which displays both primase and helicase activities. This protein appears to play a role in the initiation of discontinuous synthesis of strands during synthesis and the unwinding of parental duplex DNA of T7.

An additional observation arising from studies with in vitro systems is that there may be a hierarchy of origins of DNA replication. While the origin defines the specificity of DNA replication, there may be secondary origins which can be utilized by the organism in the face of a deficiency of a primary origin.

Eucaryotic systems are beginning to allow the definition of enzymatic activities in DNA replication. Once again it is clear that the simplest DNA genomes are going to facilitate study, with simian virus 40 being exploited as a model system. It seems clear that there will be many parallels with procaryotic systems. For example, the helicase action is apparently required for DNA replication and is supplied by the large T antigen produced from the simian virus 40 genome. On the other hand, this small genome requires numerous host functions for DNA replication.

A number of significant questions remain and new questions have been raised by studies with the in vitro systems of DNA replication.

(i) A major question regards the control of the initiation of DNA replication. This must be subdivided into the control of macroinitiation and the control of microinitiation. There does not appear to be uniformity among organisms in this regard, and whereas the control of macroinitiation in procaryotic cells appears to reflect the growth status of the cell, protein synthesis, and the availability of replicating enzymes, initiation of simpler phage genomes requires only the availability of proteins.

(ii) Likewise, the control of elongation of DNA replication has not been elucidated in procaryotic studies. However, in eucaryotes there are additional levels of control regarding both elongation and initiation. Presently, we have limited models regarding the controls preventing reinitiation during S phase in mammalian systems.

(iii) Understanding the coupling of reactions in DNA replication remains a major challenge. Little is understood of the linkage of initiation and elongation or the switch between these two processes. The model of dimerization of the replicating complex affords an idea of how the synthesis on the leading and lagging strands might be coupled, but this remains to be demonstrated in vivo.

(iv) The interaction of the termination of DNA replication with the elongation phase may vary from organism to organism, and the regulation at this point is not understood.

(v) The systems described in this part of the volume afford investigators the first real opportunity to study DNA mutagenesis in a situation where DNA replication can be controlled. Investigators can test both the contribution of the template itself to the mutagenesis process and the contribution of the individual factors in the assembly of the daughter strands.

The Editors

Chapter 8

Formation and Propagation of the Bacteriophage T7 Replication Fork

Hiroshi Nakai, Benjamin B. Beauchamp, Julie Bernstein, Hans E. Huber, Stanley Tabor, and Charles C. Richardson

Bacteriophage T7 provides a model system for studying the enzymatic mechanisms involved in the replication of a linear duplex molecule (see reference 41). It has a linear chromosome 40 kilobases (kb) in length, and its entire sequence has been determined (10). Over 100 copies of the T7 chromosome are generated within a 12-min period after infection of *Escherichia coli*.

Upon injection of phage DNA into the host cell, DNA replication initiates at the primary origin, located 15% of the distance from the genetic left end of the molecule (44, 57). DNA replication proceeds bidirectionally, and DNA synthesis is discontinuous on one strand of each replication fork, giving rise to Okazaki fragments 1 to 6 kb in length (27, 52). These fragments bear oligoribonucleotides at the 5' ends, mostly tetraribonucleotides with the sequences pppACCN and pppACAN, N being predominantly A and C (39, 49). Synthesis of these fragments initiates at the sites NGGTC and NTGTC (12). Finally, newly synthesized T7 DNA molecules accumulate in the form of linear concatemers many times the unit length of the genome (21). These concatemers are subsequently processed and packaged.

A particularly interesting feature of phage T7 is the efficiency with which it replicates its genome. T7 encodes most of its own replication proteins, and this partly accounts for the high efficiency of replication. Upon injection of phage DNA into the host cell, *E. coli* RNA polymerase transcribes the early genes (Fig. 1), the class I genes, composed of genes located at the left end of the genome through gene 1.3 (53). One of these early genes, gene 1, encodes an RNA polymerase that transcribes the remainder of the genes, those of class II (genes 1.7 through 6) and class III (genes 7 through 19).

Here we discuss enzymatic mechanisms involved in the establishment of a replication fork and the propagation of that fork. We examine the proteins at the replication fork and discuss how protein-protein interactions modulate the individual enzymatic properties of these proteins. Finally, we consider problems involved in the catalysis of leading-

---

Hiroshi Nakai, Benjamin B. Beauchamp, Julie Bernstein, Hans E. Huber, Stanley Tabor, and Charles C. Richardson • Department of Biological Chemistry and Molecular Pharmacology, Harvard Medical School, Boston, Massachusetts 02115.

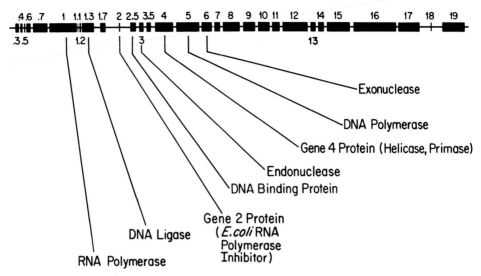

FIGURE 1. Genetic map of phage T7 (adapted from reference 10).

and lagging-strand synthesis, in particular, problems involved in recycling T7 replication proteins for multiple rounds of lagging-strand synthesis.

We focus on four proteins at the bacteriophage T7 replication fork: products of T7 genes 5, 4, and 2.5 and *E. coli* thioredoxin. The gene 5 protein is the DNA polymerase. As discussed below, a one-to-one complex of gene 5 protein with *E. coli* thioredoxin serves as a highly processive DNA polymerase at the replication fork. For simplicity, we refer to the gene 5 protein-thioredoxin complex as T7 DNA polymerase. The gene 4 protein serves as both helicase and primase at the replication fork. Finally, the gene 2.5 protein is a single-stranded DNA-binding protein.

INITIATION OF DNA REPLICATION

Two prominent features of the DNA sequence at the T7 primary origin (Fig. 2) are the presence of two tandem T7 RNA polymerase promoters and the presence of an AT-rich region just downstream of the promoters (44, 57). Both the T7 RNA and DNA polymerases are required for the initiation of DNA replication in vitro (13, 14, 43). T7 RNA polymerase initiates transcription at either of these promoters, and these transcripts are used as primers by T7 DNA polymerase (13). The initial products of DNA synthesis have 10 to 60 ribonucleotides covalently attached at the 5' termini. The transition from RNA to DNA synthesis appears to be nonrandom, taking place more readily at some sequences than at others. In particular, many transitions take place within the AT-rich region, a region that is essential for initiation at the primary origin (44, 57). The role of RNA polymerase transcripts in the initiation of leading-strand synthesis is supported by analysis of DNA replication in vivo (54).

In the model for initiation (Fig. 2), a primer transcript is laid down for rightward leading-strand synthesis, and a single-strand region is exposed on the opposite template. The next step is the recruitment of replication proteins for the catalysis of leading- and lagging-strand synthesis.

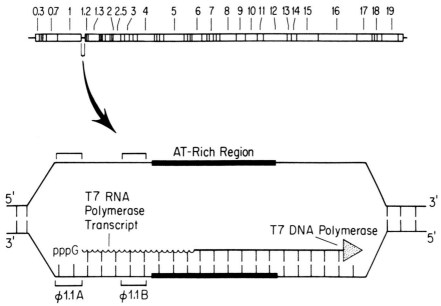

FIGURE 2. Model for initiation. The primary origin of T7 is shown. T7 RNA polymerase initiates transcription at promoter φ1.1A or φ1.1B. T7 DNA polymerase uses the transcript as primer, initiating rightward leading-strand synthesis.

PROPAGATION OF THE T7 REPLICATION FORK

Once the primer-template-DNA polymerase complex is established at the fork, additional proteins are necessary to propagate the replication fork. T7 DNA polymerase cannot by itself catalyze DNA synthesis through a duplex region. The helicase activity of gene 4 protein is needed to unwind the helix for leading-strand synthesis.

Upon generation of the replication bubble at the primary origin, gene 4 protein binds to the displaced strand. Binding of gene 4 protein to single-stranded DNA requires the presence of a nucleoside triphosphate (NTP), preferably dTTP (29). Once bound, gene 4 protein can translocate 5′ to 3′ along single-stranded DNA to the replication fork, utilizing the energy of hydrolysis of NTP to nucleoside diphosphate (NDP) and P_i (28, 30, 56). At the fork, movement of gene 4 protein unwinds duplex DNA and enables T7 DNA polymerase to catalyze leading-strand synthesis (Fig. 3A) (23, 26). This strand displacement synthesis reaction requires a specific interaction between gene 4 protein and T7 DNA polymerase. Gene 4 protein cannot promote strand displacement DNA synthesis by other DNA polymerases such as bacteriophage T4 DNA polymerase (22, 26).

The gene 4 protein has a second function at the replication fork. As a primase, it initiates multiple rounds of lagging-strand synthesis (Fig. 3B). As it translocates 5′ to 3′ along single-stranded DNA, it searches for major primase recognition sequences T/GGGTC and GTGTC, at which it catalyzes the synthesis of the tetraribonucleotides pppACCC/A and pppACAC, respectively. These primers are then extended at the site of their synthesis by T7 DNA polymerase (37, 42, 47, 56). A specific interaction between gene 4 protein and T7 DNA polymerase is also required for the transition from primer synthesis to DNA synthesis. Other DNA polymerases cannot extend

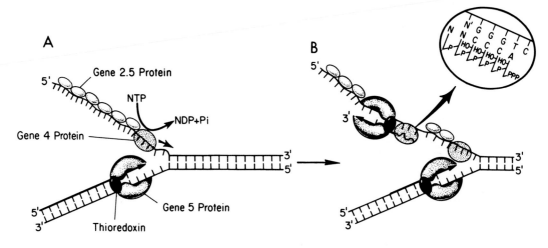

FIGURE 3. Model for leading-strand (A) and lagging-strand (B) synthesis.

primers synthesized by gene 4 protein (46, 47). The affinity of gene 4 protein for T7 DNA polymerase and the independent binding of these proteins to single-stranded DNA play an important role in the transition from primer synthesis to DNA synthesis (36, 37).

As leading-strand synthesis proceeds and displaces single-stranded DNA, the T7 gene 2.5 protein can bind to the displaced strand (Fig. 3). T7 gene 2.5 protein has been considered the T7 counterpart of the *E. coli* single-strand binding protein (SSB) (see reference 41). It has been viewed as dispensable on the basis of the observation that a T7 mutant with an amber lesion in gene 2.5, expressing a shortened polypeptide approximately 90% the length of the wild-type gene 2.5 protein, can grow on strains expressing functional *E. coli* SSB (5). However, more recent analyses with additional gene 2.5 mutations have shown that the gene 2.5 protein is required for T7 DNA replication (F. W. Studier, personal communication).

In the *E. coli* DNA replication system SSB binds cooperatively to single-stranded DNA and maintains single-stranded DNA in an extended conformation (50). It serves a number of functions at the replication fork; for example, it is required for RNA primer synthesis by the *E. coli* priming apparatus, and it enables lagging-strand DNA synthesis to proceed unimpeded by secondary structure on the displaced strand (see reference 59). It is not yet clear whether *E. coli* SSB plays a major role at the T7 replication fork in vivo. It can greatly increase the processivity of DNA synthesis catalyzed by T7 DNA polymerase on a primed single-stranded template (55). T7 gene 2.5 protein also stimulates DNA synthesis by T7 DNA polymerase on primed single-stranded DNA templates (40, 45, 48). But do T7 gene 2.5 protein and *E. coli* SSB stimulate DNA synthesis by similar mechanisms? As we discuss below, these two proteins exert very different effects on lagging-strand synthesis at the T7 replication fork.

MODULATION OF ENZYMATIC ACTIVITIES AT THE REPLICATION FORK

In this section we focus on the individual enzymatic activities of the proteins at the fork and how protein-protein interactions modulate these activities.

Interaction of T7 Gene 5 Protein and *E. coli* Thioredoxin

The T7 gene 5 protein has DNA polymerase and $3' \rightarrow 5'$ exonuclease activities. The exonuclease digests both single- and double-stranded DNA. In the absence of thioredoxin, the gene 5 protein has extremely low levels of polymerase and double-stranded DNA exonuclease activities. It is highly distributive, extending primers on single-stranded DNA templates by only 1 to 50 nucleotides before dissociating (55).

E. coli thioredoxin binds tightly to the gene 5 protein to stimulate polymerase and double-stranded DNA exonuclease activity several hundredfold (1, 17, 18). Single-stranded DNA exonuclease activity of gene 5 protein remains the same in the presence of thioredoxin. Thioredoxin, bound tightly to the gene 5 protein in a one-to-one stoichiometry (34, 35), acts as an accessory protein to increase the affinity of gene 5 protein for the primed template (20). It increases the processivity of DNA synthesis on a primed single-stranded template to greater than 10 kb (55).

An important question is how thioredoxin produces this effect. Thioredoxin is involved in many metabolic pathways in *E. coli* requiring a reducing agent (see reference 6). It has an active site with two cysteine residues that can undergo reversible oxidation to a disulfide. Only reduced thioredoxin can bind to the gene 5 protein and stimulate polymerase and double-stranded DNA exonuclease activities (2). However, the redox potential of thioredoxin is not necessary to stimulate DNA polymerase activity. A mutant thioredoxin with the two cysteine residues at the active site replaced by serine residues stimulates DNA polymerase activity of the gene 5 protein to almost the same maximal level as wild-type thioredoxin; however, the mutant thioredoxin has a 100-fold reduced affinity for gene 5 protein compared to the wild type (19).

One consequence of the increased affinity of the gene 5 protein-thioredoxin complex for the primer-template is that it recycles less efficiently than the gene 5 protein (20). The gene 5 protein efficiently dissociates from a primer-template to transfer to another primer-template. The gene 5 protein-thioredoxin complex does so at a 20- to 100-fold lower rate. It is therefore of interest to speculate whether T7 regulates the affinity of its DNA polymerase for the primer-template via the oxidation state of thioredoxin. Upon completion of an Okazaki fragment, T7 DNA polymerase would most efficiently be recycled for the next round of Okazaki fragment synthesis if the thioredoxin were oxidized and dissociated from the gene 5 protein. Gene 5 protein would then most efficiently dissociate from the 3' end of the completed Okazaki fragment. Phage T7 can grow at a reduced efficiency on a mutant *E. coli* strain with the altered thioredoxin, which has the cysteine residues at the active site replaced by serine residues (19). Although the cysteine residues are not essential for T7 growth, fine control of the affinity of DNA polymerase for the primer-template may ensure optimal T7 DNA replication.

The Two Molecular Weight Forms of the Gene 4 Protein

The gene 4 protein exists in two molecular weight species of 56,000 and 63,000. The smaller form arises as a result of a second initiation codon and ribosome binding site within gene 4 and therefore lacks 63 amino acids at the amino terminus (10). The gene 4 protein has been purified as an equimolar mixture of the two molecular weight forms (11, 28, 47) as well as a mixture that is enriched for the small form (22). The mixture has DNA-dependent nucleoside triphosphatase (NTPase), helicase, and primase activities. The large form of gene 4 protein has not yet been purified free of the small form. However, the small form has been purified by two different procedures.

In the first procedure a clone of gene 4 was constructed so that the segment encoding the amino terminus of the large form was deleted (S. Tabor and C. C. Richardson, unpublished results). The small form of gene 4 protein produced by this clone was purified, and it has DNA-dependent NTPase and helicase activities but no primase activity (6a). These results indicated that the large form of gene 4 protein is required for primase activity.

The small form of the gene 4 protein has also been purified from an equimolar mixture of both forms (H. Nakai and C. C. Richardson, *J. Biol. Chem.*, in press). As it is separated from the large form, the small form loses primase activity but not helicase activity. However, the small form prepared by this procedure is reduced in primase activity by only six- to sevenfold compared to an equimolar mixture of both forms even though the large form is present at a level of less than 0.5%. Mixing experiments reveal that submolar amounts of the large form present in a mixture of both forms are sufficient to restore high levels of primase activity. At a constant concentration of the large form, the level of primase activity is directly proportional to the concentration of small form, spanning molar ratios of small form to large form of 1:1 to 16:1. These results indicate that the small form, although devoid of primase activity, can play a role in primer synthesis. They suggest that an active primase may be an oligomer composed of both molecular weight forms of gene 4 protein.

A model for the organization of the domains on the gene 4 protein based on sequence analysis is shown in Fig. 4 (6a). The amino terminus of the large form, not present on the small form, has a potential metal-binding domain, $Cys-X_2-Cys-X_{15}-Cys-X_2-Cys$, known as a "zinc finger" (see reference 6). On the basis of recent findings that several proteins containing this domain bind to DNA at specific sequences (7, 33), we have proposed that the zinc finger recognizes the specific primase sites. The primase and helicase domains, tentatively assigned on the basis of sequence similarity to the *E. coli dnaG* primase and *dnaB* helicase, are contained in both the large and small forms. In addition, both forms contain the nucleotide binding domains, NB1 and NB2, defined by matching the sequence with a consensus topography for a mononucleotide binding fold (8).

In the model shown in Fig. 4, dTTP occupies the nucleotide binding site NB2. Its presence enables gene 4 protein to bind to single-stranded DNA, and its hydrolysis drives the movement of gene 4 protein 5' to 3' (Fig. 4A). While gene 4 protein is in this translocation mode, NB1 is unavailable for nucleotide binding. As gene 4 protein encounters a primase recognition site, the zinc finger binds at that site, inducing a conformation change and opening up NB1 so that it may be occupied by ATP (Fig. 4B). NB2 can then be occupied by CTP, aligning ATP with CTP and promoting the formation of the dinucleotide pppAC. In such a way, the binding of the zinc finger to the primase recognition site may function to switch the gene 4 protein from a translocation mode to an oligonucleotide-synthesizing mode.

Absent from this model is the role of the small form in primer synthesis. The small form, which lacks the zinc finger for recognition of primase sites, would be unable to enter the template-dependent oligonucleotide-synthesizing mode. However, it has the domain for primer synthesis. In the absence of single-stranded DNA template, the mixture of the two forms of gene 4 protein can catalyze the synthesis of ribonucleotide dimers, albeit at low levels, and the small form retains this activity (6a). This result supports the hypothesis that the primase domain is contained within the small form even though it may not enter the template-dependent primer synthesis mode.

What, then, is the role of the small form of gene 4 protein in primer synthesis? Is the active gene 4 primase composed of both

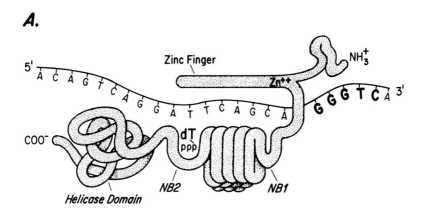

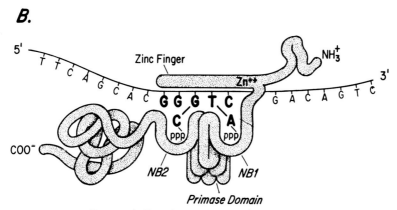

FIGURE 4. Domains on the gene 4 protein.

molecular weight forms? There is as yet no information about the physical structure assumed by gene 4 protein molecules as a helicase or primase. The close relationship between helicase and primase activities has parallels in other DNA replication systems. Both the bacteriophage T4 gene 61 primase and the *E. coli dnaG* primase interact with a helicase, gene 41 protein or *dnaB* protein, respectively, for maximal primase activity (4, 9, 15, 25, 32, 38, 58). In a later section we present a model for the role of the two molecular weight forms of gene 4 protein in the propagation of the replication fork.

RECONSTITUTION OF LEADING- AND LAGGING-STRAND SYNTHESIS ON A PREFORMED REPLICATION FORK

In examining reactions at the T7 replication fork, we can bypass the reactions that are involved in initiation by the use of a preformed replication fork (26). This template (Fig. 5A) consists of a fully duplex M13 DNA molecule with a 5′-single-stranded tail. The 3′ end of this molecule provides the primer for leading-strand synthesis. The tail provides the single-stranded DNA to which the gene 4 protein binds.

Leading-Strand Synthesis

In the presence of the four deoxyribonucleoside triphosphates (dNTPs), the gene 4 protein and DNA polymerase together catalyze strand displacement DNA synthesis via a rolling-circle mode on the preformed fork to generate a DNA product that is more than six times the length of the M13 template (Fig. 5B). The rate of leading-strand synthesis as measured on this template is 300 nucleotides per s at 30°C (26).

Leading-strand synthesis proceeds by a highly processive mechanism (Nakai and Richardson, in press). Both the gene 4 protein and T7 DNA polymerase remain bound to the replication fork as DNA synthesis proceeds more than 40 kb. A stable complex of gene 4 protein, T7 DNA polymerase, and the replication fork is formed in the presence of dTTP, and upon addition of the other three dNTPs, leading-strand synthesis can proceed uninterrupted for at least six revolutions around the M13 template in the presence of excess challenger DNA (single-stranded f1 DNA). Challenger DNA traps gene 4 protein and DNA polymerase that are not bound to the fork. In addition, although the initiation of leading-strand synthesis can be inhibited by diluting the reaction mixture, leading-strand synthesis is impervious to dilution once the stable complex is established at the fork.

Lagging-Strand Synthesis

At the replication fork the gene 4 protein acts not only as helicase but also as primase. In the previous section we examined the processivity of gene 4 protein as it translocates 5' to 3' along the DNA to unwind the helix for leading-strand synthesis. The gene 4 protein acting as helicase encounters primase recognition sites as it translocates along single-stranded DNA (Fig. 5A). An important question is whether this gene 4 protein acting as helicase can also catalyze primer synthesis. Moreover, can the gene 4 protein catalyze multiple rounds of primer synthesis to initiate multiple rounds of lagging-strand synthesis without dissociat-

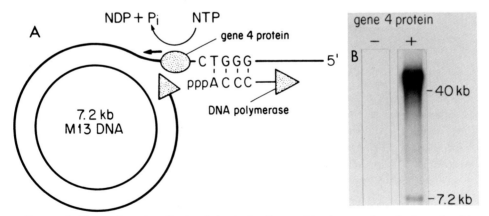

FIGURE 5. The preformed replication fork. (A) Leading- and lagging-strand synthesis catalyzed by gene 4 protein and T7 DNA polymerase on the preformed replication fork. Gene 4 protein unwinds the helix, utilizing the energy of hydrolysis of NTP to NDP and P_i, and enables T7 DNA polymerase to catalyze leading-strand synthesis. Gene 4 protein also catalyzes the synthesis of the tetranucleotide primer at the primase recognition sequence (as shown here, the tetranucleotide primer pppACCC is synthesized at the recognition sequence GGGTC). T7 DNA polymerase extends the primer, initiating lagging-strand synthesis. (B) Products of leading-strand synthesis, catalyzed by the T7 DNA polymerase in the presence or absence of gene 4 protein, are resolved on a denaturing agarose gel.

ing from the template? Polymerization of ribonucleotides for primer synthesis would proceed 5' to 3' in a direction that is opposite to the movement of gene 4 protein along single-stranded DNA.

Gene 4 protein and T7 DNA polymerase catalyze both leading- and lagging-strand synthesis on the preformed replication fork when both dNTPs and ribonucleoside triphosphates are present. Template challenge and dilution experiments analogous to the ones described for studying leading-strand synthesis indicate that the gene 4 protein acting as primase is distributive (Nakai and Richardson, in press). At a replication fork at which both leading- and lagging-strand syntheses are catalyzed, leading-strand synthesis is relatively resistant to challenge with excess single-stranded DNA whereas lagging-strand synthesis is not. Moreover, leading-strand synthesis is resistant to dilution of the reaction mixture whereas lagging-strand DNA synthesis is not, even when the diluent contains high concentrations of T7 DNA polymerase. These results indicate that the gene 4 protein molecules acting processively to promote leading-strand synthesis do not efficiently catalyze primer synthesis for lagging-strand synthesis. Moreover, the gene 4 protein molecules acting as primase are unable to catalyze multiple rounds of primer synthesis as they translocate 5' to 3' along single-stranded DNA.

Role of the T7 and *E. coli* Single-Stranded DNA-Binding Proteins at the T7 Replication Fork

Experiments in vitro indicate that the T7 gene 2.5 protein and *E. coli* SSB exert different effects at the T7 replication fork (Nakai and Richardson, in press). With limiting concentrations of gene 4 protein and in the absence of the T7 and *E. coli* DNA-binding proteins, leading-strand synthesis proceeds more efficiently than lagging-strand synthesis. Neither the T7 nor the *E. coli* DNA-binding protein stimulates leading-strand synthesis to a significant extent once it has been initiated by gene 4 protein and T7 DNA polymerase. However, they can both stimulate lagging-strand synthesis. The T7 gene 2.5 protein greatly increases the frequency with which lagging-strand synthesis is initiated. *E. coli* SSB can greatly increase the processivity of lagging-strand DNA synthesis but, unlike the T7 gene 2.5 protein, it cannot increase the initiation rate of lagging-strand synthesis. Thus, with limiting concentrations of gene 4 protein, the Okazaki fragments synthesized in the presence of T7 gene 2.5 protein (typically 1 to 6 kb in length) are smaller than fragments synthesized in the presence of *E. coli* SSB (typically more than 6 kb in length). Our results suggest that the T7 gene 2.5 protein specifically interacts with T7 replication proteins not only to stimulate gene 4 primase activity but also to increase the efficiency with which primers are utilized by the T7 DNA polymerase.

Model for the Role of the Two Molecular Weight Forms of Gene 4 Protein in the Propagation of the Replication Fork

An important consideration in the understanding of the mechanism of fork movement is the role of the two molecular weight forms of the gene 4 protein. The model for the role of the two forms presented below accounts for the highly processive nature of the gene 4 protein as a helicase and its distributive property as a primase.

In this model (Fig. 6) the small form of gene 4 protein serves as helicase to promote leading-strand synthesis. The large form binds to the small form to form a primase (Fig. 6A). As this helicase-primase complex encounters a primase recognition site, the large form dissociates from the small form and synthesizes a primer (Fig. 6B). The small form continues to serve as helicase. Its

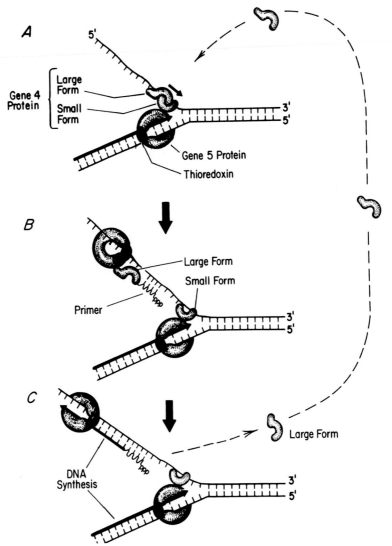

FIGURE 6. Roles of the two molecular weight forms of gene 4 protein at the replication fork.

movement along single-stranded DNA is not interrupted by primer synthesis. As T7 DNA polymerase extends the primer, the large form dissociates from the template (Fig. 6C) and is available to catalyze primer synthesis at another replication fork.

We have hypothesized that the amino terminus of the large form contains the domain for recognition of the specific primase site. The small form, which lacks that domain, is necessary for processive leading-strand synthesis. In this model the active primase complex disassembles before catalyzing primer synthesis. The rate-limiting step in primer synthesis may therefore be the reformation of the active primase, the oligomer composed of both forms, for multiple rounds of primer synthesis.

ROLE OF ADDITIONAL PROTEINS AT THE REPLICATION FORK

The above results indicate that a stable complex of gene 4 protein, T7 DNA polymerase, and the replication fork is established for processive leading-strand synthesis. However, such a stable complex is not established for the catalysis of multiple rounds of lagging-strand synthesis. This leads to the problem of how the gene 4 primase and DNA polymerase are efficiently recycled for lagging-strand synthesis. It has been proposed for the bacteriophage T4 and *E. coli* DNA replication systems that a replisome consisting of two DNA polymerase molecules, a primase, and a helicase efficiently catalyzes both leading-strand synthesis and the multiple cycles of lagging-strand synthesis without disassembling (3, 24, 51). The *E. coli* DNA polymerase III holoenzyme, for example, has been shown to be a dimeric enzyme, raising the possibility that the lagging-strand DNA polymerase is bound to the leading-strand DNA polymerase (31). Such a structure would prevent the lagging-strand DNA polymerase from dissociating from the replication fork after completion of an Okazaki fragment. It is not yet clear whether the lagging-strand DNA polymerase at the T7 replication fork may be recycled in this way.

Additional proteins may be involved in recycling primase and DNA polymerase for lagging-strand synthesis. Indeed, there are a number of class II genes of T7 (the class encoding most of the T7 replication proteins) whose functions are still undefined (see reference 10). These proteins may serve as accessory proteins which are necessary for the assembly of a replisome. Can such accessory proteins enable gene 4 protein to act processively as a primase? Can these accessory proteins regulate the processivity of the DNA polymerase via the oxidation state of thioredoxin? Do reactions involved in initiation play a role in the assembly of a putative replisome? These problems are currently under investigation.

LITERATURE CITED

1. Adler, S., and P. Modrich. 1979. T7-induced DNA polymerase. Characterization of associated exonuclease activities and resolution into biologically active subunits. *J. Biol. Chem.* **254**:11605–11614.
2. Adler, S., and P. Modrich. 1983. T7-induced DNA polymerase: requirement for thioredoxin sulfhydryl groups. *J. Biol. Chem.* **258**:6956–6962.
3. Alberts, B. M. 1984. The DNA enzymology of protein machines. *Cold Spring Harbor Symp. Quant. Biol.* **49**:1–12.
4. Arai, K., R. L. Low, and A. Kornberg. 1981. Movement and site selection for priming by the primosome in phage φX174 DNA replication. *Proc. Natl. Acad. Sci. USA* **78**:707–711.
5. Araki, H., and H. Ogawa. 1981. A T7 amber mutant defective in DNA-binding protein. *Mol. Gen. Genet.* **183**:66–73.
6. Berg, J. M. 1986. Potential metal-binding domains in nucleic acid binding proteins. *Science* **232**:485–487.
6a. Bernstein, J. A., and C. C. Richardson. 1988. A 7-kDa region of the bacteriophage T7 gene 4 protein is required for primase but not for helicase activity. *Proc. Natl. Acad. Sci. USA* **85**:396–400.
7. Blumberg, H., A. Eisen, A. Sledziewski, D. Bader, and E. T. Young. 1987. Two zinc fingers of a yeast regulatory protein shown by genetic evidence to be essential for its function. *Nature* (London) **328**:443–445.
8. Bradley, M. K., T. F. Smith, R. H. Lathrop, D. M. Livingston, and T. A. Webster. 1987. Consensus topography in the ATP binding site of the simian virus 40 and polyomavirus large tumor antigens. *Proc. Natl. Acad. Sci. USA* **84**:4026–4030.
9. Cha, T.-A., and B. M. Alberts. 1986. Studies of the DNA helicase-RNA primase unit from bacteriophage T4. *J. Biol. Chem.* **261**:7001–7010.
10. Dunn, J. J., and F. W. Studier. 1983. Complete nucleotide sequence of bacteriophage T7 DNA and the locations of T7 genetic elements. *J. Mol. Biol.* **166**:477–535.
11. Fischer, H., and D. C. Hinkle. 1980. Bacteriophage T7 DNA replication *in vitro*. Stimulation of DNA synthesis by T7 RNA polymerase. *J. Biol. Chem.* **255**:7956–7964.
12. Fujiyama, A., Y. Kohara, and T. Okazaki. 1981. Initiation sites for discontinuous DNA synthesis of bacteriophage T7. *Proc. Natl. Acad. Sci. USA* **78**:903–907.

13. Fuller, C. W., and C. C. Richardson. 1985. Initiation of DNA replication at the primary origin of bacteriophage T7 by purified proteins. Site and direction of initial DNA synthesis. *J. Biol. Chem.* 260:3185–3196.
14. Fuller, C. W., and C. C. Richardson. 1985. Initiation of DNA replication at the primary origin of bacteriophage T7 by purified proteins. Initiation of bidirectional DNA synthesis. *J. Biol. Chem.* 260:3197–3206.
15. Hinton, D. M., and N. G. Nossal. 1987. Bacteriophage T4 DNA primase-helicase. Characterization of oligomer synthesis by T4 61 protein alone and in conjunction with T4 41 protein. *J. Biol. Chem.* 262:10873–10878.
16. Holmgren, A. 1985. Thioredoxin. *Annu. Rev. Biochem.* 54:237–271.
17. Hori, K., D. F. Mark, and C. C. Richardson. 1979. Deoxyribonucleic acid polymerase of bacteriophage T7: purification and properties of the phage-encoded subunit, the gene 5 protein. *J. Biol. Chem.* 254:11591–11597.
18. Hori, K., D. F. Mark, and C. C. Richardson. 1979. Deoxyribonucleic acid polymerase of bacteriophage T7: characterization of the exonuclease activities of the gene 5 protein and the reconstituted polymerase. *J. Biol. Chem.* 254:11598–11604.
19. Huber, H. E., M. Russel, P. Model, and C. C. Richardson. 1986. Interaction of mutant thioredoxins of *Escherichia coli* with the gene 5 protein of phage T7. *J. Biol. Chem.* 261:15006–15012.
20. Huber, H. E., S. Tabor, and C. C. Richardson. 1987. *Escherichia coli* thioredoxin stabilizes complexes of bacteriophage T7 DNA polymerase and primed templates. *J. Biol. Chem.* 262:16224–16232.
21. Kelly, T. J., and C. A. Thomas. 1969. An intermediate in the replication of bacteriophage T7 DNA molecules. *J. Mol. Biol.* 44:459–475.
22. Kolodner, R., Y. Masamune, J. E. LeClerc, and C. C. Richardson. 1978. Gene 4 protein of bacteriophage T7. Purification, physical properties, and stimulation of T7 DNA polymerase during the elongation of polynucleotide chains. *J. Biol. Chem.* 253:566–573.
23. Kolodner, R., and C. C. Richardson. 1977. Replication of duplex DNA by bacteriophage T7 DNA polymerase and gene 4 protein is accompanied by hydrolysis of nucleoside 5'-triphosphates. *Proc. Natl. Acad. Sci. USA* 74:1525–1529.
24. Kornberg, A. 1982. *DNA Replication, 1982 Supplement*, p. 123–125. W. H. Freeman & Co., San Francisco.
25. LeBowitz, J. H., and R. McMacken. 1986. The *Escherichia coli* dnaB replication protein is a DNA helicase. *J. Biol. Chem.* 261:4738–4748.
26. Lechner, R. L., and C. C. Richardson. 1983. A preformed, topologically stable replication fork. *J. Biol. Chem.* 258:11185–11196.
27. Masamune, Y., G. D. Frenkel, and C. C. Richardson. 1971. A mutant of bacteriophage T7 deficient in polynucleotide ligase. *J. Biol. Chem.* 246:6874–6879.
28. Matson, S. W., and C. C. Richardson. 1983. DNA-dependent nucleoside 5'-triphosphatase activity of the gene 4 protein of bacteriophage T7. *J. Biol. Chem.* 258:14009–14016.
29. Matson, S. W., and C. C. Richardson. 1985. Nucleotide-dependent binding of the gene 4 protein of bacteriophage T7 to single-stranded DNA. *J. Biol. Chem.* 260:2281–2287.
30. Matson, S. W., S. Tabor, and C. C. Richardson. 1983. The gene 4 protein of bacteriophage T7. Characterization of helicase activity. *J. Biol. Chem.* 258:14017–14024.
31. McHenry, C. S., R. Oberfelder, K. Johanson, H. Tomasiewicz, and M. A. Franden. 1987. Structure and mechanism of the DNA polymerase III holoenzyme. *UCLA Symp. Mol. Cell. Biol. New Ser.* 47:47–61.
32. McMacken, R., K. Ueda, and A. Kornberg. 1977. Migration of *Escherichia coli* dnaB protein on the template DNA strand as a mechanism in initiating DNA replication. *Proc. Natl. Acad. Sci. USA* 74:4190–4194.
33. Miller, J., A. D. MacLachlan, and A. Klug. 1985. Repetitive zinc-binding domains in the protein transcription factor IIIA from *Xenopus* oocytes. *EMBO J.* 4:1609–1614.
34. Modrich, P., and C. C. Richardson. 1975. Bacteriophage T7 deoxyribonucleic acid replication *in vitro*. A protein of *Escherichia coli* required for bacteriophage T7 DNA polymerase activity. *J. Biol. Chem.* 250:5508–5514.
35. Modrich, P., and C. C. Richardson. 1975. Bacteriophage T7 deoxyribonucleic acid replication *in vitro*. Bacteriophage T7 DNA polymerase: an enzyme composed of phage- and host-specified subunits. *J. Biol. Chem.* 250:5515–5522.
36. Nakai, H., and C. C. Richardson. 1986. Interactions of the DNA polymerase and gene 4 protein of bacteriophage T7. *J. Biol. Chem.* 261:15208–15216.
37. Nakai, H., and C. C. Richardson. 1986. Dissection of RNA-primed DNA synthesis catalyzed by gene 4 protein and DNA polymerase of bacteriophage T7. *J. Biol. Chem.* 261:15217–15224.
38. Nossal, N. G., and D. M. Hinton. 1987. Bacteriophage T4 primase-helicase. Characterization of the DNA synthesis primed by T4 61 protein in the absence of T4 41 protein. *J. Biol. Chem.* 262:10879–10885.
39. Ogawa, T., and T. Okazaki. 1979. RNA-linked

nascent DNA pieces in phage T7-infected *Escherichia coli*. III. Detection of intact primer RNA. *Nucleic Acids Res.* 7:1621–1633.
40. Reuben, R. C., and M. L. Gefter. 1973. A DNA-binding protein induced by bacteriophage T7. *Proc. Natl. Acad. Sci. USA* 70:1846–1850.
41. Richardson, C. C. 1983. Replication of bacteriophage T7 DNA, p. 163–204. *In* Y. Becker (ed.), *Replication of Viral and Cellular Genomes*. Martinus Nijhoff Publishing, Boston.
42. Romano, L. J., and C. C. Richardson. 1979. Characterization of the ribonucleic acid primers and the deoxyribonucleic acid product synthesized by the DNA polymerase and gene 4 protein of bacteriophage T7. *J. Biol. Chem.* 254:10483–10489.
43. Romano, L. J., F. Tamanoi, and C. C. Richardson. 1981. Initiation of DNA replication at the primary origin of bacteriophage T7 by purified proteins: requirement of T7 RNA polymerase. *Proc. Natl. Acad. Sci. USA* 78:4107–4111.
44. Saito, H., S. Tabor, F. Tamanoi, and C. C. Richardson. 1980. Nucleotide sequence of the primary origin of bacteriophage T7 DNA replication: relationship to adjacent genes and regulatory elements. *Proc. Natl. Acad. Sci. USA* 77:3917–3921.
45. Scherzinger, E., and G. Klotz. 1975. Studies of bacteriophage T7 DNA synthesis *in vitro*. II. Reconstitution of the T7 replication system using purified proteins. *Mol. Gen. Genet.* 141:233–249.
46. Scherzinger, E., E. Lanka, and G. Hillenbrand. 1977. Role of bacteriophage T7 DNA primase in the initiation of DNA strand synthesis. *Nucleic Acids Res.* 4:4151–4163.
47. Scherzinger, E., E. Lanka, G. Morelli, D. Seiffert, and A. Yuki. 1977. Bacteriophage T7 induced DNA-priming protein. A novel enzyme involved in DNA replication. *Eur. J. Biochem.* 72:543–558.
48. Scherzinger, E., F. Litfin, and E. Jost. 1973. Stimulation of T7 DNA polymerase by a new phage-coded protein. *Mol. Gen. Genet.* 123:247–262.
49. Seki, T., and T. Okazaki. 1979. RNA-linked nascent DNA pieces in phage T7-infected *Escherichia coli*. II. Primary structure of the RNA portion. *Nucleic Acids Res.* 7:1603–1619.
50. Sigal, N., H. Delius, T. Kornberg, M. L. Gefter, and B. Alberts. 1972. A DNA-unwinding protein isolated from *Escherichia coli*. Its interaction with DNA and with DNA polymerases. *Proc. Natl. Acad. Sci. USA* 69:3537–3541.
51. Sinha, N. K., C. F. Morris, and B. M. Alberts. 1980. Efficient *in vitro* replication of double-stranded DNA templates by a purified T4 bacteriophage replication system. *J. Biol. Chem.* 255:4290–4303.
52. Sternglanz, R., H. F. Wang, and J. J. Donegan. 1976. Evidence that both growing DNA chains at a replication fork are synthesized discontinuously. *Biochemistry* 15:1838–1843.
53. Studier, F. W. 1972. Bacteriophage T7. *Science* 176:367–376.
54. Sugimoto, K., Y. Kohara, and T. Okazaki. 1987. Relative roles of T7 RNA polymerase and gene 4 primase for the initiation of T7 phage DNA replication *in vivo*. *Proc. Natl. Acad. Sci. USA* 84:3977–3981.
55. Tabor, S., H. E. Huber, and C. C. Richardson. 1987. *Escherichia coli* thioredoxin confers processivity on the DNA polymerase activity of the gene 5 protein of bacteriophage T7. *J. Biol. Chem.* 262:16212–16223.
56. Tabor, S., and C. C. Richardson. 1981. Template recognition sequence for RNA primer synthesis by gene 4 protein of bacteriophage T7. *Proc. Natl. Acad. Sci. USA* 78:205–209.
57. Tamanoi, F., H. Saito, and C. C. Richardson. 1980. Physical mapping of primary and secondary origins of bacteriophage T7 DNA replication. *Proc. Natl. Acad. Sci. USA* 77:2656–2660.
58. Venkatesan, M., L. L. Silver, and N. G. Nossal. 1982. Bacteriophage T4 gene 41 protein, required for the synthesis of RNA primers, is also a DNA helicase. *J. Biol. Chem.* 257:12426–12434.
59. Wickner, S. H. 1978. DNA replication proteins of *Escherichia coli*. *Annu. Rev. Biochem.* 47:1163–1191.

Chapter 9

Reverse Genetics and *Saccharomyces cerevisiae* DNA Replication and Repair

Judith L. Campbell, Martin Budd, Dave Burbee, Karen Sitney, Kevin Sweder, and Fred Heffron

While the description of the formation of the replication fork and its movement by stable complexes of proteins conserved between procaryotes and eucaryotes represents an enormous achievement in itself, in many ways it is just another beginning. For one thing, it opens up to experiment a fundamental but poorly understood aspect of DNA replication, namely the regulation of replication of eucaryotic chromosomes. The major questions are: (i) What is the molecular basis for entry into S phase? (ii) Once in S phase, what determines the spatial and temporal activation of the multiple replication units that make up the large chromosomes? (iii) What makes reinitiation within a single replication unit in a single S phase so rare? These problems are too complex to resolve easily in larger eucaryotes, but the simple fungus *Saccharomyces cerevisiae* is helping to answer some of them. Yeast is attractive because its cell cycle, chromosome organization, and replication are typical of all eucaryotes. Furthermore, the ways in which biochemistry can be combined with both classical and molecular genetics in yeast are unparalleled in any metazoan system. Of particular importance has been the ability to target exogenous DNA into the chromosomes by homologous recombination, allowing systematic mutation and genetic analysis of numerous processes, including replication. Finally, the yeast genome contains only about 14,000 kilobases of DNA, a total of about 400 replicons. This small size allowed the cloning, in 1979, of chromosomal origins of replication into plasmids that were then able to replicate outside the chromosome as minichromosomes under strict cell cycle control, which focused attention on yeast as the only eucaryotic system in which a single chromosomal replicon could be studied in detail.

A prerequisite to understanding regulation is knowing what proteins and genes are involved and how they are regulated. Thus, the initial challenge for us was to identify the yeast analogs of the replication proteins identified in other systems and to clone the genes encoding them. This was done by a

---

Judith L. Campbell, Martin Budd, Karen Sitney, and Kevin Sweder • Braun Laboratories, California Institute of Technology, Pasadena, California 91125. *Dave Burbee and Fred Heffron* • Department of Molecular Biology, Scripps Clinic and Research Foundation, La Jolla, California 92037.

combined biochemical and genetic strategy we have called "reverse genetics." Putative replication proteins are purified one by one by biochemical assays for the activities. The protein is used to clone the corresponding gene, and the gene is used to obtain mutants to verify that the protein is a bona fide constituent of the yeast replication apparatus. Thus far, we have successfully used the technique for polymerases, SSBs, origin-binding proteins, and nucleases. Others have studied primase, RNase H, and a polymerase accessory protein (4a).

DNA POLYMERASE α, ANOTHER RED HERRING?

Unexpected findings are always exciting, and the past year has brought a very important one in the replication field. The central tenet of eucaryotic DNA replication, that DNA polymerase α (pol α) was the sole polymerase responsible for DNA replication, has been challenged by new findings which have raised the possibility that DNA polymerase δ (pol δ) is also involved (7, 12).

Yeast is ideal for resolving the issue. Yeast DNA polymerase I, pol I, is the yeast counterpart of pol α. The subunit structure is identical to pol α from all other sources; pol I is sensitive to aphidicolin and to 1 μM BuPdGTP, like pol α, and is free of 3'→5' exonuclease activity. But is it required for replication? That was not so clear, because yeast also contains a second high-molecular-weight nuclear DNA polymerase, DNA polymerase II (pol II), which is sensitive to aphidicolin, contains a 3'→5' exonuclease, and is relatively insensitive to BuPdGTP. By these criteria it appears to be the analog of higher-cell pol δ, though neither pol II nor δ has been well enough characterized physically to warrant a strong statement on the subject (4, 4a, 6, 9, 10, 12, 18; M. Budd and J. L. Campbell, unpublished data).

To discover whether pol I and pol II are the same or different and whether one or both are required for replication, we set out in 1981 to isolate mutants missing pol I by reverse genetics. We began with pol I since it was easy to purify by published procedures and antibodies were already available. We used the antibodies to clone the *polI* gene from a yeast genomic DNA library in the expression vector lambda gt11 and proceeded to make mutants. The simplest kind of mutation that one can make, because it requires no knowledge of the structure of the gene, is called a gene disruption. A large deletion or insertion is made in the cloned gene and the altered gene is introduced into the chromosome by transformation and homologous gene replacement. Such mutations yield a scorable phenotype as long as the gene is a single copy. If the gene is essential, the phenotype is lethality and therefore the disruption must be carried out in a diploid, so only one copy of the gene is altered. If the gene disruption knocks out an essential function, after sporulation one sees two viable and two inviable spores. Disrupting the *polI* gene told us three things: pol I is required for viability, it is encoded by a single gene, and pol II cannot compensate for the loss of pol I. It did not tell us the basis for the loss of viability, whether pol I was the sole polymerase required, or whether pol II was also affected by the mutation.

To investigate the contribution of the two polymerases more thoroughly, conditional lethal mutants were constructed, using a technique called "plasmid shuffling" (3). First, mutagenized plasmids carrying the *polI* gene were transformed into a pol 1 knockout strain in which the only good copy of the gene was carried on an unstable plasmid. The resident plasmid was then exchanged for a mutagenized *polI* plasmid and transformants were screened to identify clones containing temperature-sensitive alleles. The mutant alleles were introduced into the chromosomes by gene replacement to produce temperature-sensitive strains. When mitotic cultures of the mutants growing at the permissive temperature are shifted to the nonpermissive temperature, DNA synthesis ceases immediately and completely. The mutants are also deficient in premeiotic DNA synthesis, commitment to recombination, and sporulation (3). The complete absence of mitotic or premeiotic DNA synthesis is an extremely important observation because it not only confirms that pol α is required for replication—that is it is not a red herring—but also tells us that if pol II has a role in replication, it must act in concert with DNA pol I and cannot act independently.

DNA repair is a phenomenon closely related to DNA replication. DNA repair in mammalian cells is aphidicolin sensitive and therefore mediated by either pol α or δ. (The findings in the early 1970s indicating that polymerase β was the exclusive repair enzyme in brain cells seem to be incomplete.) The replication-deficient pol I mutants were found to be normal for both single- and double-stranded break repair. Therefore, either pol II can compensate for loss of pol I in repair or pol I is not involved at all. Drug sensitivity of repair in permeabilized cells also supports a greater role for pol δ in repair than pol α.

ARE pol I AND II THE SAME OR DIFFERENT PROTEINS?

Since pol I is undetectable while pol II is normal in the *pol1* mutants, we infer that pol I and pol II are encoded by different genes. Furthermore, when pol I is overproduced by cloning the *pol1* gene under the control of the strong *gal* yeast promoter, pol II is not coordinately overexpressed. The surprising result is that there is a new polymerase in the *pol1* mutant. It is identical to pol II except in chromatographic properties on one final column, and since it is not overproduced in the pol I overproducer, we assume it is a new form of pol II rather than an entirely new DNA polymerase.

YEAST ORIGINS OF REPLICATION

In yeast, chromosomal sequences that might serve as replicators have been known for many years. They were identified by their ability to mediate high-frequency transformation and autonomous replication of plasmids carrying yeast selectable markers, hence the designation autonomous replication sequence (ARS) (16). Although early on we showed that ARSs served as origins of replication in an in vitro replication system (5), as recently as the past few months two clever methods have been devised that established the crucial point that ARSs are origins of replication in vivo, at least on plasmids (2, 8). That they are also chromosomal origins is implied by (i) analogy with bacteria (19), (ii) the fact that the number of ARSs in yeast (400) corresponds well with the number of replicons, and (iii) the coincidence of observed replication bubbles with functional ARSs in the spacer regions of the ribosomal DNA repeats.

The ARS is tripartite. The primary sequence of domain A is highly conserved at all ARSs; it is essential for replication and is also sufficient (5, 15). The domain B and C sequences surrounding the core consensus are more degenerate, but since their removal decreases plasmid stability, they are thought to contribute to ARS function as well. As with promoters, which often contain multiple overlapping functional sequences that make them difficult to dissect by mutation alone, the way to best assess the functions of the various domains seemed to be through a detailed study of protein-DNA interactions at an ARS sequence.

Exploiting the sensitivity of gel retardation assays, we have detected several proteins that bind to ARS1. We have thus far characterized only the most abundant of these, and it binds to domain B of ARS1. DNase I footprinting at a second ARS, the HMR-E ARS, shows that the protein protects a 24-base-pair (bp) region that is 66 bp 5′ to the T-rich strand of the ARS core sequence HMR-E. Thus it appears to be the same as the protein detected in crude extracts by Shore et al. (13), and we have followed their nomenclature and called it SBF-B. Since SBF-B binds to more than one ARS and binds to a functionally important region in both ARSs, the protein is probably essential for ARS function and its availability paves the way for a reverse genetics approach to verify this.

Although binding at HMR-E and ARS1 is clearly due to the same protein, there is limited homology between the sequences protected at the HMR-E ARS and at ARS1. It is possible that the protein is recognizing some shared DNA structure rather than the primary sequence. The region at element B of ARS1 contains a bent structure that may be an important element in binding, and enzymatic probes reveal a "non-B" structure adjacent to domain A (14, 17). Bent DNA seems to be an important feature of many origins of replication (14). Another possibility is that the recognition site is not a continuous sequence of base pairs. Of 24 bp protected at the HMR-E ARS, 12 are conserved at ARS1. The element B consensus proposed by Palzkill et al. (11) is included in the region protected at ARS1 but not at HMR-E. In fact, sequence analysis per se by us and others (1, 11) has not been able to define precisely the recognition signals in domain B. One advantage of having the purified protein in hand is that footprinting of the purified protein to additional ARSs will define the consensus unambiguously.

YEAST DEOXYRIBONUCLEASES

As the first step in the enzymatic and genetic characterization of yeast deoxyribonucleases, we have purified one of the major endonucleases of vegetatively growing diploid yeast cells. The purified nuclease degrades both double- and single-stranded DNAs in the presence of a divalent cation. We have raised antibodies against the purified enzyme and used the antibodies to select yeast fragments that encode cross-reactive epitopes on lambda gt11 clones. The cloned gene was mutagenized and fragments were used to disrupt the chromosomal gene. Loss of this nuclease, resulting from the disruption of the nuclease gene, results in a yeast cell unable to repair DNA double-strand breaks.

In summary, reverse genetics is proving to be a feasible method for studying initiation and elongation of DNA replication. It is also being extended to recombination and repair.

LITERATURE CITED

1. **Brand, A. H., G. Micklem, and K. Nasmyth.** 1987. A yeast silencer contains sequences that can promote autonomous plasmid replication and transcriptional activation. *Cell* 51:709–719.
2. **Brewer, B. S., and W. L. Fangman.** 1987. The localization of replication origins on ARS plasmids in S. cerevisiae. *Cell* 51:463–471.
3. **Budd, M., and J. L. Campbell.** 1987. Temperature sensitive mutants of yeast DNA polymerase I. *Proc. Natl. Acad. Sci. USA* 84:2838–2842.
4. **Byrnes, J. J., K. M. Downey, V. L. Black, and A. G. So.** 1976. A new mammalian DNA polymerase with 3' to 5' exonuclease activity: DNA polymerase delta. *Biochemistry* 15:2817–2823.
4a. **Campbell, J. L.** 1986. Eukaryotic DNA replication. *Annu. Rev. Biochem.* 55:733–771.
5. **Celniker, S. E., K. S. Sweder, F. Srienc, J. E. Bailey, and J. L. Campbell.** 1984. Deletion mutations affecting autonomously replicating sequence ARS1 of *Saccharomyces cerevisiae*. *Mol. Cell. Biol.* 4:2455–2466.
6. **Crute, J. J., A. F. Wahl, and R. A. Bambara.** 1986. Purification and characterization of two new high molecular weight forms of DNA polymerase delta. *Biochemistry* 25:26–36.
7. **Dresler, S. L., and M. G. Frattini.** 1986. DNA replication and UV-induced repair synthesis in human fibroblasts are much less sensitive than DNA polymerase alpha to inhibition by butylphenyl-deoxyguanosine triphosphate. *Nucleic Acids Res.* 14:7093–7102.
8. **Huberman, J. A., L. D. Spotila, K. A. Nawotka, S. M. El-Assoli, and L. R. Davis.** 1987. The in vivo replication origin of the yeast 2 mu plasmid. *Cell* 51:473–481.
9. **Johnson, L. M., M. Snyder, L. M. S. Chang,**

R. W. Davis, and J. L. Campbell. 1985. Isolation of the gene encoding yeast DNA polymerase I. *Cell* **43**:369–377.
10. Lee, M. Y. W. T., and N. L. Toomey. 1987. Human placental DNA polymerase delta: identification of a 170-kilodalton polypeptide by activity staining and immunoblotting. *Biochemistry* **26**:1076–1085.
11. Palzkill, T. G., S. G. Oliver, and C. S. Newlon. 1986. DNA sequence analysis of ARS elements from chromosome III of Saccharomyces cerevisiae: identification of a new conserved sequence. *Nucleic Acids Res.* **14**:6247–6264.
12. Prelich, G., M. Kostura, D. R. Marshak, M. B. Mathews, and B. Stillman. 1987. The cell-cycle regulated proliferating cell nuclear antigen is required for SV40 DNA replication in vitro. *Nature* (London) **326**:471–475.
13. Shore, D., D. J. Stillman, A. H. Brand, and K. A. Nasmyth. 1987. Identification of silencer binding proteins from yeast: possible roles in SIR control and DNA replication. *EMBO J.* **6**:461–467.
14. Snyder, M., A. R. Buchman, and R. W. Davis. 1986. Bent DNA at a yeast autonomously replicating sequence. *Nature* (London) **324**:87–89.
15. Srienc, F., J. E. Bailey, and J. L. Campbell. 1985. Effect of ARS1 mutations on chromosome stability in *Saccharomyces cerevisiae*. *Mol. Cell. Biol.* **5**:1676–1684.
16. Struhl, K., D. T. Stinchcomb, S. Scherer, and R. W. Davis. 1979. High-frequency transformation of yeast: autonomous replication of hybrid DNA molecules. *Proc. Natl. Acad. Sci. USA* **76**:1035–1039.
17. Umek, R. M., and D. Kowalski. 1987. Yeast regulatory sequences preferentially adopt a non-B conformation in supercoiled DNA. *Nucleic Acids Res.* **15**:4467–4480.
18. Wahl, A. F., J. J. Crute, R. D. Sabatino, J. B. Bodner, R. L. Marraccino, L. W. Harwell, E. M. Lord, and R. A. Bambara. 1986. Properties of two forms of DNA polymerase delta from calf thymus. *Biochemistry* **25**:7821–7827.
19. Zyskind, J. W., J. M. Cleary, W. S. A. Brusilow, N. E. Harding, and D. W. Smith. 1983. Chromosomal replication origin from the marine bacterium Vibrio harveyi functions in Escherichia coli: oriC consensus sequence. *Proc. Natl. Acad. Sci. USA* **80**:1164–1168.

Chapter 10

Origin Methylation in the Replication of P1 Plasmids

Ann L. Abeles and Stuart J. Austin

The study of plasmid replicons in *Escherichia coli* and other bacterial species shows that a number of individual strategies have developed for the control of replication. In only one case, that of the class of small high-copy-number plasmids typified by plasmid ColE1 (28, 43), is the control system for plasmid replication close to being fully elucidated. The plasmid-borne information for the control of these plasmids is kept to a minimum. The modest capabilities of the system are compensated by a high copy number that can absorb fluctuations in the control response. However, a number of plasmid types are maintained at very low copy numbers and therefore must make use of more stringent and accurate regulation. Such systems tend to be complicated and, perhaps as a consequence, are less well understood. One group of plasmids, the IncFII plasmids typified by plasmid R1, achieve stringent control by a system that can be regarded as an elaboration of the basic ColE1 strategy (23, 32, 45). However, the majority of low-copy-number plasmids fall into a diverse class that appears to bear some general resemblance to the host chromosomal origin *oriC*. They all encode replicon-specific proteins that are required for initiation. Some might be functional analogs of the host DnaA protein that is required for initiation at *oriC*. The plasmid origins also contain multiple direct repeats of a sequence, typically about 20 base pairs (bp) long, that is important for origin function. In some cases it has been shown that a specific protein acts by binding to these sequences (P1 [1], F [27, 42], pSC101 [24, 44], and R6K [10, 20]). The *oriC* sequence contains two series of repeated sequences of this size range. One constitutes the binding site for DnaA protein (11, 15). The other consists of 18-bp repeats present in a region thought to be important in the formation of an open complex for primosome assembly (13, 38). Several of the plasmid replicons have both replicon-specific repeat sequences and DnaA binding sites within their origins. In several cases these replicons have been shown to require both the replicon-specific initiation protein and the host DnaA protein for initiation (P1 [16], F [31], and pSC101 [9, 17]).

Another feature that several of these plasmid replicons share with *oriC* is the clustering within the origin of multiple repeats of the sequence GATC, the recogni-

Ann L. Abeles and Stuart J. Austin • Laboratory of Chromosome Biology, BRI-Basic Research Program, NCI-Frederick Cancer Research Facility, Frederick, Maryland 21701.

tion site for the host DNA adenine methylase. These sequences are often present as part of a repeated motif of about 7 bp. The methylase activity is the product of the host *dam* gene and causes the postreplicational N^6 methylation of the adenine residues of the sequences (14, 18, 21, 26). Despite the fact that *dam*-methylase-defective strains of *E. coli* are viable, several groups have presented data which suggest that *oriC* methylation plays a role in origin function and control (19, 30, 37). Recently, Russell and Zinder (35) showed that hemimethylation of *oriC* DNA blocks replication, whereas the fully methylated or unmethylated DNA is competent for replication. This lends credence to the suggestion that the transient hemimethylation of DNA that follows replication in wild-type strains might be important in providing a latent period in which reinitiation is prevented (19, 30, 37).

In summary, the origins of low-copy-number plasmids usually prove to have one or more of three "*oriC*-like" structural features: large direct repeats, one or more DnaA-binding sites, and a cluster of GATC sequences. The plasmid replication origin of P1 should prove to be an ideal subject for investigating the significance of and relationship between these features as it contains all three elements in separate but closely apposed regions of the 246-bp functional origin. Here we briefly summarize for P1 plasmids what is known of the DnaA-binding and direct-repeat regions and present information on the significance of the GATC-rich region of the origin and the role of origin methylation.

THE P1 PLASMID REPLICON

The P1 plasmid replication region consists of three major elements: the origin, an open reading frame for the RepA protein, and the *incA* locus (Fig. 1). The *repA* open reading frame produces a protein of 32 kilodaltons (kDa). Nonsense mutations and deletions which eliminate *repA* function block replication, showing that RepA is essential for replicon function (6).

The *incA* locus consists of nine consensus repeats of a 19-bp sequence that constitutes the RepA-binding site (1). *incA* is not essential for replication. However, progressive deletions of these repeat sequences show a corresponding increase in plasmid copy number (33). It has been proposed that the function of the locus is to titrate RepA protein, which in turn limits origin function (7, 33). However, alternative explanations have merit in explaining the genetic properties of the locus. It is possible that the *incA* locus in conjunction with the bound RepA protein interacts directly with the origin, by either intraplasmid or interplasmid interactions (D. K. Chattoraj, R. J. Mason, and S. H. Wickner, manuscript in preparation; A. L. Abeles, L. D. Reaves, and S. J. Austin, manuscript in preparation).

THE P1 ORIGIN

The origin region is the only locus that is essential in *cis* to the plasmid in order for specific replication to occur. Plasmids containing only the 246-bp origin and a suitable marker gene can replicate when the RepA protein is supplied in *trans* (6). The sequence of the origin is shown in Fig. 2. The leftmost part of the sequence consists of a tandem repeat of the DnaA protein-binding site as defined by Fuller et al. (11) and Hansen et al. (15). The P1 origin can function in certain *dnaA* temperature-sensitive mutants of *E. coli* that cannot support chromosomal origin function (7). However, it has recently been shown that origin function is blocked in *dnaA* null mutants (16). Thus DnaA protein is essential for plasmid replication. It is presumably required for initiation and acts by binding to the sites in the origin.

The rightmost portion of the origin sequence consists of five direct repeats of a 19-bp consensus sequence. This is the same as the sequence that is present nine times in the *incA* control locus (Fig. 1). These sequences are specific binding sites for the P1 RepA protein, and all five of these repeats can bind purified RepA protein in vitro (1). As RepA is essential for origin function in vivo and in vitro (3, 6), it is presumed that RepA acts at these sites to allow initiation Embedded in the repeat region is the promoter sequence for the RepA gene. RepA binding to the region decreases transcription, thus providing autoregulation of RepA synthesis (8). The *rep-11* deletion that removes two of the repeats, including the promoter sequence, blocks replication. However, *rep-11* plasmids fail to replicate even when suitable levels of RepA protein are supplied in *trans*. Thus, these repeats are an essential part of the origin as well as being involved in *repA* transcription (6).

THE CENTRAL REGION OF THE ORIGIN

The function of the central region of the origin is not known. It is, however, important for origin function as a single base change from C to T at position 457 blocks replication (the *rep-30* mutation [6]). It is this region that contains five GATC sequences, four of which are part of the short repeat motif C/TAGATCCC/A. A fifth related sequence, CAGGTCCA, exists between the first and second copies of the sequence, completing a compact array of these elements with spacers of 6, 2, 3, and 2 bp between them (Fig. 2). The *rep-30* mutation lies within the fourth of these repeats and, by changing a C to a T, eliminates one GATC sequence. An additional GATC sequence lies immediately to the right of the array. Thus, there are five GATC sequences in a region of 56 bp. Less than one such sequence would be expected to appear in the region if the bases were randomly distributed.

The sequence GATC is the substrate for the DNA adenine methylase of *E. coli*

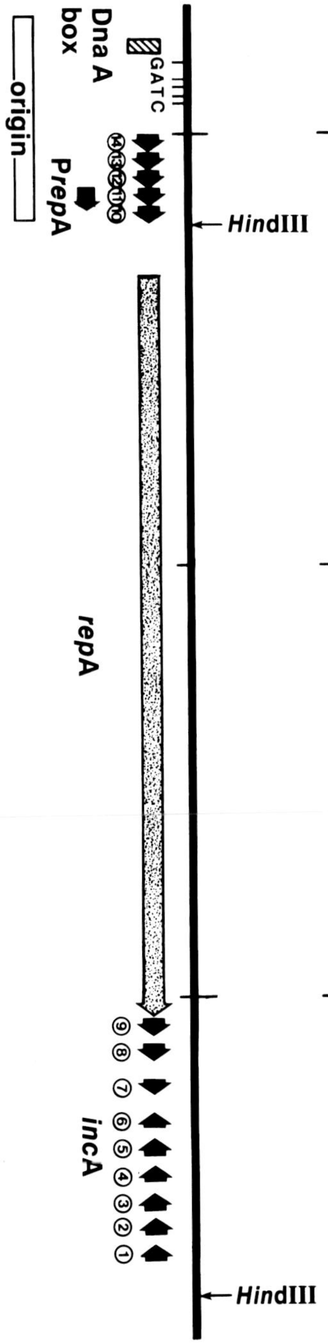

FIGURE 1. The P1 plasmid replicon. The origin region contains a direct repeat of the DnaA recognition site (DnaA box), five GATC sequences (GATC), and five 19-bp direct repeats (short arrows). The gene for the essential replication protein is labeled *repA*. The copy control element, *incA*, contains nine more of the 19-bp repeats oriented as indicated by the arrows. The scale above shows the size of the region and arrows indicate some restriction sites.

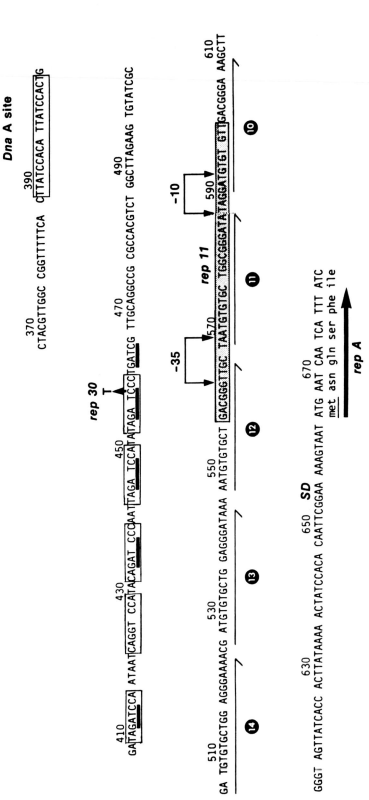

Figure 2. The DNA sequence of the minimal P1 origin cloned into M13mp10 and 11. The numbers above the sequence correspond to the numbering scheme adopted previously (2). The large box near bp 390 encloses the proposed DnaA recognition site. The small boxes mark short repeats which surround some of the underlined GA⁻C (*dam*) sites. The long arrows, numbered 14 to 10, mark the repeats which share the same consensus sequence with the *incA* region of the replicon (2). The location of the *rep-30* mutation is marked. The −35 and −10 regions of the *repA* promoter are indicated. SD marks the ribosome binding site, and the heavy arrow labeled *repA* shows the beginning of the *repA* open reading frame.

that adds methyl groups to the symmetrically placed adenine residues after DNA replication (14, 18, 21, 26). Attempts to introduce P1 miniplasmids into a *dam* strain which is deficient in the methylase were unsuccessful, suggesting that origin function might be impaired when the origin is unmethylated (6). This prompted an investigation of whether the methylation of the origin sequences was important for function.

P1 AND F PLASMID ESTABLISHMENT IN *dam* STRAINS

The F plasmid of *E. coli* contains a replication region that is nonhomologous to that of P1 at the DNA level. However, both replicons are similarly organized and both are subject to very stringent replication control (2, 36). Both P1 and F miniplasmids have been constructed by cloning fragments of the parental plasmids into a λ phage vector. λ-P1:5R (39) and λ-F:583 (5) have the complete plasmid-maintenance regions of P1 and F, respectively, introduced at the unique *Eco*RI site of the vector. The vector lacks the λ *att* site and cannot integrate into the host chromosome. Thus, both the P1 and F constructs form lysogens by replicating as low-copy-number plasmids under P1 or F control. Both constructs formed lysogens at the same frequency in wild-type strains. However, although the F construct formed lysogens of a *dam* strain at normal frequency, the P1 construct formed no lysogens at all. Introduction of plasmid pMQ3 (4), which contains a wild-type dam^+ gene, into the *dam* strain fully restored λ-P1:5R lysogeny. It was concluded that some step in the establishment or maintenance of the plasmid prophage is *dam* dependent (3).

P1 MINIPLASMID MAINTENANCE IS *dam* DEPENDENT

A plasmid, pALA1005, was constructed that contains the complete plasmid maintenance region of F and the essential plasmid replication region of P1 (3). The plasmid transformed and was stably maintained in the *dam* strain. To determine which of the two replication origins was functional under these conditions, the strain was subsequently transformed with a second plasmid, pFAMK, which carries the F *incB* and *incC* elements that switch off replication from the F origin by titrating the essential F replication initiation protein RepE. pALA1005 was displaced from the population of *dam* cells by pFAMK, but transformation by pFAMK did not have any effect on the maintenance of pALA1005 in a dam^+ strain. Thus, both the F and P1 origins are capable of function in a dam^+ strain, so if one is switched off, the other can support replication. However, in the *dam* strain only the F origin is capable of function, so the plasmid becomes sensitive to switching off of the F origin. It can be concluded that the P1 origin cannot function properly in a *dam* strain even if the plasmid is already established in the strain (3).

AN IN VITRO REPLICATION SYSTEM FOR THE P1 ORIGIN

To further investigate the function of the origin, we have developed an in vitro replication system based on that developed by Fuller et al. for the replication of *oriC* plasmids (12). The template consists of M13 replicative-form DNA containing the 246-bp P1 origin sequence. A crude extract of *E. coli* cells was used to supply the necessary host factors and was supplemented with highly purified RepA made as previously described (1). The system incorporates labeled deoxynucleotides into acid-precipitable material and is dependent on the presence of the origin-containing DNA and RepA protein (Table 1). Incorporation is approximately proportional to added RepA except at high RepA concentrations (greater than 10 pmol of RepA per 20 fmol of DNA), at which saturation of the response is ob-

TABLE 1
In Vitro Replication[a]

DNA	pmol of dTTP incorporated in the presence (+) or absence (−) of RepA	
	−	+
M13mp10	7	10
M13mp11	7	9
M13-P1ori-38[b]	7	68
M13-P1ori-49[b]	8	58
M13-P1ori-88[c]	12	11
oriC[d]	214	216

[a] Incorporation of dTTP was based on the amount of [³H]dTTP incorporated into a trichloroacetic acid precipitate under the assay conditions described by Abeles and Austin (3). Each reaction contained about 0.04 pmol of DNA and 7 pmol of RepA. A more detailed account of these results was recently published (3).
[b] 38 and 49 are M13mp10 and M13mp11 with the 246-bp P1 origin cloned in opposite orientations, respectively.
[c] This plasmid contains the rep-30 mutation (Fig. 2).
[d] The E. coli oriC origin was carried on pAL3, a pACYC184 derivative (22) kindly supplied by B. R. Munson.

served (data not shown). Without added RepA, the incorporation was low and indistinguishable from that given by controls without any exogenous DNA template or with M13 DNA that does not contain the P1 origin. Templates containing the rep-30 origin mutation also showed no RepA-stimulated incorporation (3).

METHYLATION OF THE TEMPLATE IS ESSENTIAL FOR IN VITRO ACTIVITY

The template DNA used for the successful development of the in vitro replication system was isolated from dam^+ cells. When DNA isolated from dam cells was used, only background levels of incorporation were observed in the in vitro system and added RepA had no effect. When this DNA was methylated in vitro with purified dam methylase prior to the addition of the in vitro system, its template activity was fully restored (Table 2). We conclude that replication from the P1 origin is absolutely dependent on N^6-adenine methylation of the template DNA (3). Methylation of one or more of the five clustered GATC sequences in the origin core appears to be essential, because these sequences are the only substrates for methylation present in the cloned 246-bp origin sequence.

THE P1 PLASMID AND HOST oriC ORIGINS RESPOND DIFFERENTLY TO METHYLATION

The motif of clustered GATC sequences within the origin is not unique but is found in a number of different origins, including the E. coli chromosomal origin oriC (29, 40, 46). Methylated plasmids containing the oriC transform dam strains poorly or not at all (30, 37). Recently, it has been shown that the block to oriC replication on plasmid transformation exists when the origin DNA is hemimethylated. Russell and Zinder (35) have shown that when fully methylated DNA is introduced into dam strains, a single round of replication occurs but the resulting hemimethylated DNA fails to replicate further. In contrast, unmethylated DNA transforms and continues to replicate in dam strains. It appears that the P1 plasmid origin behaves very differently. The P1-F double-replicon experiment described above generates unmethylated plasmids by growth using the F origin for many generations in a dam strain. When replication from the F origin was blocked, the P1 origin did not function under these dam conditions, showing that the unmethylated P1 origin is not functional. This difference is also reflected in the in vitro systems. Unmethylated oriC plasmids retain substantial activity in vitro (19, 30, 37), whereas unmethylated M13-P1ori plasmids have none. The hemimethylated state is the only blocked state for oriC. For the P1 plasmid, the unmethylated state is blocked. We do not know yet whether the hemimethylated P1 origin is functional.

TABLE 2
Dependence of P1 Replication on Methylation[a]

Host[b]	Methylation	dTTP incorporated into trichloroacetic acid-precipitable material (pmol/assay per h) in the presence (+) or absence (−) of RepA in vitro	
		−	+
dam^+	−	9	42
	+	9	33
dam	−	12	11
	+	10	47

[a] A more detailed account of these experiments was recently published (3).
[b] The template DNA was isolated from either a dam^+ (GM30) or a dam (GM2150) host (3).

POTENTIAL ROLES FOR ORIGIN METHYLATION

It seems probable that methylation plays some dynamic role in *oriC* control. The hemimethylated products of replication appear to persist for some time before methylation is complete (25). Thus, the generation of hemimethylated DNA may be involved in delaying the potential for reinitiation (30). It has been suggested that completion of methylation of the *oriC* sites is delayed because of secondary-structure formation in the origin (35). This allows the possibility that origin hemimethylation persists through much of the cell cycle and hence that completion of methylation constitutes an important signal for reinitiation (35).

The clustering of multiple GATC sites within the P1 plasmid origin is reminiscent of the *oriC* case. If hemimethylation proves to be a block to P1 *ori* function, it is possible that methylation at the P1 origin plays a role similar to that postulated for the *oriC* region. However, it is clear that the two systems differ, at least in detail. In addition to the fact that P1 *ori* responds very differently from *oriC* in the unmethylated state, P1 *ori* has no potential for secondary structure. Thus, if some system exists that delays the completion of methylation of the P1 origin, it must involve tertiary interactions, perhaps with protein factors, rather than the secondary interactions that have been suggested for *oriC*.

OTHER POSSIBLE ROLES FOR P1 ORIGIN METHYLATION

The fact that the unmethylated P1 *ori* DNA is blocked for initiation allows other possibilities for the P1 case. Only replicons that are subject to rapid successive rounds of replication are expected to contain a substantial proportion of unmethylated sites. Unmethylated sites can accumulate in high-copy-number plasmids such as pBR322 (41) and with phages such as λ when in the lytic cycle (34, 41). This suggests two potential roles for regulation by methylation of the P1 origin. First, it could help to prevent "runaway" plasmid replication, which might occur if the normal control systems failed. Second, it might be involved in shutdown of the plasmid origin during the lytic growth of the parental P1 phage or during infection of new host cells. Lytic growth proceeds from a different origin on the P1 phage (M. B. Yarmolinsky and N. Sternberg, *in* R. Calender, ed., *The Bacteriophages*, in press). It is conceivable that initiation from the plasmid origin could interfere with lytic replication or some other aspect of phage morphogenesis such as DNA packaging. Thus, the switching off of the plasmid origin in the rapidly replicating, undermethylated DNA pool might be a requirement for efficient phage propagation.

MECHANISMS FOR ORIGIN SWITCHING

Both the role and the mechanism for methylation control of P1 *ori* remain to be elucidated. In the case of the host *oriC* sequence, the GATC sites are found within a complicated pattern of repeats which could form different secondary structures. The actual mechanism by which methylation alters *oriC* function is unclear, but, as pointed out by Smith et al. (37), it might involve the secondary structures in the region; structure might be altered by the methylation in such a way as to switch the origin on or off. Thus, secondary structure may be involved both in the modulation of methylation and in the subsequent control of initiation.

However, in the P1 case, origin methylation is capable of switching the origin on and off without the involvement of secondary structure. It seems probable that regulation of the P1 origin occurs by modulation of the binding to the GATC-rich regions of *trans*-acting factors that are required to form an active initiation complex, a suggestion that has also been made for the *oriC* system by Messer et al. (30).

Many questions remain to be answered. The effect of hemimethylation on the P1 plasmid origin is being tested. Subsequently, attempts will be made to identify protein factors that may interact with the core region of the origin. It is hoped that an understanding of the relationship between DNA modification and the dynamic structure and function of the origin can be developed.

ACKNOWLEDGMENT. This research was sponsored by the National Cancer Institute under contract no. NO1-CO-74101 with Bionetics Research, Inc.

LITERATURE CITED

1. **Abeles, A.** 1986. P1 plasmid replication: purification and DNA-binding activity of the replication protein RepA. *J. Biol. Chem.* 261:3548–3555.
2. **Abeles, A., K. Snyder, and D. Chattoraj.** 1984. P1 plasmid replication: replicon structure. *J. Mol. Biol.* 173:307–324.
3. **Abeles, A. L., and S. J. Austin.** 1987. P1 plasmid replication requires methylated DNA. *EMBO J.* 6:3185–3189.
4. **Arraj, J. A., and M. G. Marinus.** 1983. Phenotypic reversal in *dam* mutants of *Escherichia coli* K-12 by a recombinant plasmid containing the dam^+ gene. *J. Bacteriol.* 153:562–565.
5. **Austin, S., and A. Abeles.** 1983. Partition of unit-copy miniplasmids to daughter cells. I. P1 and F miniplasmids contain discrete, interchangeable sequences sufficient to promote equipartition. *J. Mol. Biol.* 169:353–372.
6. **Austin, S. J., R. J. Mural, D. K. Chattoraj, and A. L. Abeles.** 1985. *trans*- and *cis*-acting elements for the replication of P1 miniplasmids. *J. Mol. Biol.* 183:195–202.
7. **Chattoraj, D., K. Cordes, and A. Abeles.** 1984. Plasmid P1 replication: negative control by repeated DNA sequences. *Proc. Natl. Acad. Sci. USA* 81:6456–6460.
8. **Chattoraj, D. K., K. Cordes, and A. Abeles.** 1985. P1 plasmid replication: multiple functions of RepA protein at the origin. *Proc. Natl. Acad. Sci. USA* 82:2588–2592.
9. **Felton, J., and A. Wright.** 1979. Plasmid pSC101 replication in integratively suppressed cells requires *dnaA* function. *Mol. Gen. Genet.* 175:231–233.
10. **Filutowicz, M., E. Uhlenhopp, and D. R. Helinski.** 1985. Binding of purified wild-type and mutant π initiation proteins to a replication origin region of plasmid R6K. *J. Mol. Biol.* 187:225–239.
11. **Fuller, R. S., B. Funnell, and A. Kornberg.** 1984. The *dnaA* protein complex with the *E. coli* chromosome replication origin (*oriC*) and other DNA sites. *Cell* 38:889–900.
12. **Fuller, R. S., J. M. Kaguni, and A. Kornberg.** 1981. Enzymatic replication of the origin of the *Escherichia coli* chromosome. *Proc. Natl. Acad. Sci. USA* 78:7370–7374.
13. **Funnell, B. E., T. A. Baker, and A. Kornberg.** 1987. In vitro assembly of a prepriming complex at the origin of the *Escherichia coli* chromosome. *J. Biol. Chem.* 262:10327–10334.
14. **Geier, G. E., and P. Modrich.** 1979. Recognition sequence of the *dam* methylase of *Escherichia coli* K12 and mode of cleavage of DpnI endonuclease. *J. Biol. Chem.* 254:1408–1413.
15. **Hansen, E. B., F. G. Hansen, and K. von Meyenburg.** 1982. The nucleotide sequence of the

dnaA gene and the first part of the *dnaN* gene of *Escherichia coli* K-12. *Nucleic Acids Res.* 10:7373–7385.

16. Hansen, E. B., and M. Yarmolinsky. 1986. Host participation in plasmid maintenance: dependence upon *dnaA* of replicons derived from P1 and F. *Proc. Natl. Acad. Sci. USA* 83:4423–4427.

17. Hasunuma, K., and M. Sekiguchi. 1977. Replication of plasmid pSC101 in *Escherichia coli* K-12: requirement for *dnaA* function. *Mol. Gen. Genet.* 154:225–230.

18. Hattman, S., J. E. Brooks, and M. Masurekar. 1978. Sequence specificity of the P1 modification methylase (M. *Eco* P1) and the DNA methylase (M. *Eco dam*) controlled by the *Escherichia coli dam* gene. *J. Mol. Biol.* 126:367–380.

19. Hughes, P., F. Z. Squali-Houssaini, P. Forterre, and M. Kohiyama. 1984. In vitro replication of a dam methylated and nonmethylated *oriC* plasmid. *J. Mol. Biol.* 176:155–159.

20. Kelly, W., and D. Bastia. 1985. Replication initiator protein of plasmid R6K autoregulates its own synthesis at the transcriptional step. *Proc. Natl. Acad. Sci. USA* 82:2574–2578.

21. Lacks, S., and B. Greenberg. 1977. Complementary specificity of restriction endonucleases of *Diplococcus pneumoniae* with respect to DNA methylation. *J. Mol. Biol.* 114:153–168.

22. Leonard, A. C., M. Weinburger, B. R. Munson, and C. E. Helmstetter. 1980. The effects of *oriC*-containing plasmids on host cell growth. *ICN-UCLA Symp. Mol. Cell. Biol.* 19:171–180.

23. Light, J., and S. Molin. 1982. The sites of action of two copy control functions of plasmid R1. *Mol. Gen. Genet.* 187:486–493.

24. Linder, P., G. Churchward, X. Guixian, Y. Yi-Yi, and L. Caro. 1985. An essential replication gene, *repA*, of plasmid pSC101 is autoregulated. *J. Mol. Biol.* 181:383–393.

25. Lyons, S. M., and P. F. Schendel. 1984. Kinetics of methylation in *Escherichia coli* K-12. *J. Bacteriol.* 159:421–423.

26. Marinus, M., and N. R. Morris. 1973. Isolation of deoxyribonucleic acid methylase mutants of *Escherichia coli* K-12. *J. Bacteriol.* 114:1143–1150.

27. Masson, L., and D. S. Ray. 1986. Mechanism of autonomous control of the *Escherichia coli* F plasmid: different complexes of the initiator/repressor protein are bound to its operator and to an F plasmid replication origin. *Nucleic Acids Res.* 14:5693–5711.

28. Masukata, H., and J.-I. Tomizawa. 1985. Control of ColE1 plasmid replication: conformational change of the primer transcript. *Cell* 40:527–535.

29. Meijer, M., E. Beck, F. G. Hansen, H. E. N. Bergmans, W. Messer, K. von Meyenburg, and H. Schaller. 1979. Nucleotide sequence of the origin of replication of the *Escherichia coli* K-12 chromosome. *Proc. Natl. Acad. Sci. USA* 76:580–584.

30. Messer, W., V. Bellekes, and H. Lother. 1985. Effect of *dam* methylation on the activity of the E. coli replication origin, *oriC*. *EMBO J.* 4:1327–1332.

31. Murakami, Y., H. Ohmori, T. Yura, and T. Nagata. 1987. Requirement of the *Escherichia coli dnaA* gene function for *ori-2*-dependent mini-F plasmid replication. *J. Bacteriol.* 169:1724–1730.

32. Nordstrom, M., and K. Nordstrom. 1985. Control of replication of FII plasmids: comparison of the basic replicons and of the *copB* systems of plasmids R100 and R1. *Plasmid* 13:81–87.

33. Pal, S., R. J. Mason, and D. K. Chattoraj. 1986. P1 plasmid replication. Role of initiator titration in copy number control. *J. Mol. Biol.* 192:275–285.

34. Pukkila, P. J., J. Peterson, G. Herman, P. Modrich, and M. Meselson. 1983. Effects of high levels of DNA adenine methylation of methyl-directed mismatch repair in *Escherichia coli*. *Genetics* 104:571–582.

35. Russell, D. W., and N. D. Zinder. 1987. Hemimethylation prevents DNA replication in *E. coli*. *Cell* 50:1071–1079.

36. Scott, J. R. 1984. Regulation of plasmid replication. *Microbiol. Rev.* 48:1–23.

37. Smith, D. W., A. M. Garland, R. E. Enns, T. A. Baker, and J. W. Zyskind. 1985. Importance of state of methylation of *oriC* GATC sites in initiation of DNA replication in *E. coli*. *EMBO J.* 4:1319–1326.

38. Stayton, M. M., L. Bertsch, S. Biswas, P. Burgers, N. Dixon, J. E. Flynn, Jr., R. Fuller, J. Kaguni, J. Kobori, M. Kodaira, R. Low, and A. Kornberg. 1983. Enzymatic recognition of DNA replication origins. *Cold Spring Harbor Symp. Quant. Biol.* 47:693–700.

39. Sternberg, N., and S. Austin. 1983. Isolation and characterization of P1 minireplicons, λ-P1:5R and λ-P1:5L. *J. Bacteriol.* 153:800–812.

40. Sugimoto, K., A. Oka, H. Sugisake, M. Takanami, A. Nishimura, Y. Yasuda, and Y. Hirota. 1979. Nucleotide sequence of *Escherichia coli* K-12 replication origin. *Proc. Natl. Acad. Sci. USA* 76:575–579.

41. Szyf, M., K. Avraham-Haetzni, D. Reifman, J. Shlomai, F. Kaplan, A. Oppenheim, and A. Razin. 1984. DNA methylation pattern is determined by the intracellular level of the methylase. *Proc. Natl. Acad. Sci. USA* 81:3278–3282.

42. Tokino, T., T. Murotsu, and K. Matsubara. 1986. Purification and properties of the mini-F plasmid-encoded E protein needed for autonomous replication control of the plasmid. *Proc. Natl. Acad. Sci. USA* 83:4109–4113.

43. Tomizawa, J.-I. 1985. Control of ColE1 plasmid

replication: initial interaction of RNA I and the primer transcript is reversible. *Cell* 40:527–535.
44. **Vocke, C., and D. Bastia.** 1985. The replication initiator protein of plasmid pSC101 is a transcriptional repressor ot its own cistron. *Biochemistry* 82:2252–2256.
45. **Wagner, E. G. H., J. Von Heijne, and K. Nordstrom.** 1987. Control of replication of plasmid R1: translation of the 7k reading frame in the RepA mRNA leader region counteracts the interaction between CopA RNA and CopT RNA. *EMBO J.* 6:515–522.
46. **Zyskind, J. W., J. M. Cleary, W. S. A. Brusilow, N. E. Harding, and D. W. Smith.** 1983. Chromosomal replication origin from the marine bacterium *Vibrio harveyi* functions in *Escherichia coli*: *oriC* consensus sequence. *Proc. Natl. Acad. Sci. USA* 80:1164–1168.

Chapter 11

R1 Plasmid Replication In Vitro
repA- and dnaA-Dependent Initiation at oriR

Hisao Masai and Ken-ichi Arai

REPLICATION FUNCTIONS INVOLVED IN R1 PLASMID REPLICATION

The R1 plasmid, approximately 90 kilobases in size, is maintained stably in *Escherichia coli* with a copy number of one to two per cell and confers drug resistance to host cells. Elucidation of the mechanism of R1 plasmid replication should provide insight into the host chromosome replication since the R1 plasmid relies largely on host replication functions. Furthermore, the R1 plasmid provides an attractive model system for studying regulation of DNA replication since its copy number is strictly controlled. Genetic and molecular biological analyses indicate that plasmid-encoded replication functions are clustered in a small region ("basic replicon") of the R1 plasmid genome (Fig. 1). Two *trans*-acting regulators, *copB* and *copA*, control the copy number by regulating the level of expression of *repA* which is essential for replication of the R1 plasmid (9). The *copB* gene product, an 11-kilodalton (kDa) pro-

Hisao Masai and Ken-ichi Arai • Department of Molecular Biology, DNAX Research Institute of Molecular and Cellular Biology, Palo Alto, California 94304.

tein, binds to the promoter region for *repA* and represses transcription, while *copA* encodes a small RNA which inhibits translation of the *repA* message by its ability to form a hybrid with the 5'-leader sequence of the *repA* transcript. A high-speed supernatant of gently lysed *E. coli* cells (fraction I) can initiate R1 plasmid replication when coupled to de novo protein synthesis (1a). By using this in vitro system, RepA protein, a 33-kDa polypeptide, was shown to be required for initiation (7).

oriR, REPLICATION ORIGIN OF R1 PLASMID

Analyses by electron microscopy indicated that replication initiates downstream of *repA*, within the basic replicon, and proceeds unidirectionally (Fig. 1) (10). Either mutations within *repA* (*repA ori*$^+$) or deletions extending into the presumptive origin region (*repA*$^+$ *ori*) abolish replication activity both in vivo and in vitro, suggesting that at least part of the origin sequence is present downstream of *repA* (7). However, it was not clear whether the origin sequence and *repA* could be completely separated. Furthermore, *repA* protein is *cis*-acting since plas-

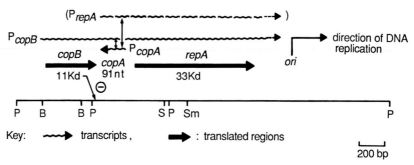

FIGURE 1. Basic replicon of R1 plasmid. The products of three structural genes, *copB*, *copA*, and *repA*, are shown according to the key. The three promoters of interest are shown as P*copB*, P*copA*, and P*repA*. Initiation of transcription from P*repA* is inhibited in the presence of the *copB* product, but transcription from P*copB* is unaffected. The *copA* RNA, a 91-nucleotide transcript, is complementary to the 5' leader region of the *repA* mRNA. *copA* RNA interacts with the *repA* transcripts (shown by double-headed arrows), thereby inhibiting their translation. *ori* indicates the replication origin mapped by electron microscopic analysis. Restriction enzyme recognition sites are indicated as follows: B, *Bgl*II; P, *Pst*I; S, *Sal*I; Sm, *Sma*I.

mids lacking a functional *repA* gene (*repA ori*$^+$) fail to replicate in vitro even in the presence of a *repA*$^+$ *ori*$^+$ helper plasmid that provides *repA* function in *trans* (Fig. 2A). However, *repA ori*$^+$ DNA is replicated efficiently when a *repA*$^+$ *ori* plasmid is used as a *repA* donor (Fig. 2B). With this in vitro *trans*-complementation assay, the minimum DNA sequence (*oriR*) that is required for efficient initiation in vitro was defined in the presence of *repA* protein provided in *trans* (Fig. 2C) (7). The *oriR* sequence is 188 base pairs (bp) long and is separated from *repA* by a nucleotide sequence of approximately 170 bp. *oriR* does not contain any characteristic repeated sequences, as are often discovered in the replication origins of other replicons (Fig. 3). The sequence TTATCCACA that is recognized by host *dnaA* protein is found at position 1427 to 1435. The sequence 1513 to 1586 is extremely AT rich (78%), and three repeats of TCNTTTAAA separated by 23-bp intervals appear in this region. The AT-rich sequence is essential for the origin function since deletion in this region abolishes replication activity. A sequence homologous to the integration host factor (IHF) binding site is also found within this region (pointed out by Gordon Guay; see Summary).

The *oriR* sequence, defined in vitro, also supports replication in vivo even when it is located upstream of *repA* or when it is in an inverted orientation. An additional sequence (*CIS*) between *repA* and *oriR* is required for efficient replication of a *repA-oriR* plasmid in vivo. In the absence of the *CIS* sequence, the efficiency of transforming bacterial cells and the plasmid copy number are decreased by factors of 50 and 3, respectively. The *CIS* sequence contains a rho-dependent transcriptional terminator, which terminates the *repA* message at position 1299 and is also required for *cis* action of RepA protein in vitro (H. Masai and K. Arai, *Fed. Proc.* 43:1669, 1984; unpublished data).

RepA PROTEIN AND HOST DnaA PROTEIN ARE REQUIRED FOR INITIATION AT *oriR*

RepA protein, purified from an overproducing strain (5), behaves as a 33-kDa monomer in a solution containing 0.5 M potassium glutamate and specifically stimu-

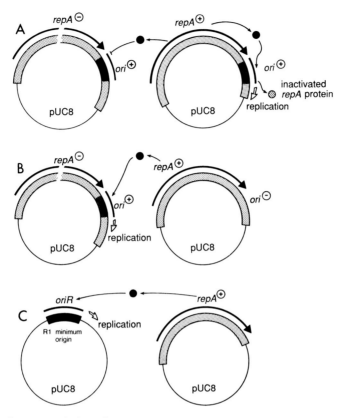

FIGURE 2. Action of *repA* protein in vitro. (A) *cis* acting when *ori* is present on the same molecule; (B) *trans* acting when *ori* is removed; (C) separation of *ori* and *repA*.

lates replication of *oriR* plasmids. The replication is dependent on DNA gyrase and other host replication functions such as *dnaB*, *dnaC*, *dnaG* (primase), single-strand DNA-binding protein, and DNA polymerase III holoenzyme but does not require the action of RNA polymerase. We found, to our surprise, that host *dnaA* function is essential for replication from *oriR* (5, 11). Extracts prepared from *dnaA204*, *dnaA205*, *dnaA46*, and *dnaA177* were inert for R1 plasmid replication and addition of purified *dnaA* protein restored the replicative activity. Although 40 to 50 molecules of *repA* proteins per template are required for initiation at *oriR*, only a couple of molecules of *dnaA* protein are required, as revealed by titration of both proteins.

PROTEIN-DNA INTERACTION AT *oriR*

Purified *repA* protein has a strong affinity for double-stranded DNA as revealed by nitrocellulose filter binding assays; it binds to DNA fragments with no sequence specificity at low salt concentrations (below 0.2 M KCl), whereas only the fragments carrying the *oriR* sequence are specifically retained on the filter at a high salt concentration (0.5 M KCl). The exact nucleotide sequences that are involved in *repA* protein binding were localized by nuclease protection experiments. Strong protection from DNase I attack was observed over nearly 90 bp of sequence within *oriR* from position

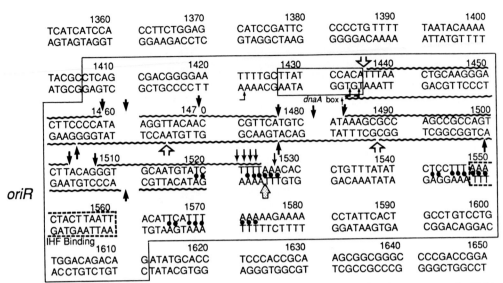

FIGURE 3. Nucleotide sequence of *oriR* and *repA* protein footprint at *oriR*. Boxed region represents the 206-bp *oriR* sequence. A small box from 1427 to 1435 and a dashed box from 1538 to 1560 represent bp of *dnaA* protein recognition sequence (*dnaA* box) and a putative IHF binding site, respectively. Dotted nucleotides represent three repeats of TCNTTTAAA in the AT-rich region. Sequences with wavy lines are protected from DNase I attack in the presence of *repA* protein. Arrows between 1436 and 1520 indicate sites that are nuclease sensitive even in the presence of *repA* protein. Big open arrows indicate the hypersensitive sites. Arrows at 1421, 1433, and 1434 represent the extent of *Exo*III digestion in the presence of *repA* (1433 and 1434) or *repA* and *dnaA* (1421) proteins. Arrows between 1521 and 1530 show the sites of P1 nuclease cleavage in the presence of *repA* and *dnaA* proteins.

1433 to 1524 (5). On the upper strand, DNase I-sensitive sites appeared with 10- to 11-bp intervals, which is reminiscent of the nuclease digestion pattern of isolated nucleosomes (Fig. 3 and 4). *dnaA* protein, which is essential for in vitro DNA replication of *oriC* plasmid (3), binds specifically to *oriC* DNA as well as to other sequences found near the origins of pBR322 and pSC101 (2). A 9-bp consensus sequence for recognition by *dnaA* protein ("*dnaA* box"), which appears four times within *oriC*, has also been discovered within *oriR* (Fig. 3). However, specific binding of *dnaA* protein to the *dnaA* box cannot be detected by DNase I footprinting. When *oriR* is already bound with *repA* protein, the 9-bp *dnaA* box is efficiently protected from nuclease attack (5) (Fig. 5). On the basis of these data, we speculate that formation of a nucleoprotein structure involving part of the *oriR* sequence and *repA* and *dnaA* proteins is an essential event for initiation of R1 plasmid replication.

Additional *repA* protein binding sites (sites II and III) have been identified in the carboxy-terminal region of the *repA* structural gene (unpublished data; see Fig. 8). Binding of *repA* protein to site II or site III is more salt sensitive (stable up to 0.3 M KCl) than binding to site I and has an affinity about 10-fold lower than site I (*oriR*). Comparison of the nucleotide sequences of sites I, II, and III has yielded the consensus recognition sequence of *repA* protein ("*repA* box," Fig. 6). At present we do not know whether sites II and III play any role in R1 plasmid replication.

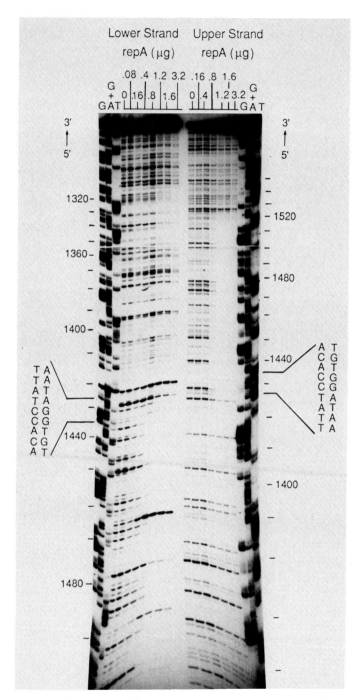

FIGURE 4. Footprint of *repA* protein at *oriR*. Five nanograms of a 5'-end-labeled fragment containing the *oriR* sequence was mixed with *repA* protein and briefly digested with DNase I. The digests were analyzed on a sequencing gel together with size markers generated by Maxam-Gilbert reactions performed with the same fragment.

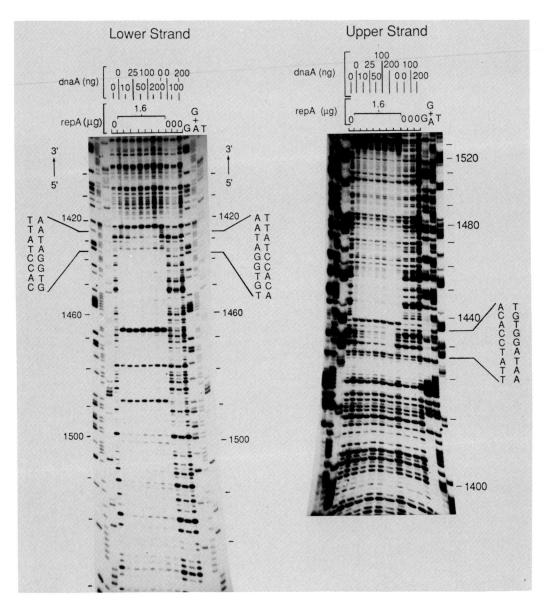

FIGURE 5. Footprint of *dnaA* protein at *oriR* with or without *repA* protein. The same fragment as in Fig. 3 was mixed with *dnaA* protein in the presence or absence of *repA* protein and was briefly digested with DNase I. The digests were analyzed in the same way.

REGULATION OF R1 PLASMID REPLICATION

The R1 plasmid has evolved several mechanisms to ensure that the copy number of the plasmid is kept constant throughout generations of host cells. The first involves strict regulation of *repA* expression (9). Mutations in either *copB* or *copA* result in increased copy number, while the loss of both functions induces uncontrolled replication (runaway phenotype) (9). The extent of replication is proportional to the level of *repA* expression both in vivo (4) and in vitro

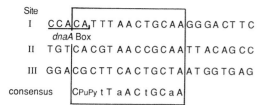

FIGURE 6. Possible recognition sequence of *repA* protein (*repA* box). Sequences that are protected in DNase I footprinting with site I (*oriR*), site II, and site III were compared and a homologous sequence was deduced. Although site I protection is nearly 90 bp long, no binding is observed if the proposed *repA* box is deleted, indicating that the primary recognition sequence lies in or near the *repA* box.

(unpublished data). The second mechanism employs *cis*-specific actions of *repA* protein. *repA* protein, newly synthesized in vitro, is so efficiently utilized by *oriR* present on the same DNA molecule that no excess free *repA* molecule is accumulated (unpublished data). Since the protein appears to be non-reusable for the next cycle of replication, overinitiation of DNA replication may be prevented. Furthermore, replication of the R1 plasmid may be coordinated with that of the host chromosome through *dnaA* protein acting as a common regulator.

SUMMARY AND FUTURE PROSPECTS

The initial event in the initiation of DNA replication of a procaryotic replicon is the formation of a specialized protein-DNA complex at a replication origin. R1 plasmid-encoded *repA* protein and host *dnaA* protein bind specifically to the *oriR* sequence, which generates a nucleosomelike protein-DNA complex. Titration of protein in a DNA replication assay and the extent of nuclease protection suggest that nearly 40 molecules of *repA* protein and only 1 or 2 monomers of *dnaA* protein are involved in the complex. Binding of *repA* and *dnaA* proteins induces P1 nuclease-sensitive sites within the AT-rich sequence which are limited to one to seven nucleotides (unpublished data; Fig. 3). They may be created by strand opening, as proposed in the *dnaA-oriC* interaction (1), or by another type of DNA conformational change such as sharp bending.

There are several unanswered questions related to R1 plasmid replication. (i) What is the role of *dnaA* protein for initiation at *oriR*? Binding of 20 to 30 *dnaA* monomers to *oriC* (2) induces P1 sensitivity in the *oriC* template (1). In R1, although binding of *repA* protein is sufficient to create a P1-sensitive region in the *oriR* template, the binding of *dnaA* protein further augments P1 sensitivity, probably by stabilizing the *repA* protein-DNA complex. *dnaA* protein may also help to incorporate the *dnaB-dnaC* complex into the initiation complex. (ii) What is the role of the AT-rich sequence? In *oriC*, the AT-rich sequence may serve as an entry site for *dnaB-dnaC* complexes. In R1, we cannot rule out the possibility that another host protein besides the *dnaB* and *dnaC* proteins recognizes the AT-rich sequence within *oriR*. Replication of pSC101 requires IHF, which binds to the AT-rich sequence and induces bending of the DNA (12). Although IHF binds to a putative IHF binding site within *oriR*, it is not essential for R1 plasmid replication, since the *himA/himD* double mutant (lacking IHF) can support replication of mini R1 (unpublished data). (iii) What determines the unidirectional mode of R1 plasmid replication? Direction of replication in vitro seems to be determined by the orientation of *oriR*. A strong termination point for lagging-strand synthesis has been identified within *oriR* near one end of the *repA* binding site (C. Miyazaki and E. Ohtsubo, personal communication). *repA* protein bound at *oriR* may physically block continuation of lagging-strand synthesis, which may otherwise provide a primer for leading-strand synthesis in the opposite direction. (iv) How and where are leading-strand and lagging-strand syntheses initiated and how are they coordinated? Inasmuch as *repA* protein has no primase

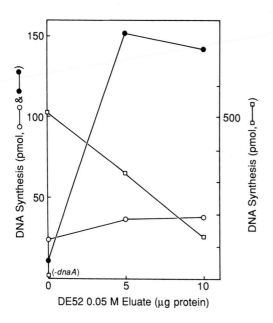

FIGURE 7. R1 plasmid replication in a partially reconstituted system. Standard *oriC* reconstitution assay, in the presence or absence of *repA* protein, was supplemented with a partially purified stimulation factor, which was prepared as follows. Fraction I was fractionated by adding 0.28 g of ammonium sulfate per ml. To the supernatant was added 0.07 g of ammonium sulfate per ml, and the precipitate was collected and resuspended. The suspension was applied to a DE52 column at 0.05 M KCl. Flowthrough fraction was concentrated by ammonium sulfate precipitation and was used for assays. Assays were at 30°C for 30 min with 600 pmol as the nucleotide of pMOB45 DNA (mini-R1 derivative) in the absence (○) or the presence (●) of *repA* protein. As a control, 600 pmol as the nucleotide of pBS*oriC* DNA (*oriC* plasmid constructed by T. Baker) was used as template (□).

activity and RNA polymerase is not essential for R1 replication, it is most likely that primase makes the primers on both strands. Since antibody against protein i does not inhibit R1 replication, lagging-strand synthesis may not be mediated by primosome (unpublished data) (6). These issues may be resolved by using the system reconstituted with purified replication proteins. Although the enzyme system composed of *dnaA* protein, HU protein, *dnaB* protein, *dnaC* protein, primase, single-strand DNA-binding protein, DNA polymerase III holoenzyme, and DNA gyrase supported *oriC* replication but failed to support *repA*-dependent *oriC* replication, a partially purified fraction derived from fraction II (0.28 to 0.35 ammo-

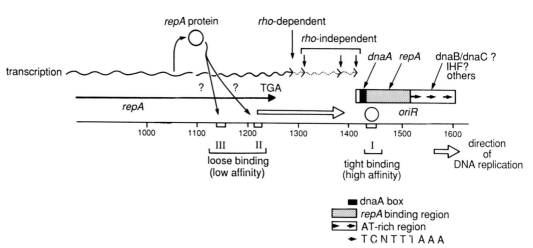

FIGURE 8. Multiple binding sites for *repA* protein within *oriR* and its vicinity and a sliding model for *cis* action of *repA* protein. Small open boxes under the scale represent "*repA* box" sequence in three binding sites. The wavy line and its arrowheads indicate the *repA* transcript and its termination sites (unpublished data). Long open arrows in the center indicate that newly synthesized *repA* protein, first binding to site II or III, slides along DNA until it locates site I, where it can form a tight complex.

nium sulfate fraction) restored *repA*-dependent R1 plasmid replication (Fig. 7). Further purification of this fraction will help to identify host components that are specifically required for R1 plasmid replication.

Another interesting aspect of R1 plasmid replication is the *cis*-specific action of *repA* protein. Although a few examples of *cis*-acting proteins have been described (8), to our knowledge, this is the first example that *cis* action of a protein can be observed in vitro. To explain the mechanism of the *cis* action, we propose that sites II and III may function as entry sites for the newly synthesized *repA* protein and that the protein slides along the DNA template until it finds the *oriR* sequence (site I) where a stable protein-DNA complex can be formed (Fig. 8). This model is in keeping with the fact that *oriR* supports replication regardless of orientation as long as it is present on the same DNA molecule. If *oriR* is absent on the same template, *repA* protein dissociates from the DNA and can activate *oriR* on a different template in vitro. In vivo, the protein that leaves the DNA may be rapidly inactivated and is unable to activate *oriR* in *trans*. We are currently designing experiments to test this model.

LITERATURE CITED

1. Bramhill, D., and A. Kornberg. 1988. Duplex opening by *dnaA* protein at novel sequences in initiation of replication at the origin of the *E. coli* chromosome. *Cell* 52:743–755.
1a. Diaz, R., K. Nordström, and W. L. Staudenbauer. 1981. Plasmid R1 DNA replication dependent on protein synthesis in cell-free extracts of *E. coli*. *Nature* (London) 289:326–328.
2. Fuller, R. S., B. E. Funnel, and A. Kornberg. 1984. The *dnaA* protein complex with the *E. coli* chromosomal replication origin (*oriC*) and other DNA sites. *Cell* 38:889–900.
3. Fuller, R. S., and A. Kornberg. 1983. Purified *dnaA* protein in initiation of replication at the *Escherichia coli* chromosome origin of replication. *Proc. Natl. Acad. Sci. USA* 80:5817–5821.
4. Light, J., and S. Molin. 1981. Replication control functions of plasmid R1 act as inhibitors of expression of a gene required for replication. *Mol. Gen. Genet.* 184:56–61.
5. Masai, H., and K. Arai. 1987. RepA and DnaA proteins are required for initiation of R1 plasmid replication *in vitro* and interact with the *oriR* sequence. *Proc. Natl. Acad. Sci. USA* 84:4781–4785.
6. Masai, H., M. W. Bond, and K. Arai. 1986. Cloning of the *Escherichia coli* gene for primosomal protein i: the relationship to *dnaT*, essential for chromosomal DNA replication. *Proc. Natl. Acad. Sci. USA* 83:1256–1260.
7. Masai, H., Y. Kaziro, and K. Arai. 1983. Definition of *oriR*, the minimum DNA segment essential for initiation of R1 plasmid replication *in vitro*. *Proc. Natl. Acad. Sci. USA* 80:6814–6818.
8. McFall, E. 1986. *cis*-Acting proteins. *J. Bacteriol.* 167:429–432.
9. Nordström, K., S. Molin, and J. Light. 1984. Control of replication of bacterial plasmids: genetics, molecular biology, and physiology of the plasmid R1 system. *Plasmid* 12:71–90.
10. Ohtsubo, E., J. Feingold, H. Ohtsubo, S. Mickel, and W. Bauer. 1977. Unidirectional replication of three small plasmids derived from R factor R12 in *Escherichia coli*. *Plasmid* 1:8–18
11. Ortega, S., E. Lanka, and R. Diaz. 1986. The involvement of host replication proteins and of specific origin sequences in the *in vitro* replication of miniplasmid R1 DNA. *Nucleic Acids Res* 14:4865–4879.
12. Stenzel, T. T., P. Patel, and D. Bastia. 1987. The integration host factor of *Escherichia coli* binds to bent DNA at the origin of replication of the plasmid pSC101. *Cell* 49:709–717.

Chapter 12

Processive DNA Synthesis In Vitro by the φ29 DNA Polymerase
Structural and Functional Comparison with Other DNA Polymerases

Luis Blanco, Antonio Bernad, and Margarita Salas

Bacteriophage φ29 contains a linear, double-stranded DNA 19,285 base pairs (bp) long (11, 36, 38) with the terminal protein p3 covalently linked at the 5' ends (31) through a phosphodiester bond between the OH group of serine residue 232 and dAMP (13). Replication of φ29 starts at either DNA end (12, 15) by a protein-priming mechanism in which the terminal protein p3 is covalently linked to dAMP in the presence of the viral DNA polymerase p2, dATP as substrate, and φ29 DNA-protein p3 as template (5, 37). By addition of the remaining dNTPs the p3-dAMP initiation complex is elongated to produce full-length φ29 DNA (6). The viral protein p6, which interacts with the φ29 DNA ends (30), stimulates both the initiation and elongation steps in φ29 DNA-protein p3 replication (4, 28).

The φ29 DNA polymerase is a unique polymerase since it not only catalyzes the covalent linkage of a nucleotide to the 3'-OH group of another nucleotide but also catalyzes the covalent linkage of dAMP to the OH group of serine residue 232 in protein p3. In addition, it has 3'→5' exonuclease activity on single-stranded DNA (7).

HIGH LEVEL OF φ29 DNA-PROTEIN p3 SYNTHESIZED IN THE IN VITRO REPLICATION SYSTEM

To find out the efficiency of φ29 DNA-protein p3 replication in vitro using the minimal system consisting of the purified terminal protein p3 (29), DNA polymerase p2 (5), and the four deoxynucleoside triphosphates, a sample of the reaction mixture, after 15 min of incubation, was diluted twofold with addition of all the components except the φ29 DNA-protein p3 template, and the same process was repeated up to six times. There was a considerable φ29 DNA-protein p3 mass increase, up to 14 times the initial value after six cycles of incubation (Fig. 1). These results indicated that the

---

Luis Blanco, Antonio Bernad, and Margarita Salas • Centro de Biología Molecular (CSIC-UAM), Universidad Autónoma, Canto Blanco, 28049 Madrid, Spain.

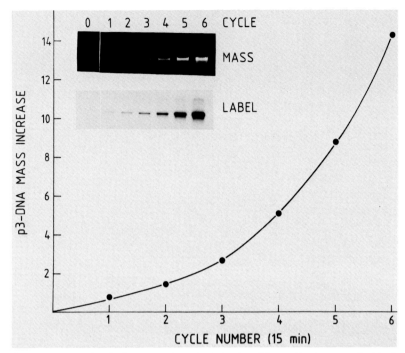

FIGURE 1. High level of φ29 DNA synthesis in vitro. The replication assay, in 50 μl, was carried out as described (6), using 21 ng of terminal protein (p3), 80 ng of φ29 DNA polymerase (p2), and 0.5 μg of φ29 DNA-protein p3 as template. After incubation for 15 min at 30°C, 25 μl of each sample was removed for further analysis and the other 25 μl was made up to 50 μl with the corresponding amount of proteins p3 and p2 and reaction mixture; the reaction was allowed to continue for another 15 min, this process being repeated six times (cycles). The DNA mass present at the end of each cycle was analyzed by agarose gel electrophoresis of the corresponding portions corrected by the dilution factor. The DNA mass was detected by ethidium bromide staining and then the gels were dried and autoradiographed. Quantitation was done by densitometry of the autoradiographs and stained bands.

minimal in vitro φ29 DNA replication system with only two proteins, the terminal protein and DNA polymerase, is very efficient.

PROCESSIVITY OF THE φ29 DNA POLYMERASE IN THE REPLICATION OF φ29 DNA-PROTEIN p3

The φ29 DNA polymerase involved in the formation of the initiation complex and in its further elongation is likely to remain attached to the replication complex after the initiation reaction and to elongate the chain without dissociation. To test this hypothesis, i.e., the processivity of the φ29 DNA polymerase, a series of dilutions of the latter were carried out in the replication assay of φ29 DNA-protein p3. The labeled DNA synthesized in vitro was subjected to alkaline agarose gel electrophoresis to assay for the synthesis of unit-length φ29 DNA. Even after a 128-fold dilution, unit-length φ29 DNA was obtained (Fig. 2), indicating that the DNA polymerase that initiates replication proceeds to elongate the DNA chain without dissociation. In addition, this result suggests that the φ29 DNA polymerase has

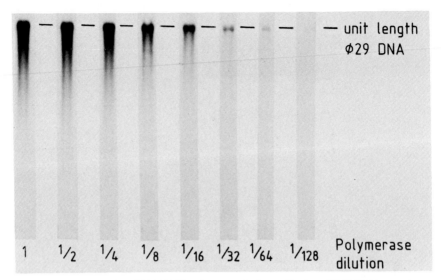

FIGURE 2. Processivity of ϕ29 DNA polymerase in the replication of ϕ29 DNA-protein p3. The replication assay was carried out as described (6) using 21 ng of terminal protein (p3), 0.5 μg of ϕ29 DNA-protein p3 as template, and the indicated amounts of ϕ29 DNA polymerase (p2); polymerase dilution 1 indicates 160 ng of ϕ29 DNA polymerase. After incubation for 10 min at 30°C, the samples were subjected to alkaline agarose gel electrophoresis. After electrophoresis, unit-length ϕ29 DNA was detected with ethidium bromide, and then the gels were dried and autoradiographed with intensifying screens at −70°C.

helicaselike activity, being able to produce strand displacement. However, since the terminal protein was present, the possibility that it was the one with the helicase activity had to be ruled out.

STRAND DISPLACEMENT BY THE ϕ29 DNA POLYMERASE IN THE REPLICATION OF PRIMED M13 DNA

To determine whether the ϕ29 DNA polymerase itself was able to produce strand displacement and was a processive enzyme in the absence of the terminal protein, replication of primed M13 DNA was analyzed by alkaline agarose gel electrophoresis. As a control, the Klenow fragment of *Escherichia coli* DNA polymerase I was also used. The ϕ29 DNA polymerase was able to replicate M13 DNA in 5 min at 30°C (Fig. 3). At longer incubation times, a length of over 70 kilobases was obtained, indicating that the ϕ29 polymerase by itself is able to produce strand displacement and is a very processive enzyme. The Klenow enzyme replicated M13 DNA in 10 min at 30°C, and no strand displacement occurred after incubation for longer times. In addition, replication by the ϕ29 DNA polymerase was resistant up to 100 mM NaCl, whereas no replication by the Klenow enzyme occurred at 100 mM NaCl.

INHIBITION OF THE ϕ29 AND T4 DNA POLYMERASES BY DRUGS AND NUCLEOTIDE ANALOGS

The replication of ϕ29 DNA-protein p3 was inhibited by aphidicolin, phosphonoacetic acid (PAA), and the nucleotide analogs butylanilino dATP (BuAdATP) and

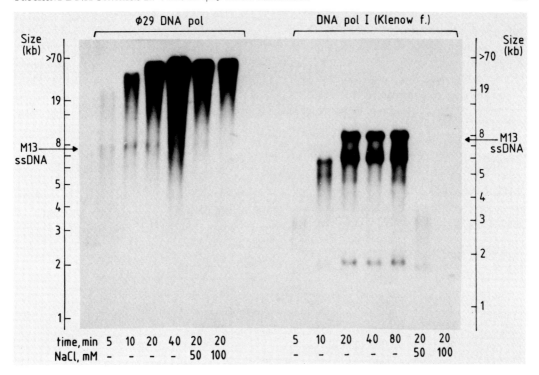

FIGURE 3. Replication of singly primed M13 single-stranded DNA (ssDNA). The incubation mixture contained, in 10 µl, 50 mM Tris hydrochloride, pH 7.5; 10 mM MgCl$_2$; 80 µM each dGTP, dCTP, dTTP, and [α-^{32}P]dATP (1 µCi); 1 µg of M13 ssDNA annealed with a 15-mer–M13 sequencing primer; and 160 ng of the φ29 DNA polymerase or 2.5 U of the Klenow fragment of E. coli DNA polymerase I. In some cases, NaCl was added at the indicated concentrations. After incubation for the indicated times at 30°C, the reaction was stopped and the samples were subjected to electrophoresis in alkaline 0.7% agarose gels alongside DNA length markers. After electrophoresis, DNA length markers were detected with ethidium bromide and the gel was dried and autoradiographed with intensifying screens at −70°C.

butylphenyl dGTP (BuPdGTP) (3, 8), which are known inhibitors of the eucaryotic DNA polymerase α (14, 18, 19) and other α-like DNA polymerases of viral origin (16, 21, 25, 26, 33). Whereas the nucleotide analogs BuAdATP and BuPdGTP inhibited both the initiation and elongation activities of the φ29 DNA polymerase, the drugs aphidicolin and PAA inhibited the elongation but essentially had no effect on the initiation step of φ29 DNA replication (3, 8). These results suggested the existence of two domains in the φ29 DNA polymerase, one for initiation and the other for elongation.

When replication of activated DNA by the T4 DNA polymerase was studied, it was found that aphidicolin and the nucleotide analogs were very strong inhibitors (3).

AMINO ACID HOMOLOGIES BETWEEN DNA POLYMERASES

Three conserved amino acid domains (SLYP/NS-YG-F/Y-DTDS) have been reported to be present in the C-terminal portion of several viral α-like DNA polymerases and in the φ29 DNA polymerase, and one of them is proposed to be the PAA binding site (22, 24). These homologous regions are also present in the phage PRD1 DNA polymerase (32) and in the T4 DNA polymerase (3), in agreement with the inhibition of the latter by aphidicolin and nucleotide analogs. In addition, we have found that the linear DNA killer plasmid pGKL1 from yeast (20) and the linear S1 mitochondrial DNA from maize (17), both of which have 5'-terminal protein, also contain the three consensus regions, suggesting that these two linear DNAs code for their own DNA polymerase.

The consensus sequences derived from the study of the amino acid sequence of the different DNA polymerases are shown (Fig. 4). It is possible to distinguish consensus sequences that account for different replication strategies, such as consensus 1 for DNA polymerases of genomes containing terminal protein and consensus 3 for the rest of the

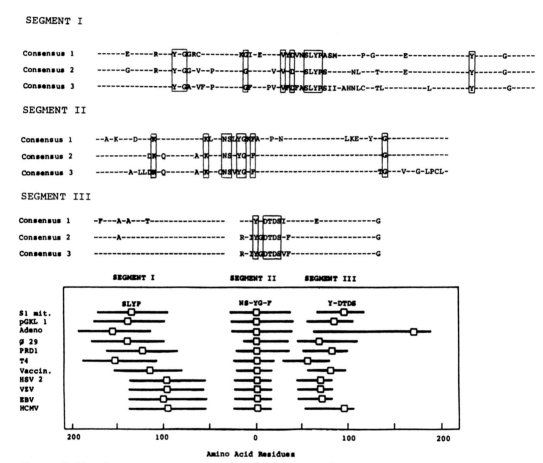

FIGURE 4. Homology segments among several DNA polymerases of procaryotic and eucaryotic origin. The DNA polymerase sequences compared were S1 mitochondrial DNA (27), pGKL1 (34), adenovirus type 2 (1), phage ɸ29 (38), phage PRD1 (32), phage T4 (E. K. Spicer, unpublished results; GenBank, 1986), vaccinia virus (10), herpes simplex virus type 2 (HSV-2) (35), varicella-zoster virus (VZV) (9), Epstein-Barr virus (EBV) (2), and human cytomegalovirus (HCMV) (23). A residue was considered to be a consensus residue if it was present in more than half of the polypeptides examined. Consensus 1 indicates the consensus residues between DNA polymerases of terminal protein-containing genomes (S1 mit, pGKL1, Adeno, ɸ29, and PRD1). Consensus 3 indicates the consensus residues between the remaining DNA polymerases (T4, vaccinia virus, HSV-2, VZV, EBV, and HCMV). Consensus 2 indicates the consensus residues between all DNA polymerases. The gap in segment III corresponds to insertions of 78 and 13 amino acid residues in the adenovirus and HCMV DNA polymerases, respectively. The consensus amino acids which are identical in the three kinds of consensus considered are boxed. The linear relationships between segments I to III (black bars) of each DNA polymerase are shown at the bottom. The tyrosine residue in the consensus motif NF-Y G-F of segment II was taken as residue 0 in each case. Boxes indicate the main consensus motif in each segment.

DNA polymerases. It is also seen (Fig. 4) that the three segments of amino acid homology are located in the same linear arrangement in each protein.

EVOLUTIONARY RELATIONSHIPS BETWEEN DNA POLYMERASES

The homologies in segment I were considered as an index of evolutionary relationship between DNA polymerases since the lengths of the polypeptides of segments II and III differed greatly. From the percentage of identical residues in segment I between each pair of DNA polymerases (Fig. 5) it can be seen that the DNA polymerases of the herpesvirus group are very proximal. Surprisingly, T4 DNA polymerase was related more to the eucaryotic viral DNA polymerases than to the procaryotic ones, especially to the vaccinia virus DNA polymerase. In fact, detailed comparison of segment I of vaccinia virus and T4 DNA polymerases suggests that they derived from a common ancestor (3).

The predicted DNA polymerases of the linear plasmids pGKL1 and S1 mitochondrial DNA have maximal homology between them and with the φ29 and PRD1 DNA polymerases, in agreement with a possible protein-priming mechanism of replication. These results suggest a phylogenetic relationship between viruses and plasmid DNAs.

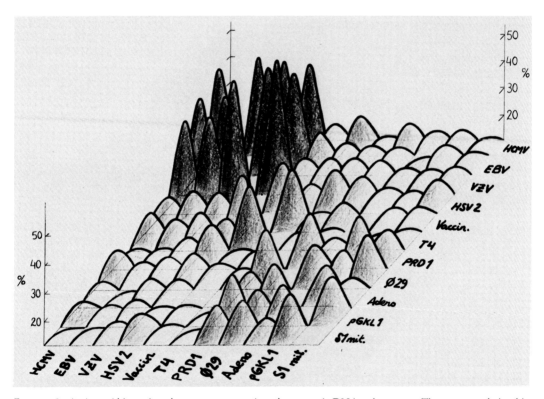

FIGURE 5. Amino acid homology between procaryotic and eucaryotic DNA polymerases. The percent relationship homologies in segment I between each pair of DNA polymerases are indicated in the third dimension of the plot. Reciprocal values are also indicated because of the different length of the sequence considered as 100% in each comparison.

STRUCTURAL AND FUNCTIONAL RELATIONSHIPS BETWEEN DNA POLYMERASES

The conservation of the three different domains among the DNA polymerases shown in Fig. 4 and 5 suggests that these regions correspond to functional domains of the enzyme. The fact that the three domains are present in the DNA polymerases from viruses that do or do not replicate by protein priming suggests that these domains may be involved in elongation or $3' \rightarrow 5'$ exonuclease activity, or both, and that different domains may be involved in the initiation activity. Site-directed mutagenesis in the three conserved domains may help to clarify the structural and functional relationships between DNA polymerases.

ACKNOWLEDGMENTS. We are grateful to A. Zaballos and G. Martin for help with the computer analysis.

This investigation was aided by Public Health Service research grant 5 R01 GM27242-08 from the National Institutes of Health, by grant 3325 from Comisión Asesora para la Investigación Científica y Técnica, and by a grant from Fondo de Investigaciones Sanitarias. A.B. was a recipient of a fellowship from Caja de Ahorros de Madrid.

LITERATURE CITED

1. Alesträm, P., G. Akusjärvi, M. Petterson, and U. Petterson. 1982. DNA sequence analysis of the region encoding the terminal protein and the hypothetical N-gene product of adenovirus type 2. *J. Biol. Chem.* 257:13492–13498.
2. Baer, R., A. T. Bankier, M. D. Biggin, P. L. Deininger, P. J. Farrel, T. J. Gibson, G. Hatfull, G. S. Hudson, S. C. Satechwell, C. Seguin, P. S. Tuffnell, and B. G. Barrell. 1984. DNA sequence and expression of the B95-8 Epstein-Barr virus genome. *Nature* (London) 263:211–214.
3. Bernad, A., A. Zaballos, M. Salas, and L. Blanco. 1987. Structural and functional relationships between prokaryotic and eukaryotic DNA polymerases. *EMBO J.* 6:4219–4225.
4. Blanco, L., J. Gutiérrez, J. M. Lázaro, A. Bernad, and M. Salas. 1986. Replication of phage ϕ29 DNA *in vitro*: role of the viral protein p6 in initiation and elongation. *Nucleic Acids Res.* 14:4923–4937.
5. Blanco, L., and M. Salas. 1984. Characterization and purification of a phage ϕ29-encoded DNA polymerase required for the initiation of replication. *Proc. Natl. Acad. Sci. USA* 81:5325–5329.
6. Blanco, L., and M. Salas. 1985. Replication of phage ϕ29 DNA with purified terminal protein and DNA polymerase: synthesis of full length ϕ29 DNA. *Proc. Natl. Acad. Sci. USA* 82:6404–6408.
7. Blanco, L., and M. Salas. 1985. Characterization of a $3' \rightarrow 5'$ exonuclease activity in the phage ϕ29-encoded DNA polymerase. *Nucleic Acids Res.* 13:1239–1249.
8. Blanco, L., and M. Salas. 1986. Effect of aphidicolin and nucleotide analogs on the phage ϕ29 DNA polymerase. *Virology* 153:179–187.
9. Davison, A. J., and J. E. Scott. 1986. The complete DNA sequence of Varicella-Zoster virus. *J. Gen. Virol.* 67:1759–1816.
10. Earl, P. L., E. V. Jones, and B. Moss. 1986. Homology between DNA polymerases of poxviruses, herpesviruses and adenoviruses: nucleotide sequence of the vaccinia virus DNA polymerase gene. *Proc. Natl. Acad. Sci. USA* 83:3659–3663.
11. Garvey, K. J., H. Yoshikawa, and J. Ito. 1985. The complete sequence of the *Bacillus* phage ϕ29 right early region. *Gene* 40:301–309.
12. Harding, N. E., and J. Ito. 1980. DNA replication of bacteriophage ϕ29: characterization of the intermediates and location of the termini of replication. *Virology* 104:323–338.
13. Hermoso, J. M., E. Méndez, F. Soriano, and M. Salas. 1985. Location of the serine residue involved in the linkage between the terminal protein and the DNA of ϕ29. *Nucleic Acids Res.* 13:7715–7728.
14. Huberman, J. A. 1981. New views of the biochemistry of eukaryotic DNA replication revealed by aphidicolin, an unusual inhibitor of DNA polymerase α. *Cell* 23:647–648.
15. Inciarte, M. R., M. Salas, and J. M. Sogo. 1980. Structure of replicating DNA molecules of *Bacillus subtilis* bacteriophage ϕ29. *J. Virol.* 34:187–199.
16. Kallin, B., L. Sternan, A. K. Saemundssen, J. Luka, H. Jornvale, B. Eriksson, P. Z. Tas, M. T. Nilsson, and G. Klein. 1985. Purification of Epstein-Barr virus DNA polymerase from P3HR-1 cells. *J. Virol.* 54:561–568.
17. Kemble, R. J., and R. D. Thompson. 1982. S1 and S2, the linear mitochondrial DNAs present in a male sterile line of maize, possess terminally attached proteins. *Nucleic Acids Res.* 10:8181–8190.

18. Khan, N. N., G. E. Wright, L. W. Dudycz, and N. C. Brown. 1984. Butylphenyl dGTP: a selective and potent inhibitor of mammalian DNA polymerase alpha. *Nucleic Acids Res.* 12:3695–3706.
19. Khan, N. N., G. E. Wright, L. W. Dudycz, and N. C. Brown. 1985. Elucidation of the mechanism of selective inhibition of mammalian DNA polymerase alpha by 2-butylanilinopurines: development and characterization of 2-(p-n-butylanilino) adenine and its deoxyribonucleotides. *Nucleic Acids Res.* 13:6331–6342.
20. Kikuchi, Y., K. Hirai, and F. Hishinuma. 1984. The yeast linear DNA killer plasmids, pGKL1 and pGKL2, possess terminally attached proteins. *Nucleic Acids Res.* 12:5685–5692.
21. Knopf, K. W. 1979. Properties of herpes simplex virus DNA polymerase and characterization of its associated exonuclease activity. *Eur. J. Biochem.* 98:231–244.
22. Knopf, K. W. 1987. The herpes simplex virus type I DNA polymerase gene: site of phosphonoacetic acid resistance mutation in strain Angelotti is highly conserved. *J. Gen. Virol.* 68:1429–1433.
23. Kouzarides, T., A. T. Bankier, S. C. Satchwell, K. Weston, P. Tomlinson, and B. G. Barrell. 1987. Sequence and transcription analysis of the human cytomegalovirus DNA polymerase gene. *J. Virol.* 61:125–133.
24. Larder, B. A., S. A. Kemp, and G. Darby. 1987. Related functional domains in virus DNA polymerases. *EMBO J.* 6:169–175.
25. Moss, B., and N. Cooper. 1982. Genetic evidence for vaccinia virus-encoded DNA polymerase: isolation of phosphonoacetic-resistant enzyme from the cytoplasm of cells infected with mutant virus. *J. Virol.* 43:673–678.
26. Nishiyama, Y., K. Maeno, and S. Yoshida. 1983. Characterization of human cytomegalovirus-induced DNA polymerase and the associated 3' to 5' exonuclease. *Virology* 124:221–231.
27. Paillard, M., R. R. Sederoff, and C. S. Levings III. 1985. Nucleotide sequence of the S-1 mitochondrial DNA from the S cytoplasm of maize. *EMBO J.* 4:1125–1128.
28. Pastrana, R., J. M. Lázaro, L. Blanco, J. A. García, E. Méndez, and M. Salas. 1985. Overproduction and purification of protein p6 of *Bacillus subtilis* phage ϕ29: role in the initiation of DNA replication. *Nucleic Acids Res.* 13:3083–3100.
29. Prieto, I., J. M. Lázaro, J. A. García, J. M. Hermoso, and M. Salas. 1984. Purification in a functional form of the terminal protein of *Bacillus subtilis* phage ϕ29. *Proc. Natl. Acad. Sci. USA* 81:1639–1643.
30. Prieto, I., M. Serrano, J. M. Lázaro, M. Salas, and J. M. Hermoso. 1988. Interaction of the bacteriophage ϕ29 protein p6 with double-stranded DNA. *Proc. Natl. Acad. Sci. USA* 85:314–318.
31. Salas, M., R. P. Mellado, E. Viñuela, and J. M. Sogo. 1978. Characterization of a protein covalently linked to the 5' termini of the DNA of *Bacillus subtilis* phage ϕ29. *J. Mol. Biol.* 119:269–291.
32. Savilahti, H., and D. H. Bamford. 1987. The complete nucleotide sequence of the left very early region of *Escherichia coli* bacteriophage PRD1 coding for the terminal protein and the DNA polymerase. *Gene* 57:121–130.
33. Sridhar, P., and R. C. Condit. 1983. Selection for temperature-sensitive mutations in specific vaccinia virus genes: isolation and characterization of a virus mutant which encodes a phosphonoacetic acid-resistant, temperature-sensitive DNA polymerase. *Virology* 128:444–457.
34. Stark, M. J. R., A. J. Mileham, M. A. Romanos, and A. Boyd. 1984. Nucleotide sequence and transcription analysis of a linear DNA plasmid associated with the killer character of the yeast *Kluyveromyces lactis*. *Nucleic Acids Res.* 12:6011–6030.
35. Tsurumi, T., K. Maeno, and Y. Nishiyama. 1987. Nucleotide sequence of the DNA polymerase gene of herpes simplex virus type 2 and comparison with the type 1 counterpart. *Gene* 52:129–137.
36. Vlček, C., and V. Pačes. 1986. Nucleotide sequence of the late region of *Bacillus* phage ϕ29 completes the 19285-bp sequence of ϕ29 genome. Comparison with the homologous sequence of phage PZA. *Gene* 46:215–225.
37. Watabe, K., M. Leusch, and J. Ito. 1984. Replication of bacteriophage ϕ29 DNA *in vitro*: the roles of terminal protein and DNA polymerase. *Proc. Natl. Acad. Sci. USA* 81:5374–5378.
38. Yoshikawa, H., and J. Ito. 1982. Nucleotide sequence of the major early region of bacteriophage ϕ29. *Gene* 17:323–335.

Chapter 13

N4 DNA Replication In Vivo and In Vitro

Gordon Lindberg, Margaret Sue Pearle, and Lucia B. Rothman-Denes

N4 virions consist of a hexagonally shaped head approximately 70 nm in diameter, a base plate, a small noncontractile tail, and a number of short tail fibers originating from the junction between the head and tail (23). The genome is a linear double-stranded DNA molecule approximately 72 kilobase pairs (kbp) in length with a G+C content of 44 mol% (24), 400- to 450-base-pair (bp) direct terminal repeats, and 3' extensions at each end (27; H. Ohmori, L. L. Haynes, and L. B. Rothman-Denes, *J. Mol. Biol.*, in press). The left end is relatively unique with a 3' protruding sequence, 3'-CATAA or 3'-CA TAAA. The right end is heterogeneous. It consists of at least six families each differing in length by 10 bp, giving rise to the variability in the length of the terminal repeats.

These ends have a 1- to 3-base 3' overhang. The protruding sequences from the left and right ends of the genome are not complementary to each other (Ohmori et al., in press). N4 DNA is not modified but is still resistant to cleavage by a large number of restriction endonucleases (27; Ohmori et al., in press).

The phage life cycle can be divided into three temporal phases: early, middle, and late (28). After adsorption to a sensitive strain of *Escherichia coli*, N4 injects a virion-associated, DNA-dependent RNA polymerase along with its DNA (22; D. Kiino and L. B. Rothman-Denes, *in* R. Calendar (ed.), *The Bacteriophages*, vol. 2, in press). This RNA polymerase, which is responsible for N4 early transcription (3), consists of a single M_r 320,000 polypeptide (4). The activities of *E. coli* DNA gyrase and *E. coli* single-stranded DNA-binding protein are absolutely required for the virion polymerase to recognize its promoters in vivo (6, 17, 18). In addition to its role in early transcription, the virion-associated, DNA-dependent RNA polymerase is directly involved in N4 DNA replication (8). Three products of N4 early transcription, proteins of M_r 40,000,

---

Gordon Lindberg • Department of Biochemistry and Molecular Biology, University of Chicago, Chicago, Illinois 60637.   *Margaret Sue Pearle* • Department of Biochemistry and Molecular Biology, University of Chicago, Chicago, Illinois 60637, and Northwestern University School of Medicine, Chicago, Illinois 60611.   *Lucia B. Rothman-Denes* • Departments of Biochemistry and Molecular Biology and of Molecular Genetics and Cell Biology, University of Chicago, Chicago, Illinois 60637.

30,000, and 15,000, are required for middle RNA synthesis, which starts 3 to 5 min after infection (26, 28). These middle mRNAs code, in part, for the proteins required for N4 DNA replication. This process begins 7 to 10 min after infection (7). Late transcription, which begins 13 to 15 min after infection, requires the activity of *E. coli* RNA polymerase holoenzyme (28) and a product of middle transcription but is not dependent on replication (N. Y. Baek, unpublished results). N4 does not code for any lysis function and up to 3,000 mature virions can accumulate during the first 3 h of infection (14, 21).

Nonhomologous 3' protrusions are unusual structures to find at the ends of a genome. Their generation is difficult to explain using models of replication that have been described for other linear bacteriophage DNAs such as T4, T7, and lambda. This unusual genomic structure and the direct involvement of the virion-associated, DNA-dependent RNA polymerase in DNA replication prompted us to investigate N4 replication.

HOST AND PHAGE FUNCTIONS REQUIRED FOR N4 DNA REPLICATION

As a first step in elucidating the mechanism of N4 DNA replication, we determined which host- and phage-coded functions are required in this process (Table 1) (8, 15).

N4 shuts off *E. coli* DNA synthesis by an unknown mechanism that requires an early protein product (8, 20). In contrast to other lytic bacteriophages, N4 does not degrade the host chromosome (19, 20). It is not surprising, therefore, that phage replication is dependent on *E. coli* ribonucleotide reductase, since the *E. coli* chromosome does not serve as a deoxynucleotide pool. We do not know, however, if N4 encodes for proteins that can synthesize deoxynucleotide precursors de novo, as has been shown for bacteriophage T4 (13).

TABLE 1
Host and Phage Functions Known To Be Required for N4 DNA Replication

Gene	Source	Function and properties
dnaF	*E. coli*	Ribonucleotide reductase
gyrA and gyrB	*E. coli*	DNA gyrase
lig	*E. coli*	Ligase
polA	*E. coli*	5'→3' exonuclease
dnp	N4	DNA polymerase with 3'→5' exonuclease, 87,000 M_r
dbp	N4	Single-stranded DNA-binding protein, 30,000 M_r
exo	N4	5'→3' exonuclease, 45,000 M_r
dns	N4	Unknown function, 78,000 M_r
vRNAP	N4	Virion RNA polymerase, 320,000 M_r

Viral DNA replication is arrested in an *E. coli* host containing a *gyrB* temperature-sensitive mutation when the infected cells are raised to the nonpermissive temperature after DNA replication has initiated (8). Therefore, DNA gyrase activity is needed for N4 DNA synthesis as well as early transcription. The phage genome is 72 kbp long, and gyrase activity may be required to facilitate unwinding and untangling of replicating DNA. In addition, the DNA may have to be in a negatively supercoiled state for assembly of a correct initiation complex.

N4 infections of *E. coli* deficient in the 5'→3' exonuclease activity of *E. coli* DNA polymerase I yield wild-type levels of DNA synthesis but no mature phage. Alkaline sucrose gradient analysis of DNA labeled in vivo shows the accumulation of 1- to 6-kb DNA fragments (8; M. Pearle, Ph.D. thesis, University of Chicago, Chicago, Ill., 1985) which do not accumulate in infections of wild-type *E. coli*. These fragments hybridize to all regions and both strands of viral DNA, indicating that there is discontinuous synthesis of both strands of the viral genome. Since the fragments can be processed to mature length DNA (7; Pearle, Ph.D. thesis), they probably represent Okazaki fragments which are processed by host proteins. In agreement, E. coli deficient in DNA ligase activity does not support N4 DNA synthesis. We do not know whether the phage DNA polymerase or E. coli DNA polymerase I fills in the gaps generated by primer excision from the Okazaki fragments.

E. coli containing temperature-sensitive alleles for the dnaA, dnaB, dnaC, dnaE, and dnaG genes is able to support N4 growth at the nonpermissive temperatures (8). Mutations in the E. coli rep (8) or uvrD (J. Gregg, unpublished data) gene do not affect N4 growth. In addition, viral DNA replication does not require the polymerizing activity of E. coli DNA polymerase I (8). Therefore, N4 infection must induce gene products required for the initiation, priming, and elongation steps of phage DNA synthesis.

We have identified five N4-coded gene products that are directly required for N4 DNA replication. Mutations in three of the genes, dns, dnp, and dbp, result in a D0 phenotype; i.e., N4 DNA replication does not occur (8, 15). The dns gene induces the synthesis of a 78,000-M_r protein of unknown function (8, 15). The dnp gene encodes a DNA polymerase (15), and the dbp gene induces the synthesis of a single-stranded DNA-binding protein (15).

Phage containing conditional lethal mutations in the dnp gene are unable to initiate DNA synthesis under nonpermissive conditions. Crude lysates of E. coli polA1 cells infected with different dnp mutants contain either no DNA polymerase activity (dnp amber alleles) or a temperature-sensitive DNA polymerase activity (dnp temperature-sensitive alleles), strongly suggesting that dnp is the structural gene for a phage-encoded DNA polymerase. This phage-encoded DNA polymerase has been purified to homogeneity by following DNA polymerase activity on activated DNA templates (G. Lindberg, J. K. Rist, T. A. Kunkel, A. Sugino, and L. B. Rothman-Denes, J. Biol. Chem., in press). The enzyme is a single polypeptide of M_r 87,000 and is a monomer in solution. The polymerase requires preexisting RNA or DNA primers to initiate DNA synthesis and synthesizes DNA with high fidelity because it has a very strong intrinsic $3' \rightarrow 5'$ exonuclease activity. It is quasi-processive, synthesizing an average of six nucleotides each time it interacts with a template-primer. The polymerase has no strand-displacing activity and is unable to initiate DNA synthesis at a nick or extend DNA synthesis past the 5' end of a gapped duplex. Furthermore, it cannot synthesize past hairpins present in the template strand (Lindberg et al., in press). Therefore, N4 must provide priming and helicase activities plus processivity factors to aid the polymerase during replication fork movement.

Mutations in the N4 dbp gene also have a D0 phenotype. This gene induces the synthesis of a single-stranded DNA-binding protein (N4 DBP) that specifically enhances the rate of synthesis by the N4 DNA polymerase (G. Lindberg, S. C. Kowalczykowski, J. K. Rist, A. Sugino, and L. B. Rothman-Denes, submitted for publication). N4 DBP has been purified to homogeneity by in vitro complementation of N4 dbp am33A7-infected cell extracts and has a monomeric molecular weight of 31,000 (Lindberg et al., submitted). Each monomer of N4 DBP covers 11 nucleotides and binding is moderately cooperative to single-stranded DNA. N4 DBP is able to lower the melting transition of poly(dA) · poly(dT) by 60°C but has no effect on the melting transition of N4 DNA. It stimulates the N4 DNA polymerase by increasing processivity of the polymerase 300-fold and removing hairpins in the template strand. It does not bestow any strand-displacing activity on the N4 polymerase and cannot be responsible for driving replication fork movement (Lindberg et al., submitted).

Since N4 encodes its own DNA polymerase and single-stranded DNA-binding protein, we expect that N4 encodes the rest of the protein activities associated with a moving replication fork (1). Although viral infections are normal in the absence of *rep* or *uvrD* activities, we cannot rule out the possibility that N4 replication requires a helicase activity from the host. Likewise, we do not know if N4 requires an unidentified host function or encodes factors that enhance the processivity of the replication machinery. The product of the *dns* gene may provide some of these functions.

The fourth N4-coded protein required for replication is the virion-associated, DNA-dependent RNA polymerase. If cultures infected with phage containing temperature-sensitive mutations in the virion polymerase are raised to the nonpermissive temperature after initiation of DNA replication, DNA synthesis immediately stops (8). Bacteriophage T7 utilizes its phage-coded RNA polymerase to prime leading-strand synthesis (5). Unfortunately, the speed at which N4 replicates its genome prevented us from determining whether the virion polymerase is required for initiation of N4 DNA synthesis or is an integral part of the replication fork. Virion RNA polymerase is able to recognize and transcribe its promoters on single-stranded phage DNA without the activities of *E. coli* DNA gyrase or *E. coli* single-stranded DNA-binding protein (9). If the virion RNA polymerase initiates DNA synthesis, then *E. coli* DNA gyrase may be required to unwind the DNA into a single-stranded form that can be used by the virion polymerase to prime leading-strand synthesis.

Finally, N4 also encodes a 5′→3′ exonuclease, the product of the *exo* gene, which is required for phage replication. The purified protein has a denatured molecular weight of 45,000 and exists as a dimer in solution (7). It degrades linear duplex DNA in a 5′→3′ direction to 5′ mononucleotides by a distributive mechanism. It has a marked preference for duplex DNA containing 3′ overhangs and is inactive on single-stranded DNA (7). The exonuclease is also inactive on nicked or gapped circular DNA (7) and therefore is unable to process Okazaki fragments formed during lagging-strand synthesis. Mutations in the gene encoding the exonuclease have a temperature-dependent effect on DNA replication (D. Guinta, S. Spellman, and L. B. Rothman-Denes, unpublished results). At 37°C phage containing amber mutations in the exonuclease gene can synthesize DNA normally until 60 min after infection, at which time replication stops. At 43°C, N4 DNA replication begins for a very brief time and stops. The phage burst size in both infections is reduced 100- to 200-fold (Guinta et al., unpublished results). In summary, exonuclease-deficient infections reveal a temperature-sensitive replication pathway that is abortive after 60 min. This may indicate two roles for the exonuclease in replication or two modes of N4 DNA replication.

ANALYSIS OF IN VIVO REPLICATING N4 DNA

Replicating N4 DNA molecules have been analyzed on sucrose gradients and examined by electron microscopy. When replicating DNA is pulse-labeled for 2 min with [^{3}H]thymidine at different times after infection and separated on agarose gels after *Xba*I restriction and the bands are visualized by autoradiography, one detects radiolabel accumulating in all the internal fragments of the genome (Fig. 1). The end fragments are missing or underrepresented. Instead, a new fragment, which is not present in mature virion DNA, is detected. Restriction analysis of this fragment, which we call the "joint fragment," indicates that it can originate by homologous recombination between the terminally repeated sequences present at the left and right ends of the genome. Radiolabel in the joint fragment can be chased into the mature ends of the genome (Pearle, Ph.D. thesis) (Fig. 1).

Sucrose gradient analysis of DNAs radiolabeled at different times after infection reveals in all cases three different migrating

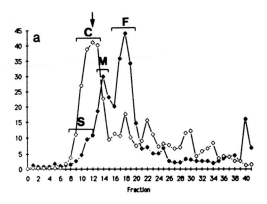
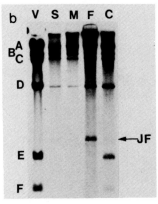

FIGURE 1. Sucrose gradient analysis of in vivo replicating N4 DNA. *E. coli* W3350 was grown at 37°C to an optical density at 620 nm of 0.5 in glucose-M9 medium. Cells were pelleted, suspended in fresh medium, and infected with N4 phage at a multiplicity of infection of 10. At 10 min after infection, the cells were labeled with 50 μCi of [^{3}H]thymine per ml. After 2 min of incubation, half the sample was added to ice-cold 25 mM NaN$_3$, while the remainder was further incubated for 5 min with 2 mg of unlabeled thymidine per ml. The cells were pelleted, suspended in 0.1 volume of 10 mM Tris hydrochloride (pH 8.0)–10 mM EDTA, and frozen in a dry ice-ethanol bath. After thawing, the cells were lysed with 0.5 mg of lysozyme per ml at 0°C for 10 min and 0.1 mg of proteinase K per ml plus 0.1% sarcosyl NL97 for 1 h at 55°C. The samples were layered on a prechilled 10-ml linear 10 to 30% (wt/vol) sucrose gradient in 20 mM Tris hydrochloride (pH 8.0)–10 mM EDTA–1 M NaCl with a 1-ml 80% sucrose cushion and spun at 4°C in a Beckman SW41 rotor for 4.5 h at 36,000 rpm. Fractions were collected by hand, and acid-insoluble radioactivity was determined in a portion of each fraction. The remainder of the samples were pooled on the basis of their sedimentation, dialyzed against 10 mM Tris hydrochloride (pH 8.0)–1 mM EDTA, and ethanol precipitated. The DNA was suspended, digested with *Xba*I, and separated on 1% agarose gels. The bands were visualized by treating the gel with En^{3}Hance before drying and exposing it to film. (a) Sucrose gradient profile. Fractions are from top to the bottom, and S, M, and F represent the pooled fractions corresponding to slow-, middle-, and fast-migrating species of DNA. C represents the migration of DNA after a 5-min chase and corresponds to the migration of virion N4 DNA. The numbers on the ordinate axis represent counts per minute (in thousands) of acid-insoluble radioactivity. (b) Agarose gel of the restriction fragments from the pooled fractions. Lane V shows the restriction pattern of mature virion N4 DNA. Lanes S, M, F, and C are as in panel a. A to F indicate N4 *Xba*I restriction fragments. JF indicates joint fragment.

species (Pearle, Ph.D. thesis). Most of the label is present in a species that migrates faster than mature N4 DNA, with the rest in species that migrate at the same or a lower speed than mature N4 DNA. The label in all these species can be chased into mature N4 DNA. When the DNA in each fraction is analyzed as described in the preceding paragraph, differences in the relative amounts of radiolabel accumulating in the end fragments and the joint fragment, as compared to internal sequences, are observed. At all times after infection, the faster-migrating species lack the mature ends but contain the joint fragment. At late times, radiolabel in the middle-migrating species appears at the left end of the genome and in the joint fragment, while the slower-migrating species accumulate radiolabel primarily at both ends of the genome. In contrast, early in infection, the middle- and slower-migrating species do not contain any label in terminal fragments. The only terminal sequences that accumulate label are present in the joint fragment, and the joint fragment appears only in the faster-migrating population of replicating DNA (Pearle, Ph.D. thesis).

These results suggest that the sequences present in the joint fragment are direct precursors of the mature ends of virion DNA. The faster-migrating species containing the joint fragment could be part of a concatemeric, circular, branched, or extensively single-stranded form of N4 DNA. N4 replicating DNA was analyzed by electron microscopy to determine more precisely what form of DNA contains the joint fragment and the overall pathway of replicating N4 DNA. Since N4 does not degrade the host chromosome, phage replicating DNA was first separated from *E. coli* DNA. To accomplish this, *E. coli thyA* was grown in medium containing bromodeoxyuridine and switched to glucose-M9 medium 10 min before infection. N4 replicating DNA was separated from *E. coli* DNA by two sequential cesium chloride density gradients and analyzed by electron microscopy (Pearle, Ph.D. thesis).

The most frequently observed replicating molecules fall in three classes. We have followed the nomenclature of Lechner and Kelley (11) and Inciarte et al. (10) in naming them. Fifty-one percent of the molecules were type II or linear molecules with one duplex and one single-stranded region (Fig. 2). The duplex regions in these molecules were generally longer than the single-stranded regions and their total length was close to unit length. Twenty-two percent of the molecules were full-length, /2-kbp, duplex, circular DNA. However, these molecules did not contain any replication bubbles and may not be true replicative intermediates. Nineteen percent of the molecules were type I or partial duplex branched molecules containing single-stranded or partially duplex branches. The remaining molecules were duplex linear molecules with internal single-stranded regions (type IV), duplex linear molecules with single-stranded regions at both ends of the genome (type III), or molecules that had characteristics of type II, III, and IV molecules. All classes of molecules contained longer than unit-length DNA at low percentages (Pearle, Ph.D. thesis).

Strikingly, there is a total absence of replication bubbles in the various forms observed. This suggests that replication begins at or near the ends of the DNA molecule.

The fact that the Okazaki fragments contain DNA sequences from all regions of the genome and hybridize to both strands implies that if replication does begin at the ends of the genome, it can begin at either end. Alternatively, extensive recombination and conversion of the recombinational intermediates to replication forks could explain the lack of polarity of the Okazaki fragments. Replication from the terminal regions of the genome is supported by in vitro studies described below.

IN VITRO N4 DNA SYNTHESIS

We have developed a crude in vitro DNA replication system from N4-infected *E. coli* cells that is specific for exogenous N4 DNA (16). Other double- or single-stranded DNAs are poor templates in this reaction. DNA synthesis initiates at both ends of the N4 genome, with a strong preference for the right end, and proceeds toward internal regions of the genome. Initiation occurs through hairpin priming (16). Furthermore, the protein products of the N4 *dnp*, *dbp*, and *exo* genes are absolutely required for in vitro DNA synthesis (15, 16). Surprisingly, only these three proteins, when combined with N4 DNA, are needed for efficient DNA synthesis (Pearle, Ph.D. thesis). Moreover, this reaction exhibits the same template specificity, mode of initiation, and directionality of N4 DNA synthesis as the reaction carried out by infected cell lysates (Pearle, Ph.D. thesis). When the reaction products are restricted with *Xba*I and analyzed by two-dimensional gel electrophoresis, the terminal fragments migrate at a position twice their expected size under denaturing conditions. Pretreatment of the DNA with S1 nuclease results in terminal fragments with similar mobilities under native and denaturing gel conditions (Fig. 3). This result indicates that DNA synthesis is initiated by hairpin priming.

This simplified reaction using the N4

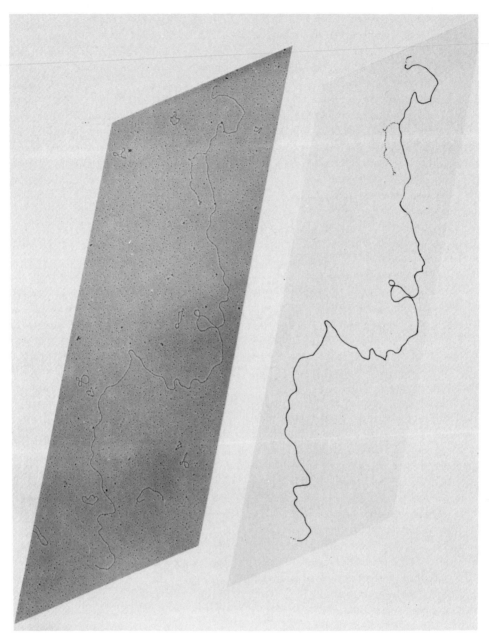

FIGURE 2. Electron micrograph of an in vivo N4 replicating DNA molecule. *E. coli thyA* was grown at 37°C in M9 medium supplemented with 50 μg of bromodeoxyuridine per ml, 3 μg of thymidine per ml, and 10 μg of trimethoprim per ml to an optical density at 620 nm of 0.6 (3×10^8 cells per ml). The cells were pelleted, suspended in M9 medium supplemented with 20 μg of thymine per ml, incubated for 10 min at 37°C, infected with N4 at a multiplicity of infection of 3 to 5, and incubated at 37°C. At 38 min after infection, the N4 replicating DNA was labeled with 50 μCi of [^{3}H]thymidine per ml for 2 min and then the culture was added to ice-cold 25 mM NaN_3. The cells were lysed as described in the text. The samples were adjusted to a density of 1.4000 g/ml with CsCl saturated Tris-EDTA and centrifuged at 4°C in a Beckman 75 Ti rotor at 25,000 rpm for 80 h. DNA samples were prepared for electron microscopy by the Kleinschmidt procedure (2).

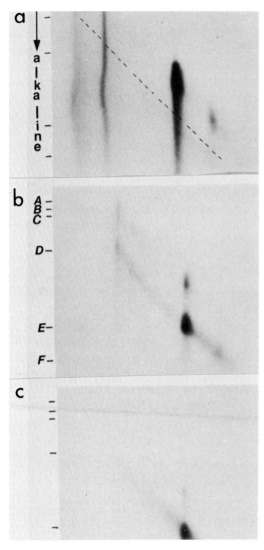

FIGURE 3. Two-dimensional gel electrophoretic analysis of in vitro replicating N4 DNA. The reaction contained N4 DNA polymerase (1.8 U/ml), N4 single-stranded DNA-binding protein (3.6 μg/ml), N4 5'→3' exonuclease (57.8 U/ml), N4 DNA (30 μg/ml), and [^{32}P]dATP (250 μCi/ml) in standard in vitro replication buffer (15). The reaction products were purified by gel filtration and ethanol precipitation before treatment with XbaI and S1 nuclease as described previously (15). (a) No S1 nuclease; (b) 10 U of S1 nuclease; (c) 50 U of S1 nuclease. A to F incdiate N4 XbaI restriction fragments.

DNA polymerase, single-stranded DNA-binding protein, and 5'→3' exonuclease has allowed us to examine the specificity of the in vitro replication reaction in more detail. The reaction is specific for both the N4 proteins and phage DNA template. *E. coli* DNA polymerase I or T4 DNA polymerase cannot substitute for the N4 DNA polymerase. Reactions containing *E. coli* DNA polymerase I initiate DNA synthesis randomly on the N4 DNA template, and reactions containing T4 DNA polymerase do not synthesize DNA efficiently (Pearle, Ph.D. thesis). Other single-stranded DNA-binding proteins cannot substitute for the N4 single-stranded DNA-binding protein (G. Lindberg, unpublished data). The specificity for the N4 DNA polymerase and N4 single-stranded DNA-binding protein is not surprising as the N4 DNA polymerase is stimulated only by the N4 single-stranded DNA-binding protein (Lindberg et al., in press). Also, the N4 single-stranded DNA-binding protein can only stimulate the N4 DNA polymerase (G. Lindberg, unpublished data). Finally, the T7 gene 6 and phage lambda 5'→3' exonucleases cannot efficiently substitute for the N4 exonuclease in initiating DNA synthesis (Pearle, Ph.D. thesis). The specificity for N4 5'→3' exonuclease is due in part to the preference of the 5'→3' exonuclease for the structures at the ends of N4 DNA. The presence of a 3' overhang at the end of a linear duplex DNA is not, however, the sole determinant of the observed template specificity. Blunt-ended N4 DNA is still a better template in this reaction than blunt-ended lambda DNA (G. Lindberg, unpublished data). This implies that internal sequences in N4 DNA, perhaps involved in hairpin priming, are required for efficient DNA synthesis. Several inverted repeats are present, especially at the right ends of the N4 genome (Ohmori et al., in press). Their involvement in in vivo initiation of N4 DNA replication remains to be seen.

CONCLUSIONS

Using an in vitro DNA replication system, we have been able to purify and characterize three N4 proteins required for phage replication. Moreover, analysis of in vivo replicative intermediates has allowed us to propose some models for N4 replication of its genome. Unfortunately, the variety of replicative intermediates observed by electron microscopy prevents us from clearly differentiating between them. The absence of replication bubbles in isolated in vivo replicative intermediates suggests that replication initiates at or near the ends of the genome, and the in vitro replication system supports this model. The hairpin priming observed in vitro may reflect the in vivo mechanism of initiating DNA synthesis. However, before drawing firm conclusions about the in vitro data, we need to clearly delineate the in vivo mode of DNA replication.

As a first step we must identify the origin of DNA replication to see whether replication initiates from the ends or in an internal region of the genome. In addition, the structure of replicating DNA must be determined. The pulse-chase experiments in Fig. 1 indicate the presence of concatemeric or circular intermediates in DNA replication. We are currently studying in vivo replicative intermediates, using pulse-field gel electrophoresis, which allows us to resolve large linear and circular DNA molecules.

Finally, the mechanism of generation of the ends of mature N4 phage DNA needs to be elucidated. The data in Fig. 3 suggest that the joint fragment may serve as a direct precursor of the mature ends of N4 DNA. If the joint fragment is part of a concatemeric structure and virion DNA is generated by simple site-specific cleavage, there would be a net loss of half of the replicated DNA. An alternative mechanism is concatemer formation followed by staggered cleavage and filling in of the 5' protruding ends, as proposed for T7 DNA (25). This could generate blunt-ended N4 DNA molecules which contain all of the sequences present in the 3' extensions. Controlled exonuclease digestion or endonucleolytic cleavage during packaging would generate the mature ends of virion DNA. Finally, the terminal sequences present in the joint fragment may be donated to just one end of the genome. The other end would then use a strand invasion mechanism, as proposed for T4 (12), to synthesize its terminal sequences. These models can be tested by repeating the pulse-chase experiments described in the legend to Fig. 1 and monitoring the processing of the unique and terminally redundant sequences present in the joint fragment during phage development.

ACKNOWLEDGMENTS. We thank A. Camacho and D. Kiino for comments on this manuscript.

This work was supported by Public Health Service grant GM 35170 from the National Institutes of Health to L.B.R.-D. G.L. and M.S.P. were trainees supported by Public Health Service grants GM 07183 and GM 07281, respectively, from the National Institutes of Health.

LITERATURE CITED

1. **Alberts, B. M.** 1984. The DNA enzymology of protein machines. *Cold Spring Harbor Symp. Quant. Biol.* 49:1–12.
2. **Davis, R. W., M. Simon, and N. Davidson.** 1971. Electronmicroscope heteroduplex methods for mapping regions of base sequence homology in nucleic acids. *Methods Enzymol.* 21:413–428.
3. **Falco, S. C., K. VanderLaan, and L. B. Rothman-Denes.** 1977. Virion-associated RNA polymerase required for bacteriophage N4 development. *Proc. Natl. Acad. Sci. USA* 74:520–523.
4. **Falco, S. C., W. Zehring, and L. B. Rothman-Denes.** 1980. DNA-dependent RNA polymerase from bacteriophage N4 virions. Purification and characterization. *J. Biol. Chem.* 255:4339–4347.
5. **Fuller, C. W., and C. C. Richardson.** 1985. Initiation of DNA replication at the primary origin of bacteriophage T7 by purified proteins: site and

direction of initial DNA synthesis. *J. Biol. Chem.* 260:3185–3196.
6. Glucksmann, A., and L. B. Rothman-Denes. 1987. N4 virion RNA polymerase-promoter interaction, p. 67–76. *In* B. Thompson and J. Papaconstantinou (ed.), *DNA-Protein Interactions.* University of Texas Press, Austin.
7. Guinta, D., G. Lindberg, and L. B. Rothman-Denes. 1986. Bacteriophage N4 coded 5'-3' exonuclease: purification and characterization. *J. Biol. Chem.* 261:10736–10743.
8. Guinta, D., J. Stambouly, S. C. Falco, J. K. Rist, and L. B. Rothman-Denes. 1986. Host and phage-coded functions required for coliphage N4 DNA replication. *Virology* 150:33–44.
9. Haynes, L. L., and L. B. Rothman-Denes. 1985. N4 virion RNA polymerase sites of transcription initiation. *Cell* 41:597–605.
10. Inciarte, M. R., M. Sala, and J. M. Sogo. 1980. Structure of replicating DNA molecules of *Bacillus subtilis* bacteriophage φ29. *J. Virol.* 34:187–199.
11. Lechner, R. L., and T. J. Kelley, Jr. 1977. The structure of replicating adenovirus 2 DNA molecules. *Cell* 12:1007–1020.
12. Luder, A., and G. Mosig. 1982. Two alternative mechanisms of initiation of DNA replication forks in bacteriophage T4: priming by RNA polymerase and by recombination. *Proc. Natl. Acad. Sci. USA* 79:1101–1105.
13. Mathews, C. K., and J. R. Allen. 1983. DNA precursor biosynthesis, p. 59–70. *In* C. K. Mathews, E. M. Kutter, G. Mosig, and P. B. Berget (ed.), *Bacteriophage T4.* American Society for Microbiology, Washington, D.C.
14. Pesce, A., G. Satta, and G. C. Schito. 1969. Factors in lysis-inhibition by N4 coliphage. *G. Microbiol.* 17:119–129.
15. Rist, J. K., D. R. Guinta, A. Sugino, J. Stambouly, S. C. Falco, and L. B. Rothman-Denes. 1983. Bacteriophage N4 DNA replication, p. 245–254. *In* N. R. Cozzarelli (ed.), *Mechanisms of DNA Replication and Recombination.* Alan R. Liss, Inc., New York.
16. Rist, J. K., M. Pearle, A. Sugino, and L. B. Rothman-Denes. 1986. Development of an *in vitro* bacteriophage N4 DNA replication system. *J. Biol. Chem.* 261:10506–10510.
17. **Rothman-Denes, L. B., K. Abravaya, A. Glucksmann, C. Malone, and P. M. Markiewicz. 1987. Bacteriophage N4-coded RNA polymerases, p. 37–46. *In* W. S. Reznikoff, R. R. Burgess, J. E. Dahlberg, C. A. Gross, M. T. Record, and M. P. Wickens (ed.), *RNA Polymerase and the Regulation of Transcription.* Elsevier Science Publishing, Inc., New York.**
18. **Rothman-Denes, L. B., L. Haynes, P. M. Markiewicz, A. Glucksmann, C. Malone, and J. Chase. 1985. Bacteriophage N4 virion RNA polymerase promoters, p. 41–53. *In* R. Calendar and L. Gold (ed.), *Sequence Specificity in Transcription and Translation.* Alan R. Liss, Inc., New York.**
19. **Rothman-Denes, L. B., and G. C. Schito. 1974. Novel transcribing activities in N4-infected *Escherichia coli*. *Virology* 60:65–72.**
20. **Schito, G. C. 1973. The genetics and physiology of coliphage N4. *Virology* 55:254–265.**
21. **Schito, G. C. 1974. Development of coliphage N4: ultrastructural studies. *J. Virol.* 13:186–196.**
22. **Schito, G. C., A. M. Molina, and A. Pesce. 1965. Un nuovo batteriofago attivo sul ceppo K12 di *E. coli*. I. Caractteristiche biologiche. *Boll. Ist. Sieroter. Milan.* 44:329–332.**
23. **Schito, G. C., G. Rialdi, and A. Pesce. 1966. Biophysical properties of N4 coliphage. *Biochim. Biophys. Acta* 129:482–490.**
24. **Schito, G. C., G. Rialdi, and A. Pesce. 1966. The physical properties of the deoxyribonucleic acid of N4 coliphage. *Biochim. Biophys. Acta* 129:491–501.**
25. **Watson, J. D. 1972. Origin of concatemeric T7 DNA. *Nature* (London) *New Biol.* 239:197–201.**
26. **Zehring, W. A., S. C. Falco, C. Malone, and L. B. Rothman-Denes. 1983. Bacteriophage N4-induced transcribing activities in *E. coli*. III. A third cistron required for N4 RNA polymerase II activity. *Virology* 126:678–687.**
27. **Zivin, R., C. Malone, and L. B. Rothman-Denes. 1980. Physical map of coliphage N4 DNA. *Virology* 104:205–218.**
28. **Zivin, R., W. Zehring, and L. B. Rothman-Denes. 1981. Transcriptional map of bacteriophage N4. Location and polarity of N4 RNAs. *J. Mol. Biol.* 152:335–356.**

Chapter 14

Repressorlike Effect of a Membrane Protein on the Initiation of *Bacillus subtilis* DNA Replication

J. Laffan and W. Firshein

Membranes have been associated with DNA replication both in theory (7, 12) and in observation (1, 4, 8–10, 12, 16, 17) for many years. Their exact role in vivo has yet to be determined. However, there is mounting evidence that membranes do play some part in DNA replication. Many replication models use the membrane as an anchor site for the replicative apparatus (12, 16). DNA-membrane complexes have been isolated from both gram-positive and gram-negative bacteria (1, 4, 8). These complexes are often enriched with DNA from the origin of replication (13). In some cases, entire functional replicative complexes can be isolated from such complexes (1, 4, 8, 9). The questions that we address here are (i) what attaches the DNA (and specifically the origin region) to the membrane and (ii) how this attachment is important in DNA replication.

DNA BINDING TO MEMBRANE PROTEINS

To identify membrane anchor protein(s) we used a survey technique modified from Bowen et al. (2). Membrane proteins were extracted and purified from *Bacillus subtilis* as described by Laffan and Firshein (10) and subjected to sodium dodecyl sulfate-polyacrylamide gel electrophoresis. The separated proteins were allowed to renature and capillary transferred to nitrocellulose. The nitrocellulose was cut into thin strips and incubated with various *Bacillus* origin and nonorigin DNA double-stranded fragments which had been radioactively labeled (Fig. 1) (for method, see reference 10). Under a low-stringency wash (50 mM NaCl) many DNA-binding proteins are seen; nevertheless, they are a small subset of the hundred or so proteins that transferred. There seem to be no consistent qualitative or quantitative differences between the different probes with the low-stringency wash. However, when similar strips were incubated with the same probes but washed with a higher salt concentration (200 mM NaCl), all the probes were dissociated except for

J. Laffan and W. Firshein • Molecular Biology and Biochemistry Department, Wesleyan University, Middletown, Connecticut 06457.

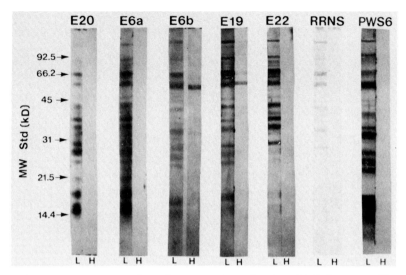

FIGURE 1. Interaction of membrane proteins and various DNA fragments. The autoradiogram shows binding of double-stranded probes, made from various origin-region (left five) and non-origin-region (right two) fragments, to Western blotted *Bacillus* membrane proteins (10). L, 50 mM NaCl wash; H, 200 mM NaCl wash. MW Std (kD), molecular size standards (kilodaltons).

two adjacent origin-region probes, which remained bound to a 63- to 64-kilodalton (kDa) protein and possibly a 67-kDa protein for one of the probes. Flanking origin-region fragments as well as non-origin-region fragments did not show this binding. Thus, these membrane proteins show specificity for a confined area in the origin region of the *Bacillus* chromosome. Similar tests with single-stranded probes show many nonspecific DNA-binding proteins (including the 64-kDa protein), but the specificity appears only with the double-stranded probes. It is possible that this protein represents an anchor that binds the chromosomal origin of replication in *B. subtilis* to the membrane. It should be noted, however, that other membrane protein anchors may also exist, e.g., the *dnaB* gene product (16) or Sargent's membrane particle (13).

ANTIBODIES TO *BACILLUS* MEMBRANE PROTEINS

Polyclonal antibodies to several *B. subtilis* membrane proteins were obtained from P. C. Tai (Boston Biomedical Research Institute; references 5, 6, and 11). Figure 2 shows the localization of the proteins recognized by the various antibodies on strips prepared similarly to the strips used in Fig. 1. We were particularly interested in an antibody that reacted with a 64-kDa protein (64-kDa antibody) as its molecular size was essentially equivalent to that of the specific DNA-binding protein described above. After preincubating the strips with the 64-kDa antibody versus a control antibody or none, it was observed that binding of one of the probes described above to the 64-kDa protein was prevented (Fig. 3). The control antibody had no such effect. This indicated that the 64-kDa antibody recognizes the 64-kDa origin-binding membrane protein.

To test the specific role of the 64-kDa membrane protein in replication, the 64-kDa antibody was added to an in vitro *B. subtilis* replication system consisting of a membrane-associated DNA complex (1, 9). By preincubating the DNA-membrane complex with the antibody, it was hoped that the function of the membrane protein might be

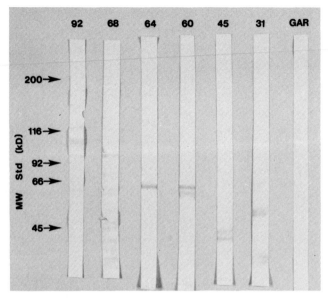

FIGURE 2. Binding of various polyclonal antibodies to *Bacillus* membrane proteins. Each interaction was visualized by peroxidase-antiperoxidase staining. The strips were prepared as described for Fig. 1. The number above each strip refers to the molecular size of the *Bacillus* membrane protein used to induce each antibody. Goat anti-rabbit (GAR) antibody, a generic antibody, was chosen because it did not bind to any of the proteins of the complex.

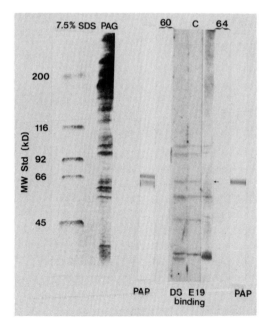

impaired, and if the 64-kDa protein was involved in DNA replication, the antibody-treated synthetic system should be affected, perhaps adversely. Pretreatment of the complex with the antibody did affect replication, but in contrast to our initial supposition, replication was enhanced significantly over control levels (Fig. 4). Nearly all of the increase in synthesis was in the form of initiation activity (as determined from total

FIGURE 3. Antibody blocking of DNA binding. Leftmost, Coomassie blue-stained molecular size markers and membrane proteins on a portion of a polyacrylamide gel, the rest of which was transferred to nitrocellulose. Strips cut from the transfer were incubated with antibody (60 or 64) or just antibody-binding buffer (C). The antibodies on the strips were then either visualized by peroxidase-antiperoxidase staining (PAP) or allowed to bind to the double-stranded origin probe (DS E19) and washed under intermediate stringency (0.1 M NaCl). The autoradiogram (center three strips) shows inhibition of binding to the 64-kDa protein only when the strip is preincubated with the 64-kDa antibody (arrow).

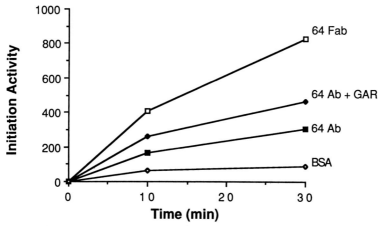

FIGURE 4. Effects of pretreatment with antibody on initiation activity of a *B. subtilis* DNA-membrane complex. Initiation activity is defined as the difference between total incorporation and incorporation in the presence of an initiation inhibitor (Streptovaricin, Upjohn Co.) (9). BSA, Bovine serum albumin; GAR, goat anti-rabbit, used as a secondary antibody to the 64-kDa antibody; Fab, the enzymatically cleaved specific recognition portion of the 64-kDa antibody.

synthetic activity minus activity in the presence of an initiation inhibitor [9]). The enhancement of initiation could be increased further by addition of a secondary antibody to the 64-kDa antibody or increased even more by using Fab fragments of the 64-kDa antibody, which, being smaller, might interact with the protein in the complex to a greater extent. Other antibodies to different membrane proteins showed little or no enhancement of synthesis over generic antibody controls and controls without antibody (Fig. 5).

MULTIPLE ROUNDS OF INITIATION

A possible mechanism of action of the 64-kDa protein could involve a repressive effect on initiation. By inhibiting this repres-

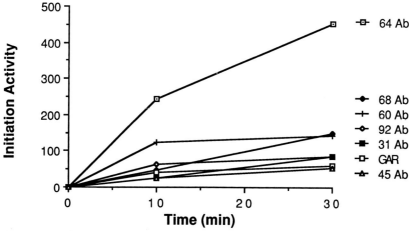

FIGURE 5. Effect of antibodies (Ab) against other membrane proteins on initiation activity of the DNA-membrane complex. The labels refer to the antibodies shown in Fig. 2.

sorlike action (via the antibody), an increase in initiation activity should occur because each initiation site would be undergoing more than one initiation event. To test this hypothesis, in vitro replication was performed in the presence of a heavy-density DNA precursor (bromodeoxyuridine triphosphate). This addition had previously been shown to produce a significant density shift of the synthesized DNA (1, 9). It was hoped that the addition of the 64-kDa antibody would allow some of the newly synthesized hybrid-density DNA to be converted further to heavy/heavy-density DNA. Such a result would suggest strongly that more than one round of initiation had occurred. Figure 6 shows the profile of DNA synthesized with and without the 64-kDa antibody in a CsCl density gradient. It can be seen that the 64-kDa antibody does indeed allow heavy/heavy DNA to be synthesized. Thus, it appears that the 64-kDa protein is involved in repression of initiation. Further support for this model comes from the observation that the origin fragments that bind specifically to the 64-kDa protein contain at least one initiation site in *B. subtilis* (9, 14). Nevertheless, it should be pointed out that if the 64-kDa protein functions as a repressor, it is a "leaky" repressor because initiation occurs in the in vitro system without the specific antibody block, but not as well. Although the mechanism of action of the 64-kDa protein is unknown, this protein binds a double-stranded initiation region DNA so strongly that it may inhibit the separation of the DNA strands required for replication to begin. Another possibility is that the conformation of the DNA is altered (e.g., bending or unbending) when the pro-

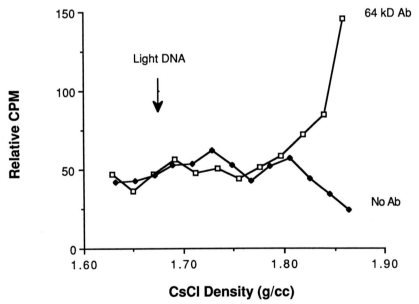

FIGURE 6. Distribution of newly synthesized DNA in a neutral CsCl density gradient. Arrow shows position of light parental DNA. DNA was synthesized in the presence of a heavy-density precursor (bromodeoxyuridine triphosphate) and with and without preincubation with the 64-kDa antibody. Both treatments resulted in similar amounts of hybrid (light/heavy) DNA, but only the system treated with 64-kDa antibody shows significant amounts of reinitiated (heavy/heavy) DNA. The total shift in DNA from light- to heavy/heavy-density DNA (~170 mg) is typical of bromouracil-substituted DNA.

tein is removed to facilitate recognition of this region by other essential initiation proteins (3).

REPLICON AS A REGULON

A case for interconnecting regulatory networks (regulons) in *B. subtilis* has been made for macromolecular synthesis (15). Something similar may occur with the 64-kDa protein. The antibodies were originally made against membrane proteins thought to be involved in protein secretion (5, 6, 11). The 64-kDa antibody in particular was important because it was found to be covered with ribosomes. The protein can be removed from the cytoplasmic (inner) surface of the membrane, but only after ribosomes have been released from the preparation. At this time, we do not know how replication and secretion are interrelated through this molecule, but there are many interesting possibilities.

ACKNOWLEDGMENTS. This work was supported by National Science Foundation grant DCB8711088.

We thank Phang C. Tai for providing the antibodies. We also thank S. Dahl and E. Laffan for their assistance.

LITERATURE CITED

1. Benjamin, P., and W. Firshein. 1983. Initiation of DNA replication *in vitro* by a DNA-membrane complex extracted from *Bacillus subtilis*. *Proc. Natl. Acad. Sci. USA* **80**:6214–6218.
2. Bowen, B., J. Steinberg, U. K. Laemmli, and H. Weintraub. 1980. The detection of DNA binding proteins by protein blotting. *Nucleic Acids Res.* **8**:1–20.
3. Echols, H. 1986. Multiple DNA-protein interactions governing high precision DNA transactions. *Science* **233**:1050–1056.
4. Firshein, W. 1972. The DNA/membrane fraction of pneumococcus contains a DNA replication complex. *J. Mol. Biol.* **70**:383–397.
5. Horiuchi, S., D. Marty-Mazars, P. C. Tai, and B. Davis. 1983. Localization and quantitation of proteins characteristic of the complexed membrane of *Bacillus subtilis*. *J. Bacteriol.* **154**:1215–1221.
6. Horiuchi, S., P. C. Tai, and B. Davis. 1983. A 64-kilodalton membrane protein of *Bacillus subtilis* covered by secreting ribosomes. *Proc. Natl. Acad. Sci. USA* **80**:3287–3291.
7. Jacob, F., S. Brenner, and F. Cuzin. 1963. On the regulation of DNA replication. *Cold Spring Harbor Symp. Quant. Biol.* **28**:289–348.
8. Kornacki, J. A., and W. Firshein. 1986. Replication of plasmid RK2 in vitro by a DNA-membrane complex: evidence for initiation of replication and its coupling to transcription and translation. *J. Bacteriol.* **167**:319–326.
9. Laffan, J., and W. Firshein. 1987. DNA replication by a DNA-membrane complex extracted from *Bacillus subtilis*: site of initiation in vitro and initiation potential of subcomplexes. *J. Bacteriol.* **169**:2819–2827.
10. Laffan, J., and W. Firshein. 1987. Membrane protein binding to the origin region of *Bacillus subtilis*. *J. Bacteriol.* **169**:4135–4140.
11. Marty-Mazars, D., S. Horiuchi, P. C. Tai, and B. Davis. 1983. Proteins of ribosome-bearing and free-membrane domains in *Bacillus subtilis*. *J. Bacteriol.* **154**:1381–1388.
12. Mosig, G., and P. Macdonald. 1986. New membrane-associated DNA replication protein, the gene 69 product of bacteriophage T4, shares a patch of homology with the *Escherichia coli dnaA* protein. *J. Mol. Biol.* **189**:243–248.
13. Sargent, M.G., and M. F. Bennett. 1986. Identification of a specific membrane-particle-associated DNA sequence in *Bacillus subtilis*. *J. Bacteriol.* **166**:38–43.
14. Seiki, M., N. Ogasawara, and H. Yoshikawa. 1981. Structure and function of the region of the replicative origin of the *Bacillus subtilis* chromosome. *Mol. Gen. Genet.* **183**:220–226.
15. Seror, S. J., F. Vannier, A. Levine, and G. Henckes. 1986. Stringent control of initiation of chromosomal replication in *Bacillus subtilis*. *Nature (London)* **321**:709–710.
16. Winston, S., and N. Sueoka. 1980. DNA-membrane association is necessary for initiation of chromosomal and plasmid replication in *Bacillus subtilis*. *Proc. Natl. Acad. Sci. USA* **77**:2834–2838.
17. Winston, S., and N. Sueoka. 1982. DNA replication in *Bacillus subtilis*, p. 35–69. *In* D. Dubnau (ed.), *The Molecular Biology of the Bacilli*, vol. 1. *B. subtilis*. Academic Press, Inc., New York.

Chapter 15

Control of Plasmid Mini-F *oriS* and Chromosomal *oriC* Replication in *Escherichia coli*

Bruce C. Kline, Malcolm S. Shields, Ross A. Aleff, and Jeffrey E. Tam

Here we compare control of mini-F and chromosome replication as a function of cell physiology, and we also relate our results to a replication control model developed for mini-F (20). Citations of many original articles about the replication and replicative control properties of F and the chromosomal *oriC* origin are omitted for the sake of brevity. They have been summarized in recent review articles (7, 8, 12, 22, 24). The present studies are the result of long-standing efforts to determine whether replication control in plasmid F can serve as a paradigm for this process in the *Escherichia coli* chromosome. The initial impetus for this goal was the observation that the numbers of F plasmids and chromosomes per cell are near parity (4). Parity indicates that each replicon is controlled by mechanisms of equal rigor. The question is whether the control mechanisms are of equivalent design.

Early study of chromosome replication

Bruce C. Kline, Malcolm S. Shields, Ross A. Aleff, and Jeffrey E. Tam • Department of Biochemistry and Molecular Biology, Mayo Clinic/Foundation, Rochester, Minnesota 55905.

led Donachie to propose that initiation was triggered when the cell mass per chromosomal origin (*oriC*) reached a critical value, a value referred to as the initiation mass (3). Furthermore, he calculated that for bacteria with growth rates between one and three doublings per hour, the mass per origin value should remain constant.

In the mid 1970s, Pritchard and colleagues observed that the ratio of F' *lac* to *oriC* genes decreased as growth rates increased in the range of the above values (16). Assuming that the initiation mass for *oriC* remained constant, they concluded that the initiation mass of F' *lac* decreased. This indicated that F plasmids and chromosomes initiated at independent times in the cell cycle and implied that control of the plasmid and chromosome replication was achieved by different mechanisms. Their conclusions severely tarnished the image of F as a paradigm for chromosomes. However, subsequent studies have provided a partial and simple explanation for this situation.

F is a complex of more than one replicon (7, 8). The simplest replicon with full

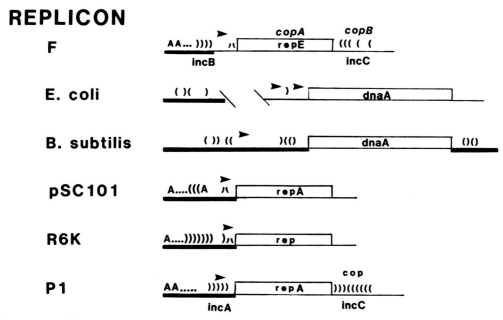

FIGURE 1. Replicons with similar molecular-genetic organization. The heavy line in each replicon represents the origin (*ori*). Symbols:) or (, repeat sequence that acts as a binding site for Rep proteins; ⋀, origin repeat sequences that act as the operator locus for transcriptional control of *rep* gene expression (in P1, origin repeats serve a dual function, acting as both origin and operator); arrowheads, *rep* gene promoters; ....., A-T-rich sequences. The letter A represents DnaA protein-binding sites. The *E. coli* and *B. subtilis* origins both use the DnaA protein as initiator, and binding sites are represented by))) marks. The *inc* gene symbolizes the fact that cloned extra copies of the direct repeat sequences interfere with timely replication of plasmids and cause loss of plasmids from the host cell—the incompatibility response. References: F, 14; *E. coli* and *B. subtilis*, 15; pSC101, 9 and 21; R6K, 11; and P1, 1.

replication control consists of *oriS*, *repE*, and *copB* sequences (Fig. 1); it exhibits a constant initiation mass at a variety of balanced growth rates (19). Hence, the variable initiation mass of F' *lac* results from its composition as a mosaic of replicons, and this property is retained in the pML31 mini-F (18). Since the molecular-genetic organization of the simplest mini-F's replication genes is similar to that of *oriC* and several other replicons (Fig. 1), there is substantial reason to view this mini-F as a putative paradigm for these replicons. Furthermore, because these replicons belong to both gram-positive and gram-negative organisms, the prospect exists that we are considering a major theme for control in procaryotes.

This theme has been articulated (20, 25) in most detail for the mini-F plasmid structured as depicted in Fig. 1 and is presented schematically in Fig. 2. Briefly, the *repE* gene product, RepE, is a protein of 29,000 M_r. It functions as an autorepressor of its own synthesis (E_r) and as an essential protein (initiator [E_i]) for replication. It functions as an initiator by binding to repeat DNA sequences of 22 base pairs (bp) (iterons) in *oriS*. The operator of *repE* is an inverted repeat of 8 bp formed from the first 8 bp of the consensus *oriS* iteron; RepE also binds to these repeats. Five copies of the *oriS* iteron are also found adjacent to the 3' terminus of *repE* in a locus called *copB*. *copB* is presumed to regulate the availability of E_i by competing with *oriS*. The primary mechanism for maintaining RepE levels is autoregulation. Two forms of RepE are hypothesized in order to prevent binding of RepE to *oriS* or

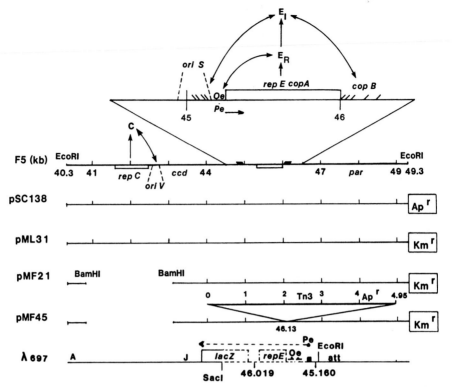

FIGURE 2. Mini-F and derived replicons. The *Eco*RI fragment of native F (F5) is 9 kilobases (kb) in length and has the F coordinates of 40.3 to 49.3 kb. Although the f5 fragment contains two origins, *oriV* and *oriS*, only *oriS* along with the RepE protein (E) can form a plasmid, as described in the text. The origins of all these plasmids are given in the reviews by Lane (8) and Kline (7). λ697 is a recombinant bacteriophage made from the *Sac*I-digested left arm of λTL25 (10) viral DNA and the *Eco*RI-digested right arm of λTL25. The "stuffer" region to make λ697 complete (the dotted lines) was the *Eco*RI-*Sac*I fragment contained in a pUC8 *repE*+ clone (45.16 to 46.019 kb) that had *lacZ* transcriptionally fused to *repE*+. λ697 was made in our laboratory by R. A. Aleff and B. C. Kline.

copB from depleting the free concentration of RepE, thereby triggering replacement synthesis. The latter scenario could result in runaway plasmid replication. An essential feature of the two-forms hypothesis is that E_i is derived irreversibly from E_r. Without this feature, the condition of replacement synthesis would prevail when E_i bound to *oriS* or *copB*. A further supposition in the model is that RepE and iterons interact in agreement with the rules of equilibrium thermodynamics.

This last supposition implies that cells of various sizes should have the same equilibrium concentration of plasmid molecules. This prediction was examined and found to be true for cells bearing pMF21 (Fig. 2) at growth rates between one and three doublings per hour (19). In these experiments the faster growing cells were about 3.0 times larger than the slowest growing cells (M. S. Shields, unpublished observation). Furthermore, in these experiments the *oriC* concentration was also found to be constant over the same range of doubling rates, as predicted by Donachie (3). These findings represented the first time any plasmid was seen to mimic chromosomal control at the physiological level. Earlier, Shields et al. (18) had found that pML31 (Fig. 2), the parent of pMF21, had a variable initiation mass just like F' *lac*.

To expand our comparison of pMF21 and chromosomes and the general applicability of our model for control, we have examined plasmid and *oriC* concentrations in cells whose sizes have been varied in a cyclical way as a function of a standard growth curve. For these experiments we cultured cells in broth (pH 7.0) composed of 10 g of tryptone, 5 g of yeast extract, and 5 g of NaCl at 37°C with aeration. For the sake of illustration, the slightly idealized behavior (6) of cells in a comparable medium is presented in Fig. 3. Relative plasmid and *oriC* concentrations were measured by a rapid and sensitive technique developed recently in our laboratory (18).

The highlights of our results are presented in Table 1; detailed results will be presented elsewhere. The most informative aspects of our observation were the responses of stationary-phase cells upon dilution into fresh broth (the stationary- and lag-phase data in Table 1). These data indicated that the *oriC* concentration in F$^-$ cells increased threefold during the period when the cell mass had increased only 50%. Conversely, the pMF21 concentration dropped in proportion to the mass increase for ap-

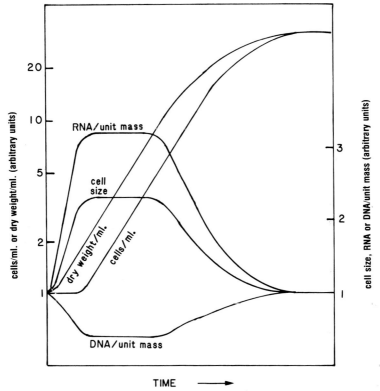

FIGURE 3. A slightly idealized growth curve of bacteria in rich broth. The presentation is replotted from the data of Herbert (6). Initial values for all variables are taken at 1.0. Values on the left axis are exponential, and on the right axis they are arithmetical. Molecules per unit of mass is essentially a concentration term since mass is related to volume by a constant factor.

TABLE 1
Concentration of *oriC* or Mini-F as a Function of Culture Growth Phase in Rich Medium[a]

Growth phase	Gene or plasmid		
	oriC	pMF21	pML31
Stationary	5,000	31,000	15,600
Lag[b]	14,600	7,850	12,000
Exponential	18,000	12,000	15,000
Early stationary	10,700	18,487	16,000

[a] *oriC* and mini-F were detected by specific, ^{32}P-labeled nucleic acid probes by the method of Shields et al. (18). The amount of hybridized probe was detected by autoradiography, and the values reported represent arbitrary densitometry units. For each column, cells from all four phases of growth were processed simultaneously on a single sheet of nitrocellulose and, therefore, exposed to identical conditions of hybridization, etc. Each value is the average of eight determinations. The standard error is ±10%.

[b] Measurements in this line represent cells that, after dilution into fresh broth, had increased their mass 50% for the F⁻ culture (*oriC* concentration), twofold for the pML31 culture, or threefold for the pMF21 culture. Culturing conditions are described in the text.

proximately two mass doublings. In contrast to these two replicons, mini-F pML31 changed its concentration little, if at all, under comparable conditions. Although not presented, compensating changes in replicon concentration were seen in the stationary phase for *oriC* and pMF21 but not for pML31. Both pMF21 and *oriC* are similar in cycling their concentrations about the same amount as cells change sizes, but the timing of their increases is 180° out of phase.

Several points can be made from the findings in Table 1 and our growth curves. First, these results indicate that our model for control is not adequate to cover all aspects of cell physiology. While Womble and Rownd (25) have found by computer simulation that our model is conceptually sound and self-correcting for cells in the steady state of growth, this fitness does not preclude deviations from model behavior. Our results show that, in non-steady states of growth, other considerations impinge upon regulation.

One of these considerations is a change in the turnover rate for E_i in the stationary phase leading to its higher concentration. At least for plasmid F, another consideration is that a cellular component is rate limiting in the exponential phase of growth. This could be the level of *E. coli* heat shock RNA polymerase, a minor polymerase that uses a novel sigma factor (sigma-32), a 32,000-M_r protein. The concentration of this polymerase is known to be elevated by the stringent growth response (5), a condition that should occur in our host cells in the stationary phase. Furthermore, sigma-32 RNA polymerase is known to be required for *repE* transcription (23).

Whether this latter speculation is correct, we have some evidence (Table 2) that more RepE is synthesized in the stationary phase of growth as compared with the exponential phase. The results in Table 2 show that lysogenic bacteria containing recombinant lambda bacteriophage (λ697 [Fig. 2]) with a *repE⁺-lacZ⁺* transcriptional fusion have higher levels of β-galactosidase in the stationary as opposed to the exponential phase. Since *repE⁺* is upstream of the *lacZ* gene, the data indicate that the RepE concentration should be comparably higher as well.

While the foregoing explanation is appealing, it may not be acceptable. Several mutations (*copA*) have been mapped in the *repE⁺* gene of pMF45. These mutations sig-

TABLE 2
Levels of β-Galactosidase in λ697 Lysogens as a Function of Culture Growth Phase

Expt	Miller units[a] of β-galactosidase sp act (A_{550}[b]) at growth phase:		Ratio of stationary- to exponential-phase β-galactosidase
	Stationary	Exponential	
I	58.6 (5.2)	7.2 (0.7)	8.1
II	29.7 (1.8)	18.2 (0.6)	1.6
III	39.3 (6.4)	10.2 (0.7)	3.9

[a] Bacteria were cultured as described for Table 1. β-Galactosidase specific activity was determined as described by Miller (13). One Miller unit of activity represents one tetrameric molecule made from LacZ monomer. This provides an estimate of the RepE protein levels as well.

[b] A_{550}, Absorbance at 550 nm of the culture at the time of assay for β-galactosidase.

nificantly reduce the autorepressor function of RepE protein, thereby increasing the concentration of mutant plasmids up to 28-fold in the exponential phase (17). In combination with *copB*, *copA-copB* double mutations in pMF45 can cause at least a 50-fold increase in plasmid amounts (R. W. Seelke, J. D. Trawick, and B. C. Kline, unpublished data). These observations are clearly at odds with the notion that a chromosomal gene product limits the extent of replication of mini-F by a factor of about fourfold, the magnitude of variation between stationary- and lag-phase cells.

Understanding the control problems in pMF21-type mini-F seems a small task compared with the problems we face in understanding control of pML31 replication. Both pML31 and pSC138 show a variable initiation mass as a function of changes in the rate of balanced cell growth (18, 19). Replication of these mini-F plasmids initiated at *oriV* requires the gene products from the *repC* (*pif*) and *ccd* operons (Fig. 2) as well as *repE*. Each of these operons and *repE* are autoregulated, and one might have expected to see the same physiological responses as were observed for pMF21. However, we do not yet know even the simplest of facts about pML31 control, such as which gene product is rate limiting for initiation. Hence, it is not possible to interpret the behavior of pML31 at present. Rather, we report its behavior to acquire a file of information about it and to serve as a counterpoint for pMF21 and *oriC* behavior.

The basis for fluctuation in *oriC* concentration is unknown. It is as if stationary-phase cells accrued or stockpiled initiator and subsequent dilution into fresh broth induced the initiation process. Presently, we do not know the exact kinetics of induction. However, since the concentration of *oriC* increased at least threefold during a period of time when the cell mass increased only 50%, the implications are clear: (i) initiation is rapid, and (ii) initiation occurred twice in the time frame needed to increase the cell mass 50%. These are unprecedented observations in the biology of *oriC* (12, 22).

Is there a physiological purpose for *oriC* amplification? Such amplification would also increase the concentration of genes proximal to *oriC*, such as the genes for protein synthesis. This amplification meets the sudden demand for protein synthesis when cells are diluted into fresh broth. It is interesting that the profile for total RNA concentration in a standard growth curve seen in Fig. 3 is strikingly similar in shape and amplitude to the profile (not presented) for *oriC* concentration as the function of a standard growth curve. There is precedence for the interpretation that *oriC* concentration could amplify if initiator were stockpiled in the stationary phase since Atlung et al. (2) induced overproduction of the DnaA protein (*oriC* initiator) and observed *oriC* amplification. Interestingly, they found that these initiations were "abortive" in that the replication forks paused after traveling partway towards the terminus of replication. This point has not been examined yet in our standard growth curve experiments, but logic suggests that this occurs here as well.

In closing, we note that F *oriS* and chromosomal *oriC* replicons share major features of structure and function with other replicons of *E. coli* and *Bacillus subtilis*. At least control of *oriS* and *oriC* appears to be similar for cells at a variety of steady-state growth rates for which a constant initiation mass is the rule. In cells cycling in and out of the steady-state growth rate, the concentration of each replicon also cycles, suggesting a variation in the initiation mass. Furthermore, the *oriS* concentration amplifies when *oriC* decreases and vice versa. By contrast, pML31, which has two origins, has a variable initiation mass at different steady-state growth rates and a constant initiation mass when its host cells cycle in and out of a steady state of growth. Cycling represents a new observation for the mini-F and *oriC* and at present remains unexplained. Our initial question in the introduction was whether the control mechanisms (for F *oriS* and chromosomal *oriC*) are of equivalent design. The results presented here show that control at the physiological level is similar but not equivalent. Given the differences in architecture (Fig. 1), undoubted origin transcriptional differences (8, 12, 22), and the physiological requirements to be satisfied by *oriC* and F *oriS* duplication, the similarities in control are remarkable.

ACKNOWLEDGMENTS. This work was supported by Public Health Service grant GM 25604 from the National Institutes of Health, General Medical Section, and by the Mayo Foundation.

LITERATURE CITED

1. **Abeles, A. L., K. M. Snyder, and D. K. Chattoraj.** 1984. P1 plasmid replication: replicon structure. *J. Mol. Biol.* 173:307–324.
2. **Atlung, T., K. V. Rasmussen, E. Clausen, and F. G. Hansen.** 1985. Role of dnaA protein in control of DNA replication, p. 282–297. *In* M. Schaechter, F. C. Neidhardt, J. L. Ingraham, and N. O. Kjeldgaard (ed.), *The Molecular Biology of Bacterial Growth*. Jones and Bartlett Publishers, Inc., Boston.
3. **Donachie, W. D.** 1968. Relationship between cell size and time of initiation of DNA replication. *Nature* (London) 219:1077–1079.
4. **Frame, R., and J. O. Bishop.** 1971. The number of sex-factors per chromosome in *Escherichia coli*. *Biochem. J.* 121:93–103.
5. **Grossman, A. D., W. E. Taylor, Z. F. Burton, R. R. Burgess, and C. A. Gross.** 1985. Stringent response in *Escherichia coli* induces expression of heat shock proteins. *J. Mol. Biol.* 186:357–365.
6. **Herbert, D.** 1961. The chemical composition of micro-organisms as a function of their environment. *Symp. Soc. Gen. Microbiol.* 11:391–416.
7. **Kline, B. C.** 1985. A review of mini-F plasmid maintenance. *Plasmid* 14:1–16.
8. **Lane, H. E. D.** 1981. Replication and incompatibility of F and plasmids in the incF1 group. *Plasmid* 5:100–126.
9. **Linder, P., G. Churchward, X. Guixian, Y.-Y. Yu, and L. Caro.** 1985. An essential replication gene, *repA*, of plasmid pSC101 is autoregulated. *J. Mol. Biol.* 181:383–393.
10. **Linn, T., and G. Ralling.** 1985. A versatile multiple- and single-copy vector system for the *in vitro* construction of transcriptional fusion to *lacZ*. *Plasmid* 14:134–142.
11. **McEachern, M. J., M. Filutowicz, S. Yang, A. Greener, P. Mukhopadhyay, and D. R. Helinski.** 1986. Elements involved in the copy number regulation of the antibiotic resistance plasmid R6K, p. 195–208. *In* S. Levy and R. Novick (ed.), *Banbury Conference on Evolution and Environmental Spread of Antibiotic Resistance Genes*. Cold Spring Harbor Laboratory, Cold Spring Harbor, N.Y.
12. **Messer, W.** 1987. Initiation of DNA replication in *Escherichia coli*. *J. Bacteriol.* 169:3395–3399.
13. **Miller, J. H.** 1972. *Experiments in Molecular Genetics*, p. 13–23. Cold Spring Harbor Laboratory, Cold Spring Harbor, N.Y.
14. **Murotsu, T., H. Tsutsui, and K. Matsubara.** 1984. Identification of the minimal essential region for the replication origin of miniF plasmid. *Mol. Gen. Genet.* 196:373–378.
15. **Ogasawara, N., S. Moriya, K. von Meyenburg, F. G. Hansen, and H. Yoshikawa.** 1985. Conservation of genes and their organization in the chromosomal replication origin region of *Bacillus subtilis* and *Escherichia coli*. *EMBO J.* 4:3345–3350.
16. **Pritchard, R. H., M. G. Chandler, and J. Collins.** 1975. Independence of F replication and chromosome replication in *Escherichia coli*. *Mol. Gen. Genet.* 138:143–155.
17. **Seelke, R. W., B. C. Kline, J. D. Trawick, and G. D. Ritts.** 1982. Genetic studies of F plasmid maintenance genes involved in copy number control, incompatibility, and partitioning. *Plasmid* 7:163–179.
18. **Shields, M. S., B. C. Kline, and J. E. Tam.** 1986. A rapid method for the quantitative measurement of gene dosage: mini-F plasmid concentration as a function of cell growth rate. *J. Microbiol. Methods* 6:33–46.
19. **Shields, M. S., B. C. Kline, and J. E. Tam.** 1987. Similarities in control of mini-F plasmid and chromosomal replication in *Escherichia coli*. *J. Bacteriol.* 169:3375–3378.
20. **Trawick, J. D., and B. C. Kline.** 1985. A two-stage molecular model for control of mini-F replication. *Plasmid* 13:59–69.
21. **Vocke, C., and D. Bastia.** 1983. Primary structure of the essential replicon of the plasmid pSC101. *Proc. Natl. Acad. Sci. USA* 80:6557–6561.
22. **von Meyenberg, K., and F. G. Hansen.** 1987. Regulation of chromosome replication, p. 1555–1577. *In* F. C. Neidhardt, J. L. Ingraham, K. B. Low, B. Magasanik, M. Schaechter, and E. Umbarger (ed.), *Escherichia coli and Salmonella typhimurium: Cellular and Molecular Biology*, vol. 2. American Society for Microbiology, Washington, D.C.
23. **Wada, C., Y. Akiyama, K. Ito, and T. Yura.**

1986. Inhibition of F plasmid replication in *htpR* mutants of *Escherichia coli* deficient in sigma 32 protein. *Mol. Gen. Genet.* **203**:208–213.

24. **Willets, N., and R. Skurray.** 1987. Structure and function of the F factor and mechanism of conjugation, p. 1110–1133. *In* F. C. Neidhardt, J. L. Ingraham, K. B. Low, B. Magasanik, M. Schaechter, and E. Umbarger (ed.), *Escherichia coli and Salmonella typhimurium: Cellular and Molecular Biology*, vol. 2. American Society for Microbiology, Washington, D.C.

25. **Womble, D. D., and R. H. Rownd.** 1987. Regulation of mini-F plasmid DNA replication. A quantitative model for control of plasmid mini-F replication in the bacterial cell division cycle. *J. Mol. Biol.* **195**:99–113.

Chapter 16

Protein-Protein Interactions of the *Escherichia coli* Single-Stranded DNA-Binding Protein

Ralph R. Meyer, Steven M. Ruben,
Suzanne E. VanDenBrink-Webb, Phyllis S. Laine,
Frederick W. Perrino, and Diane C. Rein

The *Escherichia coli* single-stranded DNA-binding protein (SSB) (10) was first isolated in 1972 (40). Initially its function in DNA metabolism was thought to be the melting of double-stranded DNA (33), leading to the name "DNA-unwinding protein" (40, 45) or "helix-destabilizing protein" (1). It has also been known as "DNA-binding protein" (26, 32, 33) or, more recently, as "single-stranded (DNA)-binding protein" (20, 28, 29).

In early in vitro studies of DNA replication with small bacteriophage templates (18, 45) (and later, using *oriC* plasmids [16]), SSB was a standard reaction component.

Ralph R. Meyer, Phyllis S. Laine, and Diane C. Rein • Department of Biological Sciences, University of Cincinnati, Cincinnati, Ohio 45221. *Steven M. Ruben* • Department of Molecular Oncology, Roche Institute of Molecular Biology, Nutley, New Jersey 07110. *Suzanne E. VanDenBrink-Webb* • Department of Rheumatology, University of Connecticut Health Sciences Center, VA Medical Center, South Newington, Connecticut 06111. *Frederick W. Perrino* • Department of Pathology, University of Washington School of Medicine, Seattle, Washington 98195.

However, under certain conditions, the requirement for SSB could be obviated by addition of spermidine or excess *dnaB* protein (2). For lack of a specific inhibitor or a mutation, the in vivo function remained a subject of speculation for many years (34). The discovery of two mutations, *ssb-1* (20, 28) and *ssb-113* (30), confirmed the essential role of SSB for phage and *E. coli* chromosome replication both in vivo and in vitro. Surprisingly, *ssb* strains were found to be UV sensitive (20, 24, 46) and defective in *recA* protein induction (3, 31), phage lambda induction (43), postirradiation DNA degradation (3, 24), Weigle reactivation (46), and mutagenesis (23, 46). Recombination deficiencies in in vivo (20, 21) and in vitro (27) experiments using mutant *ssb* and *recA* proteins have been documented.

The ability of SSB to bind tightly, specifically, and cooperatively to single-stranded DNA is undoubtedly important for functioning in these processes. However, the pleiotropic effects of these binding protein mutations, coupled with the observation that the *ssb-113* mutant is not defective in DNA binding (8), argues for a significant interaction of SSB with other proteins involved in these processes. Some direct protein-protein interactions of SSB have been demonstrated by density gradient centrifugation. These include cosedimentation with DNA polymerase II (33), exonuclease I (34), and protein n (25). Using SSB-affinity chromatography, we have demonstrated an interaction with a 25,000-M_r folded chromosomal protein (F. W. Perrino, R. R. Meyer, A. M. Bobst and D. C. Rein, submitted for publication). Functional interaction of SSB with *recA* protein in vitro is known, although recent studies indicate that this may not be specific (12). Genetic evidence suggests possible interactions between *rep* protein (42) and *rho* protein (13).

EXTRAGENIC SUPPRESSORS OF *ssb-1*

Direct approaches to the study of proteins which interact with SSB have not been especially successful. Because SSB binds strongly to single-stranded DNA, interactions with other proteins may occur only as a nucleoprotein complex. In other cases, such protein-protein interactions may be transient or too weak to withstand the rigors of centrifugation, or they may require other proteins in a higher-order complex. Thus, we have pursued a genetic approach by seeking extragenic suppressors. A deficiency resulting from a mutation in one protein may be suppressed by a compensatory mutation in another protein with which it interacts. We have generated a library of 51 independent, spontaneously arising suppressors of the temperature-sensitive phenotype of *ssb-1* in *E. coli* RM121 (31). Spontaneous revertants were chosen for two reasons: (i) we did not wish to utilize a mutagen that could produce more than one mutation in a single strain; and (ii) the spontaneous reversion frequency is very high for both the *ssb-1* and *ssb-113* mutations. These rates are approximately 1×10^{-5} for *ssb-1* and 2×10^{-4} for *ssb-113* when measured on LB plates with normal (85 mM) NaCl and using strains RM121 (*ssb-1*) and RM139 (*ssb-113*), which are C600 derivatives. The reversion frequency is influenced markedly by strain background and salt. Phage P1 transduction of each of the 51 suppressors was carried out, as we were only interested in extragenic suppressors at this time.

SUPPRESSOR 411 IS AN ALLELE OF *groEL*

The first suppressor studied in detail was designated as *sup-411* (strain RM304). By phage P1 transduction, this suppressor could be cotransduced with *ssb-1* at a very low frequency, indicating it was located within 2 min of the *ssb* gene on the *E. coli* map. This provided a convenient region to search for the identity of the suppressor gene. A genomic library from strain RM304 was prepared by *Eco*RI digestion and cloned into the *Eco*RI site in the chloramphenicol acetyltransferase gene of plasmid pBR325. This library was used to transform strain RM121, selecting for suppression of the temperature-sensitive phenotype of *ssb-1*. Restriction analysis of the resulting clones indicated that there were two unique classes of fragments of 10.0 or 8.3 kilobases (kb). Upon maxicell analysis, all of the clones carrying the 10.0-kb fragment were found to express a prominent protein of ~19,000 M_r which was subsequently shown to be the SSB-1 protein (S. R. Ruben, Ph.D. dissertation, University of Cincinnati, Cincinnati, Ohio, 1988). This was not unexpected, as Chase et al. (9) have shown that overproduction of SSB-1 protein on a multicopy plasmid suppresses the temperature-sensitive phenotype of *ssb-1*. A restriction map of the 8.3-kb fragment was constructed (Fig. 1), and the proteins carried on the fragment were identified by maxicell labeling (37). A

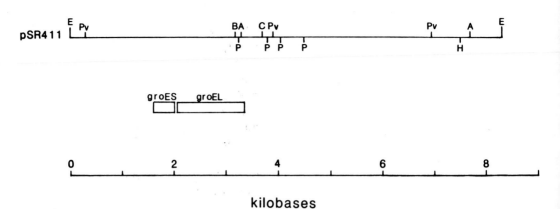

FIGURE 1. Restriction map of the cloned 8.3-kb fragment containing the *ssb-1* suppressor. Restriction endonucleases were: A, *Ava*I; B, *Bam*HI; C, *Cla*I; E, *Eco*RI; H, *Hin*dIII; P, *Pst*I; Pv, *Pvu*I. The positions of the *groES* and *groEL* genes are indicated. This restriction map is consistent with partial maps previously reported (14, 22).

major protein of M_r 65,000 was found along with two proteins of M_r 45,000 and 15,000. The *ssb* region of the *E. coli* map at 92.2 min contains several DNA replication and repair proteins, but none have corresponding molecular weights. However, the *groE* locus at 94.0 min had been reported to code for a *groEL* protein of M_r 65,000 to 68,000, a *groES* protein of M_r 15,000, and an unknown protein of M_r 45,000. Moreover, the region was contained on an 8.1-kb (14) or 8.2-kb (22) *Eco*RI fragment, and the published partial restriction map was consistent with the map for the suppressor 411. The following evidence established that the suppressor was an allele of *groEL*: (i) the Clarke-Carbon plasmid (11) pLC43-46 was known to contain the *groE* locus (35), and the restriction patterns of the 8.3-kb fragment from pLC43-46 and suppressor 411 corresponded; (ii) Southern blot analysis showed that the suppressor 411 8.3-kb fragment hybridized with the 8.3-kb fragment of pLC43-46; (iii) plasmid pSR411 carrying the suppressor 411 was able to complement the *groES30* mutation for phage lambda growth but not the *groEL140* mutation. The suppressor has, therefore, been named *groEL411*.

SPECIFICITY OF THE *groEL411* SUPPRESSOR

Two questions were addressed regarding suppression by *groEL411*. The first concerned the specificity for the *ssb-1* mutation. This mutation lies at residue 55 in the domain of SSB involved in DNA binding and subunit interaction, whereas the *ssb-113* mutation has a substitution at residue 176, which is one residue from the carboxyl terminus (10). Although both mutations produce similar (although not identical) pleiotropic deficiencies in vivo, only the *ssb-1* mutation was suppressible by *groEL411*. Since the suppressor was selected solely by its ability to suppress temperature-sensitive growth, some of the other deficiencies were also examined. The *groEL411* strain was found to suppress lethal filamentation at 42.5°C, but could not correct the defect in lambda phage growth. The UV sensitivity of the *ssb-1* mutation at 32°C was partially suppressed; however, *ssb-1 groEL411* cells would still be considered UV sensitive. Indeed, the failure to suppress UV sensitivity has proved useful in determining the presence or absence of the *ssb-1* mutation in constructed strains.

The second question concerned the specificity for the *groEL* allele. Strains carrying either *ssb-1* or *ssb-113* and either the *groEL411* or *groEL$^+$* plasmid were tested for suppression of temperature sensitivity. Suppression was restricted to the combination of *ssb-1* and *groEL411*.

WHAT IS THE MECHANISM OF SUPPRESSION OF *ssb-1* BY *groEL411*?

One unusual property of the *ssb-1* mutation is that overproduction by multicopy plasmids suppresses all of the deficiencies of the *ssb-1* mutation (9). Could *groEL411* be acting by increasing the levels of SSB-1 protein? To address this, we measured SSB or SSB-1 by a radioimmunoassay under conditions in which the presence of single-stranded DNA would not interfere with the assay (5, 36) (Table 1). Surprisingly, the relative levels of wild-type SSB decreased 30% at 42.5°C. In addition, the level of SSB-1 at 32°C in the mutant strain was reduced by about 50% as compared with the wild type. The presence of the *groEL411* suppressor, either as a single chromosome copy or on a multicopy plasmid, increased SSB-1 levels by 17 to 30% at 32°C and by about 200% at 42.5°C. Specificity for the *groEL411* allele was demonstrated, since overproduction of *groEL$^+$* protein does not significantly affect SSB-1 levels. We next looked at the effect of temperature shift on DNA synthesis (Fig. 2). Within 1 min at 42.5°C, the *ssb-1* strain rapidly ceased DNA synthesis and did not recover. The *ssb$^+$* strain showed a slightly elevated rate of DNA synthesis. In contrast, the *groEL411 ssb-1* strain showed a rapid decrease in DNA synthesis in parallel with the *ssb-1* strain. However, within 5 min, DNA synthesis recovered and increased exponentially.

If the mechanism of suppression of *ssb-1* by *groEL411* is at the level of SSB-1 induction, this raises some interesting questions. First, why are only some of the effects of the *ssb-1* mutation suppressed? Chase et al. (9) reported that all effects were reversed by an *ssb-1* plasmid. Overproduction was 25- to 30-fold, however. It is conceivable that two- to threefold overproduction can overcome the defects in growth and DNA synthesis,

TABLE 1
Relative Levels of SSB and SSB-1 in Various Strains at 32 and 42.5°C by Radioimmunoassay[a]

Strain	Genotype	Relative level of SSB[b] at temp:		Relative enhancement by *groEL411*[c] at temp:	
		32°C	42.5°C	32°C	42.5°C
RM98	*ssb$^+$*	1.00	0.70		
RM121	*ssb-1*	0.49	ND[d]		
RM304	*ssb-1 groEL411*	0.57	1.06	1.17	3.23
RM121(pSR411)	*ssb-1 groEL411*	0.64	1.18	1.30	2.76
RM121(pSR401)	*ssb-1 groEL$^+$*	0.42	ND		

[a] Radioimmunoassays were carried out in 2.0 M NaCl, using ^{125}I-labeled SSB or SSB-1 as previously described (5, 36).
[b] The values have been normalized to that for the wild-type strain RM98 at 32°C, which was 0.48 ng/µg of soluble protein.
[c] This value was calculated by the actual value of SSB-1 divided by the expected value, based upon the behavior of strain RM98.
[d] ND, Not determined, since these conditions were lethal.

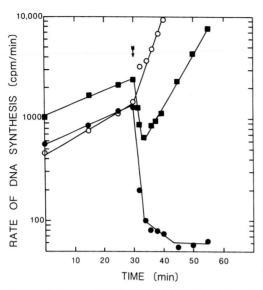

FIGURE 2. Rates of DNA synthesis in ssb^+, ssb-1, and ssb-1 $groEL411$ cells in a temperature-shift experiment. Cells were grown on LB to $A_{595} = 0.10$ at 32°C and then pipetted into flasks prewarmed to 42.5°C at 30 min (arrow). Incubation was continued, and at the times indicated in the figure, 0.5-ml samples were removed and placed in prewarmed test tubes (12 by 75 mm). The cells were pulse-labeled by addition of 0.2 μCi of [³H]thymidine for 5 min at 42.5°C or 3 min at 32°C. Reactions were stopped with 2 ml of cold 10% trichloroacetic acid on ice. The acid-precipitable DNA was collected on glass-fiber filters and counted. At the 30-min time point, labeling was carried out at 32°C. Symbols: ○, RM121 (ssb^+); ●, RM121 (ssb-1); ■, RM304 (ssb-1 $groEL411$).

but cannot suppress defects in phage growth or UV sensitivity. This can be tested by controlling levels of SSB-1 with copy-control plasmids.

If $groEL411$ is suppressing ssb-1 through regulation of SSB-1 levels, how is this accomplished? The effect is specific for the $groEL411$ allele, as $groEL^+$ does not increase SSB-1 levels. Indeed, strain RM121-(pSR401) showed about a 15% decrease in SSB-1 as compared with RM121 (Table 1). The $groEL$ gene has never been shown to regulate other genes. The regulation of ssb has yet to be elucidated. The regulatory region of ssb is quite complex. The $uvrA$ and ssb genes (38) lie close together, with transcription occurring in opposite directions. The $uvrA$ promoter lies a few nucleotides upstream from the PI promoter of ssb (6, 7). An SOS box, presumably controlling $uvrA$, terminates only one nucleotide from PI. There are reports that this promoter can be activated under SOS repair conditions (7, 36); however, this does not lead to increased cellular levels of SSB (36). The ssb gene has two noninducible promoters downstream (7). We have suggested that transcription from these promoters is decreased when the PI promoter is activated (36). At present there is no other information on possible regulation of ssb. An increase in the cellular levels of SSB-1 under the influence of $groEL411$ does not require an involvement at the level of the gene. If ssb-1 and $groEL411$ interact, the $groEL411$ protein may protect SSB-1 from degradation. Enhancement of the half-life of SSB-1 could lead to an accumulation of SSB-1 at the levels observed. This is another easily testable hypothesis.

DOES HEAT SHOCK PLAY A ROLE IN $groEL411$ SUPPRESSION OF ssb-1?

The $groEL$ protein is one of the major heat shock proteins of E. coli (35, 48). The relationship of suppression of ssb-1 by $groEL411$ to heat shock induction needs to be examined. Recovery of DNA synthesis in ssb-1 strains carrying $groEL411$ is rapid, occurring within 5 min (Fig. 2). Yamamori and Yura (48) have shown that induction of $groEL$ is very rapid, reaching a maximum rate of synthesis within 7.5 min after temperature shift. However, the presence of large amounts of $groEL411$ (introduced through a plasmid carrying $groEL411$ into an ssb-1 $groEL411$ strain) does not prevent the transient cessation of DNA synthesis upon temperature shift (R. R. Meyer and P. S. Laine, unpublished observations). Although it is possible that correction of the temperature sensitivity of ssb-1 strains for growth may be due to massive amounts of $groEL411$, the presence of large amounts of *groEL411* does not alter the transient defect in DNA synthesis.

CAN *groEL411* SUBSTITUTE FOR SSB IN DNA REPLICATION?

The *groEL411* protein has never been demonstrated to be a DNA-binding protein, although this question may never have been addressed. It seems very unlikely that *groEL* can simply substitute for SSB in replication, since overproduction of *groE*$^+$ proteins by a multicopy plasmid fails to suppress *ssb-1* temperature-sensitive DNA synthesis (S. R. Ruben, S. E. VanDenBrink-Webb, D. C. Rein, and R. R. Meyer, *J. Biol. Chem.*, in press). Moreover, as indicated above, *groEL411* plasmid-containing *ssb-1* strains still show transient inhibition of DNA synthesis at 42°C. Thus the effect of the *groEL411* allele on DNA replication must lie elsewhere.

DOES *groEL411* AFFECT THE MONOMER-TETRAMER EQUILIBRIUM OF SSB-1 PROTEIN?

In solution, SSB exists as a tetramer (32, 45), which is most likely its functional form (4, 47). Moreover, transfer of SSB from one DNA molecule to another may not involve dissociation to monomers (39). Williams et al. (47) have examined the effects of concentration and temperature on the subunit structure of SSB-1 protein. Whereas SSB tetramers are very stable, SSB-1 tetramers tend to dissociate into monomers at concentrations below 0.5 μM, which are within the physiological range for *E. coli* (47). Consequently, there may be a significant pool of monomers within cells carrying the *ssb-1* mutation. Williams et al. (47) have noted that at elevated temperatures, the SSB-1 tetramers are not particularly unstable, while the monomers seem to be non-functional with regard to tetramer formation or DNA binding. It is likely that these monomers have undergone a temporary, but reversible, change. These observations offer several possible mechanisms for the effect of the *groEL411* suppressor. As discussed above, and as suggested by Chase et al. (9), overproduction of SSB-1 protein could simply shift this equilibrium by increasing levels of SSB-1 monomers. Although we have found an increase in SSB-1 levels in *groEL411* strains (Table 1), we do not know whether its rate of induction is fast enough to account for the effect, nor whether the increased levels are high enough to shift monomers to tetramers. If, as it seems likely, SSB-1 and *groEL411* proteins interact directly, *groEL411* protein could sequester SSB-1 and raise its concentration in the microenvironment to levels favoring the tetrameric form. There are other possibilities that do not invoke effects on the concentration of SSB-1. For example, *groEL411* protein may prevent the temporary inactivation of SSB-1 monomers at restrictive temperatures. Alternatively, *groEL411* protein may promote reassociation of monomers to tetramers at 42.5°C. These models will be tested shortly.

WHAT ROLE(S) DOES *groEL* PLAY IN DNA REPLICATION?

Biochemical studies seeking to reconstitute *E. coli* chromosome replication in vitro from purified proteins by using *oriC* plasmid templates have been very successful (17). This process involves a large number of proteins, including *dnaA*, *dnaB*, *dnaC*, SSB, and HU proteins, gyrase, primase, RNA polymerase, and DNA polymerase III holoenzyme. The first step is the formation of a prepriming initiation complex. A crucial role is played by *dnaA* protein. This is a sequence-specific binding protein for which four sequences exist at *oriC* (16). Between 20 and 40 monomers of *dnaA* protein are bound. Neither *groES* or *groEL* proteins have been shown to be required or involved with initiation of replication, although one mutation, *groES131*, results in temperature-sensitive DNA synthesis (44). It is also interesting that overproduction of wild-type *groE* proteins can suppress certain *dnaA* mutations (14, 22). Thus, the *groE* proteins may serve to facilitate formation of or to stabilize the prepriming complex through interaction with *dnaA* protein and possibly SSB. This is also consistent with the known ability of *groEL* protein to organize and interact with various phage proteins in a macromolecular complex during phage morphogenesis (15). An involvement of *groEL411* protein with SSB-1 during chain elongation must also occur, as the *ssb-1* mutation has a "fast-stop" phenotype that is also suppressed by *groEL411*.

DOES SSB PLAY OTHER ROLE(S) IN PHAGE LAMBDA GROWTH?

The *groE* genes were first identified by the inability of certain *groE* mutants to support phage lambda growth. *groE* mutants are impaired in lambda head morphogenesis, and the role of *groEL* in this process is well known (see the review by Friedman et al. [15]). Mutants show defects in morphogenesis of a variety of other phages as well, including T4, T5, T6, R17, and φ80 (15). Some *groE* mutants are temperature sensitive (19), and some show temperature-sensitive filamentation (19), altered membrane permeability (41), and temperature-sensitive DNA synthesis (44). Clearly these are vital cellular functions involving *groE* proteins. While SSB would play an obvious role in phage lambda DNA replication, it may also be involved in phage morphogenesis. This suggestion comes from the following set of observations. (i) *groEL411* is not a temperature-sensitive allele. (ii) *groEL411* permits SSB-1 to function at 42.5°C, and DNA replication is no longer impaired. Presumably lambda phage replication can proceed (although this needs to be confirmed). (iii) The suppressor strain can support lambda growth at 32°C but not at 42.5°C. Since the DNA-replication defect of *ssb-1* has been corrected, the failure of lambda to grow suggests the intriguing possibility that SSB may have other functions in the lambda life cycle.

LITERATURE CITED

1. **Alberts, B., and R. Sternglanz.** 1977. Recent excitement in the DNA replication problem. *Nature* (London) 269:655–661.
2. **Arai, K.-I., and A. Kornberg.** 1979. A general priming system employing only *dnaB* protein and primase for DNA replication. *Proc. Natl. Acad. Sci. USA* 76:4308–4312.
3. **Baluch, J., J. W. Chase, and R. Sussman.** 1980. Synthesis of *recA* protein and induction of bacteriophage lambda in single-strand deoxyribonucleic acid-binding protein mutants of *Escherichia coli*. *J. Bacteriol.* 144:489–498.
4. **Bandyopadhyay, P. K., and C.-W. Wu.** 1978. Fluorescence and chemical studies on the interaction of *Escherichia coli* DNA-binding protein with single-stranded DNA. *Biochemistry* 17:4078–4085.
5. **Bobst, E. V., A. M. Bobst, F. W. Perrino, R. R. Meyer, and D. C. Rein.** 1985. Variability in the nucleic acid binding site size and the amount of single-stranded DNA-binding protein in *Escherichia coli*. *FEBS Lett.* 181:133–137.
6. **Brandsma, J. A., D. Bosch, C. Backendorf, and P. van de Putte.** 1983. A common regulatory region shared by divergently transcribed genes of the *Escherichia coli* SOS system. *Nature* (London) 305:243–245.
7. **Brandsma, J. A., D. Bosch, M. de Ruyter, and P. van de Putte.** 1985. Analysis of the regulatory region of the *ssb* gene of *Escherichia coli*. *Nucleic Acids Res.* 13:5095–5109.
8. **Chase, J. W., J. J. L'Italien, J. B. Murphy, E. K. Spicer, and K. R. Williams.** 1984. Characterization of the *Escherichia coli* SSB-113 mutant single-stranded DNA-binding protein. Cloning of the gene, DNA and protein sequence analysis, high pressure liquid chromatography peptide mapping, and DNA-binding studies. *J. Biol. Chem.* 259:805–814.
9. **Chase, J. W., J. B. Murphy, R. F. Whittier, E. Lorensen, and J. J. Sninsky.** 1983. Amplification of *ssb-1* mutant single-stranded DNA-binding protein in *Escherichia coli*. *J. Mol. Biol.* 163:193–211.

10. Chase, J. W., and K. R. Williams. 1986. Single-stranded DNA binding proteins required for DNA replication. *Annu. Rev. Biochem.* **55**:103–136.
11. Clarke, L., and J. Carbon. 1976. A colony bank containing synthetic ColE1 hybrid plasmids representative of the entire *E. coli* genome. *Cell* **9**:91–99.
12. Egner, C., E. Azhderian, S. S. Tsang, C. M. Radding, and J. W. Chase. 1987. Effects of various single-stranded-DNA-binding proteins on reactions promoted by RecA protein. *J. Bacteriol.* **169**:3422–3428.
13. Fansler, J. S., I. Tessman, and E. S. Tessman. 1985. Lethality of the double mutations *rep rho* and *rho ssb* in *Escherichia coli. J. Bacteriol.* **161**:609–614.
14. Fayet, O., J.-M. Louarn, and C. Georgopoulos. 1986. Suppression of the *Escherichia coli dnaA46* mutation by amplification of the *groES* and *groEL* genes. *Mol. Gen. Genet.* **202**:435–445.
15. Friedman, D. I., E. R. Olson, C. Georgopoulos, K. Tilly, I. Herskowitz, and F. Banuett. 1984. Interactions of bacteriophage and host macromolecules in the growth of bacteriophage λ. *Microbiol. Rev.* **48**:299–325.
16. Fuller, R. S., J. M. Kaguni, and A. Kornberg. 1981. Enzymatic replication of the origin of the *Escherichia coli* chromosome. *Proc. Natl. Acad. Sci. USA* **78**:7370–7374.
17. Funnell, B. E., T. A. Baker, and A. Kornberg. 1987. In vitro assembly of a prepriming complex at the origin of the *Escherichia coli* chromosome. *J. Biol. Chem.* **262**:10327–10334.
18. Geider, K., and A. Kornberg. 1974. Conversion of the M13 viral single strand to the double-stranded replicative forms by purified proteins. *J. Biol. Chem.* **249**:3999–4005.
19. Georgopoulos, C., and H. Eisen. 1974. Bacterial mutants which block phage assembly. *J. Supramol. Struct.* **2**:349–359.
20. Glassberg, J., R. R. Meyer, and A. Kornberg. 1979. Mutant single-stranded binding protein of *Escherichia coli*: genetic and physiological characterization. *J. Bacteriol.* **140**:14–19.
21. Golub, E. I., and K. B. Low. 1983. Indirect stimulation of genetic recombination. *Proc. Natl. Acad. Sci. USA* **80**:1401–1405.
22. Jenkins, A. J., J. B. March, I. R. Oliver, and M. Masters. 1986. A DNA fragment containing the *groE* genes can suppress mutations in the *Escherichia coli dnaA* gene. *Mol. Gen. Genet.* **202**:446–454.
23. Lieberman, H. B., and E. Witkin. 1981. Variable expression of the *ssb-1* allele in different strains of *Escherichia coli* K-12 and B: differential suppression of its effects on DNA replication, DNA repair, and ultraviolet mutagenesis. *Mol. Gen. Genet.* **183**:348–355.
24. Lieberman, H. B., and E. Witkin. 1983. DNA degradation, UV sensitivity, and SOS-mediated mutagenesis in strains of *Escherichia coli* deficient in single-strand DNA-binding protein: effects of mutations and treatments that alter levels of exonuclease V or *recA* protein. *Mol. Gen. Genet.* **190**:92–100.
25. Low, R. L., J. Shlomai, and A. Kornberg. 1982. Protein n, a primosomal DNA replication protein of *Escherichia coli*. Purification and characterization. *J. Biol. Chem.* **257**:6242–6250.
26. MacKay, V., and S. Linn. 1976. Selective inhibition of the DNase activity of the *recBC* enzyme by the DNA binding protein from *Escherichia coli. J. Biol. Chem.* **251**:3716–3719.
27. McEntee, K., G. M. Weinstock, and I. R. Lehman. 1980. *recA* protein-catalyzed strand assimilation: stimulation by *Escherichia coli* single-stranded DNA-binding protein. *Proc. Natl. Acad. Sci. USA* **77**:857–861.
28. Meyer, R. R., J. Glassberg, and A. Kornberg. 1979. An *Escherichia coli* mutant defective in single-strand binding protein is defective in DNA replication. *Proc. Natl. Acad. Sci. USA* **76**:1702–1705.
29. Meyer, R. R., J. Glassberg, J. V. Scott, and A. Kornberg. 1980. A temperature-sensitive single-stranded DNA-binding protein from *Escherichia coli. J. Biol. Chem.* **255**:2897–2901.
30. Meyer, R. R., D. C. Rein, and J. Glassberg. 1982. The product of the *lexC* gene of *Escherichia coli* is single-stranded DNA-binding protein. *J. Bacteriol.* **150**:433–435.
31. Meyer, R. R., D. W. Voegele, S. M. Ruben, D. C. Rein, and J. M. Trela. 1982. Influence of single-stranded DNA-binding protein on *recA* induction in *Escherichia coli. Mutat. Res.* **94**:299–313.
32. Molineux, I. J., S. Friedman, and M. L. Gefter. 1974. Purification and properties of the *Escherichia coli* deoxyribonucleic acid-unwinding protein. Effects on deoxyribonucleic acid synthesis *in vitro. J. Biol. Chem.* **249**:6090–6098.
33. Molineux, I. J., and M. L. Gefter. 1974. Properties of the *Escherichia coli* DNA binding (unwinding) protein: interactions with DNA polymerase and DNA. *Proc. Natl. Acad. Sci. USA* **71**:3858–3862.
34. Molineux, I. J., and M. L. Gefter. 1975. Properties of the *Escherichia coli* DNA-binding (unwinding) protein: interactions with nucleolytic enzymes and DNA. *J. Mol. Biol.* **98**:811–825.
35. Neidhardt, F. C., T. A. Phillips, R. A. VanBogelen, M. W. Smith, Y. Georgalis, and A. R. Subramanian. 1981. Identity of the B56.5 protein, the A-protein, and the *groE* gene product of *Escherichia coli. J. Bacteriol.* **145**:513–520.
36. Perrino, F. W., D. C. Rein, A. M. Bobst, and R. R. Meyer. 1987. The relative rate of synthesis and levels of single-stranded DNA binding protein

during induction of SOS repair in *Escherichia coli*. *Mol. Gen. Genet.* 202:612–614.
37. **Sancar, A., R. P. Wharton, S. Seltzer, B. M. Kacinski, N. D. Clarke, and W. D. Rupp.** 1981. Identification of the *uvrA* gene product. *J. Mol. Biol.* 148:45–62.
38. **Sancar, A., K. R. Williams, J. W. Chase, and W. D. Rupp.** 1981. Sequence of the *ssb* gene and protein. *Proc. Natl. Acad. Sci. USA* 79:4274–4278.
39. **Schneider, R. J., and J. G. Wetmur.** 1982. Kinetics of transfer of *Escherichia coli* single strand deoxyribonucleic acid binding protein between single-stranded deoxyribonucleic acid molecules. *Biochemistry* 21:608–615.
40. **Sigal, N., H. Delius, T. Kornberg, M. Gefter, and B. Alberts.** 1972. A DNA-unwinding protein isolated from *Escherichia coli*: its interaction with DNA and with DNA polymerases. *Proc. Natl. Acad. Sci. USA* 69:3537–3541.
41. **Takano, T., and T. Kakefuda.** 1972. Involvement of a bacterial factor in morphogenesis of bacteriophage capsids. *Nature* (London) *New Biol.* 239:34–37.
42. **Tessman, E. S., and P. K. Peterson.** 1982. Suppression of the *ssb-1* and *ssb-113* mutations of *Escherichia coli* by wild-type *rep* gene, NaCl, and glucose. *J. Bacteriol.* 152:572–583.
43. **Vales, L. D., J. W. Chase, and J. B. Murphy.** 1980. Effect of *ssbA1* and *lexC113* mutations on lambda prophage induction, bacteriophage growth, and cell survival. *J. Bacteriol.* 143:887–896.
44. **Wada, M., and H. Itikawa.** 1984. Participation of *Escherichia coli* K-12 *groE* gene products in the synthesis of cellular DNA and RNA. *J. Bacteriol.* 157:694–696.
45. **Weiner, J. H., L. L. Bertsch, and A. Kornberg.** 1975. The deoxyribonucleic acid unwinding protein of *Escherichia coli*. Properties and functions in replication. *J. Biol. Chem.* 250:1972–1980.
46. **Whittier, R. F., and J. W. Chase.** 1981. DNA repair in *E. coli* strains deficient in single-strand DNA binding protein. *Mol. Gen. Genet.* 183:341–347.
47. **Williams, K. R., J. B. Murphy, and J. W. Chase.** 1984. Characterization of the structural and functional defect in the *Escherichia coli* single-stranded DNA binding protein encoded by the *ssb-1* mutant gene. *J. Biol. Chem.* 259:11804–11811.
48. **Yamamori, T., and T. Yura.** 1982. Genetic control of heat-shock protein synthesis and its bearing on growth and thermal resistance in *Escherichia coli* K-12. *Proc. Natl. Acad. Sci. USA* 79:860–864.

Chapter 17

DNA Replication Mutants from Hamster V79 Cells Exhibiting Hypersensitivity to Aphidicolin

Philip K. Liu and Thomas Norwood

DNA POLYMERASES AND APHIDICOLIN

At least four major DNA-dependent DNA polymerases (pol) have been identified in mammalian cells (11, 20, 42) and designated DNA pol-α, -β, -γ, and -δ. The criteria for this classification are based on molecular weight, template-primer preference, and inhibitor sensitivity of these DNA polymerases. Since the discovery of DNA pol-α in animal cells (J. Bollum and V. R. Potter, Letter, *J. Am. Chem. Soc.* 79:3603–3604, 1957), it is believed that DNA pol-α and -δ are sensitive to aphidicolin inhibition and are primarily involved in chromosomal replication, while the β-type and γ-type polymerases are primarily involved in the repair synthesis and mitochondrial DNA synthesis, respectively. There is at present some controversy as to the existence of pol-δ as a distinct species of polymerase associated with exonuclease activity that may have an important, if not primary, role in chromosome replication. In this monograph, we shall assume that the activity sensitive to the antibiotic aphidicolin is associated with the DNA pol-α.

Specific inhibitors of polymerase activity have been very useful in defining the various species of polymerases and are of great potential value in elucidation of the functions of this class of enzymes. Of these inhibitors, the tetracyclic diterpenoid aphidicolin, synthesized naturally by the fungal species *Cephalosporium aphidicola*, is one of the most promising probes for the study of the properties of DNA polymerases (4, 17). It is a specific inhibitor of mammalian pol-α (17). In addition, it inhibits the activity of DNA polymerases I, II, and III of *Saccharomyces cerevisiae* (5a) and the DNA polymerases of certain viruses, such as that of the herpes simplex virus (29). Nevertheless, this antibiotic does not inhibit the activity of pol-β or -γ and exhibits no measurable effect on eucaryotic RNA or protein synthesis or on any of the procaryotic DNA polymerases (15). The extent to which aphidicolin inhibits its repair synthesis of UV- and chemically induced DNA damage is controversial.

Philip K. Liu • Department of Environmental Health Sciences, Case Western Reserve University School of Medicine, Cleveland, Ohio 44106. *Thomas Norwood* • Department of Pathology, University of Washington, Seattle, Washington 98195.

Some investigators have reported that aphidicolin does not inhibit repair DNA synthesis in mitotic cells after exposure to mutagenic agents (13, 32, 34). However, the weight of evidence, based on the observations from a number of laboratories, indicates that aphidicolin inhibits (up to 90%) chemically induced repair DNA synthesis (3, 7, 8, 10, 14, 27). This suggests that DNA pol-α is involved in, at least, certain types of repair synthesis.

It is generally agreed that aphidicolin is not directly mutagenic in herpesviruses and procaryotes (5), indirectly mutagenic through metabolic activation by rat liver microsomes in the Ames assay (31), or mutagenic for induction of resistance to certain drugs in animal cells (Table 1). Indirect evidence from studies of sister chromatid exchange in the CHL-V79 cell line suggests that aphidicolin acts at the replication fork (18, 19). The fact that aphidicolin induces fragile sites and endoreduplication, i.e., chromosomal duplication without cell division, in certain cell types (16) suggests that there is a relationship between cytokinesis and chromosomal DNA replication by pol-α.

Mammalian Mutants Isolated on the Basis of Resistance to Aphidicolin

In view of the apparent specificity of aphidicolin for inhibition of pol-α, selection of mutants resistant to this agent provides an obvious approach to the study of the biological and chemical properties of this pivotal enzyme. Aphidicolin has been used to isolate aphidicolin-resistant (Aph^r) polymerase mutants of DNA polymerase I from *S. cerevisiae* (35) and Aph^r DNA polymerase from cultured cells of *Drosophila melanogaster* (36),

TABLE 1
Aphidicolin Does Not Enhance UV-Induced Mutant Frequency[a]

Expt	UV (J/m^2)	Expression time (days)	Aphidicolin (nM)	Mutation frequency (10^6)[b]		
				Oua^r	6TG^r	DT^r
1	0	3	0	—	—	59
	8.4	3	0	—	—	215
	8.4	3	500	—	—	208
	0	5	0	—	—	88
	8.4	5	0	—	—	305
	8.4	5	500	—	—	432
	0	9	0	—	30	—
	8.4	9	0	—	320	—
	8.4	9	500	—	320	—
	0	13	0	—	30	—
	8.4	13	0	—	560	—
	8.4	13	500	—	340	—
2	0	8	0	4	19	55
	12.6	8	0	192	533	625
	12.6	8	500	131	329	535
	0	10	0	10	13	53
	12.6	10	0	208	523	389
	12.6	10	500	121	335	165

[a] Aphidicolin was added to Aph^r-4-2 cells 3 days after UV irradiation and was removed during selection. The drugs (ouabain [Oua], 1 mM; 6-thioguanine [6TG], 10 µg/ml; diphtheria toxin [DT], 2 floculating units per ml) were used to determine forward mutation frequency (22).
[b] —, Not tested.

hamster lung fibroblast V79 cells (6), mouse teratocarcinoma cells (2), and herpes simplex virus (29). Four mechanisms have been found to be responsible for the Aph^r phenotype: (i) amplification in the level of dCTP (or deoxynucleotide triphosphate [dNTP]); (ii) amplified activity of DNA pol-α; (iii) reduced affinity for aphidicolin; or (iv) elevated affinity for dCTP by the DNA pol-α (11). Other types of mutant DNA pol-α exhibiting temperature sensitivity (28) or hypersensitivity to aphidicolin (12) have been reported.

Rodent Mutant DNA pol-α

A mammalian DNA pol-α mutant cell was isolated (6, 21–23, 25) from an established cell line of male Chinese hamster fibroblasts (V79 cells, clone no. 743x [9]). This V79 mutant cell line originally was designated Aph^r-4 and later was recloned (6) and had been designated Aph^r-4-2 (24). Although the gene coding for human DNA pol-α appears to be located in the X chromosome (39, 44), the procedure to isolate this Aph^r mutant cell line is unique because the cell clone was isolated after combined treatments of bromodeoxyuridine photolysis and UV irradiation.

DNA pol-α purified from this mutant cell line has been found to be relatively resistant to aphidicolin as compared with the wild-type enzyme and to have a molecular weight of at least 120,000 (L. A. Loeb, P. K. Liu, M. E. Reyland, W. R. Pendergrass, and K. P. Gopinathan, Fed. Proc. 44:855, 1985; C. C. Chang, personal communication). Data in Table 2 show that this mutant enzyme could be inactivated by a monoclonal antibody against human DNA pol-α, SJK 132-20 (39). The mutant DNA pol-α exhibited a 6- to 10-fold reduction in the Michaelis constant (K_m) for dCTP and dTTP, while the K_m for dATP was virtually unaltered (Table 3). This is in agreement with the known competition between aphidicolin and dCTP, or between aphidicolin and dTTP at low concentrations of dCTP (15), and suggests a mechanism for the Aph^r phenotype. Thus, this mutant strain provides the first example of a mutation in a gene involved in the mammalian DNA replication process whose phenotypic enzyme defect has been identified. The kinetic data further suggest that the Aph^r pol-α may not be very accurate in copying DNA (22, 23). This is supported by the observation that there is a relationship between the misincorporation of dNTP and the K_m of the polymerase for that substrate in a test tube replication assay system using purified DNA polymerase (11; M. F. Goodman, J. Petruska, M. S. Boosalis, C. Bonner,

TABLE 2
Inactivation of DNA pol-α[a]

SJK 132-20 (μg/assay)	% Activity of DNA pol-α from:	
	HeLa cells	Aph^r-4-2 cells
0	100	100
0.5	40	35
1.0	26	22
4.5	15	20
9.0	15	20

[a] DNA pol-α was purified and assayed according to Liu et al. (23), using activated calf thymus DNA. The enzyme was mixed with monoclonal antibody against DNA pol-α (SJK 132-20, a generous gift from D. Korn, Stanford University, Stanford, Calif.) and incubated at 4°C for 30 min before addition of assay mixture (50 mM Tris hydrochloride [pH 8.0], 4 mM MgCl$_2$, 2 mM dithiothreitol, 5 μg of activated calf thymus DNA, 50 μM dNTP) (specific activity of [^{32}P]dTTP, 2,100 cpm/pmol in a 50-μl reaction mixture). The incorporation was carried out at 37°C for 30 min. 100% Activity = 13 pmol (HeLa) or 7 pmol (Aph^r-4-2) of dTMP incorporation.

TABLE 3
Michaelis Constants for DNA pol-α from 743x (Parental Wild Type) and Aphidicolin-Resistant Aph^r-4-2 Cells[a]

Cells	K_m (μM)		
	dATP	dCTP	dTTP
743x (wild type)	4	14	12
Aph^r-4-2	5	1.4	2

[a] Michaelis constants (K_ms) were determined by DNA synthesis in activated calf thymus, using purified DNA pol-α by the method of Liu et al. (23). The DNA polymerase was purified from either a DEAE-DE52 or a phosphocellulose P-11 column (23).

Phenotypes of the Aph^r-4-2 Mutant

The mutant cell clone Aph^r-4-2 grows more slowly (division time, 31 h) than the parental cell clone 743x does (division time, 12 h). Preliminary data from the cell cycle analysis have shown that each phase of the cycle in Aph^r-4-2 cells is increased by 2- to 3.6-fold as compared with the parental wild-type cells. Specifically, the length in time of each phase in Aph^r-4-2 cells is 10.7, 5.4, and 14.4 h for G_1, S, and G_2, respectively, whereas the length of the respective phases in the wild-type parental cells is 3, 2.6, and 5.8 h (W. Pendergrass, personal communication). We think the increased length in the S phase may be due to a decreased replication efficiency by the mutant pol-α. The reason for the lengthening of G_1 and G_2 phases of the cycle is not clear. However, we speculate that it could be attributed to the time required for postreplication repair of DNA damage incurred during replication by the mutant enzyme.

The involvement of DNA pol-α in mutagen-induced DNA repair synthesis has been suggested by several investigators. Further characterization using the mutant cells has revealed that the Aph^r-4-2 clone has an increased mutation frequency relative to the wild type after treatments with N-methyl-N'-nitro-N-nitrosoguanidine (MNNG) in addition to UV (22, 25). However, the repair capacity appears to be unchanged as measured by UV-induced unscheduled thymidine incorporation (22) or by UV dimer removal using antisera (40) against thymine dimers (P. Liu, unpublished observation). Also the cloned mutant cells display no increased sensitivity or mutability upon exposure to X-ray or dimethyl sulfate. Arrest of cell division by liquid holding after UV irradiation reduces lethality and mutation frequency at loci conferring resistance to ouabain, diphtheria toxin, and 6-thioguanine in wild-type cells. In contrast, clone Aph^r-4-2 cells display increased cell death and mutability at these loci in response to this experimental maneuver (22). According to the observations from other laboratories that DNA pol-α plays a major role in repair synthesis of UV-induced DNA damage (3, 7, 8, 10, 14, 27), our observations are consistent with the notion that the mutant enzyme carries out repair synthesis at a reduced fidelity.

Genetic Analysis of Aph^r-4-2 Mutants

Somatic cell genetic studies characterizing the Aph^r-4-2 clone have been carried out. Cell hybridization studies have revealed codominance between the Aph^r phenotype and the wild-type cells (Chang, personal communication). In these studies, a series of subclones (designated S_4) resistant to 6-thioguanine were selected from the Aph^r-4-2 cell line and fused to a variant isolated from V79 cells (30) that is unable to utilize galactose (43). Therefore, hybrids would be selected in HAT medium (supplemented with hypoxanthine, aminopterin, and thymidine) in which galactose is the only carbon source. The hybrids exhibited an intermediate level of aphidicolin resistance between that of clone Aph^r-4-2 and the wild-type V79 clone.

DNA transfection studies have provided the most direct evidence that a mutation in the locus coding for the catalytic subunit of DNA pol-α is responsible for the Aph^r phenotype of clone Aph^r-4-2 (24). The transfer of the resistance phenotype is abolished by excessive shearing or endonuclease digestion (BamHI) of the DNA prior to transfer to the recipient cells. The precise nature of the mutation in the Aph^r-4-2 clone remains to be determined. However, there is evidence that the Aph^r phenotype can result from a single amino acid substitution. This has been reported in a resistant mutant carrying a mutation in the DNA polymerase gene of herpes simplex virus type 2 (38).

While much work remains to be completed in the characterization of the Aph^r phenotype, the studies that have been completed clearly indicate that this class of mutants will extend our knowledge of the biological and chemical properties of pol-α. Another class of mutants, those exhibiting hypersensitivity to aphidicolin, could extend the utility of this drug as a probe for the function of DNA polymerases. Below we describe the isolation and initial characterization of a series of variants displaying increased sensitivity to aphidicolin, from clone 743x of the V79 hamster lung fibroblast cell line.

METHODS TO OBTAIN MUTANTS EXHIBITING HYPERSENSITIVITY TO APHIDICOLIN

Mammalian Mutant Cells Exhibiting Hypersensitivity to Aphidicolin

The preliminary data on the isolation and partial characterization of mutants from mammalian cells that exhibit hypersensitivity to aphidicolin are presented here. These mutants were isolated in the presence of 50 nM aphidicolin and tritiated thymidine (10 μCi/ml), which is selective against growing cells in a synchronized cell population (1). The lethal growth selection was first described by Lubin (26) for selection of an auxotropic mutant and was later used by others to select for DNA repair-defective mutants of CHO cells and temperature-sensitive mutants (28, 37, 41).

In our first attempt, the parental wild-type cells were mutagenized (6) and the mutants sensitive to minimal concentrations of aphidicolin were selected at 37°C. One such variant, designated BBU-1, was recovered. The sensitivity of the BBU-1 and wild-type cells to aphidicolin was examined at 34 and 37°C (Table 4). The concentration of aphidicolin to inhibit the colony-forming ability (CFA) of BBU-1 by 50% was 50 nM at 34°C and 100 nM at 37°C. At this concentration, more than 95% of the wild-type cells formed colonies. In fact, it would take more than 200 nM aphidicolin to inhibit 50% CFA in 743x cells at 37°C. The concentration was fourfold higher than that found in the BBU-1 line. The results suggested the procedure is useful for isolation of aphidicolin-hypersensitive mutants.

The mechanisms of the temperature dependence of sensitivity to aphidicolin are unclear (Table 4). It has been previously reported that 400 nM aphidicolin produces 50% inhibition of colony formation of clone 743x of the V79 cell line at 37°C (6), suggesting this then is a positive relationship between the extent of inhibition and temperature around the physiologic (34 to

TABLE 4
CFA in the Presence of Aphidicolin[a]

Temp (°C)	Cell line	% CFA at aphidicolin concn:				
		0 nM	25 nM	50 nM	100 nM	150 nM
34	743x	100	NT	97	NT	NT
	BBU-1	100	NT	48	NT	NT
37	743x	100	103	98	95	84
	BBU-1	100	88	79	58	37

[a] Actively growing cell lines were trypsinized and plated as single cells at a density of 400 cells per plate (100 mm) in medium supplemented with 5% fetal calf serum and aphidicolin. The total concentration of dimethyl sulfoxide was less than 0.5%. Medium and aphidicolin were changed every 48 h. The colonies were fixed with 95% ethanol and stained with 1% crystal violet after 6 days (743x) or 12 days (BBU-1). Colonies of greater than 16 cells were scored. The data have been corrected with the plating efficiencies without aphidicolin. NT, Not tested.

40°C) temperature range. We speculate that this temperature dependence may be due to differences in growth rate at the different temperatures, a change in the extent of induction of sensitive enzymes, or temperature-dependent conformational changes in the molecular target(s) of the antibiotic.

Isolation of Aphidicolin-Hypersensitive Mutants at 39.5°C

In a second set of studies designed to isolate variants hypersensitive to aphidicolin, the selection was carried out at 39.5°C. The rationale for this modification was based on the following observations: (i) aphidicolin has been noted to induce endoreduplication (16); (ii) since DNA pol-α is sensitive to aphidicolin inhibition, it is possible that this endoreduplication is carried out by DNA pol-β; and (iii) DNA pol-β is relatively heat sensitive (20, 23). We reasoned that selection at the higher temperature may reduce the loss of hypersensitive variants due to the incorporation of tritiated nucleotides via aphidicolin-insensitive polymerase activity.

In these studies, the parental wild-type cells were maintained at 34°C during mutation induction and the expression period. After release from the double thymidine block (1), the selection was carried out with aphidicolin and tritiated thymidine as described above except that the cultures were maintained at 39.5°C during the period of exposure to the tritium. The viable cells were enriched in HAT medium at 34°C. A relatively large number of cells survived the first selection cycle. Therefore, we subjected the surviving cells to a second round of synchronization and selection. Four isolates were obtained after the second selection and were designated aph^{hs}-1, aph^{hs}-2, aph^{hs}-3, and aph^{hs}-4.

The initial phase of characterization of these variants was to determine their CFA as compared with the parental wild-type culture. The cloning efficiency was observed to be similar or slightly reduced in comparison with that of the wild-type cells. Aphidicolin-hypersensitive (Aph^{hs}) mutants grow more slowly than the wild-type cells, as evidenced by the fact that 12 days were required to form visible colonies as compared with 7 days for the wild-type cells. The aphidicolin sensitivity was measured at three temperatures. The concentrations of aphidicolin required to reduce the CFA by 50% in 743x parental wild-type cells were 200 nM at 34°C, 400 nM at 37°C, and 600 nM at 39°C. Compared with V79 parental wild-type cells, the CFA of the mutants was reduced by at least 90% at the respective concentrations and temperatures (data not shown).

A mutant cell line carrying defective DNA pol-α reported previously (Aph^r-4-2, see above) exhibited a hypersensitivity to UV light irradiation. We wanted to know whether any of the newly isolated Aph^{hs} mutants exhibited UV hypersensitivity. At 34°C (Table 5), aph^{hs}-1, -2, and -3 were more sensitive to UV irradiation at doses of 6 or 12 J/m². The aph^{hs}-4 isolate was only slightly, but not significantly, more sensitive than wild-type cells. This UV hypersensitivity suggests that aph^{hs}-1, aph^{hs}-2, and aph^{hs}-3 could have a mutation in DNA repair enzymes such as DNA pol-α.

Possible Mechanisms for Aphidicolin-Hypersensitive Phenotypes

We described here the isolation of mutant cells that cannot grow in aphidicolin at a concentration not toxic to parental wild-type cells. Clone BBU-1 was selected at 37°C, but this variant was not stable for its Aph^{hs} phenotype. We later modified the procedure: the parental wild-type cells were mutagen treated (bromodeoxyuridine photolysis and UV irradiation [6]), were allowed expression, and were synchronized at 34°C. The cells were twice selected at 39.5°C. We obtained four Aph^{hs} mutants. The Aph^{hs} phenotype is stable for at least 30 days in the absence of selection. Three of the four mutants also exhibited hypersensitivity to UV.

TABLE 5
CFA after UV Light Irradiation[a]

UV (J/m²)	% Survival of cells:				
	743x (wild type)	aph^{hs}-1	aph^{hs}-2	aph^{hs}-3	aph^{hs}-4
0	100	100	100	100	100
6	75	ND	45	25	55
12	56	14	0.4	7.3	36

[a] Survival rate after UV irradiation was determined (26). In short, cells were trypsinized to single cells and seeded at a density of 400 cells per plate (100 mm) in growth medium supplemented with 5% fetal calf serum. At 16 h after attachment, medium was removed, and the cells were irradiated with UV (1.8 J/m² per s). The cells were grown at 34°C in the growth medium until visible colonies were formed (10 to 30 days). ND, Not determined.

We speculate that there are several possible mechanisms for the observed phenotypes, as follows.

(i) Mutations in dNTP Production

Because aphidicolin competes with dCTP or dTTP, it is possible that the Aph^{hs} phenotype is conferred by a reduced cellular dNTP pool. The growth rate in this type of mutants could be improved with an exogenous source of nucleosides. Although aph^{hs}-1 and aph^{hs}-2 grew better in the presence of HAT at 34°C (data not shown), they did not exhibit a better CFA in medium containing hypoxanthine, uridine, and thymidine (C15 medium).

(ii) Mutations Resulting in a Decreased DNA Polymerase Activity

Overproduction of the target enzyme has been known to be one of the mechanisms for drug resistance. Therefore, a down regulation of the target enzyme could also be a mechanism for a drug hypersensitivity phenotype. Such mutations could be in the regulatory protein, an auxiliary protein, or any subunit of DNA polymerase such that its activity is compromised.

(iii) Mutations in the Binding Site for Aphidicolin in the DNA Polymerase

Sugino and Nakayama (36) have reported that a mutation in the aphidicolin-binding site (perhaps a reduced affinity) confers an aphidicolin resistance. Conversely, we believe that another mutation in this site would confer an increased affinity, such that this type of mutant becomes hypersensitive to nontoxic concentrations of aphidicolin. This type of mutation may not affect dNTP binding, and such mutants may not exhibit UV hypersensitivity (36).

(iv) Mutations in the Binding Site for dCTP in the DNA Polymerase

An increased affinity for dCTP in DNA pol-α gives an aphidicolin resistance phenotype (23). This mutation, such as that which occurred in Aph^r-4-2, is believed to confer a mutagen hypersensitivity, mutator, or slow growth phenotype. It is likely that another type of mutation in this locus may also alter dCTP affinity in DNA pol-α such that the binding of dCTP or dTTP is reduced. This mutation could result in an Aph^{hs} phenotype. This type of Aph^{hs} cells (e.g., aph^{hs}-1, -2, and -3) may have some of the phenotypes shown in the Aph^r-4-2 cell clone.

(v) Mutations in Loci Coding for Enzymes Other than That for DNA pol-α

Recent developments in animal polymerase research suggest that DNA pol-α and DNA pol-δ activities could be in one enzyme. Both polymerases are similar except that DNA pol-δ contains 3′,5′-exonuclease activities. However, both polymerases can be distinguished by their respective antibodies (M. Y. W. T. Lee, N. L. Toomey, and S.-J. Zang, this volume). DNA pol-δ is more sensitive to inhibition by aphidicolin than DNA pol-α, and the former enzyme's activity on certain template-primers can be stimulated by cyclin (33). In vitro, the pol-δ is relatively resistant to N^2-(p-n-butylphenyl)-g-(2-deoxy-β-D-ribofuranosyl)guanine 5′-triphosphate and N^2-(p-n-butylanilino)-g-(2-deoxy-β-D-ribofuranosyl)adenine 5′-triphosphate. We speculate that an "overproduction" in DNA pol-δ could give rise to the Aph[hs] phenotype.

Conclusions

These initial characterization studies provide a very preliminary picture of the phenotype of this series of variants that are hypersensitive to aphidicolin. Clearly the most important immediate task is to determine whether any of these variants result from a mutation in a locus coding for the aphidicolin-sensitive DNA polymerase. A mutation in such a locus resulting in hypersensitivity to this drug would be of value for a variety of reasons. This class of mutants is of potential value in the cloning of the polymerase gene from normal (wild-type) cells. It will also be important to carry out a detailed study comparing and contrasting the phenotypes of the resistant and hypersensitive mutants. Such studies could potentially yield information relevant to the function of the polymerases (e.g., repair versus chromosome replication) and insights to the molecular level of the mechanisms of DNA binding and nucleotide selection. In addition, those variants that prove to be due to mutations in loci other than that coding the DNA pol-α may be of value in the elucidation of other cellular functions that regulate the activity of DNA polymerases.

ACKNOWLEDGMENTS. We thank Toni Leirer and Marie Walters for preparation of this manuscript.

Research was supported by a grant from the National Science Foundation (DCB 86000659), a Cancer Research Center Grant from the National Cancer Institute (P30 CA43703) to P.K.L., and a grant from the National Institutes of Health (AG-01751-05) to T.N.

LITERATURE CITED

1. **Adams, R. L. P.** 1980. *Laboratory Techniques in Biochemistry and Molecular Biology: Cell Culture for Biochemists.* Elsevier/North-Holland Biochemical Press, New York.
2. **Aizawa, S., L. A. Loeb, and G. Martin.** 1987. Aphidicolin-resistant mutator strains of mouse teratocarcinoma. *Mol. Gen. Genet.* 208:342–348.
3. **Berger, N. A., K. K. Kurohara, S. J. Petzold, and G. W. Sikorski.** 1979. Aphidicolin inhibits eukaryotic DNA replication and repair — implications for involvement of DNA polymerase in both processes. *Biochem. Biophys. Res. Commun.* 89:218–221.
4. **Brundret, K. M., W. Dalziel, B. Hesp, J. A. J. Tarvis, and S. Neidle.** 1972. X-ray crystallographic determination of the structure of the antibiotic aphidicolin: a tetracyclic diterpenoid containing a new ring system. *J. Chem. Soc. Chem. Commun.* 13:1027–1028.
5. **Bucknall, R. A., H. Moores, R. Simms, and B. Hesp.** 1973. Antiviral effects of aphidicolin, a new antibiotic produced by *Cephalosporium aphidicola*. *Antimicrob. Agents Chemother.* 4:294–304.
5a. **Burger, P. M., and G. A. Bauer.** 1988. DNA polymerase III from *Saccharomyces cerevisiae*. *J. Biol. Chem.* 263:925–930.
6. **Chang, C. C., J. A. Boezi, S. T. Warren, C. L. K. Sabourin, P. K. Liu, L. Glatzer, and J. E. Trosko.** 1981. Isolation and characterization of a UV-sensitive hypermutable aphidicolin-resistant Chinese hamster cell line. *Somatic Cell Genet.* 7:235–253.
7. **Ciarrocci, G., J. G. Jose, and S. Linn.** 1979. Further characterization of a cell-free system for measuring replicative and repair DNA synthesis with cultured human fibroblasts and evidence for the involvement of DNA polymerase-α in DNA repair. *Nucleic Acids Res.* 7:1205–1219.
8. **Dresler, S. L., J. D. Roberts, and M. W. Lieberman.** 1982. Characterization of deoxyribonucleic acid repair synthesis in permeable human fibroblasts. *Biochemistry* 21:2557–2564.

9. Ford, D. K., and G. Yerganian. 1958. Observation on the chromosomes of Chinese hamster cells in tissue culture. *J. Natl. Cancer Inst.* **21**:393–423.
10. Friedberg, E. C. 1984. *DNA Repair.* W. J. Freeman & Co., New York.
11. Fry, M., and L. A. Loeb. 1986. *Animal Cell DNA Polymerases.* CRC Press, Boca Raton, Fla.
12. Gibb, J. S., H. C. Chiou, J. D. Hall, D. W. Mount, M. J. Retondo, S. K. Weller, and D. M. Coen. 1985. Sequence and mapping analyses of the herpes simplex virus DNA polymerase gene predict a C-terminal substrate binding domain. *Proc. Natl. Acad. Sci. USA* **82**:7969–7973.
13. Giulotto, E., and C. Mondello. 1981. Aphidicolin does not inhibit the repair synthesis of mitotic chromosomes. *Biochem. Biophys. Res. Commun.* **99**:1287–1294.
14. Hanaoka, F., H. Kato, S. Ikegami, M. Ohashi, and M.-A. Yamada. 1979. Aphidicolin does inhibit repair replication in HeLa cells. *Biochem. Biophys. Res. Commun.* **87**:575–580.
15. Holmes, A. M. 1981. Studies on the inhibition of highly purified calf thymus 8S and 7.3S DNA polymerase α by aphidicolin. *Nucleic Acids Res.* **9**:161–168.
16. Huang, Y., C.-C. Chang, and E. Trosko. 1983. Aphidicolin-induced endoreduplication in Chinese hamster cells. *Cancer Res.* **43**:1361–1363.
17. Huberman, J. A. 1981. New views of the biochemistry of eukaryotic DNA replication revealed by aphidicolin, an unusual inhibitor of DNA polymerase-α. *Cell* **23**:641–648.
18. Ishi, Y., and M. A. Bender. 1980. Effects of inhibitors of DNA synthesis on spontaneous and ultraviolet light-induced sister chromatid exchanges in Chinese hamster cells. *Mutat. Res.* **79**:19–32.
19. Kato, H. 1980. Evidence that the replication point is the site of sister chromatid exchange. *Cancer Genet. Cytogenet.* **2**:69–77.
20. Kornberg, A. 1982. *DNA Replication, 1982 Supplement.* W. H. Freeman & Co., San Francisco.
21. Liu, P. K., C.-C. Chang, and J. E. Trosko. 1982. Association of mutator activity with UV sensitivity in an aphidicolin-resistant mutant of Chinese hamster V79 cells. *Mutat. Res.* **106**:317–332.
22. Liu, P. K., C.-C. Chang, and J. E. Trosko. 1984. Evidence for mutagenic repair in V79 cell mutant with aphidicolin-resistant DNA polymerase-α. *Somatic Cell Mol. Genet.* **10**:235–245.
23. Liu, P. K., C.-C. Chang, J. E. Trosko, D. K. Dube, G. M. Martin, and L. A. Loeb. 1983. Mammalian mutator mutant with an aphidicolin-resistant DNA polymerase-α. *Proc. Natl. Acad. Sci. USA* **80**:797–801.
24. Liu, P. K., and L. A. Loeb. 1984. Transfection of the DNA polymerase-α gene. *Science* **226**:833–835.
25. Liu, P. K., J. E. Trosko, and C.-C. Chang. 1982. Hypermutability of a UV-sensitive aphidicolin-resistant mutant of Chinese hamster fibroblasts. *Mutat. Res.* **106**:333–348.
26. Lubin, M. 1959. Selection of auxotrophic bacterial mutants by tritium-labeled thymidine. *Science* **129**:838–839.
27. Miller, M. R., and D. N. Chinault. 1982. The roles of DNA polymerases α, β, and γ in DNA repair synthesis induced in hamster and human cells by different DNA damaging agents. *J. Biol. Chem.* **257**:10204–10209.
28. Murakami, Y., H. Yasuda, H. Miyazawa, F. Hanaoka, and M. A. Yamada. 1985. Characterization of a temperature-sensitive mutant of mouse FM3A cells defective in DNA replication. *Proc. Natl. Acad. Sci. USA* **82**:1761–1765.
29. Nishiyama, Y., S. Suzuki, M. Yamachi, K. Maeno, and S. Yoshida. 1984. Characterization of an aphidicolin-resistant mutant of herpes simplex virus type 2 which induces an altered viral DNA polymerase. *Virology* **135**:87–96.
30. Norwood, T. H., and C. J. Zeigler. 1982. The use of dimethyl sulfoxide in mammalian cell fusion, p. 35–45. *In* J. W. Shay (ed.), *Techniques in Somatic Cell Genetics.* Plenum Publishing Corp., New York.
31. Pedrali-Noy, G., G. Mazza, F. Focher, and S. Spadari. 1980. Lack of mutagenicity and metabolic inactivation of aphidicolin by rat liver microsomes. *Biochem. Biophys. Res. Commun.* **93**:1094–1103.
32. Pedrali-Noy, G., and S. Spadari. 1980. Aphidicolin allows a rapid and simple evaluation of DNA-repair synthesis in damaged human cells. *Mutat. Res.* **70**:389–394.
33. Prelich, G., M. Kostura, D. R. Marshak, M. B. Mathews, and B. Stillman. 1987. The cell-cycle regulated proliferating cell nuclear antigen is required for SV40 DNA replication *in vitro*. *Nature* (London) **326**:471–476.
34. Seki, S., M. Ohashi, H. Oguro, and T. Oda. 1982. Possible involvement of DNA polymerases α and β in bleomycin-induced unscheduled DNA synthesis in permeable HeLa cells. *Biochem. Biophys. Res. Commun.* **104**:1502–1508.
35. Sugino, A., H. Kojo, B. D. Greenberg, P. D. Brown, and K. C. Kim. 1981. *In vitro* replication of yeast 2-μm plasmid DNA. *ICN-UCLA Symp. Mol. Cell Biol.* **22**:529–535.
36. Sugino, A., and K. Nakayama. 1980. DNA polymerase α mutants from a *Drosophila melanogaster* cell line. *Proc. Natl. Acad. Sci. USA* **77**:7049–7053.
37. Thompson, L. H. 1985. DNA repair mutants, p. 641–667. *In* M. M. Gottesman (ed.), *Molecular Cell Genetics.* John Wiley & Sons, Inc., New York.

38. Tsurumi, T., K. Naeno, and Y. Nishiyama. 1986. A single-base change within the DNA polymerase locus of herpes simplex virus type 2 can confer resistance to aphidicolin. *Virology* **61**:388–394.
39. **Wang, T. S.-F., B. E. Pearson, H. A. Suomalainen, T. Mohandas, L. J. Shapiro, J. Shroeder, and D. Korn.** 1985. Assignment of the gene for human DNA polymerase-α to the X chromosome. *Proc. Natl. Acad. Sci. USA* **82**:52–70.
40. Wani, A. A., S. M. D'Ambrosio, and N. K. Alvi. 1987. Quantitation of pyrimidine dimers by immunoslot blot following sublethal UV-irradiation of human cells. *Photochem. Photobiol.* **46**:477–482.
41. **Washmuth, J. J.** 1985. Chinese hamster cell protein synthesis mutants, p. 375–421. *In* M. M. Gottesman (ed.), *Molecular Cell Genetics*. John Wiley & Sons, Inc., New York.
42. Weissbach, A., D. Baltimore, F. J. Bollum, F. C. Gallo, and D. Korn. 1975. Nomenclature of eukaryotic DNA polymerases. *Eur. J. Biochem.* **59**:1–2.
43. **Whitfield, C. D., R. Bostedor, D. Goodrum, M. Haak, and E. H. Y. Chu.** 1981. Hamster cell mutants unable to grow on galactose and exhibiting an overlapping complementation pattern are defective in the electron transport chain. *J. Biol. Chem.* **256**:6651–6656.
44. **Wong, S. W., A. F. Wahl, P. M. Yuan, N. Arai, B. E. Pearson, K. I. Arai, D. Korn, M. W. Hunkapiller, and T. S.-F. Wang.** 1988. Human DNA polymerase-α gene expression is cell proliferation dependent and its primary structure is similar to both prokaryotic and eukaryotic replicative DNA polymerases. *EMBO J.* **7**:37–47.

Chapter 18

Viral DNA Synthesis and Recombination in the Transformation Pathway of Polyomavirus

David L. Hacker, Karen H. Friderici, Claudette Priehs, Susan Kalvonjian, and Michele M. Fluck

During infection of nonpermissive cells by papovaviruses, polyomavirus, and simian virus 40, little or no viral DNA synthesis takes place, and the infection aborts (23). Transient and stable transformation occurs in approximately 1/10 and 1/100 of the infected cell population, respectively (23). Integration of the viral genome into the host cell must occur for stable transformation (23). The patterns of viral integration have been analyzed: usually, transformants demonstrate multiple sites of viral integration, at which the viral genome is present in more than one copy, arranged in a head-to-tail manner (1, 2, 5). Viral integration appears to be under the control of large-T antigen since temperature-sensitive mutants of large-T antigen demonstrate a normal frequency of abortive transformation but a high decrease in the frequency of stable transformation (9, 13, 21). Furthermore, the pattern of viral integration is altered in the rare transformants obtained at high temperature, in which integration results in less than a complete copy of the viral genome (6). A role for viral DNA replication has been hypothesized in the process of integration based on the following considerations: (i) such a role is compatible with the requirements for large-T antigen and the viral origin of DNA replication for the normal process of integration; (ii) the frequency of transformation of infected cells is similar to the frequency of cells which produce detectable levels of viral DNA in the same population; and (iii) the topology of integration might reflect the result of integrating replication intermediates.

Recently, we analyzed the role of viral DNA replication in the transformation of Fischer rat cells by polyomavirus. We found conditions in which the amount of replication could be varied, and we monitored the effect on the frequency of transformation and the topology of integration. On the other hand, we have also analyzed the poten-

---

David L. Hacker, Karen H. Friderici, Claudette Priehs, Susan Kalvonjian, and Michele M. Fluck • Department of Microbiology and Public Health, Michigan State University, East Lansing, Michigan 48824-1101.

tial role of interviral recombination in the generation of integrated concatamers of the viral genome.

ROLE OF VIRAL DNA REPLICATION

When Fischer rat cells (here FR-3T3 [19]) are infected with wild-type polyomavirus at 37°C, very little net viral DNA synthesis is observed. When these infected cells are examined by hybridization of low-molecular-weight DNA extracted from them, at most a net doubling or tripling of the initial signal is observed within the first 2 to 3 days postinfection (Fig. 1A) or later in other cell lines (D. Hacker and M. Fluck, manuscript in preparation). In contrast, when infected cells are incubated at 33°C, a high yield of viral DNA is observed (Fig. 1B), representing at least a 10-fold increase over the 37°C yield. In FR-3T3, this synthesis peaks by 5 days postinfection.

Using in situ hybridization (16; data not shown), we were able to demonstrate that

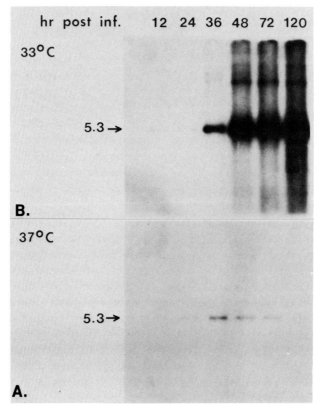

FIGURE 1. Viral DNA synthesis in polyomavirus-infected nonpermissive Fischer rat cells (FR-3T3). Fischer rat FR-3T3 cells were infected at an MOI of 10 PFU per cell. Infected cells were incubated at 37°C (A) or 33°C (B). Total DNA was extracted at the times shown. DNA was digested with EcoRI, which linearizes the polyomavirus genome, electrophoresed, transferred to nitrocellulose, and hybridized. Small increases in viral genomes are seen at 37°C, while very substantial viral DNA synthesis is seen at 33°C, at which temperature the yields approach those seen in permissive cells.

the low net increase in the amount of viral genomes observed at 37°C is contributed by a small fraction of the population in which high levels of synthesis are occurring. This observation confirms that 37°C infections undergo viral DNA replication. Similar results have been obtained previously by others for infection of BHK cells (10). These experiments also reveal that the increase at 33°C compared with that at 37°C is due, in part, to a two- to threefold-higher fraction of the population recruited to synthesize viral DNA to detectable levels. Since the increase in the number of cells engaged in synthesis does not match the total increase in viral DNA yield, we assume that the yield from positive cells is also increased. However, it is also possible that a larger fraction of cells contribute DNA than can be detected by in situ hybridization and that the number of these minimally producing cells is also increased at 33°C compared with 37°C.

The effect of increased viral DNA synthesis on transformation was assessed in two ways, by examining both transformation frequencies and integration patterns in transformants.

To assess transformation frequencies, F-111 (11) cells at a density of 10^5 cells per 60-mm-diameter culture dish, in Dulbecco modified Eagle medium, were infected with wild-type virus A2 (15) at a multiplicity of infection (MOI) of 10 PFU per cell, after which cells were passed at a ratio of 1:4 and grown at either 33 or 37°C. Transformation frequencies were the average percent transformation of three experiments and were based on the number of transformants per 10^5 cells. Transformation frequencies did increase, from 0.11 to 0.25% transformation, after incubation at 33°C in comparison with 37°C infections. However, the two- to threefold increase in transformation frequency does not match the over-10-fold increase in the yield of viral DNA. Furthermore, transformation frequencies obtained with a replication-defective strain also increase two- to threefold at 33°C in the absence of an increase in replication (D. Hacker, K. Friderici, and M. Fluck, manuscript in preparation), suggesting that the increase of wild-type transformation at 33°C can be uncoupled from viral DNA synthesis.

Integration patterns in transformants derived from infections at 37 and 33°C were also analyzed. As reviewed above, the typical pattern of integration of the polyomavirus genome consists of repeats of the viral genome integrated at multiple sites in the host chromosome. Thus, when host high-molecular-weight DNA is digested with a restriction endonuclease which cuts the viral genome at a single site, a band the size of the viral genome is generated as well as fragments of nonviral size presumed to represent the host-to-virus junction fragments. The number of such junction fragments is related to the number of viral integration sites. In addition to integrated repeats of the viral genome, free viral DNA will also yield a band of viral genome size. However, as demonstrated by Basilico's group (25), normal-size, free viral genomes can only be generated by in situ replication from integrated concatamers, so that the presence of genome-size bands does necessarily reflect the integration of concatamers of the viral genome. As illustrated in Fig. 2, wild-type transformants derived from 33 and 37°C infections do not differ in those aspects of the integration pattern. Similar results were obtained for a larger group of transformants derived from the F-111 Fischer rat cell line (Hacker and Fluck, manuscript in preparation).

SELECTION FOR INTERVIRAL RECOMBINATION CONCOMITANT WITH TRANSFORMATION

To estimate the potential role of interviral recombination in the formation of concatamers of the viral genome in the integration-transformation pathway, we undertook a study of recombination concomitant with

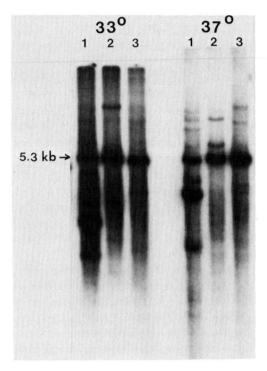

FIGURE 2. Integration patterns of polyomavirus transformation obtained under conditions of low or high DNA synthesis. Transformed foci were derived from the infections described in the legend to Fig. 1. (Three each are shown, derived from 33 or 37°C infections.) Total DNA was isolated, digested with EcoRI, and analyzed by gel electrophoresis, transfer, and hybridization to polyomavirus probes (20). The presence of genome-size (5.3 kbp) viral sequences demonstrates the presence of head-to-tail tandem integrations of the viral genome at both temperatures. The number of nonviral-size bands, presumed to represent host-viral joints, is similar at the two temperatures.

TABLE 1
Recombination between *hrt* Mutants in Infections of Fischer Rat Cells[a]

Parental genotype[b]	MOI[c]	Foci/plate[d]	
		Expt 1	Expt 2
Wild type	100	>>	>>
	10	80	80
	1	5	13
	0.1	1	3
	0.01	0	0
SD-15	12.5	0/2	
II-5	50	0/2	
SD-15 × II-5	6.3 × 6.3	>4/10	
SD-15	12.5	0/2	8/5
1387-T	200	0/2	0/5
SD-15 × 1387-T	6.3 × 6.3	>28/10	>110/10

[a] Plates containing 10^5 Fischer rat cells were infected either singly or mixedly by the wild-type or mutant strains.
[b] Wild-type strain A2 (15) and various mutant virus strains were grown on baby mouse kidney cultures from plaque-purified viruses as previously described (24). Transformation-defective *hrt* mutants II-5 and SD-15 and *mlt* mutant 1387-T were gifts from T. Benjamin. II-5 and SD-15 have deletions at 649 to 749 and 435 to 575 bp, respectively (3, 17). 1387-T contains a point mutation at nucleotide 1387, leading to premature termination of the middle-T polypeptide (4).
[c] Equal MOIs were used in the cross. Transformants were selected by focus formation (8).
[d] The total number of foci per total number of plates included in each experiment is shown. >>, Too many foci to count; >, the minimal number of foci which are clearly primary transformation events (as opposed to secondary foci migrating from primary).

transformation by selecting transformants from mixed infections with pairs of nonoverlapping, transformation-defective deletion mutants. Because the parental strains used in these crosses are nontransforming, coinfections of rat cells are expected to induce transformation only when a recombination event occurs. As shown in Table 1, such events do occur. Whereas no transformants are observed in infections with a single parent (even when these are used at multiplicities vastly exceeding those used in the cross), transformants are observed in the mixed infection. Because the parental strains used in these crosses carry different restriction endonuclease markers, it was possible to demonstrate that the viral genomes integrated in the transformants arising from the mixed infections had arisen from recombination events (S. Kalvonjian, C. Priehs, and M. Fluck, manuscript in preparation).

The following considerations can be made concerning the recombination frequency exhibited in these crosses. The transformation frequency observed in the cross between mutants SD-15 and II-5 in which the mutants are present at an MOI of 6 corresponds to approximately 1% of that expected for a wild type used at the same multiplicity. These results are compatible with a recombination rate of 2%. By the same reasoning, recombination between mutants SD-15 and 1387-T appears to be approximately 20%.

The recombination events detected in coinfections with SD-15 and II-5 occur in a region of 72 base pairs (bp), the distance which separates the two deletions. Recombination between mutants II-5 and 1387-T involves an interval of 638 bp. As seen in Table 1, an increase in distance between mutations corresponds to an increase in transformation frequency. In addition to the crosses described in this section, evidence for recombination has been obtained in crosses between the following pairs of nontransforming mutants: II-5 × dl1015, SD-15 × MOP1033, and SD-15 × 1387-T (Kalvonjian, Priehs, and Fluck, in preparation). Evidence for recombination has also been obtained for crosses performed in the hamster BHK cell line (data not shown).

NONSELECTIVE CROSSES

Since recombination appears to occur at rather high frequency in crosses in which recombinants are selected, we pursued the analysis in mixed infections in which recombination is not a selective factor for transformation. For these we selected two restriction site-minus strains, MOP1033 (22) and $ts3$ (7). MOP1033 lacks the two AvaI restriction sites located at nucleotides 672 and 1031 in the early region of the viral genome (Fig. 3A). The point mutation within the second site (nucleotide 1033) causes premature termination of the middle-T antigen; therefore, this strain is defective in cell transformation. The second strain, $ts3$, lacks the BamHI restriction site (nucleotide 4647) in the late region of the genome (Fig. 3B), presumably the cause of a decapsidation of this virus at the nonpermissive temperature. Infections with this strain were carried out at 33°C, and the infected cells were maintained at this temperature for 24 h to allow for decapsidation before shift-up to 37°C or further incubation at 33°C. When this schedule is followed, the transformation rate with $ts3$ is normal (7).

Homologous recombination between the two viral genomes can be monitored by analysis with restriction endonucleases. Digestion with a combination of BamHI and AvaI produced restriction fragments of either 3,616 bp if recombination had occurred in the large AvaI-BamHI interval or 1,317 or 1,676 bp if the two parental genomes recombined in the small AvaI-BamHI interval or between the two AvaI sites, respectively (Fig. 3C). The two parental genomes may also be detected. AvaI-BamHI digests produced a single genome-size restriction fragment (5,292 bp) from the MOP1033 genome and two fragments, 4,933 and 359 bp, from the $ts3$ genome. The 359-bp fragment was usually not detected because it was electrophoresed past the end of the gel. Thus, recombination could be followed in the unintegrated viral DNA early after infection as well as in the integrated viral genomes in transformed cells. The large AvaI-BamHI fragment was detected by hybridization to the 917-bp EcoRI-XbaI restriction fragment of polyomavirus (Fig. 3C). The 1,317- and 1,676-bp AvaI-BamHI fragments were detected by using the polyomavirus HpaII-5 restriction fragment (400 bp) as a probe (Fig. 3C).

Absence of Recombination in the Population of Unintegrated Viral Genomes

Monolayers of F-111 cells were infected with a combination of MOP1033 and $ts3$. Interviral recombination in the population of unintegrated viral genomes was monitored from the time of infection until the time when transformed foci appeared. For this purpose, viral DNA was extracted from the infected cells at times between 1 and 10 days postinfection. The DNAs were digested with a combination of AvaI and

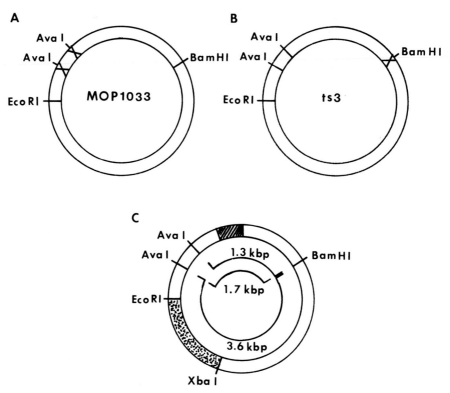

FIGURE 3. Partial restriction maps of polyomavirus strains MOP1033 and ts3. (A) MOP1033. MOP1033 (22) was derived from a wild-type strain by site-directed mutagenesis of nucleotide 1033 (numbering is as in reference 14). The point mutation introduced into the middle-T antigen reading frame produced a transformation-defective virus and eliminated the AvaI site at position 1031. The AvaI restriction site is also absent in MOP1033. X, AvaI sites at nucleotides 672 and 1031 that are absent in the MOP1033 genome. (B) ts3. The ts3 strain (7) was derived from a wild-type virus by bisulfite mutagenesis. A mutation within the VP2 gene was obtained which prevents decapsidation at the nonpermissive temperature. X, BamHI site at nucleotide 4647 that is absent in ts3. (C) Wild-type polyomavirus. The sizes of the three AvaI-BamHI fragments are given. The map also shows the location of the HpaII-5 fragment (▨) and the EcoRI-XbaI fragment (▨) which are used as hybridization probes.

BamHI. After transfer to nitrocellulose, the blot was hybridized with the HpaII-5 probe (Fig. 4A). Neither the 1.3- nor the 1.7-kilobase-pair (kbp) AvaI-BamHI restriction fragment was present at detectable levels in the population of infected cells at the times indicated. This was true even when longer film exposures were used. From a reconstruction experiment, we estimate that the wild type would have been detected if present in more than 5% of the total viral genomes (data not shown). Viral DNA replication was higher at 33 than at 37°C (Fig.

4A), a common finding in F-111 cells as shown above (Hacker and Fluck, submitted).

High Level of Recombination in Integrated Genomes

Transformed foci arising from the infections described above were visible by 10 and 17 days postinfection for the 37 and 33°C cultures, respectively. These were isolated, and total DNA was extracted from the transformants and analyzed for integrated recombinant viral genomes. Recombination in the

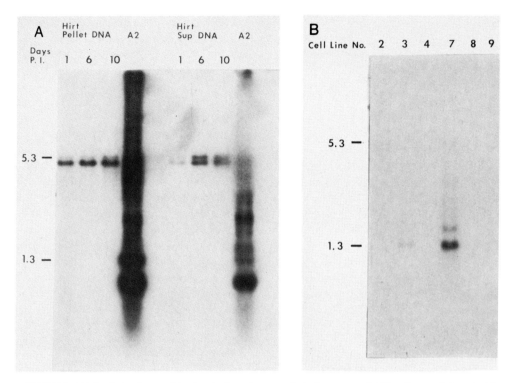

FIGURE 4. (A) Analysis of recombination in the unintegrated viral genomes. F-111 cells were infected with MOP1033 and ts3 separately at an MOI of 50 PFU per cell. Coinfections with MOP1033 and ts3 (1:1 ratio) were performed at a total MOI of 100 PFU per cell. For each time point, the viral DNA from one plate of infected cells was fractionated into high- and low- (18)-molecular-weight fractions and extracted. One-tenth of each sample was digested with a combination of AvaI and BamHI. Hybridization was to the HpaII-5 probe. The positions of the 1.3-kbp AvaI-BamHI fragments and the MOP1033 genome-size fragment (5.3 kbp) are marked. (B) Recombination in integrated viral genomes. Transformants were derived from the infections described for panel A. Total cellular DNA (10 μg) was isolated and analyzed as previously described (12).

3.6-kbp AvaI-BamHI interval was analyzed by hybridizing the blots of the AvaI-BamHI digestions with the polyomavirus EcoRI-XbaI probe (data not shown) In 23 of the 65 cell lines analyzed (35%), the 3.6-kbp AvaI-BamHI restriction fragment was present, indicating that recombination between the two parents had occurred. In addition to the recombinant fragments, the 5.0- and 5.3-kbp fragments from the two parental viral genomes were also visible in many cases. Both parental genomes were present in 14 out of 23 cell lines, while a single parental genome was present in 6 out of 23 cell lines. Recombination was also observed in the 1.3- and 1.7-kbp AvaI-BamHI intervals (Fig. 4B). The blots from the analysis above were washed and then hybridized with the HpaII-5 probe to detect evidence of recombination within the small AvaI-BamHI interval or between the AvaI sites. In many of the cell lines, both the 1.3- and the 1.7-kbp restriction fragments were present (data not shown). In all, 25 of the 65 (38%) contained at least one of the three fragments that is diagnostic of recombination between MOP 1033 and ts3. In most cases, more than one of the recombinant restriction fragments were present. Fourteen of the recombination-positive transformants had all three of the fragments, five had the 1.3- and 3.6-kbp fragments, three had the 1.7- and 3.6-kbp fragments, and three had only one of the fragments.

CONCLUSION

In experiments discussed above, we explored the role of viral DNA replication and that of interviral recombination in the generation of the integrated viral genomes present in cells transformed by polyomavirus.

The experiments demonstrate that increasing the level of viral DNA replication affects neither the frequency of transformation nor the pattern of integration of the viral genome to a detectable level. However, these experiments do not rule out more subtle differences between transformants obtained under conditions of high or low levels of DNA synthesis, such as the absolute number of integration sites (which is difficult to establish) or the number of viral genomes present in the head-to-tail concatamers of the viral genome. It further appears from our experiments that the infected cells which produce large amounts of viral DNA early postinfection may not appear among the transformants, since the unintegrated viral-DNA pools do not contain recombinant genomes at a detectable level, even in cases when the recombination frequency among the viral sequences integrated in transformants is very high.

Although these experiments do not clarify the reason for the requirement of large-T antigen in the integration pathway, they rule out one hypothesis, that large-T antigen is required simply to amplify the number of viral templates to increase the probability of integration. The requirements for large-T antigen and for the viral origin are still compatible with a viral initiation function, leading, for example, to a nicked template which may favor recombination with the host genome.

On the other hand, we have demonstrated that, at least under some circumstances, the viral genomes which end up integrated in the transformed cells have undergone very high levels of recombination, even when recombination does not appear to present a selective advantage. Concerning the potential role of interviral recombination in the generation of concatameric genome structures, it must be noted that the recombination events selected for and analyzed in the present studies consist of intragenomic events which might or might not yield concatameric structures. Thus, although the experiments discussed above clearly demonstrate that recombination occurs at high frequency in the process of integration, they do not prove that recombination has a role in the formation of the concatameric structures. Although we have previously documented some examples of tandem integration of two different parental genomes at the same site of the host cell (12), it is possible to conceive that both recombination and viral DNA replication (of the recombinant) play a role in the formation of the known integrants. If recombination alone represents the major pathway by which concatamers are formed, a role for large-T antigen in this process must be invoked. As suggested above, such a role might be linked to the role of large-T antigen in the initiation of viral DNA replication which may possibly involve a recombinogenic nick.

LITERATURE CITED

1. Basilico, C., S. Gattoni, D. Zouzias, and G. Della Valle. 1979. Loss of integrated viral DNA sequences in polyoma-transformed cells is associated with an active viral A function. *Cell* **17**:645–659.
2. Birg, F., R. Dulbecco, M. Fried, and R. Kamen. 1979. State and organization of polyoma virus DNA sequences in transformed rat cell lines. *J. Virol.* **29**:633–648.
3. Carmichael, G., and T. Benjamin. 1980. Identification of DNA sequence changes leading to loss of transforming ability in polyoma virus. *J. Biol. Chem.* **155**:230–235.

4. Carmichael, G., B. Schaffhausen, D. Dorsky, D. Oliver, and T. Benjamin. 1982. Carboxy terminus of polyoma middle-sized tumor antigen is required for attachment to membranes, associated protein kinase activities, and cell transformation. *Proc. Natl. Acad. Sci. USA* 79:3579–3583.
5. Chia, W., and P. W. J. Rigby. 1981. Fate of viral DNA in nonpermissive cells infected with simian virus 40. *Proc. Natl. Acad. Sci. USA* 78:6638–6642.
6. Della Valle, G., R. G. Fenton, and C. Basilico. 1981. Polyoma large T-antigen regulates the integration of viral DNA sequences into the genome of transformed cells. *Cell* 23:347–355.
7. Dulbecco, R., and W. Eckhart. 1970. Temperature-dependent properties of cells transformed by a thermosensitive mutant of polyoma virus. *Proc. Natl. Acad. Sci. USA* 67:1775–1781.
8. Fluck, M., and T. Benjamin. 1979. Comparisons of two early gene functions essential for transformation in polyoma virus and SV-40. *Virology* 96:205–228.
9. Fluck, M. M., R. J. Staneloni, and T. L. Benjamin. 1977. hr-t and ts-a: two early gene functions of polyoma virus. *Virology* 77:610–624.
10. Folk, W. R., J. Bancuk, and P. Vollmer. 1981. Polyoma virus replication in BHK-21 cells: semipermissiveness is due to cellular heterogeneity. *Virology* 111:165–172.
11. Freeman, A. E., R. V. Gilden, M. L. Vernon, R. G. Wolford, P. E. Hugunin, and R. J. Huebner. 1973. 5-Bromo-deoxyuridine potentiation of transformation of rat embryo cells induced *in vitro* by 3-methyl cholanthrene: induction of rat leukemia virus gs antigen in transformed cells. *Proc. Natl. Acad. Sci. USA* 70:2415–2419.
12. Friderici, K., S. Y. Oh, R. Ellis, V. Guacci, and M. M. Fluck. 1984. Recombination induces tandem repeats of integrated viral sequences in polyoma-transformed cells. *Virology* 137:67–73.
13. Fried, M. 1965. Cell transforming ability of a temperature-sensitive mutant of polyoma virus. *Proc. Natl. Acad. Sci. USA* 53:486–491.
14. Friedman, T., A. Esty, P. LaPorte, and D. Deininger. 1979. The nucleotide sequence and genome organization of the polyoma early region: extensive nucleotide and amino acid homology with SV40. *Cell* 17:715–724.
15. Griffin, B. E., M. Fried, and A. Cowie. 1974. Polyoma DNA: a physical map. *Proc. Natl. Acad. Sci. USA.* 71:2077–2081.
16. Haase, A., M. Brahic, L. Stowring, and H. Blum. 1984. Detection of viral nucleic acids by *in situ* hybridization. *Methods Virol.* 7:189–226.
17. Hattori, J., G. Carmichael, and T. Benjamin. 1979. DNA sequence alterations in hr-t deletion mutants of polyoma virus. *Cell* 16:505–513.
18. Hirt, B. 1967. Selective extraction of polyoma DNA from infected mouse cell cultures. *J. Mol. Biol.* 26:365–369.
19. Seif, R., and F. Cuzin. 1977. Temperature-sensitive growth regulation in one type of transformed rat cells induced by the *tsa* mutant of polyoma virus. *J. Virol.* 24:721–728.
20. Southern, E. M. 1975. Detection of specific sequences among DNA fragments separated by gel electrophoresis. *J. Mol. Biol.* 98:503–517.
21. Stoker, M., and R. Dulbecco. 1969. Abortive transformation by the tsa mutant of polyoma virus. *Nature* (London) 223:397–398.
22. Templeton, D., and W. Eckhart. 1982. Mutation causing premature termination of the polyoma virus medium T antigen blocks cell transformation. *J. Virol.* 41:1014–1024.
23. Tooze, J. (ed.). 1980. *DNA Tumor Viruses: Molecular Biology of Tumor Viruses*. Cold Spring Harbor Laboratory, Cold Spring Harbor, N.Y.
24. Winocour, E. 1963. Purification of polyoma virus. *Virology* 19:158–168.
25. Zouzias, D., I. Prasad, and C. Basilico. 1977. State of the viral DNA in rat cells transformed by polyoma virus. II. Identification of the cells containing nonintegrated viral DNA and the effect of viral mutations. *J. Virol.* 24:142–150.

Chapter 19

Fidelity of DNA Synthesis by Human Cell Extracts during In Vitro Replication from the Simian Virus 40 Origin

John D. Roberts and Thomas A. Kunkel

Measurements of spontaneous mutation rates in cells indicate that the genetic material is replicated and maintained with great accuracy. This accuracy, estimated to be 10^{-9} to 10^{-11} errors per base pair replicated in animal cells (9, 31, 45, 47), represents the sum of all the mutational events produced by DNA metabolic processes, including replication, repair, and recombination. While the quantitative contributions of each of these processes to overall mutation rates are not yet known, each involves the synthesis of new DNA by DNA polymerases. Logically, then, DNA polymerases are expected to have a major role in determining spontaneous mutation rates. This concept is supported by genetic and biochemical studies of procaryotic DNA polymerases, such as the mutator and antimutator T4 polymerases (10, 30, 42) and *Escherichia coli* polymerase III holoenzyme (6, 27, 39), and from investigations of mutator and antimutator phenotypes in eucaryotic systems, such as the aphidicolin-resistant polymerase α of CHO cells (22) and the more accurate mutant forms of herpes simplex virus DNA polymerase (12).

An extensive literature exists on the fidelity of purified DNA polymerases in vitro (see references 24, 25, and 37 for reviews). These studies have indicated that, although purified polymerases are significantly more accurate than is predicted by considering the free energy differences between correct and incorrect base pairs, they are inaccurate relative to the extremely low spontaneous mutation rates observed in cells. One potential explanation for this difference is that replication involves more complex and more accurate forms of the DNA polymerases than the purified preparations generally studied. There is substantial evidence in both procaryotic and eucaryotic systems that replication is the result of the concerted action of a large number of proteins in addition to the DNA polymerase itself. Eucaryotic viral model systems that primarily use host cell replication proteins are now being manipulated with good success to identify and characterize the compo-

John D. Roberts and Thomas A. Kunkel • Laboratory of Molecular Genetics, National Institute of Environmental Health Sciences, Research Triangle Park, North Carolina 27709.

nents of the mammalian replication apparatus (3).

Especially promising is the simian virus 40 (SV40) in vitro replication system recently established in several laboratories (1, 20, 43, 49). This system has a number of advantages as a model for a chromosomal replicon (3, 7). The viral genome is a circular duplex molecule that exists in vivo as a minichromosome with a nucleoprotein structure similar to that of cellular chromatin. DNA replication initiates at a unique origin and proceeds bidirectionally and semiconservatively to produce double-stranded circular daughter molecules. DNA replication depends almost entirely on host components, requiring only the SV40 T antigen from the virus. Elongation of the nascent SV40 DNA chains takes place by mechanisms that closely resemble those that operate at chromosomal replication forks.

We have taken advantage of the recent advances in the in vitro replication of SV40 DNA and combined this with an in vitro mutagenesis assay system to estimate the accuracy with which a human replication apparatus copies DNA during bidirectional, semiconservative replication. We report here some of our initial results and speculate about the implications of recent work with DNA polymerases α and δ for the fidelity of replication in eucaryotic cells.

ASSAY FOR THE FIDELITY OF DNA REPLICATION

To examine the fidelity of DNA replication, we inserted the SV40 origin of replication (on a HindIII-SphI fragment) into the unique AvaII site in M13mp2. This construct, called M13mp2SV, permits the in vitro replication of the double-stranded (replicative form [RF]) DNA by HeLa cell cytoplasmic extracts when supplemented with purified SV40 T antigen. Figure 1 illustrates the replication and mutagenesis system used. The lacZα gene on the M13mp2SV molecule serves as a mutational target that allows the detection of a wide variety of mutational events, including base substitutions, frameshifts, insertions, deletions, and more complex mutations (15). The lacZα target has been used extensively in previous studies of the accuracy of DNA synthesis by purified polymerases (15–19). It is a well-defined target for forward mutagenesis (i.e., $lacZ\alpha^+ \rightarrow lacZ\alpha$) that contains at least 110 sites at which over 200 base substitutions can be scored and at least 150 sites at which single-base frameshifts can be detected.

ANALYSIS OF M13mp2SV DNA REPLICATION PRODUCTS

Replication of the M13mp2SV DNA template by HeLa cell cytoplasmic extracts (prepared as described in reference 21) requires the presence of the SV40 T antigen in addition to the SV40 origin of replication (Table 1). The products of the reaction were analyzed by agarose gel electrophoresis and were found to be mostly nicked or covalently closed, monomer-length circles (Fig. 2, lane 1). Electrophoretic analysis in the absence of ethidium bromide indicated that the covalently closed products were partially supertwisted (data not shown). The presence of some DNA products having lower mobility in the agarose gel is consistent with data from other laboratories (1, 21, 43, 49) and probably represents replication intermediates (theta structures) or the products of rolling-circle replication.

We utilized the previously described DpnI assay (20, 33, 43, 49) to selectively monitor semiconservative replication of M13mp2SV. This assay makes use of the specificity of DpnI for DNA that is methylated on both strands at the adenine of 5'-GATC-3' sequences. The input RF DNA is methylated at the five GATC sequences present in M13mp2SV and is thus sensitive to digestion by DpnI. Upon a single round of

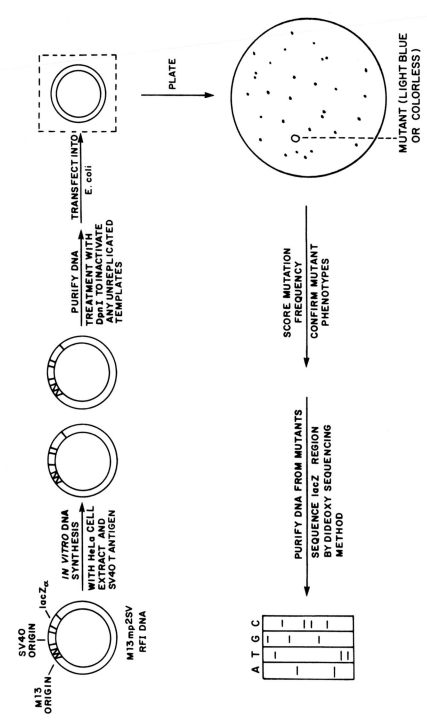

FIGURE 1. Experimental outline of the M13mp2SV mutagenesis assay. M13mp2SV, a derivative of M13mp2 that contains the SV40 origin of replication, is replicated in vitro using purified SV40 large-T antigen and a HeLa cell cytoplasmic extract. The system replicates double-stranded DNA from the SV40 origin in a bidirectional, semiconservative fashion. The DNA products are purified, treated with DpnI to eliminate any unreplicated DNA templates, and used to transfect an F'-containing E. coli strain. Mutations are scored in the lacZα gene, carried in the intergenic region of the bacteriophage M13mp2. This gene contains the information needed to obtain intracistronic complementation between the proteins produced from the lacZ information on the M13 derivative and from that carried on the F' lac in the host bacterium [CSH50 Δ(pro lac) thi ara rpsL F' proAB lacI^{q}Z ΔM15 traD36]. Such "α-complementation" restores β-galactosidase activity, which can be detected by blue-plaque formation when isopropyl-β-D-thiogalactopyranoside is used as an inducer for the lac operon and X-Gal (5-bromo-4-chloro-3-indolyl-β-D-galactopyranoside) is used as a color indicator. Plaques formed from wild-type M13mp2SV-infected cells are deep blue in color, whereas mutants with decreased or no β-galactosidase activity are lighter blue or white. Mutants can arise from a wide variety of mutational events, all of which can be determined by DNA sequence analysis.

TABLE 1
Requirements for In Vitro Replication of M13mp2SV[a]

Component omitted	DNA synthesized (pmol)
None	89
T antigen	0.4
HeLa extract	0.4
SV40 origin[b]	0.4

[a] DNA synthesis reactions (25 μl) contained 30 mM HEPES (N-2-hydroxyethylpiperazine-N'-2-ethanesulfonic acid; pH 7.8); 7 mM MgCl$_2$; 4 mM ATP; 200 μM each CTP, GTP, and UTP; 100 μM each dATP, dGTP, dTTP, and [α-^{32}P]dCTP (4,000 cpm/pmol); 40 mM creatine phosphate; 100 μg of creatine phosphokinase per ml; 15 mM sodium phosphate (pH 7.5); 40 ng of M13mp2SV RF DNA (123 pmol of nucleotide); ~1 μg of SV40 T antigen (purified as described in either reference 8 or 40); ~75 μg of protein of HeLa cytoplasmic extract (prepared as described in reference 21); and ~250 U of topoisomerase I (a gift of L. Liu, Johns Hopkins University, Baltimore, Md.). DNA synthesis was carried out for 2 h at 37°C. Acid-precipitable radioactivity was then measured as described (50).
[b] M13mp2 RF DNA was used in place of M13mp2SV.

semiconservative replication in the human cell extracts, the DNA becomes hemimethylated and is no longer sensitive to digestion by this enzyme. The finding that all the M13mp2SV in vitro replication products are completely insensitive to DpnI digestion (Fig. 2, lane 2) under conditions that allow the digestion of input M13mp2SV DNA (data not shown) indicates that the products are indeed the result of semiconservative replication and not the result of repair synthesis. The product DNA is sensitive to digestion by Sau3A1 (Fig. 2, lane 4), an isoschizomer of DpnI that incises GATC sequences regardless of their methylation, indicating that the DNA products retain the GATC sequences in a cleavable form. The circular DNA products are also partially sensitive to digestion by MboI (Fig. 2, lane 3), a DpnI isoschizomer that digests only fully unmethylated DNA, suggesting that part of the M13mp2SV templates have undergone more than one round of replication. However, even if all the DNA products arose from two rounds of replication, our estimates of error rates during replication would change by only twofold, and the data indicate that a considerable portion of the monomer circle products have been through only a single round of replication.

ANALYSIS OF MUTATION RATES DURING SV40 REPLICATION IN VITRO

We have examined the forward mutation frequency in the wild-type *lacZα* gene of M13mp2SV after replication by HeLa cell extracts. The DNA products were purified as described (50), treated with DpnI to eliminate any unreplicated templates, and used to transfect an *E. coli mutS* strain (NR9162, obtained from R. M. Schaaper, this Institute) as described (14). The frequency of mutants in the replicated DNA was only twofold above the background mutation frequency of the input M13mp2SV template (Table 2). This value can be used to calculate (see Table 2, footnote *a*) that the error rate is about 1 error for every 190,000 base pairs replicated. This is only an initial estimate since the mutation frequency of the replicated DNA is only slightly above the spontaneous background frequency.

COMPARISON OF SV40 MUTATION RATES WITH PURIFIED POLYMERASES

Replication by the SV40 system is 20- to 50-fold more accurate than DNA synthesis by purified preparations of human DNA polymerase α and is more than 100 times as accurate as human DNA polymerase β (Table 2). This suggests that either a different form of these polymerases or a different polymerase is responsible for replicative synthesis. There is strong evidence that DNA polymerase α, thought to be the primary cellular replicative enzyme (11), is absolutely required for SV40 DNA synthesis (29). Recent data also implicate polymerase δ in the replication of SV40 DNA in vitro (35). However, it is not yet clear what the precise roles of these polymerases are, or which DNA polymerase(s) is responsible for limiting the mutation rate during SV40 replication. The finding that an auxiliary protein of polymerase δ, proliferating cell nuclear antigen (2, 35), that greatly increases the

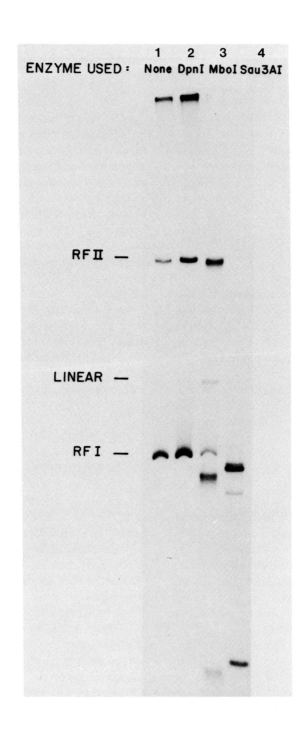

TABLE 2
Forward Mutation Frequencies of Replicated DNA

Replicative system	No. of plaques scored		Mutation frequency (10^{-4})
	Total	Mutant	
HeLa cell extract[a]			
Unreplicated M13mp2SV RF	64,975	21	3.2
M13mp2SV in vitro replication products	61,930	39	6.3
Purified enzymes[b]			
KB polymerase α	21,484	340	160
HeLa polymerase α	6,912	52	75
HeLa polymerase β	4,435	161	360

[a] M13mp2SV DNA was replicated in vitro, and the products were purified as described in the legend to Fig. 2 and used to transfect *E. coli* NR9162 [*hsdR hsdM$^+$ araD* $\triangle$(*ara leu*) $\triangle$(*lacLPOZY*) *galU galK rpsL mutS*] as described previously (14, 15). Because potential mutant replication products would be hemimethylated heteroduplexes and therefore subject to differential correction by the *E. coli* methylation-instructed mismatch repair system (26), transfections were performed using host cells deficient in this repair system (*mutS* cells). DNA replication products, generated by incubation with HeLa extract and T antigen, were treated with *Dpn*I before transfection, whereas unreplicated RF DNA was not. The average error frequency per nucleotide incorporated was determined by subtracting the background mutation frequency (3.2 × 10^{-4}), dividing by the approximate probability of expression of a mutation in the *mutS* cells (0.54), and then dividing by the approximate number of template bases at which errors coud be detected (110). The calculation yields an error rate of 1/190,000.
[b] The results with purified polymerases were obtained with gap-filling synthesis of M13mp2. The data are taken from references 16 and 18.

processivity of some preparations of the enzyme (44) is also required for SV40 DNA replication in vitro (36) has led to speculation that polymerase δ might be an efficient leading-strand polymerase, while a polymerase α-primase complex, either independent of or in physical association with polymerase δ, might carry out lagging-strand synthesis. Thus, one might expect that the mutation rate at a bidirectional replication fork would be an average of the mutation rates for polymerases α and δ alone. The precise roles of these enzymes in replication and their contributions to error rates during replication are major problems that remain to be solved.

Exonucleolytic proofreading contributes significantly to accuracy in procaryotic systems (13). Whether this mechanism operates to correct errors during eucaryotic DNA replication is not known. Interestingly, eucaryotic DNA polymerase δ is known to have a 3'→5' exonuclease activity that is capable of proofreading errors during DNA synthesis in vitro (19). Whereas most preparations of purified polymerase α have been found to lack proofreading activity (11), several reports suggest that polymerase α may also have the ability to proofread errors under appropriate conditions. Work from several laboratories indicates that a 3'→5' exonuclease is associated with multi-

FIGURE 2. Analysis of in vitro M13mp2SV DNA replication products by agarose gel electrophoresis. DNA synthesis reactions were performed as described in footnote *a* of Table 1, except that replication was carried out for 6 h at 37°C. The reaction was terminated and the DNA was purified approximately as described (50). Samples were digested with the indicated restriction enzyme and analyzed by agarose gel electrophoresis (2 to 3 V/cm) in TAE buffer (28) containing 0.2 μg of ethidium bromide per ml of buffer, followed by autoradiography of the dried gel. RF DNA standards were run in parallel lanes. The difference in mobility between the *Sau*3A1 and *Mbo*I fragments was eliminated when the products of the restriction digest were extracted with phenol before electrophoresis.

subunit forms of polymerase α (32, 41), forms that may more closely approximate those present during DNA synthesis in vivo. In addition, recent studies of the *Drosophila melanogaster* DNA polymerase α have revealed a cryptic 3'→5' exonuclease that becomes active only when the 182-kilodalton catalytic polypeptide is purified free of a tightly associated 73-kilodalton subunit (4). Thus, the possibility exists that human polymerase α may utilize a proofreading activity during replicative synthesis. Meuth and his colleagues recently demonstrated that mutation rates in cultured mammalian cells change as nucleotide pool levels are altered in a pattern consistent with the existence of a proofreading activity in vivo (34). The qualitative and quantitative effects of such an activity may be approachable using the SV40 in vitro replication system.

It seems logical that mechanisms should exist in vivo to reduce all classes of errors to approximately the same low frequency. Thus, proteins may exist in the replication complex that reduce the frequency of those errors that are most often committed by the polymerase alone. An example of such a situation probably exists in procaryotes in the case of the *E. coli* polymerase III holoenzyme. A mutation, thought to reside in an accessory subunit of the holoenzyme, has been shown to lead to a dramatic increase in a specific class of base-substitution mutations (A · T→C · G transversions) (5, 46). It appears that the polymerase alone is inaccurate for this event, but the presence of a functional accessory protein specifically reduces this mutational pathway at least 50-fold (38).

Since it would not be surprising to find similar effects in eucaryotic systems, it is instructive to consider more closely the types of mutations that have been detected during gap-filling synthesis by DNA polymerases purified from human cells and to speculate on the types of mutations that may be eliminated by a complex replication apparatus such as that anticipated to be involved in SV40 replication. A forward mutational spectrum has not yet been reported for purified polymerase δ. However, a spectrum of 175 mutants generated by human DNA polymerase α is available, as are the spectra of errors generated by polymerase α from calf thymus and chicken embryos (16–18). All show similar mutational spectra which include three major classes of errors: single-base substitutions, single-base frameshifts, and deletions. Among the base substitutions, there is a striking preference for the misinsertion of purine residues, particularly adenine bases. These base substitutions were frequent at template G residues, leading to G→T and G→C transversions. This preference is similar to the purine-insertion preference opposite apurinic sites (for review, see reference 23), a lesion that has no base hydrogen-bonding potential. This led to the speculation (18) that certain polymerase α base-substitution errors result not from miscoding but rather from misinsertion at sites where the polymerase is unable to read the pairing information of the template base, for example, at sites where the purified enzyme is prone to dissociate (partially or completely) from the DNA. Eucaryotic DNA polymerases have been shown to pause at specific template positions (48), often guanine residues. This leads to the prediction that a replication complex containing polymerase accessory proteins that increase the affinity of the polymerase for the DNA might reduce the frequency of errors generated by this process. Specifically, the frequency of (at least) the G→T and G→C transversions might be lowered.

A second striking and perhaps related feature of the polymerase α mutational spectra was the high frequency of nonrun frameshifts at template G residues (16, 17). These errors may also reflect the difficulty the polymerase has in inserting a nucleotide at these locations. Dissociation of the polymerase at certain sites may allow more conformational fluctuations at the primer terminus which could lead to the extrusion of a template base, an intermediate for a −1 frameshift mutation. In a replication complex, subunits that help the polymerase bind to the template (clamp proteins) or that confine the DNA to specific conformations during synthesis (single-stranded DNA-binding proteins) might be expected to selectively reduce the frequency of such errors.

We anticipate that the SV40 in vitro replication system will allow a detailed analysis of the proteins involved in bidirectional replication and that such an inspection will lead to insights into the mechanisms by which a DNA replication apparatus achieves fidelity.

LITERATURE CITED

1. **Ariga, H., and S. Sugano.** 1983. Initiation of simian virus 40 DNA replication in vitro. *J. Virol.* **48**:481–491.
2. **Bravo, R., R. Frank, P. A. Blundell, and H. Macdonald-Bravo.** 1987. Cyclin/PCNA is the auxiliary protein of DNA polymerase-δ. *Nature* (London) **326**:515–517.
3. **Challberg, M. D., and T. J. Kelly.** 1982. Eukaryotic DNA replication: viral and plasmid model systems. *Annu. Rev. Biochem.* **51**:901–934.
4. **Cotterill, S. M., M. E. Reyland, L. A. Loeb, and I. R. Lehman.** 1987. A cryptic 3'-5' exonuclease associated with the polymerase subunit of the DNA polymerase-primase from *Drosophila melanogaster*. *Proc. Natl. Acad. Sci. USA* **84**:5635–5639.
5. **Cox, E. C.** 1973. Mutator gene studies in *Escherichia coli*: the *mutT* gene. *Genetics* **73**(Suppl.):67–80.
6. **Cox, E. C., and D. L. Horner.** 1982. Dominant mutations in *Escherichia coli*. *Genetics* **100**:7–18.
7. **DePamphilis, M. L., and P. M. Wasserman.** 1982. Organization and replication of papovavirus DNA, p. 37–114. *In* A. S. Kaplan (ed.), *Organization and Replication of Viral DNA*. CRC Press, Inc., Boca Raton, Fla.
8. **Dixon, R. A. F., and D. Nathans.** 1985. Purification of simian virus 40 large T antigen by immunoaffinity chromatography. *J. Virol.* **53**:1001–1004.
9. **Drake, J. W.** 1969. Comparative rates of spontaneous mutation. *Nature* (London) **221**:1132.
10. **Drake, J. W., and E. F. Allen.** 1968. Antimutagenic DNA polymerases of bacteriophage T4. *Cold Spring Harbor Symp. Quant. Biol.* **33**:339–344.
11. **Fry, M., and L. A. Loeb.** 1986. *Animal Cell DNA Polymerases*. CRC Press, Inc., Boca Raton, Fla.
12. **Hall, J. D., P. A. Furman, M. H. St. Clair, and C. W. Knopf.** 1985. Reduced in vivo mutagenesis by mutant herpes simplex DNA polymerase involves improved nucleotide selection. *Proc. Natl. Acad. Sci. USA* **82**:3889–3893.
13. **Kornberg, A.** 1980. *DNA Replication, 1982 Supplement*. W. H. Freeman & Co., San Francisco.
14. **Kunkel, T. A.** 1984. Mutational specificity of depurination. *Proc. Natl. Acad. Sci. USA* **81**:1494–1498.
15. **Kunkel, T. A.** 1985. The mutational specificity of DNA polymerase-β during in vitro DNA synthesis. Production of frameshift, base substitution, and deletion mutants. *J. Biol. Chem.* **260**:5787–5796.
16. **Kunkel, T. A.** 1985. The mutational specificity of DNA polymerases-α and γ during in vitro DNA synthesis. *J. Biol. Chem.* **260**:12866–12874.
17. **Kunkel, T. A.** 1986. Frameshift mutagenesis by eukaryotic DNA polymerases in vitro. *J. Biol. Chem.* **261**:13581–13587.
18. **Kunkel, T. A., and P. S. Alexander.** 1986. The base substitution fidelity of eucaryotic DNA polymerases. Mispairing frequencies, site preferences, insertion preferences, and base substitution by dislocation. *J. Biol. Chem.* **261**:160–166.
19. **Kunkel, T. A., R. D. Sabatino, and R. A. Bambara.** 1987. Exonucleolytic proofreading by calf thymus DNA polymerase δ. *Proc. Natl. Acad. Sci. USA* **84**:4865–4869.
20. **Li, J. J., and T. J. Kelly.** 1984. Simian virus 40 DNA replication in vitro. *Proc. Natl. Acad. Sci. USA* **81**:6973–6977.
21. **Li, J. J., and T. J. Kelly.** 1985. Simian virus 40 DNA replication in vitro: specificity of initiation and evidence for bidirectional replication. *Mol. Cell. Biol.* **5**:1238–1246.
22. **Liu, P. K., C. C. Chang, J. E. Trosko, D. K. Dube, G. M. Martin, and L. A. Loeb.** 1983. Mammalian mutator mutant with an aphidicolin-resistant DNA polymerase-α. *Proc. Natl. Acad. Sci. USA* **80**:797–801.
23. **Loeb, L. A.** 1985. Apurinic sites as mutagenic intermediates. *Cell* **40**:483–484.
24. **Loeb, L. A., and T. A. Kunkel.** 1982. Fidelity of DNA synthesis. *Annu. Rev. Biochem.* **52**:429–457.
25. **Loeb, L. A., and M. E. Reyland.** 1987. Fidelity of DNA synthesis, p. 157–173. *In* F. Eckstein and D. M. J. Lilley (ed.), *Nucleic Acids and Molecular Biology*, vol. I. Springer-Verlag, Berlin.
26. **Lu, A.-L., S. Clark, and P. Modrich.** 1983. Methyl-directed repair of DNA base-pair mismatches in vitro. *Proc. Natl. Acad. Sci. USA* **80**:4639–4643.
27. **Maki, H., and A. Kornberg.** 1987. Proofreading by DNA polymerase III of *Escherichia coli* depends on cooperative interaction of the polymerase and exonuclease subunits. *Proc. Natl. Scad. Sci. USA* **84**:4389–4392.

28. Maniatis, T., E. F. Fritsch, and J. Sambrook. 1982. *Molecular Cloning: a Laboratory Manual*. Cold Spring Harbor Laboratory, Cold Spring Harbor, N.Y.
29. Murakami, Y., C. R. Wobbe, L. Weissbach, F. B. Dean, and J. Hurwitz. 1986. Role of DNA polymerase α and DNA primase in simian virus 40 DNA replication in vitro. *Proc. Natl. Acad. Sci. USA* 83:2869–2873.
30. Muzyczka, N., R. L. Poland, and M. J. Bessman. 1972. Studies on the biochemical basis of spontaneous mutation. I. A comparison of the deoxyribonucleic acid polymerases of mutator, antimutator, and wild type strains of bacteriophage T4. *J. Biol. Chem.* 247:7116–7122.
31. Nalbantoglu, J., O. Goncalves, and M. Meuth. 1983. Structure of mutant alleles of the aprt locus of Chinese hamster ovary cells. *J. Mol. Biol.* 167:575–594.
32. Ottiger, H.-P., and U. Hubscher. 1984. Mammalian DNA polymerase α holoenzymes with possible functions at the leading and lagging strand of the replication fork. *Proc. Natl. Acad. Sci. USA* 81:3993–3997.
33. Peden, K. W. C., J. M. Pipas, S. Pearson-White, and D. Nathans. 1980. Isolation of mutants of an animal virus in bacteria. *Science* 209:1392–1396.
34. Phear, G., J. Nalbantoglu, and M. Meuth. 1987. Next-nucleotide effects in mutations driven by DNA precursor pool imbalances at the aprt locus of Chinese hamster ovary cells. *Proc. Natl. Acad. Sci. USA* 84:4450–4454.
35. Prelich, G., M. Kostura, D. R. Marshak, M. B. Mathews, and B. Stillman. 1987. The cell-cycle regulated proliferating cell nuclear antigen is required for SV40 DNA replication in vitro. *Nature* (London) 326:471–474.
36. Prelich, G., C. K. Tan, M. Kostura, M. B. Mathews, A. G. So, K. M. Downey, and B. Stillman. 1987. Functional identity of proliferating cell nuclear antigen and a DNA polymerase-δ auxiliary protein. *Nature* (London) 326:517–520.
37. Roberts, J. D., and T. A. Kunkel. 1986. Mutational specificity of animal cell DNA polymerases. *Environ. Mutagen.* 8:769–789.
38. Schaaper, R. M., and R. L. Dunn. 1987. *mutT* mutator effect of *Escherichia coli* during in vitro DNA synthesis: increased occurrence of A · G replicational errors. *J. Biol. Chem.* 262:16267–16270.
39. Scheuermann, R. H., and H. Echols. 1984. A separate editing exonuclease for DNA replication: the ε subunit of *Escherichia coli* DNA polymerase III holoenzyme. *Proc. Natl. Acad. Sci. USA* 81:7747–7751.
40. Simanis, V., and D. P. Lane. 1985. An immunoaffinity purification procedure for SV40 large T antigen. *Virology* 144:88–100.
41. Skarnes, W., P. Bonin, and E. Baril. 1986. Exonuclease activity associated with a multiprotein form of HeLa cell DNA polymerase α. Purification and properties of the exonuclease. *J. Biol. Chem.* 261:6629–6636.
42. Speyer, J. F., J. D. Karam, and A. B. Lenny. 1966. On the role of DNA polymerase in base selection. *Cold Spring Harbor Symp. Quant. Biol.* 31:693–697.
43. Stillman, B. W., and Y. Gluzman. 1985. Replication and supercoiling of simian virus 40 DNA in cell extracts from human cells. *Mol. Cell. Biol.* 5:2051–2060.
44. Tan, C. K., C. Castillo, A. G. So, and K. M. Downey. 1986. An auxiliary protein for DNA polymerase-δ from fetal calf thymus. *J. Biol. Chem.* 261:12310–12316.
45. Thacker, J. 1985. The molecular nature of mutations in cultured mammalian cells: a review. *Mutat. Res.* 150:431–442.
46. Treffers, H. P., C. Spirelli, and N. O. Belser. 1954. A factor (or mutator gene) influencing mutation rates in *Escherichia coli*. *Proc. Natl. Acad. Sci. USA* 40:1064–1071.
47. Wabl, M., P. D. Burrows, A. von Gabain, and C. Steinberg. 1985. Hypermutation at the immunoglobulin heavy chain locus in a pre-B-cell line. *Proc. Natl. Acad. Sci. USA* 82:479–482.
48. Weaver, D. T., and M. L. DePamphilis. 1982. Specific sequences in native DNA that arrest synthesis by DNA polymerase-α. *J. Biol. Chem.* 257:2075–2086.
49. Wobbe, C. R., F. Dean, L. Weissbach, and J. Hurwitz. 1985. In vitro replication of duplex circular DNA containing the simian virus 40 DNA origin site. *Proc. Natl. Acad. Sci.* 82:5710–5714.
50. Yang, L., M. S. Wold, J. J. Li, T. J. Kelly, and L. F. Liu. 1987. Roles of DNA topoisomerases in simian virus 40 DNA replication in vitro. *Proc. Natl. Acad. Sci. USA* 84:950–954.

III. Mechanisms of Misincorporation

Mechanisms of Misincorporation: Introduction

Mutagenesis in the cell may be divided into several steps. The first of these is the incorporation of an "incorrect" nucleotide into the newly synthesized DNA. This part contains studies which investigate this step and which are directed to answering the question of how an incorrect nucleotide gets into DNA. Of necessity, this step must focus on the synthetic aspects of misincorporation. In both procaryotes and eucaryotes misincorporation occurs at a frequency approximately four orders of magnitude greater than the resulting frequency of mutation in the cellular system. Misincorporation frequencies have been demonstrated to vary widely, depending upon the polymerase used, and may be as high as 10^{-3} to 10^{-4} per nucleotide incorporated in viral systems to 10^{-5} to 10^{-6} when cellular DNA polymerases are used. The subsequent reactions lowering the misincorporation frequency to the appropriate mutation frequency will be dealt with in the next chapters.

It has been a major challenge to design systems to monitor misincorporation at these two ranges. Such systems must be exquisitely sensitive. Although misincorporation can be measured by identifying bases either by chemical means or with radioactive label, such an approach has limitations in sensitivities. For this reason, misincorporation studies have tended to monitor gene function by either a reporter gene or expression of selectable gene activity.

Despite the fact that misincorporation reflects a synthetic function, it is appropriate to divide misincorporation into two steps: utilization of the incorrect base in synthesis (misinsertion) and the subsequent editing step. At the synthesis step polymerases are more accurate than would be supposed on the basis of free-energy differences between correct and incorrect base pairing. While this seems appropriate and intuitively correct, models to explain this phenomenon are just being developed.

Studies with DNA polymerase I demonstrated that the intrinsic $3' \rightarrow 5'$ exonuclease activity of this enzyme removed mispaired bases at a free terminus ~1,000-fold more rapidly than it removed correctly paired bases. This led to the concept of a proofreading or "editorial" exonuclease activity which would constantly check the incorporation activity of the polymerase itself. It seems reasonable to conclude that editorial exonuclease activities contribute two orders of magnitude in improvement of the accuracy of the replication process.

While DNA polymerase I from *Esche-*

richia coli is an example of a polypeptide which contains a $3' \to 5'$ exonuclease activity intrinsic with the polymerase activity, this does not seem to be the general case. DNA polymerase III holoenzyme from *E. coli* contains the apparent $3' \to 5'$ exonuclease activity in a separate subunit (ε, product of the *dnaQ* gene). As discussed in Chapter 1, eucaryotic DNA polymerases apparently may not have $3' \to 5'$ exonuclease activity. Recent studies raise the question of whether an editorial exonuclease may be present in the synthetic subunit of a DNA polymerase but not be apparent unless the structure of the polypeptide is altered, for example, by partial proteolysis.

In the synthesis reaction the presence of the editorial exonuclease may slow the rate of synthesis past sites of misincorporation. Inactivation of this activity may result in increased processivity for the enzyme and decreased discrimination against mispairing. This yields a DNA polymerase which is useful for incorporations through regions of complex structure.

The relationship of the editorial function to the synthesis function in a physical or structural sense is also being actively investigated. In the case of DNA polymerase I these functions are at different "sites" in the enzyme. Investigators have assumed that the editorial exonuclease must have ready access to the inserted nucleotide to detect mispairing and therefore must function in apposition to synthesis. Structural data indicate that the editorial site of DNA polymerase I may not lie immediately adjacent to the synthesis site.

The in vitro replication systems described in Chapters 1 and 2 have allowed the development of mutagenesis assay systems in which synthesis is dependent upon a holoenzyme, as in the case of DNA polymerase III holoenzyme from *E. coli*. They have also allowed the addition of recognized accessory proteins of DNA replication to purified synthesis subunits. These systems promise to give us information regarding misincorporation under conditions in which replicative synthesis more closely approaches the format used in the cell.

The nature of the lesion facing the DNA polymerase in the template or the nature of the incorrect base to be incorporated influences the level of misincorporation. For example, the response of the replicative apparatus to an apurinic site might be different from its response to a pyrimidine dimer. Some lesions essentially terminate DNA replication with a complete blockade, whereas other lesions seem to represent little hindrance to the replicative apparatus. Likewise, some base analogs are readily incorporated into DNA, whereas other base analogs seem to be discriminated against by the DNA polymerases.

Thus, it appears that at the level of misincorporation, in addition to the polymerase activity itself and the editorial or proofreading function, there are several other associated important elements: the nature of the surrounding template or primer DNA (the nucleotide neighborhood), the nature of the lesion, the nature of the incoming nucleotide, and the rate of synthesis.

A number of questions remain to be investigated.

(i) We do not understand the effect of surrounding DNA on the local rate of misincorporation. Likewise, we lack sophistication with regard to subtle differences in a template; for example, do all apurinic sites, regardless of cause, give the same misincorporation at a given site in the DNA?

(ii) We do not have an explanation at the present time for the discrimination by the DNA polymerase in selecting the incoming base.

(iii) Although systems utilizing replicative complexes or DNA polymerases with accessory enzymes promise to make misincorporation rates more meaningful in terms of the intercellular environment, they add complexity to interpretation.

(iv) Accessory proteins may have more than one activity. A subset of investigation in this area will be the effect of extragenic suppression of deficiencies in accessory proteins on the rate of mutagenesis. This promises to be an exciting and fruitful area of investigation and should also allow us to determine the roles of accessory proteins. For example, the ε subunit of DNA polymerase III holoenzyme in *E. coli* seems to be strictly required and may be serving functions other than editorial proofreading.

(v) Lastly, we do not understand why DNA polymerases can vary so widely in misincorporation rates. Understanding this will require more information regarding the physical structure of a number of DNA polymerases.

The Editors

Chapter 20

Fidelity of Base Selection by DNA Polymerases
Site-Specific Incorporation of Base Analogs

Bradley D. Preston, Richard A. Zakour, B. Singer, and Lawrence A. Loeb

INTRODUCTION

The preservation of species requires an exceptionally high accuracy for the replication of the genome. Measurements of spontaneous mutation frequencies in procaryotes and eucaryotes suggest that the average frequency of errors in DNA replication is in the range of 10^{-7} to 10^{-9} and 10^{-9} to 10^{-11} errors per base pair, respectively (8). This value is the net result of errors made during DNA replication minus errors corrected by cellular DNA repair mechanisms. In both of these processes, DNA polymerases and associated enzyme activities play a central role. How can a cell exploit minor differences in chemical structures between different nucleotide pairs to generate large differences in free energy that are required for the high accuracy of DNA replication in vivo? Current evidence suggests that this high fidelity is achieved by a multistep process involving base pairing, base selection by DNA polymerases, and proofreading by an associated $3' \rightarrow 5'$ exonuclease (13). In addition, mistakes by DNA polymerases are corrected postsynthetically in bacteria by a mismatch correction system (38), and an analogous system may be present in eucaryotic cells (18).

CONTRIBUTION OF BASE PAIRINGS TO THE FIDELITY OF DNA SYNTHESIS

The free energy differences between complementary and noncomplementary base pairs are insufficient to account for the accuracy of DNA synthesis even by DNA polymerases without associated exonucleases. With few exceptions in the literature, this difference in free energy is usually estimated as 1 to 2 kcal (or 4,000 to 8,000 J) per mol. At equilibrium in an aqueous environment this small free-energy difference

---

Bradley D. Preston and Lawrence A. Loeb • Joseph Gottstein Memorial Cancer Research Laboratory, Department of Pathology, University of Washington, Seattle, Washington 98195.   *Richard A. Zakour* • Molecular and Applied Genetics Laboratory, Allied Corp., Morristown, New Jersey 07960.   *B. Singer* • Donner Laboratory, University of California, Berkeley, California 94720.

($-\Delta\Delta G$) would only account for a level of discrimination of 1 in 10 to 1 in 100 (42; Table 1). This estimate has received strong experimental support from studies on non-enzyme-mediated polymerization of activated nucleotides on synthetic polynucleotide templates (37), in which error rates of 1/100 to 1/200 were indeed obtained.

BASE SELECTION BY DNA POLYMERASES

A tabulation of experimentally determined error rates of different DNA polymerases as measured by the frequency of single base substitutions at a designated template position in a biologically active DNA is shown in Table 1. In this assay a singly primed single-stranded φX174 am3 DNA template is copied by DNA polymerase, and errors are quantitated from the reversion frequency of progeny phage after plating on *Escherichia coli* permissive or nonpermissive for the amber mutation (69). For measurement of accuracy greater than 10^{-6}, the deoxyribonucleoside triphosphate (dNTP) pool is biased to favor misincorporation, and the results are extrapolated to equal concentrations of all four dNTPs. The error rates of DNA polymerases lacking activity that excises terminal noncomplementary nucleotides varies from 10^{-3} to 10^{-4} for avian myeloblastosis virus RNA-dependent DNA polymerase (AMV-pol) (16) and DNA polymerase β (31) to 6×10^{-6} for the four-subunit DNA polymerase-primase complex purified from calf thymus by immunoaffinity chromatography (54). The fact that the accuracy of DNA polymerases in the absence of exonucleolytic activity is greater than that indicated by differences in free energy between complementary and noncomplementary base pairings indicates that these DNA polymerases enhance the selection of complementary base pairs.

One of the simplest mechanisms for enhancement of accuracy in the absence of exonuclease is a nonselective increase in $-\Delta\Delta G$ mediated by the interaction of DNA polymerase with the DNA template. Petruska et al. (47) have proposed a model in which $-\Delta\Delta G$ is magnified by the exclusion of water from the active site of DNA polymerase. In fact, the energy for dissociation of base pairs in vacuum is much greater than for those obtained in water. Also, the X-ray diffraction data obtained with *E. coli* DNA polymerase I (pol I) indicate that the DNA template is buried in a hydrophobic cleft in the enzyme that could exclude water (25, 46). With this mechanism for enhanced base selection by DNA polymerase, the error rate should be proportional to the difference in K_m between the complementary and noncomplementary nucleotides, and this can now be experimentally verified. However, this nonselective mechanism does not account for the differences in mutational spectra that have been observed in vitro with different species of DNA polymerases (28).

Alternatively, the nucleotide binding site in DNA polymerases could be versatile

TABLE 1
Mechanisms for the Fidelity of DNA Synthesis

Mechanism and representative DNA polymerases	Error frequency (reference)
Base pairing (free energy estimate)	1/10–1/100[a]
Base selection	
AMV-pol	1/1,000–1/8,000 (16)
Calf thymus pol β	1/6,700 (31)
Calf thymus pol α	
Homopolymeric form	1/30,000 (31)
Four-subunit complex	1/670,000 (54)
Base selection and exonucleolytic proofreading	
E. coli pol I	1/680,000 (13)
Bacteriophage T4 pol	$1/10^7$ (32)
Drosophila 182-kDa pol subunit	$1/10^7$ (7)

[a] Calculated as follows: $-\Delta\Delta G = -(\Delta G_{correct} - \Delta G_{incorrect}) = -RT \ln (\frac{incorrect}{correct})$. Since $-\Delta\Delta G = 1$ to 2 kcal/mol, therefore $\ln (\frac{incorrect}{correct}) = 1/10$ to $1/100$.

and change conformation at each nucleotide addition step. Support for a conformational model includes the presence of a single dNTP binding site on *E. coli* DNA pol I (27), the characteristic mutational spectrum for each species of DNA polymerase lacking exonuclease (28), and the immobilization of nucleotide substrates by DNA polymerases (12). Recent studies on the kinetics of incorporation of single noncomplementary nucleotides suggest predominantly a K_m effect but also a V_{max} effect on fidelity (4a), implying that DNA polymerases increase discrimination by both kinetic parameters.

Kinetic mechanisms for increased fidelity by DNA polymerases in the absence of exonucleolytic proofreading have been proposed (22, 23, 45). For example, discrimination could occur by rejection of the incorrect dNTP after binding to the enzyme but prior to incorporation (29, 43). Alternatively, the DNA polymerase could dissociate from the template-primer after incorporation of an incorrect nucleotide. In the φX fidelity assay, the resultant template-primers with terminal mismatches would not yield revertant progeny and thus would not be scorable (31, 69). However, in the eucaryotic cell in vivo this mechanism would permit the correction of polymerase errors by a separate $3'{\rightarrow}5'$ exonuclease. The recent demonstrations of high fidelity in the absence of detectable exonuclease by the multisubunit DNA polymerase α obtained from *Drosophila* embryos (26) and calf thymus (54) will provide attractive experimental material to test these and other hypotheses.

CONTRIBUTION OF THE $3'{\rightarrow}5'$ EXONUCLEASE TO ACCURACY

Proofreading is a major contributor to the accuracy of many DNA polymerases (27). Most procaryotic DNA polymerases and DNA polymerases from DNA viruses contain a $3'{\rightarrow}5'$ exonuclease as an integral part of the catalytic subunit (13). As a general rule, the fidelity of these polymerases is equal to or greater than that exhibited by enzymes lacking this exonuclease. Based on an increase in error rates by selective inhibition of the $3'{\rightarrow}5'$ exonuclease, it has been estimated that this activity contributes as much as three orders of magnitude to the fidelity of DNA synthesis by procaryotic DNA polymerases (32).

EXONUCLEOLYTIC PROOFREADING IN EUCARYOTES

The contribution of exonucleolytic proofreading to fidelity in eucaryotes is controversial. The power of this mechanism to achieve high accuracy in procaryotes and the even higher accuracy likely to be required for replication of the larger eucaryotic genomes suggests that an exonucleolytic proofreading step will also be operative in eucaryotes. A proofreading exonuclease need not be on the catalytic subunit of a DNA polymerase; for example, the $3'{\rightarrow}5'$ exonuclease of *E. coli* polymerase III is located on a separate subunit (57). However, studies with DNA polymerase α have uniformly demonstrated a lack of exonucleolytic activity (13), and even the multisubunit DNA polymerase-primase complex is devoid of exonucleolytic activity (7). Nevertheless, DNA polymerase δ, an enzyme closely related to DNA polymerase α, is primarily characterized by an associated $3'{\rightarrow}5'$ exonuclease (13). Moreover, although the multisubunit DNA polymerase-primase from *Drosophila melanogaster* lacks an associated exonucleolytic activity, a cryptic exonuclease is revealed after separation of the 182-kilodalton (kDa) polymerase subunit (7). Thus, it is possible that DNA polymerase α generically contains a $3'{\rightarrow}5'$ exonuclease but this activity is not apparent under assay conditions in vitro. Alternatively, a separate proofreading exonuclease could be present in eucaryotic cells and work in association with DNA polymerase α, as is the case for E. coli polymerase III.

SITE-SPECIFIC INCORPORATION OF NONCOMPLEMENTARY NUCLEOTIDES

The presentation of a single noncomplementary dNTP substrate to a DNA template-primer is handled differently by DNA polymerases with and without an associated 3'→5' exonuclease. DNA polymerases with 3'→5' exonucleases repetitively incorporate and excise noncomplementary substrates in an idling-turnover reaction (27, 43) that generates deoxyribonucleoside monophosphates. In fact, this turnover reaction has been used as a measure of proofreading (27). The yields of extended primers depend on the relative rates of the misincorporation and excision processes, which are determined in part by the DNA sequence and the availability of the next correct dNTP (15, 33). Incubation of the large fragment of E. coli DNA pol I with 15-mer-primed φX174 DNA in the presence of single dNTPs yields 16-mers only with the correct nucleotide dTTP (Fig. 1). With dCTP the idling-turnover reaction presumably occurs continuously at the 3'-terminal C residue of the primer to yield predominantly unchanged 15-mer, and with dGTP the 3'→5' exonuclease activity presumably exceeds the polymerase activity, resulting in removal of the 5'-^{32}P-labeled primer from the template.

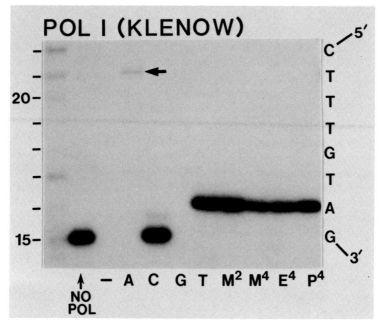

FIGURE 1. Polyacrylamide gel electrophoresis of oligonucleotides site specifically extended by the large Klenow fragment of E. coli pol I. 5'-^{32}P-labeled 15-mer was hybridized to single-stranded φX174 am3 DNA and incubated with polymerase and a single dNTP. The nucleotide sequence in the right margin indicates the sequence of the template strand. The oligonucleotide lengths are indicated in the left margin as determined from standard markers (leftmost lane). Lanes: No pol, no polymerase; −, no dNTP; A, dATP; C, dCTP; G, dGTP; T, dTTP; M^2, O^2-methyl-dTTP; M^4, O^4-methyl-dTTP; E^4, O^4-ethyl-dTTP; P^4, O^4-i-propyl-dTTP. For details, see references 49 and 50.

With dATP this hydrolysis is also the predominant reaction; however, a new band is observed at position 21 (Fig. 1, arrow). This can be accounted for by the "next nucleotide effect" (15, 33), where dATP increases the rate of polymerization relative to excision so as to favor misincorporation at both position 16 and 18. The effective ratio of polymerase to exonuclease activities of *E. coli* pol I can be increased by using α-thio-dNTPs as substrates; these are readily incorporated by pol I but are not excised by its 3'→5' exonuclease (30). Under these conditions, mismatched nucleotides formed during the idling-turnover reaction of pol I are captured as a result of blocked exonucleolytic proofreading (59).

DNA polymerases lacking 3'→5' exonucleases incorporate nucleotides without excision. Therefore, given an infinite amount of time, a DNA polymerase lacking exonucleolytic proofreading activity should, in principle, incorporate all possible noncomplementary nucleotides. If it is assumed that base selection is determined, at least in part, by the dNTP binding site(s) of a polymerase, then error-prone polymerases should be relatively nondiscriminatory in the use of dNTP substrates. Incubation of error-prone, RNA-dependent DNA polymerases with 15-mer-primed φX174 DNA and single dNTPs yields 16-mers with all possible mismatches opposite the template A residue (Fig. 2). Interestingly, the varied yields of these 16-mers do not completely coincide with the predicted stabilities of the mismatched base pairs (67) and presumably reflect differences in base selection by the polymerases. Continued incubation of the terminally mismatched products results in the addition of multiple noncomplementary (and eventually, complementary) nucleotides. However, the multiplicity of terminal additions is limited to about 3 to 4 mismatched nucleotides; this could be mediated by a progressive dissociation of the primer from the template with each noncomplementary nucleotide addition step.

The incorporation of noncomplementary nucleotides by error-prone polymerases has been exploited to engineer site-specific mutations into recombinant DNA vectors (70, 71) and should be adaptable to permit the introduction of a broad spectrum of

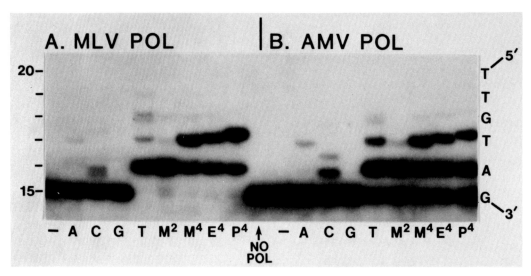

FIGURE 2. Polyacrylamide gel electrophoresis of oligonucleotides site specifically extended by (A) MLV-pol and (B) AMV-pol. The experimental protocol and abbreviations are described in the legend to Fig. 1 and in references 49 and 50.

Base Selection by DNA Polymerases

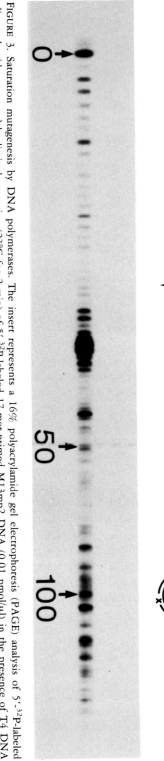

FIGURE 3. Saturation mutagenesis by DNA polymerases. The insert represents a 16% polyacrylamide gel electrophoresis (PAGE) analysis of 5′-^{32}P-labeled oligonucleotides prepared by limited extension (23°C for 2 min) of 5′-^{32}P-labeled 17-mer-primed M13mp2 DNA (0.01 pmol/μl) in the presence of T4 DNA polymerase (0.05 U/μl), the large fragment of *E. coli* pol I (0.001 U/μl), and 500 μM each of dATP, dCTP, dGTP, and dTTP. The numbers of nucleotides incorporated onto the 17-mer primer are indicated in the lower margin. POL, DNA polymerase; AMVrt, AMV-pol. For details on limited primer extension, see reference 50.

mutations throughout the entire length of a target gene (Fig. 3). This enzymatic approach to "saturation mutagenesis" is based on two properties of DNA polymerases: (i) the ability to incorporate noncomplementary nucleotides and (ii) the tendency to pause at multiple sites during polymerization in vitro. The pausing of polymerases can be accentuated under conditions that limit the rates of polymerization (e.g., by using limited substrate or polymerase concentrations), thus yielding extended primers with unit increases in length (Fig. 3). A second round of polymerization with an error-prone polymerase (e.g., AMV-pol) and single dNTPs would then yield populations of nascent strands with base substitution mutations at each 3' terminus. Finally, these noncomplementary nucleotides can be sealed into the mutant strands by continued polymerization in the presence of all four dNTPs and DNA ligase. This strategy for saturation mutagenesis is currently being developed in our laboratory to introduce phenotypically selectable mutations in the lacZα gene of bacteriophage M13mp2 (41).

SITE-SPECIFIC INCORPORATION OF BASE ANALOGS

Studies of DNA polymerization in the presence of base analogs provide a means of probing the selectivity and structural requirements for catalysis by DNA polymerases. By selectively replacing unmodified dNTPs with dNTP analogs during DNA polymerization in vitro, one can determine the apparent kinetic parameters for an analog and thus obtain a measure of its efficacy as a polymerase substrate and/or inhibitor. If polymerase fidelity is determined in part by faithful base selection, then it follows that error-prone polymerases might be more flexible in their use of dNTP analogs than highly accurate polymerases. This hypothesis is supported by studies com-

TABLE 2
O-Alkyl-dTTPs as Substrates for DNA Polymerases

dTTP Analog	Relative nucleotide incorporation[a] by polymerase:	
	AMV-pol	pol I
Control (no dTTP)	2 ± 1	2 ± 0.2
dTTP	100 ± 18	100 ± 18
O^2-Methyl-dTTP	45 ± 2	23 ± 2
O^4-Methyl-dTTP	71 ± 3	87 ± 7
O^4-Ethyl-dTTP	46 ± 2	22 ± 4
O^4-i-Propyl-dTTP	19 ± 3	3 ± 0.3

[a] Single-stranded φX174 am3 [^{3}H-dT]DNA was primed with a synthetic 15-mer and incubated at 30°C for 15 min with AMV-pol or E. coli pol I in the presence of dATP, [α-^{32}P]dCTP, dGTP, and either dTTP or an O-alkyl-dTTP. Nucleotide incorporation was calculated from acid-precipitable radioactivity and is expressed as the average incorporation for each polymerase relative to dTTP (mean plus or minus standard deviation of four determinations). For details, see references 49 and 50.

paring a series of O-alkyl-dTTPs as substrates for polymerization by AMV-pol and E. coli pol I (Table 2). In all of the reactions, with the exception of those containing O^4-methyl-dTTP, synthesis with the error-prone polymerase AMV-pol was greater than with the highly accurate enzyme E. coli pol I. The interpretation of these data is complicated by the presence of the 3'→5' exonuclease activity of pol I. However, the apparent inability of this exonuclease to excise O^4-alkyl-dTMP residues (60) suggests that the differential incorporation of the analogs reflects a more stringent base selection by pol I.

There is much literature showing that a variety of dNTP analogs, including those produced by mutagenic alkylating agents, are substrates for DNA polymerases (Table 3). Several of these studies compare the incorporation of dNTP analogs by polymerases with different fidelities (1, 17, 21, 34, 48, 52, 53, 55, 56, 65, 66). However, the association of 3'→5' exonuclease activity with many of the enzymes (13, 27, 31) and the differences in the fidelity of different polymerase preparations (13, 54) prevent a simple interpretation of the data in terms of

TABLE 3
Base Analogs Incorporated by DNA Polymerases

Analog	Polymerase incorporated[a]	Reference(s)
dATP analogs		
1-Methyl-	pol I; T4; AMV	68; Preston et al.[b]
1,N^6-Etheno-	pol I; pol III; pol β	52, 53; K. Beattie[c]
2-Aminopurine	pol I; *Salmonella typhimurium*; T4; pol α	4, 10, 48
2,6-Diaminopurine	pol I	6
N^6-Hydroxy-	pol I; pol III; T4	1
N^6-Methyl-	pol I; T4	9, 39
dCTP analogs		
N^4-Hydroxy-	pol I; T4	68
N^4-Methoxy-	pol I	51, 62
5-Bromo-	pol I; T2	3, 27; Preston et al.
5-Fluoro-	pol I; T2	27
5-Hydroxymethyl-	T2	27
5-Iodo-	AMV	Preston et al.
5-Mercuri-	AMV	Preston et al.
5-Methyl-	pol I; T2; AMV	3, 27; Preston et al.
dGTP analogs		
O^6-Methyl-	pol I; T4; T5; AMV; pol α	17, 65, 66; Preston et al.
7-Deaza-	pol I	2, 44
7-Methyl-	pol I	19
8-Aminofluorene-	pol I; AMV	Preston et al.
8-Hydroxy-	AMV	Preston et al.
Hypoxanthine	pol I	3
dTTP/dUTP analogs		
O^2-Methyl-	pol I; AMV; MLV	49, 50, 63
O^2-Ethyl-	pol I	B. Singer[d]
O^2-*i*-Propyl-	pol I	B. Singer
O^4-Ethyl-	pol I; AMV; MLV	49, 50, 63
O^4-Methyl-	pol I; AMV; MLV; pol α	17, 49, 50, 61, 63
O^4-*i*-Propyl-	pol I; AMV; MLV	49, 50, 63
5-Azido-	pol I	11
5-Biotinyl-	pol I; T4; HSV; pol α; pol β	5, 20, 34, 58
5-Br-	pol I; pol III; T2; T4; AMV; pol β	3, 21, 27, 35, 36, 53; Preston et al.; K. Beattie
5-*n*-Butyl-	pol I; pol α; pol β	56
5-*t*-Butyl-	pol I; pol α; pol β	56
5-Ethyl-	pol I; pol α; pol β	56
5-Fluoro-	pol I; T2	27
5-*n*-Hexyl-	pol I; pol α; pol β	56
5-Iodo-	pol I; AMV	21, 53; Preston et al.
5-Mercuri-	AMV	Preston et al.
5-*n*-Pentyl-	pol I; pol α; pol β	56
5-*i*-Propyl-	pol I; pol α; pol β	56
5-*n*-Propyl-	pol I; HSV; pol α; pol β	55, 56
5,6-Dihydro-	pol I	24
E-5-(2-Bromovinyl)-	HSV; pol α; pol β	55
E-5-Propenyl-	HSV; pol α; pol β	55

[a] Abbreviations: AMV, avian myeloblastosis virus polymerase; HSV, herpes simplex virus polymerase; MLV, murine leukemia virus polymerase; pol I, *E. coli* pol I; pol III, *E. coli* pol III core or holoenzyme (see each reference); pol α, eucaryotic pol α; pol β, eucaryotic pol β; *Salmonella typhimurium*, *S. typhimurium* pol I; T2, bacteriophage T2 pol; T4, bacteriophage T4 pol; T5, bacteriophage T5 pol.
[b] B. D. Preston, D. Wu, and L. A. Loeb, unpublished data.
[c] K. Beattie, personal communication.
[d] B. Singer, unpublished data.

base selection accuracy. An intriguing influence of polymerase binding sites is suggested by the recent observation that incorporation of 1,N^6-etheno-dATP occurs much more readily with *E. coli* pol III holoenzyme than with eucaryotic pol β, despite the lack of $3' \rightarrow 5'$ exonucleolytic proofreading activity in the latter enzyme (52; K. Beattie, personal communication). Thus, the stringency of base analog selection is not necessarily a simple reflection of polymerase fidelity. Studies on the recently cloned pol III α subunit (40) are required to determine the fidelity of base selection by this polymerase in the absence of exonucleolytic proofreading.

The abilities of dNTP analogs to replace dNTPs during DNA polymerization suggest that, under the appropriate conditions, polymerases might serve as general tools for the site-specific incorporation of base analogs. To determine the abilities of DNA polymerases to site-specifically incorporate O-alkyl-dTTPs, 15-mer-primed φX174 DNA was incubated in the presence of single dTTP analogs, and the extended oligonucleotides were analyzed by denaturing polyacrylamide gel electrophoresis. A comparison of three different polymerases shows that the yields of extended primers are dependent on both the source of the polymerase and the structure of the alkyl adduct (Fig. 1 and 2). With the error-prone polymerases murine leukemia virus RNA-dependent DNA polymerase (MLV-pol) (Fig. 2A) or AMV-pol (Fig. 2B) that lack $3' \rightarrow 5'$ exonuclease activity, the O-alkyl-dTTPs are incorporated onto the 3'-prime terminus to produce 16-, 17-, and 18-mers in relatively high yields. The total yield of extended primers is greatest with MLV-pol and approaches 100%. In contrast, extension of the primers by the highly accurate large fragment of pol I is limited to the insertion of single nucleotides opposite the template A residue to yield exclusively 16-mers. Under these conditions, the size or presence of the alkyl group does not appear to affect the yields of extended primers by either the RNA-dependent DNA polymerases or pol I.

CONCLUSIONS AND FUTURE DIRECTIONS

These studies on dNTPs and dNTP analogs illustrate the potential applicability of site-specific primer extension experiments for studying the stringencies and mechanisms of base selection by DNA polymerases. By comparing polymerases with different fidelities in the absence of $3' \rightarrow 5'$ exonucleolytic proofreading, it should be possible to identify the relative contributions of polymerase base selection to the fidelity of DNA synthesis and to probe the molecular environment of the dNTP binding sites with dNTP analogs. A reciprocal approach using site-specifically mismatched (64; M. E. Reyland, I. R. Lehman, and L. A. Loeb, *J. Biol. Chem.*, in press) or mutagen-modified (14, 60) primer-template structures as substrates for $3' \rightarrow 5'$ exonucleases will permit a more detailed characterization of the structural requirements for $3' \rightarrow 5'$ exonucleases.

LITERATURE CITED

1. Abdul-Masih, M. T., and M. J. Bessman. 1986. Biochemical studies on the mutagen, 6-N-hydroxylaminopurine. Synthesis of the deoxynucleoside triphosphate and its incorporation into DNA *in vitro*. *J. Biol. Chem.* 261:2020–2026.
2. Barr, P. J., R. M. Thayer, P. Laybourn, R. C. Najarian, F. Seela, and D. R. Tolan. 1986. 7-Deaza-2'-deoxyguanosine-5'-triphosphate: enhanced resolution in M13 dideoxy sequencing. *BioTechniques* 4:428–432.
3. Bessman, M. J., I. R. Lehman, J. Adler, S. B. Zimmerman, E. S. Simms, and A. Kornberg. 1958. Enzymatic synthesis of deoxyribonucleic acid. III. The incorporation of pyrimidine and purine analogues into deoxyribonucleic acid. *Proc. Natl. Acad. Sci. USA* 44:633–640.
4. Bessman, M. J., N. Muzyczka, M. F. Goodman, and R. L. Schnaar. 1974. Studies on the biochemical basis of spontaneous mutation. II. The incor-

poration of a base and its analogue into DNA by wild-type, mutator and antimutator DNA polymerases. *J. Mol. Biol.* 88:409–421.

4a. Boosalis, M. S., J. Petruska, and M. F. Goodman. 1987. DNA polymerase insertion fidelity: gel assay for site-specific kinetics. *J. Biol. Chem.* 262:14689–14696.

5. Brigati, D. J., D. Myerson, J. J. Leary, B. Spalholz, S. Z. Travis, C. K. Y. Fong, G. D. Hsiung, and D. C. Ward. 1983. Detection of viral genomes in cultured cells and paraffin-embedded tissue sections using biotin-labeled hybridization probes. *Virology* 126:32–50.

6. Cerami, A., E. Reich, D. C. Ward, and I. H. Goldberg. 1967. The interaction of actinomycin with DNA: requirement for the 2-amino group of purines. *Proc. Natl. Acad. Sci. USA* 57:1036–1042.

7. Cotterill, S. M., M. E. Reyland, L. A. Loeb, and I. R. Lehman. 1987. A cryptic proofreading 3'→5' exonuclease associated with the polymerase subunit of the DNA polymerase-primase from *Drosophila melanogaster*. *Proc. Natl. Acad. Sci. USA* 84:5635–5636.

8. Drake, J. W., E. F. Allen, S. A. Forsberg, R. M. Preparata, and E. O. Greenberg. 1969. Spontaneous mutation. *Nature* (London) 221:1128–1132.

9. Engel, J. D., and P. H. von Hippel. 1978. d(m6ATP) as a probe of the fidelity of base incorporation into polynucleotides by *Escherichia coli* DNA polymerase I. *J. Biol. Chem.* 253:935–939.

10. Engler, M. J., and M. J. Bessman. 1978. Characterization of a mutator DNA polymerase I from *Salmonella typhimurium*. Cold Spring Harbor Symp. Quant. Biol. 43:929–935.

11. Evans, R. K., J. D. Johnson, and B. E. Haley. 1986. 5-Azido-2'-deoxyuridine 5'-triphosphate: a photoaffinity-labeling reagent and tool for the enzymatic synthesis of photoactive DNA. *Proc. Natl. Acad. Sci. USA* 83:5382–5386.

12. Ferrin, L. J., R. A. Beckman, L. A. Loeb, and A. S. Mildvan. 1986. Kinetic and magnetic resonance studies of the interaction of Mn^{2+}, substrates and templates with DNA polymerases, p. 259–273. *In* F. C. Wedler and V. L. Schramm (ed.), *Manganese in Metabolism and Enzyme Function*. Academic Press, Inc., New York.

13. Fry, M., and L. A. Loeb. 1986. *Animal Cell DNA Polymerases*. CRC Press, Inc., Boca Raton, Fla.

14. Fuchs, R. P. P. 1984. DNA binding spectrum of the carcinogen N-acetoxy-N-2-acetylaminofluorene significantly differs from the mutation spectrum. *J. Mol. Biol.* 177:173–180.

15. Goodman, M. F., and E. W. Branscomb. 1986. DNA replication fidelity and base mispairing, p. 191–232. *In* T. B. L. Kirkwood, R. F. Rosenberger, and D. J. Galas (ed.), *Accuracy in Molecular Processes*. Chapman and Hall, London.

16. Gopinathan, K. P., L. A. Weymouth, T. A. Kunkel, and L. A. Loeb. 1979. Mutagenesis *in vitro* by DNA polymerase from an RNA tumor virus. *Nature* (London) 278:857–959.

17. Hall, J. A., and R. Saffhill. 1983. The incorporation of O^6-methyldeoxyguanosine and O^4-methyldeoxythymidine monophosphates into DNA by DNA polymerases I and α. *Nucleic Acids Res.* 11:4185–4193.

18. Hare, J. T., and H. Taylor. 1985. One role for DNA methylation in vertebrate cells is strand discrimination in mismatch repair. *Proc. Natl. Acad. Sci. USA* 82:7350–7354.

19. Hendler, S., E. Furer, and P. R. Srinivasan. 1970. Synthesis and chemical properties of monomers and polymers containing 7-methyguanine and an investigation of their substrate or template properties for bacterial deoxyribonucleic acid or ribonucleic acid polymerases. *Biochemistry* 9:4141–4153.

20. Herman, T. M., E. Lefever, and M. Shimkus. 1986. Affinity chromatography of DNA labeled with chemically cleavable biotinylated nucleotide analogs. *Anal. Biochem.* 156:48–55.

21. Hillebrand, G. G., A. H. McCluskey, K. A. Abbott, G. G. Revich, and K. L. Beattie. 1984. Misincorporation during DNA synthesis, analyzed by gel electrophoresis. *Nucleic Acids Res.* 12:3155–3171.

22. Hopfield, J. J. 1974. Kinetic proofreading: a new mechanism for reducing errors in biosynthetic processes requiring high specificity. *Proc. Natl. Acad. Sci. USA* 71:4135–4139.

23. Hopfield, J. J. 1980. The energy relay: a proofreading scheme based on dynamic cooperativity and lacking all characteristic symptoms of kinetic proofreading in DNA replication and protein synthesis. *Proc. Natl. Acad. Sci. USA* 77:5248–5252.

24. Ide, H., R. J. Melamede, and S. S. Wallace. 1987. Synthesis of dihydrothymidine and thymidine glycol 5'-triphosphates and their ability to serve as substrates for *Escherichia coli* DNA polymerase I. *Biochemistry* 26:964–969.

25. Joyce, C. M., and T. A. Steitz. 1987. DNA polymerase I: from crystal structure to function via genetics. *Trends Biochem. Sci.* 12:288–292.

26. Kaguni, L. S., R. A. DiFrancesco, and I. R. Lehman. 1984. The DNA polymerase-primase from *Drosophila melanogaster* embryos. Rate and fidelity of polymerization on single-stranded DNA templates. *J. Biol. Chem.* 259:9314–9319.

27. Kornberg, A. 1980. *DNA Replication*. W. H. Freeman & Co., San Francisco.

28. Kunkel, T. A., and P. S. Alexander. 1986. The base substitution fidelity of eucaryotic DNA polymerases. *J. Biol. Chem.* 261:160–166.

29. Kunkel, T. A., R. A. Beckman, and L. A. Loeb.

1986. On the fidelity of DNA synthesis: pyrophosphate-induced misincorporation allows detection of two proofreading activities. *J. Biol. Chem.* 261: 13610–13616.

30. Kunkel, T. A., F. Eckstein, A. S. Mildvan, R. M. Koplitz, and L. A. Loeb. 1981. Deoxynucleoside [1-thio]triphosphates prevent proofreading during *in vitro* DNA synthesis. *Proc. Natl. Acad. Sci. USA* 78:6734–6738.

31. Kunkel, T. A., and L. A. Loeb. 1981. Fidelity of mammalian DNA polymerases. *Science* 213:765–767.

32. Kunkel, T. A., L. A. Loeb, and M. F. Goodman. 1984. On the accuracy of DNA replication: the accuracy of T4 DNA polymerases in copying φX174 DNA *in vitro*. *J. Biol. Chem.* 259:1539–1545.

33. Kunkel, T. A., R. M. Schaaper, R. A. Beckman, and L. A. Loeb. 1981. On the fidelity of DNA replication. Effect of the next nucleotide on proofreading. *J. Biol. Chem.* 256:9883–9889.

34. Langer, P. R., A. A. Waldrop, and D. C. Ward. 1981. Enzymatic synthesis of biotin-labeled polynucleotides: novel nucleic acid affinity probes. *Proc. Natl. Acad. Sci. USA* 78:6633–6637.

35. Lasken, R. S., and M. F. Goodman. 1984. The biochemical basis of 5-bromouracil-induced mutagenesis. Heteroduplex base mispairs involving bromouracil in G·C→A·T and A·T→G·C mutational pathways. *J. Biol. Chem.* 259:11491–11495.

36. Lasken, R. S., and M. F. Goodman. 1985. A fidelity assay using dideoxy DNA sequencing: a measurement of sequence dependence and frequency of forming 5-bromouracil-guanine base mispairs. *Proc. Natl. Acad. Sci. USA* 82:1302–1305.

37. Lohrmann, R., and L. E. Orgel. 1980. Efficient catalysis of polycytidylic acid-directed oligoguanylate formation by Pb^{2+}. *J. Mol. Biol.* 142:555–567.

38. Lu, A.-L., S. Clark, and P. Modrich. 1983. Methyl-directed repair of DNA base-pair mismatches *in vitro*. *Proc. Natl. Acad. Sci. USA* 80:4639–4643.

39. Mace, D. C. 1984. N^6-Methyldeoxyadenosine 5'-triphosphate as a probe of the fidelity mechanisms of bacteriophage T4 DNA polymerase. *J. Biol. Chem.* 259:3616–3619.

40. Maki, H., and A. Kornberg. 1985. The polymerase subunit of DNA polymerase III of *Escherichia coli*. II. Purification of the α subunit, devoid of nuclease activities. *J. Biol. Chem.* 260:12987–12992.

41. Messing, J. 1983. New M13 vectors for cloning. *Methods Enzymol.* 101:20–78.

42. Mildvan, A. S., and L. A. Loeb. 1979. Role of metal ions in the mechanisms of DNA and RNA polymerases. *Crit. Rev. Biochem.* 6:219–244.

43. Mizrahi, V., P. Benkovic, and S. J. Benkovic. 1986. Mechanism of DNA polymerase I: exonuclease/polymerase activity switch and DNA sequence dependence of pyrophosphorolysis and misincorporation reactions. *Proc. Natl. Acad. Sci. USA* 83:5769–5773.

44. Mizusawa, S., S. Nishimura, and F. Seela. 1986. Improvement of the dideoxy chain termination method of DNA sequencing by use of deoxy-7-deazaguanosine triphosphate in place of dGTP. *Nucleic Acids Res.* 14:1319–1324.

45. Ninio, J. 1975. Kinetic amplification of enzyme discrimination. *Biochimie* 57:587–595.

46. Ollis, D. L., P. Brick, R. Hamlin, N. G. Xuong, and T. A. Steitz. 1985. Structure of large fragment of *Escherichia coli* DNA polymerase I complexed with dTMP. *Nature* (London) 313:762–766.

47. Petruska, J., L. C. Sowers, and M. F. Goodman. 1986. Comparison of nucleotide interactions in water, proteins, and vacuum: relevance to DNA polymerase fidelity. *Proc. Natl. Acad. Sci. USA* 83:1559–1562.

48. Pless, R. C., L. M. Levitt, and M. J. Bessman. 1981. Nonrandom substitution of 2-aminopurine for adenine during deoxyribonucleic acid synthesis *in vitro*. *Biochemistry* 20:6235–6244.

49. Preston, B. D., B. Singer, and L. A. Loeb. 1986. Mutagenic potential of O^4-methylthymine *in vivo* determined by an enzymatic approach to site-specific mutagenesis. *Proc. Natl. Acad. Sci. USA* 83:8501–8505.

50. Preston, B. D., B. Singer, and L. A. Loeb. 1987. Comparison of the relative mutagenicities of O-alkylthymines site-specifically incorporated into φX174 DNA. *J. Biol. Chem.* 262:13821–13827.

51. Reeves, S. T., and K. L. Beattie. 1985. Base-pairing properties of N^4-methoxydeoxycytidine 5'-triphosphate during DNA synthesis on natural templates, catalyzed by DNA polymerase I of *Escherichia coli*. *Biochemistry* 24:2262–2268.

52. Revich, G. G., and K. L. Beattie. 1986. Utilization of 1,N^6-etheno-2'-deoxyadenosine 5'-triphosphate during DNA synthesis on natural templates, catalyzed by DNA polymerase I of *Escherichia coli*. *Carcinogenesis* 7:1569–1576.

53. Revich, G. G., G. G. Hillebrand, and K. L. Beattie. 1984. High-performance liquid chromatographic purification of deoxynucleoside 5'-triphosphates and their use in a sensitive electrophoretic assay of misincorporation during DNA synthesis. *J. Chromatogr.* 317:283–300.

54. Reyland, M. E., and L. A. Loeb. 1987. On the fidelity of DNA replication: isolation of high fidelity DNA polymerase-primase complexes by immunoaffinity chromatography. *J. Biol. Chem.* 262:10824–10830.

55. Ruth, J. L., and Y.-C. Cheng. 1981. Nucleoside analogues with clinical potential in antivirus chemo-

56. Sagi, J., R. Nowak, B. Zmudzka, A. Szemzo, and L. Otvos. 1980. A study of substrate specificity of mammalian and bacterial DNA polymerases with 5-alkyl-2'-deoxyuridine 5'-triphosphates. *Biochim. Biophys. Acta* **606**:196–201.
57. Scheuermann, R. H., and H. Echols. 1984. A separate editing exonuclease for DNA replication: the ε subunit of *Escherichia coli* DNA polymerase III holoenzyme. *Proc. Natl. Acad. Sci. USA* **81**:7747–7751.
58. Shimkus, M., J. Levy, and T. Herman. 1985. A chemically cleavable biotinylated nucleotide: usefulness in the recovery of protein-DNA complexes from avidin affinity columns. *Proc. Natl. Acad. Sci. USA* **82**:2593–2597.
59. Shortle, D., P. Grisafi, S. J. Benkovic, and D. Botstein. 1982. Gap misrepair mutagenesis: efficient site-directed induction of transition, transversion, and frameshift mutations *in vitro*. *Proc. Natl. Acad. Sci. USA* **79**:1588–1592.
60. Singer, B. 1986. O^4-Methyldeoxythymidine replacing deoxythymidine in poly[d(A-T)] renders the polymer resistant to the 3'→5' exonuclease activity of the Klenow and T4 polymerases. *Nucleic Acids Res.* **14**:6735–6743.
61. Singer, B., F. Chavez, and S. J. Spengler. 1986. O^4-Methyl-, ethyl-, and isopropyl deoxythymidine triphosphates as analogues of deoxythymidine triphosphate: kinetics of incorporation by *Escherichia coli* DNA polymerase I. *Biochemistry* **26**:1201–1205.
62. Singer, B., H. Fraenkel-Conrat, L. G. Abbott, and S. J. Spengler. 1984. N^4-Methoxydeoxycytidine triphosphate is in the imino tautomer and substitutes for deoxythymidine triphosphate in primed poly d[A-T] synthesis with *E. coli* DNA polymerase I. *Nucleic Acids Res.* **12**:4609–4619.
63. Singer, B., J. Sagi, and J. T. Kusmierek. 1983. *Escherichia coli* polymerase I can use O^2-methyldeoxythymidine or O^4-methyldeoxythymidine in place of deoxythymidine in primed poly(dA-dT) · poly(dA-dT) synthesis. *Proc. Natl. Acad. Sci. USA* **80**:4884–4888.
64. Sinha, N. K. 1987. Specificity and efficiency of editing of mismatches involved in the formation of base-substitution mutations by the 3'→5' exonuclease activity of phage T4 DNA polymerase. *Proc. Natl. Acad. Sci. USA* **84**:915–919.
65. Snow, E. T., R. S. Foote, and S. Mitra. 1984. Kinetics of incorporation of O^6-methyldeoxyguanosine monophosphate during *in vitro* DNA synthesis. *Biochemistry* **23**:4289–4294.
66. Toorchen, D., and M. D. Topal. 1983. Mechanisms of chemical mutagenesis and carcinogenesis: effects on DNA replication of methylation at the O^6-guanine position of dGTP. *Carcinogenesis* **4**:1591–1597.
67. Topal, M. D., and J. R. Fresco. 1976. Complementary base pairing and the origin of substitution mutations. *Nature* (London) **263**:285–293.
68. Topal, M. D., C. A. Hutchinson III, and M. S. Baker. 1982. DNA precursors in chemical mutagenesis: a novel application of DNA sequencing. *Nature* (London) **298**:863–865.
69. Weymouth, L. A., and L. A. Loeb. 1978. Mutagenesis during *in vitro* DNA synthesis. *Proc. Natl. Acad. Sci. USA* **75**:1924–1928.
70. Zakour, R. A., E. A. James, and L. A. Loeb. 1984. Site specific mutagenesis: insertion of single noncomplementary nucleotides at specified sites by error-directed DNA polymerization. *Nucleic Acids Res.* **12**:6615–6628.
71. Zakour, R. A., and L. A. Loeb. 1982. Site-specific mutagenesis by error-directed DNA synthesis. *Nature* (London) **295**:708–710.

Chapter 21

Oligonucleotides with Site-Specific Structural Anomalies as Probes of Mutagenesis Mechanisms and DNA Polymerase Function

G. Peter Beardsley, Thomas Mikita, Alton B. Kremer, and James M. Clark

Structural anomalies or "lesions" may arise in DNA through chemical or radiation damage and also through misincorporation of abnormal precursors. Knowledge of the resulting structural perturbations in DNA and the attendant alterations in DNA physicochemical and biochemical properties is critical for eventual understanding of the mechanisms by which DNA damage produces profound effects on living cells (Fig. 1). The investigation of such mechanisms is complex at the cellular level as a result of low lesion frequency, multiplicity of lesion types, wide variation in lesion sites, and, often, ongoing repair processes. DNA oligomers containing a single lesion at 100% frequency can be prepared using chemical or enzymatic methodology. These allow the study of variations in DNA physicochemical and biochemical properties which result from the presence of the structural lesions. Precise control is possible over the critical factors of lesion frequency, specificity, site, and sequence context. The structurally modified DNA oligomers may also be valuable as probes for the mechanisms of enzymes which act on DNA. Work from our laboratory on the 5-fluorouracil (U^F), arabinosyl cytosine (araC), and thymine glycol lesions, using this approach, will be described.

OLIGOMER PREPARATION

The choice between chemical or enzymatic methodology for preparation of lesion-containing oligomers is dictated by several factors. Chemical synthesis has the primary advantage of scale and is favored when large quantities of oligomers are necessary, such as for physicochemical studies, particularly those using nuclear magnetic

---

G. Peter Beardsley and Thomas Mikita • Departments of Pediatrics and Pharmacology, Yale University School of Medicine, New Haven, Connecticut 06510. *Alton B. Kremer* • Ortho Pharmaceuticals, Raritan, New Jersey 08869. *James M. Clark* • National Institute of Environmental Health Sciences, Research Triangle Park, North Carolina 27709.

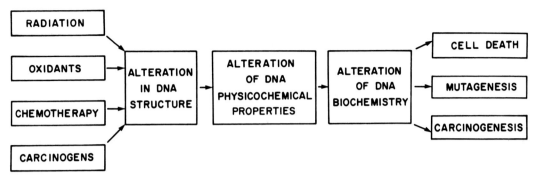

FIGURE 1. Origins and consequences of lesions in DNA.

resonance (NMR). Solid-phase, sequential DNA oligomer synthesis using either phosphoramidite, phosphotriester (2), or cyanoethyl phosphite (18) chemistry is the usual method. This requires the chemical synthesis of a suitably protected and activated precursor of the abnormal nucleotide residue. The ability of the precursor to introduce the desired lesion should be checked by preparation of a short test oligomer whose structure can be proven by physical measurements such as NMR spectroscopy and also by enzymatic degradation to the expected mixture of natural and modified nucleotides or nucleosides. Some structural abnormalities in DNA introduce chemical instability not compatible with the relatively vigorous conditions necessary for precursor or oligomer synthesis. DNA oligomers themselves undergo low levels of chemical damage during synthesis, and precursor purification can never be absolute. For these reasons, chemically synthesized oligonucleotides are probably not suitable for study of very-low-frequency biologic events. Certainly, no valid conclusion can be drawn regarding events which occur at a frequency lower than the demonstrable purity of the oligomer.

Alternative methods for synthesis of lesion-containing oligomers include introduction of lesions enzymatically using the 5'-triphosphate of the anomalous nucleotide and a tolerant or "promiscuous" polymerase (B. D. Preston, R. A. Zakour, B. Singer, and L. A. Loeb, this volume). Another alternative which is sometimes possible is direct chemical introduction of lesions into preformed oligomers. This requires a chemical reagent with high specificity for producing only one type of structural lesion, as well as conditions in which the reaction proceeds essentially to completion. To obtain the necessary site specificity, it is usually necessary to arrange the sequence of the target oligomers so that there is only one residue susceptible to reaction with the modification reagent. It may be necessary to introduce the lesion into a short oligomer which can more easily be separated from other reaction or nonreaction products. For biologic studies, these short oligomers must be subsequently ligated into longer pieces of DNA. The same general cautions relating to purity apply to oligomers prepared by any of these methods, including enzymatic synthesis.

THE U^F LESION

The pyrimidine analog 5-fluorouracil, U^F, and several of its derivatives are important antineoplastic drugs (7). Misincorporation of the analogs into RNA and DNA occurs as a consequence of treatment with the drug and may play a role in its antiproliferative effects (12, 22). Perturbations of DNA structure and function by U^F are of particular interest as it represents an extreme among the 5-halouracils, fluorine being the most electronegative of all the elements. Figure 2A shows the ionization through loss of the N-3 proton of U^F. In DNA this occurs with pK_a 7.9, in comparison to thymidine, which has a corresponding pK_a above 9 (8). Thus a significant fraction of U^F in DNA and RNA will be ionized in the physiologic pH range. Figure 2B shows the hypothetical base mispair which might be formed between ionized U^F and guanosine.

We prepared the DNA oligomer duplex containing the following site-specific U^F to test the base-pairing properties of U^F in DNA:

GTAAAACGACGGCC
CATTTTGCTGCCGGGU^{F}AC

The 14-mer primer was extended to full length over the lesion-containing template by using [α-^{32}P]dGTP along with the other unlabeled deoxynucleotide triphosphates (dNTPs). In every full-length extension product, [^{32}P]dGMP was added once opposite the terminal deoxycytosine (dC) of the template. A second [^{32}P]dGMP was added opposite U^F with a frequency dependent on the occurrence of U^F-G base pairs. The relative amounts of incorporation opposite dC and U^F in the template can be determined by exhaustive 3' phosphodiesterase digestion of the extended products, followed by separation and quantitation of the 3'-[^{32}P]dCMP (resulting from incorporation of [^{32}P]dGMP opposite U^F) and 3'-[^{32}P]dTMP (resulting from incorporation of [^{32}P]dGMP opposite C). The ratio of 3'-[^{32}P]dCMP to 3'-[^{32}P]dTMP gives the misincorporation frequency opposite U^F. The sensitivity, or minimally detectable misincorporation level, is 0.5 to 1%. None of the polymerases tested, including those with and without 3'→5' exonuclease activity, showed misincorporation frequencies above the minimum sensitivity level (T. Mikita, A. B. Kremer, and G. P. Beardsley, unpublished data). This was true even when reactions were carried

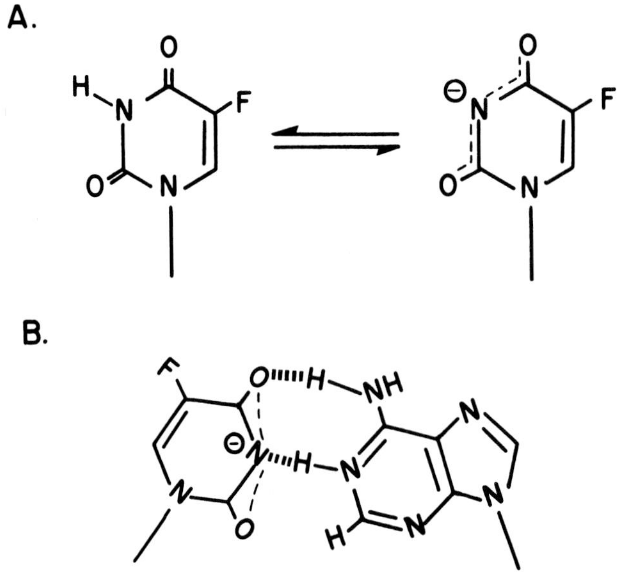

FIGURE 2. (A) Ionization of U^F. (B) Hypothetical ionized U^F-G base pair.

out at pH 8.2, where the majority of U^F in DNA is ionized. The conclusion is that U^F-G base pairs, if they are formed at all, occur at a frequency less than the sensitivity of the assay. Detection of the true frequency of U^F-G mispairing will require a more sensitive assay.

We next assessed the physicochemical stability of the U^F-A base pair by NMR spectroscopy (8). Figure 3 shows a portion of the ^{1}H-NMR spectrum of the self-complementary duplex dodecamer analog of the Pribnow box:

```
1 2 3 4 5  6 6  5 4 3 2 1
CGATUF AT A ATCG
GCTA A TAUFTAGC
```

at various temperatures. Spectra were run in the presence of 1 mM cacodylate, 1 mM EDTA, and 100 mM NaCl to avoid the possibility of buffer catalysis of imino proton exchange. The peaks shown are the resonance signals of the hydrogen-bonded imino protons of the six unique base pairs. Broadening of the signal with increasing temperature is due to proton exchange with solvent

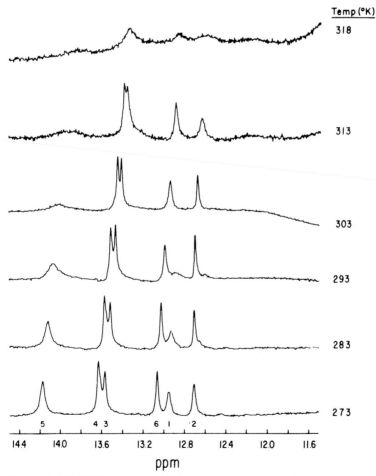

FIGURE 3. ^{1}H-NMR spectrum of U^F-A-containing oligomer duplex at various temperatures.

and is a function of the base-pair opening rate. The signal for the U^F-A (position 5) base pair broadens at a lower temperature than the signals for the other base pairs and at a lower temperature than for the corresponding A-T base pair in the control Pribnow duplex:

```
1 2 3 4 5 6 6 5 4 3 2 1
C G A T T A T A A T C G
G C T A A T A T T A G C
```

The overall melting temperatures for the helices did not significantly differ (data not shown).

Base-pair opening rates were derived by determining the exchange rate at various buffer concentrations and then extrapolating to infinite buffer. When this was done for the above dodecamers, the opening rate for the U^F-A base pair was faster than that for the analogous base pair in the control helix. Comparison of the opening rates for the other base pairs in the U^F-containing helix yielded values comparable to those for the analogous base pairs in the control helix (8).

These results indicate that the U^F-A base pair is somewhat less stable than the natural A-T base pair. However, only minimal perturbation of overall duplex stability is incurred by substitution of U^F for T. The use of U^F as a nondisruptive "reporter" nucleus to monitor DNA-protein interactions (13) appears justified.

THE THYMINE GLYCOL LESION

Figure 4 shows the structure of one of two diastereomers of cis-thymine glycol which arise as a consequence of ionizing radiation or oxidant damage to thymine base in DNA. The major chemical differences between thymine glycol and the thymine from which it is derived are (i) loss of π electron aromaticity and, consequently, loss of the capacity for positive stacking interactions with adjacent bases; (ii) loss of pyrimidine ring planarity, with major increase in steric bulk on one side of the ring as the methyl group of the thymine glycol assumes a quasi-equatorial position; and (iii) addition of two hydroxyl groups in the glycol with consequent increase in potential for inter- and intramolecular H bonding.

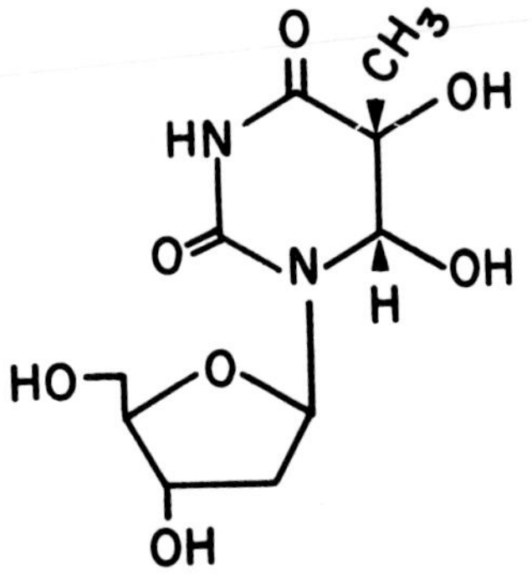

FIGURE 4. Structure of cis-($5R,6S$)-thymine glycol.

The thymine glycol lesion is unstable to strongly basic conditions and therefore cannot be introduced into DNA oligomers by the usual solid-phase oligomer synthesis method. Fortunately, osmium tetroxide, a chemical oxidant, can convert a thymine base in DNA to the cis-thymine glycol lesion with excellent selectivity and in high yield (1). We were able to find conditions under which a single thymine residue in a DNA oligomer could be quantitatively converted to cis-thymine glycol (3). Of the two possible diastereomers possible, stereochemical considerations predict that the $5R,6S$ diastereomer would be the predominant product.

The major functional consequence of the thymine glycol lesion in template DNA is inhibition of replication past the lesion site (3). Surprisingly, polymerases are inhibited from progression after inserting a base opposite the lesion, whereas most other lesions, e.g., bulky adducts or abasic sites, result in stop sites one nucleotide before the lesion (15). The nucleotide added opposite the thymine glycol lesion is primarily the "correct" one, dAMP, since omission of dATP from the polymerase reaction mixture causes primer extension to stop one nucleotide before the thymine glycol lesion. The methodology used would not detect incorporation of incorrect bases opposite thymine glycol at a frequency below 1 to 2%. Finally, by varying the sequence context of the thymine glycol lesion, we were able to establish that bypass of the thymine glycol lesion is favored by a 5'-following pyrimidine, rather than a purine, and by longer-length templates 5' to the lesion.

We constructed an energy-minimized computer graphic model of the *cis*-thymine glycol lesion in duplex DNA to understand the structural basis for the functional effects of the lesion (5). In the case where a purine follows the glycol lesion on the 5' side, quite severe distortions of the duplex occur, and there is resulting destabilization relative to the unmodified duplex. The flanking purine is displaced by the bulky quasi-equatorial methyl group of the thymine glycol, and destabilization results primarily from diminished stacking interactions. H-bonding base-pair interactions between thymine glycol and adenine are not significantly perturbed. The steric disturbances of the thymine glycol methyl group can be better accommodated by a 5'-flanking pyrimidine. These properties of the model are consistent with the biochemical observations and suggest that the origin of the effects lies primarily in the thermodynamics of the structurally altered DNA rather than in perturbed DNA-protein interactions between the modified template-primer and DNA polymerases.

The overall prediction of our modeling and in vitro biochemical studies is that the *cis*-thymine glycol lesion may inhibit overall DNA replication, but clearly can be bypassed under some conditions. Thymine glycol base pairs primarily with adenine, and thus the lesion in vivo is not likely to be highly mutagenic. Data from other laboratories have borne out these predictions. Basu et al. (A. K. Basu, S. A. Leadon, and J. M. Essigmann, Abstr. ASM Conference on DNA Replication and Mutagenesis, 1987, abstr. no. 48, p. 17) have site-specifically inserted the *cis*-thymine glycol lesion into a plasmid which was then transfected into mammalian cells. They detected T→C mutations at the lesion site with a frequency of 0.3%. Also, LeClerc and co-workers showed that mutations produced by osmium tetroxide were primarily at cytosine rather than thymidine sites (R. C. Hayes, L. A. Petrullo, H. Huang, S. S. Wallace, and J. E. LeClerc, submitted for publication).

THE AraC LESION

AraC is a nucleoside analog with potent antileukemic activity. In the course of its action it is misincorporated into DNA, and the amount of misincorporation correlates strongly with cytotoxicity (9, 11). The misincorporation of araC constitutes a structural lesion in DNA in which the abnormality is in the sugar-phosphate backbone rather than in the base.

Using chemical methodology, we developed reagents for the introduction of araC into either chain-terminal or internucleotide positions in synthetic DNA oligomers (G. P. Beardsley, T. Mikita, M. M. Klaus, S. C. Srivastava, and A. L. Nussbaum, *Biochemistry*, in press). These were used to construct araC-containing template-primer oligomer duplexes to study the specific consequences of the araC lesion on DNA replication processes (T. Mikita and G. P. Beardsley, submitted for publication).

Using template-primers such as

GTAAAACGAGGGCC*
CATTTTGCTCCCGGGTAC

where C* indicates araC or dC, we were able to investigate the effects of araC at 3' chain termini on DNA polymerase-catalyzed chain

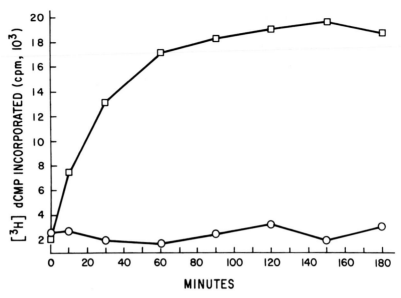

FIGURE 5. Kinetics of next-nucleotide addition to araC (○) and dC (□) primers by HeLa cell DNA polymerase α_2.

elongation. Figure 5 shows the comparative rates of addition of [^{3}H]dCTP to araC- and dC-terminated primers by HeLa cell polymerase α_2 (21). The dramatic difference in template primer utilization (Fig. 5) was quantitated by repeating the experiment using [α-^{32}P]dCTP, stopping the reactions at an early time point when the incorporation rate was approximately linear, electrophoretically separating the products, and counting the relative amounts of radiolabel in the full-length extension products (20-mers). By this method it was determined that the rate of elongation of the araC-terminated primer is at least 100-fold less than that of the control dC primer.

The araC primer extension product material was then exhaustively digested to 3' mononucleotides to determine the 5'-nearest neighbor of the incorporated [α-^{32}P]dCMP. Analysis of this material showed that the ^{32}P label was exclusively associated with 3' dCMP rather than with 3' araC monophosphate. Thus, the small amount of extension product obtained with the araC primer is formed only after araC monophosphate is excised from the primer terminus by the 3'→5' exonuclease activity associated with this enzyme (19). Therefore, the 100-fold difference in primer extension product formation determined above greatly underestimates the actual difference between the rate of addition of the next nucleotide to araC and of the similar addition to dC.

Similar results were obtained with the large (Klenow) fragment of *Escherichia coli* DNA polymerase I and T4 polymerase, both of which have associated 3'→5' exonuclease activity. Initial experiments with mammalian DNA polymerase β, an enzyme lacking an editing exonuclease, indicate that whereas extension of the araC-terminated primer is approximately 100-fold slower than that with the control primer, extension occurs via addition to araC. Thus, DNA polymerase β appears more tolerant of the aberrant sugar moiety than the α polymerase.

In view of the profound inhibitory effects on chain elongation when araC is situated at the primer terminus, the possibility existed that araC similarly situated might affect the activity of DNA ligase when it forms a phosphodiester bond between the 3' terminus of one DNA fragment and the 5' terminus of an adjacent fragment. The effect on ligation, as catalyzed by T4 DNA ligase, was observed to be much less inhibitory than for the chain elongation reaction; a rate difference of only three- to fourfold was found. The reason for this large difference between the effects of araC on ligation and its effects on chain elongation is not known, but may be related to the processivity of the chain elongation reaction as opposed to the ligase reaction, which is distributive.

The effect araC had on chain elongation when it was situated at an internucleotide site in the template was surprising. The lesion produced a partial block to replication with an overall 3- to 15-fold reduction (depending on the polymerase used) in the rate of primer extension past the lesion site (Fig. 6). Analysis of the products formed showed the site of the block to be opposite the template lesion site, with insertion of the correct nucleotide, dGMP, opposite the lesion. This latter observation is not surprising as the base of araC is structurally normal.

Whereas the structural anomalies in the sugar portion of 3'-terminal araC are located in close proximity to the sites of bond formation and breaking in the chain elongation and ligase reactions, they are spatially much further removed from those activities when the lesion occurs in the template. Furthermore, crystallographic and NMR studies on araC mononucleosides show only very minimal deviance from the structure of dC (6, 17). Although we have not yet constructed an energy-minimized model of the araC lesion in DNA, there seems little reason to expect that it will show distortions of the same magnitude as were found for the *cis*-thymine glycol lesion, and the functional consequences of the araC lesion are unlikely to be explained by perturbed stability of the DNA substrate. In fact, we have found that

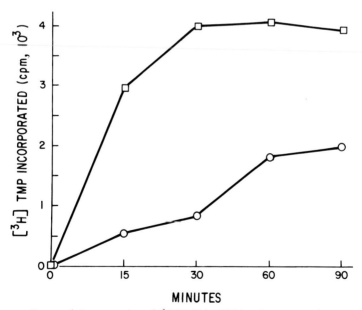

FIGURE 6. Incorporation of [^{3}H]TMP by DNA polymerase α_2 into

GTAAAACGAGGGCC
CATTT TGCT CCCGGGC*AC

where C* is araC (○) and dC (□).

the melting temperature of araC-containing oligomers is only slightly different from that of controls.

The origin of araC-induced effects on DNA replication processes is likely to reside in perturbed DNA-protein interactions between the altered DNA substrates and the relevant enzymes. Physical models of araC in B-form duplex DNA show that the additional hydroxyl group is located at one edge of the major groove. Until more is known about polymerase structure and the detailed mechanism of its catalysis, however, the ways in which this structural anomaly might perturb the interactions of this enzyme with its DNA substrate will remain unknown.

BLUNT-END ADDITION REACTIONS

In experiments in which short oligonucleotide template-primers were extended with various polymerases, we often detected small amounts of product having a length one nucleotide longer than the primer. When a mutant of the large (Klenow) fragment of *E. coli* DNA polymerase I, lacking $3' \rightarrow 5'$ exonuclease activity (C. M. Joyce, J. M. Friedman, L. Beese, P. S. Fremont, and T. A. Steitz, this volume), was used, "+1 addition" became the predominant reaction (Fig. 7). Determination of the sequence of the +1 product showed it to result from addition of dAMP to the 3' terminus of the blunt-end duplex first formed by extension of the primer to the full length of the template. This unexpected reaction was further investigated by utilizing the blunt-end duplex

5'^{32}P-GTCCGTCTCTGCCTC
3'   CAGGCAGAGACGGAG

The +1 and smaller amounts of similar +2 addition product were again found, and sequencing showed these to be products of

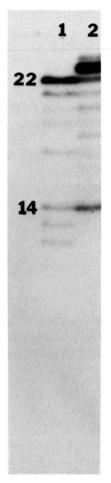

FIGURE 7. The +1 addition reaction. Autoradiogram of products of elongation of a 5'-end-labeled 14-mer primer annealed to a 22-mer template by wild-type (lane 1) and *exo*-mutant (lane 2) Klenow fragment. The +1 and +2 products are seen with the mutant enzyme.

dAMP addition (4). These could not have arisen as a result of any primer-template slippage, since the "template" (lower) strand contains no T residues, and "fold-back" structures of the "primer" (upper) strand would have low stability.

Further investigation showed that this non-template-directed addition (Fig. 8) does not occur on single-strand DNA, and with equal dNTP pool sizes, there is a strong preference for dAMP addition. Non-tem-

NON-TEMPLATE-DIRECTED END ADDITION

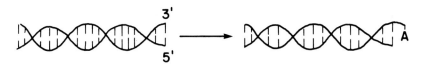

PSEUDO-TEMPLATE-DIRECTED END ADDITION

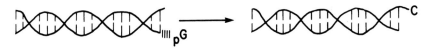

FIGURE 8. Blunt-end addition phenomena.

plate-directed addition may be a general property of DNA polymerases, as we have also detected it by using wild-type Klenow fragment, mammalian DNA polymerases α and β, yeast DNA polymerase I, and DNA polymerase from *Thermis aquaticus*.

Comparative rates of non-template- versus template-directed addition have not been directly measured. However, the observation that +1 products are only minor products when 3'→5' exonuclease activity is present implies that their rate of formation must be slower than their removal by the 3'→5' exonuclease. Since 3'→5' exonuclease rates are roughly 2 orders of magnitude slower than template-directed addition, it can be estimated that non-template-directed addition is approximately 3 orders of magnitude slower than template-directed addition.

The phenomenon which we term pseudo-template-directed addition (Fig. 8) was discovered when high concentrations of a deoxynucleotide monophosphate (dNMP) were added to blunt-end addition reaction mixtures (4). Under these conditions, in addition to the preference for dAMP addition noted earlier, further preference for addition of the base-pairing partner of the added monophosphate was found. Thus, for example, in the presence of dGMP and equal concentrations of all four dNTPs, the product of +1 addition of dCMP is found, in addition to the product of +1 dAMP addition. The pseudo-template-directed +1 addition reaction may involve noncovalent association of the dNMP with the 5' terminus at the blunt end, as could occur through positive stacking interactions. This might then serve as a pseudo-template to direct addition of the appropriate dNTPs to the 3' terminus of the blunt end. The process does not seem to involve dNMP binding to the monophosphate-binding site of the Klenow fragment, as pseudo-template-directed addition was seen with the mutant enzyme which does not bind dNMPs (Joyce et al., this volume).

Whether either of these processes has any evolutionary significance is a matter for speculation. It is conceivable that these phenomena represent functional vestiges of prebiotic polymerases, less stringent in their template requirements. In a sense, blunt ends represent the ultimate in noninstructional lesions, and the preference for addition of dAMP is reminiscent of the similar reference seen with noninstructional lesions located within template strands (16).

It is possible that these activities may play some role in mutagenesis, producing frameshift mutations. Polymerases would only encounter blunt ends infrequently at double-strand breaks or at chromosomal telomeres. The frequency of +1 frameshift mutations is very low, and blunt-end addition reactions could account for at least some of them.

SUMMARY

The examples discussed above represent only a small part of an emerging use of oligonucleotides containing structural anomalies as probes of biochemical and biological processes. A large number of lesion types have been studied, including alkylation products (10), bulky adducts (14), and models of abasic sites (20). Such lesions have been used for studies of mutagenesis, DNA-protein interactions, and DNA stability and dynamics. Utilization of this methodology will undoubtedly increase with expansion to additional lesions and other DNA-related phenomena.

LITERATURE CITED

1. Beer, M., S. Stern, D. Carmalt, and K. H. Mohlenrich. 1966. Determination of base sequence in nucleic acids with the electron microscope. V. The thymine-specific reactions of osmium tetroxide with deoxyribonucleic acid and its components. *Biochemistry* 5:2283–2288.
2. Caruthers, M. H. 1985. Gene synthesis machines; DNA chemistry and its uses. *Science* 230:281–285.
3. Clark, J. M., and G. P. Beardsley. 1987. Functional effects of cis-thymine glycol lesions on DNA synthesis in vitro. *Biochemistry* 26:5398–5403.
4. Clark, J. M., C. M. Joyce, and G. P. Beardsley. 1987. Novel blunt-end addition reactions catalyzed by DNA polymerase of *Escherichia coli*. *J. Mol. Biol.* 198:123–127.
5. Clark, J. M., N. Pattabiraman, W. Jarvis, and G. P. Beardsley. 1987. Modeling and molecular mechanical studies of the cis-thymine glycol radiation damage lesion in DNA. *Biochemistry* 26:5404–5409.
6. Dalton, J. G., A. L. George, F. E. Hruska, T. N. McGaig, K. K. Ogilvie, J. Peeling, and D. J. Wood. 1977. Comparison of arabinose and ribose nucleosides conformation in aqueous and dimethylsulfoxide solution. *Biochim. Biophys. Acta* 478:261–273.
7. Heidelberger, C., L. Griesbach, B. J. Montag, D. Mooren, O. Cruz, R. J. Schnitzer, and E. Gruenberg. 1958. Studies on fluorinated pyrimidines. II. Effects on transplant tumors. *Cancer Res.* 18:305–317.
8. Kremer, A. B., T. Mikita, and G. P. Beardsley. 1987. Chemical consequences of incorporation of 5-fluorouracil into DNA as studied by NMR. *Biochemistry* 26:391–397.
9. Kufe, D. W., P. P. Major, E. M. Egan, and G. P. Beardsley. 1980. Correlation of cytotoxicity with incorporation of araC into DNA. *J. Biol. Chem.* 19:8997–9000.
10. Loechler, E. L., C. L. Green, and J. M. Essigmann. 1984. In vivo mutagenesis by O^6-methylguanine built into a unique site in a viral genome. *Proc. Natl. Acad. Sci. USA* 81:6271–6275.
11. Major, P. P., E. M. Egan, G. P. Beardsley, M. D. Minden, and D. W. Kufe. 1981. Lethality of human myeloblasts correlates with the incorporation of arabinofuranosyl cytosine into DNA. *Proc. Natl. Acad. Sci. USA* 78:3235–3239.
12. Major, P. P., E. M. Egan, D. Herrick, and D. W. Kufe. 1982. 5-Fluorouracil incorporation in DNA of human breast carcinoma. *Cancer Res.* 42:3005–3009.
13. Metzler, W. J., K. Arndt, E. Tecza, J. Wasilewski, and P. Lu. 1985. Lambda phage cro repressor interaction with its operator DNA: 2'-deoxy-5-fluorouracil O_R3 analogues. *Biochemistry* 24:1418–1424.
14. Mitchel, N., and G. Stohrer. 1979. Mutagenesis originating in site-specific DNA damage. *J. Mol. Biol.* 191:177–180.
15. Moore, P. D., and B. Strauss. 1979. Sites of inhibition of in vitro DNA synthesis in carcinogen and UV tested OX174 DNA. *Nature* (London) 278:664–666.
16. Sagher, D., and B. Strauss. 1983. Insertion of nucleotides opposite apurinic/apyrimidic sites in deoxyribonucleic acid during in vitro synthesis: uniqueness of adenine nucleotides. *Biochemistry* 22:4518–4526.
17. Sherfinski, J. S., and R. E. Marsh. 1973. The crystal structure and absolute configuration of 1-(β-D-arabinofuranosyl)cytosine hydrochloride. *Acta Crystallogr. Sect. B Struct. Sci.* 29:192–198.
18. Sinha, N. D., J. Biernat, and H. Koster. 1983. β-Cyanoethyl N, N-dialkylamino/N-morpholinomonochloro phosphoamidites, new phosphitylation agents facilitating ease of deprotection and work-up of synthesized oligonucleotides. *Tetrahedron Lett.* 24:5843–5846.
19. Skarnes, W., P. Bonin, and E. Barel. 1986. Exonuclease activity associated with a multiprotein form of HeLa cell DNA polymerase α. *J. Biol. Chem.* 261:6629–6636.

20. Takeshita, M., C.-N. Chang, F. Johnson, S. Will, and A. P. Grollman. 1987. Oligodeoxyribonucleotides containing synthetic abasic sites. *J. Biol. Chem.* **262**:10171–10179.
21. Vishwantha, J. K., S. A. Coughlin, M. Wesolowski-Owen, and E. F. Barel. 1986. A multiprotein form of DNA polymerase α from HeLa cells. *J. Biol. Chem.* **261**:6619–6628.
22. Wilkinson, D. W., and J. Crumley. 1977. Metabolism of 5-fluorouracil in sensitive and resistant Novikoff hepatoma cells. *J. Biol. Chem.* **252**:1051–1056.

Chapter 22

Structural Model for the Editing Decision of *Escherichia coli* DNA Polymerase I

Catherine M. Joyce, Jonathan M. Friedman, Lorena Beese, Paul S. Freemont, and Thomas A. Steitz

An important question in the enzymology of DNA replication is how DNA polymerases achieve high fidelity in copying genetic information. The accuracy of DNA polymerases far exceeds the discrimination that would be provided merely by the free energy difference between correct and incorrect base pairs in DNA (11). It is generally accepted that, at least for procaryotic polymerases, accuracy is determined by a combination of enzymatic mechanisms that (i) enhance the discrimination between correct and incorrect nucleotides in the synthesis step and (ii) allow for subsequent removal (editing) of incorrect nucleotides incorporated into DNA.

For a simple procaryotic polymerase, such as DNA polymerase I (pol I) of *Escherichia coli*, three processes are important in determining the overall accuracy of the enzyme (Fig. 1). These are the selection and incorporation of deoxynucleotide triphosphates (dNTPs) into DNA (reaction 1), the hydrolytic removal of the 3'-terminal residue by the 3'→5' exonuclease (reaction 2), and the commitment to further rounds of polymerization (reaction 3). Of crucial importance in polymerase editing is the branch point between reactions 2 and 3, which we shall refer to as the "editing decision." The choice made by the polymerase at this branch point determines whether a mismatched DNA 3' terminus will escape exonucleolytic editing and gain the chance to become fixed as a mutation in DNA.

In this paper we shall ignore the synthesis stage of the polymerase reaction and focus instead on the editing decision. Our model system is the large fragment (Klenow fragment) of DNA pol I, where the availability of high-resolution structural information allows us to attempt a molecular description of editing.

GENERAL KINETIC DESCRIPTION OF EDITING

Any editing process must involve a branch in the reaction pathway, and the

---

Catherine M. Joyce • Department of Molecular Biophysics and Biochemistry, Yale University Medical School, New Haven, Connecticut 06510. *Jonathan M. Friedman, Lorena Beese, and Thomas A. Steitz* • Department of Molecular Biophysics and Biochemistry and Howard Hughes Medical Institute, Yale University, New Haven, Connecticut 06511. *Paul S. Freemont* • Imperial Cancer Research Fund Laboratories, London, United Kingdom.

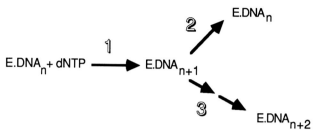

FIGURE 1. Simplified reaction scheme for a DNA polymerase. The numbered arrows indicate reactions (which may involve more than one step) that contribute to fidelity, as described in the text. The order of substrate addition (DNA before dNTP) was established by Bryant et al. (4).

choice made by the enzyme at that branch point will be determined by the balance between the rates of the competing reactions. Perturbations of the relevant reaction rates can change the relative flux down the two (or more) alternative pathways and thus influence the extent of editing. For example, the fidelity of pol I is lower either when reaction 3 (Fig. 1) is favored by high concentrations of dNTPs (the "next nucleotide effect" [10]) or when reaction 2 is slowed either by the $3' \rightarrow 5'$ exonuclease inhibitor deoxynucleoside monophosphate (dNMP) or by mutations at the exonuclease active site (10; K. Bebenek, C. M. Joyce, and T. A. Kunkel, unpublished data). In either case, the net effect is the same: an increase in the flux down pathway 3 relative to pathway 2. Conversely, a decrease in the rate of reaction 3 will increase the efficiency of editing and increase fidelity.

Since the editing decision depends on the balance between reaction rates, effective editing does not require absolute discrimination between correctly and incorrectly base-paired DNA termini. Thus it is not surprising to find that pol I is capable both of extending incorrectly base-paired termini and of hydrolyzing correctly base-paired molecules, resulting in the turnover of dNTPs to dNMPs. All that is required of an "intelligent" editing system is that the relevant reaction rates for correct and incorrect DNA termini should be biased such that a mispaired terminus would be less likely to persist in DNA and become fixed as a mutation.

In the preceding discussion we have ignored another process that may contribute to the editing decision. This is the dissociation of the enzyme-DNA complex ($E.DNA_{n+1}$; see Fig. 1). Currently we can only speculate that dissociation could contribute to fidelity if DNA molecules with mismatched termini were preferentially released and could be processed by other repair or degradative pathways in vivo.

What aspects of protein and DNA structure control the relevant reaction rates that influence the editing decision? To answer this question we need to understand how a DNA polymerase can sense the difference between a normal Watson-Crick base pair and a mismatched base pair, and how the protein-DNA interaction is translated into an enzymatic response that influences reaction rates. Although such a detailed description is beyond our reach at present, it is certainly a realistic goal, given the availability of high-resolution crystallographic data on Klenow fragment (13) and on several DNA duplexes containing mismatched base pairs (1, 2, 7, 9, 16).

THE KLENOW FRAGMENT STRUCTURE: SEPARATE DOMAINS CORRESPOND TO SEPARATE ENZYMATIC ACTIVITIES

The Klenow fragment of DNA pol I contains polymerase and $3' \rightarrow 5'$ (editing) exonuclease functions and thus, for our present purposes, provides a convenient-sized derivative having the important attributes of the parent molecule. X-ray crystallography shows that Klenow fragment has two structural domains, a large C-terminal domain and a smaller N-terminal domain (13). The position of the exonuclease active site was determined crystallographically as the binding site of the exonuclease inhibitor dNMP. The location of the polymerase active site has been inferred from a number of DNA footprinting and protein chemistry experiments designed to locate the binding sites for the DNA primer terminus and the dNTP substrate. The two active sites are separated by about 30 Å (3 nm) with the polymerase active site on the large domain and the exonuclease active site on the small domain. This conclusion is supported by a large body of experimental evidence (reviewed in reference 8). Perhaps the most convincing data supporting the assignment of the active sites and their spatial separation come from site-directed mutagenesis. Mutations in residues at the proposed exonuclease active site gave proteins deficient in exonuclease activity but with normal polymerase activity (4a). Conversely, mutations in the region thought to contain the polymerase active site gave proteins with altered polymerase activity but essentially normal exonuclease activity (A. Polesky and C. M. Joyce, unpublished data).

DNA-BINDING SITES OF KLENOW FRAGMENT

Model building suggests that the polymerase domain of Klenow fragment has a binding site for duplex DNA (13). The domain contains a large cleft of the appropriate dimensions to bind a B-DNA duplex. Theoretical calculations of electrostatic surface potential indicate that virtually all of the positive electrostatic charge of the Klenow fragment molecule is located on the interior surface of the binding cleft, describing an approximately spiral path with a pitch similar to that of the DNA helix (20). Thus a duplex DNA molecule bound to the polymerase domain would be almost completely surrounded by protein, with positively charged amino acid side chains neutralizing the negative charge of the phosphodiester backbone. There are some portions of the Klenow fragment molecule at the periphery of the binding cleft that are partially disordered in the native structure. These regions are in the appropriate position to close off the top of the binding cleft, raising the possibility that the polymerase may completely envelop its duplex DNA substrate.

Recent cocrystallization experiments have yielded a wealth of information on the binding of single-stranded DNA oligonucleotides to the $3' \rightarrow 5'$ exonuclease active site (T. A. Steitz, L. Beese, P. Freemont, J. Friedman, and M. Sanderson, *Cold Spring Harbor Symp. Quant. Biol.*, in press; P. S. Freemont, L. Beese, J. M. Friedman, and T. A. Steitz, unpublished data). The binding of three bases of DNA can clearly be seen in the electron density maps, with the 3'-terminal residue positioned at the exonuclease active site. Amino acid side chains provide polar interactions with the DNA phosphates as well as hydrophobic stacking interactions with the exposed nucleotide bases. The latter interactions in particular suggest that this region of the protein is specifically designed to bind single-stranded (as opposed to duplex) DNA; this idea is consistent with the observed preference of the exonuclease for a single-stranded DNA substrate (3).

The path of the single-stranded DNA seen in these cocrystals can fairly easily be connected to the trajectory of the DNA primer strand modeled into the duplex DNA-binding cleft (8; Steitz et al., in press). The model so derived suggests a fairly simple relationship between a DNA primer terminus in the polymerase active site and that same terminus in the exonuclease active site (Fig. 2). For the DNA terminus to move from polymerase to exonuclease site would require the protein and DNA to move relative to one another over a distance equivalent to 8 base pairs; at the same time the terminal 3 or 4 base pairs would become melted. (The movement would of course follow a spiral path so as to maintain the favorable protein-DNA contacts in the binding cleft on the polymerase domain.)

MECHANISTIC IMPLICATIONS OF THE SEPARATION OF POLYMERASE AND EXONUCLEASE ACTIVE SITES

The model illustrated in Fig. 2 is of necessity naive in the absence of definitive structural information from a Klenow fragment complex with duplex DNA. However, it provides a useful starting point for considering the relationship between the polymerase and exonuclease active sites and the mechanistic implications of their separation.

As we have pointed out previously, the distance between the active sites is unlikely to cause a kinetic problem (5). A rough calculation indicates that pol I catalyzes about 10 rounds of dNTP incorporation and one exonucleolytic hydrolysis per s at 37°C. Thus the rate of movement between active sites would merely need to be fast relative to these (not particularly rapid) reactions. Although no rate data are available for any process precisely analogous to the one under consideration, the rate of sliding calculated for lac repressor on DNA (10^6 nucleotides per s [21]) is orders of magnitude faster than would be required.

It is less easy to decide whether the model is feasible energetically. Clearly, the need to melt several base pairs at the primer terminus means that a trip from the polymerase active site to the exonuclease active site could be energetically costly. Maybe there is sufficient compensation from the interactions, described above, between the single-stranded DNA terminus and the exonuclease region. Moreover, a fine balance between the energetics of binding the substrate DNA as a duplex to the polymerase

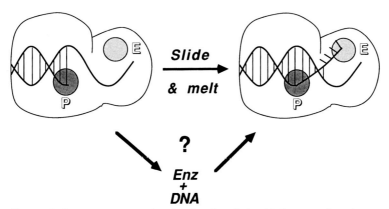

FIGURE 2. Cartoon representation showing the relationship between the polymerase (P) and 3'→5' exonuclease (E) active sites of Klenow fragment. The left-hand molecule represents a Klenow fragment-DNA complex with the 3' end of the primer strand at the polymerase active site. The right-hand molecule shows this complex with the same primer terminus at the exonuclease active site. (The length of template strand shown is arbitrary.) The two alternative pathways reflect our level of ignorance of the mechanism by which the primer terminus moves from one active site to the other.

active site versus binding that same substrate in a partially melted form to the exonuclease active site could provide the physical basis for the editing decision. In this case, the presence of a mispair at the primer terminus could tip the balance by decreasing the energy cost of melting the terminal base pairs and therefore increasing the likelihood that a mispaired terminus will be edited. Note that this model for the editing decision is a thermodynamic one, based on the difference in DNA-binding affinity at the two active sites. Whether the DNA substrate remains associated with the enzyme when moving from one site to the other, or whether it dissociates (see Fig. 2), is unimportant provided that the rate of switching between the sites is compatible with the overall kinetics of DNA synthesis.

STRUCTURAL MODEL FOR EDITING

If one accepts that the editing decision could be based largely on the energetics of DNA base pairing, then it is easy to rationalize the structural organization of the Klenow fragment molecule. Far from being a mechanistic inconvenience, the separation of the two active sites becomes an important component of the editing decision. The editing process outlined above would demand the DNA-binding preference of the polymerase active site to be the opposite of that of the exonuclease active site. This requirement clearly could not be satisfied if the two catalytic sites were juxtaposed and shared a DNA-binding site. As discussed above, the properties of the two Klenow fragment domains fit in with this idea. The polymerase domain appears optimized for binding a duplex DNA substrate, with an almost cylindrical binding cleft lined with positive charge. Conversely, the exonuclease region, with its appropriately positioned combination of polar and hydrophobic side chains, seems optimized for binding single-stranded

DNA. We therefore suggest that the presence of a terminal mispair destabilizing the base pairing at the primer terminus would both facilitate movement of the primer terminus to the exonuclease active site (for the energetic reasons indicated above) and also weaken the interaction of that DNA terminus with the polymerase active site. The latter could decrease the rate of subsequent polymerization, thus providing an additional bias in favor of editing. Experiments with exonuclease-deficient derivatives of Klenow fragment clearly demonstrate a substantial block to polymerization caused by a mismatched primer terminus that cannot be removed (C. M. Joyce, unpublished data). Since a slow rate of dNTP addition to a mismatched end will increase the probability of exonucleolytic editing, these results emphasize the importance of the rate of reaction 3 (Fig. 1) in the editing decision.

The arguments outlined above are essentially a restatement of the "frayed-unfrayed" model of Brutlag and Kornberg (3), modified so as to take account of the physical separation of the two catalytic sites in the Klenow fragment structure. Moreover, we have suggested that, for successful discrimination at the level of base-pairing stability, the two active sites must necessarily be physically separate because their DNA-binding properties will be mutually incompatible. We are encouraged in this view by the obvious analogy between Klenow fragment and *E. coli* DNA pol III, which has polymerase and exonuclease activities on separate subunits within a complex multisubunit assembly (12, 17). The many differences between pol I and pol III make it unlikely that the analogous separation of polymerase and exonuclease active sites is merely an accident of evolutionary history; they suggest a functional significance.

An attractive feature of a simple frayed-unfrayed model is that it exploits a property common to all incorrect base pairs, namely, their lower base-pairing stability, to explain the editing decision of Klenow fragment.

Crystallographic studies of a number of DNA duplexes containing mismatched base pairs indicate that all the mismatches studied to date can be accommodated within a relatively undistorted B-DNA structure (1, 2, 7, 9, 16). This argues against any model for editing based on the hypothesis that mismatched base pairs in DNA cause a gross structural distortion. There are, however, a number of subtle structural differences that distinguish incorrect from correct base pairs (7). As we have pointed out previously (18, 19), such differences could be detected by an enzymatic scanning mechanism operating in the duplex DNA-binding site. The model presented above certainly does not rule out such additional contributions to fidelity which might serve to augment the discrimination against further dNTP addition to a mismatched primer terminus.

IS THIS MODEL REASONABLE?

A frayed-unfrayed model for editing predicts that changes in the rate of exonucleolytic degradation, dNTP addition, or both should be interpretable in a fairly straightforward manner in terms of the base-pairing stability of the DNA substrates under consideration. Unfortunately there is too small a data base of rate constants for the relevant reactions in different DNA sequence contexts to permit any meaningful correlations to be made at present. The results from other studies, while not conclusive, are at least encouraging. Studies on Klenow fragment fidelity using the substrates poly(dI-dC) and poly(dG-dC) suggest an important role for DNA duplex stability (14). Petruska and Goodman have achieved considerable success in interpreting misincorporation frequency on a natural DNA template in terms of base-pairing and base-stacking interactions (15). As a model for polymerase-catalyzed misincorporation these authors used 2-aminopurine incorporation by a mutant form of T4 DNA polymerase. It is clear from numerous in vitro experiments (e.g., reference 6) that the reaction rates determining polymerase accuracy are exquisitely sensitive to local sequence context. A major challenge for any model for polymerase accuracy will be to understand this variation in terms of the protein-DNA interactions involved.

ACKNOWLEDGMENTS. We are grateful to Nigel Grindley for helpful discussions and a critical reading of the manuscript.

Work in our laboratories is supported by Public Health Service grants GM-28550 (to N. D. F. Grindley) and GM-22778 (to T.A.S.) from the National Institutes of Health and by American Cancer Society grant NP-421 (to T.A.S.). J.M.F. acknowledges a postdoctoral fellowship from the Jane Coffin Childs Memorial Fund.

LITERATURE CITED

1. **Brown, T., W. N. Hunter, G. Kneale, and O. Kennard.** 1986. Molecular structure of the G · A base pair in DNA and its implications for the mechanism of transversion mutations. *Proc. Natl. Acad. Sci. USA* **83:**2402–2406.
2. **Brown, T., O. Kennard, G. Kneale, and D. Rabinovitch.** 1985. High resolution structure of a DNA helix containing mismatched base pairs. *Nature* (London) **315:**604–606.
3. **Brutlag, D., and A. Kornberg.** 1972. Enzymatic synthesis of deoxyribonucleic acid. A proofreading function for the 3′-5′ exonuclease activity in deoxyribonucleic acid polymerases. *J. Biol. Chem.* **247:**241–248.
4. **Bryant, F. R., K. A. Johnson, and S. J. Benkovic.** 1983. Elementary steps in the DNA polymerase I reaction pathway. *Biochemistry* **22:**3537–3546.
4a. **Derbyshire, V., P. S. Freemont, M. R. Sanderson, L. Beese, J. M. Friedman, C. M. Joyce, and T. A. Steitz.** 1988. Genetic and crystallographic studies of the 3′-5′ exonucleolytic site of DNA polymerase I. *Science* **240:**199–201.
5. **Freemont, P. S., D. L. Ollis, T. A. Steitz, and C. M. Joyce.** 1986. A domain of the Klenow fragment of *Escherichia coli* DNA polymerase I has polymerase but no exonuclease activity. *Proteins* **1:**66–73.
6. **Hillebrand, G. G., A. H. McCluskey, K. A.**

Abbott, G. G. Revich, and K. L. Beattie. 1984. Misincorporation during DNA synthesis, analyzed by gel electrophoresis. *Nucleic Acids Res.* 12:3155–3171.
7. Hunter, W. N., T. Brown, N. N. Anand, and O. Kennard. 1986. Structure of an adenine · cytosine base pair in DNA and its implications for mismatch repair. *Nature* (London) 320:552–555.
8. Joyce, C. M., and T. A. Steitz. 1987. DNA polymerase I: from crystal structure to function via genetics. *Trends Biochem. Sci.* 12:288–292.
9. Kennard, O. 1985. Structural studies of DNA fragments: the G · T wobble base pair in A, B and Z-DNA; the G · A base pair in B-DNA. *J. Biomol. Struct. Dyn.* 3:205–226.
10. Kunkel, T. A., R. M. Schaaper, R. A. Beckman, and L. A. Loeb. 1981. On the fidelity of DNA replication. Effect of the next nucleotide on proofreading. *J. Biol. Chem.* 256:9883–9887.
11. Loeb, L. A., and T. A. Kunkel. 1982. Fidelity of DNA synthesis. *Annu. Rev. Biochem.* 52:429–457.
12. Maki, H., and A. Kornberg. 1985. The polymerase subunit of DNA polymerase III of *Escherichia coli*. Purification of the α subunit, devoid of nuclease activities. *J. Biol. Chem.* 260:12987–12992.
13. Ollis, D. L., P. Brick, R. Hamlin, N. G. Xuong, and T. A. Steitz. 1985. Structure of large fragment of *Escherichia coli* DNA polymerase I complexed with dTMP. *Nature* (London) 313:762–766.
14. Patten, J. E., A. G. So, and K. M. Downey. 1984. Effect of base-pair stability of nearest-neighbor nucleotides on the fidelity of deoxyribonucleic acid synthesis. *Biochemistry* 23:1613–1618.
15. Petruska, J., and M. F. Goodman. 1985. Influence of neighboring bases on DNA polymerase insertion and proofreading fidelity. *J. Biol. Chem.* 260:7533–7539.
16. Prive, G. G., U. Heinemann, S. Chandrasegaran, L.-S. Kan, M. L. Kopka, and R. E. Dickerson. 1987. Helix geometry, hydration, and G · A mismatch in a B-DNA decamer. *Science* 238:498–504.
17. Scheuermann, R. H., and H. Echols. 1984. A separate editing exonuclease for DNA replication: the ε subunit of *Escherichia coli* DNA polymerase III holoenzyme. *Proc. Natl. Acad. Sci. USA* 81:7747–7751.
18. Steitz, T. A. 1987. The Klenow fragment structure suggests mechanisms for fidelity and processivity of DNA polymerase I, p. 45–55. *In* R. M. Burnett and H. J. Vogel (ed.), *Biological Organization: Macromolecular Interactions at High Resolution*. Academic Press, Inc., New York.
19. Steitz, T. A., and C. M. Joyce. 1987. Exploring DNA polymerase I of *E. coli* using genetics and X-ray crystallography, p. 227–235. *In* D. L. Oxender and C. F. Fox (ed.), *Protein Engineering*. Alan R. Liss, Inc., New York.
20. Warwicker, J., D. L. Ollis, T. A. Steitz, and F. M. Richards. 1985. Electrostatic field of the large fragment of *Escherichia coli* DNA polymerase I. *J. Mol. Biol.* 186:645–649.
21. Winter, R. B., O. G. Berg, and P. H. von Hippel. 1981. Diffusion-driven mechanisms of protein translocation on nucleic acids. 3. The *Escherichia coli lac* repressor-operator interaction. Kinetic measurements and calculations. *Biochemistry* 20:6961–6977.

Chapter 23

DNA Replication Errors
Frameshift Errors Produced by *Escherichia coli* Polymerase I

Lynn S. Ripley and Catherine Papanicolaou

Our ongoing interest in defining the mutational mechanisms that contribute to spontaneous frameshift mutagenesis led us to examine the nature of frameshift errors characteristic of DNA polymerization. Within the cell, DNA replication occurs as a consequence of highly regulated interactions among multiprotein complexes, the DNA genome, and nucleotide precursors. Such replication occurs within an overall context of regulated maintenance of DNA integrity through metabolic events involving DNA recombination and repair. It is not known which metabolic process(es) contributes most frequently to spontaneous frameshift mutagenesis.

An important role for DNA polymerase in influencing the frequency and specificity of spontaneous frameshifts has been demonstrated in bacteriophage T4 (8, 9). Examination of frameshift frequencies in many different mutant DNA polymerase backgrounds demonstrated that most polymerase muta-

Lynn S. Ripley and Catherine Papanicolaou • Department of Microbiology and Molecular Genetics, New Jersey Medical School, University of Medicine and Dentistry of New Jersey, Newark, New Jersey 07103.

tions exert a detectable quantitative effect on the frequency of frameshift mutation. These effects are not exhibited uniformly within the DNA, suggesting contributions by different frameshift mechanisms and/or DNA polymerase-specific DNA context effects. We previously observed that some mutant polymerases reduce the frequencies of some frameshift mutations while increasing the frequencies of others (8). This suggested that multiple mechanisms contribute in quantitatively important ways to spontaneous frameshift mutation and that these mechanisms can be differentially influenced by genetic perturbation of the DNA polymerase enzyme.

Our approaches to understanding the potential role of polymerization errors in spontaneous frameshift mutagenesis have proceeded from both an in vivo and an in vitro perspective. In vivo we have sequenced spontaneous frameshifts to examine the characteristics of the mutations whose origin we wish to determine. In vitro we have begun to examine the DNA sequences of frameshift mutations that arise as the consequence of DNA polymerization errors.

The in vivo approach has yielded a

description of the types of DNA sequence changes and their relative frequency in the N-terminal portion of the *rII*B gene of T4 (6). These studies revealed that a number of different classes of frameshift mutations occur spontaneously. In some cases the DNA sequences of the mutations suggest that the DNA context in which frameshift mutations arise is a direct determinant of frameshift specificity. That is, the mutant DNA sequences are consistent with accurately templated synthesis involving transient local misalignments of the primer and template DNA strands during DNA metabolism (1, 6; C. Papanicolaou and L. S. Ripley, submitted for publication). Several different classes of these events have been distinguished. A major class is consistent with the class of misalignment originally suggested by Streisinger et al. (10) involving locally misaligned pairing between repeated DNA sequences. Two other classes that account for certain complex frameshift mutations consisting of changes in base sequence as well as base number involve misaligned pairing mediated within quasipalindromic or quasirepeated DNA sequences. Although the general outlines of the misalignments that could account for these mutations have been discussed, the enzymes actually responsible for creating these mutations in vivo are unknown (5, 6).

In addition, there is a substantial fraction of spontaneous frameshift sequences whose DNA context does not suggest that misaligned but accurate DNA synthesis can account for the mutations. Some of these represent major contributors to spontaneous frameshift mutation in vivo (6). For example, 50% of base additions and deletions in the *rII*B gene fall into this category as do all of the 2-base-pair (bp) deletions and duplications. In particular, we are eager to identify the sources of the major mutation groups for which no molecular paradigms have been identified. Our examination of the specificity of errors associated with DNA polymerization in vitro has been undertaken to identify the molecular characteristics of frameshift errors most frequently produced by polymerization reactions as a guide to recognizing such mutations in vivo.

EXAMINATION OF FRAMESHIFT SPECIFICITY IN BACTERIOPHAGE T4 DNA SEQUENCES AFTER IN VITRO DNA REPLICATION

In vitro we have pursued the characterization of polymerization errors in some of the same *rII*B DNA sequences that we have examined in vivo. We constructed DNA templates from bacteriophage M13mp8 by incorporating a 74-bp fragment of the *rII*B DNA sequence into the N terminus of the alpha-complementing, β-galactosidase gene fragment. The incorporation produces a frameshift mutation in the β-galactosidase gene, thereby disrupting its ability to complement. Mutations that restore the wild-type reading frame restore complementation despite the presence of the extra T4 DNA. Thus, mutants can be readily recognized by their blue-plaque phenotype in a colorless plaque background when the M13mp8 *rII*B fusion-bearing phage are plated on an appropriate host and in the presence of the indicator dye X-Gal (5-bromo-4-chloro-3-indolyl-β-D-galactopyranoside) and the inducer IPTG (isopropyl-β-D-thiogalactopyranoside) (2).

We began studies of frameshift errors introduced in vitro during DNA synthesis by using the large fragment of DNA polymerase I of *Escherichia coli* (pol-LF). These studies used a gapped DNA substrate that limited in vitro synthesis to the region in which mutations were to be detected. Following DNA synthesis in vitro, DNA was transfected and blue plaques were isolated. The mutant frequency was increased as a consequence of DNA synthesis by 10- to 100-fold over the background, depending on the template (2).

Characterization of the mutant DNA sequences recovered after transfection of the replicated DNA and recovery of blue plaques revealed that the most frequent frameshift mutations were 1-bp deletion mutations. At substantially lower frequencies, 2-bp deletion mutations were found and at even lower frequencies were complex frameshifts and large deletions (2).

COMPARISON OF ERRORS PRODUCED BY *E. COLI* POL I AND ITS LARGE-FRAGMENT DERIVATIVE

In our studies we have extended the characterization of DNA polymerization errors to a comparison between pol-LF and *E. coli* polymerase I (pol I). We have used a single-stranded DNA template (M74) primed with an oligonucleotide in these studies. The DNA sequence of the template in the N-terminal region of the β-galactosidase gene fragment is shown in Fig. 1.

In vitro DNA synthesis was carried out as follows. The oligonucleotide primers were annealed to the template at a template concentration of 80 µg/ml of 50 mM Tris hydrochloride (pH 7.5)–0.1 M NaCl buffer. The molar ratio of primer to template was 2. After 5 min at 55°C, the annealing mixture was incubated at 37°C for 15 min and then

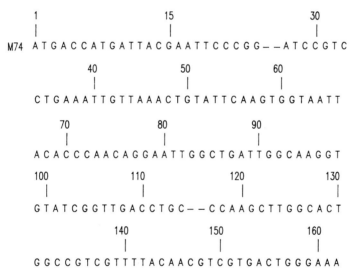

FIGURE 1. The DNA sequence of the M74 template at the N terminus of the β-galactosidase gene. Numbering begins at the ATG codon that begins the coding region for the alpha-complementing fragment of the β-galactosidase gene fragment. The DNA sequence from positions 34 to 110 is from the T4 *rIIB* gene. Dashes shown at positions 25 and 26 and 115 and 116 represent positions at which bases were deleted from M13mp8 DNA sequences to adjust the reading frame for detecting −1 reading frameshifts (2). The oligonucleotide primers used to initiate DNA synthesis had 3' termini at 142 (15-mer) or at 148 (17-mer). Mutations that create +1 reading frameshifts are detected in the M74 template when they produce termination near TTG or GTG codons that are in the −1 reading frame. The very light blue phenotype of these mutants probably reflects its dependence upon translational reinitiation (2). Base pair substitution mutations have also been detected using this template (see Table 1). Most of these substitutions produce nonsense codons near the same potential reinitiation sites, and they also have a very light blue phenotype. The detailed description of these mutants is presented elsewhere (Papanicolaou and Ripley, submitted).

chilled on ice. DNA synthesis was carried out at 20 μg of primed template per ml of 50 mM Tris hydrochloride (pH 7.6)–10 mM $MgCl_2$–1 mM dithiothreitol–0.4 mg of bovine serum albumin per ml–250 μM each dNTP. DNA polymerase (0.1 U/ml) was added, equivalent to a 20-fold molar excess of pol-LF to template and a 10-fold molar excess of pol I to template. The reaction was incubated at 37°C for 30 min and stopped by the addition of 50 mM EDTA. Mutants were assayed after transfections of 50 to 100 ng of copied template per ml of $CaCl_2$-treated cells. Plating conditions have been previously described (2). Mutants were identified by their blue-plaque phenotype. After plaque purification, the DNA was purified and sequenced. The results for the experiments reported here are summarized in Table 1. Overall, there was a significant increase in mutation. Blue-plaque frequencies were less than 10^{-6} in unreplicated templates. In the sections that follow, the characteristics of 1-bp deletions, large deletions, and complex frameshifts will be discussed in more detail. A more complete account of these mutations has been submitted elsewhere (Papanicolaou and Ripley, submitted).

DELETIONS OF 1 bp: IDENTIFICATION OF A DNA SEQUENCE DETERMINANT COMMON TO BOTH ENZYMES AND IDENTIFICATION OF A HOT SPOT SPECIFIC FOR POLYMERASE I

As described in our earlier work (2) and confirmed in these studies, 1-bp deletions occurring as a consequence of in vitro replication by the pol-LF enzyme occur most frequently in a consensus DNA sequence. The deletion occurs opposite template G when that G is followed on the 5' side by TT or CT. These sites, when examined within the pol I spectrum, are still among the most frequent sites of mutation. A comparison of the most frequent sites of pol-LF and pol I 1-bp deletions is given in Table 2. Approximately 50% of the mutants in each spectrum lie at four sites. All of the sites lie in the consensus sequence except site 15. Mutants occurred at this site 7 times in the pol I

TABLE 1
Summary of Mutations Produced as a Consequence of In Vitro Replication by pol-LF or pol I of the M74 DNA Template

Polymerase (no. of expts)[a]	Total plaques (10^6)[b]	Frequency (10^{-4}) of mutants (range)[c]	Sequences[d]	No. of mutations
pol-LF (6)	8.6	0.5 (0.2–1.3)	1-bp deletion	193
			1-bp substitution	42
			Complex	71
pol I (3)	5.6	0.5 (0.45–0.90)	1-bp deletion	67
			1-bp substitution	11
			Complex	84
			Large deletion	50

[a] Two of the pol I experiments were carried out in parallel with two of the pol-LF experiments, using the same preparation of template-primer and the same batches of cells for transfection.
[b] Total plaques examined for blue-plaque phenotypes after transfection.
[c] Frequency of blue plaques averaged from all experiments.
[d] A random sample of mutants from the experiments was sequenced. There were four major classes of mutations: deletions of 1 bp, base pair substitutions (transitions and transversions), complex frameshifts that change both the number of bases and the base sequence, and deletions of multiple bases. An additional 31 mutants sequenced after pol I replication and 73 mutants sequenced after pol-LF replication have no DNA sequence change in the portion of the alpha-complementing fragment expected to be able to revert by reading-frame restoration, nor do they have changes in the upstream lac regulatory region. The nature of these mutations is unknown.

spectrum but only once in the much larger pol-LF spectrum. (Mutations at site 131 or 132 occurred 4 times in the pol I spectrum.) The presence of a new hot spot in the pol I spectrum is significant not only because it occurs at a much higher frequency than it does during pol-LF synthesis but because it occurs in a different DNA context than do the other frequent DNA mutations in either the pol-LF or pol I spectra. The existence of the position-15 hot spot in the pol I spectrum suggests that some additional or different molecular perturbation to which only pol I is highly sensitive is responsible for this mutation. In contrast, both polymerases respond in a similar manner to the molecular perturbation that produces 1-bp deletions opposite G in when the 5' context is TT or CT.

Finding a novel 1-bp deletion hot spot in vitro due to a perturbation of the DNA polymerase may be relevant to some of the in vivo specificity differences that we have observed. When we compared the spectrum of spontaneous frameshifts produced in T4 with a wild-type polymerase to the spectrum produced in T4 with the mutant DNA polymerase $tsL141$, the $tsL141$ allele produced a striking new hot spot in $rIIB$ DNA (8). A major contributor to that hot spot is a single base deletion that occurs rarely in the wild-type polymerase background (unpublished data). We are currently examining the characteristics of frameshifts produced by the T4 DNA polymerase on the M74 DNA template.

LARGE DELETIONS BETWEEN DIRECTLY REPEATED DNA SEQUENCES: A CLASS OF FRAMESHIFTS PROMOTED BY DNA POL I

A striking difference between the spectrum of the pol I polymerase and that of pol-LF is the frequency of large deletions. None were found in the pol-LF spectrum, while large deletions represented 20% (50/255) of the pol I spectrum. The deletion sequences are illustrated in Table 3. Each deletion has its endpoints in DNA sequences that are direct repeats. Thus, these sequences resemble a major class of deletion

TABLE 2
Hot Spots for 1-bp Deletions in the Spectra of Frameshifts Arising as a Consequence of Polymerization by pol I and pol-LF

Polymerase and no. of mutations[a]	Template sequence and deletion sites			Nucleotide deleted[c]
	5' Context	Deletion site	3' Context[b]	
pol-LF				
37	CT	G	CCCAAG	113
33	TT	G	ACCTGC	108
16	CT	G	GCCGTC	131 or 132
13	TT	G	GCACTG	125 or 126
pol I				
9	TT	G	ACCTGC	108
8	CT	G	CCCAAG	113
7	TT	G	GCACTG	125 or 126
7	AC	G	AATTCC	15

[a] The total number of 1-bp deletions sequenced was 193 for pol-LF and 67 for pol I.
[b] The 3' context of the most frequent deletion sites has 4/6 G · C pairs (2). The most frequent sites also have a C in the template 2 bases before the deletion site.
[c] Numbering of nucleotides is shown in Fig. 1.

TABLE 3
DNA Sequences of Deletion Mutations Produced as a Consequence for pol I DNA Synthesis

Mutation	No. of occurrences	Initial/mutated sequences[a]		
LD1	37	48 AAC AAC	TGTATTCAAG//GTTGGCAAGG	98 TGTATCGGTT TGTATCGGTT
LD2	7	44 GTT GTT	AAACTGTATT//GCTGATTGGC	94 AAGGTGTATC AAGGTGTATC
LD3	3	37 TGA TGA	AATTGTTAAA//ACCCAACAGG	79 AATTGCCTGA AATTGGCTGA
LD4	2	54 ATT ATT	CAAGTGGTAA//GGCTGATTGG	93 CAAGGTGTAT CAAGGTGTAT
LD5	1	7 CCA CAA	TGATTACGAA//AGGAATTCCG	86 TGATTGGCAA TGATTGGCAA

[a] The initial DNA sequences are shown above; the deleted sequence below. // indicates that not all of the sequence is illustrated; for the complete sequence refer to Fig. 1. Underlining illustrates the portions of the repeated sequences believed to participate in producing these mutations as described in the text.

sequences seen frequently in vivo that also have endpoints in such repeats (3). These repeats may mediate the misalignments between the complement of one copy and the second copy. Such misalignments have been proposed as intermediates that could account for either deletion or duplication of one copy of the repeat and the intervening DNA sequence. The production of these mutations as a consequence of in vitro DNA polymerization suggests that these particular mutations can arise as a consequence of misalignment between the primer and the template, thereby producing the deletions. The absence of duplications is notable.

As can be seen in Table 3, deletions were observed consistent with the misalignment of primers extended to position 49, 45, 38, 55, or 8. The maximum length of the repeated sequences to mediate this misaligned pairing was 5 bp for the LD1, LD3, and LD5 mutation sequences and 7 out of 9 bp for the LD2 and LD4 sequences. The LD1 sequence dominates the pol I spectrum, accounting for 37 of 50 deletions. LD3 and LD5, which depend on similar misalignments and repeats, occur about 10-fold less frequently. This specificity is likely to reflect some characteristic of synthesis by the pol I enzyme since deletions occur more rarely during pol-LF synthesis. However, the fact that not all similar repeats lead to deletion with similar frequencies suggests a role for specific DNA substrate determinants in the specificity of this mutagenesis as well.

COMPLEX FRAMESHIFTS: PALINDROMIC COMPLEMENTARITY AND DNA POLYMERASE SPECIFICITY

A major class of frameshifts isolated in these experiments with both *E. coli* polymerases are complex changes in DNA sequence that change both the number and sequence of bases at the mutant site. All but 4 out of 114 mutant sequences examined can be readily explained by misaligned DNA synthesis mediated by misalignment of the primer in a palindromic DNA sequence, followed ultimately by return of the newly synthesized DNA to the template, mediated by homology to the DNA template (Papanicolaou and Ripley, submitted). An example of the hypothesized molecular intermediates whose formation can account for the complex DNA sequence changes is shown in Fig. 2. During the production of the mutation, the DNA primer is misaligned by folding back on itself for a transient period of DNA synthesis. This misalignment is mediated by palindromic complementarity. Realignment of the primer to the template is accomplished in a manner similar to that believed to be responsible for the pol I-mediated deletions.

In the illustrated example, the DNA primer, extended to position 93, realigns to position 125. This specific realignment of the primer was also involved in producing four other complex frameshift sequences. The illustrated sequence was found only after replication by pol I, but the remaining four were found in both the pol I and pol-LF spectra. The sequences of these complex frameshifts and the frequency of their occurrence among the sequenced mutants are shown in Table 4.

While the overall frequency differences between pol-LF and pol I for complex frameshift mutations were modest, the site-specific frequencies for particular sequences were very different. The CPX1 sequence was not seen at all among mutants produced by pol-LF, but it was the most frequent complex frameshift among all complex mutants produced by pol I. Conversely, the CPX3 sequence dominated the pol-LF spectrum and was rarely seen among pol-I-produced complex frameshifts. These five complex sequences do not exhaust the possibility for misalignment frameshifts initiated from this fold-back position. For example, DNA synthesis for 14, 17, and 18 bp could realign to the template by homologies of 4/4, 5/7, and 5/5, respectively, to produce complex frameshifts that produce a −1 shift in the reading frame. It appears that the DNA polymerases each interact with these unusual DNA substrates in a unique manner. Previous in vitro studies using homopolymer templates have demonstrated distinct differences between the pol I and pol-LF enzymes in their template interactions (4). Overall DNA context and topology are likely to play a role in a manner we do not yet understand. Further studies will be required to dissect the contributing factors.

SUMMARY

The sequencing of large samples of frameshift mutations arising as a consequence of in vitro DNA replication by the DNA pol I of *E. coli* and its large fragment reveals substantial similarities among some groups of mutations and substantial differences among others. The most frequent class of frameshifts, deletions of 1 bp, occur with a similar distribution and frequency in both spectra. Although the mechanism(s) responsible for this class of frameshifts is not known, its inability to be predicted by slippage models, while retaining a strong sequence-specific characteristic, offers a new paradigm for the examination of in vivo frameshift specificity. DNA synthesis by *E. coli* DNA pol I produced a new −1-bp hot spot. This hot spot is also not explained by a slippage model. The high frequency of this hot spot in the pol I spectrum and its low frequency in the pol-LF spectrum demonstrate that specific polymerase perturbations can strikingly influence one class of replication-produced mutation in the absence of major effects on other classes just as they do in vivo. The observation that the large-deletion mutations are more frequently produced by pol I than by pol-LF coupled with the observation that the most frequent complex frameshift for each enzyme is different suggests that major specificity determinants

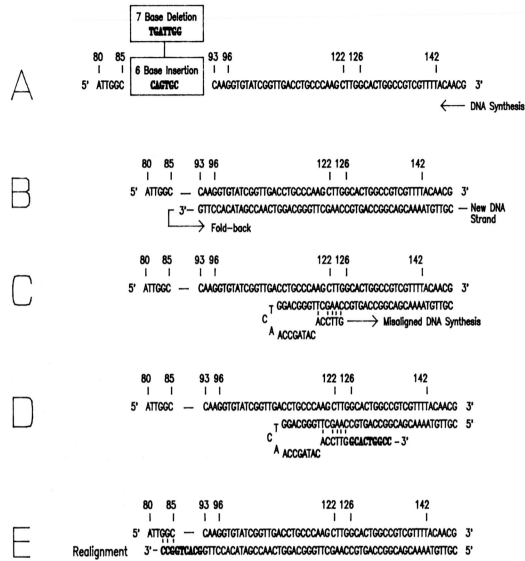

FIGURE 2. The primer fold-back hypothesis, a molecular proposal that accounts for most complex frameshifts produced as a consequence of in vitro DNA synthesis by pol I or pol-LF. (A) DNA sequence change of a complex frameshift isolated frequently after pol I synthesis but not represented among pol-LF mutants (see Table 3). (B) DNA synthesis proceeds normally on the template up to position 93. (C) Realignment of the extended primer in B through palindromic complementarity to position 125. This realignment is characteristic of all of the mutations described in Table 3 and represents the most frequent misalignment accounting for this class of mutations in either the pol I or the pol-LF spectra. (D) DNA synthesis in the misaligned position. To account for the illustrated mutation, DNA synthesis extends for 9 nucleotides. The other mutations described in Table 3 require additional synthesis. (E) Realignment of the DNA to the template. In this case there are only 3 nucleotides of complementarity to the DNA template after the site of the mutation. The realignments to the DNA template for the other mutant sequences in Table 3 depend on more extensive homologies.

TABLE 4
Complex Frameshifts Initiated by a Misalignment of the Primer as a
Consequence of Palindromic Complementarity

Mutation	No. of mutations[a] with:		Original/mutated DNA sequence			No. of bases added by misaligned synthesis	Realignment homology
	pol-LF	pol-I					
CPX1	0	11	GGC GGC	86 92 TGATTGG CAGTGC	CAA CAA	9	3/3
CPX2	5	1	AGT AGT	60 92 GGTAATTACA//GGCTGATTGG TGTAAAACGACGGCCAGTGC	CAA CAA	22	6/7
CPX3	35	2	TGT TGT	44 92 TAAACTGTAT//GGCTGATTGG AAAACGACGGCCAGTGC	CAA CAA	21	8/9
CPX4	7	4	GAA GAA	39 92 ATTGTTAAAC//GGCTGATTGG GTTGTAAAACGACGGCCAGTGC	CAA CAA	22	8/9
CPX5	1	1	TGA TGA	38 92 AATTGTTAAA//GGCTGATTGG CGTTGTAAAACGACGGCCAGTGC	CAA CAA	25	10/13

[a] The number of mutants having the indicated sequence among 55 sequenced complex frameshifts produced by pol-LF and 59 produced by pol I.

in these categories of mutations are likely to involve sequence-specific differences in polymerase-template interactions and not only the characteristics of DNA misalignment probabilities.

ACKNOWLEDGMENT. This work was supported by Public Health Service grant AG06171 from the National Institutes of Health to L.S.R.

LITERATURE CITED

1. deBoer, J. G., and L. S. Ripley. 1984. Demonstration of the production of frameshift and base-substitution mutations by quasipalindromic DNA sequences. Proc. Natl. Acad. Sci. USA 81:5528–5531.
2. deBoer, J. G., and L. S. Ripley. 1988. An in vitro assay for frameshift mutations: hotspots for deletions of 1 bp by Klenow fragment polymerase show a consensus DNA sequence. Genetics 118:181–191.
3. Farabaugh, P. J., U. Schmeissner, M. Hofer, and J. H. Miller. 1978. Genetic studies of the lac repressor. VII. On the molecular nature of spontaneous hotspots in the lacI gene of Escherichia coli. J. Mol. Biol. 126:847–863.
4. Papanicolaou, C., P. Lecomte, and J. Ninio. 1986. Mnemonic aspects of E. coli DNA polymerase I: interaction with one template influences the next interaction with another template. J. Mol. Biol. 189:435–448.
5. Ripley, L. S. 1982. Model for the participation of quasipalindromic DNA sequences in frameshift mutation. Proc. Natl. Acad. Sci. USA 79:4128–4132.
6. Ripley, L. S., A. Clark, and J. G. deBoer. 1986. Spectrum of spontaneous frameshift mutations: sequences of bacteriophage T4 rII gene frameshifts. J. Mol. Biol. 19:601–613.
7. Ripley, L. S., and B. W. Glickman. 1983. DNA

secondary structures and mutation. *UCLA Symp. Mol. Cell. Biol.* 11:521–540.

8. Ripley, L. S., B. W. Glickman, and N. B. Shoemaker. 1983. Mutator versus antimutator activity of a T4 DNA polymerase mutant distinguishes two different frameshifting mechanisms. *Mol. Gen. Genet.* 189:113–117.

9. Ripley, L. S., and N. B. Shoemaker. 1983. A major role for bacteriophage T4 DNA polymerase in frameshift mutagenesis. *Genetics* 103:353–366.

10. Streisinger, G., Y. Okada, J. Emrich, J. Newton, A. Tsugita, E. Terzaghi, and M. Inouye. 1966. Frameshift mutations and the genetic code. *Cold Spring Harbor Symp. Quant. Biol.* 31:77–84.

Chapter 24

Effect of Accessory Replication Proteins on the Misincorporation of a Carcinogen-Modified Nucleotide by Bacteriophage T4 DNA Polymerase during DNA Replication In Vitro

MaryClaire Shiber and Navin K. Sinha

Ever since the realization that most carcinogens are mutagens (1), much effort has been spent in trying to elucidate how carcinogens alter the structure and function of DNA (for a review, see reference 20). One of the lines of investigation has been to examine the pairing properties of carcinogen-modified DNA or DNA precursors (deoxynucleotides) during DNA synthesis in vitro (2, 6, 7, 13, 21, 23). However, these studies so far have used a DNA polymerase alone, quite often *Escherichia coli* DNA polymerase I. DNA replication in vivo, in contrast, always involves a multienzyme complex consisting of a DNA polymerase along with several accessory proteins—DNA-binding protein, helicase, primase, ATP-driven factors which increase processivity, etc. (12). Here we describe the results of studies examining the pairing properties of a model carcinogen-modified nucleotide (1,N^6-etheno deoxyadenosine triphosphate [ε-dATP]) during DNA replication in vitro with bacteriophage T4 DNA polymerase, alone or in combination with its accessory proteins.

Original studies on the interaction of chloroacetaldehyde (a metabolite of vinyl chloride) with bases (9), nucleosides and nucleotides (18, 19), and homopolymeric nucleic acids (2, 10) confirmed the presence of etheno-modified derivatives. The in vivo formation of these derivatives was demonstrated in the hydrolytic products of liver DNA purified from rats exposed to vinyl chloride (5). Screening by a *Salmonella typhimurium* (*his* to *his*[+]) reversion assay with microsomal extract helped to further establish the metabolic cause-and-effect relationship between vinyl chloride metabolites in mutagenesis and possibly carcinogenesis (3).

MaryClaire Shiber and Navin K. Sinha • Waksman Institute of Microbiology, Rutgers University, Piscataway, New Jersey 08854.

UTILIZATION OF $1,N^6$-ε-dATP BY PHAGE T4 DNA POLYMERASE DURING DNA SYNTHESIS IN VITRO

Sequencing reactions provide an extremely sensitive assay for examining the pairing behavior of a modified nucleotide in DNA synthesis (7, 11, 23). We have used an assay (7) in which a set of four DNA synthesis reactions are performed with the modified nucleotide substituting for one of the four normal nucleotides. In addition, three control reactions are done—a "minus" reaction containing only the three normal deoxynucleotide triphosphates (dNTPs), a "plus" reaction containing all four normal dNTPs, and a "dideoxy" reaction (17) containing the four normal dNTPs with the addition of a chain-terminating ddNTP. The incorporation of the modified nucleotide can be detected either as chain elongation beyond the termination site for synthesis with minus reactions or as enhanced synthesis up to the termination site seen in minus reactions. These DNA synthesis reactions were carried out with T4 DNA polymerase on φX174 DNA template primed with a 17-mer primer (complementary to nucleotides 603 to 619 [16, 22]).

Enhanced synthesis to the −dNTP termination site is detectable in the presence of 2.5×10^{-3} to 5×10^{-3} μM of a normal dNTP and chain elongation, beyond that obtained with only three dNTPs, at 2.5×10^{-3} to 10×10^{-3} μM (Fig. 1A and 2A). An examination of the ability of ε-dATP to substitute for the normal nucleotides revealed that it functions best as dATP (visible primer elongation at 5 μM [Fig. 1B]) and more poorly as dCTP and dTTP (visible primer extension at 250 μM [Fig. 2B; Table 1]). However, ε-dATP substituted very poorly for dGTP (little primer elongation even at 1,000 μM). Therefore, T4 DNA polymerase uses ε-dATP as a substitute for normal nucleotides in the order A > C > T > G. ε-dATP works as dATP at a frequency of 1×10^{-3}, dCTP at 1×10^{-4}, dTTP at 5×10^{-5}, and dGTP at $<5 \times 10^{-6}$ (Table 1). Since pairing as adenine should not be mutagenic, these results suggest that ε-dATP should primarily produce transversion mutations. It is worth pointing out that we have examined only the pairing properties of ε-dATP as a nucleotide substrate. Since most mutations caused by this modified nucleotide are likely to result from its mispairing while it is in DNA, the above prediction would hold only if the pairing properties as a DNA template nucleotide are the same as those as a deoxynucleotide precursor.

MECHANISM OF BASE PAIRING BY ε-dAMP

We have shown that $1,N^6$-ε-dATP can serve as a substrate for DNA synthesis with

TABLE 1
Incorporation Efficiency of ε-dATP with T4 DNA Polymerase during In Vitro DNA Synthesis

−dNTP reaction	Chain elongation beyond −dNTP reaction			Increased synthesis relative to −dNTP reaction		
	Minimum [ε-dATP] (μM)	Minimum [dNTP] (μM)	Incorporation frequency	Minimum [ε-dATP] (μM)	Minimum [dNTP] (μM)	Incorporation frequency
−dATP	5	5×10^{-3}	1×10^{-3}	5	5×10^{-3}	1×10^{-3}
−dCTP	250	1×10^{-2}	4×10^{-5}	25	2.5×10^{-3}	1×10^{-4}
−dGTP	—[a]	5×10^{-3}	$<5 \times 10^{-6}$	—[a]	5×10^{-3}	$<5 \times 10^{-6}$
−dTTP	250	2.5×10^{-3}	1×10^{-5}	50	2.5×10^{-3}	5×10^{-5}

[a] —, No increase in chain elongation beyond the −dNTP reaction or no increase in synthesis relative to the −dNTP reaction at dNTP concentrations up to 1,000 μM.

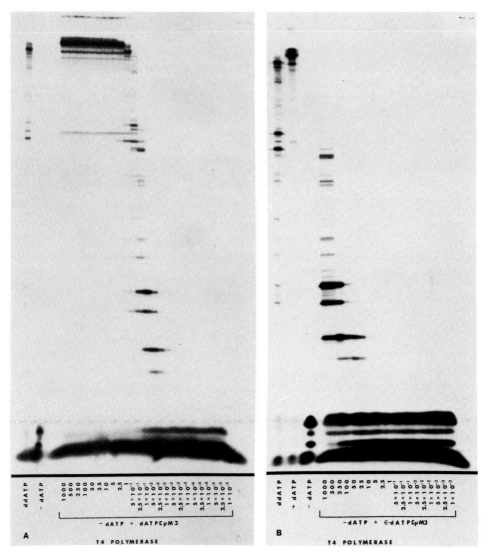

FIGURE 1. (A) Titration of the minimal dATP concentration needed to yield the extension of the 17-mer primer chain. (B) Titration of the minimal ε-dATP concentration needed as a substitute for dATP to yield detectable chain elongation with T4 DNA polymerase. The method used for DNA synthesis on synthetic oligodeoxynucleotide primed-φX174 single-stranded DNA templates was as described previously (22) with the following exceptions: (i) minus reactions contained only the three normal dNTPs at 30 μM; (ii) plus reactions contained three normal dNTPs at 25 μM with the addition of the fourth normal nucleotide at 100 μM unless otherwise indicated; (iii) ε-dATP pairing reactions contained three normal dNTPs at 25 μM with the addition of ε-dATP as indicated; (iv) dideoxy reactions contained 100 μM ddATP, 50 μM ddCTP, 50 μM ddGTP, or 200 μM ddTTP each in separate reactions. The dNTP corresponding to the ddNTP was included at 1.2 μM, whereas the three remaining normal dNTPs were present at 25 μM. All reactions (5 μl) contained 3×10^{10} molecules of φX174 am3 DNA primer hybrid. Synthesis reactions were stopped by the addition of formamide sample buffer. Samples were heated to 90°C prior to loading on an 8% acrylamide sequencing gel.

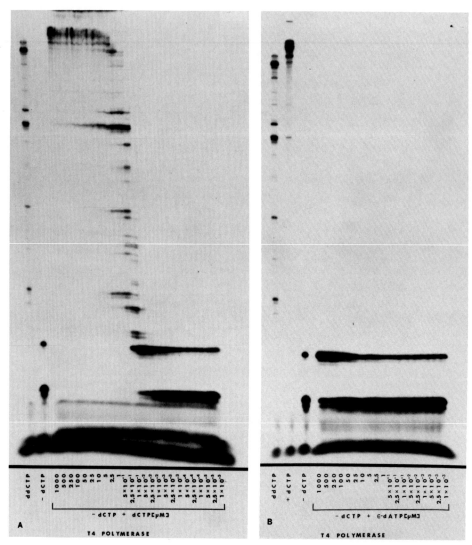

FIGURE 2. (A) Extent of primer extension upon the addition of different concentrations of dCTP to −dCTP reactions with T4 DNA polymerase. (B) Titration of the ε-dATP concentration needed to substitute for dCTP in −dCTP reactions with T4 DNA polymerase.

T4 DNA polymerase. Our findings confirm the results of Revich and Beattie, using *E. coli* DNA polymerase I (13), that ε-dATP functioned most readily as an analog of dATP (misincorporation frequency of 10^{-3}). We also have evidence for ambiguous behavior of this compound, the order of base pairing being ε-A · T > ε-A · G > ε-A · A > ε-A · C. The misincorporation frequency of ε-dAMP in place of dAMP was previously explained (13) by assuming that the base pairing of ε-dAMP with a normal nucleotide involved a protonated species of the analog (pK_a 4.3 [19]). At a physiological pH, it was calculated that 1 in every 1,000 ε-dATPs would be protonated. Since we have observed that ε-dATP substitutes for dATP at a frequency of 10^{-3}, if a protonated species is used then it must be accepted by the DNA polymerase as well as dATP, the normal nucleotide. Incorporation of ε-dATP involves rotation of the glycosyl bond to a *syn* conformation and is going to cause some strain, at least, due to a change in the angle of this bond and due to a size difference compared with adenine. Therefore, we felt that it was unlikely that the DNA polymerase would accept protonated ε-dATP as well as the normal nucleotide. DNA synthesis kinetics studies confirm this expectation (data not shown). The ambiguous pairing properties of ε-adenine were explained previously (13) by using minor tautomer and wobbling of the protonated ε-dAMP. An examination of these protonated pairs suggests that the stabilities should be as follows: $T \cdot \varepsilon\text{-A}^+_{syn} = G \cdot \varepsilon\text{-A}^+_{syn} > C \cdot \varepsilon\text{-A}^+_{syn} = A \cdot \varepsilon\text{-A}^+_{syn}$. The presence of an exocyclic hydrogen bond involving a carbonyl oxygen and an endocyclic nitrogen, in addition to a hydrogen bond involving two endocyclic nitrogens in the $T \cdot \varepsilon\text{-A}^+_{syn}$ or $G \cdot \varepsilon\text{-A}^+_{syn}$ pair, could contribute to higher stability than $C \cdot \varepsilon\text{-A}^+_{syn}$ or $A \cdot \varepsilon\text{-A}^+_{syn}$. The latter base pairs exhibit a hydrogen bond scheme involving an exocyclic amino group and an endocyclic nitrogen in addition to a hydrogen bond involving two endocyclic nitrogens. The results obtained here disagree with these predictions. We propose that perhaps the nonprotonated form of ε-dAMP is involved in ambiguous pairing. Using these, it is possible to explain the observed order of base pairing (ε-A · T > ε-A · G > ε-A · A > ε-A · C). Even though no information is available for the arrangement of the glycosyl bond of ε-dATP in solution, the crystal structure for a derivative of 1,N^6-etheno adenosine hydrochloride (ethyl-etheno-adenosine hydrochloride) shows the glycosyl bond to be in the *syn* conformation (24). The steric hindrance obtained in the *syn* conformation may be

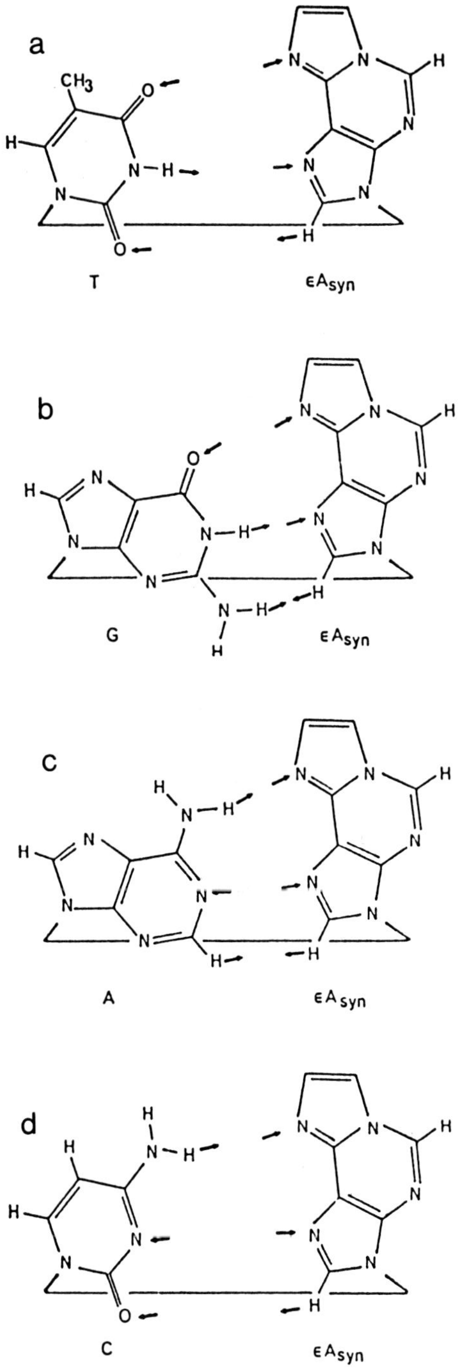

FIGURE 3. Suggested base-pairing schemes involving ε-dATP and the four normal nucleotides. (a) $T \cdot \varepsilon\text{-A}_{syn}$; (b) $G \cdot \varepsilon\text{-A}_{syn}$; (c) $A \cdot \varepsilon\text{-A}_{syn}$; (d) $C \cdot \varepsilon\text{-A}_{syn}$. An arrow pointing toward an atom indicates an H-bond acceptor. An arrow pointing away from an atom indicates an H-bond donor.

minimized if the sugar was oriented in a $C_{2'}$-endo conformation (15). The crystal structure of the ethyl-etheno-adenosine hydrochloride supports this expectation.

The process by which ε-dAMP might pair with each of the four normal nucleotides is shown in Fig. 3. The following parameters were assumed: (i) ε-dAMP is in a nonprotonated form; (ii) normal nucleotides and ε-dAMP exhibit $C_{1'}\ldots C_{1'}$-N angles of approximately 56 and 44°, respectively; (iii) $C_{1'}\ldots C_{1'}$ separation is the same in all four pairs; and (iv) ε-dAMP is arranged about the glycosyl bond in the *syn* conformation. Upon examination of the T · ε-A_{syn} pair or the G · ε-A_{syn} pair, a possible hydrogen bond may be drawn between the two endocyclic nitrogens in both cases. The T · ε-A_{syn} pair shows an additional possible hydrogen bond between the carbonyl oxygen of T and the hydrogen at position C-8 of ε-A_{syn}. In contrast, the two endocyclic nitrogens constitute a repulsive force in the A · ε-A_{syn} and C · εA_{syn} pairs. Another possibility exists between exocyclic amino groups on either A or C and endocyclic N^6 atoms in ε-A_{syn}. The distance in the A · ε-A_{syn} hydrogen bond appears to be shorter than that in the C · ε-A_{syn} case. The base-pairing possibilities discussed here can account for the relative order of misincorporation frequencies of ε-dATP as a substitute for each of the four normal dNTPs observed during DNA synthesis. It should be noted that, in the absence of any experimental data (such as those provided by nuclear magnetic resonance or X-ray crystallography), possible hydrogen bond schemes discussed are only a hypothetical attempt to explain the observed incorporation frequencies.

EFFECT OF THE ACCESSORY REPLICATION PROTEINS ON THE UTILIZATION OF ε-dATP BY T4 DNA POLYMERASE

Nothing is known about the contribution of the accessory proteins of a replication apparatus to the incorporation of nucleotide analogs by a DNA polymerase during DNA synthesis in vitro. To examine the effects of the accessory proteins, primer extension experiments similar to those described above were carried out in the presence of phage T4 DNA polymerase alone and in various combinations with the other six replication proteins. The ε-dATP concentration was held constant at 1,000 μM in each of the four −dNTP reactions. For example, each −dATP reaction contained 1,000 μM ε-dATP and 25 μM each dCTP, dGTP, and dTTP. To this was added T4 DNA polymerase, alone or with other proteins, to determine the overall effects on incorporation of ε-dATP residues as a substitute for dATP residues.

At this high concentration (1,000 μM), ε-dATP was readily utilized by phage T4 DNA polymerase as a substitute for dATP, dCTP, and dTTP and was poorly utilized for dGTP (see above). Addition of gene 32 protein (32p, a helix-destabilizing protein) decreased the extent of utilization of ε-dATP as any of the three correct nucleotides (C, G, or T), as evidenced by a decrease in the extent of primer elongation compared with that seen with the polymerase alone (Fig. 4 and 5). Addition of 45p, 44p-62p complex, or a mixture of 41p-61p each somewhat increased the misincorporation of ε-dATP in place of any of the three correct

FIGURE 4. Effect of the addition of phage T4 accessory replication proteins in the indicated combinations on the ability of T4 DNA polymerase to utilize ε-dATP as a substitute for dCTP. The method used for DNA synthesis is as described in the legend to Fig. 1 except that all reactions (10 μl) contained 1,000 μM rATP with 4×10^{10} molecules of φX174 am3 DNA primer hybrid. The T4 replication protein concentrations were as follows: 43p, 6 μg/ml; 32p, 300 μg/ml; 41p-61p, 20 μg of 41p per ml with 1 μg of 61p per ml; 44p-62p complex, 72 μg/ml; 45p, 60 μg/ml.

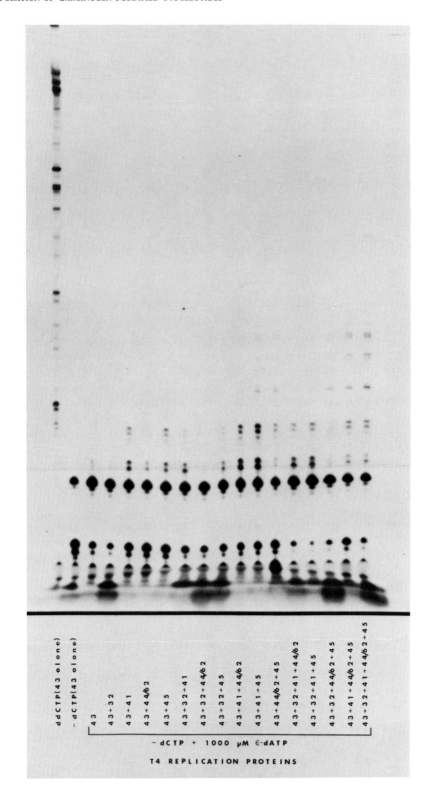

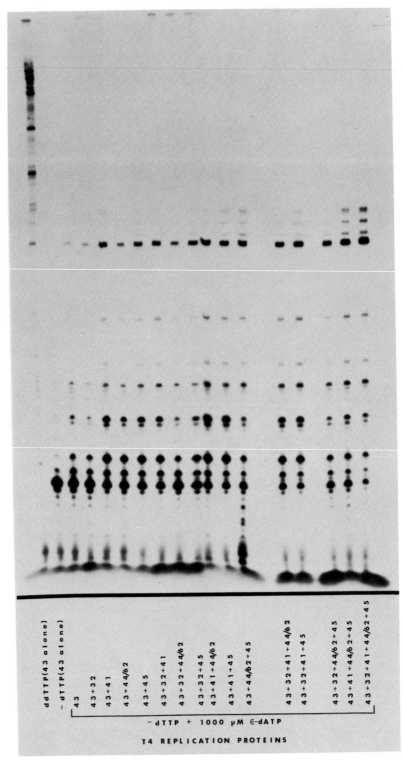

FIGURE 5. Effect of accessory replication proteins on the utilization of ε-dATP by T4 DNA polymerase as a substitute for TTP in -dTTP reactions. See the legend to Fig. 4 for more details.

nucleotides. The addition of 32p somewhat moderated this misincorporation (i.e., a decrease was seen for some but not all nucleotide substitutions). Addition of the combinations 41p-61p plus 44p-62p or 41p-61p plus 45p generally resulted in more misincorporation than either accessory-protein combination alone. The addition of 44p-62p plus 45p always increased misincorporations significantly (Fig. 3 and 4). The addition of 32p to reactions containing 41p-61p plus 44p-62p, 41p-61p plus 45p, or 44p-62p plus 45p slightly decreased the utilization of ε-dATP as dCTP (Fig. 4) or dGTP, and sometimes as dTTP (Fig. 5). The highest misincorporations of ε-dATP were observed when the polymerase was supplemented with 41p-61p plus 44p-62p plus 45p. Interestingly, the addition of 32p to this combination had very little effect on the degree of misincorporation of ε-dATP as dATP, dCTP, or dTTP (Fig. 4 and 5; data not shown).

32p binds very tightly to single-stranded DNA and, even though incapable of fully melting natural double-stranded DNAs, can impart partly single-stranded character to double-stranded DNA (12). It also interacts specifically with T4 DNA polymerase. The reason for enhanced accuracy (i.e., reduced incorporation of ε-dATP) in the presence of 32p, we believe, is due to a greater destabilization of the primer terminus after the incorporation of the analog (also see references 4 and 14). 44p-62p and 45p together make a sliding clamp for the T4 DNA polymerase (12), and this greatly increases the processivity of T4 DNA polymerase and reduces dissociation when difficulties in chain elongation are encountered (for example, at hairpins in the DNA template [8, 14]). We suggest that there is likely to be an increased tendency for the T4 polymerase to dissociate from the primer terminus upon the insertion of a less favorable nucleotide like ε-dAMP. The presence of the sliding clamp of 44p-62p plus 45p or the 41p helicase (which is known to be stimulated by the 61p primase [12]) might decrease the dissociation after the insertion of the ε-dAMP. Addition of 32p partially offsets the increased misincorporation seen by the addition of 44p-62p plus 45p or 41p-61p plus 45p. However, 32p is unable to prevent increased incorporation of ε-dAMP once the T4 polymerase is supplemented with all five accessory proteins (44p-62p plus 45p plus 41p-61p). This is most likely because the polymerase is held on the template-primer very tightly and 32p is now not able to destabilize the primer terminus and cause the dissociation of the polymerase.

ACKNOWLEDGMENTS. This work was supported by Public Health Service grant GM24391 from the National Institutes of Health and by a grant from the Busch Foundation.

LITERATURE CITED

1. **Ames, B. N., W. E. Durston, E. Yamasaki, and F. D. Lee.** 1973. Carcinogens are mutagens: a simple test system combining liver homogenates for activation and bacteria for detection. *Proc. Natl. Acad. Sci. USA* 70:2281–2285.
2. **Barbin, A., H. Bartsch, P. Leconte, and M. Radman.** 1981. Studies on the miscoding properties of 1,N^6-ethenoadenine and 3,N^4-ethenocytosine, DNA reaction products of vinyl chloride metabolites, during *in vitro* DNA synthesis. *Nucleic Acids Res.* 9:375–387.
3. **Bartsch, H., C. Malaveille, and R. Montesano.** 1975. Human, rat, and mouse liver-mediated mutagenicity of vinyl chloride in *S. typhimurium* strains. *Int. J. Cancer* 15:429–437.
4. **Bedinger, P., and B. M. Alberts.** 1983. The 3'-5' proofreading exonuclease of bacteriophage T4 DNA polymerase is stimulated by other T4 DNA replication proteins. *J. Biol. Chem.* 258:9649–9656.
5. **Green, T., and D. E. Hathway.** 1978. Interactions of vinyl chloride with rat-liver DNA *in vivo*. *Chem.-Biol. Interactions* 22:211–224.
6. **Hall, J. A., R. Saffhill, T. Green, and D. E. Hathway.** 1981. The induction of errors during *in vitro* DNA synthesis following chloroacetaldehyde treatment of poly(dA-dT) and poly(dC-dG) templates. *Carcinogenesis* 2:141–146.
7. **Hillebrand, G. G., A. H. McCluskey, K. A.**

Abbott, G. G. Revich, and K. L. Beattie. 1984. Misincorporation during DNA synthesis, analyzed by gel electrophoresis. *Nucleic Acids Res.* 12:3155–3171.

8. Huang, C.-C., J. E. Hearst, and B. M. Alberts. 1981. Two types of replication proteins increase the rate at which T4 DNA polymerase traverses the helical regions in a single-stranded DNA template. *J. Biol. Chem.* 256:4087–4094.

9. Kotchetkov, N. K., V. N. Shibaev, and A. A. Kost. 1971. New reaction of adenine and cytosine derivatives, potentially useful for nucleic acids modification. *Tetrahedron Lett.* 22:1993–1996.

10. Kusmierek, J. T., and B. Singer. 1982. Chloroacetaldehyde-treated ribo- and deoxyribopolynucleotides. 1. Reaction products. *Biochemistry* 21:5717–5722.

11. Lasken, R. S., and M. F. Goodman. 1985. A fidelity assay using "dideoxy" DNA sequencing: a measurement of sequence dependence and frequency of forming 5-bromouracil·guanine base mispairs. *Proc. Natl. Acad. Sci. USA* 82:1301–1305.

12. Nossal, N. G., and B. M. Alberts. 1983. Mechanism of DNA replication catalyzed by purified T4 replication proteins, p. 71–81. *In* C. K. Mathews, E. M. Kutter, G. Mosig, and P. B. Berget (ed.), *Bacteriophage T4*. American Society for Microbiology, Washington, D.C.

13. Revich, G. G., and K. L. Beattie. 1986. Utilization of 1,N^6-etheno-2'-deoxyadenosine 5'-triphosphate during DNA synthesis on natural templates, catalyzed by DNA polymerase I of *Escherichia coli*. *Carcinogenesis* 7:1569–1576.

14. Roth, A. C., N. G. Nossal, and P. T. Englund. 1982. Rapid hydrolysis of deoxynucleoside triphosphates accompanies DNA synthesis by T4 DNA polymerase and T4 accessory proteins 44/62 and 45. *J. Biol. Chem.* 257:1267–1273.

15. Saenger, W. 1984. *Principles of Nucleic Acid Structure*, p. 103. Springer-Verlag, Inc., New York.

16. Sanger, F., A. R. Coulson, T. Friedmann, G. M. Air, B. G. Barrell, N. L. Brown, J. C. Fiddes, C. A. Hutchison, P. M. Slocombe, and M. Smith. 1978. The nucleotide sequence of bacteriophage ϕX174. *J. Mol. Biol.* 125:225–246.

17. Sanger, F., S. Nicklen, and A. R. Coulson. 1977. DNA sequencing with chain-terminating inhibitors. *Proc. Natl. Acad. Sci. USA* 74:5463–5467.

18. Sattsangi, P. D., J. R. Barrio, and N. J. Leonard. 1980. 1,N^6-etheno-bridged adenines and adenosines. Alkyl substitution, fluorescence properties, and synthetic applications. *J. Am. Chem. Soc.* 102:770–774.

19. Secrist, J. A., J. R. Barrio, N. J. Leonard, and G. Weber. 1972. Fluorescent modification of adenosine-containing coenzymes. Biological activities and spectroscopic properties. *Biochemistry* 11:3499–3506.

20. Singer, B., and D. Grunberger. 1983. *Molecular Biology of Mutagens and Carcinogens*. Plenum Publishing Corp., New York.

21. Singer, B., J. Sagi, and J. T. Kusmierek. 1983. *Escherichia coli* polymerase I can use O^2-methyldeoxythymidine or O^4-methyldeoxythymidine in place of deoxythymidine in primed poly(dA-dT)·poly(dA-dT) synthesis. *Proc. Natl. Acad. Sci. USA* 80:4884–4888.

22. Sinha, N. K. 1987. Specificity and efficiency of editing of mismatches involved in the formation of base-substitution mutations by the 3'→5' exonuclease activity of phage T4 DNA polymerase. *Proc. Natl. Acad. Sci. USA* 84:915–919.

23. Topal, M. D., C. A. Hutchison, and M. S. Baker. 1982. DNA precursors in chemical mutagenesis: a novel application of DNA sequencing. *Nature* (London) 298:863–865.

24. Wang, A. H.-J., L. G. Dammann, J. R. Barrio, and I. C. Paul. 1974. The crystal and molecular structure of a derivative of 1,N^6-ethenoadenosine hydrochloride. Dimensions and molecular interactions of the fluorescent ε-adenosine (εAdo) system. *J. Am. Chem. Soc.* 96:1205–1213.

Chapter 25

Essential Function of the Editing Subunit of DNA Polymerase III in *Salmonella typhimurium*

Russell Maurer, Edward Lancy, Miriam Lifsics, and Patricia Munson

Most proteins involved in bacterial DNA replication are essential for viability of the bacterial cell. Exceptions to this statement arise in the case of many proteins that affect the fidelity of the replication process, because the generalized increase in spontaneous mutation frequency observed when such "mutator" genes are inactive is not itself a lethal event for the population (even if an occasional cell dies after acquiring a lethal mutation) (2). The editing (ϵ) subunit of DNA polymerase III (pol III) seems to occupy an ambiguous middle ground: in its known role as an editing exonuclease (15, 16) it does not appear to be essential in principle, and viable mutants deficient in exonuclease activity are known (3–5). On the other hand, as a core component of an essential enzyme (11), ϵ modulates the polymerase activity of the α subunit of pol III (9). It is not known whether this function is essential.

Russell Maurer, Edward Lancy, Miriam Lifsics, and Patricia Munson • Department of Molecular Biology and Microbiology, Case Western Reserve University, Cleveland, Ohio 44106.

We have attempted to address the cellular requirement for ϵ by making defined chromosomal null mutations in its gene (*dnaQ*). Our experiments show that, at least in *Salmonella typhimurium* where these experiments were carried out, *dnaQ* is essential. Perhaps more surprisingly, the lethality of *dnaQ* null mutations is rather easily suppressed by second-site mutations. In this article we will give a brief summary of our experimental system and the results of our investigation to date. We will also discuss possible interpretations of our results and indicate future directions.

dnaQ MUTATIONS

Defined null mutations were constructed by Tn*10* insertion into the cloned *S. typhimurium dnaQ* gene (10) and subsequent modification of the insertion mutations. In one such modification, two Tn*10* insertions, one located near the middle of *dnaQ* and the other located near the 3' end of the gene, were recombined (that is, by an unequal crossover between the two Tn*10*s) to form a deletion-substitution mutation in which Tn10 exactly replaces the 3' portion of the gene. Further modifications to this construct extended the substituted region to include increasing 5' portions of *dnaQ*. The joint point between Tn10 and *dnaQ* in several such constructs was sequenced, and in the course of this sequencing, a substantial portion of the *dnaQ* gene sequence was obtained. The sequence data show convincingly that the *dnaQ* gene isolated by genetic complementation, and deleted in these experiments, is authentic on the basis of extensive similarity to the *Escherichia coli dnaQ* gene (and derived protein) sequence.

The deletion-substitution alleles of *dnaQ* were transferred by homologous recombination into the *S. typhimurium* chromosome, where they could be selected by their associated tetracycline resistance phenotype. The line of experimentation described in this paper derives from our initial observation (10) that the tetracycline-resistant colonies that formed upon introduction of the *dnaQ* null allele exhibited extensive lethal sectoring; in fact, most of the visible mass of a colony consisted of dead cells. Nonetheless, from each such colony we could recover a few viable cells that had become immortalized, that is, they would go on to form colonies without lethal sectors. Since we had taken some care to construct *dnaQ* alleles that were incapable of reversion to *dnaQ*$^+$, and since Southern blots showed the expected structure of the mutated *dnaQ* gene in the chromosome of the immortalized cells (and the absence of wild-type *dnaQ*$^+$), we guessed that the immortalized cells contained a second-site suppressor of the *dnaQ* mutation. This hypothesis has proved correct.

Genetic Analysis

Full details of the analysis of immortalized cells will appear elsewhere. Here, we will summarize our analysis. Each immortalized cell line carries a mutation, which we have provisionally named *spq* (suppressor of Q). The *spq* mutants are viable in a *dnaQ*$^+$ background as well as with the *dnaQ* null mutations. The properties of wild-type, *spq*, and *dnaQ*::Tn10 *spq* strains are summarized in Table 1.

All *spq* mutations examined to date map in close proximity to the *dnaE* gene (encoding the α, or polymerase, subunit of DNA pol III). We do not know how many distinct sites of mutation are represented in our group of *spq* mutants. Nor do we know with certainty whether the *spq* mutations all define a single locus, but we argue (below) that one *spq* mutation is an allele of *dnaE*, and we think it likely that others are *dnaE* alleles as well.

From an *spq-2* strain we isolated a 7-

TABLE 1
Properties of Wild-Type, *spq*, and *dnaQ*::Tn10 *spq* Strains

Strain	Criteria		
	Growth rate[a] (min)	Lethal sectoring[b]	Mutation rate[c]
Wild type	50	Yes	1
spq mutant	50	No	1
dnaQ:Tn10 *spq* mutant	50	NA[d]	200–5,000

[a] Doubling time in L broth at 37°C.
[b] Colony morphology upon fresh transduction to *dnaQ*::Tn10.
[c] Forward mutation frequencies to rifampin resistance and nalidixate resistance, normalized to wild type. The range is shown for 10 different *spq* alleles, each assayed in no fewer than three separate cultures. The frequencies measured for wild type were as follows: Nal^r, 9.3 × 10^{-9}; Rif^r, 2.1 × 10^{-8}.
[d] NA, Not applicable.

kilobase fragment of DNA which was capable of complementing a *dnaE*(Ts) mutant. This fragment also conferred the suppressor phenotype on otherwise wild-type cells. We were able to locate the *dnaE* coding sequence, as well as other coding sequences, within this fragment by a combination of restriction mapping, complementation tests, and expression of proteins (under T7 RNA polymerase control; 17) from this DNA fragment and from mutated derivative fragments. As a point of reference we located a *Bgl*II restriction site roughly in the middle of *dnaE*. Our analysis showed that DNA from both sides of this *Bgl*II site was necessary if the fragment was to confer the suppressor phenotype. On one hand, if we deleted material to one side of *Bgl*II (deleting the 3' portion of *dnaE*), then suppression was lost; therefore material on the 3' side was necessary. On the other hand, if we replaced the 3' deleted material with a comparable segment from wild type, suppression was restored, but a fragment consisting entirely of wild-type sequences was unable to suppress; therefore, material from *spq-2* on the 5' side of the *Bgl*II site was also necessary for suppression. The simplest interpretation of these data is that the suppressor gene is *dnaE* and the *spq-2* mutation is located in the 5' half of *dnaE*. We note, however, that we cannot rule out a more complicated possibility, i.e., that one of the coding sequences located 5' to *dnaE* collaborates with *dnaE* (in *cis*) to produce the suppressor phenotype.

Biochemical Analysis of DNA Polymerase

Guided by the results of the genetic analysis, we have begun to analyze the properties of DNA pol III from wild-type and *dnaQ*::Tn*10 spq-2* cells. Both enzymes have been purified to 10 to 20% purity by a gap-filling assay. Both enzymes are similarly sensitive to N-ethylmaleimide and KCl. The major difference noted has been in the differential appearance of the ε subunit and its associated exonuclease activity. This subunit and its activity were present in the preparation from wild-type cells, but neither was detected in the preparation from mutant cells. In summary, then, the polymerase preparations clearly reflect the difference in *dnaQ* genotype of the cell source, but to date, using rather nonspecific tests, we have not detected any property of *spq-2* polymerase that could explain the cellular phenotype.

MODELS FOR *dnaQ* LETHALITY AND ITS SUPPRESSION

One of the most striking features of the experimental system we have described is the ability of freshly formed *dnaQ*::Tn*10* mutants to continue growing and dividing for many generations (enough to form a visible colony on a plate) before succumbing to the lethal effect. This feature is not the result of residual ε activity, since mutants in which the Tn*10* replaces virtually the entire *dnaQ* produce this phenotype. We consider several possible explanations of this feature.

Phenotypic Lag

According to the model of phenotypic lag, ε protein existing in the cell at the time of elimination of the *dnaQ* gene supports further rounds of DNA replication and cell division. Only after ε falls to inadequate levels do cells die. To push this model to its extreme, one could even imagine that after the preexisting ε has been distributed into the maximum possible number of daughter cells, these cells will continue to grow arithmetically, producing one viable daughter cell (containing ε) and one inviable cell (lacking ε) at each division. This model requires no particular assumption about the nature of the lethal effect caused by the absence of ε. For example, this model would be compatible with lethality caused by a gradual loss of polymerase activity as the ε/α ratio declines

(9), as well as with lethality related to lack of editing (see below). The main difficulty with this model is the great disparity between the likely amount of ε protein initially present (tens or hundreds of molecules per cell) and the number of progeny cells that are produced (on the order of a million).

Excessive Mutagenesis

The total contribution of the ε subunit to replication fidelity is unknown. Fersht and Knill-Jones (6) have argued that editing contributes a factor of some 10^2. The *mutD5* mutation in *E. coli dnaQ* increases mutation frequency by about 10^4 (4); this mutation drastically curtails exonuclease activity (5), but because of its dominance over $dnaQ^+$ its precise mode of action is unclear (3, 4, 18). The model of excessive mutagenesis is founded on the notion that the contribution of ε to fidelity (through editing and perhaps other mechanisms, additively) is greater than previously estimated; if so, complete loss of ε might increase mutation frequency to an intolerable level (i.e., above that observed in *mutD5 E. coli*, which does not exhibit lethal sectoring). Lethal sectoring would occur as cells randomly develop lethal mutations. This model implies that the suppressor mutations in *dnaE* would compensate for lethal mutagenesis by exerting an antimutator effect, e.g., by improving the fidelity of the initial base selection or by recruiting an editing function alternative to the replication fork.

The available data tend to argue against this model. Most importantly, the *spq* mutations, in an otherwise wild-type background, do not appear to be significant antimutators. In addition, the absence of detectable exonuclease in the pol III preparation from *dnaQ*::Tn*10 spq-2* bacteria argues against the activation of a latent exonuclease in the α subunit and indicates that if a hypothetical novel editing function is recruited by the *spq-2* polymerase, it either does not copurify with pol III activity or is not revealed by the assays used.

Replication Blockage Resulting from Misinsertion

At a frequency generally estimated (8) at once per 10^4 to 10^5 base insertions, DNA pol III inserts an incorrect base. In ordinary cells, almost all such misinsertions are removed by editing. Those few misinsertions that are not edited and hence become stable misincorporations are revealed as an increment of mutation frequency when postreplication repair (i.e., methyl-directed mismatch repair) is blocked. It is not known in detail how a misinsertion is converted to a misincorporation; mechanistic studies on DNA pol I and pol III suggest that addition of nucleotides onto the mispaired base is a difficult step (1, 6, 7, 12, 13). Since it is clear that normal bacterial cells do occasionally misincorporate, cells must have at least limited capacity to convert misinsertions into misincorporations. A mutant cell devoid of the editing subunit will face misinsertions at a much increased frequency and may find its capacity to convert them to misincorporations overwhelmed. Confronted repeatedly with nonextendable (misinsertion) substrates, replication forks may be able to progress in one of two ways suggested by other studies. If the polymerase can dissociate from the primer terminus and reinitiate downstream (as is thought to occur, for example, on UV-treated DNA [14]), a molecule with single-strand gaps is produced; the gaps can later be filled in by DNA pol I with its associated exonuclease activity, or by homologous recombination. In either case, the potential mutation at the site of the misinsertion would be lost. Alternatively, the pol III can undergo a modification that enables it to extend the mispaired primer terminus, resulting in fixation of a mutation (19). Either of these mechanisms might be aided by proteins induced under SOS control, since stalled replication forks are thought to act as SOS-inducing signals (19).

In the replication blockage model, we suppose that one or both of the secondary mechanisms described above operate to prolong the viability of dnaQ::Tn10 cells during lethal sectoring and may continue to operate in the immortalized spq mutants as well. In accord with this idea, we have noted that DNA pol I is required both for lethal sectoring and for indefinite propagation of dnaQ::Tn10 spq-2 strains.

The replication blockage model for lethality implies that suppressor mutations ought to facilitate either the removal of misinsertions or their conversion to misincorporations. The mutation frequency data argue that the suppressor mutations allow many misinsertions to be converted to misincorporations. Therefore, we favor the idea that the suppressor mutations in dnaE alter the intrinsic ability of pol III to extend misinsertions. It is not difficult to imagine that different spq mutations might alter this property to various degrees, a factor that, in combination with the continuing operation of a nonmutagenic system for processing some misinsertions (e.g., the pol I system of gap filling), could account for some of the variation in mutation frequency that we see in different spq alleles.

ε AS THE MISINCORPORATION FUNCTION IN NORMAL CELLS?

What accounts for the capacity of normal cells to convert misinsertions to misincorporations? A hint may come from consideration of the mutD5 mutation of E. coli. This mutation in dnaQ results in severe reduction in editing capacity (5), yet the cells are viable without the necessity of a suppressor mutation (4). In terms of the misinsertion-replication blockage model, we can account for this fact if the mutD5 cells retain (or possibly amplify) their ability to convert misinsertions. Unlike the S. typhimurium dnaQ::Tn10 mutations, the mutD5 mutation is dominant over wild type and the mutant ε protein is presumably present in the pol III complex. This difference between the two types of mutation in dnaQ and their differential lethality lead to the hypothesis that ε itself promotes misincorporation in normal cells. In this view, ε actively partitions misinsertions between excision and extension, with excision normally the favored pathway. In editing-defective mutants such as mutD5, there is no lethality because ε partitions sufficient misinsertions into the extension pathway to permit viability. If, however, the ε subunit is eliminated, then the effect is lethal because misinsertions cannot be processed efficiently and replication blocks or unfilled gaps accumulate.

DEVELOPMENT OF spq MUTATIONS DURING LETHAL SECTORING

It is clear that the spq mutations we have detected are not preexisting in the parental bacterial strains. Had this been the case, we would have observed normal colony formation rather than lethal sectoring upon introduction of the dnaQ::Tn10 mutation (see Table 1). The fact that essentially every sectored colony gives rise to surviving spq mutants may be explained if the dying cells are subjected to extensive mutagenesis. This would certainly be an expected consequence of loss of editing function.

FUTURE DIRECTIONS

Even though our experimental results thus far do not compel us to accept replication blockage due to misinsertion as the mechanism of lethality and misinsertion-extension as the mechanism of suppression, our attention is focused in these directions because these possibilities have the most interesting implications and also are subject to explicit test. At present, biochemical tests of extension are in progress using purified α and defined, misprimed template-primers. The postulated role of ε in promoting extension can also be tested using such primers, provided that exonuclease-deficient ε can be prepared.

In another context, the idea of a permissive polymerase has been offered to explain the ability of SOS-induced cells to copy UV-damaged DNA. This situation differs from the misinsertion-extension problem in two significant ways. First, in the UV case, the problem is a damaged template strand, whereas in the misinsertion case, the problem is in the primer strand. This difference is significant because editing can operate to remove the source of trouble only in the misinsertion case. Second, in the UV case successful copying depends on several auxiliary proteins which are thought to modify the polymerase to make it more permissive: UmuC and UmuD (or plasmid analogs of these) and RecA (19). In the misinsertion-extension case, both in normal cells and in our $dnaQ$::Tn10 strains, the necessary adjustment appears to be intrinsic to pol III. Nonetheless, the two situations both involve a mispaired template-primer that must be extended, and a potentially fruitful avenue of research will be to examine whether the adjustment made in one situation will apply in the other. Along these lines, we have begun to investigate whether the spq mutations confer on $S.$ $typhimurium$ improved ability to copy UV-damaged DNA, as assayed by reactivation of UV-treated P22 phage. Thus far, we have detected tantalizing suggestions that this hypothesis may be correct for some of the spq alleles. The experiments, however, suffer from two major problems. One is that the strains in which an effect has been seen, being $dnaQ$::Tn10, are strong mutator strains, making it difficult to rule out an unsuspected mutation elsewhere as the source of our observations. The second problem is that the UvrABC repair system is active in $S.$ $typhimurium$, providing a source of P22 reactivation that limits the sensitivity of detection for an spq effect. The latter problem, at least, is solvable by suitable strain constructions.

ACKNOWLEDGMENTS. R.M. gratefully acknowledges support from the National Institutes of Health (Public Health Service grant AI19942-01) and from the G. Harold and Leila Y. Mathers Charitable Foundation. M.L. was supported during part of this work through the Genetic Toxicology training program at Case Western Reserve University, supported by Public Health Service grant T32-ES 07080 from the National Institutes of Health.

LITERATURE CITED

1. Brutlag, D., and A. Kornberg. 1972. Enzymatic synthesis of deoxyribonucleic acid. XXXVI. A proofreading function for the 3'→5' exonuclease activity in deoxyribonucleic acid polymerases. *J. Biol. Chem.* 247:241–248.
2. Cox, E. C. 1976. Bacterial mutator genes and the control of spontaneous mutation. *Annu. Rev. Genet.* 10:135–156.
3. Cox, E. C., and D. L. Horner. 1982. Dominant mutators in *E. coli*. *Genetics* 100:7–18.
4. Degnen, G. E., and E. C. Cox. 1974. Conditional mutator gene in *Escherichia coli*: isolation, mapping, and effector studies. *J. Bacteriol.* 117:477–487.
5. Echols, H., C. Lu, and P. M. J. Burgers. 1983. Mutator strains of *Escherichia coli*, *mutD* and *dnaQ*, with defective exonucleolytic editing by DNA polymerase III holoenzyme. *Proc. Natl. Acad. Sci. USA* 80:2189–2192.
6. Fersht, A. R., and J. W. Knill-Jones. 1983. Contribution of 3'→5' exonuclease activity of DNA polymerase III holoenzyme from *Escherichia coli* to specificity. *J. Mol. Biol.* 165:669–682.
7. Hillebrand, G. G., A. H. McCluskey, K. A. Abbott, G. G. Revich, and K. L. Beattie. 1984. Misincorporation during DNA synthesis, analyzed by gel electrophoresis. *Nucleic Acids Res.* 12:3155–3171.
8. Loeb, L. A., and T. A. Kunkel. 1982. Fidelity of DNA synthesis. *Annu. Rev. Biochem.* 52:429–457.
9. Maki, H., and A. Kornberg. 1987. Proofreading by DNA polymerase III of *Escherichia coli* depends on cooperative interaction of the polymerase and exonuclease subunits. *Proc. Natl. Acad. Sci. USA* 84:4389–4392.

10. Maurer, R., B. C. Osmond, E. Shekhtman, A. Wong, and D. Botstein. 1984. Functional interchangeability of DNA replication genes in *Salmonella typhimurium* and *Escherichia coli* demonstrated by a general complementation procedure. *Genetics* 108:1–23.
11. McHenry, C. S. 1985. DNA polymerase III holoenzyme of *Escherichia coli*: components and function of a true replicative complex. *J. Mol. Cell. Biochem.* 108:1–23.
12. Mizrahi, V., P. Benkovic, and S. J. Benkovic. 1986. Mechanism of DNA polymerase I: exonuclease/polymerase activity switch and DNA sequence dependence of phosphorolysis and misincorporation reactions. *Proc. Natl. Acad. Sci. USA* 83:5769–5773.
13. Mizrahi, V., P. A. Benkovic, and S. J. Benkovic. 1986. Mechanism of the idling-turnover of the large (Klenow) fragment of *Escherichia coli* DNA polymerase I. *Proc. Natl. Acad. Sci. USA* 83:231–235.
14. Rupp, W. D., and P. Howard-Flanders. 1968. Discontinuities in the DNA synthesized in an excision-defective strain of *Escherichia coli* following ultraviolet irradiation. *J. Mol. Biol.* 31:291–304.
15. Scheuermann, R., T. Schuman, P. M. J. Burgers, C. Lu, and H. Echols. 1983. Identification of the ε-subunit of *Escherichia coli* DNA polymerase III holoenzyme as the *dnaQ* gene product: a fidelity subunit for DNA replication. *Proc. Natl. Acad. Sci. USA* 80:7085–7089.
16. Scheuermann, R. H., and H. Echols. 1984. A separate editing exonuclease for DNA replication: the ε subunit of *Escherichia coli* DNA polymerase III holoenzyme. *Proc. Natl. Acad. Sci. USA* 81:7747–7751.
17. Tabor, S., and C. C. Richardson. 1985. A bacteriophage T7 RNA polymerase/promoter system for controlled exclusive expression of specific genes. *Proc. Natl. Acad. Sci. USA* 82:1074–1078.
18. Takano, K., Y. Nakabeppu, H. Maki, T. Horiuchi, and M. Sekiguchi. 1986. Structure and function of *dnaQ* and *mutD* mutators of *Escherichia coli*. *Mol. Gen. Genet.* 205:9–13.
19. Walker, G. C. 1984. Mutagenesis and inducible responses to deoxyribonucleic acid damage in *Escherichia coli*. *Microbiol. Rev.* 48:60–93.

Chapter 26

Very-High-Frequency Mutagenesis Induced by 5-Bromo-2'-Deoxyuridine and Deoxynucleotide Pool Imbalance in Syrian Hamster Cells

Elliot R. Kaufman

The thymidine (dThd) analog 5-bromodeoxyuridine (BrdUrd) has been shown to be an effective mutagen in procaryotic (2, 9, 21, 25, 30) and eucaryotic (1, 11, 17, 29) systems. A model (9) for the mutagenic action of this base analog suggested that 5-bromouracil (BrUra) would undergo a tautomeric shift to the rare enol state, allowing it to mispair with guanine more frequently than would thymine (Thy). Thus, two possible mechanisms for BrUra mutagenesis were proposed, errors of incorporation and errors of replication. Errors of incorporation were thought to occur when BrdUrd triphosphate (BrdUTP) mispaired with a guanine residue in replicating DNA. Errors of replication were thought to occur when a BrUra residue in replicating DNA mispaired with dGTP.

Studies in mammalian systems initially suggested that BrdUrd mutagenesis was exclusively due to errors of incorporation, was determined by the concentration of BrdUrd to which the cells were exposed, and was independent of the amount of BrUra incorporated into DNA in place of Thy residues (17). These studies utilized a mutagenesis protocol called INC mutagenesis (Fig. 1). The INC mutagenesis protocol involved the exposure of cells to supplemented culture medium which allowed the BrdUrd concentration in the medium and the level of BrUra substituted for Thy in DNA to be varied independently. The FBT medium contained BrdUrd, dThd, and 5-fluorodeoxyuridine, an inhibitor of de novo thymidylate biosynthesis (5). Under these culture conditions, the ratio of BrUra to Thy residues in newly synthesized DNA was determined by the BrdUrd/dThd ratio in the culture medium (18). In this medium, the substitution of BrUra residues for Thy residues in DNA was varied, at a constant BrdUrd concentration, by varying only the dThd concentration. Conversely, the level of BrUra substituted for Thy was held constant while the BrdUrd concentration in the medium was varied, by maintaining a constant ratio of

---

Elliot R. Kaufman • Department of Genetics, University of Illinois College of Medicine at Chicago, Chicago, Illinois 60612.

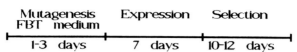

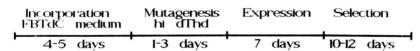

FIGURE 1. BrdUrd mutagenesis protocols. The 2E cell line was derived from a Syrian hamster melanoma line, RPMI 3460 (22), by its ability to grow with all of the Thy residues in DNA replaced by BrUra (3). The basic culture medium for the cells was Dulbecco modified Eagle medium (DMEM) supplemented with 10% fetal bovine serum. All cells exposed to BrdUrd were protected from wavelengths of light below 550 nm (16). The FBTdC medium used in the REP mutagenesis experiments was DMEM supplemented with 10 μM 5-fluorodeoxyuridine, an inhibitor of de novo thymidylate biosynthesis (5), 10 μM dCyd, and the indicated levels of BrdUrd and dThd. To generate the various levels of BrUra substituted for Thy residues in DNA, FBTdC medium contained the following: 0 μM BrdUrd plus 20 μM dThd (0% substitution); 10 μM BrdUrd plus 30 μM dThd (25% substitution); 10 μM BrdUrd plus 10 μM dThd (50% substitution); or 10 μM BrdUrd plus 3.3 μM dThd (75% substitution). The ratio of BrdUrd to dThd in the culture medium has been shown to determine the ratio of BrUra to Thy in DNA when de novo thymidylate biosynthesis is inhibited (18).

BrdUrd to dThd. Studies using the INC protocol (7, 8, 12, 17, 19) provided evidence that (i) BrdUrd mutagenesis in mammalian cells was driven by high concentrations of BrdUrd which caused an increase in the intracellular ratio of BrdUTP to dCTP, thus driving the mispairing of BrdUTP with guanine residues in replicating DNA; (ii) BrdUrd mutagenesis occurred only during incorporation of BrdUTP into DNA; and (iii) mutations induced by the replication of BrUra residues in DNA, errors of replication, did not occur.

The results of the studies with the INC mutagenesis protocol suggested that mutations caused by the replication of BrUra residues in DNA did not play an important role in BrdUrd mutagenesis in mammalian cells. However, the deoxynucleoside triphosphate (dNTP) precursor pool imbalance demonstrated to be required for INC mutagenesis (a high BrdUTP/dCTP ratio, i.e., incorrect/correct) suggested that the unbalanced precursor environment necessary to induce the mispairing of replicating BrUra residues may not have been present under the conditions of the INC protocol. By analogy, a high dGTP/dATP ratio (incorrect/correct) might serve to increase the probability of dGTP mispairing with BrUra residues in replicating DNA.

In more recent studies with mammalian cells (12–15), a new BrdUrd mutagenesis protocol was developed which has allowed the unambiguous demonstration of mutations induced by the replication of BrUra residues in DNA, leading to errors of replication. This new protocol, called REP mutagenesis, is also shown in Fig. 1. The REP mutagenesis protocol involves the incorporation of BrUra into DNA under nonmutagenic conditions, the removal of the BrdUrd from the culture medium, and the subsequent replication of the BrUra-containing DNA in the presence of high intracellular levels of dTTP and dGTP. The incorporation of BrUra into DNA under nonmutagenic conditions was accomplished through the use of FBTdC medium (12), which utilized both deoxycytidine (dCyd) and low concentrations of BrdUrd to prevent the occurrence of any errors of incorporation during the incorporation of BrUra into DNA (7, 17). The BrdUrd/dThd ratio in FBTdC medium was varied at a constant BrdUrd concentration to control the level of substitution of BrUra for Thy in the DNA. After 4 to 5 days of growth in FBTdC medium, the cells, which had incorporated BrUra into their DNA, were washed free of BrdUrd and transferred into medium lacking BrdUrd but supplemented with high concentrations of dThd and dCyd. The cells were allowed to replicate their BrUra-containing DNA under these conditions for 1 to 3 days, followed by time for expression of induced mutations and selection. The presence of high concentrations (1 to 3 mM) of dThd plus 0.1 mM dCyd caused large increases in dTTP levels, which in turn caused significant increases in dGTP levels. It has been shown that dTTP is a positive effector of the ribonucleotide reductase-catalyzed reduction of GDP to dGDP (31). Although the major effect on the dNTP pools was the increased levels of dTTP, a similarly large increase in the dGTP/dATP ratio was observed due to increased dGTP levels without any significant change in dATP levels.

In this report, a line of Syrian hamster melanoma cells, which had been previously selected for its ability to grow with all the Thy residues in DNA replaced by BrUra (3), was subjected to the REP protocol under optimized conditions. Mutation frequencies that were 1 to 2 orders of magnitude higher than those previously described with classical mutagenesis protocols were observed at a number of loci.

VERY-HIGH-FREQUENCY REP MUTAGENESIS

The REP BrdUrd mutagenesis protocol has been used to induce high levels of mutations in the Syrian hamster melanoma cell line 2E (12). It has also been shown that mutation frequencies at the hypoxanthine-guanine phosphoribosyltransferase locus and the Na^+-K^+ ATPase locus induced by the REP protocol were 1 to 2 orders of magnitude greater in 2E cells than in other cell lines, such as Chinese hamster ovary (CHO) cells (15). It was, therefore, of interest to attempt to optimize the REP protocol for 2E cells and thereby increase the frequency of mutations induced.

In previous studies involving REP mutagenesis, the highest concentration of dThd that the cells were exposed to was 3 mM. Therefore, 2E cells, whose DNA was 50% substituted with BrUra for Thy, were allowed to replicate this DNA in the presence of from 3 to 15 mM dThd. Exposure of cells to 3 mM dThd resulted in a 100-fold increase in the frequency of thioguanine-resistant (Sgu^r) and ouabain-resistant (Oua^r) mutants above those observed at a low (10 μM) dThd concentration (Table 1). However, increasing the dThd concentration above 3 mM to 6, 9, 12, or 15 mM was not found to have any further significant effect on the mutation frequencies induced.

Mutagenesis induced by the REP protocol has been shown to be correlated with the intracellular dGTP/dATP ratio generated during the exposure of the cells to high concentrations of dThd (12). Exposure of 2E cells to 3 mM dThd resulted in a 30-fold increase in the dGTP/dATP ratio above that observed at a low (10 μM) dThd concentration (Table 1). However, increasing the concentration of dThd above 3 mM, as high as 15 mM, was not found to significantly increase the dGTP/dATP ratio. It was found that although both the dTTP and dGTP levels increased two- to threefold with increasing dThd concentration, the level of

TABLE 1
Effect of dThd Concentration on REP Mutagenesis[a]

Medium additives (mM)		Mutation frequency[b]		dNTP[c] (pmol/10^6 cells)				Ratio (dGTP/dATP)
dThd	dCyd	Sgu^r	Oua^r	dCTP	dTTP	dATP	dGTP	
0.01	0.001	5 ± 2	2 ± 1	26	84	41	14	0.3
3	0.3	842 ± 191	195 ± 55	157	3,673	41	392	9.6
6	0.6	764 ± 97	250 ± 66	383	6,545	68	629	9.3
9	0.9	710 ± 106	148 ± 56	508	7,890	76	775	10.2
12	1.2	872 ± 104	204 ± 50	843	10,362	81	921	11.4
15	1.5	689 ± 108	190 ± 65	833	10,299	83	921	11.1

[a] REP mutagenesis protocol was as follows. Culture dishes (diameter, 100 mm) containing FBTdC medium were inoculated with 10^5 cells. After 4 to 5 days of growth, during which time the cells were allowed to incorporate BrUra into their DNA under nonmutagenic conditions, i.e., low concentrations of BrdUrd in the presence of dCyd, the cells were trypsinized, and culture dishes (diameter, 100 mm), containing DMEM supplemented with various concentrations of dThd and dCyd and no BrdUrd, were each inoculated with 5 × 10^5 cells. After 3 days, the dishes were rinsed three times and renewed with DMEM. The cells were grown for 7 days, with renewal and transfer, in this nonselective medium to allow for the expression of mutations before the selection of mutants. 2E cells (50% BrUra substituted) were grown in the indicated concentrations of dThd plus dCyd for 3 days to induce mutations or for 6 h to determine dNTP pools. Cultures were then processed to assay for mutants and to determine the dNTP pools as described below.
[b] For the selection of Sgu^r mutants, 10 100-mm culture dishes containing DMEM supplemented with 36 μM thioguanine were each inoculated with 10^5 cells (controls) or 10^4 cells (treated). For the selection of Oua^r mutants, 10 100-mm culture dishes containing DMEM supplemented with 1 mM ouabain were each inoculated with 5 × 10^5 cells (control) or 10^5 cells (treated). The plating efficiency of the cells was determined by inoculating 100-mm culture dishes containing unsupplemented DMEM with 10^2 cells. After 7 to 9 days, the dishes to be scored were stained and colonies of 50 or more cells were counted. Mutation frequencies are expressed as resistant colonies per 10^5 cells for both Sgu^r or Oua^r mutants, after corrections for plating efficiency in the absence of selection. The data shown are the mean ± standard deviation calculated from 10 selection dishes.
[c] The dNTP pools were determined as previously described (12). Values shown are the averages of at least two determinations.

dATP also increased twofold. This increase in dATP levels, which was not observed at dThd concentrations up to 3 mM (15), prevented any significant increase in the dGTP/dATP ratio. The observed increase in dATP levels was probably due to the stimulation of the ribonucleotide reductase-catalyzed reduction of ADP to dADP caused by the increased levels of dGTP (31).

The REP BrdUrd mutagenesis protocol has been shown to be dependent on the level of incorporation of BrUra into DNA (12). Therefore, the effects of BrUra levels greater than 50% on the mutation frequencies induced in 2E cells were determined. The mutation frequencies induced for both Sgu^r and Oua^r mutants increased twofold when the level of BrUra substitution was increased from 25 to 50% (Table 2), as would be expected from a twofold increase in substitution if the mutation frequencies induced were proportional to the amount of BrUra in DNA. However, when the level of BrUra substitution was increased from 50 to 75%, a 1.5-fold increase, the mutation frequencies induced were only increased by 10% or 1.1-fold. Thus, it appears that at these very high levels of BrUra substitution, the mutation frequencies induced were not strictly proportional to the level of BrUra in DNA. These results may be due to a decrease in growth rate observed with the 75% BrUra-substituted cells (data not shown).

The above results indicated that the optimal conditions for the REP protocol, which would induce very high mutation frequencies without affecting growth and viability, were 50% BrUra substitution followed by replication in the presence of 3 mM dThd plus 0.3 mM dCyd. These mutagenesis conditions were therefore used to induce very high mutation frequencies in 2E cells in an attempt to detect very rare events, such as double mutants, which were not detectable when other mutagens were used. When 2E

TABLE 2
Effect of BrUra Substitution on REP Mutagenesis[a]

BrUra substitution (%)	Medium additives (mM)		Mutation frequency[b]	
	dThd	dCyd	Sgu[r]	Oua[r]
25	0.01	0.001	2 ± 1	1 ± 2
	3	0.3	498 ± 93	29 ± 10
50	0.01	0.001	5 ± 2	1 ± 1
	3	0.3	836 ± 101	122 ± 34
75	0.01	0.001	26 ± 16	17 ± 9
	3	0.3	917 ± 120	136 ± 18

[a] REP mutagenesis was performed as described in footnote a of Table 1. 2E cells were grown in the appropriate FBTdC medium and then allowed to replicate their DNA in the indicated concentrations of dThd plus dCyd for 3 days to induce mutations.
[b] Mutation frequencies were determined as described in footnote b of Table 1 and are expressed as resistant colonies per 10^5 cells for both Sgu[r] or Oua[r] mutants after correction for plating efficiency in the absence of selection. The data shown are the mean ± standard deviation calculated from 10 selection dishes.

cells were exposed to these REP mutagenesis conditions, very high frequencies of Sgu[r] (8.5×10^{-3}, 7.7×10^{-3}) and Oua[r] (2.7×10^{-3}, 1.7×10^{-3}) mutants were induced (Table 3). These results predict that Sgu[r] Oua[r] double mutants would be induced at frequencies of from 1×10^{-5} to 3×10^{-5}, a value which was calculated as the product of the individual, observed mutation frequencies. The actual observed frequencies of Sgu[r] Oua[r] double mutants were 1.9×10^{-5} and 2.0×10^{-5} (Table 3), values well within the expected range.

Two mutational events might also be expected to be required to generate mutants resistant to the toxic effects of high concentrations of 2,6-diaminopurine and of BrdUrd. These two analogs select for the loss of adenine phosphoribosyltransferase (4) and thymidine kinase (20) activities, respectively. As the loci for both of these enzymes have two functional alleles, both copies of the gene must be inactivated to result in the selected phenotype. If one makes the assumption that the frequency of inactivation of a single allele at the adenine phosphoribosyltransferase locus or at the thymidine kinase locus is approximately equivalent to the frequency of inactivation of the hypoxanthine-guanine phosphoribosyltransferase

TABLE 3
Very-High-Frequency REP Mutagenesis[a]

Expt	Selection[b]	Mutation frequency[c]	
		Expected[d]	Observed
1	Sgu		8.5×10^{-3}
	Oua		2.7×10^{-3}
	Sgu + Oua	2.3×10^{-5}	1.9×10^{-5}
	DAP	7.2×10^{-5}	5.3×10^{-5}
	BrdUrd	7.2×10^{-5}	1.2×10^{-5}
2	Sgu		7.7×10^{-3}
	Oua		1.7×10^{-3}
	Sgu + Oua	1.3×10^{-5}	2.0×10^{-5}
	DAP	5.9×10^{-5}	3.9×10^{-5}
	BrdUrd	5.9×10^{-5}	1.1×10^{-5}

[a] REP mutagenesis was performed as described in footnote a of Table 1. 2E cells were grown in the appropriate FBTdC medium to attain 50% substitution of BrUra for Thy in their DNA and were then allowed to replicate their DNA in 3 mM dThd plus 0.3 mM dCyd for 3 days to induce mutations.
[b] Selections for Sgu[r] and Oua[r] mutants were as described in footnote b of Table 1. For the selection of mutants resistant to both thioguanine and ouabain, DMEM was supplemented with 36 μM thioguanine plus 1 mM ouabain; for the selection of DAP[r] mutants, DMEM was supplemented with 150 μM 2,6-diaminopurine; and for the selection of BrdUrd[r] mutants, DMEM was supplemented with 100 μM BrdUrd. For all selections, 10 150-mm culture dishes containing the indicated medium were each inoculated with 5×10^5 cells (control and treated).
[c] All mutation frequencies were corrected for plating efficiency in the absence of selection. The data shown are the means calculated from 10 selection dishes.
[d] The expected frequencies for Sgu[r] Oua[r] mutants were calculated as the products of the observed individual frequencies. The expected frequencies for both DAP[r] mutants and BrdUrd[r] mutants were calculated as the square of the observed frequency of Sgu[r] mutants.

locus, then the expected frequency of inactivation of both alleles at either of the former loci could be approximated by calculating the square of the observed frequency of Sgu^r mutants. The observed frequencies of Sgu^r mutants were used to predict that 2,6-diaminopurine-resistant (DAP^r) and BrdUrd^r mutants would be induced at frequencies ranging from 6×10^{-5} to 7×10^{-5}. Frequencies of 3.9×10^{-5} and 5.3×10^{-5} for DAP^r mutants and 1.1×10^{-5} and 1.2×10^{-5} for BrdUrd^r mutants were observed (Table 3). The observed frequencies were lower than expected, but the assumptions made to calculate the expected frequencies were rather broad.

The detection, at these frequencies in a single-step selection, of mutants resistant to high concentrations of DAP and BrdUrd was apparently not due to the high-frequency occurrence of recessive drug resistance phenotypes occasionally observed in other systems (10, 24, 26, 32). This conclusion was based on two observations, the inability to detect any spontaneously arising DAP^r or BrdUrd^r mutants (frequencies of $<10^{-8}$) and the inability to induce any such mutants with a chemical mutagen such as ethyl methanesulfonate. When 2E cells were exposed to increasing concentrations of ethyl methanesulfonate, from 1 to 20 mM for 4 h, a dose-dependent decrease in survival was observed ranging from >90% survival at 1 mM to <10% survival at 20 mM. Under the same expression and selection conditions as were used for the REP protocol, the maximum mutation frequencies of Sgu^r (2.0×10^{-3}) and Oua^r (4.9×10^{-4}) mutants were obtained from the cells exposed to 20 mM ethyl methanesulfonate. These observed frequencies were used to calculate the expected frequencies of Sgu^r Oua^r double mutants (9.6×10^{-7}) and DAP^r mutants (3.8×10^{-6}), using the assumptions discussed above. However, no Sgu^r Oua^r double mutants or DAP^r mutants were observed even though 5×10^6 cells were subjected to each selection. This resulted in mutation frequencies of $<5 \times 10^{-7}$ for both selections after correction for plating efficiency.

These results suggested that the toxicity caused by the concentrations of ethyl methanesulfonate necessary to induce high frequencies of Sgu^r and Oua^r mutants had a negative effect on the ability to recover double mutants at the expected frequencies. However, these results also indicated that the double mutants observed with the REP protocol were probably not due to allelic instability effects as has been suggested for CHO cells (27). It should also be noted that the REP protocol has no effect on the growth or survival of the treated cells, and this characteristic allowed the recovery of potentially every mutant induced. The induction of double mutations (at relatively high frequencies in 2E cells) at any locus for which either a selection scheme or a screening protocol is available may provide a valuable source of null mutants involving genes of special interest.

MECHANISM OF BrdUrd MUTAGENESIS

The studies presented in this report have continued the analysis of the mechanism by which BrdUrd induces mutations in mammalian cells. Previous work had indicated that the levels of mutations (12) and sister chromatid exchanges (SCEs) (14) induced were proportional to the amount of BrUra substituted for Thy in DNA and dependent upon the concentration of dThd present in the culture medium during the replication of this BrUra-substituted DNA. When the dNTP pools generated by the conditions of the REP protocol were analyzed, it was found that mutations and SCEs increased along with the intracellular levels of dTTP and dGTP. It was suggested that the resulting increased intracellular ratio of dGTP to dATP might serve to drive the mispairing of dGTP with BrUra residues in replicating DNA. The replicaton of such mismatched bases would result in base substitution mutations, namely, $A \cdot T$-to-$G \cdot C$ transitions, while the action of a mismatch repair function might generate an intermediate that could lead to the generation of an SCE. Recent findings in support of this model (14) indicated that 3-aminobenzamide, a potent inhibitor of poly(ADP-ribose) synthesis (23, 28), increased the level of induction of SCEs but had no effect on mutations induced by the REP protocol. As 3-aminobenzamide has been reported to inhibit the ligation of DNA strand breaks (6), the frequency of SCEs observed might be increased due to the increased lifetime of the break in the presence of this inhibitor. However, 3-aminobenzamide should have had no effect, and was observed to have no effect, on mutations induced via a mechanism that does not involve DNA strand breaks as an intermediate, i.e., the replication of a mismatched base pair.

Further evidence to support this model came from the comparison of 2E Syrian hamster cells and CHO Chinese hamster cells, which has provided the observation of the apparent uncoupling of mutagenesis and SCE induction by the REP protocol (15). When 2E and CHO cells were exposed to the REP mutagenesis protocol under identical conditions, mutations leading to Sgu^r and Oua^r cells were detected at a 10- to 50-fold higher frequency in 2E cells than in CHO cells. This difference was shown not to be due to any cellular difference in the ability of these two loci to be mutagenized, but was shown to correlate with the degree of dNTP pool perturbation induced by the REP protocol in each cell line; i.e., the increase in the dGTP/dATP ratio caused by exposure to high concentrations of dThd was five- to sixfold greater in 2E cells than in CHO cells. This was another indication that this pool imbalance is directly involved in the mechanism of REP mutagenesis. In contrast to their greater sensitivity to the induction of mutations by the REP protocol, 2E cells were found to be much less sensitive than CHO cells to the induction of SCEs under identical experimental conditions. The difference observed in the ability of the REP protocol to induce SCEs in these two cell lines was not due to any difference in the ability of the cells to generate SCEs, as shown by their very similar response to the induction of SCEs by mitomycin C. The difference, therefore, seems to be a function of the manner in which the two cell types deal with the base pair mismatches thought to be generated by the REP protocol. The ability of dNTP pool imbalance to modulate the level of SCE induction by REP mutagenesis suggests that the mere presence of BrUra in DNA may be a necessary, but not a sufficient, condition to induce SCEs. Rather, the mismatched base pairs resulting from the replication of BrUra-substituted DNA in the presence of specific dNTP pool imbalance may be the necessary substrate of a mismatch repair system that serves to generate DNA strand breaks. CHO cells could be thought to respond to mismatched bases with a mismatch repair function that generates an intermediate involving a DNA strand break, resulting in a potential substrate for the generation of SCEs. This mismatch repair system would not be mutagenic. In contrast, 2E cells could be thought not to repair these mismatched bases to a significant degree, but rather to allow them to persist until they were replicated. This would result in the fixation of mutations without involving significant numbers of DNA strand breaks to serve as intermediates for SCEs. Thus, the replication of the mismatched bases would result in mutations but not SCEs, while the repair of the mismatched bases could lead to SCEs but not mutations.

ACKNOWLEDGMENTS. This work was supported by Public Health Service grants CA31777 from the National Cancer Institute and GM33156 from the National Institute of General Medical Sciences.

LITERATURE CITED

1. **Aebersold, P. M.** 1976. Mutagenic mechanism of 5-bromodeoxyuridine in Chinese hamster cells. *Mutat. Res.* 36:357–362.
2. **Benzer, S., and E. Freese.** 1958. Induction of specific mutations with 5-bromouracil. *Proc. Natl. Acad. Sci. USA* 44:112–119.
3. **Bick, M. D., and R. L. Davidson.** 1976. Nucleotide analysis of DNA and RNA in cells with thymidine totally replaced by bromodeoxyuridine. *Somatic Cell Genet.* 2:63–76.
4. **Chasin, L. A.** 1974. Mutations affecting adenine phosphoribosyl transferase activity in Chinese hamster cells. *Cell* 2:37–41.
5. **Cohen, S. S., J. G. Flaks, H. D. Barner, M. R. Loeb, and J. Lichtenstein.** 1958. The mode of action of 5-fluorouracil and its derivatives. *Proc. Natl. Acad. Sci. USA* 44:1004–1012.
6. **Creissen, D., and S. Shall.** 1982. Regulation of DNA ligase activity by poly(ADP-ribose). *Nature* (London) 296:271–272.
7. **Davidson, R. L., and E. R. Kaufman.** 1978. Bromodeoxyuridine mutagenesis in mammalian cells is stimulated by thymidine and suppressed by deoxycytidine. *Nature* (London) 276:722–723.
8. **Davidson, R. L., and E. R. Kaufman.** 1979. Resistance to bromodeoxyuridine mutagenesis and toxicity in mammalian cells selected for resistance to hydroxyurea. *Somatic Cell Genet.* 5:873–885.
9. **Freese, E.** 1959. The specific mutagenic effect of base analogues on phage T_4. *J. Mol. Biol.* 1:87–105.
10. **Gupta, R. S., and L. Siminovitch.** 1978. Genetic and biochemical studies with the adenosine analogs toyocamycin and tubercidin: mutation at the adenosine kinase locus in Chinese hamster cells. *Somatic Cell Genet.* 4:715–735.
11. **Huberman, E., and C. Heidelberger.** 1972. The mutagenicity to mammalian cells of pyrimidine nucleoside analogs. *Mutat. Res.* 14:130–132.
12. **Kaufman, E. R.** 1984. Replication of DNA containing 5-bromouracil can be mutagenic in Syrian hamster cells. *Mol. Cell. Biol.* 4:2449–2454.
13. **Kaufman, E. R.** 1985. Reversion analysis of mutations induced by 5-bromodeoxyuridine mutagenesis in mammalian cells. *Mol. Cell. Biol.* 5:3092–3096.
14. **Kaufman, E. R.** 1986. Induction of sister-chromatid exchanges by the replication of 5-bromouracil-substituted DNA under conditions of nucleotide-pool imbalance. *Mutat. Res.* 163:41–50.
15. **Kaufman, E. R.** 1987. Uncoupling of the induction of mutations and sister-chromatid exchanges by the replication of 5-bromouracil-substituted DNA. *Mutat. Res.* 176:133–141.
16. **Kaufman, E. R., and R. L. Davidson.** 1977. Novel phenotypes arising from selection of hamster melanoma cells for resistance to BUdR. *Exp. Cell Res.* 107:15–24.
17. **Kaufman, E. R., and R. L. Davidson.** 1978. Bromodeoxyuridine mutagenesis in mammalian cells: mutagenesis is independent of the amount of bromouracil in DNA. *Proc. Natl. Acad. Sci. USA* 75:4982–4986.
18. **Kaufman, E. R., and R. L. Davidson.** 1978. Biological and biochemical effects of bromodeoxyuridine and deoxycytidine in Syrian hamster melanoma cells. *Somatic Cell Genet.* 4:587–601.
19. **Kaufman, E. R., and R. L. Davidson.** 1979. Bromodeoxyuridine mutagenesis in mammalian cells is stimulated by purine deoxyribonucleosides. *Somatic Cell Genet.* 5:653–663.
20. **Kit, S., D. R. Dubbs, L. J. Piekarski, and T. C. Hsu.** 1963. Deletion of thymidine kinase activity from L cells resistant to bromodeoxyuridine. *Exp. Cell Res.* 31:297–312.
21. **Litman, R., and A. B. Pardee.** 1956. Production of bacteriophage mutants by a disturbance of deoxyribonucleic acid metabolism. *Nature* (London) 178:529–531.
22. **Moore, G.** 1964. In vitro cultures of a pigmented hamster melanoma cell line. *Exp. Cell Res.* 36:422–423.
23. **Oikawa, A., H. Tohda, M. Kanai, M. Miwa, and T. Sugimura.** 1980. Inhibitors of poly(adenosine diphosphate ribose) polymerase induce sister chromatid exchanges. *Biochem. Biophys. Res. Commun.* 97:1311–1316.
24. **Rabin, M. S., and M. M. Gottesman.** 1979. High frequency of mutation of tubercidin resistance in CHO cells. *Somatic Cell Genet.* 5:571–583.
25. **Rudner, R.** 1961. Mutation as an error in base pairing. I. The mutagenicity of base analogs and their incorporation into the DNA of *Salmonella typhimurium*. *Z. Vererbungsl.* 92:336–360.
26. **Siminovitch, L.** 1976. On the nature of heritable variation in cultured somatic cells. *Cell* 7:1–11.
27. **Simon, A. E., M. W. Taylor, W. E. C. Bradley, and L. H. Thompson.** 1982. Model involving gene inactivation in the generation of autosomal recessive mutants in mammalian cells in culture. *Mol. Cell. Biol.* 2:1126–1133.
28. **Sims, J. L., G. W. Sikorski, D. M. Catino, S. J. Berger, and N. Berger.** 1982. Poly (adenosinediphosphoribose) polymerase inhibitors stimulate unscheduled deoxyribonucleic acid synthesis in normal human lymphocytes. *Biochemistry* 21:1813–1821.
29. **Stark, R. M., and J. W. Littlefield.** 1974. Muta-

genic effect of BUdR in diploid human fibroblasts. *Mutat. Res.* 22:281–286.

30. **Strelzoff, E.** 1962. DNA synthesis and induced mutations in the presence of 5-bromouracil. II. Induction of mutations. *Z. Vererbungsl.* 93:301–318.

31. **Thelander, L., and P. Reichard.** 1979. Reduction of ribonucleotides. *Annu. Rev. Biochem.* 48:133–158.

32. **Tischfield, J. A., J. J. Trill, Y. I. Lee, K. Coy, and M. W. Taylor.** 1982. Genetic instability at the adenine phosphoribosyltransferase locus in mouse L cells. *Mol. Cell. Biol.* 2:250–257.

Chapter 27

DNA Structure and Mutation Hot Spots

Robert P. P. Fuchs, Anne-Marie Freund, Marc Bichara, and Nicole Koffel-Schwartz

The determination of mutation spectra by sequencing independent mutants in forward mutation assays made it clear that mutations do not arise with equal probability at all bases within a target DNA sequence. There are clearly sites where mutations occur at high frequency, the so-called mutation hot spots. It should be stressed that a site can be referred to as a mutation hot spot only if there is no bias in the selection procedure of the mutants. Such a situation is achieved when dealing with frameshift mutation in a forward mutation assay (for example, an assay involving the inactivation of a gene responsible for the resistance to an antibiotic). All frameshift mutations occurring in the early part of the gene have equal chances to inactivate the gene product (no silent mutations). On the other hand, in terms of mechanisms of mutagenesis, it is misleading to refer to mutation hot spots when describing sites where base substitutions occur at higher frequency in a mutation assay that relies on phenotypic detection of mutants. Indeed, such sites are likely to reflect selection biases due to the biology of the product of the target gene. The study of base substitution hot spots requires determination of the mutation spectrum by using an assay that can detect all the mutations, including the silent mutations.

Using a forward mutation assay based on the inactivation of the tetracycline resistance gene, we have determined the mutation spectra induced by several DNA-damaging agents including chemical carcinogens (derivatives of N-2-acetylaminofluorene [AAF]) (1, 7, 8) and chemotherapeutic agents (*cis*-DDP) (2). The covalent binding of AAF residues to position C-8 of guanine residues induces mostly frameshift mutations at specific sites (Fig. 1). We suggest that mutation hot spots result from the processing of unusual DNA structures that either are induced by chemicals or occur naturally within specific DNA sequences.

TWO PATHWAYS FOR AAF-INDUCED FRAMESHIFT MUTATION

Using a forward mutation assay based on the inactivation of the tetracycline resis-

---

Robert P. P. Fuchs, Anne-Marie Freund, Marc Bichara, and Nicole Koffel-Schwartz • Groupe de Cancérogénèse et de Mutagénèse Moléculaire et Structurale, Institut de Biologie Moleculaire et Cellulaire, Centre National de la Recherche Scientifique, 67084 Strasbourg, France.

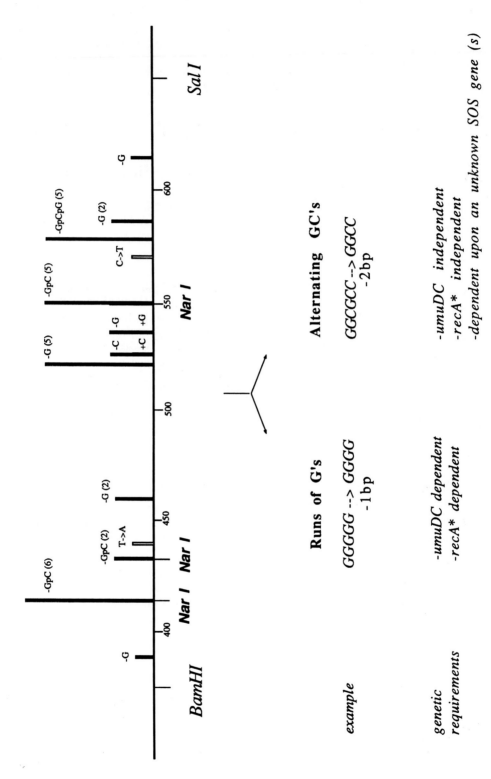

FIGURE 1. AAF-induced mutation spectrum in wild-type and *uvrA E. coli* strains (8). AAF-induced frameshift mutations can be grouped in two distinct groups of mutation hot spots, runs of Gs and alternating GC sequences which have specific genetics requirements (Koffel-Schwartz and Fuchs, unpublished results). RecA* represents the activated form of the *recA* protein.

tance gene located on plasmid pBR322, we have shown that AAF adducts induce primarily frameshift mutations within two kinds of DNA sequences (8; Fig. 1): (i) runs of guanines in which AAF triggers −1 frameshift mutations, and (ii) alternating GC sequences, in which AAF induces −2 frameshift mutations. These two kinds of mutation hot spots appear to belong to two distinct mutagenic pathways since they have distinct genetic requirements. Both require the induction of functions that belong to the SOS regulon. The former pathway requires a functional *umuDC* locus and the activated form of RecA (RecA*), while the second has neither of these two requirements but depends on an unidentified SOS function(s) (Koffel-Schwartz and Fuchs, unpublished results).

Mutations occurring within runs of a repeated unit (a unit being a single base or a group of bases) have been described as being hot spots for spontaneous frameshift mutations and are referred to as strand slippage mutants (Streisinger model [15]). In this model, the frameshift mutation involves the addition or deletion of the repeated unit within the run. It appears that AAF adducts specifically increase the frequence of the −1 frameshift mutation events within short runs of G residues.

On the other hand, mutations that occur within small stretches of alternating GC sequences have first been discovered within the recognition sequence of the restriction enzyme *Nar*I (GGCGCC) (8). These mutations involve the loss of 2 base pairs. We have subsequently shown that sequences containing more than two alternating GC pairs are also hot spots for AAF-induced mutations. It should be stressed that we consider these hot spots to belong to a pathway that is different from the one described above (i.e., mutations within a run of a repeated unit) on the basis of the different genetic requirements that are involved (see above).

RANDOM BINDING OF AAF ADDUCTS TO THE TARGET DNA SEQUENCE

Mutation hot spots can be accounted for in at least two hypothetical ways: (i) hot spots result from the preferential binding of the mutagen to specific DNA sequences, or (ii) hot spots result from the processing of lesions at specific DNA sequences. To investigate the first possibility we have established the binding spectrum of the AAF adducts to the DNA sequence that is used in the mutation assay (5). This was made possible by the observation that the $3' \rightarrow 5'$ exonuclease activity that is associated with the T4 DNA polymerase is blocked in the vicinity of an AAF adduct. Using the sequencing gel technology, we were able to show that the binding of the AAF adducts is in fact random in the sense that there is no preferential reactivity (toward N-AcO-AAF) of the guanine residues that belong to the mutation hot spots. In other words, the average reactivity of a guanine within a mutation hot spots is equal to the average reactivity of a guanine residue that does not belong to the sites of mutation hot spots (Fig. 2).

FRAMESHIFT MUTATION HOT SPOTS RESULT FROM THE PROCESSING OF UNUSUAL DNA STRUCTURES

If mutation hot spots cannot be explained by the binding spectrum of the mutagen to the target DNA sequence, one must consider two possibilities: (i) mutations arise as errors made when replication proceeds through a DNA sequence that can adopt an unusual structure, or (ii) mutations result prereplicatively from the processing of unusual DNA structures. For the two frameshift mutation pathways that we have described for AAF, what are the potential unusual DNA structures that lead to mutagenesis?

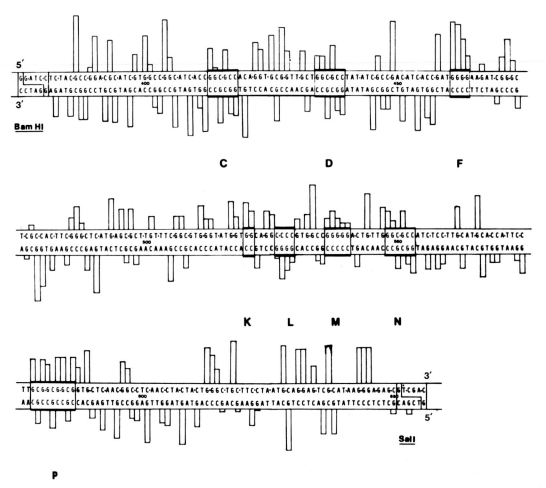

FIGURE 2. Distribution of the AAF adducts along the BamHI-SalI fragment of pBR322. The vertical bars represent the reactivity index of a given guanine residue. The mutation hot spots are framed with bold lines (7).

Runs of Gs

The interaction of an AAF adduct within a DNA sequence has been described by the so-called insertion-denaturation model in which the binding of the AAF adduct to the C-8 position of a guanine residue triggers the insertion of the fluorene ring between the base pairs (3, 4, 6). In this model the guanine residue is rotated outside the helix. Such a local geometry is likely to favor the slipped-strand mispairing intermediate that is required in the Streisinger model (15). This intermediate would represent the unusual DNA structure that is triggered by the mutagen and that is eventually processed into a −1 frameshift mutation by the mismatch repair apparatus. This mutation pathway was found to have several genetic requirements, namely that it involves the *umuDC* operon (8) and the activated form of RecA (Koffel-Schwartz and Fuchs, unpublished results). It should be stressed that the insertion-denaturation model (3, 4, 6) predicts −1 frameshift mutations as opposed to +1 mutations as it is actually observed.

Alternating GC Sequences

The binding of AAF to alternating purine-pyrimidine containing polynucleotides $[(GC)_n$ or $(GT)_n]$ favors the B-to-Z transition of the helix (11–14, 16). We suggest that AAF is able to locally induce a Z-like DNA structure within a sequence such as the *Nar*I sequence (GGCGCC). We hypothesize that the processing of this unusual DNA structure will lead to the -2 frameshift mutations observed at these sites. This hypothesis implies that a protein(s) recognizes and processes such structures. We are currently involved in the isolation of *Escherichia coli* mutants that are defective in this frameshift mutation pathway. A second implication of this hypothesis is that sequences that can undergo a B-to-Z transition under natural conditions might be subject to high mutation rates.

IS Z-DNA MUTAGENIC?

To test the hypothesis that Z-DNA is mutagenic, we have constructed a series of plasmids that contain stretches of defined lengths of alternating GC sequences inserted in the early part of the *lacZ* gene (*lacZ'*) and have made use of the β-galactosidase alpha complementation assay to follow the frameshift mutation frequency within the inserted sequence (Fig. 3). The parent plasmid (pUC8) contains the *lacZ'* gene that codes for the alpha-complementing β-galactosidase polypeptide and therefore gives rise to blue colonies when grown on plates containing X-Gal (5-bromo-4-chloro-3-indolyl-β-D-galactopyranoside). If the inserted sequence contains $3n+1$ or $3n+2$ base pairs, then the colony is white since the *lacZ'* gene is out of frame. The frameshift mutation assay scores colonies that have restored the reading frame and therefore appear as blue colonies (Lac$^-$ → Lac$^+$). Plasmids containing an insert that is $3n$ base pairs long have been constructed. With such constructions the mutation assay will score all the frameshift events that will set the target gene out of frame (Lac$^+$ → Lac$^-$). These assays were used to monitor the frequency at which spontaneous mutations occur within $(GC)_n$ sequences (with n ranging from 8 to 13).

INCREASE IN FREQUENCY OF SPONTANEOUS FRAMESHIFT MUTATIONS WITH THE SIZE OF THE INSERTED $(GC)_n$ SEQUENCE

The frequency of spontaneous mutations increases dramatically with the size of the $(GC)_n$ insert as exemplified by plasmids containing 8, 9, 12, and 13 GC repeats, respectively (Table 1). For all these plasmids, the mutation assay scores the passage from the Lac$^-$ to the Lac$^+$ phenotype [except for the $(GC)_{13}$ plasmid which scores the passage from the Lac$^+$ to the Lac$^-$ phenotype]. The extremely high spontaneous mutation frequency (0.5%) that is found with the $(GC)_{12}$ plasmid should be stressed. Moreover, this represents only a fraction of the actual mutation frequency since the selection procedure only scores the Lac$^-$ to Lac$^+$ events. Indeed, when we use the $(GC)_{13}$ plasmid that scores the Lac$^+$ to Lac$^-$ events, which are theoretically twice as frequent as the Lac$^-$ to Lac$^+$ events, we find a spontaneous mutation frequency that is equal to ≈3%. Taken together, this means that the total spontaneous frameshift muta-

TABLE 1
Spontaneous-Mutation Frequency in a Wild-Type Strain with Plasmids Containing Different $(GC)_n$ Inserts

Plasmid and insert length	Mutation	Frequency (10^{-4})
pUC-$(GC)_8$	White to blue	0.7
pUC-$(GC)_9$	White to blue	2.3
pUC-$(GC)_{12}$	White to blue	48.6
pUC-$(GC)_{13}$	Blue to white	300.0

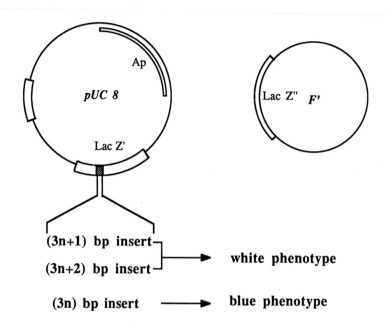

FIGURE 3. Strategy of the frameshift mutation assay. Plasmids containing stretches of defined lengths of alternating GC sequences inserted in the early part of the *lacZ* gene are used in the β-galactosidase alpha complementation assay to monitor the frameshift mutation frequency within the inserted sequence.

tion frequency for a plasmid that contains a $(GC)_{12}$ is in the order of 1.5%.

TENDENCY OF SPONTANEOUS MUTAGENESIS TO REDUCE THE LENGTHS OF THE INSERTED SEQUENCE, FROM $(GC)_n$ TO $(GC)_6$ OR $(GC)_7$

In a typical experiment, a collection of 50 independent mutants was analyzed to determine the nature of the events that give rise to the observed phenotype. The plasmid DNA isolated from a collection of 50 independently isolated mutants was analyzed in the vicinity of the inserted $(GC)_n$ insert by determining, on a sequencing gel, the length of a small restriction fragment that contains the inserted sequence.

In a typical experiment involving the $(GC)_{12}$ plasmid (a white-colony-forming plasmid), when 50 independent blue-colony-forming mutants were analyzed as described above, a set of bands corresponding to plasmids containing $(GC)_4$, $(GC)_7$, $(GC)_{10}$, and $(GC)_{13}$ inserts is found. The events that can theoretically give rise to blue colonies when starting with the $(GC)_{12}$ plasmid are either insertions of $3p+2$ base pairs or deletions of $3p+1$ base pairs. From the data, it is seen that only one insertion event is observed, corresponding to the insertion of 2 base pairs [$(GC)_{12} \rightarrow (GC)_{13}$]; this event represents 8% of the observed mutants. Among the deletions, only the events corresponding to the deletion of an *even* base pair number are found: 4-, 10-, and 16-base-pair deletions for plasmids containing inserts of $(GC)_{10}$, $(GC)_7$, and $(GC)_4$, respectively (the relative frequency being 10, 68, and 14%, respectively). It should be stressed that the mutation event leading to the $(GC)_7$ plasmid is by far the most frequent event (67%). None of the possible *odd* numbers of base-pair deletions has been found (i.e., base-pair deletions of 1, 7, 13, etc.).

PARALLEL BETWEEN SPONTANEOUS-MUTAGENESIS FREQUENCY AND POTENTIAL OF THE $(GC)_n$ INSERT TO ADOPT A Z-DNA CONFORMATION

We made use of two-dimensional agarose gel electrophoresis technology (9, 10) to analyze the presence of Z-DNA in the $(GC)_n$-containing plasmids (Fig. 4). In such gels, the different topoisomers of a given plasmid migrate along a monotonous curve if the plasmid contains no Z-DNA or inverted repeats (hairpin formation). On the other hand, there is a break in this curve when some of the topoisomers of the plasmid DNA contain a portion of their DNA in the Z conformation. From the magnitude of the break one can derive the length of the DNA portion that has adopted the Z conformation (9, 10). With the $(GC)_n$-containing plasmids isolated from the wild-type strain in which we performed the mutagenesis experiments, we observed that the $(GC)_6$-containing plasmid presents a family of topoisomers that line up along a monotonous curve. For the $(GC)_{12}$-containing plasmid, one can see a discontinuity in the migration pattern of the different topoisomers showing that about half of them contain a stretch of Z-DNA. The magnitude of the break is in agreement with the entire $(GC)_{12}$ sequence being flipped into the Z conformation. Therefore, it should be stressed that the high spontaneous-mutation frequency in the $(GC)_n$-containing plasmids parallels the potential of the $(GC)_n$ sequence to adopt the Z conformation in vitro.

CONCLUSION

The study of the frameshift mutations induced by a chemical carcinogen, namely AAF, led to the discovery of strong mutation hot spots. Neither a bias in the binding of the chemical nor in the selection procedure of the mutants can explain these mutation

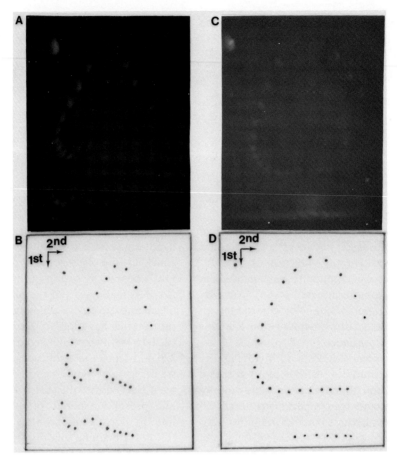

FIGURE 4. Analysis of the topoisomers of plasmid pUC-(GC)$_{12}$ (panels A and B) or pUC-(GC)$_6$ (panels C and D) by two-dimensional agarose gel electrophoresis (9, 10). Panels A and C are actual photographs of the gels, whereas panels B and D represent respective schematic drawings. The upper part of each panel represents the different topoisomers that were generated in vitro by using topoisomerase I in the presence of different concentrations of ethidium bromide (for details, see references 9 and 10). The lower part of each panel represents the distribution of the topoisomers as they were isolated from the wild-type bacteria that were used in the mutation experiments. From these data it is seen that none of the topoisomers of the pUC-(GC)$_6$ plasmid have undergone a B → Z transition. On the other hand, there is a break in the migration pattern of the different topoisomers of plasmid pUC-(GC)$_{12}$ showing that about half of the topoisomers that are isolated from the wild-type strain contain their (GC)$_{12}$ stretch in the Z conformation.

hot spots. Therefore, we propose that the mutation hot spots result from the processing of unusual DNA structures that are induced by the carcinogen within specific DNA sequences.

Within runs of identical bases (for example, GGGG) the AAF adducts induce an insertion-denaturation type of conformational change (3, 4, 6) that will stabilize the slipped-strand mispairing intermediate that is postulated to be important in the Streisinger model for frameshift mutations (15).

On the other hand, small stretches of alternating GC sequence (i.e., the NarI site contains two alternating GC units) appear to be strong mutation hot spots for 2-base-pair deletions. We propose that the unusual DNA structure is a local Z-like DNA conformation that is induced by the binding of AAF within such sequences.

To further investigate the potential mutagenicity of Z-DNA sequences, we have constructed a series of plasmids containing defined numbers of alternating GC units within the *lacZ* gene. These plasmids provide a sensitive tool to measure the stability of such sequences. Our results show that sequences that can adopt a Z conformation are subject to high rates of spontaneous deletions that reduce the length of the stretch of DNA that has adopted the Z conformation to a segment that is now too short to be in the Z conformation any longer. We are presently investigating the genetics of this mutation pathway.

LITERATURE CITED

1. Bichara, M., and R. P. P. Fuchs. 1985. DNA binding and mutation spectra of the carcinogen N 2 aminofluorene in *Escherichia coli*. *J. Mol. Biol.* 183:341–351.
2. Burnouf, D., M. Daune, and R. P. P. Fuchs. 1987. Spectrum of cisplatin-induced mutations in *E. coli*. *Proc. Natl. Acad. Sci. USA* 84:3758–3762.
3. Fuchs, R., and M. Daune. 1972. Physical studies on deoxyribonucleic acid after covalent binding of a carcinogen. *Biochemistry* 11:2659–2666.
4. Fuchs, R. P. P. 1975. In vitro recognition of carcinogen-induced local denaturation sites in native DNA by S1 endonuclease from *Aspergillus oryzae*. *Nature* (London) 257:151–152.
5. Fuchs, R. P. P. 1984. DNA binding spectrum of the carcinogen N-acetoxy-N-2-acetyl aminofluorene significantly differs from the mutation spectrum. *J. Mol. Biol.* 177:173–180.
6. Fuchs, R. P. P., J. F. Lefèvre, J. Pouyet, and M. P. Daune. 1976. Comparative orientation of the fluorene residue in native DNA modified by N-aco-N-2-acetylaminofluorene and two 7-halogeno derivatives. *Biochemistry* 15:3347–3351.
7. Fuchs, R. P. P., N. Schwartz, and M. P. Daune. 1981. Hot spots of frameshift mutations induced by the ultimate carcinogen N-acetoxy-N-2-acetylaminoflurone. *Nature* (London) 294:657–659.
8. Koffel-Schwartz, N., J. M. Verdier, M. Bichara, A. M. Freund, M. P. Daune, and R. P. P. Fuchs. 1984. Carcinogen-induced mutation spectrum in wild-type, *uvrA* and *umuC* strains of *Escherichia coli*. *J. Mol. Biol.* 177:33–51.
9. Lee, C. H., H. Mizusawa, and T. Kakefuda. 1981. Unwinding of double-stranded DNA helix by dehydration. *Proc. Natl. Acad. Sci. USA* 78:2838–2842.
10. Peck, L. J., and J. C. Wang. 1983. Energetics of B to Z transition in DNA. *Proc. Natl. Acad. Sci. USA* 80:6206–6210.
11. Sage, E., and M. Leng. 1980. Conformation of poly(dG-dC) · poly(dG-dC) modified by the carcinogen N-acetoxy-N-acetyl-2-aminofluorene and N-hydroxy-N-2-aminofluorene. *Proc. Natl. Acad. Sci. USA* 77:4597–4601.
12. Sage, E., and M. Leng. 1981. Conformational changes of poly(dG-dC) · poly(dG-dC) modified by the carcinogen N-acetoxy-N-acetyl-2-aminofluorene. *Nucleic Acids Res.* 9:1241–1250.
13. Santella, R. M., D. Grunberger, S. Broyde, and B. E. Hingerty. 1981. Z DNA conformation of N-2-acetylaminofluorene modified poly(dG-dC) · poly(dG-dC) determined by reactivity with anticytidine antibodies and minimized potential energy calculation. *Nucleic Acids Res.* 9:5459–5467.
14. Santella, R. M., D. Grunberger, I. B. Weinstein, and A. Rich. 1981. Induction of the Z conformation in poly(dG-dC) · poly(dG-dC) by binding of N-2-acetylaminofluorene to guanine residues. *Proc. Natl. Acad. Sci. USA* 78:1451–1455.
15. Streisinger, G., Y. Okada, J. Emrich, J. Newton, A. Tsugita, E. Terzaghi, and M. Inouye. 1966. Frameshift mutations and genetic code. Cold Spring Harbor Symp. Quant. Biol. 31:77–84.
16. Wells, R. D., J. J. Mighetta, J. Klysik, J. E. Larson, S. M. Stirdivant, and W. Zacharias. 1982. Spectroscopic studies on acetylaminofluorene-modified $(dT-dG)_n · (dC-dA)_n$ suggest a left-handed conformation. *J. Biol. Chem.* 257:10166–10171.

IV. Bypass Synthesis

Bypass Synthesis: Introduction

The fixing of misincorporation into the genome is required for the mutation to be recognized. This is a broad phase in mutagenesis which must come after the misincorporation phase. Mutation fixation requires a failure of the proofreading function as well as failure of mismatch repair and subsequent replication. The goal of the next chapters is to present information regarding replication past the incorrect base pairing. We may term this bypass synthesis. Bypass synthesis must have two broad categories: (i) synthesis past an adduct which inhibits DNA synthesis and favors misincorporation or (ii) synthesis past a noninhibitory lesion, but involving misincorporation. In the second of these categories the misincorporation and fixation of the mutation would be a reflection of the synthesis and editorial functions discussed in the preceding chapters. Therefore, in this part of the volume we deal primarily with synthesis past DNA damage in the template.

The fact that misincorporation is not equal to mutagenesis argues that there must be a later stage(s) in the process required to fix the mutation. An inference of this conclusion is that there may be a relaxed phase of DNA replication to allow incorporation with less-than-optimal base pairing at the site of DNA damage.

DNA polymerases might be expected to have some intrinsic rate of synthesis past DNA damage or tolerance for mispairing. This would reflect the sum of misincorporation and failure of proofreading functions. It appears that there is a marked difference among the polymerases in this regard. We might also expect that interaction with other proteins, notably the accessory proteins of replication or the proteins involved in holoenzyme complex formation, might significantly influence the rate of bypass synthesis.

A question from a functional standpoint is whether or not synthesis past mutagenic sites is actually continued synthesis past a site or a restart of replication beyond the mutagenic site, with subsequent patch-filling reactions occurring at the mutagenic site and therefore required for fixation of mutation. In procaryotes and eucaryotes single-stranded gaps can be observed in replicated DNA after damage. This argues for a restart mechanism being involved in such replication. However, this observation does not inform us regarding the nature of the synthesis at the mutagenic site, and it does not rule out continuous synthesis past the mutagenic site.

The concept of the SOS response linked damage, induction, and mutagenesis. An important question is whether altered cellular processes (or proteins) are required for at least some types of mutagenesis. The suggestion that DNA damage induces a response in which replication accepts a high rate of errors (mutations) to allow survival raised the question of whether or not there are alterations in proteins which increase the rate of errors. The possibility of an error-prone DNA polymerase secondary to induction processes has been suggested. Polymerases are required to fix mutations in any event, but the question of whether or not there are alterations to DNA polymerases which increase the mutation rate is important.

The question also arose as to whether or not new proteins which favored mutagenesis were induced. For the purposes of these chapters, the focus of these questions remains on the cellular proteins involved in allowing bypass synthesis to occur or favoring an increased rate of bypass synthesis. The *recA* gene is pivotal to the SOS induction process, and evidence has been presented that the RecA protein plays a direct role in mutagenesis by interacting with the replisome. This role may be viewed as distinct from the protein's other roles in induction or activation of other genes involved in mutagenesis. It has also been suggested that UmuCD may interact with the replisome. In the following chapters evidence is also presented that DNA polymerase III is directly required in at least some cases of damage-directed mutagenesis. Thus we now have direct evidence arguing that mutagenesis is not a passive event, but the fixing of mutations in the genome is an active event requiring several defined functions, some of which participate in normal DNA replication and some of which appear to be damage induced.

This is an area of investigation laden with potential. A number of questions can be foreseen.

(i) Investigators must determine whether proteins interacting with the replication apparatus simply accelerate mutation fixation or whether they are strictly required for fixation of mutations and for bypass synthesis.

(ii) Which proteins interact with the replication apparatus to allow bypass remains a primary question, as does whether existing proteins are modified to favor bypass replication.

(iii) In the immediate future the questions of how the RecA protein and the UmuCD proteins interact directly with the replisome must be addressed. Whether or not there is an altered DNA polymerase within the replisome must be determined.

(iv) Lastly, whether the mechanism of the DNA synthesis itself is altered remains to be determined.

The Editors

Chapter 28

Possible Roles of RecA Protein and DNA Polymerase III Holoenzyme in UV Mutagenesis in *Escherichia coli*

B. A. Bridges, C. Kelly, U. Hübscher, and S. G. Sedgwick

The induction of base-pair substitution mutations in *Escherichia coli* exposed to UV light is largely dependent upon genes under the control of the LexA repressor and constitutes part of a phenomenon known as the SOS response. The interest generated by the elegant work involved in the elucidation of the SOS control system tended for some time to distract attention from the nature of the mutagenic process itself. The SOS work established that RecA protein had a direct role in UV mutagenesis in addition to its role in cleaving the Lex repressor; it also established that elevated levels of UmuD and UmuC proteins were required. It is only recently, however, that attention has swung back to a consideration of what these proteins might do and how they might interact with DNA polymerases.

B. A. Bridges and C. Kelly • MRC Cell Mutation Unit, University of Sussex, Falmer, Brighton BN1 9RR, United Kingdom.   *U. Hübscher* • Institute for Pharmacology and Biochemistry, University of Zurich-Irchel, Zurich, Switzerland.   *S. G. Sedgwick* • National Institute for Medical Research, The Ridgeway, Mill Hill, London, United Kingdom.

IS DNA pol III INVOLVED?

The involvement of DNA polymerase III (pol III) in UV mutagenesis was first suggested by the observation that an excision-defective strain carrying a temperature-sensitive *dnaE* mutation ceased fixing UV-induced mutations (as determined by their loss of photoreversibility) immediately upon transfer to restrictive temperature (3). It was proposed that a $lexA^+$-dependent inducible cofactor exists which enables DNA pol III, when sealing certain daughter-strand and excision gaps, to override its normal high fidelity and to insert bases opposite pyrimidine dimers, perhaps at random. With hindsight one could argue that other explanations might exist, for example, that *dnaE* activity was necessary for the induction of the SOS system but not for the actual mutagenic event. This argument, however, seems unlikely in view of the fact that the 20-min period in growth conditions between UV irradiation and transfer to the restrictive temperature should have been sufficient for SOS induction to occur. Moreover, Hagensee et al. (9) have shown that DNA pol III activity is not needed for SOS induction to occur after UV.

Other data obtained by Hagensee et al. (9) also appear to indicate involvement of pol III in the mutagenic process. They employed *dnaE*(Ts) strains carrying a mutation, *pcbA*, which enables DNA pol I to carry out chromosomal replication at the restrictive temperature for the *dnaE* polymerase. UV mutagenesis was found to occur at 32 but not at 43°C. Two problems arise in connection with these experiments, however. First, with neither of the mutation systems employed was adequate time allowed for expression of newly induced mutations before the bacteria were plated on selective media. Thus only a very small (and possibly unrepresentative) proportion of induced mutations could appear, and one cannot exclude the possibility that the dysfunction of *dnaE* protein at 43°C may have altered that proportion. A second problem arises from the fact that the strains used were excision proficient. If replication after UV were very much slower (relative to other cellular processes) at 43 than at 32°C, then additional time would be available for excision repair of potentially mutagenic lesions, which might lead to a greatly reduced induced mutation frequency. This possibility is supported by the observations of Bridges et al. (3), who noticed a progressive reduction in UV-induced mutation frequency during a 60-min incubation of excision-proficient bacteria at restrictive temperature, which rendered their conclusion about the involvement of pol III less clear-cut than for excision-deficient bacteria. In the experiments of Hagensee et al., however, the growth rate of the *pcbA* strains was similar whether they carried *dnaE* or *dnaE*(Ts) alleles (R. Moses, personal communication), so this particular worry may be groundless.

Notwithstanding, the evidence for the involvement of pol III in UV mutagenesis in our opinion is not yet definitive. It is true that there are some alleles or revertants of *dnaE* strains in which UV mutability is altered (2, 7, 14a), two of which will be mentioned below, but such evidence is necessarily circumstantial.

ROLE OF RecA PROTEIN

Several years ago we began to think of how RecA and UmuDC proteins might act. We were impressed by the evidence of Fersht and Knill-Jones (8) that RecA protein could inhibit the proofreading activity of pol III holoenzyme (HE). Since UmuC protein appeared to be required for Weigle reactivation of bacteriophage, and as this seemed to involve DNA synthesis past blocking lesions, we wondered whether UV mutagenesis might not be a two-step process with insertion of bases opposite photoproducts ("misincorporation") due to inhibition of pol III HE proofreading by RecA protein, followed by a chain elongation step ("bypass") mediated by UmuD and UmuC proteins. If this were so, we argued that in a *umuC* strain a misincorporation event opposite a pyrimidine dimer would not be seen as a mutation since bypass could not occur. If, however, the dimer were subsequently removed by photoreversal, the misincorporation would be rescued and be seen as a mutation. The prediction was fulfilled; delayed photoreversal mutagenesis was demonstrated in *lexA* (Ind$^-$), *umuC*, and *umuD* bacteria. This phenomenon has proved to have most of the characteristics of normal UV mutagenesis in *umu*$^+$ bacteria (see references 4–5). We are therefore currently working on the assumption that delayed photoreversal mutagenesis reflects the first step in normal UV mutagenesis.

The demonstration by Lu et al. (11) that RecA protein can inhibit the epsilon proofreading subunit of pol III HE in isolation also led those authors to suppose that the effect is an (if not *the*) important component of UV mutagenesis. Although in vivo the misincorporation step is affected by different alleles of *recA* (6), it now seems that the RecA protein is not essential for misincor-

TABLE 1
Frequencies of Trp$^+$ Mutations Induced in Various recA-Defective Strains Derived from E. coli WP2 uvrA155 trpE65^a

Strain	Mutation frequency after PR at:			
	0 min	40 min	80 min	102 min
WP2$_s$ [Δ(srlR-recA)306]	6.60×10^{-9}	1.18×10^{-7}	3.92×10^{-7}	6.75×10^{-7}
CM871 (recA56 lexA102)	5.00×10^{-9}	1.20×10^{-7}	2.87×10^{-7}	7.54×10^{-7}
WP50-100 (recA1 lexA102)	0	1.19×10^{-7}	4.26×10^{-7}	3.43×10^{-7}

^{a} Frequencies expressed as a function of time of incubation between exposure to UV (0.3 J m^{-2}) and photoreversing light (PR; 40 min) (from data of Bridges [1]).

poration. Carrying out successful delayed photoreversal mutagenesis experiments with recA-defective bacteria is not easy because lethality is high and the numbers of mutant colonies per plate are low. Nevertheless, we have been able to demonstrate the induction of mutations when delayed photoreversal was given to a UV-irradiated derivative of E. coli WP2 uvrA carrying a deletion through recA (Table 1). A similar result was obtained with bacteria carrying two other recA-defective mutations, recA1 and recA56, but only in the presence of the lexA102(Ind$^-$) mutation, which prevents induction of any SOS genes and presumably also lowers the basal level of RecA protein. The frequency of delayed photoreversal mutagenesis with all these three strains was as high as (or possibly higher than) that in recA441 bacteria, and one could make an arguable case that RecA$^+$, RecA430, RecA1, and RecA56 proteins actually inhibit rather than promote misincorporation, a point that will be picked up below in connection with our in vitro studies.

MUTANT pol III HE

We have recently studied strains with pol III HE containing mutant components in the hope that this might throw some light on the problem. The mutD gene specifies the ε proofreading exonuclease activity of pol III HE, and mutD mutants have defective proofreading which results in a powerful spontaneous mutator phenotype. They do not, however, show any detectable difference from mutD$^+$ strains as regards normal UV mutagenesis or delayed photoreversal mutagenesis in the presence of the umuC122 allele (16). It might be argued that this could be due to mutD5 being too leaky for any effect to be detectable in our experiments. In view of the fact that the mutD5 allele resulted in dramatic increases in spontaneous mutation rate, however, this seems an unlikely explanation. The result is unfortunately less than satisfactory since it is compatible with both of the major competing hypotheses, (i) that pol III HE is not involved in either normal or delayed photoreversal UV mutagenesis and (ii) that pol III HE is involved but that the ε subunit is excluded or its activity is completely inhibited when mutagenic repair occurs.

The latter hypothesis is quite plausible, but if we are to consider RecA protein as being responsible for that inhibition then it should be pointed out that, in all the in vitro studies so far published, the inhibition of proofreading by RecA protein is never complete (6a, 8, 11).

Another type of mutant pol III HE is found in cells with the temprature-sensitive alleles dnaE486, dnaE511, and polC74. At permissive temperature these all confer a very modest spontaneous mutator activity

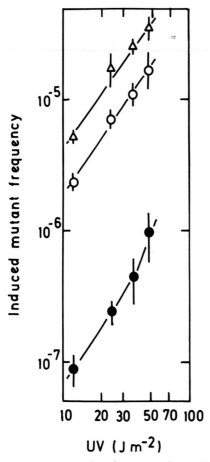

FIGURE 1. Induction of Trp⁺ mutations by UV light in *E. coli* K-12 *trpA3* carrying *dnaE*⁺ (strain 366; ●), *dnaE486* (strain 370; △), or *dnaE511* (strain 374; ○). Growth was at 34°C throughout.

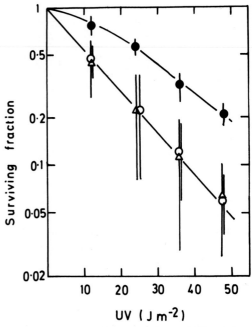

FIGURE 2. Survival of strains carrying different *dnaE* mutations following exposure to UV light. Growth was at 34°C throughout. Symbols as for Fig. 1.

and have properties consistent with defective base selection. Unlike the *mutD5* mutation, however, they all confer an elevated mutagenic response after UV (1a, 7, 14a); at some loci this is quite dramatic (e.g., Fig. 1). The alleles also confer some UV sensitivity (Fig. 2). These results do not prove that pol III HE is involved in UV mutagenesis, but assuming that it is, one can make a satisfying explanation of the results along the following lines (14a). Since both proofreading and mismatch correction operate during normal DNA replication, then one may expect that the modest spontaneous mutator effect observed (10, 15) is actually the tip of an iceberg of incorporation errors, most of which will have been detected and corrected. It then seems reasonable to assume that the elevated mutation frequency seen in UV-irradiated cells may reflect the operation of the mutant polymerase at a time when error correction is inoperative. This, put another way, suggests that proofreading and mismatch correction (either or both) may be abrogated or ineffective when mutagenic repair occurs. The inhibition or exclusion of the ε proofreading subunit of pol III HE has been discussed above. If this situation should persist, the DNA synthesized beyond the photoproduct would have a high frequency of errors made by the α subunit that would not have been corrected by proofreading. There is also a good chance that the errors would also escape mismatch correction since error-prone repair is a relatively slow process and methylation of the daughter strand could well have occurred beforehand. The outcome would be a region of newly synthesized DNA with an elevated frequency of untargeted errors. Since these would be triggered by a bypass event but would not be opposite the photoproduct itself, they could be termed hitchhiking mutations.

Despite the fairly strong UV mutator effect of *dnaE486* and *polC74*, there was no evidence for any enhancement of delayed photoreversal mutagenesis attributable to these alleles in a *umuC* background (14a). As with the *mutD* result, this is consistent with pol III HE having no role in this misincorporation process, but if pol III HE is involved, its infidelity is not rate limiting.

The information from the sort of studies described above may be summarized as follows. There is evidence that a misincorporation process occurs in bacteria that cannot perform the complete mutagenic process [e.g., *umuDC*, *lexA*(Ind$^-$), *recA*(Def)]. This may represent a step in the normal pathway, possibly the insertion of a wrong base opposite a photoproduct. The genetic requirements for the misincorporation process have not been established. RecA protein is not essential, and it may even restrict the extent of the process. There is no conclusive evidence that pol III is required, although one might argue that the loss-of-photoreversibility experiments of Bridges et al. (3) which implicated pol III reflected misincorporation rather than the whole mutagenic process. The process can occur in a number of *polA umuC* strains (F. Sharif, unpublished results). Completion of the mutagenic process, which presumably involves further chain elongation beyond the photoproduct (bypass), requires UmuC and UmuD proteins. RecA protein is required to cleave UmuD protein to an active form (G. Walker, personal communication; R. Woodgate and H. Echolls, personal communication), but whether there is any further role for RecA protein is not yet clear. The evidence implicating pol III HE in the bypass step is strong but not definitive.

STUDIES WITH AN IN VITRO SYSTEM

It may be some time before the complete mutagenic process is reconstructed in vitro. We have sought to model only the misincorporation step, which we envisage to be the laying down of a base opposite a photoproduct such as a pyrimidine dimer. Our system is similar to that pioneered by Moore and Strauss (14) with φX174 DNA. UV-irradiated single-stranded M13mp8 DNA was primed with universal primer and replicated by either pol I or pol III HE. Preliminary results have been described elsewhere (6a). Synthesis was found to terminate generally one base before pyrimidine-pyrimidine sites (putative photoproduct sites), in agreement with the results of others (13, 14). The presence of inhibitors of $3'\rightarrow 5'$ exonuclease function such as dAMP (with pol I) or dGMP or RecA protein (with pol III HE) did not cause any additional bases to be added to these termination sites. Even when the fidelity of pol I was relaxed by substituting Mn^{2+} for Mg^{2+}, an alteration which did result in some bases being inserted opposite putative pyrimidine dimers, there was no further insertion in the presence of RecA protein. We have not, therefore, found any evidence supporting the hypothesis that inhibition of proofreading by RecA protein is responsible for incorporation opposite photoproducts. This is entirely consistent with the dispensability of RecA protein for delayed photoreversal mutagenesis in vivo.

Since RecA protein could not be shown to have any effect on the ability of pol III HE to insert bases opposite putative dimers, it has to be assumed that if pol III mediates this step in vivo, some factor other than RecA protein is likely to be responsible for making it do so. It is possible, however, that RecA protein may have some effect on the proofreading of a base that has already been inserted opposite a dimer by some unknown enzyme. Such a substrate was therefore constructed using pol I in an Mn^{2+} buffer with newly synthesized DNA terminated opposite the first position of a thymine dimer. This DNA was purified and used as a substrate to examine the ability of pol I and pol III HE either to extend from or to remove the extra base. It was found that neither polymerase had any effect on this substrate. In the presence of deoxynucleotide triphosphates, pol III HE, and RecA protein, however, the band corresponding to the base opposite a pyrimidine dimer disappeared (Fig. 3). A further series of experiments indicated that exonucleolytic action occurs rather than chain extension.

This appears to be the first report of any effect of RecA protein on the interaction of pol III HE with photoproducts and further argues against RecA protein being responsible for misincorporations opposite photoproducts. At first sight it might be taken to indicate that RecA protein may actually inhibit such misincorporation, a notion already suggested by some of the in vivo data. It must be recognized, however, that our in vitro model system is imperfect in that we have been looking at the incorporation of an A opposite a TT dimer, which is a correct incorporation and not a misincorporation. We are currently entertaining the possibility that the action of RecA protein is more subtle and consists in changing the specificity of pol III proofreading (it does not appear to affect pol I proofreading). Could A become recognized as "wrong" opposite a TT dimer and be excised, yet be recognized as correct opposite a C in a dimer and not be excised? This would lead to transitions at GC sites, which are the commonest UV-induced mutations in the bacterial chromosome (12, 14b; C. W. Lawrence, personal communication). The possible alteration of pol III HE proofreading specificity is experimentally approachable and deserves attention. Its relevance, however, depends on whether genetic experiments (as yet unreported) allow any role for RecA protein in UV mutagene-

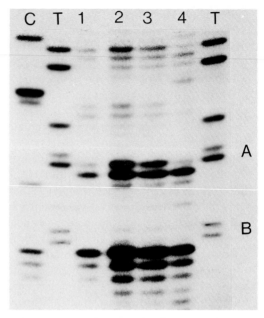

FIGURE 3. Sites of termination of synthesis opposite a UV-irradiated template. C and T are sequencing lanes. Lane 1, Termination sites with pol I in an Mg^{2+} buffer, showing stops immediately before potential TT dimer sites at A and B. Lane 2, Addition of a base opposite the dimer at A but not at B when synthesis was carried out in an Mn^{2+} buffer. This extra base was removed when the DNA was purified and reincubated in a second-stage reaction with pol III and RecA protein (lane 4), but not when incubation was with pol III alone (lane 3) or RecA protein alone (not shown).

sis other than cleavage of LexA protein and UmuD protein.

LITERATURE CITED

1. Bridges, B. 1988. Mutagenic DNA repair in *Escherichia coli*. XVI. Mutagenesis by ultraviolet light plus delayed photoreversal in *recA* strains. *Mutat. Res.* 198:343–350.
1a. Bridges, B. A. 1980. Ultraviolet light mutagenesis in bacteria: a result of the failure of normal error-correcting mechanisms?, p. 131–139. *In* M. Alacevic (ed.), *Progress in Environmental Mutagenesis*. Elsevier, Amsterdam.
2. Bridges, B. A., and R. P. Mottershead. 1978. Mutagenic DNA repair in *Escherichia coli*. VIII. Involvement of DNA polymerase III in constitutive and inducible DNA repair after ultraviolet and gamma irradiation. *Mol. Gen. Genet.* 162:35–41.
3. Bridges, B. A., R. P. Mottershead, and S. G.

Sedgwick. 1976. Mutagenic repair in *Escherichia coli*. III. Requirement for a function of DNA polymerase III in ultraviolet light mutagenesis. *Mol. Gen. Genet.* 144:53–58.

4. Bridges, B. A., and F. Sharif. 1986. Mutagenic DNA repair in *Escherichia coli*. XII. Ultraviolet mutagenesis in excision proficient *umuC* and *lexA(ind⁻)* bacteria as revealed by delayed photoreversal. *Mutagenesis* 1:111–117.

4a. Bridges, B. A., and F. Sharif. 1988. Mutagenic DNA repair in *Escherichia coli*. XV. Mutation frequency decline of ochre suppressor mutations in *umuC* and *lexA* bacteria occurring between ultraviolet irradiation and delayed photoreversal. *Mutat. Res.* 197:15–22.

5. Bridges, B. A., and R. Woodgate. 1985. The two-step model of bacterial UV mutagenesis. *Mutat. Res.* 150:133–139.

6. Bridges, B. A., and R. Woodgate. 1985. Mutagenic repair in *Escherichia coli*: products of the *recA* gene and of the *umuD* and *umuC* genes act at different steps in UV-induced mutagenesis. *Proc. Natl. Acad. Sci. USA* 82:4193–4197.

6a. Bridges, B. A., R. Woodgate, M. Ruiz-Rubio, F. Sharif, S. G. Sedgwick, and U. Hübscher. 1987. Current understanding of UV-induced base pair substitution mutation in *E. coli* with particular reference to the DNA polymerase III complex. *Mutat. Res.* 181:219–226.

7. Brotcorne-Lannoye, A., G. Maenhaut-Michel, and M. Radman. 1985. Involvement of DNA polymerase III in UV-induced mutagenesis of bacteriophage lambda. *Mol. Gen. Genet.* 199:64–69.

8. Fersht, A. R., and J. W. Knill-Jones. 1983. Contribution of 3′-5′ exonuclease activity of DNA polymerase III holoenzyme from *Escherichia coli* to specificity. *J. Mol. Biol.* 165:669–682.

9. Hagensee, M. E., T. L. Timme, S. K. Bryan, and R. E. Moses. 1987. DNA polymerase III of *Escherichia coli* is required for UV and ethyl methanesulfonate mutagenesis. *Proc. Natl. Acad. Sci. USA* 84:4195–4199.

10. Hall, R. M., and W. J. Brammer. 1973. Increased spontaneous mutation rates in mutants of *E. coli* with altered DNA polymerase III. *Mol. Gen. Genet.* 121:271–276.

11. Lu, C., R. H. Scheuermann, and H. Echols. 1986. Capacity of RecA protein to bind preferentially to UV lesions and inhibit the editing subunit of DNA polymerase III: a possible mechanism for SOS-induced targeted mutagenesis. *Proc. Natl. Acad. Sci. USA* 83:619–623.

12. Miller, J. H. 1985. Mutagenic specificity of ultraviolet light. *J. Mol. Biol.* 182:45–68.

13. Moore, P. D., K. K. Bose, S. D. Rabkin, and B. S. Strauss. 1981. Sites of termination of *in vitro* DNA synthesis on ultraviolet- and N-acetylaminofluorene-treated ϕX174 templates by prokaryotic and eukaryotic DNA polymerases. *Proc. Natl. Acad. Sci. USA* 78:110–114.

14. Moore, P. D., and B. S. Strauss. 1979. Sites of inhibition of *in vitro* DNA synthesis in carcinogen and UV treated ϕX174 DNA. *Nature* (London) 278:664–666.

14a. Ruiz-Rubio, M., and B. A. Bridges. 1987. Mutagenic DNA repair in *Escherichia coli*. XIV. Influence of two DNA polymerase III mutator alleles on spontaneous and UV mutagenesis. *Mol. Gen. Genet.* 208:542–548.

14b. Schaaper, R. M., R. L. Dunn, and B. W. Glickman. 1987. Mechanisms of ultraviolet-induced mutation. Mutational spectra in the *Escherichia coli lacI* gene for a wild-type and an excision-repair-deficient strain. *J. Mol. Biol.* 198:187–202.

15. Sevastopoulos, C. G., and D. A. Glaser. 1977. Mutator action by *E. coli* strains carrying *dnaE* mutations. *Proc. Natl. Acad. Sci. USA* 74:3947–3950.

16. Woodgate, R., B. A. Bridges, G. Herrera, and M. Blanco. 1987. Mutagenic DNA repair in *Escherichia coli*. XIII. Proofreading exonuclease of DNA polymerase III holoenzyme is not operational during UV mutagenesis. *Mutat. Res.* 183:31–37.

Chapter 29

Molecular Mechanisms of DNA Synthesis Fidelity and Isolation of a Possible SOS-Induced Polymerase

Myron F. Goodman, John Petruska, Michael S. Boosalis, Cynthia Bonner, Sandra K. Randall, Lawrence C. Sowers, and Lynn Mendelman

BASE SUBSTITUTION MUTAGENESIS

Profound biological effects can result from single-base changes in genes. Lesch-Nyhan syndrome, a debilitating and incurable childhood disease, is caused by single amino acid substitutions in the purine salvage enzyme hypoxanthine-guanine phosphoribosyltransferase (14). Substitution for glutamic acid by valine at a unique hemoglobin site results in sickle cell anemia (12). Single-base changes have also been implicated in oncogene activation resulting in malignant transformation of a normal cell (18). In this paper, we discuss our current efforts to elucidate several important components that are likely to affect spontaneous base substitution mutations and mutations which might result from SOS-induced error-prone repair (23, 24).

Based on genetic data amassed since the early 1960s, it is clear that specific DNA sites in an organism can mutate at much higher or much lower rates than average. There are some exceptional examples where mutational hot spots can be attributed to specific events such as deamination of 5-methylcytosine to thymine, leading to G→A transitions (6). However, the molecular causes underlying most mutational hot spots remain a mystery.

One of our main goals is to describe the kinetics of base substitution mutagenesis in defined DNA sequences. We would like to determine how DNA polymerase insertion and incorporation errors can be influenced by variations in neighboring DNA sequences. To address the effects of base context on fidelity in a quantitative fashion, we have recently devised an assay in which incorporation and misincorporation of single nucleotides are visualized directly by gel electrophoresis and autoradiography (2, 22).

Myron F. Goodman, John Petruska, Michael S. Boosalis, Cynthia Bonner, Sandra K. Randall, Lawrence C. Sowers, and Lynn Mendelman • Department of Biological Sciences, Molecular Biology Section, University of Southern California, Los Angeles, California 90089-1340.

The assay can be used to measure K_m and V_{max} values for insertion and incorporation of right and wrong nucleotides and to obtain the fidelity of DNA synthesis at any site within an arbitrarily chosen nucleotide sequence.

Questions to be discussed which are relevant to understanding fidelity include the following. How does DNA polymerase, acting either alone or as part of a replication complex, determine the specificity of nucleotide insertion at individual template sites? What are the enzyme kinetic mechanisms governing nucleotide selection at individual template sites? What are the enzyme kinetic mechanisms for inserting a correct nucleotide after a mispair? How do free energy differences between right and wrong base pairs, deduced from DNA melting temperature data in aqueous solution, relate to base selection by DNA polymerase?

The gel assay has recently been used to investigate the kinetics of insertion opposite abasic template sites (22). The copying of abasic (apurinic/apyrimidinic) sites in DNA is of great interest since bypass of such lesions may be a property of enzymes involved in SOS error-prone repair. The final section of this paper is devoted to a description of our use of the gel assay to isolate a DNA polymerase fraction, seemingly enriched following SOS induction, which exhibits the property of inserting nucleotides opposite and extending beyond a template abasic site.

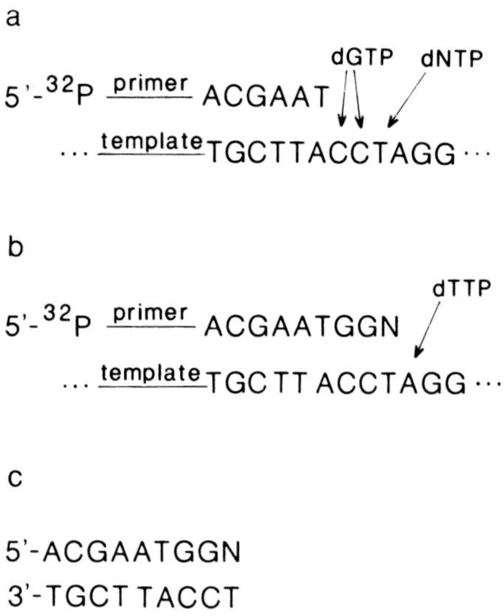

NUCLEOTIDE MISINCORPORATION KINETICS AND FIDELITY

Polyacrylamide Gel Assay to Measure Site-Specific Kinetics

Our objective is to measure the kinetics of nucleotide incorporation by DNA polymerase at any selected site along a DNA template strand (2, 22). A template-primer

a

5'-^{32}P $\underline{\text{primer}}$ ACGAAT $\overset{\text{dGTP}}{\downarrow}$ $\overset{\text{dNTP}}{\downarrow}$
... $\underline{\text{template}}$TGCTTACCTAGG ...

b

5'-^{32}P $\underline{\text{primer}}$ ACGAATGGN $\overset{\text{dTTP}}{\swarrow}$
... $\underline{\text{template}}$TGCTT ACCTAGG ...

c

5'-ACGAATGGN
3'-TGCT TACCT

FIGURE 1. DNA configurations used to measure DNA polymerase insertion and extension kinetics and primer-template thermal stability. (a) Primer-template used to measure enzymatic rates of insertion of matched and mismatched dNTP substrates opposite base T in template. (b) Primer-template used to measure enzymatic rates of extending matched and mismatched primer 3' termini by addition of T opposite A. (c) Synthetic duplex 9-mer used to measure melting temperatures for matched and mismatched primer 3' termini. In panels a and b, the synthetic primer strands are 23 and 20 nucleotides long, respectively, labeled at the 5' end with ^{32}P, and annealed to their complementary sections of circular M13 DNA template.

molecule containing an arbitrary sequence of template bases is shown in Fig. 1a. The primer strand, which must be long enough to hybridize to a unique region of the template strand, is labeled with ^{32}P at its 5' end. By choosing a proper combination of unlabeled deoxynucleotide triphosphate (dNTP) substrates, it is possible to extend the primer by one or, at most, a few bases.

For the example shown in the figure, two Gs are added opposite template C sites before each of the four nucleotides can be added opposite the template T site. To measure the kinetics of correct insertion of A opposite T, the reaction contains dGTP at saturating concentration and dATP at a series of variable concentrations. To measure

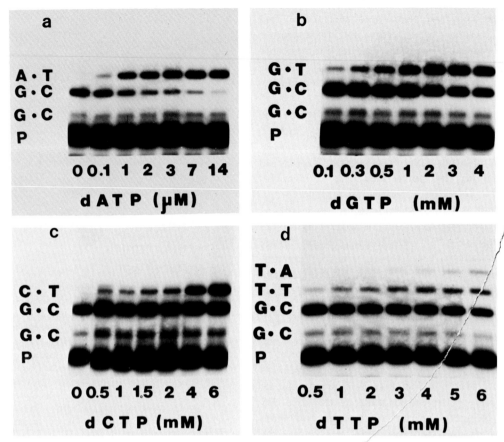

FIGURE 2. Gel autoradiogram showing band intensities as a function of added dNTP substrate. *Drosophila* DNA pol α reactions were run at various concentrations (indicated) of (a) dATP, (b) dGTP, (c) dCTP, and (d) dTTP as described in reference 2. In each case except panel b, 50 μM dGTP was present in the reaction mixture to obtain close to maximum rates of G insertion opposite C (G · C) in the first two sites without significant insertion opposite T in the third site.

the kinetics of incorrect insertion of C opposite T, saturating dGTP is again present, while dCTP concentrations are varied.

^{32}P-labeled primers elongated by one or a few nucleotides are clearly resolved on autoradiograms of polyacrylamide gels (Fig. 2). The correct insertion of A or incorrect insertions of either G, T, or C at the target T site are each detected as bands whose integrated intensity increases as a function of dNTP concentration (Fig. 2). The target bands exhibit saturation behavior (Fig. 3), and the data can be analyzed according to standard Michaelis-Menten kinetics (Fig. 4) provided that polymerase reaction times are reasonably short, typically 30 s to 4 min for correct insertions and 2 to 16 min for incorrect insertions (2).

The nucleotide incorporation velocity measured at time t in the target site (n) is related to the integrated intensities of target band (I_n) and previous band (I_{n-1}) by (2, 22):

$$v = (I_n/I_{n-1})(I_n + I_{n-1})/t \qquad (1)$$

The intensities are easily measured on a scanning densitometer having the appropriate integration software. We have written a

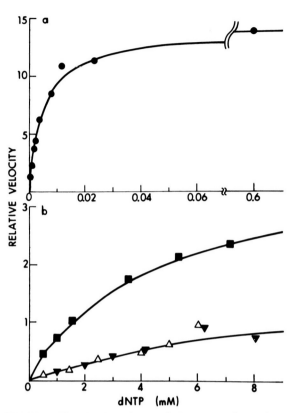

FIGURE 3. Plot showing relative velocity measured on gels as a function of dNTP concentration. Nucleotide substrates used for insertion opposite T (Fig. 1a): (a) dATP (●); (b) dGTP (■), dCTP (▼), and dTTP (△). The relative velocity, $v = I_3/I_2$ (see equation 2), was obtained by integrating the band intensities obtained from Fig. 2 at template T (I_3) and adjacent C (I_2) sites. Technical details of the gel assay have been described (2).

computer program to carry out a Michaelis-Menten analysis directly on data files generated from a Hoefer scanning densitometer.

Absolute velocities are not required to determine fidelity values at the target site. We have shown (2, 22) that a relative velocity is obtained by taking the ratio of target to previous band intensities, as follows:

$$v = (I_n/I_{n-1}) = V_{max}S/(K_m + S) \quad (2)$$

and this ratio, measured as a function of S = dNTP concentration for either right (r) or wrong (w) dNTPs, is all that is required to determine the fidelity. The nucleotide misincorporation frequency at template site n is

$$f = \frac{K_m(r)}{K_m(w)} \frac{V_{max}(w)}{V_{max}(r)} \quad (3)$$

The value of K_m/V_{max} is given by the slope of a $1/v$ versus $1/[dNTP]$ plot or, equivalently, by the intercept on the ordinate of a $[dNTP]/v$ versus $[dNTP]$ plot. Since the fidelity, $1/f$, depends only on the ratios of K_m to V_{max} for right and wrong dNTPs, there is

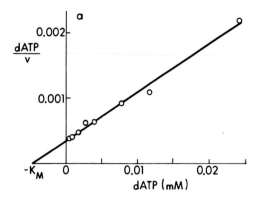

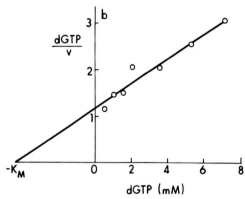

FIGURE 4. Hanes-Woolf plot to determine K_m and V_{max} for nucleotide insertion opposite T. Variable substrates were (a) dATP and (b) dGTP. The lines were obtained by a linear least-squares fit to data obtained by evaluating $v = I_3/I_2$ from gel band intensities shown in Fig. 2.

no need to obtain K_m and V_{max} values for either substrate.

The ability to evaluate fidelity by simply measuring K_m/V_{max} ratios is important since these ratios can be determined accurately at relatively low dNTP concentrations, even for the wrong dNTP (≤ 1 to 2 mM). To measure V_{max} and K_m individually, it may be necessary to use very high dNTP concentrations in the case of the wrong nucleotide, since V_{max} and K_m can only be measured when sufficient substrate is present to approach a plateau in the v versus [dNTP] plot. We have found that substrate inhibition can begin to become significant for dNTP concentrations above 6 mM, possibly caused by chelation of Mg^{2+}.

Fidelity of Insertion Opposite a Template T Site

Initially, we chose to investigate the fidelity of nucleotide insertion opposite T on an M13 DNA template, using purified *Drosophila melanogaster* DNA polymerase α (pol α) (13), a gift from I. R. Lehman (Stanford University, Stanford, Calif.). The intact *Drosophila* α holoenzyme complex contains no detectable 3'-exonuclease activity (5, 13), so that nucleotide incorporation rates are determined solely by insertion rates.

Nucleotide misinsertion frequencies opposite template T were obtained from data of the type shown in Fig. 4, using equation 3. Insertion of G occurred at a frequency of about 2×10^{-4}, whereas C and T insertions occurred fourfold less frequently, 5×10^{-5} (Table 1). As compared with correct insertion of A opposite T, K_m increased 1,100-fold for insertion of G opposite T and 2,600-fold for insertion of C or T opposite T. In contrast to the large increase in K_m, the reduction in V_{max} was only fourfold for G · T and eightfold for C · T and T · T mispairs.

Based on the kinetic data, a simple K_m discrimination mechanism can be proposed to account for the high nucleotide insertion fidelity of *Drosophila* pol α. Here, the insertion of any given nucleotide, either right or wrong, is dictated primarily by the relative dNTP residence times on the polymerase-DNA template-primer complex. When a wrong dNTP is bound to the complex, its dissociation rate is much larger than when a right dNTP is bound, resulting in a proportionally larger K_m for the wrong nucleotide.

It appears from the data that V_{max} discrimination makes only a relatively minor contribution to fidelity at the template T site under investigation. V_{max} discrimination will occur whenever the enzyme can reduce its catalytic rate constant, k_{cat}, to decrease the

TABLE 1
Kinetics of Nucleotide Misinsertion and Extension of Mismatched Primer Termini[a]

Operation	K_m (μM)	Relative V_{max}	f^b
dNTP insertion[c]			
A ↓ T	3.7	1	1
G ↓ T	4,200	0.24	2.1×10^{-4}
C ↓ T	10,000	0.13	4.9×10^{-5}
T ↓ T	10,000	0.13	4.9×10^{-5}
dTTP extension[d]			
A T ↓ T A	3.2	1	1
G T ↓ T A	178	0.29	5.2×10^{-3}
C T ↓ T A	1,290	0.36	8.9×10^{-4}
T T ↓ T A	980	0.16	5.2×10^{-4}

[a] *Drosophila* DNA pol α (13) was used in the gel assay.
[b] For dNTP insertion, these values represent relative frequency of inserting N opposite T, compared with correct insertion of A opposite T, as obtained from relative V_{max}/K_m values by equation 3. For dTTP extension, these values represent relative frequency of extending mismatched terminus (N · T) by inserting T opposite A, compared with correctly matched (A · T) terminus, as obtained from relative V_{max}/K_m values by equation 3.
[c] Misinsertion:

```
                                    dGTP  dNTP
Primer    5'-32P_____   ↓  ↓  ↓
Template  ..._____    C  C  T A G G ...
```

[d] Extension:

```
                                        dTTP
          5'-32P_____  G G N  ↓
          ..._____   C C T A G G...
```

insertion rate for any bound mispaired nucleotide. In the future, it will be interesting to investigate a variety of template sites to determine how different neighboring bases affect K_m and V_{max} discrimination. It will also be important to see whether different polymerases control fidelity by similar or different mechanisms at the same template-primer site.

A K_m Discrimination Model for Nucleotide Insertion Fidelity

In the spirit of Occam's razor, it has been suggested that nucleotide insertion fidelity can be attributed to the polymerase's ability to exploit free energy differences ($\Delta\Delta G$) between correct and incorrect base pairs when synthesizing DNA (9, 10). In the

0), the inser-
...s assumed to
bility between
...irs at the poly-
the polymerase
d from the mea-
cy, f, as

$$\cdots \ln f \quad (4)$$

Equation 4 can ~~ ~~ed as an operational definition of the free energy difference for inserting right versus wrong base pairs at the polymerase active site.

The important question is this: is there any reasonable way to account for polymerase insertion fidelity without having to resort to explanations which invoke conformational changes or other complex enzyme properties? Put in more quantitative terms, can free energy differences obtained by polymerase fidelity measurements be understood primarily in terms of base-pairing interactions, hydrogen bonding, and base stacking, or are there significant contributions made by the enzyme to discrimination at the insertion step? Our initial attempt to address this question is discussed below.

Insertion, Extension, and Thermal Melting of DNA

Before a detailed physical and chemical picture can emerge to account for nucleotide insertion specificity, it is necessary to acquire a basic understanding of the interactions between a template base and an incoming dNTP on the polymerase active site surface. Mechanistically, hydrogen bonding between complementary bases is commonly assumed to provide the specificity for preferential formation of A · T and G · C base pairs. Two critical issues yet to be treated adequately are: (i) the role of DNA polymerase in determining the specificity of nucleotide insertion and (ii) to what extent free energy differences between correct and incorrect base pairs account for nucleotide insertion fidelity.

In an attempt to address these issues, we have analyzed melting thermodynamics for matched and mismatched template-primer termini in defined oligomer sequences and compared the melting data with enzyme kinetic data for right and wrong nucleotide insertion and elongation in similar DNA sequences (J. Petruska, M. F. Goodman, M. S. Boosalis, L. C. Sowers, C. Cheong, and I. Tinoco, Jr., *Proc. Natl. Acad. Sci. USA*, in press).

Relating Melting Temperature Differences to DNA Polymerase Fidelity

It has been recognized that free energy differences derived from melting temperature differences for polymers containing matched and mismatched base pairs are much too small to explain insertion fidelity (see, e.g., reference 17). For example, substituting misinsertion frequencies from Table 1 into equation 4 gives values of $\Delta\Delta G$ = 5.1 kcal/mol (ca. 21.3 kJ/mol) for G · T mispairs and 6.0 kcal/mol (ca. 25.1 kJ/mol) for C · T and T · T mispairs relative to A · T base pairs, corresponding to nucleotide insertion discrimination factors of about 5,000 and 20,000, respectively. In contrast, typical $\Delta\Delta G$ values obtained from T_m measurements are only about 1 to 3 kcal/mol (ca. 4.2 to 12.5 kJ/mol), corresponding to discrimination values of 5 to 150.

We are interested in asking whether there might be any basis for reconciling thermal melting and insertion fidelity data. In collaboration with C. Cheong and I. Tinoco (University of California, Berkeley), we are investigating the thermodynamics of matched and mismatched primer-template 3' termini (Fig. 1c) having a 9-base-pair sequence identical to the one used in the gel misinsertion assay (Fig. 1a). Preliminary data show that $\Delta\Delta G$ values based on T_m differences, comparing A · T with G · T, C · T, and T · T termini, are between about 0.3 and 0.5 kcal/mol (ca. 1.3 to 2.1 kJ/mol) (Petruska et al., in press). However, the respective differences in enthalpy for these base mispairs range between 1.3 and 4.5 kcal/mol (ca. 5.4 to 18.8 kJ/mol). What was not recognized previously, to our knowledge, is that enthalpy and entropy changes appear to be strongly correlated (Petruska et al., in press). Since $dG = dH - TdS$ and $\Delta\Delta G = \Delta\Delta H - T\Delta\Delta S$ (see below), large changes in enthalpy are compensated by large comparable changes in entropy to give relatively small changes in free energy.

Since the enthalpy differences ($\Delta\Delta H$) between right and wrong base pairs are quite large, it is of interest to speculate as to how the polymerase might be able to exploit $\Delta\Delta H$ to discriminate between right and wrong dNTPs. One possibility might be that a dNTP interacting with a template in the polymerase active site has its degrees of freedom severely restricted by geometrical constraints. A reduction in the available degrees of freedom would lead to a concomitant decrease in entropy differences ($\Delta\Delta S$), while the magnitude of $\Delta\Delta H$ might not be changed significantly. This would make free energy differences ($\Delta\Delta G = \Delta\Delta H - T\Delta\Delta S$) on the surface of the enzyme closer to $\Delta\Delta H$ than to $\Delta\Delta G$ in aqueous solution (Petruska et al., in press).

Extension Kinetics for Mismatched Primer Termini

A gel assay to measure site-specific nucleotide misinsertion kinetics was described above, and data were presented for insertion of A, G, C, and T opposite template T (Fig. 3 and Table 1). The same assay can be used to measure the kinetics of correct nucleotide extension (Petruska et al., in press) either from a preexisting mismatched primer terminus (Fig. 1b) or from a mismatched primer terminus generated by the polymerase while in the process of copying the template.

There is a good reason for investigating the kinetics of extension past a mismatch. Extension of a mismatched primer terminus is an important component in the fidelity of DNA synthesis. A polymerase that dissociates following an insertion error or has difficulty reassociating at a melted 3' terminus would be less likely to synthesize aberrant DNA molecules than one which maintains processivity following a misinsertion or binds efficiently to a melted terminus.

We found that G · T mismatches, which are known to form relatively stable wobble structures, are easily extended. K_m for adding T opposite A following G · T (Table 1) was 178 μM, less than 60 times larger than for addition following A · T (3.2 μM). It was considerably more difficult to add onto C · T (K_m = 1.3 mM) or T · T (K_m = 0.98 mM). We noted that extension K_ms were about 10- to 20-fold lower than mismatch K_ms (Table 1). Thus, enzymatic reactions involving extension from mismatched primer termini are 10 to 20 times more efficient than the nucleotide misinsertion reactions with *Drosophila* pol α.

The two DNA polymerase reactions, mismatch extension and nucleotide misinsertion, appear to utilize similar kinetic discrimination mechanisms. Both are governed primarily by K_m rather than V_{max} discrimination. A reduction in V_{max} of only four- to eightfold was observed for the insertion of wrong compared with right nucleotides. A similar reduction in V_{max} was also observed for extension from mismatched compared to matched primer termini.

MISINSERTION AND EXTENSION AT ABASIC SITES: POSSIBLE ASSAY FOR PROTEINS INVOLVED IN SOS-INDUCED MUTAGENESIS

A series of kinetic measurements were made, using *Drosophila* pol α, to determine the "fidelity" for nucleotide insertion opposite abasic (apurinic/apyrimidinic) template lesions (22). Depending on nearest-neighbor nucleotide base, V_{max}/K_m was 6 to 11 times greater for A than for G and 20 to 50 times greater for A than for C or T. The abasic site

```
5'-³²P TCCCACTCACGACGT
       AGGGTGAGTGCTGCAAGAXTTTTTTTT
```

FIGURE 5. Synthetic primer-template DNA used to isolate and purify enzymes involved in nucleotide misinsertion and bypass of the abasic template lesion, X.

contained a synthetically prepared reduced deoxyribose (7), suitable for incorporation into any site in a synthetic DNA template, using a standard automatic DNA synthesizer. One important feature of the reduced abasic site is that it cannot undergo a β-elimination reaction and is therefore exceptionally stable.

It occurred to us that a synthetic template containing an abasic site might be a sensitive probe to assay for enzymes that are capable of inserting nucleotides opposite the abasic site and of continuing polymerization beyond the lesion (Fig. 5). It is our hope that this assay can be used to isolate and purify enzymes and auxiliary proteins involved in SOS-induced error prone repair (23, 24). The remainder of this paper is devoted to a summary of our initial findings concerning the ability of DNA polymerases, purified from uninduced and SOS-induced *Escherichia coli*, to insert opposite and to copy past a single abasic site. Contributions from Bruce Kaplan and Ramon Eritja (City of Hope Hospital, Duarte, Calif.), Kevin McEntee (University of California, Los Angeles), and Christiene Rayssiguier and Miroslav Radman (University of Paris VII, Paris, France) have been important in enabling us to pursue this project; a manuscript containing the evidence for the results summarized below is in preparation.

DNA polymerases were partially purified from parental *E. coli* GC4510 (uninduced) and GC4510 *recA730*; the latter is constitutive for SOS functions and exhibits a high level of mutagenesis (100-fold) compared with GC4510. Polymerases were also purified from CJ229 having a deletion of the pol I gene on the chromosome, but which contains the 5'→3' exonuclease fragment located on an F' episome. Strains containing the pol I gene deletion were kindly given to us by Catherine Joyce and Nigel Grindly (Yale University, New Haven, Conn.). SOS induction of pol I gene-deleted strains were carried out with nalidixic acid; successful induction was verified by measuring a 10-fold increase in the level of RecA protein, and SOS-induced mutagenesis was assayed by measuring the frequency of producing valine-resistant (Val^r) from valine-sensitive (Val^s) cells.

There are several lines of genetic evidence which suggest that pol III is involved directly in causing mutations in SOS-induced cells (3, 11). We therefore decided to purify pol III* from induced and uninduced cells. pol III* is believed to contain the holoenzyme, possibly as a dimer, minus β protein, a dissociable subunit required for processivity (19). Unless stated otherwise, pol III* was purified through the Sephacryl S300 stage by the methods of Fay et al (8).

DNA pol III*, purified from strain GC4510 *recA730* (induced) and the parental strain GC4510 (uninduced), inserts opposite the abasic site X (Fig. 6). In the presence of N-ethylamaleimide (NEM), an inhibitor of DNA pol III*, insertion opposite the abasic site persisted in the induced fraction even though normal insertion of T opposite A at the template site prior to X was virtually abolished. The addition of DNA pol I antibody prevented insertion opposite the abasic site with no discernible effect on normal primer elongation (Fig. 6).

Nucleotide insertion at the abasic site is attributed to DNA pol I that copurifies with pol III*. Integration of abasic band intensities revealed a three- to fivefold-greater insertion opposite template X for the induced polymerase fraction. Insertion at X is attributable to a higher level of DNA pol I in the induced polymerase fraction. Additional evidence that DNA pol I is present at a higher level in the induced fraction was found by examining normal nucleotide insertions prior to the abasic site. In the presence of

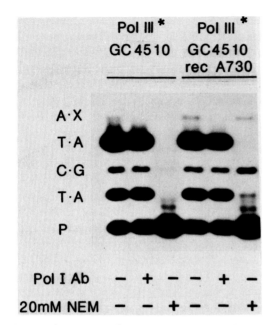

FIGURE 6. Gel autoradiogram showing nucleotide insertion activity of pol III* on a template containing an abasic site, X. pol III* represents the main phosphocellulose peak purified from uninduced and SOS-induced E. coli containing a wild-type polA gene. Lanes 1 to 3, pol III* from strain GC4510 (uninduced; lane 1) with pol I antibody added (lane 2) or with 20 mM NEM (lane 3). Lanes 4 to 6, pol III* from GC4510 recA730 (SOS induced) (lane 4) with pol I antibody added (lane 5) or with 20 mM NEM (lane 6).

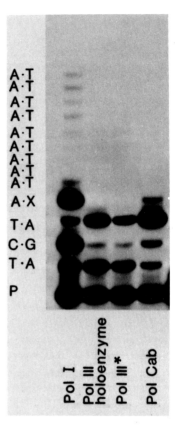

FIGURE 7. Gel autoradiogram showing synthesis by E. coli proteins pol I, pol III, and pol Cab on a template containing an abasic site, X (Fig. 5). pol III* and pol Cab were from E. coli CJ229 induced with nalidixic acid. This strain contains a deletion of the polA gene.

NEM, normal insertion of C opposite template G was clearly observed in the induced fraction, whereas only trace insertion of C occurred in the uninduced fraction.

Since it appears from these early data that pol I might be able to associate with pol III*, an interesting possibility, or perhaps pol III* may simply be contaminated with pol I which might be present at higher levels in SOS-induced cells, we decided to purify pol III* in the absence of pol I, using E. coli CJ229 deleted for the pol I gene. The major peak of pol III*, purified in the absence of pol I from induced and uninduced cells, was unable to insert nucleotides opposite the abasic site (Fig. 7). An inability to insert opposite abasic lesions was shared by other pol III forms, the holoenzyme complex (Fig. 7) and core pol III (not shown), kindly provided by Arthur Kornberg (Stanford University). In contrast, pol I (Fig. 7) and Klenow fragment (not shown) were exceedingly active in inserting opposite the abasic site and extending beyond.

However, the absence of pol I enabled us to uncover a potentially very interesting pol III-like fraction (pol Cab) which was also able to insert nucleotides opposite the abasic lesion (Fig. 7). It is also noteworthy that the Cab fraction appeared to be about 10-fold more active in induced compared with uninduced cells.

pol Cab was originally detected by assaying all fractions from a phosphocellulose column (8) by the gel assay, using the abasic site template (Fig. 5). pol Cab represented about 10% of the total activity of the main phosphocellulose peak in strain CJ229 cells induced by nalidixic acid; in uninduced cells, this peak was less than 3% of the main peak. What is especially interesting is that pol Cab insertion opposite the abasic site is estimated to occur with at least a 100-fold-higher rate than any of the pol III fractions tested. pol Cab is completely inhibited by NEM and its activity is unaffected by antibody to pol I. We cannot rule out the possibility that pol Cab may be DNA pol II, and McHenry (19) has in fact suggested the possibility that pol II may be equivalent to pol III' (pol III core plus τ).

In attempting to reconstitute a seemingly complex biochemical process such as SOS-induced mutagenesis in vitro, it is obviously important to avoid overinterpreting suggestive data. What we have specifically found is: (i) there is a polymerase activity (pol Cab) present in pol I gene-deleted cells which is able to insert dATP opposite an abasic template site in vitro and is able to add on to the misinserted A to extend beyond the lesion; (ii) the Cab activity is apparently about an order of magnitude higher in pol I gene-deleted cells induced for SOS by nalidixic acid than in uninduced cells; (iii) the misinsertion reaction appears to be much more efficient than the extension (lesion bypass) reaction; (iv) purified pol III proteins (core, III*, holoenzyme) show minimal insertion and bypass of the abasic site; and (v) pol I and Klenow fragment can also insert opposite and continue beyond the abasic site.

Genetic evidence demonstrates the necessity for *umuCD* gene products in SOS mutagenesis (4). Fixation of a mutation at an abasic site requires misinsertion at the noncoding lesion and extension beyond the lesion. The gel assay makes it possible to separate misinsertion from extension (see above and Table 1). It has been suggested that UmuC protein is not required for misincorporation of bases opposite UV-induced lesions, but is required for subsequent nucleotide addition to the misincorporated base (4). We are currently investigating whether pol Cab is present in UmuCD mutants. Also, since Cab is able to insert opposite abasic sites much more easily than it extends, we are attempting to fractionate induced crude extracts for components that enhance lesion bypass by pol Cab.

Although the genetic evidence to date implies a clear role for pol III in SOS-induced mutagenesis (3, 11), we are not aware of any reason to suppose that pol I might not also play an important role. The biochemical data on pol I* from Lackey et al. (15, 16) suggest that another form of pol I is found in SOS-induced cells showing lower fidelity than normal pol I. Recent studies by Ahmad and van Sluis (1) now show the DNA pol I operon to be under negative control (inducible) and that induction of this operon leads to increased UV resistance (1). We have measured the level of induction of RecA protein and forward mutation frequency going from Val^s to Val^r for each parental and mutant *E. coli* strain used in this study (20). One possibly relevant observation with regard to a role for pol I in SOS-induced mutagenesis is that strain CJ261 (deleted for pol I) is partially induced for RecA protein in the absence of nalidixic acid. Although this pol I deletion mutant has a 10-fold-higher mutation frequency (5×10^{-7}) than its parent strain (5×10^{-8}) under uninduced conditions, it actually shows a 10-fold-lower mutation frequency (5×10^{-6}) than its parent strain (5×10^{-5}) in the presence of UV treatment. Perhaps the inability of the pol I deletion strain to become fully induced with respect to forward mutagenesis indicates that both pol I and pol III (pol Cab) are involved in SOS-induced error prone repair. Our biochemical data obtained using the gel assay and abasic site template seem consistent with the idea that both polymerases could have meaningful roles to play in inserting nucleotides and copying past abasic lesions.

ACKNOWLEDGMENTS. This research was supported by Public Health Service grants GM21422 and GM33863 from the National Institutes of Health.

LITERATURE CITED

1. **Ahmad, S. I., and C. A. van Sluis.** 1987. Inducible DNA polymerase I synthesis in a UV hyperresistant mutant of *Escherichia coli*. *Mutat. Res.* 190:77–81.
2. **Boosalis, M. S., J. Petruska, and M. F. Goodman.** 1987. DNA polymerase insertion fidelity: gel assay for site-specific kinetics. *J. Biol. Chem.* 262:14689–14696.
3. **Bridges, B. A., and R. P. Mottershead.** 1976. Mutagenic DNA repair in *Escherichia coli*. III. Requirement for a function of DNA polymerase III in ultraviolet-light mutagenesis. *Mol. Gen. Genet.* 144:53–58.
4. **Bridges, B. A., and R. Woodgate.** 1984. Mutagenic repair in *Escherichia coli*. *Mol. Gen. Genet.* 196:364–366.
5. **Cotterill, S. M., M. E. Reyland, L. A. Loeb, and I. R. Lehman.** 1987. A cryptic proofreading $3' \rightarrow 5'$ exonuclease associated with the polymerase subunit of the DNA polymerase-primase from *Drosophila melanogaster*. *Proc. Natl. Acad. Sci. USA* 84:5635–5639.
6. **Coulondre, C., J. Miller, P. J. Farabaugh, and W. Gilbert.** 1978. Molecular basis of base substitution hotspots in *Escherichia coli*. *Nature* (London) 274:775–780.
7. **Eritja, R., P. A. Walker, S. K. Randall, M. F. Goodman, and B. E. Kaplan.** 1987. Synthesis of oligonucleotides containing the abasic site model compound 1,4-anhydro-2-deoxy-D-ribitol. *Nucleosides Nucleotides* 6:803–814.
8. **Fay, P. J., K. O. Johanson, C. S. McHenry, and R. A. Bambara.** 1982. Size classes of products synthesized processively by two subassemblies of *Escherichia coli* DNA polymerase III holoenzyme. *J. Biol. Chem.* 257:5692–5699.
9. **Galas, D. J., and E. W. Branscomb.** 1978. Enzymatic determinants of DNA polymerase accuracy: theory of coliphage T4 polymerase mechanisms. *J. Mol. Biol.* 124:653–687.
10. **Goodman, M. F., and E. W. Branscomb.** 1986. DNA replication fidelity and base mispairing mutagenesis, p. 191–232. *In* T. B. L. Kirkwood, R. F. Rosenberger, and D. J. Galas (ed.), *Accuracy in Molecular Processes: Its Control and Relevance to Living Systems*. Chapman and Hall, New York.
11. **Hagensee, M. E., T. L. Timme, S. K. Bryan, and R. E. Moses.** 1987. DNA polymerase III of *Escherichia coli* is required for UV and ethyl methanesulfonate mutagenesis. *Proc. Natl. Acad. Sci. USA* 84:4195–4199.
12. **Ingram, V. M.** 1957. Gene mutations in normal haemoglobin: the chemical difference between normal and sickle cell haemoglobin. *Nature* (London) 180:326–328.
13. **Kaguni, L. S., J. M. Rossignol, R. C. Conway, and I. R. Lehman.** 1983. Isolation of an intact DNA polymerase-primase from embryos of *Drosophila melanogaster*. *Proc. Natl. Acad. Sci. USA* 80:2221–2225.
14. **Kelley, W. N., and J. B. Wyngaarden.** 1972. The Lesch-Nyhan syndrome, p. 969–991. *In* J. B. Stanbury, J. B. Wyngaarden, and D. S. Fredrickson (ed.), *The Metabolic Basis of Inherited Diseases*, 3rd ed. McGraw-Hill Book Co., New York.
15. **Lackey, D., S. W. Krauss, and S. Linn.** 1982. Isolation of an altered form of DNA polymerase I from *Escherichia coli* cells induced for recA/lexA functions. *Proc. Natl. Acad. Sci. USA* 79:330–334.
16. **Lackey, D., S. W. Krauss, and S. Linn.** 1985. Characterization of DNA polymerase I*, a form of DNA polymerase I found in *Escherichia coli* expressing SOS functions. *J. Biol. Chem.* 260:3178–3184.
17. **Loeb, L. A., and T. A. Kunkel.** 1982. Fidelity of DNA synthesis. *Annu. Rev. Biochem.* 52:429–457.
18. **Marshall, C. J., K. H. Vousden, and D. H. Phillips.** 1984. Activation of c-Ha-ras 1 protooncogene by in vitro modification with a chemical carcinogen, benzo(a)pyrene diol-epoxide. *Nature* (London) 310:586–589.
19. **McHenry, C. S.** 1985. DNA polymerase III holoenzyme of *Escherichia coli*: components and function of a true replicative complex. *Mol. Cell. Biochem.* 66:71–85.
20. **Miller, J. H.** 1972. *Experiments in Molecular Genetics*, p. 221–229. Cold Spring Harbor Laboratory, Cold Spring Harbor, N.Y.
21. **Petruska, J., L. C. Sowers, and M. F. Goodman.** 1986. Comparison of nucleotide interactions in water, proteins, and vacuum: model for DNA polymerase fidelity. *Proc. Natl. Acad. Sci. USA* 83:1559–1562.
22. **Randall, S. K., R. Eritja, B. E. Kaplan, J. Petruska, and M. F. Goodman.** 1987. Nucleotide insertion kinetics opposite abasic lesions in DNA. *J. Biol. Chem.* 262:6864–6870.
23. **Walker, G. C.** 1984. Mutagenesis and inducible responses to deoxyribonucleic acid damage in *Escherichia coli*. *Microbiol. Rev.* 48:60–93.
24. **Witkin, E. M.** 1976. Ultraviolet mutagenesis and inducible DNA repair in *Escherichia coli*. *Bacteriol. Rev.* 40:869–907.

Chapter 30

Bypass and Termination at Lesions during In Vitro DNA Replication
Implication for SOS Mutagenesis

Zvi Livneh, Hasia Shwartz, Dana Hevroni, Orna Shavitt, Yaakov Tadmor, and Orna Cohen

The consequences of encounters of the replicative polymerase with lesions during replication are of major importance in determining the subsequent response of *Escherichia coli* to damage in its DNA. Most often replication terminates at bulky lesions, leading to the exposure of single-stranded DNA (ssDNA) regions. These serve as cofactors for the RecA protein, which becomes activated and cleaves the LexA repressor, thereby derepressing the entire array of SOS genes. It is believed that some of the SOS-induced proteins then interact with the polymerase and/or with the lesion-blocked primer-template to enable translesion synthesis. This process is likely to be mutagenic, owing to the absence of coding information or to the miscoding nature of the lesions, and is assumed to be the mechanism of SOS mutagenesis (11, 21, 22).

Zvi Livneh, Hasia Shwartz, Dana Hevroni, Orna Shavitt, Yaakov Tadmor, and Orna Cohen • Department of Biochemistry, The Weizmann Institute of Science, Rehovot 76100, Israel.

In an effort to elucidate the molecular mechanism of SOS mutagenesis, we have undertaken a biochemical approach and have begun by investigating the replication of damaged ssDNA by using the purified SS→RF replication system (8). The ssDNA was primed with a synthetic oligonucleotide and replicated with DNA polymerase III holoenzyme (pol III HE) in the presence of ssDNA-binding protein (SSB) to yield a nicked duplex (RF II). The pol III HE used in these studies is the major replicative polymerase primarily responsible for replicating the *E. coli* chromosome (8). It is likely to be the polymerase which encounters lesions in DNA during replication and seems to be essential for SOS mutagenesis, based on a variety of genetic evidence (2, 4, 7). It is a multisubunit complex, composed of at least seven subunits, and acts apparently as a dimer of two asymmetrical assemblies. In vitro it exhibits accurate, rapid, and processive DNA synthesis, properties essential for its in vivo function.

DNA pol III HE BYPASSES BULKY LESIONS DURING IN VITRO REPLICATION TO A LOW BUT SIGNIFICANT EXTENT

We recently presented evidence that pol III HE can bypass pyrimidine photodimers during replication in vitro by 20 to 30% (12, 13). Those data and those from more recent experiments provided indirect evidence, and an effort has been made to obtain direct evidence by using engineered DNA substrates which contain a single photodimer at a unique site. This unassisted replicative bypass of photodimers is likely to be the mechanism of SOS mutagenesis observed in UV-irradiated phage S13 (19).

During replication of depurinated ssDNA, the polymerase was found to bypass also apurinic (AP) sites by 10 to 20%, despite the different chemical nature of AP sites and photodimers (D. Hevroni and Z. Livneh, *Proc. Natl. Acad. Sci. USA*, in press). In a complete SS→RF replication reaction containing pol III HE, SSB, DNA polymerase I, and T4 DNA ligase, we were able to demonstrate directly the replicative bypass of AP sites by showing that covalently closed, fully replicated, depurinated ssDNA was converted to the nicked duplex form by the AP endonuclease activity of exonuclease III, proving the existence of AP sites in the fully replicated molecules (Fig. 1).

We proposed that the high processivity

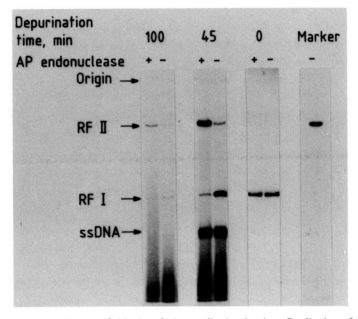

FIGURE 1. Bypass of AP sites during replication in vitro. Replication of oligonucleotide-primed depurinated M13 ssDNA was performed under standard conditions (12, 13) in the presence of SSB, DNA pol III HE, DNA polymerase I, and T4 DNA ligase. This reaction produces covalently closed duplex products. Following replication the DNA was precipitated, and the pellet was suspended in a buffer containing $CaCl_2$. The samples were boiled for 5 min, cooled on ice, and treated with exonuclease III. Under these conditions, only RF I DNA containing AP sites is converted into a nicked duplex (RF II) by the AP endonuclease activity of exonuclease III. Mixtures were then fractionated by electrophoresis in an 0.8% agarose gel. The photograph is an autoradiogram of the dried gel after electrophoresis. (From Hevroni and Livneh, in press.)

of pol III HE is one of the major factors responsible for its ability to bypass lesions during replication (12, 13). As the polymerase encounters the lesion it pauses and can take one of two pathways, bypass or termination. We argued that the higher the processivity of the polymerase, the higher its affinity for DNA, causing it to stay bound longer and increasing its chances to bypass the lesion. Supporting this view is the reduced extent of bypass of photodimers observed in the absence of SSB (13), a condition under which the polymerase shows a reduced processivity (5).

The bypass observed in vitro clearly shows that the polymerase has the intrinsic ability to replicate through lesions. However, the extent of unassisted bypass may be modulated in vivo by the presence of damaged-DNA-binding proteins, which may destabilize the polymerase-DNA complex at the damaged site, thus favoring dissociation and decreasing unassisted bypass.

DNA pol III HE UNDERGOES REPEATED CYCLES OF DISSOCIATION AND BINDING AT PHOTODIMER-BLOCKED PRIMER-TEMPLATE TERMINI DURING TERMINATION

Most encounters of pol III HE and pyrimidine photodimers or AP sites result in termination (12, 13). We previously showed that termination involved dissociation of the polymerase from the lesion-blocked primer-template and that it could replicate challenging primed ssDNA, indicating that the dissociation step does not inactivate the polymerase (13). It should be noted that a lesion-blocked primer-template is basically a structure that promotes tight binding of the polymerase when not blocked by a lesion (Fig. 2). Usually the signal for termination and dissociation is a nick on a duplex, a structure obtained at the end of synthesis of phage ssDNA (Fig. 2). It is thus remarkable

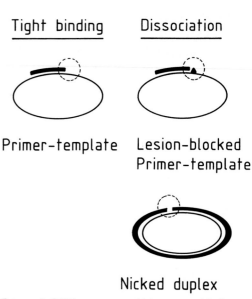

FIGURE 2. DNA structures which promote binding or dissociation of pol III HE.

that the presence of the lesion at a primer-template junction signals termination, similarly to a nick on a duplex, and causes the polymerase to dissociate from DNA.

Once the polymerase has dissociated it cannot resume synthesis at the terminated DNA strand (12, 13, 18). Why is resumption of DNA synthesis prevented? A possible reason might be the inability of pol III HE to rebind the lesion-blocked primer-template. This, however, does not seem to be the case. We previously showed that during termination on UV-irradiated DNA, when no net synthesis of DNA occurred, there was an active turnover of deoxynucleoside triphosphates (dNTPs) into deoxynucleoside monophosphates, accompanied by extensive ATP hydrolysis for an extended period of time (18). The turnover must have occurred via repeated polymerization-excision reactions while the polymerase was bound to the DNA. Since the polymerase has been shown to dissociate from the blocked termini, this implies that it did rebind the blocked primer-template following dissociation and that, while bound, it performed the polymerization-excision reactions.

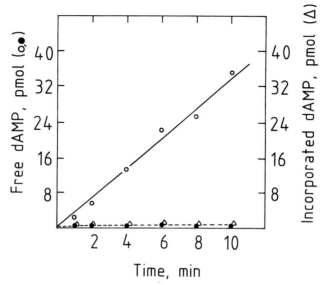

FIGURE 3. Promotion by isolated replication products of UV-irradiated DNA of the turnover of dATP to dAMP by DNA pol III HE. UV-irradiated ssDNA was replicated with pol III HE and purified on a Bio-Gel A-5m column. A fresh sample of pol III HE was then added, and the release of radiolabeled dAMP or incorporation into DNA was measured under standard replication conditions as described previously (18). Symbols: ○, free dAMP released; ●, free dAMP released in the absence of DNA; △, dAMP incorporated in DNA.

To directly demonstrate the ability of pol III HE to bind a lesion-blocked primer-template, we purified replication products and added to them a fresh sample of pol III HE. The addition of the polymerase did not promote any net DNA synthesis but caused the turnover of dATP to dAMP (Fig. 3), implying that the polymerase did bind the photodimer-blocked primer-template terminus and that, while bound, it was able to polymerize and excise nucleotides opposite the lesion.

THE TURNOVER OF dNTPs OCCURRED PRIMARILY OPPOSITE THE LESIONS

The following considerations argue that at termination the turnover of dNTPs occurred primarily opposite the blocking lesion. (i) The rate of turnover on UV-irradiated or depurinated DNA (Table 1) was much higher than that at the end of replication of untreated DNA, when reactions occurred at a normally base-paired DNA region. (ii) Similar rates of turnover were observed for UV-irradiated or depurinated ssDNA (Table 1). Taking into account the different chemical and stereochemical natures of pyrimidine photodimers and AP sites and the different nucleotide sequences in which they were preferentially formed, i.e., polypyrimidine runs and purine-containing sequences, respectively, it seems very likely that turnover occurred opposite the lesion, with little dependence on its exact chemical nature.

It should be emphasized that all dNTPs turn over, with a slight preference for dATP.

TABLE 1
Rates of Turnover of dNTPs to dNMPs during Termination of Replication of Damaged DNA Templates[a]

Nucleotide	Turnover rate (pmol of dNMP/pmol of DNA per s) for:		
	Untreated DNA[b]	Depurinated DNA[c]	UV-irradiated DNA[b]
dATP	1.03	4.10	3.74
dGTP	0.28	0.95	0.61
dCTP	0.15	0.48	0.31
dTTP	0.50	1.29	1.52

[a] Untreated or damaged ssDNA was replicated in the presence of SSB with DNA pol III HE. For each dNTP, a set of experiments was performed with the appropriate α-^{32}P-labeled dNTP. The kinetics of release of the dNMPs was measured via fractionation of samples from the reaction mixtures by thin-layer chromatography on polyethyleneimine-cellulose plates, followed by assaying of the radiolabel in free dNMP and in DNA as described previously (18).
[b] Data are from reference 18.
[c] Data are from Hevroni and Livneh, in press.

This fact most likely represents a slight preference of pol III HE for inserting a dAMP residue opposite photodimers or AP sites (18). A much more pronounced preference for inserting a dAMP residue opposite AP sites was found for DNA polymerase I (10, 17) and for DNA polymerase α from *Drosophila melanogaster* (16).

INHIBITION OF THE 3'→5' EXONUCLEASE ACTIVITY OF pol III HE IS NOT SUFFICIENT TO ENABLE BYPASS OF LESIONS

It has been argued that the 3'→5' exonuclease activity of the polymerase prevents the bypass of lesions. According to this view, as soon as the polymerase inserts a nucleotide opposite the lesion, it is recognized as a "mismatch" by the proofreading activity of its 3'→5' exonuclease, which excises out the nucleotide, thus preventing the next polymerization step. It has been further suggested that the bypass of lesions and thus mutagenesis occur via inhibition of the 3'→5' exonuclease activity of the polymerase (1, 20). The recent finding that the RecA protein, known to be involved in SOS mutagenesis, inhibited the 3'→5' exonuclease activity of the polymerase further supports this view (6, 14; H. Shwartz and Z. Livneh, submitted for publication).

We tested this hypothesis by inhibiting the 3'→5' exonuclease activity of pol III HE. If indeed this exonuclease activity is the only factor preventing bypass, its inhibition should cause an increase in the length of replication products. Using a variety of inhibitory conditions, we could not detect any increase in the length of products. For inhibition we used dGMP, purified RecA protein, and a mutant polymerase obtained from a *mutD* strain which is deficient in 3'→5' exonuclease activity (13). More recently we performed experiments with the nucleotide analogs deoxyadenosine 5'-O-(1-thiotriphosphate) and deoxythymidine 5'-O-(1-thiotriphosphate), which were efficiently incorporated into DNA by pol III but poorly excised by its 3'→5' exonuclease (O. Shavitt and Z. Livneh, unpublished data). The same results were obtained, implying that inhibition of the 3'→5' exonuclease activity is not sufficient to prevent the bypass of lesions. Inhibition of the 3'→5' exonuclease activity may still be needed in conjunction with the action of other SOS-induced proteins, such as the *umuD* and *umuC* gene products, to enable replicative translesion DNA synthesis.

WHY DOES THE POLYMERASE TERMINATE REPLICATION AT PHOTODIMERS OR AP SITES?

Each polymerization reaction consists of three distinct kinetic steps: translocation to the next nucleotide to be copied (e.g., step 1 in Fig. 4), polymerization, and proofreading, resulting from the 3'→5' exonuclease activity of the polymerase (e.g., step 2 in Fig. 4). As the polymerase encounters a lesion it pauses and can take one of two possible

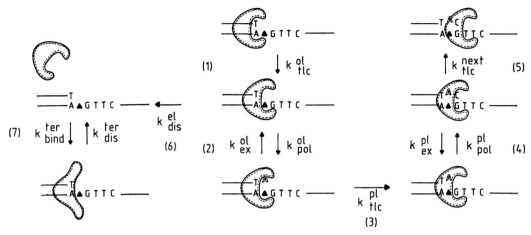

FIGURE 4. Model describing bypass and termination at lesions by DNA pol III HE. The polymerase is represented as a "C"-like structure, except for step 7, in which a change in the polymerase structure is suggested. Abbreviations: ol, opposite lesion; pl, past lesion; tlc, translocation; pol, polymerization; ex, excision; el, elongation; ter, termination; dis, dissociation; bind, binding. See text for explanations.

pathways: bypass, which is a complete translesion replication route (steps 1 to 5 in Fig. 4), or termination, in which the polymerase dissociates from the lesion-blocked primer-template and ceases elongation (steps 1, 2, 6, and 7 in Fig. 4). During replication of undamaged regions of DNA, when normal base pairing occurs, the rate of accumulation of deoxynucleotide monophosphates during the turnover reaction is determined by the rate of excision, which is slower than the rate of polymerization. What happens opposite a lesion? The rate of polymerization is likely to decrease, owing to the difficulty in copying a distorted region in DNA, whereas the rate of excision is likely to increase, for the same reason. Furthermore, we propose that the net result is that polymerization is the slower reaction opposite photodimers or AP sites and is therefore the major factor which determines the rate of turnover opposite lesions. In support of this proposition we found that purified RecA protein did not influence the rate of turnover during termination (unpublished results). Previous studies revealed that in the presence of SSB, the RecA protein did not inhibit polymerization but did inhibit the exonuclease activity of the polymerase (Shwartz and Livneh, submitted). These results suggest that the RecA protein inhibited the faster reaction during termination, the excision reaction, and thus had no effect on turnover rates.

The extensive turnover observed opposite lesions during termination, which reflects the rate of polymerization opposite the lesions, suggests that the polymerase has no difficulty in polymerizing the nucleotides opposite the lesions (step 2 in Fig. 4). We thus suggest that one of the subsequent steps is extremely slow (steps 3 and 4 in Fig. 4). Translocation of the polymerase to the nucleotide past the lesion (step 3 in Fig. 4) might be very slow, owing to steric hindrance by the lesion. Polymerization of the nucleotide past the lesion (step 4 in Fig. 4) might be very slow, owing to the need to extend a partially frayed primer-template terminus formed by the mispairing of the lesion and the nucleotide inserted opposite the lesion.

If polymerization past the lesion is extremely slow, how can the polymerase perform complete unassisted translesion replication at all? We suggested that it could do so because of the stability of the polymerase-

	ELONGATION COMPLEX	DEFECTIVE INITIATION COMPLEX
Formed at:	Primer-template	Blocked primer-template
Stability:	Very high	Low
ATPase activity:	Low	High
Bypass of photodimers:	Some	No
Polymerization opposite photodimer: First nucleotide / Next nucleotides	Yes / Yes	Yes / No

FIGURE 5. Schematic representation of a polymerase-DNA elongation complex and a defective initiation complex and their biochemical properties.

DNA complex during elongation (18). In the elongation mode, the polymerase is in a highly processive form, with a very high affinity for the DNA. When a bulky lesion such as an AP site or a pyrimidine photodimer is met and the polymerase pauses, this tight binding to the DNA permits some of the polymerase molecules to overcome the kinetic barrier of the very slow polymerization past the lesion (steps 1 to 5 in Fig. 4). Once dissociated, however, rebinding to the lesion-blocked primer-template leads to the formation of a defective initiation complex of lower stability (step 7 in Fig. 4). This complex is stable enough to allow the relatively easy first polymerization step opposite the lesion but not the next very slow steps (Fig. 5). In fact, the instability of the defective initiation complex is self-evident from our data demonstrating the dissociation of pol III HE from lesion-blocked termini (13), whereas the polymerase was stably bound to a primer-template, the elongation of which was arrested by the omission of a single dNTP (13, 15).

Thus, the formation of an unstable defective initiation complex, together with slow translocation and/or polymerization steps past the lesion, is the major factor which prevents the bypass of lesions during termination.

SPECTRUM OF AP SITE MUTAGENESIS PREDICTED FROM TURNOVER RATES DURING TERMINATION

Our model proposes that the major function of the SOS-induced proteins is to assist polymerization past the lesion, implying that the "mutagenic" step is performed by the unassisted polymerase. It is a slight modification of the two-step model proposed recently (3). This means that the spectrum of mutations produced by an AP site, for example, will be determined primarily by the identity of the nucleotide polymerized opposite the AP site. These data can be obtained from the turnover rates of the various nucleotides at the termination stage of replication of depurinated ssDNA in vitro, since these values are determined primarily by the rates of polymerization opposite the AP sites, as reasoned above. Table 2 shows a prediction of the spectrum of in vivo mutagenesis of AP sites based on the relative turnover rates of the various nucleotides in

TABLE 2
Prediction of AP Site Mutagenesis Specificity from Turnover Rates during Termination of Replication of Depurinated DNA In Vitro[a]

Depurinated site	Inserted nucleotide	Turnover rate[b]	Predicted misinsertion (%)[c]	No. of mutations in depurinated M13mp2[d]	
				Predicted	Found
Adenine	dAMP	4.10	74	12	13
	dGMP	0.95	17	3	2
	dCMP	0.48	9	1	1
Guanine	dAMP	4.10	65	31	27
	dGMP	0.95	15	7	7
	dTMP	1.29	20	10	14

[a] From Hevroni and Levneh, in press.
[b] From Table 1. The turnover of dTTP was not included for the prediction of mutations at adenine AP sites, since its insertion will not lead to a mutational event. The turnover of dCTP was excluded from the calculation for guanine AP sites for the same reason.
[c] Predicted percentage of appropriate mutational events based on the ratios of the turnover rates [e.g., the predicted misinsertion of dAMP opposite an adenine AP site is as follows: $(4.10 \times 100)/(4.10 + 0.95 + 0.48) = 74\%$].
[d] In vivo mutagenesis specificity was taken from Table 4 in reference 9, in which 16 mutations at putative adenine AP sites and 48 mutations at putative guanine AP sites were listed. The predicted occurrence of the various mutants was obtained by multiplying the predicted misinsertion percentage by the total number of mutations available (16 at adenine AP sites and 48 at guanine AP sites).

vitro, as compared with experimental data on mutations produced in vivo in depurinated ssDNA from phage M13mp2, as reported by T. A. Kunkel (9). The biochemical data taken from the termination stage predict the correct spectrum of mutations with striking accuracy and support the views that it is pol III HE which carries out the mutagenic step in AP site mutagenesis and that the specificity of mutagenesis is determined principally by the polymerase itself. Furthermore, it seems that the "misinsertion" mutagenic step does not require any SOS functions. These are needed, presumably, for performing the subsequent steps of synthesis past the lesion, as suggested by Bridges and Woodgate (3). The verification of this mechanism has to await the complete in vitro reconstitution of SOS mutagenesis.

ACKNOWLEDGMENTS. This study was supported by Public Health Service grant AG05469 from the National Institutes of Health and by grant 1743 from the Council for Tobacco Research USA Inc. Z.L. is a recipient of the Yigal Alon Scholarship.

LITERATURE CITED

1. **Bridges, B.** 1978. DNA polymerase and mutation. *Nature* (London) 275:591–592.
2. **Bridges, B. A., R. P. Motershead, and S. G. Sedgewick.** 1976. Mutagenic DNA repair in *E. coli*. III. Requirement for a function of DNA polymerase III in ultraviolet light mutagenesis. *Mol. Gen. Genet.* 144:53–58.
3. **Bridges, B. A., and R. Woodgate.** 1985. Mutagenic repair in *Escherichia coli*: products of the *recA* gene and of the *umuD* and *umuC* genes act at different steps in UV-induced mutagenesis. *Proc. Natl. Acad. Sci. USA* 82:4193–4197.
4. **Brotcorne-Lannoye, A., G. Maenhaut-Michel, and M. Radman.** 1985. Involvement of DNA polymerase III in UV-induced mutagenesis of bacteriophage lambda. *Mol. Gen. Genet.* 199:64–69.
5. **Fay, P. J., K. O. Johanson, C. S. McHenry, and R. A. Bambara.** 1981. Size classes of products synthesized processively by DNA polymerase III and DNA polymerase III holoenzyme of *Escherichia coli. J. Biol. Chem.* 256:976–983.

6. Fersht, A. R., and J. W. Knill-Jones. 1983. Contribution of 3'-5'exonuclease activity of DNA polymerase III holoenzyme from *Escherichia coli* to specificity. *J. Mol. Biol.* 165:669–682.
7. Hagensee, M. E., T. L. Timme, S. K. Bryan, and R. E. Moses. 1987. DNA polymerase III of *Escherichia coli* is required for UV and ethyl methanesulfonate mutagenesis. *Proc. Natl. Acad. Sci. USA* 84:4195–4199.
8. Kornberg, A. 1980. *DNA Replication.* W. H. Freeman & Co., San Francisco.
9. Kunkel, T. A. 1984. Mutational specificity of depurination. *Proc. Natl. Acad. Sci. USA* 81:1494–1498.
10. Kunkel, T. A., R. M. Schaaper, and L. A. Loeb. 1983. Depurination-induced infidelity of deoxyribonucleic acid synthesis with purified deoxyribonucleic acid replication proteins *in vitro. Biochemistry* 22:2378–2384.
11. Little, J. W., and D. W. Mount. 1982. The SOS regulatory system of *Escherichia coli. Cell* 29:11–22.
12. Livneh, Z. 1986. Replication of UV-irradiated single-stranded DNA by DNA polymerase III holoenzyme of *Escherichia coli*: evidence for bypass of pyrimidine photodimers. *Proc. Natl. Acad. Sci. USA* 83:4599–4603.
13. Livneh, Z. 1986. Mechanism of replication of ultraviolet-irradiated single-stranded DNA by DNA-polymerase III holoenzyme of *Escherichia coli*: implications for SOS mutagenesis. *J. Biol. Chem.* 261:9526–9533.
14. Lu, C., R. H. Scheuermann, and H. Echols. 1986. Capacity of RecA protein to bind preferentially to UV lesions and inhibit the editing subunit (ε) of DNA polymerase III: a possible mechanism for SOS-induced targeted mutagenesis. *Proc. Natl. Acad. Sci. USA* 83:619–623.
15. O'Donnell, M. E., and A. Kornberg. 1985. Dynamics of DNA polymerase III holoenzyme of *Escherichia coli* in replication of a multiprimed template. *J. Biol. Chem.* 260:12875–12883.
16. Randall, S. K., R. J. Eritja, B. E. Kaplan, J. Petruska, and M. F. Goodman. 1987. Nucleotide insertion kinetics opposite abasic lesions in DNA. *J. Biol. Chem.* 262:6864–6870.
17. Sagher, D., and B. Strauss. 1983. Insertion of nucleotides opposite apurinic/apyrimidinic sites in deoxyribonucleic acid during *in vitro* synthesis: uniqueness of adenine nucleotides. *Biochemistry* 22:4518–4526.
18. Shwartz, H., and Z. Livneh. 1987. Dynamics of termination during *in vitro* replication of ultraviolet-irradiated DNA with DNA polymerase III holoenzyme of *Escherichia coli. J. Biol. Chem.* 262:10518–10523.
19. Tessman, I. 1985. UV-induced mutagenesis of phage S13 can occur in the absence of the RecA and UmuC proteins of *Escherichia coli. Proc. Natl. Acad. Sci. USA* 82:6614–6618.
20. Villani, G., S. Boiteux, and M. Radman. 1978. Mechanism of ultraviolet-induced mutagenesis: extent and fidelity of *in vitro* DNA synthesis on irradiated templates. *Proc. Natl. Acad. Sci. USA* 75:3037–3041.
21. Walker, G. C. 1984. Mutagenesis and inducible responses to deoxyribonucleic acid damage in *Escherichia coli. Microbiol. Rev.* 48:60–93.
22. Witkin, E. M. 1976. Ultraviolet mutagenesis and inducible DNA repair in *Escherichia coli. Bacteriol. Rev.* 40:869–907.

Chapter 31

DNA Polymerase III Is Required for Mutagenesis

Sharon Bryan, Michael E. Hagensee, and Robb E. Moses

Escherichia coli has three DNA polymerases (11). The roles of DNA polymerases I (pol I) and III (pol III) have been investigated in depth. DNA pol I appears to be required for optimal DNA replication. Although mutations thermolabile in the $5' \rightarrow 3'$ exonuclease activity of this enzyme are conditionally lethal, a deletion mutant can be constructed in which the mutation is lethal only under conditions of rich medium (10). On the other hand, mutants which are conditionally defective in DNA pol III activity are nonviable. DNA pol III is a holoenzyme complex of a number of peptide subunits. The α subunit is responsible for the synthetic activity and is the product of the *dnaE* (*polC*) gene (15, 20). It was temperature-sensitive mutations in this gene that enabled identification of the gene product as being required for DNA replication (7). Mutants conditionally defective in other proteins of the DNA pol III holoenzyme complex also indicate that DNA pol III is required for DNA replication. These include *dnaZ*, *dnaQ* (*mutD*), *dnaX*, and *dnaN* mutants (16).

We have reported a mutation, *pcbA1*, which phenotypically suppresses the temperature-dependent lethality of temperature-sensitive mutations in the *dnaE* gene encoding the α subunit activity of DNA pol III holoenzyme (5, 17, 18). This mutation allows some critical step(s) of DNA replication to be performed by DNA pol I. In the presence of a functional DNA pol I activity, the cell survives at 42°C even though it contains a *dnaE*(Ts) allele. The phenotypic suppression by the *pcbA1* mutation is not allele specific, and we have observed that the mutation can phenotypically suppress four *dnaE*(Ts) alleles. Thus, these cells define a *pcbA1*-pol I replication pathway. In the wild-type cell a critical step in replication would normally be performed by DNA pol III, but in cells with *pcbA1* it would be performed by pol I. Alternatively, there may be multiple steps at which DNA pol I substitutes for DNA pol III.

The *pcbA1*-pol I replication pathway is dependent upon the normal function of other recognized DNA replication gene products. We have tested *dnaA*, *dnaB*, *dnaC*, *dnaJ*, *dnaK*, *dnaN*, and *dnaZ* temperature-sensitive mutant alleles. None of these temperature-sensitive mutations is phenotypically suppressed by the *pcbA1* allele. Using P1 transduction, we have mapped the *pcbA1*

Sharon Bryan, Michael E. Hagensee, and Robb E. Moses • Department of Cell Biology, Baylor College of Medicine, Houston, Texas 77030.

allele by tagging with Tn10 to 82.5 min on the E. coli genome (5) in the region of dnaA, dnaN, and gyrB. We know that DNA replication is strictly dependent upon the synthetic activity of DNA pol I because temperature-resistant pcbA1 strains into which a temperature-sensitive DNA pol I gene (polA12) was introduced became temperature sensitive once again. DNA replication can also be supported in the presence of the pcbA1 allele by the Klenow fragment either on an F' or another plasmid (6). Therefore the pcbA1 allele allows us to specifically define an altered polymerase role in replication.

Cells containing the pcbA1 allele give us a unique opportunity to investigate the physiology of DNA replication because replication occurs in the absence of an effective α subunit of DNA pol III. Cells containing the pcbA1 allele also afford an excellent opportunity to investigate the processes of DNA repair and mutagenesis in the absence of a functional DNA pol III holoenzyme.

DNA REPAIR DEFECTS

We have observed increased sensitivity to certain DNA-damaging agents in strains which utilize the pcbA1-pol I replication pathway (8). That is, cells which are deficient in the function of the α subunit of DNA pol III holoenzyme show an increased sensitivity to certain agents. These cells are more sensitive than normal to methyl methanesulfonate (MMS) (Fig. 1) and hydrogen peroxide (data not shown). On the other hand, these cells do not show a marked increase in sensitivity to other DNA-damaging agents such as UV irradiation, bleomycin, or psoralen. The increase in sensitivity to MMS and H_2O_2 is reflected in a decreased DNA repair response as monitored by reformation of high-molecular-weight DNA.

The increase in sensitivity to MMS and H_2O_2 is not so marked as in the case of a DNA pol I deficiency, but it is significant.

The sensitivities can be complemented by transformation or transduction of dnaE into these strains (Fig. 1). This leads us to conclude that both DNA pol I and pol III play responsible roles in the repair of DNA damage following exposure to DNA-damaging agents. Our data are compatible with a major DNA repair pathway involving DNA pol I and a secondary DNA repair pathway involving both DNA pol I and pol III.

The relatively specific repair defect or sensitivity to DNA damage expressed in cells which are defective in the α subunit might at first glance seem surprising. Perhaps these results indicate that the DNA pol III functions in the repair of "nonbulky" lesions in the DNA, but not in the repair of "bulky" lesions.

MUTAGENESIS

We have observed that cells utilizing the pcbA1-pol I DNA replication pathway do not show damage-directed mutagenesis (9). For these studies, we have used two recognized mutagens, UV irradiation and ethyl methanesulfonate (EMS). Exposure to UV in a strain containing a dnaE486 allele, DNA pol I, and the pcbA1 allele resulted in normal mutagenesis at 32°C, at which temperature the α subunit was active. However, at 43°C, with the restrictive temperature inactivating the α subunit, no damage-directed mutagenesis could be observed (Fig. 2a).

The defect in mutagenesis can be assigned specifically to the α subunit because supplying a dnaE gene by transduction or by transformation with a plasmid containing the gene established normal mutagenesis at the restrictive temperature (Fig. 2b). Complementation in trans by the structural gene indicates that the α subunit is strictly required for damage-directed mutagenesis.

The defect in mutagenesis seemingly cannot be assigned to a failure of induction of SOS response (19) since we found normal inducibility for recA-dependent processes at

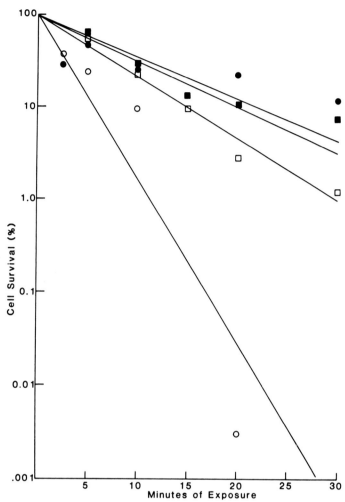

FIGURE 1. Cell survival for strains ER11 and ER11(pSB5). Cultures were grown at 32 and 43°C to mid-log phase and subsequently exposed to 38 mM MMS at the respective temperatures as previously described (8). Samples were withdrawn at various intervals for exposure during growth. Point zero is based on cells incubated at 32 or 43°C, except with no MMS. Symbols: ● and ○, strain ER11 (*dnaE486 pcbA1 zic-1*::Tn*10*) temperature resistant at 32 and 43°C, respectively; ■ and □, strain ER11(psB5) at 32 and 43°C, respectively. psB5 has *dnaE1026* cloned in pBR322.

43°C (9). It is worth noting that the background levels of mutagenesis observed in strains operating by the *pcbA1*-pol I DNA replication pathway are normal at 43°C. Thus, the strains constructed did not seem to represent mutator strains.

The role of the α subunit in DNA mutagenesis would seem to be in fixing the mutation in the cell. There are indications that DNA pol III may interact with the *recA* protein (4, 13) or may interact with additional proteins such as UmuC or UmuD (3) to allow replication past the lesion, thereby fixing the mutation. If this were the case, we might conclude that cells utilizing DNA pol I in replication do not allow passage at the site of the lesion. If replication cannot pass the lesion by synthesis, it would seem that this might be a lethal event for the cell. On the other hand, since recombination is func-

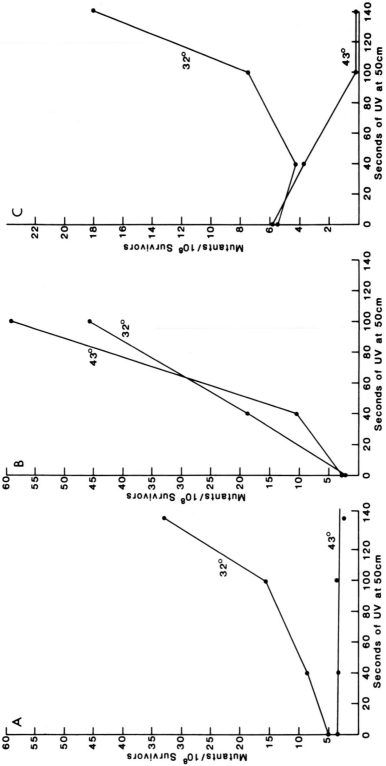

FIGURE 2. UV mutagenesis of strains ER11 (A), ER11(pSB5) (B), and CSM61(pSB5) (C). Cells were grown to mid-log phase and then exposed to a UV flux of 1 J/m² per s for various times as previously described (9). Samples were withdrawn and diluted for cell survival, and also concentrated and plated on L broth rifampin (100 µg/ml) plates to score for mutants. Cells were grown and plated at 32 and 43°C. D37 values (the dose necessary to reduce the surviving fractions to e^{-1} or 0.37) were determined graphically and were 32 to 35 s at both temperatures for all strains.

TABLE 1
Mapping *pcbA1* with λ p*lac* Mu9[a]

Lysogen[b]	% Kan[r] transductants which reversed phenotype for linkage tested[c]:			
	E486 *polC*(Ts)	CRT46 *dnaA*(Ts)	HC194 *dnaN*(Ts)	N4177 *gyrB*(Ts)
T1	78% TR (18/23)[d]	ND	<1% TR (0/140)	10% TR (4/40)
T4	83% TR (24/29)	6% TR (3/51)	ND	5% TR (4/73)
T8	60% TR (17/28)	6% TR (4/63)	14% TR (5/36)	3% TR (1/36)
T25	25% TR (2/8)	ND	<1% TR (0/100)	18% TR (2/11)

[a] Mapping *pcbA1* with λ p*lac* Mu9 was done by coinfecting strain CSM14 *pcbA1* with λ p*lac* Mu9 Kan[r] and helper phage λ pMu507 (2) and selecting for Kan[r] colonies. These lysogens were induced by UV irradiation, and the lysate was used for transduction into strain E486 *polC486 pcbA*[+](Ts). Kan[r] transductants that were temperature resistant indicated λ p*lac* Mu9 carried *pcbA1*.
[b] Temperature-resistant Kan[r] transductants (lysogens) which carried *pcbA1* on λ p*lac* Mu9 were numbered T1 through T25. These lysogens were UV induced and used for subsequent transductions into various strains.
[c] Linkage was determined by the percentage of Kan[r] transductants which reversed their phenotype from the total number of Kan[r] colonies. TR, Temperature resistant; ND, not done.
[d] Numbers in parentheses indicate actual numbers of Kan[r] transductants which reversed their phenotype over the total number of Kan[r] colonies picked.

tional under the conditions in which the *pcbA1*-pol I pathway of replication operates (9), it may be that recombination is involved in allowing the cell to reestablish chromosomal integrity. We do not have any experimental basis for arguing whether the DNA pol I-dependent replication will pass the lesion in the DNA.

CLONING OF THE *pcbA1* ALLELE

We have not succeeded in cloning the *pcbA1* allele in multicopy plasmids. However, we have been successful in cloning the allele by using λ p*lac* Mu9 (2). This specialized λ bacteriophage, which was constructed for protein fusions and monitoring expression of β-galactosidase activity, can also be used as a specialized cloning vector. With this vector we have been able to transduce *pcbA1* from strains CSM61 and HS432 of *E. coli*. These in vivo constructs were also used to locate the map position of the *pcbA1* allele, as shown in Table 1. The data generated by this approach are coincident with those reported by us for P1 mapping (5) and locate the *pcbA1* allele at ~82.5' on the *E. coli* chromosome. There are a number of genes of potential interest in DNA replication in this region, including *dnaA*, *dnaN*, *gyrB*, and *recF*.

Walker and his collaborators have noted that the *dinD* gene lies near this region (19). *dinD* is defined as a gene which is induced following DNA damage. One method of inducing the SOS system is to stall DNA replication. Since DNA replication would stall in *dnaE*(Ts) alleles at the restrictive temperature, we reasoned that perhaps the *dinD* gene product was being induced in response to hesitation in DNA replication and that the *pcbA1* allele might be an allele of the *dinD* gene. However, we were able to obtain recombinants between *dinD* and the *pcbA1* locus (Table 2). Thus, we conclude

TABLE 2
dinD Lies Near *gyrB*[a]

Marker (strain)	Linkage[b]
tnaA (BE280)	<5% (0/18)
dnaA (CRT46)	30% (3/10)
dnaN (HC194)	50% (15/30)
recF (JC9239)	69% (9/13)
gyrB (N4177)	80% (24/30)
pcbA1 (CSM61)	12% (12/97)

[a] P1 was grown on GW1040 *dinD*::Mu d Amp[r] and used to infect strains of interest. Selection was for Amp[r], and linkage was determined from the total number of Amp[r] colonies picked. For BE280 *tnaA*, a colorimetric indol test was used (5), and for JC9239 *recF*, UV irradiation on plates containing 3.5 μg of nitrofurantoin per ml was used to score UV[s] → UV[r] colonies. Transductants of the other strains were evaluated by conversion to the temperature resistance phenotype.
[b] Numbers in parentheses show actual numbers of Amp[r] colonies which reversed their phenotype over total number of Amp[r] colonies picked.

TABLE 3
Interallelic Complementation of *dnaE*(Ts) Alleles In Vivo

Strain and plasmid	Growth[a] at indicated temperature		Mutagenesis[b] after indicated treatment at temperature:				Survival after MMS damage[c] at temperature:	
			UV		EMS			
	32°C	43°C	32°C	43°C	32°C	43°C	32°C	43°C
CSM61 *pcbA1 dnaE1026*	+	+	+	−	+	−	+	−
pDS4-26	+	+	+	+	+	+	+	+
pSB5	+	+	+	−	+	−	+	−
HM10 *polA12 pcbA1 dnaE1026*	+	−						
pDS4-26	+	+	ND	ND	ND	ND	ND	ND
pSB5	+	−						
E511 *dnaE511*	+	−					+	−
pDS4-26	+	+	ND	ND	ND	ND	ND	ND
pSB5	+	+					+	+
E486 *dnaE486*	+	−	+	−				
pDS4-26	+	+	ND	ND	ND	ND	ND	ND
pSB5	+	+	+	+				
ER11 *pcbA1 dnaE486*	+	+	+	−	+	−	+	−
pDS4-26	+	+	+	+	+	+	+	+
pSB5	+	+	+	+	+	+	+	+

[a] Growth was scored on L-broth plates at 32 and 43°C.
[b] +, Increase of ≥20 mutants per 10^8 survivors over the spontaneous mutation frequency after 140 s of UV (1 J/m^2 per s) or 30 min with 0.2 M EMS; −, increase of ≤2 mutants per 10^8 survivors over the spontaneous mutation frequency after 140 s of UV or 30 min with 0.2 M EMS; ND, not done.
[c] +, Normal survival; −, survival decreased by at least 3 orders of magnitude after 30 min of incubation with 38 mM MMS at 43°C.

that, at this level of sensitivity, *dinD* and *pcbA1* are not isoallelic.

INTERALLELIC *dnaE*(Ts) COMPLEMENTATION

We have cloned the *dnaE* gene from strain HS432 which contains the *dnaE1026* allele. The behavior of cells with this clone (pSB5) is summarized in Table 3. It is clear that the *dnaE1026* allele complemented the *dnaE486* and *dnaE511* alleles for growth, DNA repair after damage by MMS or hydrogen peroxide, and mutagenesis. On the other hand, in all situations, it failed to complement a genomic copy of the *dnaE1026* allele.

These observations suggest interallelic complementation of the α subunits. Alberts et al. (1), Kornberg (12), and McHenry (14) have suggested that DNA polymerase holoenzyme may function in the cell in a dimeric form in replication. Our results might best be explained by a dimeric interaction of the α subunits permitting DNA synthesis to occur under the conditions identified. The results noted in Table 3 are not dependent on the *pcbA1* allele and were observed in strains which were wild type at that locus. The failure of isoallelic complementation is quite evident in mutagenesis trials where there was no increase in mutagenesis in strains containing the *dnaE1026* allele on plasmid pSB5 and a chromosomal copy of the *dnaE1026* allele (Fig. 2c). These results indicate that the apparent complementation is not due to overproduction of the *dnaE*(Ts) allele. One might suspect that a multiple-copy effect would produce a high concentration of a partially inactive protein, for example the *dnaE1026* product, and thus restore function on the basis of overproduction. However, the failure to complement in cells containing a chromosomal copy of *dnaE1026* clearly excludes this possibility.

If interallelic complementation is occurring, we would expect that, in strains containing the *dnaE511* and *dnaE486* allele in the chromosome and plasmid pSB5 with the *dnaE1026* allele, there would be a number of inactive dimeric forms of DNA pol III produced by interaction of two isoallelic α subunits. Nevertheless, it appears that the number of heteroallelic enzymes resulting in these cells is sufficient for normal maintenance of mutagenesis and repair.

We have also been able to observe interallelic complementation in crude DNA pol III holoenzyme preparations. Preparations made from strains containing the *dnaE1026* or the *dnaE486* allele showed a decrease in activity on holding at 43.5°C, whereas the wild-type preparation (made from strain HMS83) was activated approximately twofold (Table 4). In extracts prepared from strains containing plasmid pSB5 and a chromosomal copy of *dnaE1026* we saw the normal temperature sensitivity. On the other hand, when a strain containing the chromosomal copy of *dnaE486* and plasmid pSB5 was used for preparation of the extract, then we saw the normal twofold increase in activity, indicating temperature resistance. Thus, it appears that the in vivo observations of interallelic complementation can be substantiated in in vitro extracts.

SUPPRESSION OF NONSENSE *dnaE* DEFECTS

The results reported above for the *pcbA1*-pol I replication pathway were ob-

TABLE 4
In Vitro Complementation of *polC*(Ts) Alleles

Strain/genotype[a]	% Synthesis after incubation at 43.5°C[b] for indicated minutes:		
	0	15	30
E486 *dnaE486*	100	50	40
E486 *dnaE486*(pSB5[c])	100	152	239
CSM61 *dnaE1026*	100	65	71
CSM61 *dnaE1026*(pSB5)	100	80	78
HMS83 *polC*⁺	100	160	280

[a] DNA pol III holoenzyme was isolated from cells and held at 43.5°C for times indicated before adding to reaction mixture (17). The reaction was incubated for 30 min at 30°C. All activity was N-ethylmaleimide sensitive. 100% synthesis equals 130 to 210 pmol.
[b] Strain E486 contains the *dnaE486* allele; strain CSM61 contains the *dnaE1026* allele and HM583 carries the *polC*⁺ gene.
[c] Plasmid pSB5 was constructed by cloning gene *dnaE1026* (from HS432) into pBR322.

tained with temperature-sensitive alleles of the *dnaE* gene. The results suggested that it may be possible to eliminate DNA pol III activity from the cell and yet have the cell survive. H. Maki (Kyushu University, Japan) has isolated nonsense mutations in the *dnaE* gene. These are viable because the cell lines contain suppressor mutations, but the suppressor mutations are temperature sensitive, so the strains are temperature sensitive. The temperature resistance phenotype could be restored by transformation of these cell lines by *polC* plasmid pDS4-26 as described by McHenry (14). Thus, the defect seemed specifically an α-subunit defect. We demonstrated by immunoblotting that there was no α subunit detectable at 43°C.

These cells also provided another test of the *pcbA1* allele bypass for *dnaE*. We found that *pcbA1* phenotypically suppressed three strains tested, MK503, MK531, and MK519. Each strain became temperature resistant when *pcbA1* was introduced via transduction. Thus, *pcbA1* does allow the cell to bypass a complete defect in the α subunit of DNA pol III.

The evidence that cells can live in the absence of a functional DNA pol III holoenzyme raises the question of what, under such circumstances, the functional replicating enzyme is. The requirement for DNA pol I synthetic activity in this situation leads to the speculation that DNA pol I is incorporated into the holoenzyme to allow replication, but this has not yet been demonstrated. Whether assembly of DNA pol I into the replisome requires modification of DNA pol I or modification of a holoenzyme component is a point of speculation. Linn and co-workers (personal communication) have described modified forms of DNA pol I. In any event we can assume that the synthetic activity resulting from the interaction of DNA pol I and the *pcbA1* gene product is very processive since processivity is a feature of replisomes. Whether this results from modifications of DNA pol I or a persistent interaction of the *pcbA1* gene production in the replisome remains an open question.

ACKNOWLEDGMENTS. This work was supported by Public Health Service grants GM19122 and GM24711 and Robert A. Welch Foundation grant Q-1027. M.E.H. is the recipient of an Achievement Reward for College Scientists.

LITERATURE CITED

1. **Alberts, B., J. Barry, P. Bedinger, T. Formosa, C. V. Jongeneel, and K. N. Kreuzer.** 1983. Studies on DNA replication in the T4 bacteriophage *in vitro* system. *Cold Spring Harbor Symp. Quant. Biol.* **47:**655–668.
2. **Bremer E., T. Silhavy, and G. Weinstock.** 1985. Transposable λ*plac*Mu bacteriophages for creating *lacZ* operon fusions and kanamycin resistance insertions in *Escherichia coli*. *J. Bacteriol.* **162:**1092–1099.
3. **Bridges, B. A., and R. Woodgate.** 1984. Mutagenic repair in *E. coli*. X. The *umuC* gene product may be required for replication past pyrimidine dimers but not for the coding error in UV-mutagenesis. *Mol. Gen. Genet.* **196:**364–366.
4. **Bridges, B. A., and R. Woodgate.** 1985. Mutagenic repair in *Escherichia coli*: products of the *recA* gene and the *umuD* and *umuC* genes act at different steps in UV-induced mutagenesis. *Proc. Natl. Acad. Sci. USA* **82:**4193–4197.
5. **Bryan, S. K., and R. E. Moses.** 1984. Map location of the *pcbA* mutation and physiology of the mutant. *J. Bacteriol.* **158:**216–221.
6. **Bryan, S. K., and R. E. Moses.** 1988. Sufficiency of the Klenow fragment for survival of *polC*(Ts) *pcbA1 Escherichia coli* at 43°C. *J. Bacteriol.* **170:**456–458.
7. **Gefter, M., Y. Hirota, T. Kornberg, J. A. Wechsler, and C. Barnoux.** 1971. Analysis of DNA polymerase II and III in mutants of *Escherichia coli* thermosensitive for DNA synthesis. *Proc. Natl. Acad. Sci. USA* **68:**3150–3153.
8. **Hagensee, M. E., S. K. Bryan, and R. E. Moses.** 1987. DNA polymerase III requirement for repair of DNA damage caused by methyl methanesulfonate and hydrogen peroxide. *J. Bacteriol.* **169:**4608–4613.
9. **Hagenesee, M. E., T. L. Timme, S. K. Bryan, and R. E. Moses.** 1987. DNA polymerase III of *Escherichia coli* is required for UV and ethylmethane sulfonate mutagenesis. *Proc. Natl. Acad. Sci. USA* **84:**4195–4199.
10. **Joyce, C., and N. D. F. Grindley.** 1984. Method for determining whether a gene of *Escherichia coli* is essential: application to the *polA* gene. *J. Bacteriol.* **158:**636–643.
11. **Kornberg, A.** 1980. *DNA Replication*, p. 347–414. W. H. Freeman & Co., San Francisco.
12. **Kornberg, A.** 1982. *DNA Replication 1982 Supplement.* W. H. Freeman & Co., San Francisco.
13. **Lu, C., R. Scheuermann, and H. Echols.** 1986. Capacity of RecA protein to bind preferentially to UV lesions and inhibit the editing subunit (ε) of DNA polymerase III: a possible mechanism for SOS-induced targeted mutagenesis. *Proc. Natl. Acad. Sci. USA* **83:**619–623.
14. **McHenry, C.** 1985. DNA polymerase III holoenzyme of *E. coli*: components and function of a true replicative complex. *Mol. Cell. Biochem.* **66:**71–85.
15. **McHenry, C., and A. Kornberg.** 1981. DNA polymerase III, p. 39–50. *In* P. Boyer (ed.), *The Enzymes*, vol. 14. Academic Press, Inc., New York.
16. **McMacken, R., L. Silver, and C. Georgopoulos.** 1987. DNA replication in *Escherichia coli* and *Salmonella typhimurium*, p. 564–612. *In* F. C. Neidhardt, J. L. Ingraham, K. B. Low, B. Magasanik, M. Schaecter, and H. E. Umbarger (ed.), *Escherichia coli and Salmonella typhimurium: Cellular and Molecular Biology*. American Society for Microbiology, Washington, D.C.
17. **Niwa, O., S. K. Bryan, and R. E. Moses.** 1979. Replication at restrictive temperatures in *Esche-*

richia coli containing a *polC*$_{ts}$ mutation. *Proc. Natl. Acad. Sci. USA* **76**:5572–5576.
18. **Niwa, O., S. K. Bryan, and R. E. Moses.** 1981. Alternate pathways of DNA replication: DNA polymerase I-dependent replication. *Proc. Natl. Acad. Sci. USA* **78**:7024–7027.
19. **Walker, G.** 1984. Mutagenesis and inducible responses to deoxyribonucleic acid damage in *Escherichia coli. Microbiol. Rev.* **48**:60–93.
20. **Welch, M., and C. McHenry.** 1982. Cloning and identification of the product of the *dnaE* gene of *Escherichia coli. J. Bacteriol.* **152**:351–356.

Chapter 32

In Vitro Model of Mutagenesis

Janet Sahm, Edith Turkington, Diane LaPointe, and Bernard Strauss

It is a hypothesis that mutation occurs primarily in replicating systems. Although one can argue that the definition of mutation, as a sudden heritable change, requires the test of replication to determine heritability, it is certainly possible to envisage nonreplicating systems in which nucleic acid bases are substituted or in which repair reactions proceeding with a minimum of repair synthesis do result in changes in DNA sequence, without the necessity of replicating the whole genome (23). For the most part this does not happen. We suppose that most mutations are fixed at the time of replication. Damaged DNA is not yet mutant DNA until it has been fixed by replication. So, for example, an abasic site or the site of a cyclobutane dimer may have lost its informational character, but if the abasic site is due to the loss of a T or U and if the pyrimidine dimer is a TT dimer, then it is likely that no mutation will occur, due to the propensity of the replication apparatus to insert A's opposite noninformational sites (13, 25). Only the insertion of an incorrect base and its protection from proofreading and mismatch repair by replication results in a mutation.

It is therefore at or near the replication fork that most mutations occur. This may also be true for transposition-induced insertion mutation since transposition itself involves replication (2). What is the role of the polymerases and the accessory proteins in determining the specificity of mutation? An older view is that the role is limited. Mutation occurs by mispairing, and the mispairing is governed by the physical structure of the bases. This approach takes into account the effect of the neighboring sequence on mutation since the stability of base pairing is determined in part by stacking interactions (24) and even the production of small additions or deletions (frameshifts) can often be ascribed to the local secondary structure resulting from particular repeated base sequences as in the Streisinger model (3). The isolation of mutator genes indicates that physical factors are not the only ones involved in mutation, just as the isolation of mutants sensitive to the killing action of radiation demonstrates that physical target interactions do not alone account for the inactivating effects. Some of these mutator genes affect the $3' \rightarrow 5'$ proofreading nuclease activity of the replicative polymerase(s)

Janet Sahm, Edith Turkington, Diane LaPointe, and Bernard Strauss • Department of Molecular Genetics and Cell Biology, University of Chicago, Chicago, Illinois 60637.

(14, 26), and this finding indicates that kinetic factors of proofreading determine the fidelity of particular enzymes. Even in such cases, however, the effect of the enzyme is kinetic rather than selective. To use the terminology of Goodman and Branscomb, the enzyme is illiterate, and the lapses in fidelity are likely to be due to incorrect base pairings frozen by the kinetic behavior of the polymerase and its associated proofreading exonuclease (4).

The situation with respect to lesions which can block DNA synthesis is different, with the enzyme playing more of a determinative role in base insertion. For some lesions, particularly for the abasic site, different polymerases all make the same decision—the preferential insertion of adenine nucleotides (25, 29). This appears to be the case for some but certainly not all lesions. In the case of angelicin, for example, the insertion of T opposite a noninstructional C seems most prevalent (19). It is, therefore, a question as to what the detailed factors are which affect the behavior of the replicative apparatus when confronted with damaged sites within DNA. Various factors clearly play a role. The nature of the altered site must be of importance. Recent studies indicate that the surrounding bases are also important (e.g., reference 6). Genetic studies suggest that the structure of the polymerase plays a role, as do the associated binding proteins which form the replication complex. Working out the role of the different components requires an in vitro system that permits one to control the factors involved.

The requirements of such a system include an easily scored and easily sequenced gene, since understanding of mutational specificity requires precise definition of the base changes. We use the *lacZ* β-galactosidase-complementing peptide cloned into bacteriophage M13. Mutations at numerous sites in this gene produce mutants which are recognizable as light blue or colorless plaques when progeny phage are plated on chromogenic media. In general, colorless mutants can be nonsense, frameshifts, or large deletions. Mutants retaining enzyme activity are light blue and tend to represent base substitutions. We prepare plus (or positive) viral strands containing the insert and hybridize these to viral replicative form DNA from which the complementing insert has been removed by restriction with *Pvu*I and *Pvu*II followed by electrophoretic separation of the large fragment. Hybridization produces a molecule with a single-stranded gapped region of 392 nucleotides (Fig. 1). The plus strand is mutagenized before hybridization. In vitro DNA synthesis to fill the gap is likely to encounter lesions and therefore to generate mutants. To make sure that the mutants we obtain are generated in vitro rather than after infection of the tester strain, we used two strategems. First, we use a *rec* mutant host in which the RecA protein is not present and the host is not induced for SOS functions. Second, we use as substrate a viral plus strand synthesized in an *ung dut* mutant host so that about 5% of the thymines are substituted by uracil. This substrate was first used by Kunkel (10) for the efficient isolation of site-directed mutants. We have found that in vitro synthesis by *Escherichia coli* polymerase I (pol I) Klenow fragment (Kf) increases the plaque-forming efficiency of a gapped uracil-containing template by about 20- to 30-fold (Table 1). Viable phage are recovered only from the minus (or negative) strand, as can be deduced from Table 1; the efficiency of this transfection is between about 1 and 10% of the non-uracil-containing plus strand. We think this is partly the result of "strand bias" resulting from the biology of M13 replication (8), in which it is the plus strand which is naturally infectious. It may also be that the difference relates to the ability of both plus and minus strands to yield progeny in nonuracil hybrids, whereas with the uracil-containing hybrid, virus can come only from the minus strand.

We used either *E. coli* DNA pol I or a commercial preparation of phage T7 DNA

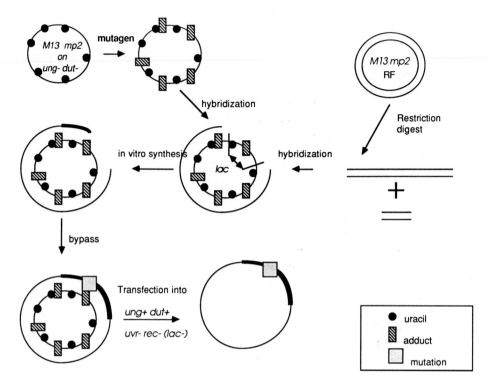

FIGURE 1. Scheme for the detection of mutants produced by in vitro synthesis. Details are given in the text.

polymerase ("Sequenase"). This polymerase preparation contains thioredoxin but has been oxidized in the presence of Fe^{2+} ion, a procedure reported to drastically reduce the $3' \rightarrow 5'$ endonuclease activity of the protein

TABLE 1
Transfection Efficiency of Different Substrates in
E. coli S90c recA56

Substrate	Avg PFU/ng of DNA
1. Replicative form I DNA (wild type)	866.0
2. Single-strand DNA (wild type)	93.0
3. Single-strand DNA (uracil)	0.03
4. Linear fragment (wild type) (PvuI-PvuII)	0.22[a]
5. Uracil hybrid (no. 3 and 4)	0.31
6. No. 5 after DNA synthesis (Kf)	10.9
7. Wild-type hybrid (no. 2 and 4)	169.0
8. No. 7 after DNA synthesis (Kf)	145.0

[a] Data from a different experiment in which replicative form I DNA had a transfection efficiency of 825 PFU/ng.

(28). Sequenase is highly processive, a property suggested by Livneh (12) as important for the bypass of noninstructional lesions. Synthesis through the gapped region of a uracil-containing, nonreacted template was highly mutagenic (Table 2). Almost 1% of the plaques obtained were mutant. When the synthesis was carried out at 37 instead of 10°C the "spontaneous" mutation frequency was 3%, with the major portion being light blue mutants. We observed fewer mutations in control experiments with E. coli Kf at either 10 or 37°C than with Sequenase. Addition of dGMP doubled the observed frequency when the experiment was carried out with E. coli Kf at 37°C. Since dGMP is an inhibitor of $3' \rightarrow 5'$ exonuclease activity (22), we suggest that the lower rate of synthesis at 10°C permits kinetic editing even in the absence of an exonuclease. At the higher temperatures, mutations may be fixed by

TABLE 2
Effect of Temperature on Spontaneous Mutations Induced by In Vitro Synthesis[a]

Sample (temp)	Total PFU	PFU/ng of DNA	No. of mutant plaques		Spontaneous mutation frequency (%)	
			Total[b]	Blue	Total	Blue
No synthesis	121	0.017	0	0	0	0
T7 Sequenase						
10°C	3,021	1.05	30	22	0.99	0.73
37°C	2,095	0.97	72	58	3.4	2.8
E. coli Kf						
10°C	2,985	1.38	10	10	0.34	0.34
37°C	1,294	0.6	8	8	0.62	0.62
+dGMP, 10°C	1,170	0.54	5	5	0.43	0.43
+dGMP, 37°C	983	0.46	12	12	1.2	1.2

[a] A gapped M13mp2 substrate was constructed by hybridization of M13mp2 plus strands with a 6,804-base-pair fragment generated by double digestion of M13mp2 with the restriction enzymes PvuI and PvuII. The double-stranded DNA template containing a 392-base single-stranded gapped region was incubated with E. coli pol I Kf or Sequenase. Synthesis was carried out in 20 mM HEPES (N-2-hydroxyethylpiperazine-N'-2-ethanesulfonic acid) (pH 7.8)–5 mM dithiothreitol–10 mM MgCl$_2$ along with 300 µM each dATP, dCTP, dGTP, and dTTP and 500 µM ATP. Incubation was for 30 min at 10 or 37°C with 1 to 2.5 U of pol I KF per µg of DNA or 0.8 U of Sequenase per µg of DNA. When used, dGMP was added to a final concentration of 10.3 mM. The reaction was stopped by the addition of EDTA to 15 mM, and the mixture was held on ice until it was transfected into E. coli S90c recA56 (11). Mutant plaques were picked, replated, and purified.
[b] Total mutants = blue + colorless.

elongation before the mispairing can be detected. These results led us to carry out the following experiments at 10°C.

We used N-acetoxy acetylaminofluorene and N-hydroxy-2-aminofluorene as model compounds for in vitro mutagenesis studies. These compounds both react with DNA to form primarily a C-8 derivative differing by an acetyl group. The presence of this acetyl group is structurally of great importance since the aminofluorene (AF) adduct is able to fit into the minor groove of DNA without producing major distortion whereas the acetylaminofluorene (AAF) results in a rotation of the base from the *anti* to the *syn* configuration to accommodate the adduct (5). As a result, the mutagenic properties of the adducts differ considerably. The AAF adduct is a major if not absolute block to DNA synthesis and is associated in vivo with deletion mutations. There is intriguing evidence from the laboratory of Fuchs et al. that the replication apparatus of E. coli is not able to bypass this lesion (at least, not easily), so that the effect of single lesions is to produce strand loss (7). AF is a less damaging lesion. In contrast to single AAF lesions which are lethal for single-stranded phage ɸX174 DNA, seven AF "hits" are required in both wild-type and *uvr rec* mutant strains (15, 16). In contrast to the mutations obtained with AAF, AF-induced mutations are reported to be transitions (1). This pair of lesions therefore provide a good system for studies of in vitro mutagenesis.

As might be expected from the in vivo results, we have been unable to obtain evidence for bypass of the AAF lesion by using either E. coli Kf or Sequenase as the enzyme. We treated a uracil-containing M13mp2 plus strand with N-acetoxy-N-acetylaminofluorene to obtain a series of substrates with different numbers of adducts and assayed the gapped, AAF-containing molecules on an

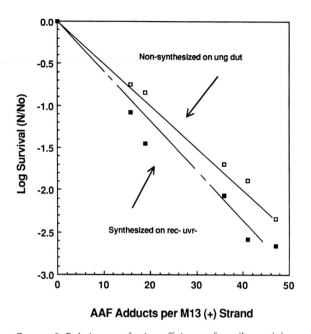

FIGURE 2. Relative transfection efficiency of uracil-containing M13mp2 DNA after treatment with N-acetoxyacetylaminofluorene. The number of adducts fixed is calculated from the radioactivity precipitated by trichloroacetic acid and the specific activity of the [^{14}C]AAF sample used. Samples were analyzed after hybridization but before synthesis on an *ung dut* mutant strain and after synthesis with *E. coli* pol I Kf on an $ung^+ \, dut^+ \, rec \, uvr$ strain provided by Wolf Epstein.

ung dut mutant strain (Fig. 2). We then carried out a polymerase synthesis reaction with *E. coli* pol I Kf and assayed the product on an $ung^+ \, dut^+ \, rec \, uvr$ strain (prepared for us by Wolf Epstein). Synthesis did not affect the lethal effect of adducts in the template strand, indicating an absence of bypass (Fig. 2). We also analyzed the results of synthesis on plus strands containing different numbers of adducts on the production of mutants and on the overall plaque-forming efficiency (Table 3). The greater the number of AAF lesions, the lower was the increase in plaque-forming efficiency due to synthesis. The absolute number of mutants decreased with N-acetoxy aminofluorene treatment, and in addition, all of the mutants in the treated series were colorless. The absolute number of these mutants did not increase with increasing number of AAF adducts, and synthesis decreased rather than increased the number recovered. DNA sequence analysis of the mutants recovered after synthesis on AAF-containing templates showed that the recovered colorless mutants did not contain any trace of the *lac* insert. We conclude that neither *E. coli* pol I Kf nor Sequenase is able to bypass the AAF lesion under our conditions. The increase in the frequency (but not the absolute number) of mutants is due to an artifact first pointed out to us by T. Kunkel. Our experiments are done using a hybridization mixture in which gapped DNA, unhybridized plus circles, and residual large linear fragments (which have been purified by electrophoresis away from the restricted replicative-form DNA) are present. Control experiments show that the mixture is a poor DNA primer before hybridization but a good primer afterwards. Since removal of the *lac-*

TABLE 3
Results of Synthesis on an AAF-Uracil-Containing Template[a]

Sample	Total PFU	PFU/ng of DNA	No. of mutant plaques		Frequency of mutation (%)	
			Total	Blue	Total	Blue
Control						
No synthesis	120	0.03	0	0	0	0
T7 Sequenase, 10°C	1,587	1.1	13	10	0.8	0.6
AAF (3.8)						
No synthesis	61	0.0038	4	0	6.6	0
T7 Sequenase	422	0.05	7	6	1.7	1.4
AAF (5.9)						
No synthesis	36	0.002	3	0	8.3	0
T7 Sequenase	64	0.008	2	0	3.1	0
Non-uracil DNA						
No synthesis	4,998	83	6	6	0.12	0.12
T7 Sequenase	5,606	93	10	7	0.18	0.12

[a] Reactions were carried out as described in Table 2, footnote a, except that, where indicated, hybridization was to an N-acetoxy-N-acetylaminofluorine-treated plus strand containing the calculated number of lesions shown in parentheses.

complementing fragment does not remove vital genes, ligation of the two ends of the large linear fragment produces viable phage. These viruses are not able to complement the *lacZ* mutation of the bacterial host and therefore produce colorless plaques. The frequency of such events is very small, but as the total number of viable phage molecules becomes smaller as a result of AAF treatment of plus strands, such illegitimate ligations result in a larger fraction of the recovered progeny.

In contrast to our results with AAF adducts, the data indicate that AF lesions can be bypassed in vitro with the production of mutations (Table 4). We observed a large absolute increase in the number of mutants as well as an increase in the relative frequency of mutants with Sequenase as the polymerizing enzyme. The mutants were in both blue and colorless categories (all putative mutants were picked, purified, and replated alongside bona fide wild types to confirm their mutant character before sequencing). The data were less compelling with pol I Kf, although there may also be mutagenesis in this case. Bypass of AF-induced lesions as a result of pol I Kf-catalyzed synthesis in vitro has been reported by two laboratories (18, 21).

The high background mutation rate produced by in vitro synthesis on nonreacted templates makes more definitive demonstration of lesion-induced mutation necessary. We therefore proceeded to determine the sequence changes in the mutants we obtained from control (uracil-containing) and AF-reacted templates. We tried to eliminate the sequencing of "sib" clones that derive from the same mutational event. These results indicated that the AF-containing class of templates generate mutants with different specificity than the controls (Table 5). The mutations produced in the controls included both single-base deletions and base substitutions divided between transitions and transversions. Most of the transitions were C→T changes of the type that would be expected as a result of cytosine deamination during storage of the substrate. Deletions seemed equally likely at the site of any base. In contrast, the mutations synthe-

TABLE 4
In Vitro Mutagenesis by Synthesis on an AF-Uracil-Containing Template[a]

Sample	Enzyme	Total PFU	PFU/ng of DNA	No. of mutant plaques		Frequency of mutation (%)	
				Total	Blue	Total	Blue
No synthesis	None	1,625	0.16	26	6	1.6	0.37
Control	Sequenase	9,057	2.1	87	55	0.96	0.60
AF (2.4)	None	495	0.07	3	1	0.61	0.20
AF (2.4)	Sequenase	15,937	1.9	414	209	2.6	1.3
AF (6)	None	280	0.04	5	3	1.8	1.1
AF (6)	Sequenase	3,312	0.40	178	74	5.4	2.2
No synthesis	None	832	0.18	14	3	1.7	0.36
Control	Kf	22,827	5.28	64	56	0.28	0.25
AF (2.4)	None	585	0.083	3	0	0.51	0
AF (2.4)	Kf	19,642	1.58	89	69	0.45	0.35
AF (6)	None	367	0.052	2	1	0.54	0.27
AF (6)	Kf	1,998	0.16	26	8	1.3	0.4

[a] Sequenase and *E. coli* Kf reactions were run at 10°C using the protocol given in Table 2, footnote *a*. Control templates were not treated. Numbers in parentheses refer to average adducts per molecule calculated from the radioactivity bound to DNA after treatment with N-hydroxy-2-aminofluorene. The Sequenase and *E. coli* polymerase experiments were done separately.

sized off the AF substrate had a different specificity. Half represented deletions of one or more bases. The majority had lost a single G. All of the transversions were G→T, and these outnumbered C→T transitions by 4:1. We suppose many of the C→T transitions may well represent the contribution of the spontaneous mutations. The difference in the type and distribution of mutation produced from the two substrates lends credence to the supposition that induced mutation has occurred in vitro. The mutations that were produced were very like those reported by Bichara and Fuchs (1) for AF-induced mutation except that in this in vitro system we observed a number of single-base deletions of the sort observed in vivo only with AAF. The location of the AF-induced mutants within the gene is also interesting since it is clear that there are particular hot spots for mutation (Fig. 4). One of these hot spots occurs at position 90 (11) adjacent to the sequence 3' CCCAAAGGG 5'. This particular region

TABLE 5
Summary of Sequencing Results[a]

Synthesis	Total Mutants	Multiple deletions	Single-base deletions				Base substitutions						
							Transversions			Transitions			
			G	C	T	A	G→T	A→T	G→C	C→T	G→A	T→C	A→G
Control (uracil)	54	5[b]	5	1	3	5	19	4	3	12	0	1	2
AF	58	9[c]	21	0	0	0	23	0	0	5	2	0	0

[a] Changes observed in mutants obtained by transfection of uracil-containing M13mp2 gapped templates following synthesis with Sequenase (control) or after synthesis with Sequenase on a gapped template containing an average of six AF adducts per molecule (AF).
[b] Two GC; one each CA, AC, and AT.
[c] Four GC; one each GT, GCA, CA, ACA, and TGCCGGTCAC.

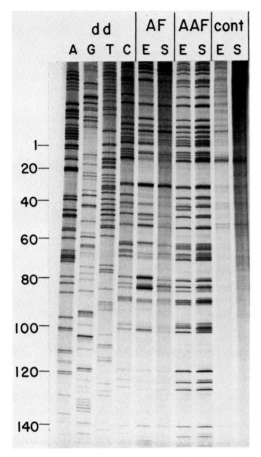

FIGURE 3. Polyacrylamide gel analysis of M13mp2 (uracil-containing) DNA templates either untreated (cont) or treated with N-hydroxy-2-aminofluorene or N-acetoxyacetylaminofluorene. The treated M13mp2 DNA plus strand was primed with a 6,804-base-pair fragment generated by double digestion of M13mp2 with the restriction enzymes PvuI and PvuII. The double-stranded DNA template, containing a 392-base single-stranded gapped region, was incubated with E. coli pol I Kf (lanes E) or Sequenase (lanes S). The A, G, T, and C lanes are dideoxy sequence standards obtained by incubating untreated DNA templates with avian myeloblastosis virus reverse transcriptase and dideoxynucleotides. Synthesis was carried out in 20 mM HEPES (pH 7.8)–5 mM dithiothreitol–10 mM MgCl$_2$. [^{35}S]dATP was added to a final concentration of 1.2 μM along with 300 μM each dGTP, dCTP, and dTTP and 500 μM ATP. Incubation was for 15 min at 10°C with either 0.4 U of pol I Kf or 0.4 U of Sequenase, followed by a chase with dATP added to give 300 μM final concentration. After synthesis on all templates, the DNA was digested with BglI, which cuts 29 bases 3' to the PvuI site of the primer. The DNA was denatured in formamide, loaded on an 8% acrylamide-urea buffer gradient gel, and run at 38 W constant for 5 h. The amounts of trichloroacetic acid-precipitable radioactivity added to each of the lanes were as follows: lane AF (E), 5.4×10^5 dpm; lane AF (S), 5.6×10^5 dpm; lane AAF (E), 3.9×10^5 dpm; lane AAF (S), 3.4×10^5 dpm; lane cont (E), 4.6×10^5 dpm; lane cont (S), 5.4×10^5 dpm. The numbers on the left side of the gel indicate the nucleotide positions as given by Kunkel (11). AF lanes: DNA treated with N-hydroxy-2-aminofluorene to give aminofluorene adducts. AAF lanes: DNA treated with N-acetoxyacetylaminofluorene to give acetylaminofluorene adducts.

has been reported by Kunkel (9) as a hot spot for depurination mutagenesis.

The AF lesion probably requires bypass DNA synthesis since an activated SOS system is required for in vivo mutagenesis. We therefore asked whether the hot spots for mutation correspond to the hot spots for termination of DNA synthesis. Termination experiments were carried out as previously described (27), using the identical substrates employed in the mutation experiments (Tables 2 through 5; see Fig. 4) and comparing substrates containing either AAF or AF residues after synthesis carried out with both E. coli pol I Kf and Sequenase. The results with the AAF-containing template were what might be expected (Fig. 3). Each G appeared to react about equally. There were differences in detail when Sequenase was compared with pol I Kf. Such differences are consistent with the assumption that Sequenase synthesis more often terminates opposite the lesion, whereas pol I Kf tends to terminate 5' to the lesion on the template strand. The results with AF-containing templates were more complex and very different from those obtained with AAF-containing substrates (Fig. 3). Both AF and AAF termination patterns indicated reaction with G's in the templates, as predicted from the known reaction patterns. However, with the AF template there were clearly hot spots for termination, for example, near positions 30, 80, and 90 with the E. coli enzyme. In contrast to the identical behavior of the two

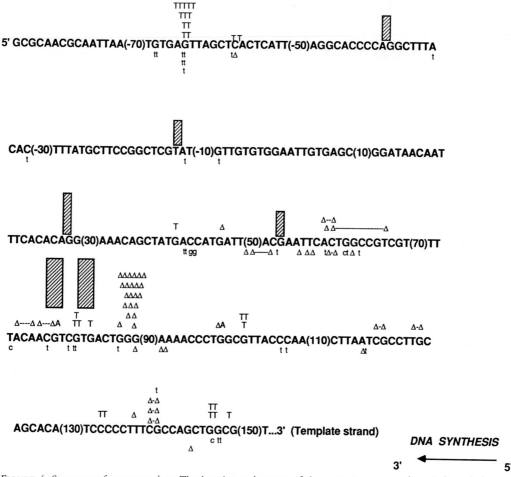

FIGURE 4. Summary of sequence data. The location and nature of the mutants sequenced are indicated above or below the sequence. Lowercase letters or symbols below the sequence indicate spontaneous mutants, i.e., mutants obtained by synthesis on a nontreated, uracil-containing, gapped template. Uppercase letters and symbols above the line indicate results with mutants obtained after synthesis on an AF- and uracil-containing, gapped template. The hatched bars indicate major termination sites on the AF template, using *E. coli* pol I Kf (see Fig. 3). The numbers in parentheses indicate the nucleotide positions as given by Kunkel (11). Δ, Deletion; Δ---Δ, deletion spanning a sequence.

enzymes on the AAF substrate, there were marked differences between the pattern obtained with the AF substrate when *E. coli* Kf and Sequenase reaction products were compared. Although there are hot spots for termination in common, for example at position 29 to 30, it appears that Sequenase passes through sites that are clear blocks for the *E. coli* enzyme. The termination results obtained with the *E. coli* enzyme on AF-containing templates indicate nonuniformity in either reactivity or termination ability at different sites within the sequence, as compared with the relative uniformity of AAF-containing templates. These results differ from similar studies reported from this laboratory, using another sequence and a more highly reacted substrate (20). We wondered whether this difference might relate to different amounts of ring opening (17) in the substrates used in 1982 and at present. Analysis of our substrate indicated the adducts to be mainly AF with only about 2.2% of the ring opened form (M. Tang, personal communication), and so an explanation for the difference is not yet available. However, it is clear that the same substrate after treatment with N-hydroxy-2-aminofluorene responds differently as a template with different polymerases.

We quantitated our results by scanning gels and determined the proportion of termination that occurred at particular sites. To normalize the results, we expressed the termination percentages as a proportion of the total counts added to the gel. We then compared the termination sites with the sites of mutation as determined in the sequencing experiments (Fig. 3 and 4). Mutations occurred in regions of termination, but hot spots for mutation were not hot spots for termination in these experiments with this substrate.

We conclude from this set of experiments that mutations are induced by Sequenase-catalyzed in vitro synthesis on an AF-containing template, since (i) the number and frequency of mutants are increased by synthesis on an AF-containing template, and (ii) the distribution of mutant types is different when spontaneous and AF-template-induced mutants are compared. In contrast to their behavior on an AAF-containing template, Sequenase and *E. coli* pol I Kf recognize the AF-containing template differently: termination occurs at or near G's in both cases, but not all G's show equivalent terminations. Some sites seem to be hot spots for mutation, but these are not necessarily the sites of strongest DNA synthesis terminations. The hot spots are not explained as multiple occurrences of sibs. Single-base G deletions and G→T transversions are the major sources of mutation on AF-containing templates in vitro. The sequence around position 90, 5′ GGGAAAACCC 3′, may promote slippage when an AF adduct blocks DNA synthesis, leading to loss of a base in the newly synthesized strand. Some secondary DNA structure is likely to be involved, and the sequence does permit formation of a hairpin. It is not obvious what is special about the structure around position −64 (Fig. 4) leading to G→T transversions, and we do not yet have termination data for this part of the sequence.

More generally, we suggest that the specificity of mutation in organisms, including induced mutation, can be accounted for by the properties of the polymerases. Insofar as the mutations we obtain really do occur in vitro rather than in the test organism, we have been able to simulate the mutational specificity obtained in vivo by these in vitro experiments. It should be possible to duplicate the process with greater precision in vitro by control of reaction conditions and by use of the complete replication apparatus. This technology then should permit us to determine the rules which determine the role of the surrounding nucleotide sequence. Since it is clear that the nature of the lesion also affects specificity, we will also be able to investigate the interactions between the inherent specificity of the polymerases and the instructive nature of an altered nucleotide site.

ACKNOWLEDGMENTS. This work was supported in part by grants from the National Institute of General Medical Sciences (GM 07816) and the National Cancer Institute (CA32436) and by a contract with the Department of Energy (DE-AC02-76ER 02040).

We thank Wolf Epstein for his preparation of the *uvr rec lac* strain and Moon-shong Tang for his analysis of our substrate.

LITERATURE CITED

1. **Bichara, M., and R. Fuchs.** 1985. DNA binding and mutation spectra of the carcinogen N-2-aminofluorene in *Escherichia coli*. A correlation between the conformation of the premutagenic lesion and the mutation specificity. *J. Mol. Biol.* **183**:341–351.
2. **Craigie, R., and K. Mizuuchi.** 1985. Mechanism

of transposition of bacteriophage Mu: structure of a transposition intermediate. *Cell* 41:867–876.

3. **Drake, J., B. Glickman, and L. Ripley.** 1983. Updating the theory of mutation. *Am. Sci.* 71:621–630.

4. **Goodman, M., and E. Branscomb.** 1986. DNA replication fidelity and base mispairing mutagenesis, p. 191–232. *In* T. Kirkwood, R. Rosenberger, and D. Galas (ed.), *Accuracy in Molecular Processes.* Chapman & Hall, London.

5. **Grunberger, D., and R. Santella.** 1981. Alternative conformations of DNA modified by N-2-acetylaminofluorene. *J. Supramol. Struct. Cell. Biochem.* 17:231–244.

6. **Hayes, R., and J. LeClerc.** 1986. Sequence dependence for bypass of thymine glycols in DNA by DNA polymerase I. *Nucleic Acids Res.* 14:1045–1061.

7. **Koffel-Schwartz, N., G. Maenhaut-Michel, and R. Fuchs.** 1987. Specific strand loss in N-2-acetylaminofluorene-modified DNA. *J. Mol. Biol.* 193:651–659.

8. **Kornberg, A.** 1980. *DNA Replication.* W. H. Freeman & Co., San Francisco.

9. **Kunkel, T.** 1984. Mutational specificity of depurination. *Proc. Natl. Acad. Sci. USA* 81:1494–1498.

10. **Kunkel, T.** 1985. Rapid and efficient site-specific mutagenesis without phenotypic selection. *Proc. Natl. Acad. Sci. USA* 82:488–492.

11. **Kunkel, T.** 1985. The mutational specificity of DNA polymerase β during *in vitro* DNA synthesis. Production of frameshift, base substitution, and deletion mutations. *J. Biol. Chem.* 260:5787–5796.

12. **Livneh, Z.** 1986. Mechanism of replication of ultraviolet-irradiated single-stranded DNA by DNA polymerase III holoenzyme of *Escherichia coli. J. Biol. Chem.* 261:9526–9533.

13. **Loeb, L.** 1985. Apurinic sites as mutagenic intermediates. *Cell* 40:483–484.

14. **Lu, C., R. Scheuermann, and H. Echols.** 1986. Capacity of RecA protein to bind preferentially to UV lesions and inhibit the editing subunit (ε) of DNA polymerase III: a possible mechanism for SOS-induced targeted mutagenesis. *Proc. Natl. Acad. Sci. USA* 83:619–623.

15. **Lutgerink, J., J. Retel, and H. Loman.** 1984. Effects of adduct formation on the biological activity of single- and double-stranded φX174 DNA, modified by N-acetoxy-N-acetyl-2-aminofluorene. *Biochim. Biophys. Acta* 781:81–91.

16. **Lutgerink, J., J. Retel, J. Westra, M. Welling, H. Loman, and E. Kriek.** 1985. By-pass of the major aminofluorene-DNA adduct during *in vivo* replication of single- and double-stranded φX174 DNA treated with N-hydroxy-2-aminofluorene. *Carcinogenesis* 6:1501–1506.

17. **Lutgerink, J., J. Retel, J. Westra, M. Welling, H. Loman, and E. Kriek.** 1986. The biological activity of single stranded φX174 DNA, modified by N-hydroxy-2-aminofluorene, is inhibited by guanine imidazole ring-opening of the major, non-lethal aminofluorene-DNA adduct. *Carcinogenesis* 7:1359–1364.

18. **Michaels, M., M. Lee and L. Romano.** 1986. Contrasting effects of *Escherichia coli* single-stranded DNA binding protein on synthesis by T7 DNA polymerase and *Escherichia coli* DNA polymerase I (large fragment). *J. Biol. Chem.* 261:4847–4854.

19. **Miller, S., and E. Eisenstadt.** 1987. Suppressible base substitution mutations induced by angelicin (isopsoralen) in the *Escherichia coli lacI* gene: implications for the mechanism of SOS mutagenesis. *J. Bacteriol.* 169:2724–2729.

20. **Moore, P., S. Rabkin, A. Osborn, C. King, and B. Strauss.** 1982. Effect of acetylated and deacetylated 2-aminofluorene adducts on *in vitro* DNA synthesis. *Proc. Natl. Acad. Sci. USA* 79:7166–7170.

21. **O'Connor, D., and G. Stohrer.** 1985. Site-specifically modified oligodeoxyribonucleotides as templates for *Escherichia coli* DNA polymerase I. *Proc. Natl. Acad. Sci. USA* 82:2325–2329.

22. **Rabkin, S., and B. Strauss.** 1984. A role for DNA polymerase in the specificity of nucleotide incorporation opposite N-acetyl-2-aminofluorene adducts. *J. Mol. Biol.* 178:569–594.

23. **Ryan, F., R. Rudner, T. Nagata, and Y. Kitani.** 1959. Bacterial mutation and the synthesis of macromolecules. *Z. Vererbungsl.* 90:148–158.

24. **Saenger, W.** 1984. *Principles of Nucleic Acid Structure.* Springer-Verlag, New York.

25. **Sagher, D., and B. Strauss.** 1983. Insertion of nucleotides opposite apurinic/apyrimidinic sites in deoxyribonucleic acid during *in vitro* synthesis: uniqueness of adenine nucleotides. *Biochemistry* 22:4518–4526.

26. **Scheuermann, R., and H. Echols.** 1984. A separate editing exonuclease for DNA replication: the ε subunit of *Escherichia coli* DNA polymerase III holoenzyme. *Proc. Natl. Acad. Sci. USA* 81:7747–7751.

27. **Strauss, B.** 1985. Translesion DNA synthesis: polymerase response to altered nucleotides. *Cancer Surveys* 4:493–516.

28. **Tabor, S., and C. Richardson.** 1987. DNA sequence analysis with a modified bacteriophage T7 DNA polymerase. *Proc. Natl. Acad. Sci. USA* 84:4767–4771.

29. **Takeshita, M., C. Chang, F. Johnson, S. Will, and A. Grollman.** 1987. Oligodeoxynucleotides containing synthetic abasic sites. Model substrates for DNA polymerases and apurinic/apyrimidinic endonucleases. *J. Biol. Chem.* 262:10171–10179.

Chapter 33

Escherichia coli mutT Mutator
Increased dGTP Misincorporation Opposite Template Adenines during DNA Replication

Roel M. Schaaper and Ronnie L. Dunn

ESCHERICHIA COLI MUTATOR STRAINS

E. coli mutator strains (i.e., strains that display elevated spontaneous mutation rates) are useful tools in studying the enzymatic pathways used in the control of mutagenesis (6). Most strong mutators are connected to the process of DNA replication. For instance, mutations in the *dnaE* and *dnaQ* genes produce strong mutator phenotypes (19, 20). These genes encode, respectively, the polymerase (α-subunit) and the 3'→5' exonuclease (ε-subunit) of the polymerase III (pol III) holoenzyme complex, the enzyme primarily responsible for the replication of the bacterial chromosome. Related to the replication process in a more general context are mutators *mutH*, *mutL*, and *mutS*, which are defective in the postreplicative process of DNA mismatch correction (10). The mechanism for another strong mutator, *mutT* (5, 6, 22), is as yet unknown. We have undertaken a study of the possible mechanisms by which *mutT* exerts its effects, in the expectation that these studies will provide further, fundamental insights into the control of mutation.

E. COLI mutT

The striking property of *mutT* which sets it apart from all other mutators is its unique specificity; only A·T→C·G base substitutions are produced (5, 24). For these changes it is a strong mutator; a 100- to 10,000-fold increase in spontaneous mutation is commonly observed (5, 6). *mutT* is recessive (7), and all known *mutT* alleles appear identical in both effectiveness and specificity (3, 6). It is likely, therefore, that the phenotype results from the absence of a functional gene product. Two observations have linked *mutT* to DNA replication. First, experiments with bacteriophage lambda, either replicating or nonreplicating, have shown that only replicated phage are mutated by *mutT* (4, 5). Second, certain temperature-sensitive *dnaE* alleles have proven more temperature resistant in the presence of the *mutT1* mutation (5; U. Ray, Fed. Proc. 37:141, 1978). A physical interaction might

---

Roel M. Schaaper and Ronnie L. Dunn • Laboratory of Molecular Genetics, National Institute of Environmental Health Sciences, Research Triangle Park, North Carolina 27709.

therefore exist between the *mutT* and *dnaE* gene products.

While a role for the *mutT* gene product in DNA replication seems reasonable, many questions still need to be answered, for example: (i) is the *mutT* gene product a component of the DNA replication complex, preventing A·T→C·G base substitutions at the polymerization step, or does it operate in a (postreplicative) repair process reversing A·T→C·G base substitutions after they have been made; (ii) does the *mutT* pathway involve the processing of A·G mispairs (at template adenines) or T·C mispairs (at template thymines), either of which would lead to A·T→C·G base substitutions; and (iii) what are the special properties of A·T→C·G base substitutions such that a separate prevention system is required?

The *mutT* gene was recently cloned (1; D.-Y. Chang, V. Desiraju, and A.-L. Liu, *Abstr. Annu. Meet. Am. Soc. Microbiol.* 1987, H1, p. 139), and the availability of the clones will likely play an important role in answering some of these questions. Below we will describe our efforts at defining a mechanism for the mutator effect of *mutT* by using single-stranded phage as a probe.

mutT ACTION ON SINGLE-STRANDED PHAGE

Early experiments demonstrated that *mutT* could exert a mutator effect on genes on the *E. coli* chromosome and F' episomes, as well as bacteriophage lambda, but not on T phages such as T1, T3, T4, and T5 (7). These phages, in contrast to phage lambda, provide their own DNA replication complexes. Interestingly, an absence of the mutator effect was also reported for phage S13 (7), a single-stranded phage related to phage φX174, which relies almost entirely on the host replication machinery. We investigated this matter further by using another single-stranded phage, M13, in the form of its derivative, M13mp2.

M13mp2 carries the *lacZα* gene and is capable of α-complementation, measured as blue-plaque formation in the presence of X-Gal (5-bromo-4-chloro-3-indolyl-β-D-galactopyranoside) and IPTG (isopropyl-β-D-thiogalactopyranoside) on certain bacterial hosts such as strain CSH50 or JM103 (13). A series of M13mp2 derivatives have been isolated which no longer display α-complementation (colorless plaques) due to single-base substitutions in the *lacZα* gene (13). We used the reversion of such *lacZα* mutants (restoration of the blue-plaque phenotype) as a measure of mutation in single-stranded phage.

We tested a total of four different *lacZα* mutants, propagating them on both wild-type and *mutT1* hosts, followed by scoring for *lacZα*$^+$ revertants (blue plaques) in an appropriate α-complementation host. For two, each carrying an ochre (TAA) mutation, a moderate but significant increase in reversion frequency was observed, five- to sevenfold (data not shown). We concluded that *mutT* can indeed be effective on a single-stranded phage such as phage M13.

DNA sequence analysis showed that the specificity of the *mutT* effect for single-stranded phage is identical to that for *E. coli* or phage lambda genes, A·T→C·G substitutions, since more than 80% of all mutants contained a change of the TAA codon to a GAA, TCA, or TAC codon. The specific increase for A·T→C·G substitutions amounted to more than 65-fold (data not shown).

mutT DURING DNA REPLICATION IN VITRO

The observation that single-stranded phage are a substrate for the *mutT* mutator opened a way for using the phage for in vitro studies of *mutT*. We designed a system for measuring the accuracy of a single round of M13 DNA synthesis by cell-free extracts of *E. coli*, as presented in Fig. 1. The approach

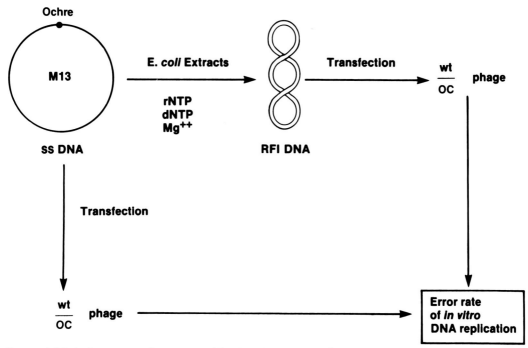

FIGURE 1. Method to measure the accuracy of DNA replication in cell-free extracts of *E. coli* by conversion of M13 single-stranded DNA to double-stranded form (RF DNA). When an M13 nonsense mutant is used, an increase in reversion frequency to wild type or pseudo wild type following DNA replication may be taken as a measure of the accuracy of a single round of in vitro DNA synthesis. In the present case a nonsense mutation in the *lacZα* portion of M13mp2 was used, monitoring the reversion frequency through the change from colorless- to blue-plaque phenotype in an α-complementation assay (13).

is based on early observations (23) that extracts of *E. coli* efficiently convert single-stranded M13 DNA into its double-stranded replicative-form (RF) DNA. The components responsible for this replication are well known (9, 12) and include RNA polymerase for priming and pol III holoenzyme and single-stranded binding protein for subsequent synthesis. The product is initially relaxed double-stranded RFII DNA, which, by DNA pol I and ligase present in the extract, may be converted into the closed RFI form. In our hands, efficient conversion of single-stranded DNA to a mixture of RFI and RFII was indeed observed, as shown in Fig. 2. Wild-type and *mutT* extracts proved equally proficient in this respect.

The reversion frequencies obtained in two separate experiments after replication of M13mp2 ochre mutant T90 by either wild-type or *mutT* extracts are presented in Table 1. DNA replication is very accurate; error rates for the chosen condition (1 mM for each of the four dNTPs) are equal to, or barely exceed, the background frequency of about 10^{-6}. Significant increases above the background can be obtained by manipulation of the dNTP concentrations, a generally useful approach in fidelity studies (14, 15). Lowering dATP 100-fold relative to the other dNTPs (an imbalance expected to induce mutations at the first position of the TAA codon) readily produced revertants at a frequency of 10×10^{-6} to 20×10^{-6}. Under these conditions a similar response was obtained in wild-type and *mutT* extracts. However, lowering dTTP 100-fold (expected to induce mutations at the two ade-

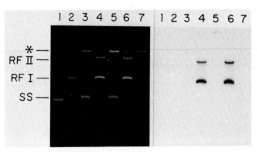

FIGURE 2. DNA synthesis in cell-free extracts of *E. coli* (M13 single stranded→RF conversion). Shown is a 0.8% agarose gel with the reaction products of DNA synthesis reactions by either wild-type- or *mutT1*-derived extracts. Left, ethidium-bromide-stained gel; right, autoradiogram of the same. Reactions were performed as in reference 23, with some modifications. dNTPs were provided at 1,000 μM each. Lanes: 1 and 2, marker DNAs (M13 single-stranded and RF DNA); 3 and 4, wild-type extracts after 0 and 30 min of synthesis at 30°C, respectively; 5 and 6, same as lanes 3 and 4, but using extracts from *mutT* cells; 7, wild-type extract but no M13 DNA added and without incubation, showing the origin of the starred (*) band. This band disappears upon incubation.

nine positions of the TAA codon) produced a response unique to the *mutT* extract, 37×10^{-6} to 62×10^{-6} versus essentially background for the wild-type extract.

DNA sequence analysis of the in vitro revertants (Table 2) showed that the in vitro mutator effect produced exclusively TCA and TAC codons, examples of A·T→C·G transversion mutation. These data clearly demonstrate that the *mutT* mutator effect can be reproduced in vitro.

NATURE OF THE *mutT*-INDUCED DNA REPLICATION ERRORS

An important conclusion can be drawn from the present results. The frequency data of Table 1 and the specificity data of Table 2 show that the *mutT* effect is specific for template adenine positions. In *mutT* extracts, TAA→(TCA or TAC) changes are increased at least 50-fold over the wild type, whereas TAA→GAA changes do not seem to be affected. From this we deduce that *mutT* cells are specifically defective in the processing of (template A)·G mismatches, as previously suggested (6, 21), rather than of the mutationally equivalent (template T)·C mismatches.

mutT AND A POSSIBLE ROLE IN THE DNA REPLICATION COMPLEX

Our observations of the *mutT* mutator effect during DNA replication in vitro confirm and strengthen previous suggestions of the *mutT* gene acting in conjunction with DNA replication. Since our experiments use extracts of *E. coli* rather than purified components, the precise step of the *mutT* involvement remains to be determined. However, we see two general possibilities. First, *mutT* may be a component of the DNA replication complex. Support for this comes from the observation that the *mutT1* mutation thermostabilizes certain *dnaE*(Ts) alleles

TABLE 1
Reversion Frequencies of M13mp2 T90[a] after Replication in *E. coli* Extracts

Strain	dNTP[b]	Reversion frequency (10^6) for expt:	
		1	2
KA796 (wild type)	1,000 μM	1.6	0.5
	GCT/A = 100	11	20
	GCA/T = 100	2.6	ND[c]
NR9082 (*mutT*)	1,000 μM	4.3	3.5
	GCT/A = 100	10	20
	GCA/T = 100	37	62

[a] M13mp2 T90 contains a G→T base substitution at position 90 in *lacZ*α (13). This mp2 derivative is incapable of α-complementation (colorless plaque); however, by single-base substitution, the blue-plaque phenotype can be restored. The reversion frequency indicates the frequency of such blue-plaque revertants among total phage.
[b] Deoxynucleoside triphosphate concentrations were 1,000 μM, except for the indicated pool bias conditions in which either dATP or dTTP was lowered to 10 μM to provide a 100-fold imbalance with respect to the other three dNTPs.
[c] ND, Not determined.

TABLE 2
DNA Sequence Changes in M13mp2 T90 Revertants Produced during In Vitro DNA Synthesis

Strain	dNTP[a]	No. of revertants with indicated revertant codon[b]:						
		CAA	GAA	AAA	TCA	TTA	TAC	TAT
KA796 (wild type)	GCT/A=100	9	3	3	0	0	0	0
NR9082 *mutT*	GCT/A=100	3	0	0	1	0	1	0
	GCA/T=100	0	0	0	31	0	10	0

[a] dNTP concentrations were as described in footnote *b* of Table 1. Sequence changes for GCA/T = 100 were not determined for the wild-type strain.
[b] All seven indicated revertant codons can be detected as blue plaques. Only A·T→G·C changes at positions 2 and 3 of the ochre codon cannot be scored since they produce TGA or TAG nonsense codons.

(5; Ray, *Fed. Proc.*, 1978), suggesting the existence of a physical interaction of the *mutT* and *dnaE* gene products. Such an interaction or association would provide a plausible basis to explain the *mutT* effect; through this association, *mutT* might directly influence the base selection process of the DNA polymerase, preventing the otherwise frequent occurrence of A·G replication errors. The recent cloning of the *mutT* gene (1; Chang et al., *Abstr. Annu. Meet. Am. Soc. Microbiol.* 1987) has allowed the molecular weight determination of the gene product. On this basis it was suggested that the *mutT* gene product might be identical to the θ-subunit of the holoenzyme complex (1). This subunit composes, with α- and ε-subunits, the polymerase core (16). Of the seven subunits characterized unequivocally as part of the holoenzyme complex, only θ has not been identified as the product of a known *E. coli* gene (16). The tight association of α, ε, and θ within the polymerase core (12, 16) would indeed provide a solid basis to explain direct involvement of *mutT* with pol III-induced replication errors. However, other, less defined subunits often associated with pol III holoenzyme preparations (16) must be considered as well.

A second possibility, which seems less likely but cannot be excluded, is that *mutT* might specify a postreplicative repair system that removes A·G mismatches after they have been made.

SPECIAL NATURE OF A·G MISMATCHES

The A·G mismatch is only 1 of 12 possible mismatches that can be produced as a DNA replication error. The question might be raised, why does *E. coli* possess a separate prevention pathway for this one particular mismatch? One reasonable assumption might be that A·G mispairs are substantially different from the other mispairs. Therefore, their prevention through other pathways, such as DNA polymerase fidelity (insertion and proofreading step) or postreplicative mismatch repair, might have been less than required. In this light, the development of a separate prevention pathway for A·G mismatches might not only have directly benefited the prevention of A·T→C·G substitutions, but might, in addition, have enhanced the efficiency of the alternative systems, former restraints on these systems being relaxed.

The A·G mismatch has been the subject of several recent investigations by physical techniques. Two different nuclear magnetic resonance studies concluded that the A·G base pairs in question were in a wobble configuration with the bases in their normal (*anti*) conformation and normal tautomeric forms, with two hydrogen bonds maintained between them (11, 17). In contrast, X-ray diffraction studies indicated the presence of a $G_{anti} \cdot A_{syn}$ pair. In this opposition also, two hydrogen bonds were maintained (2). An example of an A·G opposition producing an extrahelical structure has also been reported (8). It thus appears that multiple conformations are possible, with perhaps the neighboring sequences playing a determining role.

It is an important question how relevant these structural clues are for the conformational arrangement of the two bases during the formation of the mismatch in the active site of the DNA polymerase. The distinction between the (template A)·G and (template G)·A oppositions by *mutT* (A·T→C·G, but not the reciprocal G·C→T·A, are induced) already indicates that the simple consideration of the mismatch itself is not sufficient. Topal and Fresco (21) used model-building studies to propose that the major pathway for transversion mutagenesis (i.e., a purine· pyrimidine to pyrimidine·purine change) involved the opposition of two purines (as opposed to two pyrimidines) with one purine in a *syn*- and the other in an *anti*-orientation. The *syn*-oriented base was assumed to be the incoming base since stacking forces in the template strand would prevent the *anti→syn* rotation of the template residue. If such modeling is correct, one possible mechanism for *mutT* action may be to prevent the insertion of a G_{syn} opposite the template A. It is likely that a *syn*-oriented base would make a mismatch sufficiently different from other mismatches that a separate repair or prevention pathway would be required. One problem with such a model is to explain why this would work for $A \cdot G_{syn}$ but not other purine·purine oppositions that might involve a *syn* isomer: $G \cdot A_{syn}$, $A \cdot A_{syn}$, and $G \cdot G_{syn}$. Perhaps these other oppositions occur less frequently and/or are corrected efficiently by other, possibly unknown, correction pathways. It is worth noting that, of the four nucleotides, guanine possesses an exceptional tendency to be in the *syn* conformation (18).

LITERATURE CITED

1. **Akiyama, M., T. Horiuchi, and M. Sekiguchi.** 1987. Molecular cloning and nucleotide sequence of the *mutT* mutator of *Escherichia coli* that causes A:T to C:G transversion. *Mol. Gen. Genet.* **206**:9–16.
2. **Brown, T., W. N. Hunter, G. Kneale, and O. Kennard.** 1986. Molecular structure of the G·A base pair in DNA and its implications for the mechanism of transversion mutations. *Proc. Natl. Acad. Sci. USA* **83**:2402–2406.
3. **Conrad, S. E., K. T. Dussik, and E. C. Siegel.** 1976. Bacteriophage Mu-1-induced mutation to *mutT* in *Escherichia coli*. *J. Bacteriol.* **125**:1018–1025.
4. **Cox, E. C.** 1970. Mutator gene action and the replication of bacteriophage λ DNA. *J. Mol. Biol.* **50**:129–135.
5. **Cox, E. C.** 1973. Mutator gene studies in *Escherichia coli*: the *mutT* gene. *Genetics* **73**(Suppl.):67–80.
6. **Cox, E. C.** 1976. Bacterial mutator genes and the control of spontaneous mutation. *Annu. Rev. Genet.* **10**:135–156.
7. **Cox, E. C., and C. Yanofsky.** 1969. Mutator gene studies in *Escherichia coli*. *J. Bacteriol.* **100**:390–397.
8. **Fazakerley, G. V., E. Quignard, A. Woisard, W. Guschlbauer, G. A. van der Marel, J. H. van Boom, M. Jones, and M. Radman.** 1986. Structures of mismatched base pairs in DNA and their recognition by the *Escherichia coli* mismatch repair system. *EMBO J.* **5**:3697–3703.
9. **Geider, K., and A. Kornberg.** 1974. Conversion of the M13 viral single strand to the double-stranded replicative forms by purified proteins. *J. Biol. Chem.* **249**:3999–4005.
10. **Glickman, B. W., and M. Radman.** 1980. *Escherichia coli* mutator mutants deficient in methylation-instructed DNA mismatch correction. *Proc. Natl. Acad. Sci. USA* **77**:1063–1067.
11. **Kan, L., S. Chandrasegaran, S. M. Pulford, and P. S. Miller.** 1983. Detection of a guanine·adenine base pair in a decadeoxyribonucleotide by proton magnetic resonance spectroscopy. *Proc. Natl. Acad. Sci. USA* **80**:4263–4265.
12. **Kornberg, A.** 1980. *DNA Replication*. W. H. Freeman & Co., San Francisco.
13. **Kunkel, T. A.** 1984. Mutational specificity of depurination. *Proc. Natl. Acad. Sci. USA* **81**:1494–1498.

14. **Kunkel, T. A., and L. A. Loeb.** 1980. On the fidelity of DNA replication: the accuracy of *Escherichia coli* DNA polymerase I in copying natural DNA *in vitro*. *J. Biol. Chem.* **255**:9961–9966.
15. **Loeb, L. A., and T. A. Kunkel.** 1982. Fidelity of DNA synthesis. *Annu. Rev. Biochem.* **52**:429–457.
16. **McHenry, C. S.** 1985. DNA polymerase III holoenzyme of *Escherichia coli*: components and function of a true replicative complex. *Mol. Cell. Biochem.* **66**:71–85.
17. **Patel, D. J., S. A. Kozlowski, S. Ikuta, and K. Itakura.** 1984. Deoxyguanosine-deoxyadenosine pairing in the d(C-G-A-G-A-A-T-T-C-G-C-G) duplex: conformation and dynamics at and adjacent to the dG·dA mismatch site. *Biochemistry* **23**:3207–3217.
18. **Saenger, W.** 1984. *Principles of Nucleic Acid Structure*, p. 76–78. Springer-Verlag, New York.
19. **Scheuermann, R., S. Tam, P. M. J. Burgers, C. Lu, and H. Echols.** 1983. Identification of the ε-subunit of *Escherichia coli* DNA polymerase III holoenzyme as the *dnaQ* gene product: a fidelity subunit for DNA replication. *Proc. Natl. Acad. Sci. USA* **80**:7085–7089.
20. **Sevastopoulos, C. G., and D. A. Glaser.** 1977. Mutator action by *Escherichia coli* strains carrying *dnaE* mutations. *Proc. Natl. Acad. Sci. USA* **74**:3947–3950.
21. **Topal, M. D., and J. R. Fresco.** 1976. Complementary base pairing and the origin of substitution mutations. *Nature* (London) **263**:285–289.
22. **Treffers, H. P., C. Spinelli, and N. O. Belser.** 1954. A factor (or mutator gene) influencing mutation rates in *Escherichia coli*. *Proc. Natl. Acad. Sci. USA* **40**:1064–1071.
23. **Wickner, W., D. Brutlag, R. Schekman, and A. Kornberg.** 1972. RNA synthesis initiates in vitro conversion of M13 DNA to its replicative form. *J. Biol. Chem.* **69**:965–969.
24. **Yanofsky, C., E. C. Cox, and V. Horn.** 1966. The unusual mutagenic specificity of an *E. coli* mutator gene. *Proc. Natl. Acad. Sci. USA* **55**:274–281.

Chapter 34

Genetic Characterization of the SOS Mutator Effect in *Escherichia coli*

G. Maenhaut-Michel

Induction of the SOS functions in *Escherichia coli* promotes a mutator effect that generates mutations in undamaged DNA. The SOS hypothesis (23) postulates that the SOS mutator effect is a direct result of the mechanism of targeted mutagenesis (mutagenesis of damaged DNA; 4, 30). SOS mutagenesis of DNA damaged by UV light or radiomimetic chemicals requires *umuCD* function and activated RecA protein (RecA*) for its expression (for review, see reference 27). RecA protein has at least two different roles in SOS mutagenesis: (i) a regulatory role by stimulating the cleavage of the LexA repressor, allowing the expression of a set of functions involved in the repair and mutagenesis of damaged DNA, and (ii) another role that is necessary for the expression of mutagenesis, even in bacteria that constitutively express the SOS functions, i.e., that are deficient in the LexA repressor [*lexA*(Def)]. This second role may be the cleavage of the UmuD protein (21). The SOS mutator effect was shown either on cellular DNA in *recA441* and *recA730* mutants, in which RecA protein is activated by the *recA* mutation in the absence of DNA-damaging treatments (7, 29), or on bacteriophage DNA when undamaged phages are replicated in UV-irradiated host bacteria (UV-induced untargeted mutagenesis; 14).

GENETIC REQUIREMENTS FOR *recA441*- OR *recA730*-INDUCED MUTAGENESIS AND FOR UV-INDUCED UNTARGETED MUTAGENESIS OF PHAGES

Although in both *recA441*- or *recA730*-induced mutagenesis and UV-induced untargeted mutagenesis the expression of the mutator effect is under the control of the *lexA* and *recA* genes, many differences have been found in the other genetic requirements and in the spectra of generated mutations. Mainly, *recA441*- and *recA730*-induced mutagenesis is *umuCD* dependent, whereas UV-induced untargeted mutagenesis of phage λ is *umuC* independent. Moreover, in phage λ, untargeted mutagenesis does not absolutely require RecA* protein as does targeted mutagenesis (2, 6, 12). Indeed, we have shown that mutagenesis of undamaged phage λ could occur in the absence of RecA

G. Maenhaut-Michel • Laboratoire de Biophysique et Radiobiologie, Département de Biologie Moléculaire, Université libre de Bruxelles, B-1640 Rhode St. Genèse, Belgium.

protein in UV-irradiated *lexA*(Def) bacteria. In addition, in phage λ, the expression of untargeted mutagenesis requires a functional excision repair and the *dinB* gene product (2, 9). This would suggest that the SOS mutator effect responsible for phage untargeted mutagenesis is different from the one involved in *recA441*- or *recA730*-mediated mutagenesis. However, we have recently shown that untargeted mutagenesis of phages M13 and φX174, which both are single-stranded DNA phages, does require the *umuCD* function and the RecA* protein even in a *lexA*(Def) mutant. Moreover, neither functional excision repair nor the *dinB* gene product is required (G. Maenhaut-Michel and P. Caillet-Fauquet, submitted for publication). The differences observed in the genetic control of untargeted mutagenesis between single- and double-stranded DNA phages could reflect the importance of the nature of the DNA substrate for the function of the SOS mutator effect. This also suggests that the mutator effect functioning on cellular DNA is related to the one functioning on single-stranded DNA phages rather than to the one acting on the double-stranded DNA phage λ. These results encouraged us to consider as a working hypothesis that UV-induced phage untargeted mutagenesis is the result of the same mutator mechanism as the *recA441*- or *recA730*-mediated mutator effect on cellular DNA and that some proteins involved in this mechanism are dispensable depending on the nature of the DNA substrate.

NATURE OF THE MOLECULAR EVENT GENERATING THE SOS MUTATOR EFFECT

Current knowledge of the mechanisms controlling base substitution and frameshift mutagenesis suggests that a mutator effect could result from (i) an increased error rate of the DNA replication machinery copying intact templates, (ii) a decreased efficiency of the postreplicative mismatch correction, or (iii) chemical modifications of the DNA resulting in miscoding DNA lesions. A hypothesis has been proposed for the molecular basis of the UV-induced SOS mutator effect: blocking of the replication fork by DNA lesions could generate a new, transitory replication complex of lower fidelity that allows error-prone *trans*-lesion synthesis (4, 23). This mutagenic bypass of blocking lesions was proposed to be achieved in two steps: a first, *umuC*-independent step generating misincorporation opposite damaged bases, followed by a second, *umuC*-dependent step allowing DNA chain elongation past the lesions (1). Modification of the proofreading and/or the base selection functions of preexisting DNA polymerases or the generation of a new DNA polymerase with less accuracy could account for this process (3, 17, 25, 26, 33). The existence of UV-induced untargeted mutagenesis in phage suggests that, once induced, the error-prone replicative complex could operate in *trans* on a substrate that has not been damaged (5). The role of RecA* and UmuCD proteins in such an error-prone replication has not yet been clearly demonstrated. The requirement of *umuCD* gene products for phage untargeted mutagenesis in single-stranded DNA phages and for *recA441*- or *recA730*-mediated mutagenesis suggests that the role of *umuCD* proteins is not exclusively related to DNA chain elongation past the lesions, except if one considers the possibility that both *recA441*- or *recA730*- and UV-induced "untargeted" mutations are in fact targeted on cryptic DNA lesions. Sequencing of forward mutations has shown that a large proportion of the untargeted mutagenic events both in phage λ (31) and in phage M13 (15) are frameshifts. On the contrary, *recA441*-induced mutations are mainly base substitutions with a very high proportion of GC→TA and AT→TA transversions (19). Both DNA mutation spectra are very different from those of spontaneous mutations.

The apparent modification of the spectrum of spontaneous mutations by the SOS-induced mutator effect could result just as well from a modification of the specificity of base insertion errors by the replicative complex or by the polymerase itself (e.g., the presence of *mutT* or *dnaQ49* alleles alters the specificity of base substitution errors [8, 22]) as from the presence of specific miscoding cryptic DNA lesions (e.g., apurinic sites [16]).

THE *recA441*- OR *recA730*-MEDIATED MUTATOR EFFECT

The Untargeted-Mutation Hypothesis

The methyl-directed mismatch repair which depends in *E. coli* on *mutH*, *mutL*, *mutS*, and *mutU* genes does not recognize (and/or does not correct) all possible mispaired and unpaired bases with the same efficiency. In particular, the A·G, C·T, and C·C mismatches appear to be less repaired than any others, indicating that mismatched bases involved in transversion events are less repaired than those involved in transition events (10). Therefore, the specificity of mutations produced by the *recA441* mutant could reflect replication errors not recognized by mismatch repair.

The Targeted-Mutation Hypothesis

The strong specificity of *recA441*-induced base substitution mutations (19) and the requirement of the *umuCD* gene products for this mutator effect have suggested that *recA441*-induced mutations are actually targeted on cryptic lesions (e.g., apurinic sites) (19, 28). Such cryptic lesions were proposed to be caused by cellular metabolism; they would not induce error-prone replication by themselves but would be converted into mutations by the mutagenic synthesis involving UmuCD and activated RecA proteins. The specificity of transversion mutations induced in *recA441* mutants is in agreement with the preferential insertion of A opposite apurinic sites (16).

recA730-INDUCED MUTATIONS ELIMINATED BY THE METHYL-DIRECTED MISMATCH REPAIR

Thus, two hypotheses can be proposed for the *recA441*- or *recA730* mutator effect: (i) it causes a decrease of DNA replication fidelity that increases the rate of untargeted replication errors and could also modify the specificity of base insertion or (ii) it causes a modification of the replication complex that allows misinsertion of bases targeted opposite noncoding cryptic lesions. The expected results for both hypotheses are shown schematically in Fig. 1. Both hypotheses imply that *recA441*- or *recA730*-induced mutations are produced during DNA replication. However, up to now, there has been no direct evidence to rule out a nonreplicative effect. In the first hypothesis premutational mismatched (or unmatched) bases could be recognized and corrected by postreplicative, methyl-directed mismatch repair. Current knowledge of methyl-directed mismatch repair indicates that it operates at or near the replication fork during the short time during which the newly synthesized daughter strand contains unmethylated GATC sequences (for a review, see references 20 and 24). This repair mechanism functions to remove mispaired (or unpaired) bases from the newly synthesized undermethylated DNA strand by excision and resynthesis of tracts with the parental, intact strand as a model for daughter strand resynthesis. Methylation of adenine in GATC sequences depends on the *dam* gene (18). To be recognized and repaired, mismatched bases would have to be in an intrahelical conformation (13) and have efficient (mis)-pairing properties (32).

To validate one of the above hypotheses, we analyzed spontaneous-mutation fre-

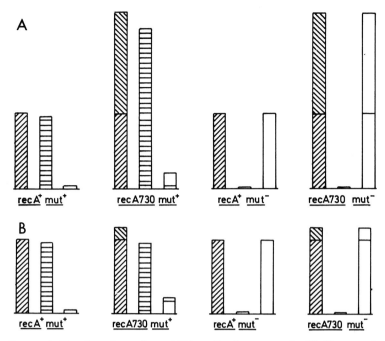

FIGURE 1. Two hypotheses for *recA730*-mediated mutagenesis. (A) Untargeted-mutation hypothesis: synergistic effects of *recA730* and *mut* mutations. (B) Targeted-mutation hypothesis: additive effects of *recA730* and *mut* mutations. Symbols: ▨, spontaneous mutations; ◧, *recA730*-induced mutations; ▤, mutations eliminated by mismatch repair (*mut*⁺); ▫, observed mutations, i.e., the difference between the induced mutations and mutations eliminated by mismatch repair.

quencies in *recA730* mutants deficient in postreplicative mismatch repair (*mut*). We found that the frequencies of mutations towards rifampin resistance in *recA730 mutL* or *recA730 mutS* double mutants are much higher than the sum of mutation frequencies observed in single mutants *recA730* and *mutL* or *mutS*, respectively (P. Caillet-Fauquet and G. Maenhaut-Michel, *Mol. Gen. Genet.*, in press). Mutations towards rifampin resistance result from base substitutions in the gene coding for the β-subunit of RNA polymerase (34). Therefore, our results favor the first hypothesis (Fig. 1A), i.e., an increase of the error rate at the replication fork, since results indicate that the majority of *recA730*-induced premutational events are corrected by the postreplicative *mutH*, *mutL*, or *mutS* mismatch repair, with only a weak proportion of these events (5 to 10%) not being corrected. These noncorrected events could have arisen either via mismatched bases poorly recognized by mismatch repair enzymes (e.g., GA) or, as proposed in the second hypothesis, opposite noncoding cryptic lesions. On the other hand it cannot be excluded that this fraction of mutations is not produced at the replication fork. Both correctable and noncorrectable *recA730*-induced mutations are suppressed by a mutation in the *umuC* gene, suggesting that UmuCD and RecA* proteins cooperate to produce replication errors in the *recA730* mutant (P. Caillet-Fauquet and G. Maenhaut-Michel, in press). We found also that *dam recA730* double mutants are unstable segregating clones that have lost the *dam* or the *recA* mutations or that have acquired a new mutation, probably in one of the genes involved in mismatch repair. We suggest that the genetic instability of the *dam recA730* mutants is caused by the high level of replication errors resulting from the *recA730* mutation, which generates DNA double-strand breaks by coincident mismatch repair on the two unmethylated DNA strands. Mismatch-stimulated killing has been shown to occur in unmethylated phage λ heteroduplex DNA containing at least one repairable mismatch (11).

We conclude (i) that the major part of *recA730*-mediated mutagenesis occurs at or near the replication fork in the narrow window where newly synthesized DNA is undermethylated and mismatch repair is operating and (ii) that *recA730*-induced mutations are very likely untargeted replication errors introduced in the newly synthesized DNA.

Although our results strongly suggest that the *recA730* mutator effect induces mainly untargeted mutations, we cannot exclude the possibility of base alterations in the nascent daughter strand that create correctable mismatched base pairs near the replication fork.

THE UV-INDUCED MUTATOR EFFECT

We have shown that untargeted mutagenesis of phage λ requires both the expression of one or more SOS genes normally repressed by LexA (*uvrAB*, *dinB*, etc.) and some effector(s) provided by the presence of UV lesions in the host DNA (2). The *recA730* mutation hardly increases spontaneous mutagenesis of phage λ, and this small increase is suppressed by a *umuC* mutation. UV irradiation of *recA730* host bacteria increases phage λ untargeted mutagenesis to the level observed in UV-irradiated *recA$^+$* strains. This UV-induced mutator effect in *recA730* mutants is not suppressed by a *umuC* mutation. Therefore, the UV-induced mutator effect on undamaged phage λ is apparently different from the *recA730* mutator effect (P. Caillet-Fauquet and G. Maen-

haut-Michel, in press). It is suggested that the role of UV-induced DNA damage in the expression of phage λ untargeted mutagenesis is not only to activate the RecA protein; alternatively, it could be that UV irradiation modifies the RecA protein in a different way than do the *recA441* and *recA730* mutations, at least as far as the role of RecA in mutagenesis is concerned.

UV-induced, untargeted mutagenesis could be observed also on bacterial DNA in UV-irradiated *recA730 mutL* or *recA730 mutS* mutants. Indeed, irradiation of these mutant strains increases the amount of mutations susceptible to elimination by mismatch repair (P. Caillet-Fauquet and G. Maenhaut-Michel, in press). Therefore, either UV-irradiation and the *recA730* mutation seem to induce different SOS mutator replication activities, both generating untargeted mutations, or UV irradiation further activates the *recA730* mutator effect, e.g., by creating newly available RecA* binding sites (A. Bailone, A. Backman, S. Sommer, J. Célérier, M. Bagdasarian, and R. Devoret, *Mol. Gen. Genet.*, in press). The UV-stimulated, error-prone replication could be achieved either through further UV-dependent modification of RecA protein or through a RecA*-independent modification of the replication fidelity triggered by UV damage in the DNA. RecA*-independent but UV-dependent, untargeted mutagenesis was observed in *lexA*(Def) mutants with unirradiated phage λ (2). The necessity of proficient excision repair for phage λ untargeted mutagenesis suggests that some processing of UV-induced lesions generates an effector for the SOS mutator activity. We have suggested that the role of the *dinB* gene product may also be at that level (2). The different involvement of RecA* and UmuC and UmuD proteins in phage untargeted mutagenesis, depending on whether the DNA is single or double stranded, might be due to the higher affinity of both proteins for single-stranded than for undamaged double-stranded DNA. This is indeed the case for

RecA and as yet unknown for UmuC and UmuD proteins (17).

CONCLUSION

Our results favor the hypothesis of an SOS-induced, error-prone replication triggered by the activation of the RecA protein and of the *umuCD* function that increases the load of untargeted mutations. These untargeted mutations are eliminated by the postreplicative mismatch repair with a high efficiency in the cellular DNA and with a low efficiency in phage DNA. Moreover, our results indicate that the *recA730* mutation does not fully activate the SOS error-prone function(s) and that an increased involvement of the RecA730 protein and/or some other SOS protein(s) is stimulated by UV lesions in the DNA.

ACKNOWLEDGMENTS. I thank my co-worker, P. Caillet-Fauquet, for her invaluable collaboration in the work reported here. I thank M. Errera, M. Radman, and A. Brandenburger for critically reviewing this manuscript.

The work reported here was supported by a research contract from the Commission of European Communities (ULB-CEE BI6-0155-B).

LITERATURE CITED

1. Bridges, B. A., and R. Woodgate. 1985. Mutagenic repair in *Escherichia coli*: products of the *recA* gene and of the *umuD* and *umuC* genes act at different steps in UV-induced mutagenesis. *Proc. Natl. Acad. Sci. USA* **82**:4193–4197.
2. Brotcorne-Lannoye, A., and G. Maenhaut-Michel. 1986. Role of RecA protein in untargeted UV mutagenesis of bacteriophage λ: evidence for requirement for the *dinB* gene. *Proc. Natl. Acad. Sci. USA* **83**:3904–3908.
3. Brotcorne-Lannoye, A., G. Maenhaut-Michel, and M. Radman. 1985. Involvement of DNA polymerase III in UV-induced mutagenesis of bacteriophage lambda. *Mol. Gen. Genet.* **199**:64–69.
4. Caillet-Fauquet, P., M. Defais, and M. Radman. 1977. Molecular mechanisms of induced mutagenesis. Replication *in vivo* of bacteriophage φX174 single-stranded, ultraviolet light-irradiated DNA in intact and irradiated host cells. *J. Mol. Biol.* **117**:95–112.
5. Caillet-Fauquet, P., G. Maenhaut-Michel, and M. Radman. 1984. SOS mutator effect in *E. coli* mutants deficient in mismatch correction. *EMBO J.* **3**:707–712.
6. Calsou, P., and M. Defais. 1985. Weigle reactivation and mutagenesis of bacteriophage λ in *lexA* (Def) mutants of *E. coli* K12. *Mol. Gen. Genet.* **201**:329–333.
7. Castellazzi, M., J. George, and G. Buttin. 1972. Prophage induction and cell division in *E. coli*. I. Further characterization of the thermosensitive mutation *tif-1* whose expression mimics the effect of UV irradiation. *Mol. Gen. Genet.* **119**:139–152.
8. Cox, E. C. 1973. Mutator gene studies in *E. coli*: the *mutT* gene. *Genetics* **73**(Suppl):67–80.
9. Devoret, R. 1965. Influence du génotype de la bactérie hôte sur la mutation du phage λ produite par le rayonnement ultraviolet. *C.R. Acad. Sci. Ser. D* **260**:1510–1513.
10. Dohet, C., R. Wagner, and M. Radman. 1985. Repair of defined single base-pair mismatches in *E. coli*. *Proc. Natl. Acad. Sci. USA* **82**:503–505.
11. Doutriaux, M. P., R. Wagner, and M. Radman. 1986. Mismatch-stimulated killing. *Proc. Natl. Acad. Sci. USA* **83**:2576–2578.
12. Ennis, D. G., B. Fisher, S. Edniston, and D. W. Mount. 1985. Dual role for *Escherichia coli* RecA protein in SOS mutagenesis. *Proc. Natl. Acad. Sci. USA* **82**:3325–3329.
13. Fazakerley, G. V., E. Quignard, A. Woisard, W. Guschlbauer, G. A. van der Marel, J. H. van Boom, M. Jones, and M. Radman. 1986. Structures of mismatched base pairs in DNA and their recognition by the *E. coli* mismatch repair system. *EMBO J.* **5**:3697–3703.
14. Ichikawa-Ryo, H., and S. Kondo. 1975. Indirect mutagenesis in phage λ by ultraviolet preirradiation of host bacteria. *J. Mol. Biol.* **97**:77–92.
15. LeClerc, J. E., N. L. Istock, B. R. Saran, and R. Allen, Jr. 1984. Sequence analysis of ultraviolet-induced mutations in M13*lacZ* hybrid phage DNA. *J. Mol. Biol.* **180**:217–237.
16. Loeb, L. A., and B. D. Preston. 1986. Mutagenesis by apurinic/apyrimidinic sites. *Annu. Rev. Genet.* **20**:201–203.
17. Lu, C., R. H. Scheuermann, and H. Echols. 1986. Capacity of RecA protein to bind preferentially to UV lesions and inhibit the editing subunit (ε) of DNA polymerase III: a possible mechanism for SOS-induced targeted mutagenesis. *Proc. Natl. Acad. Sci. USA* **83**:619–623.

18. Marinus, M. G. 1973. Location of DNA methylation genes on the *E. coli* K12 genetic map. *Mol. Gen. Genet.* **127**:47–55.
19. Miller, J. H., and K. B. Low. 1984. Specificity of mutagenesis resulting from the induction of the SOS system in the absence of mutagenic treatment. *Cell* **37**:675–682.
20. Modrich, P. 1987. DNA mismatch correction. *Annu. Rev. Biochem.* **56**:435–466.
21. Perry, K. L., S. J. Elledge, B. B. Mitchel, L. Marsh, and G. C. Walker. 1985. *umu*DC and *muc*AB operons whose products are required for UV light- and chemical-induced mutagenesis: UmuD, MucA, and LexA proteins share homology. *Proc. Natl. Acad. Sci. USA* **82**:4331–4335.
22. Piechocki, R., D. Kupper, A. Quinones, and R. Langhammer. 1986. Mutational specificity of a proof-reading defective *Escherichia coli dnaQ49* mutator. *Mol. Gen. Genet.* **202**:162.
23. Radman, M. 1974. Phenomenology of an inducible mutagenic DNA repair pathway in *Escherichia coli*: SOS repair hypothesis, p. 128–142. *In* L. Prakash, F. Sherman, M. Miller, C. Lawrence, and H. W. Taber (ed.), *Molecular and Environmental Aspects of Mutagenesis*. Charles C Thomas, Publisher, Springfield, Ill.
24. Radman, M., and R. Wagner. 1986. Mismatch repair in *E. coli. Annu. Rev. Genet.* **20**:523–538.
25. Ruiz-Rubio, M., and B. A. Bridges. 1987. Mutagenic DNA repair in *Escherichia coli*. XIV. Influence of two DNA polymerase III mutator alleles on spontaneous and UV mutagenesis. *Mol. Gen. Genet.* **208**:542–548.
26. Villani, G., S. Boiteux, and M. Radman. 1978. Mechanism of ultraviolet-induced mutagenesis extent and fidelity of *in vitro* DNA synthesis on irradiated templates. *Proc. Natl. Acad. Sci. USA* **75**:3037–3041.
27. Walker, G. C. 1984. Mutagenesis and inducible responses to deoxyribonucleic acid damage in *Escherichia coli*. *Microbiol. Rev.* **48**:60–93.
28. Witkin, E. M. 1976. Ultraviolet mutagenesis and inducible DNA repair in *Escherichia coli*. *Bacteriol. Rev.* **40**:869–907.
29. Witkin, E. M., J. O. McCall, M. R. Volkert, and I. E. Wermundsen. 1982. Constitutive expression of SOS functions and modulation of mutagenesis resulting from resolution of genetic instability at or near the *recA* locus of *E. coli*. *Mol. Gen. Genet.* **185**:43–50.
30. Witkin, E. M., and I. E. Wermundsen. 1978. Targeted and untargeted mutagenesis by various inducers of SOS functions in *E. coli. Cold Spring Harbor Symp. Quant. Biol.* **43**:881–886.
31. Wood, R. P., and F. Hutchinson. 1984. Nontargeted mutagenesis of unirradiated λ phage in *E. coli* host cells irradiated with UV light. *J. Mol. Biol.* **173**:293–305.
32. Wood, S. G., A. Ubasawa, D. Martin, and J. Jiricny. 1986. Guanine and adenine analogues as tools in the investigation of the mechanisms of mismatch repair in *E. coli*. *Nucleic Acids Res.* **14**:6591–6601.
33. Woodgate, R., B. A. Bridges, G. Herrera, and M. Blanco. 1987. Mutagenic DNA repair in *E. coli*. XIII. Proofreading exonuclease of DNA polymerase III holoenzyme is not operational during UV mutagenesis. *Mutat. Res.* **183**:31–37.
34. Zillig, W., K. Zechel, D. Rabussay, M. Schachna, V. S. Sethi, P. Palm, A. Hal, and W. Seifert. 1976. On the role of different subunits of DNA-dependent RNA polymerase from *E. coli* in the transcription process. *Cold Spring Harbor Symp. Quant. Biol.* **35**:47–58.

Chapter 35

Blocking and Recovery of DNA Polymerase Readthrough on Templates Bearing Chemically Modified Cytosines

Tadayoshi Bessho, Kazuo Negishi, and Hikoya Hayatsu

Much interest has been focused on the biological consequences of specific chemical lesions in DNA. Particularly, effects of structurally modified bases in DNA templates on DNA polymerase-mediated replication in vitro have been studied extensively.

A method to chemically modify cytosine is illustrated in Fig. 1. Cytosine (compound *a*) can be converted into N^4-amino-5,6-dihydrocytosine 6-sulfonate (*b*) on treatment with a mixture of sodium bisulfite and hydrazine. The 5,6-dihydro-6-sulfonate structure in *b* can be reverted to a 5,6-double bond structure (*c*) on treatment with phosphate buffer. Since both of these conversions, *a* to *b* and *b* to *c*, can be carried out at pH 7 and 37°C, conditions sufficiently mild to treat DNA, it is attractive to use this method of cytosine modification to ask whether residues *b* and *c* in DNA can act as blocks or mutagenic lesions for DNA replication. We have already shown that N^4-aminocytosine residues in templates can be read efficiently by bacterial and mammalian DNA polymerases without blockage (10, 13). Furthermore, N^4-aminocytosine residues in template are probably mutagenic, because treatment of phages with bisulfite-hydrazine and subsequently with phosphate buffer causes mutations (3; unpublished results).

We have prepared DNA templates bearing residues *b* and have demonstrated that DNA polymerase I-catalyzed polynucleotide synthesis stops at one base before residue *b*. Furthermore, we show that treatment of the template with phosphate buffer to form compound *c* from *b* results in readthrough of the template by the polymerase. It is also shown that similar readthrough takes place in the presence of RecA or single-strand binding protein in polymerization mixtures.

TOXICITY STUDIES ON BACTERIOPHAGES

In earlier studies, we have shown that phage lambda can be inactivated by treatment with a mixture of bisulfite and hydrazine and that the inactivated phage can be

---

Tadayoshi Bessho, Kazuo Negishi, and Hikoya Hayatsu • Faculty of Pharmaceutical Sciences, Okayama University, Tsushima, Okayama 700, Japan.

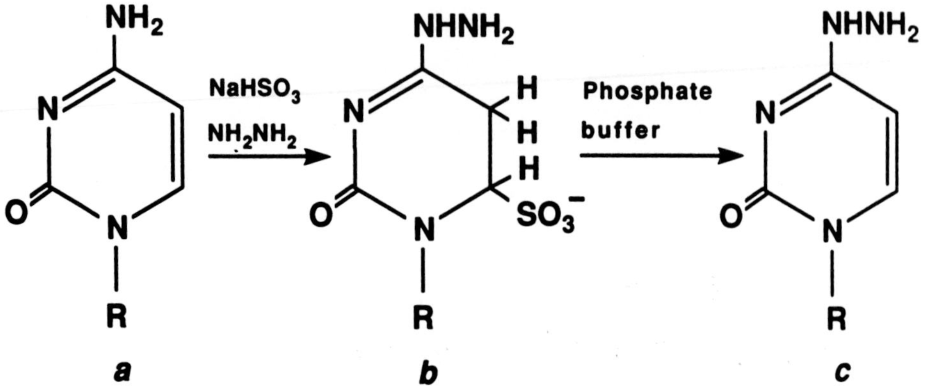

FIGURE 1. Chemical modification of cytosine.

reactivated by subsequent treatment with phosphate buffer (3). We have now treated the single-strand DNA phage M13mp2 in a similar manner and have observed the same phenomenon. Thus, when M13mp2 was treated with 0.5 M sodium bisulfite–0.5 M hydrazine at pH 7.0 and 37°C for 1 h, the phage titer, as measured on *Escherichia coli* CSH50, became 0.015 of the original, and the titer increased to 0.15 on subsequent treatment with 1 M sodium phosphate at pH 7.0 and 37°C for 7 h. Bisulfite alone or hydrazine alone did not inactivate the phage. These results suggest that the bisulfite adducts (*b*) formed in the DNA are strongly cytotoxic but the N^4-aminocytosines (*c*) are much less so.

CHAIN TERMINATION ON DNA TEMPLATES BEARING BISULFITE ADDUCTS *b* AND RESTORATION OF READTHROUGH BY TREATMENT WITH PHOSPHATE

M13 DNA was treated with 0.25 M bisulfite–0.25 M hydrazine at pH 7.0 and 37°C for 1 h. Under these conditions, every cytosine in the DNA was expected to be partially modified. The solution was filtered through Sephadex G-25 to remove the reagents, and the DNA was precipitated with ethanol. This modified DNA was annealed with a synthetic oligonucleotide primer (18 nucleotides long), and DNA polymerization was carried out with DNA polymerase I large fragment (Klenow). The polymerization of a mixture of dGTP, dATP, dTTP, and [α-^{32}P]dCTP was done with 7 mM Tris hydrochloride (pH 7.5) as the medium (also, 0.1 mM EDTA, 20 mM NaCl, and 7 mM $MgCl_2$ were present). It is known that Tris base is a poor catalyst for removal of HSO_3^- across the 5,6-bond of dihydropyrimidine 6-sulfonates (2). As a reference, elongation of the primer with mixtures of deoxyribonucleoside 5′-triphosphates and single dideoxynucleoside 5′-triphosphates (12) was carried out with unmodified DNA as a template. The polynucleotides produced were analyzed by electrophoresis on denaturing polyacrylamide gels.

Bands were found corresponding to all the guanine sites (i.e., cytosine sites in the template), and the chain lengths of all these bands were 1 nucleotide shorter than the guanine-band fragments. This observation suggests that chain termination takes place one nucleotide before the bisulfite adducts *b* in the template.

When the template DNA was first treated with phosphate buffer to convert residues *b* into *c* and then used for DNA synthesis, chain terminations were no longer observable. This phenomenon was consistent with the previous finding that N^4-aminocytosine (*c*) can serve as a template for guanine (10).

These results are also consistent with observations in toxicity studies of the whole phage particles (see above).

We conclude that the bisulfite adduct *b* is a block for DNA polymerase and that the restoration of readthrough can be obtained by removal of the bisulfite across the 5,6-dihydro linkage of the pyrimidine ring.

RECOVERY OF READTHROUGH BY RecA PROTEIN

The stop of chain elongation at one base before modified cytosines (*b*) on template may occur simply by a polymerase halt at that position. On the other hand, it may also occur in a more complicated manner; namely, the polymerase can incorporate a nucleotide opposite *b*, but the 3'→5' exonuclease activity of the polymerase molecule removes it efficiently due to the absence of base pairing between *b* and the incorporated nucleotide, and the net result is the stop at one base before *b*. To explore this possibility, we examined the effect of adding RecA protein to the assay system. RecA is known to inhibit the proofreading of DNA polymerase (1, 8), and therefore, if this mechanism is operating, the presence of RecA is expected to make the chain elongation proceed past *b*. When RecA was added to the polymerization mixture in excess of the polymerase and of the DNA, complete readthrough of the template by the polymerase resulted.

However, a contradictory observation was also obtained: when deoxyguanosine 5'-phosphate was used as an inhibitor for 3'→5' exonuclease activity of the polymerase (5), no readthrough took place. Therefore, the readthrough caused by adding RecA might have been due to the DNA-binding nature of RecA but not due to inhibition of the exonucleolytic activity. In fact, single-strand DNA-binding protein (SSB), which is known to cause *E. coli* DNA polymerase III holoenzyme to bypass pyrimidine photodimers on the template (7), can bring about readthrough when added to our reaction mixture.

These results suggest that in RecA- or SSB-mediated readthrough, binding to DNA strands is important and that proofreading exonucleolytic action is not directly responsible for the stopping of DNA chain elongation at residue *b* on the template.

PROPOSED SCHEME FOR BLOCKING AND READTHROUGH RECOVERY

Figure 2 shows our views on the mechanism of blocking and readthrough recovery. The presence of a bulky sulfonate anion at position 6 of the bisulfite adduct *b* must be important in blocking the DNA synthesis. Ionic repulsion between the sulfonate and the nearby phosphate groups may also serve to distort the structure of this portion of the template. The stop at one base before the lesion has been found for damages caused by benzo[*a*]pyrene (6), acetylaminofluorene (9), and 4-nitroquinoline 1-oxide (14). It is possible that these lesions, with large molecules attached to DNA bases, cause the polymerase to fall off when it encounters them. It is also known that a thymine diol residue on the template, which may be a "small lesion" compared with these others, can cause stops opposite to it rather than at one base before (4, 11). In this case, the polymerase falls off after it is positioned opposite to the lesion. It would be important to elucidate how these differences in positioning arise.

The actions of RecA protein and single-strand binding protein to restore readthrough may be due to covering up the sulfonate group of adduct *b* in the template, letting the polymerase pass through the

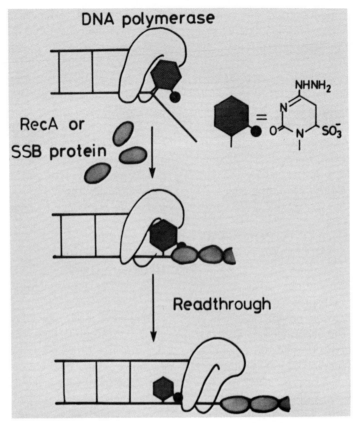

FIGURE 2. Stop and readthrough on chemically modified template in DNA polymerase I-catalyzed DNA synthesis.

damaged part. This proposition will have to be experimentally explored.

ACKNOWLEDGMENTS. This work was supported by a Grant-in-Aid for Scientific Research by the Ministry of Education, Science and Culture, Japan (61480430).

LITERATURE CITED

1. **Fersht, A. R., and J. W. Knill-Jones.** 1983. Contribution of 3'→5' exonuclease activity of DNA polymerase III holoenzyme from *Escherichia coli* to specificity. *J. Mol. Biol.* 165:669–682.
2. **Hayatsu, H.** 1976. Reaction of cytidine with semicarbazide in the presence of bisulfite. A rapid modification specific for single-stranded polynucleotide. *Biochemistry* 15:2677–2682.
3. **Hayatsu, H., and A. Kitajo.** 1977. Cooperations of bisulfite and nitrogenous compounds in mutagenesis, p. 285–292. *In* D. Scott, B. A. Bridges, and F. H. Sobels (ed.), *Progress in Genetic Toxicology*. Elsevier/North-Holland Biomedical Press, Amsterdam.
4. **Ide, H., Y. W. Kow, and S. S. Wallace.** 1985. Thymine glycols and urea residues in M13 DNA constitute replicative blocks *in vitro*. *Nucleic Acids Res.* 13:8035–8052.
5. **Kunkel, T. A., R. M. Schaaper, R. A. Beckman, and L. A. Loeb.** 1981. On the fidelity of DNA replication. *J. Biol. Chem.* 256:9883–9889.
6. **Larson, K. L., K. Angelis, and B. S. Strauss.** 1984. Replication of carcinogen damaged double and single stranded DNA templates *in vitro*. *J. Cell. Biochem.* 109:43.
7. **Livneh, Z.** 1986. Replication of UV-irradiated single-stranded DNA by DNA polymerase III holoenzyme of *Escherichia coli*: evidence for bypass of pyrimidine photodimers. *Proc. Natl. Acad. Sci. USA* 83:4599–4603.
8. **Lu, C., R. H. Scheuermann, and H. Echols.**

1986. Capacity of RecA protein to bind preferentially to UV lesions and inhibit the editing subunit (ε) of DNA polymerase III: a possible mechanism for SOS-induced targeted mutagenesis. *Proc. Natl. Acad. Sci. USA* **83**:619–623.
9. **Moore, P. D., S. D. Rabkin, A. L. Osborn, C. M. King, and B. S. Strauss.** 1982. Effect of acetylated and deacetylated 2-aminofluorene adducts on *in vitro* DNA synthesis. *Proc. Natl. Acad. Sci. USA* **79**:7166–7170.
10. **Negishi, K., M. Takahashi, Y. Yamashita, M. Nishizawa, and H. Hayatsu.** 1985. Mutagenesis by N^4-aminocytidine: induction of AT to GC transition and its molecular mechanism. *Biochemistry* **24**:7273–7278.
11. **Rouet, P., and J. M. Essigmann.** 1985. Possible role for thymine glycol in the selective inhibition of DNA synthesis on oxidized DNA templates. *Cancer Res.* **45**:6113–6118.
12. **Sanger, F., S. Nicklen, and A. R. Coulson.** 1977. DNA sequencing with chain-terminating inhibitors. *Proc. Natl. Acad. Sci. USA* **74**:5463–5467.
13. **Takahashi, M., M. Nishizawa, K. Negishi, F. Hanaoka, M.-A. Yamada, and H. Hayatsu.** 1988. Induction of mutation in mouse FM3A cells by N^4-aminocytidine-mediated replicational errors. *Mol. Cell. Biol.* **8**:347–352.
14. **Yoshida, S., O. Koiwai, R. Suzuki, and M. Tada.** 1984. Arrest of DNA elongation by DNA polymerases at guanine adducts on 4-hydroxyaminoquinoline 1-oxide-modified DNA template. *Cancer Res.* **44**:1867–1870.

V. Genetic Control of Mutagenesis

Genetic Control of Mutagenesis: Introduction

It is the goal of the chapters that follow this introduction to review and summarize current information about the role of host genes in controlling the mutagenic process. Naively, one might think that mutations, which historically have been thought of as accidents of nature, would just happen. The genotype of the cell in which the mutation occurs might be expected to have little influence on such a random, accidental process, or if it did, it might increase mutagenesis by interference with repair pathways. Indeed, if mutations are in general "bad," why has the cell evolved and retained functions that seem to promote mutagenesis?

It is now quite clear that mutagenesis is a complex process and that multiple genes affect the final outcome. Some genes such as *umuC* and *umuD* are required in order to observe any UV mutagenesis. The *recA* gene is also required. *umuCD* genes are under the negative control of a repressor encoded by the *lexA* gene. Cleavage of the LexA protein is an early step in the induction of the mutagenic pathway in *Escherichia coli* and is mediated by the proteolytic activity of so-called "activated RecA" protein. Indeed, constitutive induction of the SOS pathway is mutagenic even in the absence of specific DNA damage. One step in this pathway now appears to be the cleavage of the UmuD protein by RecA, another example of the ubiquitous role of limited proteolysis as a key biological regulatory mechanism.

A critical question in the genetic control of the SOS pathway is the nature of the inducing signal. Somehow, the RecA protease activity is regulated so that UV damage results in activated RecA. RecA protease can be activated in vitro by single-stranded DNA, so it has been proposed that single-stranded oligonucleotides, UV-treated DNA, or gapped duplex molecules may be the physiological signal that activates RecA. Although direct experimental tests of these ideas have not been devised, there is strong circumstantial evidence in favor of some form of single-stranded DNA, formed in vivo, as the activator of RecA.

In addition to genes such as the *umuCD* system that are required for mutation, genes are known that suppress or antagonize mutagenesis. Mutations in these genes result in a mutator phenotype. In *E. coli* one well-studied group of such mutators is known to regulate the specific handling of base-pair mismatches. Recent work in *Saccharomyces cerevisiae* has identified genes that may be functionally analogous to the mismatch repair genes in *E. coli*. Biochemical approaches to this problem, while not the subject of the following chapters, have also suggested that such pathways may exist in mammalian cells.

The ability to study the genetic control of the mutagenic process depends on having systems that allow identification of different classes of mutagenic events so that they can be examined in isolation. Systems using bacteriophage and plasmids have been developed that can be used to score for specific classes of mutations and to analyze chromosomal genes that control the specific mutagenic process. This approach has been particularly fruitful in studies of the process by which deletion mutagenesis is controlled.

Mutation results from the interplay of replication bypass (which is important for survival of the organism, mutant or not), fidelity of the replication system, repair of mismatched duplex DNA, specificity of the damage detection and removal systems, and DNA breaking and joining reactions (e.g., recombination and transposition).

Once these complexities are admitted, it is not surprising that many cellular genes will be found that modulate, positively and negatively, the mutagenic outcome of DNA damage, as well as the level of spontaneously occurring mutations. The following papers address the current state of research on genes that influence the mutagenic responses in procaryotes and yeasts. In these systems sophisticated genetic analysis has been brought to bear on this problem and has made possible the current level of understanding. Still, a number of problems remain unresolved, as follows.

(i) Although RecA protein functions in mutagenesis by cleavage of LexA protein and UmuD protein, it is still unclear what specific molecule activates RecA protease.

(ii) Specific correction systems seem to recognize specific base mismatches in *E. coli*, but it is unknown whether such systems function in eucaryotes.

(iii) To what extent is the regulation of replication involved in the generation of mutations, especially deletions?

(iv) Although genetic rearrangements are of critical importance in mammalian organisms (because essential genetic information may be retained intact but relocated to a different regulatory environment), little is known of their mechanism of occurrence even in procaryotes, with the exception of the special case of transposons.

The Editors

Chapter 36

RecA-Mediated Cleavage Activates UmuD for UV and Chemical Mutagenesis

Takehiko Nohmi, John R. Battista, and Graham C. Walker

Most mutagenesis of *Escherichia coli* by UV irradiation and a variety of chemicals requires the UmuD and UmuC proteins (5, 24, 29, 30) or their plasmid-derived analogs MucA and MucB. Both *umuD* and *umuC* mutants are virtually nonmutable with UV and many chemicals (5, 10, 26). The *umuD* and *umuC* genes are organized in an operon (5, 24) and encode proteins of 15.0 and 47.7 kilodaltons, respectively (12, 17). The *mucA* and *mucB* genes of the plasmid pKM101 are also organized in an operon and encode similarly sized products (17, 18).

Both the *umuDC* operon and the plasmid-derived *mucAB* operon are repressed by the LexA protein (1, 5, 24) and regulated as part of the SOS response of *E. coli* (6, 15, 29, 30, 32). SOS induction occurs by activated RecA mediating the proteolytic cleavage of LexA at its Ala-84–Gly-85 bond (14), apparently by facilitating a specific autodigestion of LexA (13). Slilaty and Little (25) have recently suggested that hydrolysis of the LexA Ala-Gly bond proceeds by a mechanism similar to that of serine proteases, with Ser-119 acting as a nucleophile and Lys-156

as an activator. LexA shares homology with the repressors of bacteriophages lambda, 434, P22, and ϕ80 (23; Y. Eguchi, T. Ogawa, and H. Ogawa, *J. Mol. Biol.*, in press), and cleavage of these proteins appears to occur by an analogous mechanism (20, 22). The cleavage site of all these proteins is an Ala-Gly bond except for ϕ80 repressor, which has a Cys-Gly cleavage site.

UmuD SHARES HOMOLOGY WITH LexA AND SEVERAL PHAGE REPRESSORS

We recently reported that UmuD and MucA share homology with the carboxyl-terminal regions of LexA and these phage repressors (17). This led us to hypothesize that UmuD and MucA might interact with activated RecA and that this interaction could result in a proteolytic cleavage of these proteins that would activate or unmask their function required for mutagenesis (17). The putative cleavage site of UmuD is the Cys-24–Gly-25 bond. A RecA-mediated cleavage of UmuD has now been shown to occur both in vivo (23a) and in vitro (3a), and work from our laboratory, discussed below, has shown that the purpose of this cleavage is to

---

Takehiko Nohmi, John R. Battista, and Graham C. Walker • Biology Department, Massachusetts Institute of Technology, Cambridge, Massachusetts 02139.

activate UmuD for its role in mutagenesis (16a).

RecA HAS AN ADDITIONAL ROLE IN MUTAGENESIS BESIDES MEDIATING LexA CLEAVAGE

The possibility that RecA mediates the cleavage of UmuD and MucA has taken on added significance in the face of a growing body of evidence that RecA is required for at least one other role in mutagenesis besides mediating the cleavage of LexA (29). This was first suggested by the observation that *lexA*(Def) *recA*(Def) strains are nonmutable despite the lack of functional LexA (1, 3, 15). A need for RecA to be activated for an additional role besides cleaving LexA has been suggested by analyses of mutagenesis in strains carrying various *recA* mutations that alter the ability of RecA to be activated (3, 7, 27–29) and by various physiological experiments (21, 33). We have recently presented evidence suggesting that RecA may play at least two additional roles in mutagenesis besides mediating the cleavage of LexA (16).

HOMOLOGY OF UmuD TO LexA AND PHAGE REPRESSORS HAS FUNCTIONAL SIGNIFICANCE

To test the significance of the homology of UmuD to LexA and certain phage repressors (17), we used site-directed mutagenesis of a $umuD^+C^+$ plasmid to create certain *umuD* mutations that were analogous to *lexA* or lambda repressor mutations that block both RecA-mediated cleavage and autodigestion. We found that all these *umuD* mutations caused major reductions in the ability of UmuD to function in UV mutagenesis. We found that changing the Gly-25 residue of the putative Cys-24–Gly-25 UmuD cleavage site to Glu or Lys largely abolished the ability of UmuD to function in UV mutagenesis. A Gly→Glu change at the Ala-Gly cleavage site of lambda repressor completely blocks cleavage (8), while a corresponding Gly→Asp change in LexA largely blocks cleavage (14, 15). In addition, we found that changing either Ser-60 or Lys-97 to Ala also greatly reduced the ability of UmuD to function in UV mutagenesis. Slilaty and Little (25) have shown that changes of the corresponding Ser-119 and Lys-156 residues of LexA to Ala completely block cleavage. They also reported that changing Ser-119 to Cys, another residue with a nucleophilic side chain, did not block LexA cleavage as severely as the change to Ala. Interestingly, we found that changing the Ser-60 of UmuD to Cys caused a less severe reduction in the ability of UmuD to function in mutagenesis than the change to Ala.

CLEAVED UmuD IS FUNCTIONAL IN MUTAGENESIS

To test more directly the hypothesis that cleavage of UmuD is important for mutagenesis, we constructed a *umuD* mutant of a $umuD^+$ $\Delta umuC$ plasmid in which overlapping termination (TGA) and initiation (ATG) codons were introduced at the site in the *umuD* sequence that corresponds to the putative cleavage site. The plasmid carrying this engineered form of UmuD encodes two polypeptides rather than one. These two polypeptides are virtually the same as those that would result from cleavage at the Cys-24–Gly-25 bond of UmuD. The smaller (NH_2-terminal) polypeptide differs from the corresponding cleavage product in lacking the COOH-terminal Cys. The larger (COOH-terminal) polypeptide is probably identical to the corresponding cleavage product. Although it would have an NH_2-terminal formylmethionine when first synthesized, it is likely to be removed since NH_2-terminal methionines adjacent to glycines are removed by *E. coli* methionine aminopeptidase with high efficiency (2). Despite the absence of an obvious Shine-Delgarno sequence, we have been able to detect the synthesis of the larger polypeptide corresponding to the COOH terminus of UmuD by the use of the maxicell technique. When a plasmid carrying this engineered *umuD* encoding two polypeptides was introduced into a nonmutable *umuD44* strain, it restored the UV mutability of the cell to that of a *umuD*$^+$ strain. This result strongly indicates that at least one of the products resulting from cleavage of the UmuD at its Cys-24–Gly-25 bond is capable of carrying out the role of the *umuD* gene product in mutagenesis. Furthermore, it rules out the possibility that the purpose of UmuD cleavage is to inactivate UmuD.

THE COOH-TERMINAL CLEAVAGE PRODUCT IS NECESSARY AND SUFFICIENT FOR THE ROLE OF UmuD IN MUTAGENESIS

A plasmid that encoded only the polypeptide corresponding to the small NH$_2$-terminal fragment of UmuD failed to complement the UV nonmutability of a *umuD44* strain, whereas a plasmid that encoded only the large COOH-terminal polypeptide made the strain more UV mutable than a plasmid carrying *umuD*$^+$. These results strongly suggest that the COOH-terminal cleavage product of UmuD is both necessary and sufficient for the role of UmuD in UV mutagenesis. The COOH-terminal polypeptide of UmuD is expressed at a higher level from the plasmid encoding only this polypeptide than from the plasmid encoding both the NH$_2$- and COOH-terminal polypeptides. This probably accounts for the higher UV mutability of the strain carrying the plasmid encoding the COOH-terminal polypeptide alone.

THE COOH-TERMINAL POLYPEPTIDE OF UmuD RESTORES UV MUTABILITY TO *recA430* STRAINS

To test the physiological significance of UmuD cleavage, we introduced plasmids carrying either the engineered *umuD* encoding two polypeptides or the COOH-terminal polypeptide of UmuD into a *lexA71*::Tn5 (Def) *recA430* strain (29). The *recA430* mutation has differential effects on the ability of RecA to mediate proteolytic cleavage. The RecA430 protein fails to mediate the cleavage of lambda repressor (19), mediates the cleavage of LexA with reduced efficiency (4, 5, 7, 11), and mediates the cleavage of φ80 repressor normally (4; Eguchi et al., in press). *recA430* strains are UV nonmutable. The introduction of the plasmid encoding the two engineered UmuD polypeptides partially restored the UV mutability of this strain, whereas the plasmid encoding only the COOH-terminal UmuD polypeptide restored the UV mutability of the strain to that of a *lexA71*::Tn5(Def) *recA*$^+$ strain carrying a *umuD*$^+$ plasmid. We suggest that the more efficient restoration of mutability by the latter plasmid is due to a higher level of expression of the COOH-terminal polypeptide. The two plasmids also restored some UV mutability to a *lexA*$^+$ *recA430* strain. The restoration of UV mutability to the *recA430* strains observed when we circumvented the need for UmuD cleavage strongly indicates that the primary cause for the UV nonmutability of *recA430* derivatives is an inability to mediate the cleavage of UmuD. It furthermore implies that the purpose of RecA-mediated cleavage of UmuD is to activate UmuD for its role in mutagenesis. Thus, it appears that RecA carries out two mechanistically related roles in UV and chemical mutagenesis: (i) transcriptional derepression of the *umuDC* operon by mediating the cleavage of LexA and (ii) posttranslation activation of UmuD by mediating its cleavage. With respect to future research in other organisms, it is worth noting that RecA-mediated cleavage of UmuD provides an example of an activating event induced by DNA damage that does not involve regulation at the transcriptional level (31).

A THIRD ROLE FOR RecA IN MUTAGENESIS

In contrast to the situation with the corresponding $lexA71$::Tn5(Def) $recA430$ strain, introduction of a plasmid encoding only the COOH-terminal polypeptide of UmuD into a $lexA71$::Tn5(Def) $\Delta recA306$ strain did not suppress the UV nonmutability of this strain. This observation strongly suggests that RecA carries out a third role in mutagenesis besides mediating the cleavage of LexA and UmuD. The nature of that requirement for RecA for mutagenesis remains to be established. However, the observation of Tessman et al. (27, 28) that particular *RecA* mutants, which are constitutively activated for mediating proteolysis but are recombination defective, are UV mutable suggests that the third role cannot require all of the functions of RecA required for homologous recombination. It might, however, require a subset of these activities.

DIFFERENTIAL EFFECTS OF CHANGING Ser-60 AND Lys-97 IN THE COOH-TERMINAL POLYPEPTIDE OF UmuD

To explore further the roles of Ser-60 and Lys-97 in UmuD function, we introduced Ser-60→Ala and Lys-97→Ala changes into the engineered *umuD* encoding two polypeptides. The Ser→Ala change only reduced the ability of the COOH-terminal polypeptide of UmuD to function in mutagenesis by 50%. This contrasts with the major reduction in mutagenesis observed when the same change was introduced into a plasmid encoding an intact UmuD protein and suggests that the primary requirement for Ser-60 in UmuD function is for cleavage rather than for the subsequent role of the COOH-terminal fragment in UV mutagenesis. In contrast, the Lys→Ala change caused a substantial reduction in the amount of UV mutagenesis observed, indicating that this change in some way disrupts the function of the COOH-terminal fragment of UmuD in UV mutagenesis.

Our results also suggest that the primary role of Ser-60 of UmuD is to function as a nucleophile in RecA-mediated cleavage. As noted previously for LexA and lambda repressor (25), the sequences surrounding this serine, Gly–Asp–Ser-60–Gly for UmuD and Gly–Ser–Ser-61–Met for MucA, closely resemble the consensus sequences around the active serine in mammalian serine proteases (Gly-Asp-Ser-Gly) and in microbial proteases such as subtilisin (Gly-Thr-Ser-Met). It is possible that the mechanism of cleavage is even more closely related to that of serine proteases than postulated by Slilaty and Little (25). In particular, it will be interesting to test whether the acidic residue conserved between UmuD (Asp-68), MucA (Asp-69), LexA (Asp-127), and the phage repressors (Asp in lambda and φ80; Glu in P22 and 434) functions as a base to accept a proton from Lys during RecA-mediated cleavage just as Asp-102 of chymotrypsin accepts a proton from His-57 during catalysis (9).

ACKNOWLEDGMENTS. We thank Lorraine Marsh and Lori Dodson and the members of our research group for many helpful discussions.

This work was supported by Public Health Service grant CA21615 awarded by the National Cancer Institute. J.R.B. was supported by a postdoctoral fellowship from the American Cancer Society, Massachusetts Division, Inc.

LITERATURE CITED

1. Bagg, A., C. J. Kenyon, and G. C. Walker. 1981. Inducibility of a gene product required for UV and chemical mutagenesis in *Escherichia coli*. *Proc. Natl. Acad. Sci. USA* 78:5749–5753.
2. Ben-Bassat, A., K. Bauer, S. Chang, K. Myambo, A. Boosman, and C. Chang. 1987. Processing of the initiation methionine from proteins: properties of the *Escherichia coli* methionine aminopeptidase and its gene structure. *J. Bacteriol.* 169:751–757.
3. Blanco, M., G. Herrera, P. Collado, J. E. Rebollo, and L. M. Botella. 1982. Influence of RecA protein on induced mutagenesis. *Biochimie* 64:633–636.
3a. Burckhardt, S. E., R. Woodgate, R. H. Scheuermann, and H. Echols. 1988. UmuD mutagenesis protein of *Escherichia coli*: overproduction, purification, and cleavage by RecA. *Proc. Natl. Acad. Sci. USA* 85:1811–1815.
4. Devoret, R., M. Pierre, and P. L. Moreau. 1983. Prophage φ80 is induced in *Escherichia coli recA430*. *Mol. Gen. Genet.* 189:199–206.
5. Elledge, S. J., and G. C. Walker. 1983. Proteins required for ultraviolet light and chemical mutagenesis: identification of the products of the *umuC* locus of *E. coli*. *J. Mol. Biol.* 164:175–192.
6. Elledge, S. J., and G. C. Walker. 1983. The *muc* genes of pKM101 are induced by DNA damage. *J. Bacteriol.* 155:1306–1315.
7. Ennis, D. G., B. Fisher, S. Edmiston, and D. W. Mount. 1985. Dual role for *Escherichia coli* RecA protein in SOS mutagenesis. *Proc. Natl. Acad. Sci. USA* 82:3325–3329.
8. Gimble, F. S., and R. T. Sauer. 1986. λ repressor inactivation: properties of purified ind^- proteins in the autodigestion and RecA-mediated cleavage reactions. *J. Mol. Biol.* 192:39–47.
9. Hunkapiller, M. W., S. H. Smallcombe, D. R. Whitaker, and J. H. Richards. 1973. Carbon nuclear magnetic resonance studies of the histidine residue in α-lytic protease: implications for the catalytic mechanism of serine proteases. *Biochemistry* 12:4732–4742.
10. Kato, T., and Y. Shinoura. 1977. Isolation and characterization of mutants of *Escherichia coli* deficient in induction of mutations by ultraviolet light. *Mol. Gen. Genet.* 156:121–131.
11. Kawashima, H., T. Horii, T. Ogawa, and H. Ogawa. 1984. Functional domains of *Escherichia coli* recA protein deduced from the mutational sites in the gene. *Mol. Gen. Genet.* 193:288–292.
12. Kitagawa, Y., E. Akaboshi, H. Shinagawa, T. Horii, H. Ogawa, and T. Kato. 1985. Structural analysis of the *umu* operon required for inducible mutagenesis in *Escherichia coli*. *Proc. Natl. Acad. Sci. USA* 82:4336–4340.
13. Little, J. W. 1984. Autodigestion of LexA and phage λ repressors. *Proc. Natl. Acad. Sci. USA* 81:1375–1379.
14. Little, J. W., S. H. Edmiston, L. Z. Pacelli, and D. W. Mount. 1980. Cleavage of the *Escherichia coli lexA* protein by the *recA* protease. *Proc. Natl. Acad. Sci. USA* 77:3225–3229.
15. Little, J. W., and D. W. Mount. 1982. The SOS regulatory system of *Escherichia coli*. *Cell* 29:11–22.
16. Marsh, L., and G. C. Walker. 1987. New phenotypes associated with *mucAB*: alteration of a MucA sequence homologous to the LexA cleavage site. *J. Bacteriol.* 169:1818–1823.
16a. Nohmi, T., J. R. Battista, L. A. Dodson, and G. C. Walker. 1988. RecA-mediated cleavage activates UmuD for mutagenesis: mechanistic relationship between transcriptional depression and posttranslational activation. *Proc. Natl. Acad. Sci. USA* 85:1816–1820.
17. Perry, K. L., S. J. Elledge, B. B. Mitchell, L. Marsh, and G. C. Walker. 1985. *umuDC* and *mucAB* operons whose products are required for UV light- and chemical-induced mutagenesis: UmuD, MucA, and LexA proteins share homology. *Proc. Natl. Acad. Sci. USA* 82:4331–4335.
18. Perry, K. L., and G. C. Walker. 1982. Identification of plasmid(pKM101)-coded proteins involved in mutagenesis and UV resistance. *Nature* (London) 300:278–281.
19. Roberts, J. W., and C. W. Roberts. 1981. Two mutations that alter the regulatory activity of the *E. coli recA* protein. *Nature* (London) 290:422–424.
20. Roberts, J. W., C. W. Roberts, and N. L. Craig. 1978. *Escherichia coli recA* gene product inactivates phage lambda repressor. *Proc. Natl. Acad. Sci. USA* 75:4714–4718.
21. Ruiz-Rubio, M., R. Woodgate, B. A. Bridges, G. Herrera, and M. Blanco. 1986. New role for photoreversible pyrimidine dimers in induction of prototrophic mutations in excision-deficient *Escherichia coli* by UV light. *J. Bacteriol.* 166:1141–1143.
22. Sauer, R. T., M. J. Ross, and M. Ptashne. 1982. Cleavage of the λ and P22 repressors by *recA* protein. *J. Biol. Chem.* 257:4458–4462.
23. Sauer, R. T., R. R. Yocum, R. F. Doolittle, M. Lewis, and C. O. Pabo. 1982. Homology among DNA-binding proteins suggests use of a conserved super-secondary structure. *Nature* (London) 298:447–451.
23a. Shinagawa, H., H. Iwasaki, T. Kato, and A. Nakata. 1988. RecA protein-dependent cleavage of UmuD protein and SOS mutagenesis. *Proc. Natl. Acad. Sci.* 85:1806–1810.
24. Shinagawa, H., T. Kato, T. Ise, K. Makino, and A. Nakata. 1983. Cloning and characterization of

the *umu* operon responsible for inducible mutagenesis in *Escherichia coli. Gene* **23:**167–174.
25. **Slilaty, S. N., and J. W. Little.** 1987. Lysine-156 and serine-119 are required for LexA repressor cleavage: a possible mechanism. *Proc. Natl. Acad. Sci. USA* **84:**3987–3991.
26. **Steinborn, G.** 1978. Uvm mutants of *Escherichia coli* deficient in UV mutagenesis. I. Isolation of *uvm* mutants and their phenotypical characterization in DNA repair and mutagenesis. *Mol. Gen. Genet.* **165:**87–93.
27. **Tessman, E. S., and P. Peterson.** 1985. Isolation of protease-proficient, recombinase-deficient *recA* mutants of *Escherichia coli* K-12. *J. Bacteriol.* **163:**688–695.
28. **Tessman, E. S., I. Tessman, P. K. Peterson, and J. D. Forestal.** 1986. Role of RecA protease and recombinase activities of *Escherichia coli* in spontaneous and UV-induced mutagenesis and in Weigle repair. *J. Bacteriol.* **168:**1159–1164.
29. **Walker, G. C.** 1984. Mutagenesis and inducible responses to deoxyribonucleic acid damage in *Escherichia coli. Microbiol. Rev.* **48:**60–93.
30. **Walker, G. C.** 1985. Inducible DNA repair systems. *Annu. Rev. Biochem.* **54:**425–457.
31. **Walker, G. C., L. Marsh, and L. A. Dodson.** 1985. Genetic analyses of DNA repair: inference and extrapolation. *Annu. Rev. Genet.* **19:**103–126.
32. **Witkin, E. M.** 1976. Ultraviolet mutagenesis and inducible DNA repair in *Escherichia coli. Bacteriol. Rev.* **40:**869–907.
33. **Witkin, E. M., and T. Kogoma.** 1984. Involvement of the activated form of RecA protein in SOS mutagenesis and stable DNA replication in *Escherichia coli. Proc. Natl. Acad. Sci. USA* **81:**7539–7543.

Chapter 37

Is Mismatch Repair Involved in UV-Induced Mutagenesis in *Saccharomyces cerevisiae*?

Friederike Eckardt-Schupp, Fred Ahne, Eva-Maria Geigl, and Wolfram Siede

During the past few years there has been increasing interest regarding functions inducible by DNA damage being involved in mutagenic processes in *Saccharomyces cerevisiae* (24, 29). From careful study of the dose dependence of mutation induction and from analysis of mutants known to be defective in induced mutagenesis, it is obvious that the problem of inducible mutagenic processes is closely related to further questions concerning the molecular mechanisms of mutation induction and repair, which can be formulated as follows. Is the fixation of mutations a pre- or postreplicational event, or is it replication dependent? Is excision repair completely error free, and if not, which role does it play in the mutagenic process? And further, does mismatch repair play a role in the processes controlling spontaneous and induced mutagenesis in *S. cerevisiae*?

Friederike Eckardt-Schupp, Fred Ahne, and Eva-Maria Geigl • Gesellschaft für Strahlen- und Umweltforschung, Institut für Strahlenbiologie, D-8042 Neuherberg, Federal Republic of Germany.   *Wolfram Siede* • Department of Pathology, Stanford University, Stanford, California 94305.

EVIDENCE FOR INDUCIBLE MUTAGENIC FUNCTIONS

Mutation Kinetics Studies

Evidence for the existence of inducible mutagenic functions in *S. cerevisiae* was derived from the interpretation of dose-effect curves (kinetics) of various mutation induction systems in excision repair-competent (EXC^+) and excision-deficient (exc^-) yeast strains. Mathematical analysis of mutation kinetics proposed that linearity reflects a dose-independent (constitutive) mutagenic process and that a quadratic branch is an indication for an inducible component (5, 9, 10). From this extended analysis the inducible mutagenic process in *S. cerevisiae* can be characterized as follows.

Gene or Allele Specificity

Under selective conditions of omission media, inducible processes are involved in the reversion of numerous ochre alleles (*ade2-1*, *lys2-1*, *lys1-1*, *arg4-17*, *his5-2*) (9, 25, 27) and a few tested missense alleles (*ilv1-92*, *his3*), but for unknown reasons not in the mutation of tRNA genes, for example, supersuppressors of the *ade2-1* and *lys2-1* alleles (unpublished data).

Influence of Conditions of Replication

Under nonselective conditions of complete or supplemented synthetic media, the mutation kinetics are linear, as for instance for forward mutants in various genes, *ADEx*, preceding *ADE1* and *ADE2* in adenine biosynthesis (6), or for locus revertants of the *ade2-1* allele if scored nonselectively on supplemented synthetic medium (unpublished data). Obviously, under conditions which allow replication, the UV inducibility of mutagenic functions is not detectable.

Dependence on Excision Competence

For all mutation systems, both reverse and forward tested, the mutation kinetics in excision-deficient mutants are always linear (5, 26). Deviations from linearity could be attributed to differences in the probabilities of survival of mutant as compared with non-mutant cells (5, 9). Hence, excision competence is required for the inducibility of some component of the mutagenic process.

Molecular Evidence for UV-Inducible Genes

Some genes which are involved in the repair of DNA damage, but probably not in mutagenic processes, were proven to be UV inducible (7). Furthermore, some yeast genes were isolated which possess UV- or chemically enhanced levels of transcription, the *DIN* (23) and *DDR* (21) genes. One of the *DDR* genes lacks UV-inducible transcription activity in an excision-deficient *rad3-2* background (19). Although the function of this gene is unknown, it supports the notion that excision repair seems to be involved in the regulation of UV-inducible processes in *S. cerevisiae*.

EVIDENCE FOR MISMATCH REPAIR IN *S. CEREVISIAE*

Mismatch repair has been characterized genetically and biochemically rather well in *Escherichia coli* (4, 22). In *S. cerevisiae*, the formation of heteroduplex DNA was regarded as a crucial intermediate of genetic recombination and gene conversion (12). The occurrence of postmeiotic segregation was considered the most direct genetic evidence for the existence of heteroduplex DNA possibly containing mismatched base pairs. A mismatch process was postulated to explain gene conversion and the normally low levels of postmeiotic segregation. The isolation and characterization of *pms* mutants which have considerably enhanced levels of postmeiotic segregation confirmed the role of mismatch correction of heteroduplex DNA arising during meiotic recombination. The *pms* mutants exhibit a mitotic mutator phenotype which argues for a role of the *PMS1* gene in error correction during DNA replication (32). For the process of induced mutagenesis, an involvement of mismatch repair was postulated to explain the amazingly high percentage of UV-induced pure mutant clones in stationary, excision-proficient *S. cerevisiae* (6).

A capacity of *S. cerevisiae* to repair artificially produced heteroduplex DNA was proven by transformation experiments with plasmids containing heteroduplex DNA at restriction sites (1). By this approach, the role of *PMS1* in correction of mismatch bases could be substantiated (2). In *E. coli*, the dam^+ gene controls methylation of the adenine in the sequence GATC, which serves as a signal for the mismatch repair complex in that unmethylated DNA will be incised and mismatch repair will occur according to the methylated template (4, 22). However, most organisms, including *S. cerevisiae*, do not use GATC methylation to direct mismatch repair, and it has been suggested that mismatch repair in eucaryotes acts at transient nicks in the newly synthesized DNA strands (22).

THE *REV2* GENE, A POSSIBLE CANDIDATE FOR AN INDUCIBLE GENE CONTROLLING MISMATCH REPAIR AND A MUTAGENIC PROCESS IN *S. CEREVISIAE*

One of the several mutants impaired in UV induction of reverse mutation (*rev2* mutant) shows decreases in the reversion frequencies of certain ochre and frameshift alleles only (16, 18). For this mutant, purely linear reversion kinetics have been reported (17), which we interpreted as possible deficiency in an inducible mutagenic process. Using a temperature-sensitive *rev2* allele, we performed kinetic studies (25, 26) and temperature-shift experiments and measured the biological endpoints of survival and mutation induction in various alleles, with prolonged times at the permissive temperature (23°C) before shifting to the nonpermissive temperature (36°C). This allowed us to deduce the temporal course of the repair of prelethal damage and of mutation fixation, both being controlled by the *REV2* process (27, 28). The results are briefly summarized as follows.

(i) The *REV2* gene product plays different roles in spontaneous and induced mutagenesis: a *rev2* mutant is a spontaneous mutator, whereas after UV and X-ray irradiation and various chemicals *rev2* mutants are mutation deficient for certain loci.

(ii) The *REV2* gene activity is UV inducible and growth dependent. The constitutive level of the *REV2* gene product is detectable in replicating cells. In contrast, the constitutive level in stationary cells is undetectably low and the UV inducibility is pronounced.

(iii) The inducibility of the *REV2* gene requires excision competence of the cells, as has been deduced from detailed kinetic studies of the *rev2*(Ts) mutation in an EXC^+ and an exc^- background at 23 and 36°C.

(iv) The timing of mutation fixation differs in EXC^+ and exc^- cells. In EXC^+ stationary cells, the mutagenic process is finished within 2 to 4 h, clearly before replication starts. In exc^- cells more than 12 h is required, beyond the onset of postirradiation DNA replication. However, mutation fixation is obviously not replication dependent in the *rev2 rad3-2* double mutant, since hydroxyurea, which blocks the ribonucleotide reductase, and aphidicolin, which blocks the α-like polymerase I, do not block mutation induction.

From these biological data we propose our model concerning the possible mechanism of the *REV2*-controlled process in the following points (29).

(i) *REV2* controls mismatch repair in connection with semiconservative DNA replication; e.g., its main function is an error-avoiding one.

(ii) The onset of excision repair gives signals for the enhanced transcription of the *REV2* gene to control repair of mispaired bases arising during repair replication.

(iii) The *REV2* gene product, possibly as part of a complex, recognizes disturbances in base pairing at the sites of DNA damage and handles these sites like a mismatch. This "misled" mismatch repair causes insertion of wrong bases, possibly yielding mutations. These mutations are fixed in the next round of replication, or, if the excision complex recognizes the DNA lesion and performs late repair, the mutation is transferred into the other strand and a pure mutant clone should arise. In support of this point, a high percentage of pure mutant clones have been found in stationary, excision-competent yeast strains (6, 13).

(iv) exc^- strains do not allow induction of the *REV2* gene product. Instead, they mutate and repair at the constitutive level. The postreplicative mutation fixation is due to the slowly acting *REV2* process, caused by the low level of the *REV2* protein. In exc^-

strains, misled mismatch repair can take place before or after replication, but due to the excision deficiency no transfer of the mutation into the other strand occurs. This explains the high percentage of sectored mutant clones found in exc⁻ mutants (6, 14). The increase in mutation frequencies found in replicating exc⁻ cells if the scoring conditions permit DNA replication (15) is explicable in our model by the replication-related high constitutive level of the *REV2* gene product leading to an increase of misled mismatch repair, as compared with stationary exc⁻ cells.

MOLECULAR ANALYSIS OF THE *REV2* PROCESS

We pursued two different approaches to verify our working hypothesis. First, we wanted to clone and sequence the *REV2* gene and initiate various molecular studies, for example, to investigate the phenotypic effect of a deletion and to study the UV inducibility and the growth phase and possibly cell cycle dependence of the transcription of the *REV2* gene. Second, we wanted to examine the assumed mismatch repair function of the *REV2* gene product.

Cloning of the *REV2* Gene

The *REV2* gene seemed to belong to that group of eukaryotic genes which are not easy to clone. Transformants of a suitable *rev2* recipient could be isolated only by directly ligating partially digested yeast genomic DNA into a centromeric vector (YCP50) without previous propagation in *E. coli*. Plasmids isolated from those transformants successfully retransformed *S. cerevisiae* again, but not *E. coli* (30). Just recently we were able to recover *E. coli* transformants in a low frequency by purifying the plasmids isolated from yeast on a Nucleogen column (Diagen, Düsseldorf, Federal Republic of Germany) by high-pressure liquid chromatography. Restriction analysis of the plasmids isolated from 15 *E. coli* transformants revealed that 12 of them had identical restriction patterns, but 3 of them showed severe structural changes. One of those plasmids with an identical restriction pattern complemented a *rev2* mutant yielding a UV resistance identical to the resistance of the *REV2* wild type. Experiments are in progress to verify that the *rev2*-complementing plasmid carries the *REV2* gene and no unspecific suppressor.

Evidence for Repair of Mispaired Bases in *S. cerevisiae*

We regard the multiply damaged bases induced by ionizing irradiation under anoxic conditions (31) as a model lesion for mispaired bases. These induced sites of denatured DNA, the so-called S1-sensitive sites (SSS), are substrates for endonuclease S1. This enzyme recognizes single-stranded, unpaired DNA regions and cuts both strands, yielding double-strand breaks (DSB; 20). To quantify DSB and SSS we used pulsed-field gel electrophoresis (8). This technique allows the separation of full-length chromosomal DNA of *S. cerevisiae* (3) and is sufficiently sensitive to identify as few as one DSB per five chromosomes. The induction of the lesions and their repair were quantified in each chromosome individually.

So far, we have investigated the sensitivity to ^{60}Co gamma rays under anoxic conditions of a repair-competent diploid (BK0) and the repair-deficient *rad18* diploid (BK18). A *rev2*(Ts) diploid showed a sensitivity comparable to that of BK0 if tested at 23°C and a sensitivity identical to that of BK18 at 36°C (unpublished data). The *RAD18* and *REV2* genes belong to the *RAD6* epistasis group, which deals with the repair of X-ray- and UV-induced damage. There are indications for a possible function of the *RAD18* gene in the repair of mispaired bases, since the *rad18* mutant is a

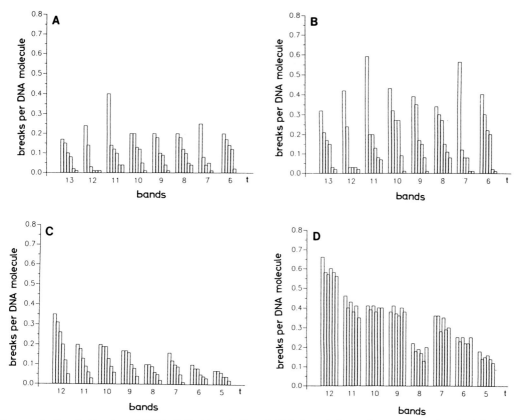

FIGURE 1. Repair kinetics of DNA DSB (A and C) and SSS (B and D) as determined with pulsed-field gel electrophoresis for various chromosomes (depicted by the band numbers 5/6 to 12/13) in the repair-competent diploid strain BK0 (A and B) and the repair-deficient diploid strain BK18 (C and D) after exposure to ^{60}Co gamma rays (200 Gy) followed by liquid holding recovery. For each band the breaks per DNA molecule were evaluated by densitometry of the intensity of the bands in the ethidium bromide-stained gel and plotted against the increasing time of liquid holding recovery (0, 2, 4, 8, 12, and 24 h). (The strains BK0 and BK18 carry various auxotrophic markers; they were kindly donated by B. A. Kunz, University of Manitoba, Winnipeg, Manitoba, Canada.)

spontaneous mutator and shows increased levels of gene conversion (11).

Using pulsed-field gel electrophoresis, we measured the induction of DSB and SSS and analyzed their repair under liquid holding conditions. We showed that both SSS and DSB are repaired in the repair-competent diploid, whereas the *rad18* diploid is competent in DSB repair, but deficient in the repair of SSS (Fig. 1). According to preliminary results, a comparable effect could be observed in the *rev2*(Ts) diploid if tested at the nonpermissive temperature of 36°C. So far it seems to be a promising experimental approach to regard SSS as a model for unpaired bases and to screen mutants possibly deficient in mismatch repair by measuring their repair capacity for SSS.

SUMMARY

We postulate that the *REV2* gene of *S. cerevisiae* controls correction of mispaired bases, e.g., mismatch repair. The *REV2* process is supposed to have error-avoiding functions for semiconservative DNA replication and repair replication during the course of excision repair, but simultaneously error-prone, mutation-creating functions at sites of DNA damage. The activity of mismatch repair is regulated by the growth phase of the cells and by excision repair dealing with DNA damage. The deficiency in mismatch repair attributed to the *rev2* mutation will be studied by following the repair kinetics of SSS, lesions induced by directly acting ionizing irradiation which contain one or more unpaired base pairs. Using pulsed-field gel electrophoresis it was shown that the *rad18* diploid strain is competent in the repair of DSB but deficient in the repair of SSS, which we take as an indication that SSS are a useful model lesion for studying the repair of unpaired or mispaired base pairs.

LITERATURE CITED

1. **Bishop, D. K., and R. D. Kolodner.** 1986. Repair of heteroduplex plasmid DNA after transformation into *Saccharomyces cerevisiae*. *Mol. Cell. Biol.* 6:3401–3409.
2. **Bishop, D. K., M. S. Williamson, S. Fogel, and R. D. Kolodner.** 1987. The role of heteroduplex correction in gene conversion in *Saccharomyces cerevisiae*. *Nature* (London) 328:362–364.
3. **Carle, G. R., and M. Olson.** 1984. Separation of chromosomal DNA molecules from yeast by orthogonal field alternation gel electrophoresis. *Nucleic Acids Res.* 12:5647–5664.
4. **Claverys, J.-P., and S. A. Lacks.** 1986. Heteroduplex deoxyribonucleic acid base mismatch repair in bacteria. *Microbiol. Rev.* 50:133–165.
5. **Eckardt, F., and R. H. Haynes.** 1977. Kinetics of mutation induction by ultraviolet light in excision deficient yeast. *Genetics* 86:225–247.
6. **Eckardt, F., S. J. Teh, and R. H. Haynes.** 1980. Heteroduplex repair as intermediate step of error-prone repair in yeast. *Genetics* 95:63–80.
7. **Friedberg, E.** 1987. The molecular biology of nucleotide excision repair of DNA: recent progress. *J. Cell Sci.* 6(Suppl.):1–23.
8. **Geigl, E.-M., F. Eckardt-Schupp, and U. Hagen.** 1986. Analysis of ^{60}Co-gamma-induced bulky lesions in yeast chromatin by orthogonal field alteration gel electrophoresis. *Yeast* 2:126.
9. **Haynes, R. H., and F. Eckardt.** 1980. Mathematical analysis of mutation induction kinetics. *Chem. Mutagens* 6:271–307.
10. **Haynes, R. H., F. Eckardt, and B. A. Kunz.** 1985. Analysis of non-linearities in mutation frequency curves. *Mutat. Res.* 15:51–59.
11. **Haynes, R. H., and B. A. Kunz.** 1981. DNA repair and mutagenesis in yeast, p. 371–414. *In* J. N. Strathern, E. W. Jones, and J. R. Broach (ed.), *The Molecular Biology of the Yeast Saccharomyces: Life Cycle and Inheritance*. Cold Spring Harbor Laboratory, Cold Spring Harbor, N.Y.
12. **Holliday, R.** 1964. A mechanism for gene conversion in fungi. *Genet. Res.* 5:282–304.
13. **James, A. P., and B. J. Kilbey.** 1977. The timing of UV mutagenesis in yeast: a pedigree analysis of induced recessive mutation. *Genetics* 87:237–248.
14. **James, A. P., B. J. Kilbey, and G. J. Prefontaine.** 1978. The timing of UV mutagenesis in yeast: continuing mutation in an excision-defective (*rad1-1*) strain. *Mol. Gen. Genet.* 165:207–212.
15. **Kilbey, B. J., T. Brychy, and A. Nasim.** 1978. Initiation of UV mutagenesis in *Saccharomyces cerevisiae*. *Nature* (London) 274:889–891.
16. **Lawrence, C. W.** 1982. Mutagenesis in *Saccharomyces cerevisiae*. *Adv. Genet.* 21:173–254.
17. **Lawrence, C. W., and R. B. Christensen.** 1978. Ultraviolet-induced reversion of *cyc1* alleles in radiation sensitive strains of yeast. II. *rev2* mutant strains. *Genetics* 90:213–226.
18. **Lemontt, J. F.** 1971. Mutants of yeast defective in mutations induced by ultraviolet light. *Genetics* 68:21–33.
19. **Maga, J. A., T. A. McClanahan, and K. McEntee.** 1986. Transcriptional regulation of DNA damage responsive (*DDR*) genes in different *rad* mutant strains of *Saccharomyces cerevisiae*. *Mol. Gen. Genet.* 205:276–284.
20. **Martin-Bertram, H., P. Hartl, and C. Winkler.** 1984. Unpaired bases in phage DNA after gamma-irradiation in-situ and in-vitro. *Radiat. Environ. Biophys.* 23:95–105.
21. **McClanahan, T., and K. McEntee.** 1984. Specific transcripts are elevated in *Saccharomyces cerevisiae* in response to DNA damage. *Mol. Cell. Biol.* 4:2346–2363.
22. **Radman, M., and R. Wagner.** 1986. Mismatch repair in *Escherichia coli*. *Annu. Rev. Genet.* 20:523–538.
23. **Ruby, S. W., and J. W. Szostak.** 1985. Specific *Saccharomyces cerevisiae* genes are expressed in response to DNA-damaging agents. *Mol. Cell. Biol.* 5:75–84.
24. **Siede, W., and F. Eckardt.** 1984. Inducibility of error-prone DNA repair in yeast? *Mutat. Res.* 129:3–11.
25. **Siede, W., and F. Eckardt.** 1986. Analysis of mutagenic DNA repair in a thermoconditional

mutant of *Saccharomyces cerevisiae*. III. Dose-response pattern of mutation induction in UV-irradiated *rev2*^{ts} cells. *Mol. Gen. Genet.* 202:68–74.
26. Siede, W., and F. Eckardt. 1986. Analysis of mutagenic DNA repair in a thermoconditional mutant of *Saccharomyces cerevisiae*. IV. Influence of DNA replication and excision repair on *REV2* dependent mutagenesis and repair. *Curr. Genet.* 10:871–878.
27. Siede, W., F. Eckardt, and M. Brendel. 1983. Analysis of mutagenic DNA repair in a thermoconditional mutant of *Saccharomyces cerevisiae*. I. Influence of cycloheximide of UV irradiated stationary phase *rev2*^{ts} cells. *Mol. Gen. Genet.* 190:406–412.
28. Siede, W., F. Eckardt, and M. Brendel. 1983. Analysis of mutagenic DNA repair in a thermoconditional mutant of *Saccharomyces cerevisiae*. II. Influence of cycloheximide on UV-irradiated exponentially growing *rev2*^{ts} cells. *Mol. Gen. Genet.* 190:413–416.
29. Siede, W., and F. Eckardt-Schupp. 1986. A mismatch repair based model can explain some features of UV mutagenesis in yeast. *Mutagenesis* 1:471–474.
30. Siede, W., and F. Eckardt-Schupp. 1986. DNA repair genes of *Saccharomyces cerevisiae*: complementing *rad4* and *rev2* mutations by plasmids which cannot be propagated in *Escherichia coli*. *Curr. Genet.* 11:205–210.
31. Ward, J. F. 1985. Biochemistry of DNA lesions. *Radiat. Res.* 103:103–111.
32. **Williamson, M. S., J. C. Game, and S. Fogel.** 1985. Meiotic gene conversion mutants in *Saccharomyces cerevisiae*. I. Isolation and characterization of *pms1-1* and *pms1-2*. *Genetics* 110:609–646.

Chapter 38

Is Exonuclease I Involved in the Regulation of the SOS Response in *Escherichia coli*?

Sidney R. Kushner and Gregory J. Phillips

The SOS response is a multigene network that is derepressed when *Escherichia coli* is exposed to conditions that either interfere with normal DNA metabolism or disrupt the integrity of the DNA (11, 17, 18). It is triggered through cleavage of the LexA repressor protein by an induced proteolytic activity associated with the RecA protein (10, 11). Inactivation of LexA causes increased transcription of 20 or more genes whose products are involved in diverse processes such as DNA repair, inhibition of cell division, and induced mutagenesis (17).

Despite an excellent understanding of the SOS response after LexA proteolysis, the exact nature of the inducing signal(s) for the activation of the RecA protein is unknown. In vitro, activation of RecA protein occurs when it forms a ternary complex with single-stranded DNA and a nucleoside triphosphate (2, 3). Accordingly, it has been presumed that the binding of RecA protein to single-strand gaps in duplex DNA could serve to activate its proteolytic activity. Recently, Lu et al. (13) demonstrated that RecA protein preferentially binds to UV-irradiated duplex DNA and that this binding is capable of inducing RecA proteolytic activity.

A potential problem with having single-strand gaps being the inducing signal relates to the accessibility of the activated RecA protein molecules to LexA repressor molecules. Induction of the system would require the gradual reduction of the LexA protein pool size through inactivation of free LexA proteins. This would lead to the eventual dissociation of more LexA repressor proteins and the eventual full induction of the system.

Another alternative is that other forms of single-stranded DNA may serve as the inducing signal(s). These could be either single-stranded regions in double-stranded DNA that have either 3' or 5' termini, or possibly small single-stranded oligonucleotides. In both cases these single-stranded molecules would have termini that would be susceptible to exonucleolytic degradation by enzymes such as exonuclease I or exonuclease VII. The fact that interruption of DNA replication is sufficient to induce the SOS response suggests the possibility that

---

Sidney R. Kushner • Department of Genetics, University of Georgia, Athens, Georgia 30602. *Gregory J. Phillips* • Department of Molecular Biology, Princeton University, Princeton, New Jersey 08544.

DNA damage is not necessary to activate the RecA proteolytic activity. In addition, the finding that RecA protein levels are elevated approximately fourfold in *recB21 recC22 sbcB15* strains (4) could be explained by the possibly higher intracellular concentrations of single-stranded DNA in the absence of exonuclease I.

EXONUCLEASE I OF *ESCHERICHIA COLI*

Exonuclease I was first shown by Lehman (8) and Lehman and Nussbaum (9) to specifically degrade single-stranded DNA in the $3' \rightarrow 5'$ direction, releasing mononucleotides. Inhibition of exonucleolytic activity occurs when the enzyme approaches within 6 to 8 nucleotides of a duplex region. Ray et al. (16) demonstrated that exonucleolytic activity is stimulated by the presence of the SSB protein. Purified exonuclease I has a monomeric molecular weight of 53,000 on the basis of DNA sequence analysis (14) and sodium dodecyl sulfate-polyacrylamide gel electrophoresis (15).

The in vivo function of exonuclease I is still not clear. Exonuclease I-deficient mutants of *E. coli* were first characterized by Kushner et al. (7) through the analysis of suppressors of *recB* and *recC* mutations. It was shown that the absence of exonuclease I (*sbcB* mutations) led to the suppression of both the repair and recombination deficiencies associated with the absence of exonuclease V (*recB recC*). Subsequently, a second type of exonuclease I mutation was identified (6) which suppressed only the repair deficiencies associated with *recB* and *recC* mutations. These so-called *xonA* mutations, however, were also deficient in the $3' \rightarrow 5'$ exonucleolytic activity. Although at this time it is not clear how to explain the differences between the two types of mutations, it is apparent that the loss of exonuclease I permits the full establishment of the RecF pathway for DNA repair and genetic recombination (1).

EXONUCLEASE I AND THE SOS RESPONSE

In our work with the cloned structural gene for exonuclease I, we noticed that strains carrying ColE1-derived recombinant plasmids with the *sbcB* coding sequence appeared more sensitive to UV light. The presence of approximately a 15-fold increase in exonuclease I activity resulted in a marked increased in the UV sensitivity of a wild-type control strain (Fig. 1). An intermediate level of UV sensitivity was observed with a fivefold increase in enzyme activity (Fig. 1). Additional experiments showed that strains carrying pDPK20 also were more sensitive to mitomycin C and nalidixic acid and showed reduced levels of Weigle reactivation (data not shown). Taken together, these results suggested a possible relationship between the level of exonuclease I and the extent of induction of the SOS system.

To examine this possibility in more detail, the plasmid pDPK20 was transformed into strain GW1030 (*dinB1* [5]). In this strain a damage-inducible gene is fused to β-galactosidase and it is possible to determine the extent of induction by measuring β-galactosidase activity. In the presence of a 15-fold increase in exonuclease I activity, the level of induction of the *dinB* gene was significantly reduced (Fig. 2). Similar results were obtained with *dinD1* and a *recA-lacZ* fusion (data not shown). Control experiments have demonstrated that high levels of exonuclease I do not nonspecifically inhibit the induction of β-galactosidase (data not shown).

CONCLUSIONS AND IMPLICATIONS

By increasing the intracellular concentration of exonuclease I, we have potentially exaggerated the normal role of this protein in the cell. Specifically, the enzyme may normally serve to reduce the intracellular

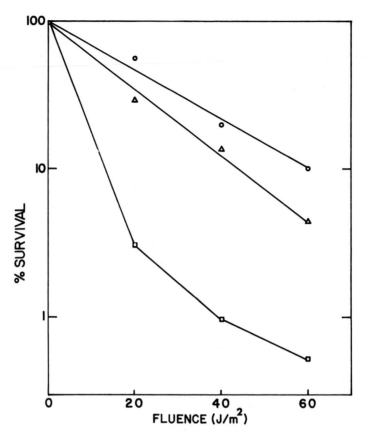

FIGURE 1. UV survival of strains as a function of intracellular exonuclease I concentration: survival curves of E. coli strains carrying various recombinant DNA plasmids. Cells were grown in Luria broth to approximately 10^8/ml and subsequently irradiated in a minimal medium as described by Kushner et al. (7). All of the strains were transformants of AB1157 (7). Symbols: ○, AB1157 containing the plasmid pBR322; □, AB1157 containing the recombinant plasmid pDPK20 (the $sbcB^+$ structural gene in pBR322); △, AB1157 containing the recombinant plasmid pDK12 (the $sbcB^+$ structural gene in a low-copy cloning vehicle). Exonuclease I activity was also measured in these strains. Cells were grown in Luria broth and assayed for exonuclease I activity as described by Prasher et al. (15). Activity (units per milligram of protein) was as follows: strain AB1157(pBR322), 92; AB1157(pDPK20), 1,500; AB1157(pDK12), 511.

concentrations of the single-stranded DNA molecules that are capable of activating the proteolytic activity of the RecA protein. These inducer substrates may be produced as by-products of normal DNA metabolism within the cell from reactions such as replication, recombination, or mismatch repair. Thus, exonuclease I could aid in maintaining full repression of the damage-inducible genes during normal growth periods. Additionally, the activity of the RecF pathway of recombination may be kept at a minimum by the additional role of exonuclease I in degrading a recombinational intermediate recognized by the RecF pathway enzymes.

After SOS induction, however, exonuclease I may immediately become saturated with substrate either from the SOS-inducing

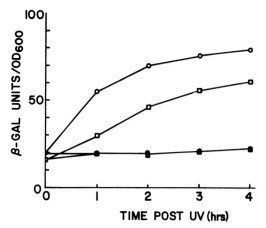

FIGURE 2. Induction of the *dinB1* locus as a function of exonuclease I concentration. Strain GW1030 was induced with UV light at a dose of 50 J/m^2. After induction, cells were removed at the times indicated and β-galactosidase activity was measured. Half of each culture was not induced. Symbols: GW1030(pBR322), uninduced (●) and induced (○); GW1030(pDPK20), uninduced (■) and induced (□).

signal(s) or from other DNA products generated by DNA repair processes (12), thereby allowing full expression of the SOS-induced genes. However, as the repair of damaged DNA nears completion, it is essential to shut off the system so that normal DNA replication and cell division can resume. This may in part be accomplished by degrading the inducing signal. Exonuclease I may also play a role at this step.

A serious problem with this hypothesis is that although increasing the level of exonuclease I inhibits full induction of the SOS system, it is necessary to have a 15-fold increase in activity to see an appreciable effect. This suggests that the substrate being acted on by exonuclease I under these circumstances is not a preferred one. Since the preferred substrate is single-stranded DNA and not short oligonucleotides, it may be assumed that the inducing signal is most likely either short oligonucleotides or short single-stranded regions with 3'-OH termini in otherwise double-stranded DNA. If either of these substrates is in fact the inducing signal, then either a different exonuclease is the primary enzyme involved in the degradation of the inducing signal, or an auxiliary protein may aid in altering the substrate preference of exonuclease I.

This hypothesis can be tested by examining an *E. coli* recombinant library for plasmids which cause increased repair deficiencies in a wild-type genetic background or, alternatively, which accentuate the repair deficiencies associated with increased exonuclease I activity. Although there is no direct evidence at this time regarding how the SOS response is controlled, it seems reasonable to assume that the turnover of the inducing signal is an important feature of the regulation. In addition, it also seems apparent that at least part of the induction results from some form of single-stranded DNA which contains 3'-OH termini. This would tend to rule out single-stranded gaps in double-stranded DNA and direct binding of the RecA protein to UV-induced photoproducts as the primary inducing signals.

ACKNOWLEDGMENTS. This work was supported by a Public Health Service grant from the National Institutes of Health (GM27997) to S.R.K.

LITERATURE CITED

1. **Clark, A. J.** 1973. Recombination deficient mutants of *E. coli* and other bacteria. *Annu. Rev. Genet.* 7:67–86.
2. **Craig, N. L., and J. W. Roberts.** 1980. *E. coli* recA protein-directed cleavage of phage lambda repressor requires polynucleotide. *Nature* (London) 283:26–30.
3. **Craig, N. L., and J. W. Roberts.** 1981. Function of the nucleoside triphosphate and polynucleotide in *Escherichia coli* recA protein-directed cleavage of phage lambda repressor. *J. Biol. Chem.* 256:8039–8044.
4. **Karu, A. E., and E. D. Belk.** 1982. Induction of *E. coli* recA protein via recBC and alternate pathways: quantitation by enzyme-linked immunosorbent assay (ELISA). *Mol. Gen. Genet.* 185:275–282.
5. **Kenyon, C. J., and G. C. Walker.** 1981. DNA-damaging agents stimulate gene expression at spe-

cific loci in *Escherichia coli*. *Proc. Natl. Acad. Sci. USA* 77:2819–2823.
6. **Kushner, S. R., H. Nagaishi, and A. J. Clark.** 1972. Indirect suppression of *recB* and *recC* mutations by exonuclease I deficiency. *Proc. Natl. Acad. Sci. USA* 69:1366–1370.
7. **Kushner, S. R., H. Nagaishi, A. Templin, and A. J. Clark.** 1971. Genetic recombination in *Escherichia coli*: the role of exonuclease I. *Proc. Natl. Acad. Sci. USA* 68:824–827.
8. **Lehman, I. R.** 1960. The deoxyribonucleases of *Escherichia coli*. I. Purification and properties of a phosphodiesterase. *J. Biol. Chem.* 235:1479–1487.
9. **Lehman, I. R., and A. L. Nussbaum.** 1964. The deoxyribonucleases of *Escherichia coli*. V. On the specificity of exonuclease I (phosphodiesterase). *J. Biol. Chem.* 239:2626–2636.
10. **Little, J. W.** 1983. The SOS regulatory system: control of its state by the level of RecA protease. *J. Mol. Biol.* 167:791–808.
11. **Little, J. W., and D. W. Mount.** 1982. The SOS regulatory system of *Escherichia coli*. *Cell* 29:11–22.
12. **Lloyd, R. G., and A. Thomas.** 1984. A molecular model for conjugation recombination in *Escherichia coli*. *Mol. Gen. Genet.* 197:328–336.
13. **Lu, C., C. R. H. Scheuermann, and H. Echols.** 1986. Capacity of RecA protein to bind preferentially to UV lesions and inhibit the editing subunit (epsilon) of DNA polymerase III: a possible mechanism for SOS-targeted mutagenesis. *Proc. Natl. Acad. Sci. USA* 83:619–623.
14. **Phillips, G. J., and S. R. Kushner.** 1987. Determination of the nucleotide sequence for the exonuclease I structural gene (*sbcB*) of *Escherichia coli* K-12. *J. Biol. Chem.* 262:455–459.
15. **Prasher, D. C., L. Conarro, and S. R. Kushner.** 1983. Amplification and purification of exonuclease I from *Escherichia coli*. *J. Biol. Chem.* 258:6340–6343.
16. **Ray, R. K., R. Reuben, I. Molineux, and M. Gefter.** 1974. The purification of exonuclease I from *Escherichia coli* by affinity chromatography. *J. Biol. Chem.* 249:5379–5381.
17. **Walker, G. C.** 1984. Mutagenesis and inducible responses to deoxyribonucleic acid damage in *Escherichia coli*. *Microbiol. Rev.* 48:60–93.
18. **Walker, G. C.** 1985. Inducible DNA repair systems. *Annu. Rev. Biochem.* 54:425–457.

Chapter 39

Mutations Suppressing Loss of Replication Control
Genetic Analysis of Bacteriophage λ-Dependent Replicative Killing, Replication Initiation, and Mechanism of Mutagenesis

Sidney Hayes

The control of λ replication initiation involves an activity of the λ repressor encoded by gene cI (Fig. 1). Replicative inhibition (39) of λ results from repressor binding at operator sites o_L and o_R. The repressor bound at o_R prevents rightward transcription from the major rightward promoter, p_R, and inhibits the expression of downstream genes O and P. The O and P gene products (gpO, gpP) are required for the initiation of λ replication. The binding of repressor at o_L prevents transcription from p_L. This action inhibits the expression of N, required for the antitermination of p_R-initiated transcription at the downstream site, t_{R1}, and other sites such as t_{L1} and t_{R2}. The product of gene N (gpN) is required for maximal depression of O and P transcription. By inhibiting transcription from p_R, the repressor also blocks a *cis* requirement for replication initiation described as transcriptional activation. This requirement is understood because it can be suppressed by ri^c (replication-inhibition-constitutive) mutations which lie outside and downstream from the assigned λ replicator region (10, 29). The need for transcriptional activation apparently represents a requirement for transcription across (or near) the origin, which may serve to denature the transcribed DNA strands.

REPLICATIVE-KILLING SELECTION

The examination of cell survival following infection or induction of λ mutants (4, 27, 37) led to the finding (30) that mutations preventing the initiation of λ replication from nonexcising, N-defective prophage also suppress killing of the cell (described as replicative killing [6]) by an induced N cI ts prophage. Among 50 mutants selected by Pereira da Silva et al. (30) from a thermally induced λ Nam7am53 cI ts857 prophage, 5 carried mutations within λ region x (likely within p_R), and the remainder carried muta-

---

Sidney Hayes • Department of Microbiology, College of Medicine, University of Saskatchewan, Saskatoon, Saskatchewan S7N 0W0, Canada.

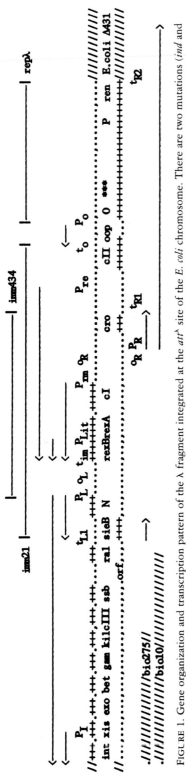

FIGURE 1. Gene organization and transcription pattern of the λ fragment integrated at the att^λ site of the E. coli chromosome. There are two mutations (ind and cI857) within gene cI in each of the RK⁺ constructs (see text). Mutation ind prevents spontaneous λ induction, making the repressor resistant to activated cellular RecA protease. Mutation cI857 confers temperature sensitivity at 39°C. Cro regulates repressor transcription. Gene cII is required for p_{re} transcription. Symbols: + + −, sense reading frame of gene; → and ←, transcription; ///, E. coli DNA; ***, O-iterons (four repeated target sites for O product binding).

tions that inactivated genes O or P. The Pereira da Silva (PDS) selection incorporated the N mutation to prevent prophage excision, dependent upon λ genes int and xis (Fig. 1). These genes, positioned downstream of t_{L1}, require an activity of gpN for their expression. In addition, the N mutation reduced late gene expression and cell lysis, which require the antitermination of rightward transcription at t_{R2}.

Rambach (32) utilized the PDS selection to isolate mutants in the cis-acting target site (termed the replicator) proposed to be recognized by the initiator protein in the replicon model (22). He reasoned that replicator mutations would represent O^+ P^+-complementable survivor clones. His replicator mutations r99, r96, and r93 (12-, 15-, and 24-base-pair [bp] deletions) have been used to define the origin region for initiation of bidirectional λ replication. They were localized within a DNA stretch joining the N-terminal and C-terminal domains of gene O (5, 9, 36). Rambach's study (32) required the following (unstated) assumptions: replication initiation will occur from an induced N mutant prophage with reduced transcription of genes O and P; sufficient rightward transcription across (near) the replicator will occur in an induced N mutant prophage to provide the requirement for transcriptional activation; the constitutive expression of the λ replication initiation proteins gpO and gpP and the other derepressed λ gene products would not be lethal to the host cell. These assumptions, still mostly unverified, are relevant to the isolation of replicative-killing-defective (RK⁻) survivor clones by replicative-killing-dependent selective pressure. Cells encoding the capacity for replicative killing are designated as having an RK⁺ (replicative-killing-competent) phenotype (19).

Brachet et al. (1) used the PDS selection to obtain RK⁻ survivors that were unable to express O and P in the absence of gpN, but could express these genes in its presence. The N gene product was found capable of suppressing polar insertions between t_{R1} and O that were not otherwise suppressible by suppressors of amber, ochre, or the UGA codon. One such mutation, $r32$, was shown to represent the insertion of IS2 (8). By the same selection method, mutations within y were shown to produce the Hyp (hyperimmune) phenotype resulting from cII-independent cI gene product synthesis, which inhibited the plating of λ v_1 v_2 v_3 (λ vir) (7, 12). Castellazzi et al. (2) combined the PDS selection with a selection for chlorate mutations to obtain very large deletions which included chromosomal gene $chlA$ through to O P in the λ prophage. The only published breakdown of survivor mutations from the PDS selection (6) reported the combined results for three selection variations and pooled data for 20 spontaneous and 20 nitrosoguanidine-induced mutations. These RK$^-$ survivor clones arose at a frequency of about 10^{-6}. Among 40 independent mutations conferring the RK$^-$ phenotype, 4 were recessive O mutations, 6 were recessive P mutations, 10 were x mutations (would not recombine with imm^{434}), 15 were N-suppressible polars, and 5 remained unclassified. One point mutation ($ti12$) rescued from this last group was subsequently sequenced within the assigned replicator region (5).

In summary, these studies suggested that replicative killing in the absence of N activity was suppressed by null mutations of O and P, inactivation of p_R, polar insertions between p_R and O, or mutations in the replicator within O. Do these mutations specify all the essential target sites and genes necessary for the bidirectional initiation of λ replication? For example, can a single mutation affect both the origin and O function? Can survivor clones carry mutations that negatively complement O and P or that are defective for participating *Escherichia coli* replication genes? An alternative selection, dependent upon obtaining clones that prevent vegetative phage growth, has been used to recognize *E. coli* genes whose products participate in initiation of λ replication (35).

RK MUTATEST SYSTEM

An attempt was made to modify the PDS selection on the assumption that variations in replicative-killing selective pressure could generate a variety of new suppressor mutations and contribute to an understanding of the mechanism and regulation of replication initiation. A defective prophage strain was selected containing deletion Δ431 (38), which removed $chlA$, the intervening *E. coli uvrB* and bio operon genes, and prophage DNA (including all late functions as Q, lysis functions, head and tail genes, and b region)—ending somewhere to the right of P. The Δ431 deletion eliminated a problem of selection interference caused by reduced expression of late prophage genes. Selector strains were made N^+ to avoid obtaining N-suppressible polar blocks (1) (observed in 37.5% of RK$^-$ survivor clones obtained by Dove et al. [6]); to ensure that transcriptional activation occurred during the selection; and to assure that the level of O and P gene expression was not limiting. The int-ral or int-kil regions of the prophage (Fig. 1) were substituted with $bio10$ or $bio275$ DNA from λ bio transducing phage to permit N^+ expression without the expression of gene kil. Both cro^+ and cro variations of the defective prophage were constructed to permit an evaluation of the involvement of Cro regulation in replication initiation and to compare the kinetics of oop and repressor establishment transcription from induced replication-competent or -defective prophage with earlier results obtained from induced intact prophage (14, 17, 18, 21). N mutant variants of some of the N^+ RK$^+$ constructs were made for assessing the influence of N activity on the selection for RK$^-$ survivors.

The prepared RK$^+$ selector strains (Fig. 1) contain gene cassettes responsible for the establishment and maintenance of immunity (imm [Fig. 1]) and the initiation of λ early replication (rep [Fig. 1]). Derepression of the λ fragment only arises by thermal inactivation of the repressor, accomplished by mechanically shifting cells grown at 30 to 42°C. This action synchronously derepresses the λ genes, resulting in initiation of replication from the nonexcisable λ fragment. The constitutive initiation of λ replication from this fragment is a lethal event to the host cell, which we continue to describe as replicative killing. Mutations that suppress the loss of replication control in these constructs arise spontaneously at frequencies of somewhat more than 10^{-6} to 10^{-8}, depending on the λ genes within the fragment. Mutant RK^- cells are selected by their ability to form a colony on an agar plate incubated at 42°C.

MODELS FOR APPEARANCE OF SPONTANEOUS AND INDUCED RK^- MUTATIONS

Spontaneous RK^- mutations can be imagined to arise by at least two mechanisms. (i) A spontaneous RK^- mutation is generated prior to λ induction. Such mutations could result from the misrepair of a DNA synthesis error occurring within the *rep* region of the λ fragment during *E. coli* replication. For example, an occasional error in strand identification during mismatch repair would result in a mutation (3). (ii) An RK^- mutation is formed after initiation of replication from the λ fragment by misrepair of a DNA synthesis error introduced while the λ replication fork moves across rep^λ.

Induced RK^- mutations might arise by model 2 if the cell treatment can elicit a greater frequency of replication errors, e.g., due to a perturbation of the chromosome or replication apparatus, perhaps even if the perturbing agent is not itself genotoxic. Induced mutagenesis could also arise by the following models. (iii) An RK^- mutation is generated in response to a DNA lesion (methylation, intercalation, cross-linking, or addition of a bulky adduct) produced by a genotoxic agent, which provokes misreplication. Adducts that block movement of the replication fork will induce an SOS response with the potential for error-prone repair. (iv) Macromutational events (e.g., large deletions, gene rearrangements) may arise by aberrant recombination between the arms of the λ replication forks or by the activation of an insertion sequence or target site upon the movement of a λ replication fork beyond the λ fragment into the contiguous *E. coli* DNA, termed herein escape replication.

RK MUTATEST

The short-term RK mutatest assay system was developed to examine whether replicative-killing selection could be used to measure the mutagenic potential of environmental substances. The RK test was validated for the prediction that the frequency with which RK^- survivor colonies appear after selection at 42°C will increase for mutagen-treated compared with untreated RK^+ cells (16, 20). In the assay, RK^+ cells are transiently exposed to the agent before being placed at 42°C. The assay endpoint is a mutation, detected by the appearance of an RK^- clone. Equivalent mutagenic potencies were observed (15) for the RK mutatest system and the SOS chromotest (31) (in which the endpoint represents a measurement of SOS response). Although the RK system has a much larger target than many mutagenesis assay systems and the induced RK^- mutations can be directly selected, the mutated DNA is not readily amplified or sequenced because it is contained within cell chromosome. Each mutation must be cloned out. This is a serious drawback of the system for rapid determination of the spectrum of RK^- mutations. This limitation may change if direct genomic-sequencing techniques can be used to characterize selected RK^- mutations. Since the target for mutagenesis in the RK system is a large, highly regulated DNA replication region, many potential mutant phenotypes are possible and numerous criteria are required to localize and characterize the divergent RK^- mutations. The determination of mutagenic spectrum is made accordingly more difficult. The RK mutatest could have an advantage over simpler systems in being able to identify agents which induce large chromosomal deletions and gene rearrangements or for studying gene amplification. The system also was found useful in identifying a conditionally genotoxic agent that stimulated the occurrence of large deletions in RK mutatest cells. We have proposed that such agents might perturb the activity of enzymes involved with the topological manipulation of DNA and thereby generate DNA lesions (S. Hayes, D. Duncan, C. Hayes, J. Blushke, V. Bennett, and S. Lal, submitted for publication).

What is the interdependence between the RK mutatest selection, cellular mechanisms for mutagenesis, and the initiation of λ replication? The forward selection of RK$^-$ survivor clones derived from genotoxin-exposed RK$^+$ cells requires that a premutagenic DNA lesion introduced by the mutagenic agent into the targeted λ fragment be either erroneously repaired or misreplicated in a manner that causes a heritable cessation of λ-initiated replication. Accordingly, the measured mutational endpoint must depend (in some part) on cellular (error-prone) DNA repair and replication functions. The understanding of cellular functions that process (fix) DNA lesions and give rise to mutations could be further advanced through attention to the interconnection between the powerful forward RK selection and mechanisms for inducible cellular repair.

CATEGORIZATION OF RK$^-$ MUTATIONS

The genetic mapping of RK$^-$ mutations has been carried out by complementation analysis, phage exclusion, phage marker rescue, and marker rescue from plasmids carrying a fragment of λ. Physical mapping and sizing of large insertions, deletions, and gene rearrangements is by DNA blotting, using fragments of λ DNA as probes against a restriction endonuclease digest of chromosomal DNA from the RK$^-$ survivor isolate. The selected and recloned RK$^-$ survivors are initially screened by a series of stab tests to plates spread with λ *vir* and λ *c*I and to an agar overlay (functional immunity [FI] test) containing a lysogen of λ *imm*^{434} T plus free λ *imm*^{434} *c*I (with plate incubated at 39°C). The FI test is a quick, sensitive technique to determine whether *imm*^{λ} can be rescued by a double recombinational event (17). The RK$^-$ clones not lysed by λ *vir* are designated RK$^-$ Hd$^-$ (host defect). These are arbitrarily considered to have either a mutation in a host gene whose product participates in vegetative λ growth or a λ-fragment mutation whose effect is to negatively complement for the growth of λ *vir*. The various Hd$^-$ survivors represent about 4 to 6% of selected RK$^-$ clones. The RK$^-$ clones that are lysed by λ *vir*, retain λ immunity at 30°C (Imm$^+$ phenotype), and are FI$^+$ are designated RK$^-$ Ilr$^-$ (defective for initiation of lambda replication). These clones tentatively carry mutations within the λ fragment which prevent replication initiation. This is confirmed by an inability to rescue *imm*^{λ} or *rep*^{λ} phage after infection of the clone with hybrid λ (*imm-rep*)^{P22} phage. The simplest interpretation for mutations conferring both FI$^-$ and Imm$^-$ phenotypes is that they represent deletions into *c*I and p_R. In a recent examination of spontaneous RK$^-$ mutations from four N^+ RK$^+$ selector strains, 256 of 650 RK$^-$ isolates were of this class. The physically mapped RK$^-$ FI$^-$ Imm$^-$ isolates have had large deletions (greater than 10 kilobases [kb]), which included the entire integrated λ fragment (Hayes et al., submitted). These clones may be analogous to survivor mutations assigned as x^- (6, 30).

SELECTION BY COMPLEMENTATION

To achieve the simple goal of obtaining more replicator mutations, RK$^-$ survivor clones from RK$^+$ selector strains (Fig. 1) were examined for isolates that would complement for O and P activity at 42°C. A total of 51 clones were identified as $O^+ P^+$ among 800 RK$^-$ isolates from four N^+ RK$^+$ selector strains. In subsequent isolations $N^+ P^+$ clones were first identified and then screened for O^+ activity; alternatively, O^+ clones were identified and then screened for P^+ or $N^+ P^+$ activity. Many of the initial isolates were mapped by phage crosses. Fifteen $O^+ P^+$ isolates were characterized from an Nam7am53 RK$^+$ strain. However, the selected $O^+ P^+$ isolates, obtained from either the N^+ RK$^+$ or N RK$^+$ strains, with time, unaccountably lost their ability to effectively complement for gene O (Table 1).

The complementation results are compared (Table 2) for the RK$^-$ isolate ilr-2-11e (which had lost O^+ activity) and the r96 isolate obtained by Rambach (32). The complementation pattern of isolate ilr-2-11e illustrates a frequent observation; the immunity of the amber infecting phage influenced the degree of complementation for O and P (see also Table 1). Is ilr-2-11e O^+ or O? How had Rambach obtained replicator-defective isolates that retained their $O^+ P^+$ phenotype? What strain differences contribute to the instability of the $O^+ P^+$ isolates we obtained? Had each $O^+ P^+$ isolate acquired an additional mutation?

In considering the last two questions, it is possible that the O complementation data were misleading, since an analogous complementation pattern to the r mutation (Table 2) was expected. Using a complementation method very similar to the one described here (only the host cells contained either lambda N to cII or N to P fragments cloned into pBR322), Rao and Rogers (33) were unable to demonstrate that cells carrying the O^+ plasmid could complement (i.e., support the efficient plating of) an infecting λ imm^{21} Oam29 phage. These cells carried an estimated 50 plasmid copies per cell at 32°C and 260 copies after inactivation of cI ts857 at 42°C (i.e., the cells with N to P-carrying plasmid). In contrast, they had no similar

TABLE 1
Loss of Ability To Complement for O

Infecting phage	% EOP on RK$^-$ isolate from N RK$^+$ construct at 42°C[a]		
	line	ilr-4RT	ilr-47RT
Initial complementation			
λ imm^{434} cI Oam8	1	64	20
λ imm^{434} cI Pam3	2	31	33
Subsequent complementation[b]			
λ imm^{434} cI Oam8	3	6.4	1.1
λ imm^λ cI ts857 Oam8	4	<0.1	<0.1
λ imm^{21} cI Oam8	5	<0.1	<0.1
λ imm^{434} cI Pam3	6	0.1	0.6
λ imm^λ cI ts857 Pam3	7	<0.1	<0.1
λ imm^{21} cI Pam3	8	18	20

[a] Tryptone agar plates were overlaid with a mixture containing an overnight culture prepared either from an RK$^-$ clone, strain W3350 sup^0, or strain TC600 $supE$ (0.25 ml), and a phage carrying an amber mutation in N, O, or P (each diluted to give 2×10^2, 2×10^4, or 2×10^6 to 2×10^8 phage per plate), plus 2.5 ml of top agar. Parallel plates were incubated at 30 and 39 to 42°C. The efficiency of plating (EOP) on the RK$^-$ cells was compared to phage titer on strain TC600 and to amber-reversion frequency (PFU on W3350 divided by PFU on TC600). The induced RK$^-$ clones were assigned as not complementing for the marker if the EOP of the plated amber phage was below 0.1% of its EOP on TC600 (R. Tuer, M.S. thesis, University of Saskatchewan, Saskatoon, 1984; R. Tuer and S. Hayes, unpublished data).
[b] Measured about 1 year later by using culture isolates that had been stored on tryptone slants at ambient temperature.

difficulty demonstrating complementation by the plasmid genes for Nam7am53 or Pam3 mutations carried by infecting imm^{434} or imm^{21} phage. Very weak evidence was provided by a burst size experiment, that some O complementation had in fact occurred. Although the products of genes O and P have both been considered to be diffusible because O and P mutant phage complement each other (11, 23), other studies suggest difficulties in demonstrating complementation for gene O. Rambach (32) indicated that certain ori O^+ phage weakly complement ori^+ O phage. Kleckner (24) suggested that the λ O product might act in cis (or be poorly complemented) under N mutation conditions. Hayes and Szybalski (21) examined the requirement of O and P gene products for lit (late immunity tran-

TABLE 2
Complementation by Induced Replicator-Defective Prophage

Infecting phage	% EOP on RK⁻ survivor isolate at 42°C[a]	
	r96	ilr-2-11e
Controls		
λ $imm^λ$ Eam68	13	0
λ $(imm$-$rep12$amN14$)^{P22}$	0	0
λ Nam7am53 imm^{434}	0	130
λ $imm^λ$ wild type	8	42
Pam complementation		
λ imm^{21} cI Pam3	54	45
λ imm^{434} cI Pam3	54	44
λ $imm^λ$ cI ts857 Pam3	40	3
Oam complementation		
λ imm^{434} Oam29	38	4.5
λ $imm^λ$ cI ts857 Oam29	21	0.08
λ imm^{434} Oam8	46	1.6
λ $imm^λ$ cI ts857 Oam8	40	0.08

[a] The complementation results were obtained by the method described in footnote a of Table 1. ilr-2-11e was obtained by C. Hayes as an RK⁻ survivor isolate that was initially selected because it complemented well for O and in subsequent screens was found to complement for N and P. It was from the N⁺ RK⁺ construct SA431 bio275 illustrated in Fig. 1 and in text. The replicator-defective r96 strain was isolated by Rambach (32) from strain M72 (λ Nam7am53 cI ts857) and obtained from M. Furth and W. F. Dove as a sup^+ variant (9,10) that has lost (line 2) its ability to suppress amber mutations.

scription) RNA (rexB) synthesis and found gpP complemented in *trans*, whereas gpO would not complement. Since O mutants prevented lit RNA synthesis, gpO was suggested to be required in *cis*. In light of our experience of having many initial O⁺ P⁺ RK⁻ isolates become phenotypically O mutants, it would appear that some nonmutational cellular event can influence the complementation properties of gene O. It is possible that we have been detecting a normal event, perhaps a regulatory switch between the replicator-dependent mode of bidirectional replication and the late-rolling-circle mode of λ replication.

NONIMMUNE EXCLUSION

An initial approach of using complementation analysis to characterize the mutations responsible for the RK⁻ phenotype of survivor isolates revealed that the predominant class of survivors was phenotypically N O P and permitted the plating of λ *vir*. In determining whether these isolates had acquired a pleiotropic mutation with the effect of inhibiting complementation, we identified among general RK⁻ Ilr⁻ isolates a new phenotype described as nonimmune exclusion (Nie). Many of the Nie clones complemented as being N O P mutants. Among the clones with a Nie phenotype, two classes (A and B) were identified. Survivors from each class exhibited typical λ immunity at 30°C, preventing the plating of λ wild type but not λ imm^{434} phage. Neither class A nor B isolates supported plaque formation by these phage at 42°C, suggesting the production of an inducible inhibitor of λ gene expression or activity. The class B isolates were distinguished because they permitted the formation of rare plaques at a somewhat higher frequency than did type A survivors. To pursue the basis for nonimmune exclusion we randomly chose seven class B RK⁻ Ilr⁻ isolates for use as hosts to isolate λ mutations which suppressed Nie. Spontaneous mutants of λ wild type were obtained (frequency about 10^{-6}) that overcame Nie and plated at 42°C on cell lawns made from the cells of each RK⁻ clone. The acquired λ se (suppress exclusion) mutations were genetically mapped and sequenced within three sites in the rightward λ operator, o_R (19).

Not all of the phenotypically N O P mutant isolates exhibited the Nie phenotype. The clones lacking Nie efficiently permitted the plating of wild-type imm^{434} and $imm^λ$ phages at 42 and 30°C. These clones were also FI⁻ (i.e., no marker rescue of $imm^λ$). DNA blotting and hybridization studies on representative isolates indicated that they had acquired large deletions encompassing the λ fragment.

The effect of nonimmune exclusion was examined by plating 60 modified or substituted λ phage on class-B isolates at 42°C. Most were inhibited. We have found that a few point mutations in cI were able to partially suppress Nie and may negatively complement for repressor activity. Coincidentally, the selected se mutations within o_R confer to λ se phage a p_{rm} mutant phenotype, making them defective for repressor maintenance transcription (Fig. 1). Although much work remains to elucidate the mechanism of nonimmune exclusion, my present working model is that Nie may reflect the binding of inactive cI ts857 repressor or inactive Cro protein to a phage target site (e.g., o_R) and that this action blocks replication initiation. Determination of the mechanism for Nie may help explain replicative inhibition.

The RK^- mutations in the seven class-B Nie isolates were mapped by genetic and physical techniques. Four of the mutants complement as N O P mutants, two as N^+ O P, and one as N O^+ P^+. The first six mutations mapped to the right as Oam205, and the remaining mutation mapped within y-cII-oop-O (N-terminal end). (This last mutant, which appears to be duplicated for 2.5 kb of the λ sequence, will complement Oam8 on imm^{21} phage but will not complement Oam8 carried on λ cI ts857 Oam8, even though λ cI ts857 will plate these RK^- cells at 42°C. Similar results were seen for complementation of the Pam3 mutation.) Four of the isolates have 1.4-kb insertions in P. Two of these insertions are not simple and involve gene rearrangement. These four insertions appear to be IS2 by parallel blotting analysis with an IS2 probe.

ESCAPE REPLICATION FROM THE λ FRAGMENT

The presumptive transposition of IS2 into the λ P gene in four of seven RK^- isolates appeared to be more than a coincidence. Does escape replication moving from the induced λ fragment into the E. coli chromosome serve as an inducer of IS2 transposition? Southern blots probed with IS2 were compared for BstEII-digested chromosomal DNA extracted from noninduced and 90-min-induced RK^+ cultures. These strains have more than 10 IS2 bands per chromosome. Among five isolates of one RK^+ strain, one isolate had one IS2 band that was clearly amplified upon λ induction (and apparently was inserted near to, but not within, the λ fragment). The remaining bands were of the same intensity as the bands from the parallel noninduced sample. This experiment revealed two interesting observations. Escape replication from the λ fragment into the E. coli chromosome does occur. However, the escape replication from the λ fragment does not continue around the E. coli chromosome since the remaining IS2 bands were not amplified after 90 min of λ-fragment induction. This observation was unexpected, but it may help to explain why λ has not been shown to integratively suppress E. coli replication. During these studies we also noted that IS2 transposition was elevated in RK^+ constructs with the Δ431 deletion, but not in the parent of this strain with an intact λ prophage. Could differences in escape replication between an induced, intact λ N prophage and the induced λ fragment (Fig. 1) account for the difference in RK^- survivor spectra, particularly the preponderance of macromutations in survivors from RK^+ strains? Does λ have sequences at its prophage ends which limit or prevent escape replication so as to limit the induction of insertion sequence transposition upon prophage induction?

REPLICATIVE-KILLING EFFECT

The lethal early effects of replicative killing differ from the effect produced by the expression of λ kil gene. Greer (13) noted that replication-defective kil^+ and kil cultures, which are thermally induced by shifting from 30 to 42°C, continue to increase in optical density at the same rate for 3.5 h and form filaments (resulting from the continuation of cellular metabolism and cell growth, while DNA synthesis is prevented). In contrast, a similar temperature shift for either int-to-kil deleted or competent RK$^+$ cultures results in complete inhibition of cell growth (A_{575}) within about 30 min for *cro* derivatives and by 60 min for *cro*$^+$ derivatives, along with a 50- to 100-fold drop in cell viability for cultures induced 1 h (20; S. Hayes and C. Hayes, unpublished data). No cellular filamentation was observed through 20 h of incubation (42°C, in culture) for these *kil*$^+$ or *kil* RK$^+$ strains. In contrast, filament formation was readily observed after thermal induction of an intact *O* mutant prophage (S. Hayes, C. Hayes, and H. Bull, unpublished data). The induction of high levels of *P* protein in cells harboring a plasmid encoding this gene (26) has also been reported to produce cellular filamentation and to interfere with overall bacterial DNA synthesis, but the high level of gpP had no lethal effect on the bacterial cells (26). Alternative studies (40) have suggested that the overproduction of λ initiators from a plasmid seems to depend on the simultaneous expression of both *O* and *P* genes and does not occur when either *O* or *P* is separately cloned.

Both genetic and biochemical studies have indicated that λ gpP interacts with the *E. coli dnaB* gene product (25) involved both in prepriming and in DNA synthesis elongation (11). Excessive gene *P* expression could stall or slow down an *E. coli* replication fork. *P*$^+$ RK$^-$ isolates containing a single integrated copy of the λ fragment apparently do not produce sufficient gpP to inhibit cell growth, since these isolates plate with an efficiency of plating of 1.0 at 42°C. However, an induced single-copy λ fragment, which remains integrated and cannot be excised from the chromosome, kills its host cell about 10^4- to 10^7-fold more efficiently than if the same λ fragment is induced from a multicopy plasmid (Hayes and Hayes, unpublished data). It would appear that the initiation of λ replication in *cis* from a nonexcisable λ fragment within the chromosome is an even more important component of replicative killing than the elevated expression of initiator proteins.

In examining replicative-killing selection, we have identified induction-resistant (IR) cells. If RK$^+$ selector cells are directly spread (10^6 to 10^8 cells) on tryptone agar plates and the plates are incubated at 42°C for 24 to 48 h, only rare RK$^-$ cells form colonies at a frequency of 10^{-6} to 10^{-8}. If these same plates are reincubated at 30°C, additional colonies appear. The new colonies arise from IR cells that remained viable (but could not form a colony) on the tryptone plate during incubation at 42°C. The IR colonies appear at about two or more times the frequency of the RK$^-$ survivor clones. Many retain an RK$^+$ phenotype and are killed if returned to 42°C. Surprisingly, some IR clones acquired large λ-fragment deletions (Hayes and Hayes, unpublished data). Some were temperature sensitive for growth at 42°C. The frequency for appearance of IR clones was substantially lower when they were selected from RK$^+$ cultures grown in rich medium, compared with selection from RK$^+$ cells grown in minimal (low-phosphate) medium (fivefold-longer culture generation time). This observation suggests that replicative killing of the host involves at least two effects: (i) it inhibits the initiation of *E. coli* replication from resting cells, and (ii) it kills cells that are undergoing cellular replication or cells that initiate replication before the build-up of initiation inhibitor. The accumulation of gpP is a likely candidate for retarding *E. coli* replication initiation.

The mechanism of replicative killing is of relevance to mutagenesis testing. Agents which do not themselves induce an SOS response could be identified as mutagenic if the premutagenic fixing capacity associated with SOS functions was induced in response to an effect of replicative killing. Untargeted mutagenesis, resulting from the induction of SOS functions (41), could enhance the appearance of RK$^-$ survivors obtained from RK$^+$ cells that were pretreated with a mutagen. It is unclear whether initiation of replication from the λ fragment in RK$^+$ cells will trigger induction of the SOS response (34,

41). This might occur either if the build-up of gpP interferes with the progression of the *E. coli* replication fork, if the initiation of λ-fragment replication creates a steric block to migration of the *E. coli* replication fork, or if a modified *E. coli* replication complex loses fidelity resulting in the introduction of multiple mismatches in the replicated *E. coli* chromosome. The filamentation by cells that were overproducing gpP from a multicopy plasmid (26) is explained if the high level of gpP triggers an SOS response, resulting in synthesis of the *sulA* product (28, 41), a known inhibitor of cellular septation.

CONCLUSION

The integrated λ fragment in RK$^+$ cells represents a form of cryptic DNA that is tolerated if its replication remains repressed, but is highly stressful to the cell when the λ fragment is derepressed and induced to replicate. The induced replicative stress to the cell results from the loss of λ replication control. New categories of mutations which suppress replicative killing have been recognized among the RK$^-$ isolates. The variety and complexity of these mutations are far greater than was initially proposed from examination of survivors for induced N cI ts prophage. The characterized map positions, DNA alterations, and metabolic effects of the RK$^-$ mutations are of interest because they serve to define the involvement of host and phage genetic elements in the regulation of λ replication initiation.

ACKNOWLEDGMENTS. This work was supported by Public Health Service grant GM 30077 from the National Institutes of Health and grant MA 9277 from the Medical Research Council of Canada.

LITERATURE CITED

1. Brachet, P., H. Eisen, and A. Rambach. 1970. Mutations of coliphage λ affecting the expression of replicative functions O and P. *Mol. Gen. Genet.* 108:266–276.
2. Castellazzi, M., P. Brachet, and H. Eisen. 1972. Isolation and characterization of deletions in bacteriophage λ residing as prophage in *E. coli* K12. *Mol. Gen Genet.* 117:211–218.
3. Claverys, J.-P., and S. A. Lacks. 1986. Heteroduplex deoxyribonucleic acid base mismatch repair in bacteria. *Microbiol. Rev.* 50:133–165.
4. Cross, R. A., and M. Lieb. 1970. Heat-sensitive early functions in induced λ N*sus* lysogens. *Virology* 6:33–41.
5. Denniston-Thompson, K., D. D. Moore, K. E. Kruger, M. E. Furth, and F. R. Blattner. 1977. Physical structure of the replication origin of bacteriophage lambda. *Science* 198:1051–1056.
6. Dove, W. F., H. Inokuchi, and W. Stevens. 1971. Replication control in phage lambda, p. 747–777. *In* A. D. Hershey (ed.), *The Bacteriophage Lambda*. Cold Spring Harbor Laboratory, Cold Spring Harbor, N.Y.
7. Eisen, H., P. Barrand, W. Spiegelman, L. F. Reichardt, S. Heinemann, and C. Georgopoulos. 1982. Mutants in the y region of bacteriophage λ constitutive for repressor synthesis: their isolation and the characterization of the Hyp phenotype. *Gene* 20:71–81.
8. Fiandt, M., W. Szybalski, and M. H. Malamy. 1972. Polar mutations in *lac, gal,* and phage λ consist of a few IS-DNA sequences inserted with either orientation. *Mol. Gen. Genet.* 119:223–231.
9. Furth, M. E., F. R. Blattner, C. McLester, and W. F. Dove. 1977. Genetic structure of the replication origin of bacteriophage lambda. *Science* 198:1046–1051.
10. Furth, M. E., W. F. Dove, and B. J. Meyer. 1982. Specificity determinants for bacteriophage λ DNA replication. III. Activation of replication in λri^c mutants by transcription outside of *ori*. *J. Mol. Biol.* 154:65–83.
11. Furth, M. E., and S. H. Wickner. 1983. Lambda DNA replication, p. 145–173. *In* R. W. Hendrix, J. W. Roberts, F. W. Stahl, and R. A. Weisberg (ed.), *Lambda II*. Cold Spring Harbor Laboratory, Cold Spring Harbor, N.Y.
12. Georgopoulos, C., N. McKittrick, G. Herrick, and H. Eisen. 1982. An IS4 transposition causes a 13-bp duplication of phage λ DNA and results in the constitutive expression of the *cI* and *cro* gene products. *Gene* 20:83–90.
13. Greer, H. 1975. The *kil* gene of bacteriophage λ. *Virology* 66:589–604.
14. Hayes S. 1979. Initiation of coliphage lambda replication, *lit, oop* RNA synthesis, and effect of gene dosage on transcription from promoters P_L, P_R, and P_R'. *Virology* 97:415–438.
15. Hayes, S., and A. Gordon. 1984. Validating RK

test: correlation with Salmonella mutatest and SOS chromotest assay results for reference compounds and influence of pH and dose response on measured toxic and mutagenic effects. *Mutat. Res.* 103:107–111.

16. Hayes, S., A. Gordon, I. Sadowski, and C. Hayes. 1984. RK bacterial test for independently measuring chemical toxicity and mutagenicity: short-term forward selection assay. *Mutat. Res.* 130:97–106.

17. Hayes, S., and C. Hayes. 1978. Control of λ repressor prophage and establishment transcription by the product of gene *tof*. *Mol. Gen. Genet.* 164:63–76.

18. Hayes, S., and C. Hayes. 1979. Control of bacteriophage λ repressor establishment transcription. Kinetics of *l*-strand transcription from the *y-c*II-*oop*-O-P region. *Mol. Gen. Genet.* 170:75–88.

19. Hayes, S., and C. Hayes. 1986. Spontaneous λ o_R mutations suppress inhibition of bacteriophage growth by nonimmune exclusion phenotype of defective λ prophage. *J. Virol.* 58:835–842.

20. Hayes, S., C. Hayes, E. Taitt, and M. Talbert. 1983. A simple, forward selection scheme for independently determining the toxicity and mutagenic effect of environmental chemicals: measuring replicative killing of *Escherichia coli* by an integrated fragment of bacteriophage lambda DNA, p. 61–77. *In* A. R. Kolber, T. K. Wong, L. D. Grant, R. S. DeWoskin, and T. J. Hughes (ed.), *In Vitro Toxicity Testing of Environmental Agents, Part A*. Plenum Publishing Corp., New York.

21. Hayes, S., and W. Szybalski. 1975. Role of *oop* RNA primer in initiation of coliphage lambda DNA replication, p. 486–512. *In* M. Goulian and P. Hanawalt (ed.), *DNA Synthesis and Its Regulation*. Benjamin-Cummings Publishing Co., Menlo Park, Calif.

22. Jacob, F., S. Brenner, and F. Cuzin. 1963. On the regulation of DNA replication in bacteria. *Cold Spring Harbor Symp. Quant. Biol.* 28:329–348.

23. Kaiser, A. D. 1971. Lambda DNA replication, p. 195–210. *In* A. D. Hershey (ed.), *The Bacteriophage Lambda*. Cold Spring Harbor Laboratory, Cold Spring Harbor, N.Y.

24. Kleckner, N. 1977. Amber mutants in the O gene of bacteriophage λ are not efficiently complemented in the absence of phage N function. *Virology* 79:174–182.

25. Klein, A., E. Lanka, and H. Schuster. 1980. Isolation of a complex between the P protein of phage λ and the *dnaB* protein of *Escherichia coli*. *Eur. J. Biochem.* 105:1–6.

26. Klinkert, J., and A. Klein. 1979. Cloning of the replication gene P of bacteriophage lambda: effects of increased P-protein synthesis on cellular and phage DNA replication. *Mol. Gen. Genet.* 171:219–227.

27. Lieb, M. 1966. Studies of heat-inducible λ prophage. III. Mutations in cistron N affecting heat induction. *Genetics* 54:835–844.

28. Mizusawa, S., and S. Gottesman. 1983. Protein degradation in *Escherichia coli:* the *lon* gene controls the stability of *sulA* protein. *Proc. Natl. Acad. Sci. USA* 80:358–362.

29. Moore, D. D., and F. R. Blattner. 1982. Sequence of λri^c5b. *J. Mol. Biol.* 154:81–83.

30. Pereira da Silva, L., H. Eisen, and F. Jacob. 1968. Sur la replication du bacteriophage λ. *C. R. Acad. Sci.* 266:926–928.

31. Quilardet, P., O. Huisman, R. D'ari, and M. Hofnung. 1982. SOS chromotest, a direct assay of induction of an SOS function in *Escherichia coli* K-12 to measure genotoxicity. *Proc. Natl. Acad. Sci. USA* 79:5971–5975.

32. Rambach, A. 1973. Replicator mutants of bacteriophage lambda: characterization of two subclasses. *Virology* 54:270–277.

33. Rao, R. N., and S. G. Rogers. 1978. A thermoinducible λ phage-colE1 plasmid chimera for the overproduction of gene products from cloned DNA segments. *Gene* 3:247–263.

34. Roberts, J. W., and R. Devoret. 1983. Lysogenic induction, p. 123–144. *In* R. W. Hendrix, J. W. Roberts, F. W. Stahl, and R. A. Weisberg (ed.), *Lambda II*. Cold Spring Harbor Laboratory, Cold Spring Harbor, N.Y.

35. Saito, H., and H. Uchida. 1977. Initiation of the DNA replication of bacteriophage lambda in *Escherichia coli* K12. *J. Mol. Biol.* 113:1–25.

36. Scherer, G. 1978. Nucleotide sequence of the O gene and of the origin of replication of bacteriophage lambda DNA. *Nucleic Acids Res.* 5:3141–3156.

37. Sly, W. S., H. A. Eisen, and L. Siminovitch. 1968. Host survival following infection with or induction of bacteriophage lambda mutants. *Virology* 34:112–127.

38. Stevens, W. F., S. Adhya, and W. Szybalski. 1971. Origin and bidirectional orientation of DNA replication of coliphage lambda, p. 515–533. *In* A. D. Hershey (ed.), *The Bacteriophage Lambda*. Cold Spring Harbor Laboratory, Cold Spring Harbor, N.Y.

39. Thomas R., and L. E. Bertani. 1964. On the control of the replication of temperate bacteriophages superinfecting immune hosts. *Virology* 24:241–253.

40. Tsurimoto, T., T. Hase, H. Matsubara, and K. Matsubara. 1982. Bacteriophage lambda initiators: preparation from a strain that overproduces the O and P proteins. *Mol. Gen. Genet.* 187:79–86.

41. Walker, G. C. 1984. Mutagenesis and inducible responses to deoxyribonucleic acid damage in *Escherichia coli*. *Microbiol. Rev.* 48:60–93.

Chapter 40

Use of Predesigned Plasmids To Study Deletions
Strategies for Dealing with a Complex Problem

Elias Balbinder

Although the existence of large genetic rearrangements such as deletions, duplications, inversions, and translocations has been known for close to 50 years, the mechanisms that bring them about are not yet understood. Interest in their study has been increasing in recent years since genetic rearrangements in both procaryotes and eucaryotes are at the base of important biological processes (7, 42) and also constitute a major class of disease-related genetic events (39, 57). The complexity of eucaryotic systems has made it difficult to study genetic rearrangements at the molecular level. Most of what we know today comes from studies using procaryotic systems. These have identified some of the important parameters in rearrangement formation and also brought about the realization that the mechanisms responsible for them are exceedingly complex. Major genetic rearrangements can occur either as the result of the movement of transposable elements (transposition; 44) or naturally from the resolution of transient secondary structures formed in the course of DNA metabolism (1, 17, 22, 43). We are concerned exclusively with the latter.

Deletions are the most extensively studied rearrangements, and most of them are explained by misalignment mutagenesis models. These are extensions of the model advanced by Streisinger et al. to explain the origin of frameshift mutations (51) and assign important roles to direct and inverted repeats. These models are based on extensive sequencing of deletion mutations and propose either that deletions take place between direct terminal repeats and can be facilitated by the presence of intervening inverted repeats (palindromes) (1), or they can occur at the end of palindromes in the absence of direct end repeats (22, 43). Palindromes are believed to stabilize misalignments resulting from slippage of transiently single-stranded regions during DNA replication (1, 22, 43). Transposon excision is a deletion event taking place between direct repeats at the end of internal inverted repeats (1). There is considerable support for these models (1, 11, 17, 24, 27, 37, 45, 50, 55), and they explain most of the deletions reported to date. A certain number of dele-

---

Elias Balbinder • Department of Biochemistry, Biophysics and Genetics, University of Colorado Health Sciences Center, Denver, Colorado 80262.

tions can be explained by errors in the action of enzymes which normally break and join DNA, such as topoisomerases, and occur on sequences that share little or no homology (13, 32, 36).

While the role of sequence homologies in the formation of deletions is well established, we know nothing about the mechanisms which bring them about. The misalignment models leave this question entirely open. Most of what has been published to date is mainly concerned with the roles of legitimate versus illegitimate recombination (4, 19, 20). Legitimate (homologous) recombination requires extensive sequence homology and the participation of the $recA^+$ protein (10), while illegitimate recombination (4, 19, 20) refers to a number of recombinatory processes which do not require large sequence homology or the participation of the $recA^+$ protein (9, 13, 32, 36, 49, 54). In many bacteria and bacteriophage systems the same deletion frequencies were observed in $recA^+$ and $recA$ backgrounds (3, 11, 12, 14, 18, 26, 27). In two cases, $recA^+$ increased deletion frequency considerably but was not essential (1, 50). These deletions were large and occurred between direct repeats, and the regions deleted could form hairpin structures. The participation of $recA^+$ is generally interpreted as indicating that homologous recombination plays a role, and independence from $recA^+$ has been explained as the result of either illegitimate recombination (4, 20) or replication errors (1, 14). It is known, however, that the $recA^+$ protein performs a number of functions; i.e., it is essential for the initiation of homologous recombination (10), plays a major role in regulating the SOS response (53), and is also needed in mutagenesis (16). There is only one published report in which the possible participation of the SOS response in deletion production was considered (38), with negative results. The role of $recA^+$ in producing deletions still remains unresolved, and we are still unclear about which mechanisms (homologous recombination, SOS processing, or other) are defined by its participation. As for whether $recA^+$-independent rearrangements are caused by errors in replication or processes related to illegitimate recombination, we have no information at this time.

General recombination and the various specialized recombination systems appear to be mediated by separate overall processes, but may share common components of DNA metabolism such as winding/unwinding enzymes, ligase, polymerases, various nucleases, and DNA-binding proteins, which also participate in DNA repair and replication (31). This sharing of major functions between different processes for DNA metabolism is one component of the complexity of the problem. The other is the multiplicity of transient secondary structures that can form on DNA and which are themselves subject to modification by such factors as location on a replicon (14) and degree of superhelical tension (48). Clearly the identification of functions which give rise to deletions is a major priority if we hope to understand how these originate. The best way to do this is by isolating and studying mutants which affect deletion frequency (4). However, in devising strategies to obtain these, we must keep in mind the inherent complexity of the problem. We know that there are at least two pathways for Tn10 excision-associated processes (33, 34) and, by analogy, also for deletions. Thus, selecting mutants for their effect on a given event will probably not uncover all functions affecting a very similar event. It is also possible that, while there are probably several pathways for deletions, there may also be overlaps; i.e., the same deletion can be brought about by more than one mechanism. This is strongly indicated by the observations that the frequencies of the same deletions which occurred in the absence of $recA^+$ were substantially increased by its presence (1, 50). Keeping such facts in mind, we have devised strategies for studing deletions by following the advice of Drake (15), i.e., trying to imagine "all the ways" in which deletions might occur and then designing "experimental attacks . . . powerful enough not only to discredit the inappropriate answers but also to ferret out the as yet unimagined possibilities." We have developed a two-pronged attack based on (i) control of all the structural variables for deletions through predesigned plasmids and (ii) the use of these plasmids to study the effect of host mutants on the frequency of structurally defined classes of deletions, and as genetic screens to isolate new mutants which increase deletion frequency. In the following sections we will show how this approach works.

USE OF PREDESIGNED HIGH-COPY-NUMBER PLASMIDS TO STUDY DELETIONS

The slipped mispairing mutagenesis models provide a good point of departure since they identify important structural parameters for deletions and provide a conceptual framework within which we can make precise predictions to be tested experimentally. We describe here the use of derivatives of the high-copy-number plasmid pBR325 (6) to test an important prediction of the slipped-mispairing model of Albertini et al. (1), namely, that a palindrome will delete more frequently than a nonpalindrome of comparable size between the same direct repeats, at the same position on the same replicon. This plasmid has a number of unique restriction sites within the genes determining resistance to the drugs tetracycline, ampicillin, and chloramphenicol (6). By the simple expedient of cloning any fragment of desired size and sequence into one of these sites, we inactivate the gene and revertants will result exclusively from the deletion of the insert. Such a system is extremely attractive for a number of reasons. (i) It allows us to manipulate one variable at a time while keeping the others constant, thus making interpretation of the results straightforward and unequivocal. (ii) By selecting for reversion from drug sensitivity to resistance we recover only deletions, making their frequency easy to quantitate. (iii) Selection for drug resistance should be very stringent for the restoration of the normal sequence of the gene into which the insert was cloned, making extensive sequencing of revertants unnecessary. (iv) The sequence and size of any DNA to be inserted can be chosen at will. (v) Plasmids can be easily purified for biochemical analysis (sequencing, restriction enzyme mapping, etc.). (vi) Plasmids are portable and can be easily introduced into strains of any desired genetic background.

This work was carried out with derivatives of pBR325 obtained by inserting two fragments of the same approximate size (66 to 68 base pairs [bp] but different sequence into the unique *Eco*RI site of the gene coding for the enzyme chloramphenicol acetyltransferase, CAT, which is responsible for resistance to chloramphenicol. The CAT gene has been completely sequenced (2) and has a unique *Eco*RI site located between bp 437 and 442 within an *Alu*I fragment of 129 bp (Fig. 1A and 2). Plasmid pOCE15 was constructed by Betz and Sadler (5) and contains a 66-bp inverted repeat of a *lac* operator fragment. This is a palindromic sequence and is capable of forming a hairpin structure (47) as illustrated in Fig. 1B-1. Plasmids pRS1 and pRS4 were obtained by R. Sinden (unpublished data) as independent transformants in the same experiment when pBR325, opened at the unique *Eco*RI site in CAT, was cloned to the 66-bp *Hae*II fragment of pBR322 to which *Eco*RI linkers had been added. This fragment is not palindromic (Fig. 1B-2). Digestion with *Alu*I, followed by electrophoresis on acrylamide gels, showed that the 129-bp *Alu*I fragment of pBR325 had been replaced by larger fragments in the derived plasmids: about 195 bp in pOCE15 and pRS1 and 202 bp in pRS4 (Fig. 2). Although both pRS1 and pRS4 contain the same nonpalindromic insert,

PREDESIGNED PLASMIDS TO STUDY DELETIONS 381

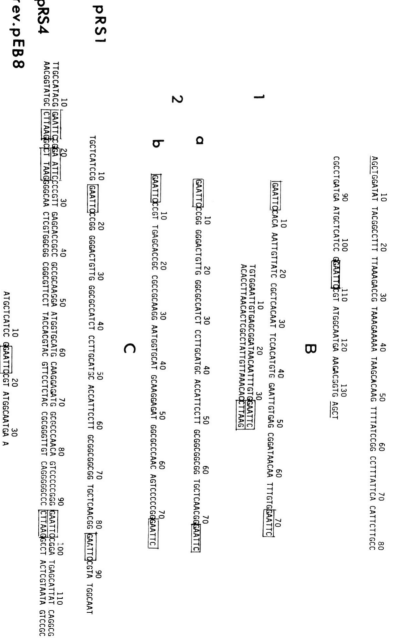

FIGURE 1. (A) Sequence of the 129-bp AluI fragment of pBR325 containing the unique EcoRI site (boxed). AluI sites are underlined. (B) Sequences of fragments cloned into the EcoRI site of PBR325 generating at least one such site (boxed) at each end. (1) pOCE15 was constructed and sequenced by Betz and Sadler (5) and contains one fragment with lac operator sequences forming an inverted repeat 66 bp long. This constitutes a perfect palindrome with a 36-bp stem, as shown. (2) Sequence of the 64-bp HaeIII fragment of pBR322 plus added terminal EcoRI linkers, used to construct plasmids pRS1 and pRS4. Both orientations are shown (a and b). This fragment has little twofold rotational symmetry. (C) Relevant sequences of plasmids pRS1, pRS4, and the revertant of pOCE15, pEB8, showing the presence of the inserted fragment (pRS1 and pRS4) and its absence (pEB8). Note that in pRS1 the insert is flanked by a single EcoRI site at each end, while in pRS4 there are two EcoRI sites in tandem at the 3' end of the insert and one at the 5' end.

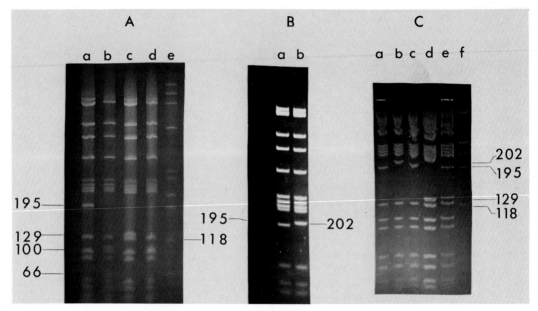

FIGURE 2. Acrylamide gel electrophoresis of plasmid digests. (A) Single (AluI) and double (AluI-EcoRI) digests of plasmids pOCE15 (a and b) and pBR325 (c and d), run on a 5% acrylamide gel (17 by 22 cm). Lanes: (a) pOCE15, AluI digest; (b) pOCE15, AluI-EcoRI digest; (c) pBR325, AluI digest; (d) pBR325, AluI-EcoRI digest; (e) φX174 HaeIII molecular size ladder. Note that EcoRI digestion of AluI-digested pOCE15 leads to the disappearance of the 195-bp replacement fragment and the appearance of fragments of 100 and 66 bp. The former corresponds to the segment between the first AluI site and EcoRI site (Fig. 1A), and the second is the 66-bp insert. Double digests of pBR325 (lane d) show the disappearance of the 129-bp AluI fragment and its replacement by the 100-bp AluI-EcoRI fragment also found with pOCE15, but the 66-bp insert cannot be seen. (B) AluI digests of plasmid pRS1 (lane a) and pRS4 (lane b) run alongside each other on a 5% acrylamide gel (17 by 22 cm). The insert-containing fragment of pRS4 runs behind the one from pRS1. The difference in size between these fragments is one EcoRI linker (8 bp, since octamers were used). (C) AluI digests of Cm^s plasmids and two Cm^r revertants run on a 15% acrylamide gel (17 by 22 cm). Lanes: (a) pOCE15; (b) pRS4; (c) pRS1; (d) pOCE15 revertant pEB1; (e) pOCE15 revertant pEB8 (dimer); (f) φX174 molecular size ladder. This gel increases resolution for the smaller fragments over the 5% gel (compare separation of the 118- and 129-bp fragments with that in panel A). Both revertants of pOCE15 have lost the insert and recovered the 129-bp fragment, but pEB8 is a dimer consisting of one Cm^r and one Cm^s plasmid and thus also shows the presence of the 195-bp fragment. (From Balbinder et al., submitted).

they differ from each other in one interesting respect, as shown by the sequences of the insert termini (Fig. 1C): while pRS1 has a single EcoRI site at each end of the insert, pRS4 shows an asymmetry since at the 5' end it has one while at the 3' end of the insert it has two EcoRI sites in tandem configuration. This size difference of about 8 bp is clearly observed on 5% and even more clearly on 15% acrylamide gels (Fig. 2B and C). A detailed analysis of revertants of all three plasmids, which included phenotype tests, restriction enzyme analysis, and sequencing across the deletion site and is described in a separate publication (E. Balbinder, C. Mac Vean, and R. E. Williams, submitted for publication), established that the selective procedure was specific for the restoration of the single, original EcoRI site. This permits quantitative measurements of deletion frequency by a simple reversion test. Deletion rates and frequencies of all three plasmids were compared in $recA^+$ and $recA$ isogenic backgrounds. The following results (Table 1) were obtained. (i) Deletion was more frequent on pOCE15, which car-

TABLE 1
Cm^s→Cm^r Reversion Rates and Frequencies[a]

Strain	Genotype	No. of cultures	No. of cultures with no Cm^r reversions	Avg no. of cells/culture (10^8)	Cm^s→Cm^r reversion rate (10^{-9})	Reversion frequency (10^{-9})		
						Low	High	Avg
EB252 [HB101(pRS1)]	recA	85	78	2.6	0.3	4	15	0.6
EB244 [HB101(pOCE15)]	recA	85	5	6	4.7	1.7	111	10
EB250 [HB101(pRS4)]	recA	85	0	1.2	>37	25	2,116	260
EB259 [D1206(pRS1)]	recA$^+$	106	79	2.4	1.2	4	40	2.4
EB246 [D1206(pOCE15)]	recA$^+$	100	11	2.1	10.5	4.7	460	33
EB261 [D1206(pRS4)]	recA$^+$	96	0	2.5	>20	24	2,076	362

[a] The deletion rates were calculated as described by Albertini et al. (1) from the fraction of cultures having no revertants (P_0, the zero term of the Poisson distribution) and represent the frequency of Cm^r reversion per cell per generation. Reversion frequencies are expressed as the number of Cm^r revertants per number of viable cells plated and were calculated from the cultures containing Cm^r revertants; the low value is from the culture showing the lowest and the high value is from the culture showing the highest number of revertants. The average frequencies represent the average number of revertants in an entire sample of 85 to 106 cultures over the average number of viable cells per culture.

ries a palindromic insert, than on pRS1, which does not, confirming the prediction of the model of Albertini et al. (1). (ii) Surprisingly, deletion on pRS4, which carries the same nonpalindromic insert as pRS1 but has a tandem duplication of an EcoRI site at the 3' end, was 5- to 10-fold higher than on pOCE15, a result totally unpredicted by slipped mispairing models and with some important implications which will be discussed below. (iii) Although these deletions occurred independently of recA$^+$, the deletion rates as well as the average deletion frequencies were two- to fourfold higher in recA$^+$ than recA cells (Table 1). We will return to this below.

IDENTIFICATION OF FUNCTIONS WHICH AFFECT DELETION FREQUENCY

As mentioned above, the role of recA$^+$ in producing deletions still remains unresolved. We need more information about what recA$^+$ and some of its alleles do to different deletions before we can understand its role in these processes. recA730 is an allele of recA$^+$ which results in constitutivity for the SOS response at all temperatures and also increases recombination frequency (56). The plasmids described in the preceding section were introduced into three isogenic strains differing only at the recA gene: one was recA$^+$, another was recA, and the third had the recA730 mutation. All these strains were derived from strain SC30 (56), which carries the mutation trpE65. Reversion of this allele to prototrophy is known to be increased by recA730, particularly in the presence of adenine, thus providing us with a marker to monitor the mutator response of this allele (56). We found that the deletion frequency on all plasmids was increased an average of three- to fourfold by recA730 (Table 2). This increase in deletion frequency can be explained by enhanced participation of the SOS response or by enhanced homologous recombination. Regardless of which mechanism is finally shown to be at work, we can say on the basis of these results and those in Table 1 that some alleles of recA$^+$, and perhaps recA$^+$ itself, increase by a small factor the frequency of a class of small (66- to 68-bp) deletions between 6-bp direct repeats represented by the ones on these constructed plasmids.

TABLE 2
Effects of Different recA Alleles on Deletion Frequency (Cm^s→Cm^r)[a]

Strain	Plasmid	recA genotype	Cm^s→Cm^r reversion frequency (10^{-9})	No. of trpE→trpE$^+$ revertants/plate[b]	
				SEM	SEM+A
EB306	pOCE15	recA730	146 (4.4)	104 (12)	202 (25)
EB309	pOCE15	recA$^+$	46 (1.4)	10 (1.2)	16 (2)
EB312	pOCE15	recA	33 (1)	8.6 (1)	7.8 (1)
EB496	pRS1	recA730	3.9 (4.2)	89 (19)	196 (30)
EB497	pRS1	recA$^+$	1.2 (1.2)	15 (3.2)	25 (3.8)
EB498	pRS1	recA	0.93 (1)	4.7 (1)	6.5 (1)
EB499	pRS4	recA730	1,212 (3)	32 (3.2)	57 (6.3)
EB500	pRS4	recA$^+$	773 (1.9)	11 (1)	20 (2)
EB501	pRS4	recA	402 (1)	10 (1)	9 (1)

[a] Overnight L-broth cultures were started from single-colony inoculation, and 0.1 to 0.5 ml was plated on L-agar containing 25 μg of chloramphenicol per ml; 0.1-ml samples of appropriate dilutions were plated on L-agar to obtain a viable count. Reversion frequency is expressed as the number of Cm^r revertants per number of viable cells plated. These results are the average of four independent determinations.
[b] Media: SEM, Bonner + Vogel minimal medium enriched with L-broth (20 ml/liter); SEM+A, SEM + adenine (75 μg/ml) (56).

PLASMIDS AS GENETIC SCREENS: ISOLATION OF *dli* (DELETION INCREASE) MUTATIONS

Plasmid pMC874 was constructed in the laboratory of M. Casadaban (8) and was an available genetic screen to select mutants which increased deletion frequency. This plasmid (Fig. 3A) has most of the *lac* operon, except for the promoter-operator region and the first few bases in *lacZ*, cloned between a *Bam*HI and a *Hinc*II-*Sal*I site about 1 to 2 kilobases downstream and in the same orientation as the Km^r gene. It is phenotypically Lac$^-$ Km^r, but it gives rise to Lac$^+$ papillae at very low frequency (1.6×10^{-9}) on MacConkey agar. Restriction enzyme analysis of Lac$^+$ plasmids showed that the Lac$^+$ phenotype was caused by the deletion of a 1-kilobase fragment between km^r and *lacZ* which eliminates the *Bam*HI site and fuses the *lac* operon to the km^r promoter (plasmid pEB7, Fig. 3B). Sequencing of the deletion termini and intervening fragment is in progress. We have isolated 31 mutants for their ability to stimulate the occurrence of the 1-kilobase deletion on plasmid pMC874 by increase in Lac$^+$ papillation (18, 28). Thirteen of these mutants were extensively tested for stimulation of Lac$^-$→Lac$^+$ deletion frequency and several characteristics typical of mutants for DNA-handling functions such as sensitivity to UV and methyl-methanesulfonate, increase in spontaneous mutation to nalidixic acid (Nal^r) and rifampin (Rif^r) resistance (mutator phenotype; 34), and growth at various temperatures (Table 3). With two exceptions (strains EB323 and EB335), the increases in Lac$^-$→Lac$^+$ deletion frequency were low (less than 10-fold) (Table 3). Most of the mutants, however, showed phenotypes consistent with alterations in genes for DNA metabolism such as mutator (strains EB325, -330, -335, -350, -359, and -360) or antimutator (EB323) effects, as well as increased sensitivity to mutagenic agents and impaired growth. The mutants were cured of pMC874 and transformed with pOCE15, pRS1, and pRS4, and the frequencies of Cm^s→Cm^r deletions were determined for each strain thus obtained. Only 4 of 13 *dli* mutants

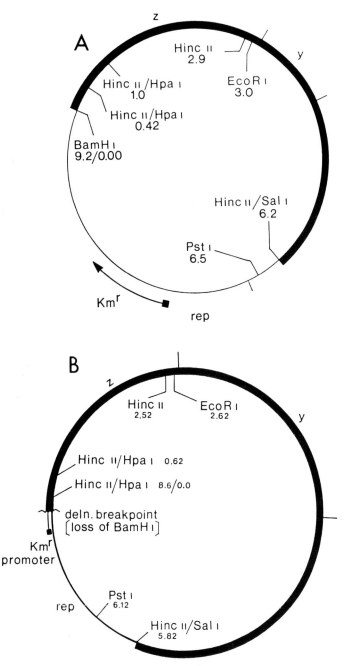

FIGURE 3. Diagrams of plasmid pMC874 (A) and its derivative pEB7 (B), which carries a 0.6-kilobase deletion joining the *lac* operon to the km^r promoter. The diagram of pMC874 has been adapted from Casadaban et al. (8).

TABLE 3
Characteristics of *dli* Mutants

Strain	Relative deletion frequency				Mutator phenotype[a]		Mutagen sensitivity[b]		Growth at temp:	
	Lac$^-$→Lac$^+$ (pMC874)	Cm^s→Cm^r			Nal^s→Nal^r	Rif^s→Rif^r	UV	MMS	30°C	42°C
		pOCE15	pRS1	pRS4						
EB265(WT)[c]	1	1	1	1	1	1	R	R	WT	WT
Eb323	30	0.4	19.3	2.2	0.5	0.16	VS	VS	Slow	Slow
EB325	5.5	1.85	12.9	3	3.5	4.7	PS	VS	Slow	Slow
EB328	1.6	0.55	1.4	0.85	1.8	2.3	WT	S	Slow	Slow
EB330	1.4	0.7	0.6	1.1	6.3	2	WT	WT	Slow	Slow
EB335	20	0.34	17	2.6	32	24.5	WT	WT	Slow	Slow
EB337	8.6	0.8	1	0.7	0.2	3.2	S	VS	Slow	Slow
EB339	3.2	1	0.5	1.2	1	1.7	WT	S	WT	WT
EB347	1	0.98	0.7	1.01	3.2	4.2	WT	WT	Slow	Slow
EB348	7.2	0.8	1.04	0.8	0.6	0.9	WT	WT	Slow	Slow
EB350	1.5	ND[d]	0.6	ND	99.2	63	WT	WT	WT	WT
EB355	1	1	0.8	0.9	1.3	2.7	WT	WT	WT	WT
EB359	1	0.7	0.4	0.5	4.3	2.8	WT	WT	WT	WT
EB360	3.7	0.25	1.9	1.3	3.3	1.04	S	S	Slow	NG[e]

[a] Defined by the increase in spontaneous mutation frequency of nalidixic acid sensitivity to resistance (Nal^s→Nal^r) and rifampin sensitivity to resistance (Rif^s→Rif^r).
[b] Abbreviations: MMS, methyl methanesulfonate; VS, very sensitive; PS, partially sensitive; S, sensitive; R, resistant (WT).
[c] WT, Wild type. Wild-type values as follows. Deletion frequency: Lac$^-$→Lac$^+$ (pMC874), 1.6 × 10^{-9}; Cm^s→Cm^r, 4 × 10^{-7} (pOCE15), 2 × 10^{-8} (pRS1), and 1.5 × 10^{-6} (pRS4). Mutation frequency: Nal^s→Nal^r, 2 × 10^{-9}; Rif^s→Rif^r, 66 × 10^{-9}.
[d] ND, Not done.
[e] NG, No growth.

tested (strains EB323, -325, -335, and -360) increased Cm^r deletion frequency on pRS1 and pRS4 and actually decreased (with the exception of strain EB325) deletion frequency on pOCE15. This suggests a preference in such mutants for events taking place between direct terminal repeats in the absence of palindromes.

Our *dli* mutants have very different phenotypes from the *tex* mutants selected by Lundblad and co-workers (33–35) for increasing the frequency of precise excision of Tn*10*. In general, *dli* mutants were more sensitive to UV and methyl methanesulfonate than *tex* mutants, but showed lower stimulation of deletion frequency and weaker mutator phenotypes. Also, of six *dli* mutants tested (Table 4), only two stimulated Tn*10* excision and the remaining four showed no effect or an actual inhibition of this deletion event. Thus, on the basis of these preliminary data we see no obvious similarity between *dli* and *tex* mutants. On the other hand, there are some superficial similarities in phenotype between some of our mutants and others described in the literature which affect different DNA metabolism functions. For example, EB360 (Table 4) resembles a conditional lethal *polA* mutant (30); slow growers with a mutator phenotype (EB325, -328, -330, and -347, Table 3) resemble mutants affected in genes for components of the DNA polymerase III replication complex such as *dnaQ* and *dnaE* (25, 29, 46). These results support the proposition that different deletions originate through different pathways and that a genetic dissection of these pathways will require specific genetic screens.

CONCLUSIONS AND FUTURE DIRECTIONS

The results we have presented illustrate how a strategy based on the use of prede-

TABLE 4
Effects of Several *dli* Mutations on Excision of Tn*10*[a]

Original *dli* mutant	Relative frequency of Tn*10* excision					
	gal::Tn*10*	*srl*::Tn*10*	*leu-82*::Tn*10*	*leu-63*::Tn*10*	*thr-34*::Tn*10*	*pro-81*::Tn*10*
EB265[b]	1	1	1	1	1	1
EB323	ND[c]	1	<0.4	0.1	<0.6	9
EB325	322	22	61	134	<0.16	173
EB335	17	12	58	2	0.1	71
EB337	ND	<0.1	0.4	<0.3	0.3	23
EB348	0.54	0.4	1.1	0.5	0.54	<0.12
EB360	ND	<0.1	1.3	<0.8	<0.36	ND

[a] Averages of four independent experiments.
[b] Wild type. Actual values: *gal*::Tn*10*, 2.24×10^{-9}; *srl*::Tn*10*, 0.1×10^{-9}; *leu-82*::Tn*10*, 1.13×10^{-9}; *leu-63*::Tn*10*, 1×10^{-9}; *thr-34*::Tn*10*, 1×10^{-9}; *pro-81*::Tn*10*, 1.5×10^{-9}.
[c] ND, Not done.

signed plasmids for the systematic study of deletions works. With the pBR325-derived plasmids, deletion frequency can be determined by a simple reversion test: selection for Cm^r only yields revertants that have the original sequence at and adjacent to the *Eco*RI site, making extensive sequencing of revertants unnecessary and the interpretation of results clear and unambiguous. In each of the plasmids we have changed one single parameter, i.e., the sequence of the insert (pRS1 and pRS4 versus pOCE15) or the number and arrangement of the direct terminal repeats (pRS1 and pOCE15 versus pRS4). Thus, the differences in deletion frequency for each plasmid can result only from the difference in the single controlled variable. In a general way, our results agree with a major postulate of slipped-mispairing models for deletions: direct and inverted repeats promote deletion events. More specifically, our observations that deletions between direct repeats are facilitated by an intervening palindrome (pOCE15) are a clear confirmation of the prediction of Albertini et al. (1), which was based on the sequencing of large *lacI-Z* deletions, and agree with many reports showing that palindromes are highly deletion prone (11, 14, 22, 24, 43, 45, 55). Our data also show that inverted repeats are not absolutely necessary, however: 66-bp inserts still delete between *Eco*RI sites, although at a decreased frequency (pRS1 versus pOCE15), and, more dramatically, a tandem duplication of one *Eco*RI site actually increases the deletion frequency of a nonpalindrome over that of a palindrome (pRS4 versus pOCE15). This observation has several important implications. First, the stability of a given stretch of DNA is not determined exclusively by its sequence, but can be strongly influenced by surrounding sequences as well, a conclusion also reached by Das Gupta et al. (14) from the study of deletions in a different system. Second, not all deletions which can be explained by slipped-mispairing models occur of necessity by the elimination of palindromes or resemble the excision of Tn*10* (1, 14, 18, 22, 43). The latter represents one class of deletions, and our data suggest that there are more deletion-prone sequences in procaryotic genomes that we recognize today. Finally, we have to consider whether structures such as the highly deletion-prone one on plasmid pRS4 can be found in contemporary procaryotic genomes. While some palindromic sequences have been retained in the course of evolution (21, 40, 41), it is not clear whether other potentially unstable structures have been preserved as well or have been entirely lost. With pBR325-derived plasmid systems we can study any sequence of interest, regardless of whether it exists in nature or has been eliminated in the course of evolution. Thus, such systems can be very useful in studies of the evolution of the genome.

We have also shown how the portability of plasmids provides a versatile approach to the problem of mechanisms, either by studying the effect of well-identified mutations for repair-recombination functions (such as recA and recA730) or by using them as genetic screens to isolate mutants for their effect on specific deletions. The preliminary results we have presented in this report are of some interest. Table 1 shows that deletion rates for all pBR325-derived plasmids are about twofold higher in $recA^+$ than in recA cells. Although differences of this magnitude are difficult to establish conclusively, the fact that they are consistent for all three plasmids is difficult to dismiss. These results suggest the possibility that $recA^+$ plays a role in causing these deletions which, although barely detectable by the fairly insensitive method of measuring rates, may not be insignificant.

As mentioned above, deletions occurring with the same frequency in recA and $recA^+$ cells are interpreted as being $recA^+$ independent and resulting from either replication errors, illegitimate recombination, or some sort of DNA cleavage which may or may not be part of illegitimate recombination processes (1, 4, 14, 19, 20, 22, 43). These different mechanisms are not necessarily mutually exclusive. In fact, the evidence available suggests, as we have mentioned earlier, that more than one mechanism may be responsible for causing the same deletions (see references 1 and 50 and above). Our observations (Tables 1 and 2) are congruent with that evidence in showing that $recA^+$ participates, together with other unknown functions, in producing deletions. The difference is that the deletions Albertini et al. (1) and Sommer et al. (50) were studying could be dramatically increased by $recA^+$, whereas ours were only slightly enhanced. Regardless of whether it does this by promoting homologous recombination, by allowing a certain level of induction of the SOS response, or by a combination of these two mechanisms, $recA^+$ seems to favor certain deletions over others. In most cases its contribution could be undetectable while in others (1, 50) it can be substantial. Certain mutant alleles of $recA^+$ such as recA730 are more efficient than $recA^+$ in causing certain deletions and thus magnify the effect of $recA^+$ to a detectable level. What this interpretation implies is that $recA^+$ is one of many intracellular functions participating in DNA metabolism which are normally active at low levels in *Escherichia coli* cells and can cause, accidentally, spontaneous deletions at very low frequency. If so, we would expect that some mutants selected exclusively for their ability to increase deletion frequency may turn out to be alleles of genetic functions which, like recA730, magnify the barely detectable effect of the wild type on the deletion process. Mutants selected for increasing the frequency of excision of Tn10 (tex mutants; 33–35) fit this prediction. Tn10 excision is normally independent of the *E. coli* recABC homologous recombination pathway, yet several tex mutations are alleles of recB and recC. In these mutants, Tn10 excision appears to depend on an altered form of the recBC nuclease, and one of them actually makes Tn10 excision dependent on $recA^+$ function (35). We are in the process of characterizing the dli mutants, and if the prediction above is correct, we expect that some of them will be alleles of already identified recombination-repair-replication genes which thus far have had no demonstrable participation in producing deletions.

Although a final understanding of the dli mutants we have isolated awaits their complete genetic characterization and the sequencing of the endpoints of the 1-kb deletion on plasmid pMC874, both of which are in progress, the preliminary data we have presented (Tables 3 and 4) contain some interesting findings which deserve some comment. First, most of the *dli* mutants had no effect on deletions on the pBR325-derived plasmids or on excision of Tn*10*. Thus, selecting for mutants which increase the frequency of one class of deletions will not uncover functions involved in other, different deletions. Lundblad and Kleckner (33, 34) reported a similar finding for *tex* mutants. These were selected for stimulating the precise excision of Tn*10* but will not affect the excision of nearly precise excision remnants of the same transposon. Second, the four mutants which affected deletions on pBR325-derived plasmids stimulated preferentially those taking place in the absence of intervening palindromes (pRS1 and pRS4) and either inhibited or had no effect on the deletion of a palindrome (pOCE15). Interestingly, EB323, the mutant which showed the largest stimulatory effect on Lac$^-$→Lac$^+$ deletions on plasmid pMC874, also had the strongest inhibiting effect on Cm^s→Cm^r deletions on pOCE15 as well as on excision of Tn*10*, both events involving deletions of palindromes. A comparison between EB323 and *texA* mutants is of interest, since the latter seem to stimulate Tn*10* excision through the interaction of the inverted repeats (35) whereas the mutation in EB323 does exactly the opposite. This suggests that certain genetic functions may code for polypeptides capable of recognizing either the palindrome or the hairpin structures capable of forming at palindromic sites. If so, presence or absence of palindromes may define two mechanistically different groups of deletions.

The purpose of this presentation was to describe a strategy for the study of a complex problem, namely, how spontaneous deletions originate. Plasmids can be designed in different ways to study not only deletions (14) but other rearrangements as well (23, 52), and the general strategy we have outlined here can also be used with other systems, such as bacteriophages (55). At this time it seems a most promising approach to deal with the complex processes causing major genetic rearrangements.

LITERATURE CITED

1. **Albertini, A. M., M. Hofer, M. P. Calos, and J. H. Miller.** 1982. On the formation of spontaneous deletions: the importance of short sequence homologies in the generation of large deletions. *Cell* **29**:319–328.
2. **Alton, N. K., and D. Vapnek.** 1979. Nucleotide sequence analysis of the chloramphenicol resistance transposon Tn9. *Nature* (London) **282**:864–869.
3. **Anderson, C. W.** 1970. Spontaneous deletion formation in several classes of *Escherichia coli* mutants deficient in recombination ability. *Mutat. Res.* **9**:155–165.
4. **Anderson, R. P.** 1987. Twenty years of illegitimate recombination. *Genetics* **115**:581–584.
5. **Betz, J. L., and J. R. Sadler.** 1981. Variants of a cloned synthetic lactose operator. I. A palindromic dimer lactose operator derived from one strand of the cloned 40-base pair operator. *Gene* **13**:1–12.
6. **Bolivar, F.** 1978. Construction and characterization of new cloning vehicles. III. Derivatives of plasmid pBR322 carrying unique *Eco*RI sites for selection of *Eco*RI generated recombinant molecules. *Gene* **4**:121–136.
7. **Borst, P., and D. R. Greaves.** 1987. Programmed gene rearrangements altering gene expression. *Science* **235**:658–667.
8. **Casadaban, M. J., J. Chou, and S. N. Cohen.** 1980. In vitro gene fusions that join an enzymatically active β-galactosidase segment to amino-terminal fragments of exogenous proteins: *Escherichia coli* plasmid vectors for the detection and cloning of translational signals. *J. Bacteriol.* **143**:971–980.
9. **Chang, S., and S. N. Cohen.** 1977. In vivo site-specific genetic recombination promoted by the *Eco*RI restriction endonuclease. *Proc. Natl. Acad. Sci. USA* **74**:1811–1815.
10. **Clark, A. J.** 1973. Recombinant deficient mutants of *E. coli. Annu. Rev. Genet.* **7**:67–86.
11. **Collins, J., G. Volckaert, and P. Nevers.** 1982. Precise and nearly-precise excision of the symmetrical inverted repeats of Tn5; common features of *recA*-independent deletion events in *Escherichia coli. Gene* **19**:139–146.
12. **Coukell, M. B., and C. Yanofsky.** 1970. Increased frequency of deletions in DNA polymerase mutants of *Escherichia coli. Nature* (London) **228**:633–635.

13. **Cozzarelli, N. R.** 1980. DNA topoisomerases. *Cell* **22**:327–328.
14. **Das Gupta, U., K. Weston-Hafer, and D. E. Berg.** 1987. Local DNA sequence control of deletion formation in *Escherichia coli* plasmid pBR322. *Genetics* **115**:41–49.
15. **Drake, J. W.** 1982. Perspectives in molecular mutagenesis, p. 361–378. *In* J. F. Lemontt and W. M. Generoso (ed.), *Molecular and Cellular Mechanisms of Mutagenesis*. Plenum Publishing Corp., New York.
16. **Ennis, D. G., B. Fisher, S. Edmiston, and D. W. Mount.** 1985. Dual role for *Escherichia coli* RecA protein in SOS mutagenesis. *Proc. Natl. Acad. Sci. USA* **82**:3325–3329.
17. **Farabaugh, P. J., U. Schmeissner, M. Hofer, and J. H. Miller.** 1978. Genetic studies of the *lac* repressor. VII. On the molecular nature of spontaneous hotspots in the *lacI* gene of *Escherichia coli*. *J Mol. Biol.* **126**:847–863.
18. **Foster, T. J., V. Lundblad, S. Hanley-Way, S. M. Halling, and N. Kleckner.** 1981. Three Tn10-associated excision events: relationship to transposition and role of direct and inverted repeats. *Cell* **23**:215–227.
19. **Franklin, N. C.** 1971. Illegitimate recombination, p. 175–194. *In* A. D. Hershey (ed.), *The Bacteriophage Lambda*. Cold Spring Harbor Laboratory, Cold Spring Harbor, N.Y.
20. **Franklin, N. C.** 1967. Extraordinary recombinational events in *Escherichia coli:* their independence of the rec+ function. *Genetics* **55**:699–707.
21. **Gilson, E., J. M. Clement, D. Perrin, and M. Hofnung.** 1987. Palindromic units: a case of highly repetitive DNA sequences in bacteria. *Trends Genet.* **3**:226–230.
22. **Glickman, B. W., and L. S. Ripley.** 1984. Structural intermediates of deletion mutagenesis: a role for palindromic DNA. *Proc. Natl. Acad. Sci. USA* **81**:512–516.
23. **Goldberg, I., and J. J. Mekalanos.** 1986. Effect of a *recA* mutation on cholera toxin gene amplification and deletion events. *J. Bacteriol.* **165**:723–731.
24. **Hagan, C. E., and G. J. Warren.** 1983. Viability of palindromic DNA is restored by deletions occurring at low but variable frequency in plasmids of *Escherichia coli*. *Gene* **24**:317–326.
25. **Horiuchi, T., H. Maki, and M. Sekiguchi.** 1981. Conditional lethality of *Escherichia coli* strains carrying *dnaE* and *dnaQ* mutations. *Mol. Gen. Genet.* **181**:24–28.
26. **Inselburg, J.** 1967. Formation of deletion mutations in recombination-deficient mutants of *Escherichia coli*. *J. Bacteriol.* **94**:1266–1267.
27. **Jones, I. M., S. B. Primrose, and S. D. Ehrlich.** 1982. Recombination between short direct repeats in a RecA host. *Mol. Gen. Genet.* **188**:486–489.
28. **Konrad, E. B.** 1977. Method for the isolation of *Escherichia coli* mutants with enhanced recombination between chromosomal duplications. *J. Bacteriol.* **130**:167–172.
29. **Konrad, E. B.** 1978. Isolation of an *Escherichia coli* K-12 *dnaE* mutation as a mutator. *J. Bacteriol.* **133**:1197–1202.
30. **Konrad, E. B., and I. R. Lehman.** 1974. A conditional lethal mutant of *E. coli* K-12 defective in the $5' \rightarrow 3'$ exonuclease associated with DNA polymerase I. *Proc. Natl. Acad. Sci. USA* **71**:2048–2051.
31. **Kopecko, D. J.** 1980. Specialized genetic recombination systems in bacteria: their involvement in gene expression and evolution. *Prog. Mol. Subcell. Biol.* **7**:135–234.
32. **Lopez, P., M. Espinosa, B. Greenberg, and S. A. Lacks.** 1984. Generation of deletions in pneumococcal *mal* genes cloned in *Bacillus subtilis*. *Proc. Natl. Acad. Sci. USA* **81**:5189–5193.
33. **Lundblad, V., and N. Kleckner.** 1982. Mutants of *Escherichia coli* K12 which affect excision of transposon Tn10, p. 245–258. *In* J. F. Lemontt and W. M. Generoso (ed.), *Molecular and Cellular Mechanisms of Mutagenesis*. Plenum Publishing Corp., New York.
34. **Lundblad, V., and N. Kleckner.** 1985. Mismatch repair mutations of *Escherichia coli* K12 enhance transposon excision. *Genetics* **109**:3–19.
35. **Lundblad, V., A. F. Taylor, G. R. Smith, and N. Kleckner.** 1984. Unusual alleles of RecB and RecC stimulate excision of inverted repeat transposons Tn10 and Tn5. *Proc. Natl. Acad. Sci. USA* **81**:824–828.
36. **Marvo, S. L., S. R. King, and S. R. Jaskunas.** 1983. Role of short regions of homology in intermolecular illegitimate recombination events. *Proc. Natl. Acad. Sci. USA* **80**:2452–2456.
37. **Meulien, P., R. G. Downing, and P. Broda.** 1981. Excision of the 40kb segment of the TOL plasmid from *Pseudomonas putida* mt-2 involves direct repeats. *Mol. Gen. Genet.* **184**:97–101.
38. **Miller, J. H., and K. B. Low.** 1984. Specificity of mutagenesis resulting from the induction of the SOS system in the absence of mutagenic treatment. *Cell* **37**:675–682.
39. **Monaco, A. P., C. J. Bertelson, W. Middlesworth, C. A. Colletti, J. Aldridge, K. H. Fischbeck, R. Bartlett, M. A. Pericak-Vance, A. D. Roses, and L. M. Kunkel.** 1985. Detection of deletions spanning the Duchenne muscular dystrophy locus using a tightly linked DNA segment. *Nature* (London) **316**:842–845.
40. **Noller, H. F.** 1984. Structure of ribosomal RNA. *Annu. Rev. Biochem.* **53**:119–162.
41. **Platt, T.** 1981. Termination of transcription and its

regulation in the tryptophan operon of *E. coli. Cell* 24:10–23.
42. **Riley, M., and A. Anilionis.** 1978. Evolution of the bacterial genome. *Annu. Rev. Microbiol.* 32:519–560.
43. **Ripley, L. S., and B. W. Glickman.** 1983. Unique self-complementarity of palindromic sequences provides DNA structural intermediates for mutation. *Cold Spring Harbor Symp. Quant. Biol.* 47:851–861.
44. **Ross, D. G., J. Swan, and N. Kleckner.** 1979. Physical structures of Tn10-promoted deletions and inversions: role of 1400 bp inverted repetitions. *Cell* 16:721–731.
45. **Sadler, J. R., and M. Tecklenburg.** 1981. Cloning and characterization of the natural lactose operator. *Gene* 13:13–23.
46. **Sevastopoulos, C. G., and D. A. Glaser.** 1977. Mutator action of *E. coli* strains carrying *dnaE* mutations. *Proc. Natl. Acad. Sci. USA* 74:3947–3950.
47. **Sinden, R. R., S. S. Broyles, and D. E. Pettijohn.** 1983. Perfect palindromic *lac* operator DNA sequence exists as a stable cruciform structure in supercoiled DNA *in vitro* but not *in vivo. Proc. Natl. Acad. Sci. USA* 80:1797–1801.
48. **Sinden, R. R, and D. E. Pettijohn.** 1984. Cruciform transitions in DNA. *J. Biol. Chem.* 259:6593–6600.
49. **Smith, G. R.** 1985. Site specific recombination, p. 147–163. *In* J. Scaife, D. Leach, and A. Galizzi (ed.), *Genetics of Bacteria.* Academic Press, Inc., New York.
50. **Sommer, H., B. Schumacher, and H. Saedler.** 1981. A new type of IS1-mediated deletion. *Mol. Gen. Genet.* 184:300–307.
51. **Streisinger, G., Y. Okada, J. Emrich, J. Newton, A. Tsugita, E. Terzaghi, and M. Inouye.** 1966. Frameshift mutations and the genetic code. *Cold Spring Harbor Symp. Quant. Biol.* 31:77–84.
52. **Tlsty, T. D., A. M. Albertini, and J. H. Miller.** 1984. Gene amplification in the *lac* region of *E. coli. Cell* 37:217–224.
53. **Walker, G. C.** 1984. Mutagenesis and inducible responses to deoxyribonucleic acid damage in *Escherichia coli. Microbiol. Rev.* 48:60–93.
54. **Weisberg, R. A., and A. Landy.** 1983. Site-specific recombination in phage Lambda, p. 211–250. *In* R. W. Hendrix, J. W. Roberts, F. W. Stahl, and R. A. Weisberg (ed.), *Lambda II.* Cold Spring Harbor Laboratory, Cold Spring Harbor, N.Y.
55. **Williams, W. L., and U. R. Muller.** 1987. Effects of palindrome size and sequence on genetic stability of bacteriophage φX174 genome. *J. Mol. Biol.* 196:743–755.
56. **Witkin, E. M., J. O. McCall, M. R. Volkert, and I. H. Wermundsen.** 1982. Constitutive expression of SOS functions and modulation of mutagenesis resulting from resolution of genetic instability at or near the *recA* locus of *Escherichia coli. Mol. Gen. Genet.* 185:43–50.
57. **Yamamoto, T., R. W. Bishop, M. S. Brown, J. L. Goldstein, and D. W. Russell.** 1986. Deletion in cysteine-rich region of LDL receptor impedes transport to cell surface in WHHL rabbits. *Science* 232:1230–1237.

VI. Damage-Directed Mutagenesis

Damage-Directed Mutagenesis: Introduction

The goal of the chapters in this section is to review and summarize recent progress in understanding the process of mutagenesis through specific analysis of the mutations caused by known DNA-damaging agents. This approach has relied on the development of specific assays for mutagenicity, systems that allow molecular analysis of the DNA lesions as well as the mutations at the DNA or protein level, and understanding of the pathways by which DNA with damage is repaired.

Current research on damage-directed mutagenesis has illuminated the mechanisms of action of some of the cellular genes that control mutagenic responses. A key question is whether the site of the mutation is restricted to the site of the DNA damage (so-called targeted mutations) or not (untargeted mutations). Although untargeted mutagenesis, controlled by the SOS system in *Escherichia coli,* can occur, it seems to be quantitatively obscured by the high amount of targeted mutagenesis in UV-treated cells.

One generality that has emerged from this work in both procaryotes and eucaryotes is the propensity of cells to insert adenine residues opposite "noninstructive" lesions in DNA (this is the so-called "A-rule"). Such a response makes evolutionary sense, since UV photodimers might be thought to be the most frequent lesion encountered by cells exposed to the sun. Since the most frequent lesion in UV-treated DNA is the T-T dimer, even if this lesion is noninstructional to the replication apparatus, under the A-rule the sequence is maintained intact.

The question of the particular instructional nature of specific lesions is of current interest. The problem here is to identify the biologically significant lesion or lesions from among all the possible lesions caused by a mutagen. Are some lesions silent, i.e., bypassed with no change in the coding capacity, and others bypassed with consistent miscoding, while still others constitute absolute blocks to replication and are therefore genetically dead? These questions are being approached by in vitro and in vivo assays using templates with defined lesions in defined sequences.

In mammalian cells, experiments on damage-directed mutagenesis have depended on the use of shuttle vector systems to recover the mutations for analysis. These vectors are of two types, those that replicate as plasmids in the nucleus and are under the control of viral replicons, for example, simian virus 40, bovine papillomavirus, or Epstein-Barr virus, and those vectors that integrate into the cellular genome but can be excised and rescued for analysis of the mutations. These latter vectors have employed both retroviruses and bacteriophage lambda. Reassuringly, the results in mammalian cells, both for episomal vectors and chromosomal vectors, seem to be quite similar to those in procaryotes: the nature of the mutation produced by a given mutagenic lesion seems the same in mammalian cells as in procaryotes.

In mammalian cells the pathways that modify DNA damage are not as fully understood as those in *E. coli*. One important reaction that removes DNA damage is the hydrolytic cleavage of the base-sugar glycosidic bond of deoxyuridylic acid residues in DNA (caused by deamination of cytosine or by misincorporation of dUMP from dUTP substrate). In *E. coli* this enzyme is encoded by the *ung* gene. The cell further attempts to reduce incorporation of uracil by destruction of dUTP produced by the various nucleotide kinases in the cell. The dUTP hydrolase in *E. coli* is the product of the *dut* gene. The uracil-N-glycosylase pathway seems to be ubiquitous and is representative of a whole class of base-removal glycosidases. Even viral genomes and the smallest free-living cells (the class *Mollicutes*) have recently been shown to encode such activities. Another very interesting mechanism for modification of the level of DNA damage is the DNA-alkyltransferase pathway, in which the alkyl group on DNA is removed by transfer to an acceptor protein (the transferase itself).

Mutagens with previously unrecognized mechanisms or public health consequences are also of obvious importance, and two examples of current research in these areas are included here.

With the recent advances in methodology for determination of mutagen specificity, both in procaryotes and in eucaryotes, rapid progress is expected. Some of the currently intriguing problem areas still to be resolved are the following.

(i) What are the interactions between the type of DNA lesion and the local sequence context that determine the outcome of the mutagenic process?

(ii) What are the rules that allow deduction of the miscoding or mutagenic mechanism of a given chemical adduct on DNA?

(iii) What is the basis for and mechanism of so-called spontaneous mutation, and what is the mechanism of spontaneous mutagenesis in cells with abnormally high mutation rates such as those from patients with xeroderma pigmentosum or Bloom syndrome?

(iv) What intracellular mutagens are important in producing DNA lesions and how are they formed? Are there endogenous mutagens responsible for some of the observed spontaneous mutations?

The Editors

Chapter 41

UV Mutagenesis in *Escherichia coli*
UmuC-Independent Targeted Mutations, Altered Spectrum in Exconjugants, and Mutagenesis Resulting from a Single T-T Cyclobutane Dimer

J. E. LeClerc, J. R. Christensen, R. B. Christensen, P. V. Tata, S. K. Banerjee, and C. W. Lawrence

Current ideas regarding the mechanisms of UV mutagenesis tend to revolve around a translesion synthesis/targeted mutation model for the process, in which DNA polymerase III holoenzyme-directed synthesis past sites of template damage is assisted by additional proteins and results in misincorporation opposite the template lesions. Aspects of this model that require further investigation include the function of the additional proteins, the identity of the mutagenic lesions, and the consequences of replicating past these lesions, that is, the probability of error and the spectrum of resulting sequence changes.

At least three proteins, in addition to the DNA polymerase III holoenzyme, are thought to play a direct, nonregulatory role in UV mutagenesis: the products of the *umuC*, *umuD*, and *recA* genes (2, 7–9, 17). RecA protein has been shown to bind to UV-irradiated duplex DNA and inhibit polymerase editing (13), but the role of the UmuC and UmuD proteins is less clear. Bridges and Woodgate (4, 5) have suggested that these proteins may be concerned with the resumption of DNA chain elongation at sites of template damage, rather than with misincorporation, and other evidence even raises the question of whether the UmuC protein always has an essential function in the production of targeted mutations. Maenhaut-Michel and Caillet-Fauquet (14), using lambda bacteriophage, and Tessman (19), using S13 phage, found comparable yields of mutations in $umuC^+$ and $umuC122$::Tn5 strains. In neither case does it seem likely that the mutations are untargeted, though further evidence is required on this point.

Much of the information concerning the misincorporation step of translesion synthesis has been obtained from sequence analysis of UV-induced mutations. The necessity of using a mutation selection or detection procedure based on change in gene function

---

J. L. LeClerc • Department of Biochemistry; *J. R. Christensen* • Department of Microbiology and Immunology; and *R. B. Christensen, P. V. Tata, S. K. Banerjee, and C. W. Lawrence* • Department of Biophysics, University of Rochester School of Medicine and Dentistry, Rochester, New York 14642.

prevents recovery of more than a small and biased sample of the mutations, however, and precludes estimates of error frequency per replication. Sequence analysis, using a variety of genetic systems (11, 12, 15, 21), shows that 80% or more of substitutions can be assigned to putative bipyrimidine target sites, and independent evidence (6) indicates that about 10% of all *lacI* mutations induced by UV are untargeted. Differences are evident, however, between the various sets of sequence data regarding the relative proportions of mutations at each of the four types of damage sites, and probably with respect to other features as well. In principle, such disparities might result from differences in photochemistry, differences in mutation detection, or differences in the mode of DNA synthesis or repair. Which of these causes, or which combination of them, is actually responsible is not yet known. In no case, however, does the proportion of mutation at the four bipyrimidine sites correspond well to the relative yields, as determined by Brash and Haseltine (3), of either cyclobutane or 6,4-pyrimidine-pyrimidone (6,4-PyO) dimers at these sites. This implies not only that factors other than photochemistry may be important in determining mutation spectra, but also that inferences regarding identity of causal lesions may not be possible from sequence data. Lastly, almost all of the UV-induced substitutions are single-base-pair changes, even though some kind of bipyrimidine lesion is thought to cause them. This observation, among others, led to the conclusion (10) that cyclobutane dimers might retain a significant fraction of normal base-pairing specificity. For thymine-containing dimers, polymerase bias in favor of adenine incorporation (18) may also reduce the frequency of tandem double substitutions. In addition, Wood et al. (21) have argued that the structure of the 6,4-PyO lesion is likely to ensure a single misincorporation opposite the 3′ nucleotide.

In the following sections of this chapter we describe the results of experiments designed to examine some of the questions and problems outlined above. Issues discussed include the properties and role of UmuC protein in mutagenesis, the influence of varying conditions of replication and repair on the spectrum of UV-induced sequence changes, and mutagenic properties of a single defined T-T cyclobutane dimer.

UmuC FUNCTION

When assayed by commonly used methods, the frequencies of UV-induced mutations in *umuC* mutants are generally very low, leading to the conclusion that UmuC function is essential for SOS-dependent targeted mutagenesis (see reference 20). As discussed above, however, observations from experiments with phage question whether this conclusion is correct. Further, in keeping with the phage work, we (J. R. Christensen, J. E. LeClerc, P. V. Tata, R. B. Christensen, and C. W. Lawrence, unpublished data) found that up to 25% of the normal yield of *lacI* mutations could be induced by a UV dose of 7 J m^{-2} in an F′ *lac uvrA6 umuC122*::Tn5 strain, if the mutations were detected nonselectively by plating on X-Gal (5-bromo-4-chloro-3-indolyl-β-D-galactopyranoside)-minimal glucose medium. Selection of *lacI* mutants by plating the *umuC*$^+$ and *umuC* mutant cells on P-Gal (phenyl-β-D-galactoside) medium reduced their frequency in the mutant to about 6% of that in the wild type, a result much more typical of experiments using reversion test systems. The chief difference between these two procedures is that all viable cells form colonies on X-Gal medium, and *lacI* mutants are detected as blue colonies or, more commonly, as blue sectors in colonies. On P-Gal medium, only cells that quickly express a *lacI* phenotype are capable of prolonged growth. It seems likely, therefore, that the chief difference between the two methods is the much greater time available for postirradiation recovery when cells are grown on X-Gal medium. If this interpretation is correct, the UmuC protein facilitates rapid recovery from UV damage, but is not always essential for this process, given enough time.

Sequence analysis of 134 spontaneous and 145 induced mutations showed that broadly similar kinds of mutations were induced in $umuC^+$ and $umuC122$ strains. More particularly, it is unlikely that more than a small proportion of the mutations induced in the $umuC$ mutant were untargeted events, since we (J. E. LeClerc, P. V. Tata, R. B. Christensen, J. R. Christensen, and C. W. Lawrence, unpublished data) have shown that authentic mutations of this variety are predominantly transversions and single-base-pair deletions. Overall, about 91% of the substitutions induced in $umuC^+$ cells could be assigned to putative bipyrimidine targets, compared with about 84% of those in $umuC$ mutants. A higher proportion of transitions than transversions could be assigned in this way (29/31 transitions versus 24/27 transversions in $umuC^+$; 23/26 versus 19/24 in $umuC$). By far the largest class of substitutions were located in C-C sequences in both strains ($umuC^+$, 22/58; $umuC$, 22/50), and the proportion of transitions was very similar in the two strains (53% and 52%). The two sets of data differ chiefly in the overall proportion of substitutions ($umuC^+$, 83%; $umuC$, 66%); a significant proportion of the additions and deletions are likely to arise by UmuC-independent processes. In addition, the two strains differ with respect to the proportion of substitutions located at the 5' as opposed to the 3' nucleotide of the probable target ($umuC^+$, 16/33; $umuC$, 24/30). The higher proportion of 3' changes in the wild type may imply that UmuC function is required particularly for recovery when the first nucleotide of the lesion to be encountered by the polymerase becomes mispaired.

In summary, we suggest that UmuC protein facilitates rapid recovery from UV damage, but is not always essential for this process. In agreement with Bridges and Woodgate (4, 5), we believe such recovery may involve the resumption of DNA chain elongation at blocked replication forks. UmuC function seems to be required particularly when the 3' nucleotide of the lesion is mispaired. It is not yet clear whether this reflects properties of the mispaired site itself or the presence of different kinds of lesion.

ALTERED MUTATION SPECTRUM IN EXCONJUGANT AND VEGETATIVE CELLS

Comparison of nucleotide sequence data from 74 induced and 64 spontaneous $lacI$ mutations isolated from exconjugants with the data from vegetative cells, discussed in the previous section, suggests that they differ in several ways. Induced mutations were isolated after irradiation of the F' lac $uvrA6$ donor with 7 J m^{-2} and the conjugational transfer of the plasmid to an SOS-induced (7 J m^{-2}) $uvrA6$ recipient; spontaneous mutations were isolated using the same procedure, but with unirradiated donor and recipient.

The two sets of data for induced mutations are similar with respect to the overall proportion of substitutions (vegetative cells, 50/70; exconjugants, 66/74) and in showing a high proportion of G·C changes (vegetative cells, 43/58; exconjugants, 54/66). They differ, however, with regard to the proportion of transitions among the substitution (vegetative cells, 31/58; exconjugants, 49/66; $P = 0.03$ to 0.02). They may also differ with regard to the target sequence: in vegetative cells most mutations are located in C-C sequences, but in mating cells the proportion assignable to T-C target sequences is greater (vegetative cells, 22 C-C, 10 T-C; exconjugants, 16 C-C, 19 T-C). This difference is not quite significant, but since $umuC$ vegetative cells give results very similar to their $umuC^+$ counterpart (22 C-C, 3 T-C) the difference may nevertheless be real. Finally, the two sets of data differ regarding the proportion of substitutions at the 3', as opposed to the 5', site of the probable target sequence (vegetative cells, 17/33; exconjugants, 30/36; P = 0.01 to 0.001).

Since the state of the F' plasmid in vegetative and donor cells is virtually identical at the time of irradiation, these differences must presumably arise from varying conditions of replication or repair, but the important variable is not yet known. All strains are $uvrA6$, and thus excision repair does not take place. During conjugation, a single F' strand is transferred to the recipient, where it becomes template for an exclusive lagging-strand synthesis (16), whereas both leading- and lagging-strand syntheses occur in vegetative cells. Partitioning vegetative cell mutations according to strand shows that mutations induced by replicating the antisense strand tend to resemble those in exconjugants, but the resemblance is not marked, and data from the two strands are not significantly heterogeneous. A significant difference between the two sets of data for spontaneous mutations from vegetative and exconjugant cells nevertheless emphasizes that some alteration in the condition of replication must occur: there is a high proportion of transitions among spontaneous substitutions in vegetative cells (26/38), but a much lower one in exconjugants (12/43; $P < 0.001$).

In addition to suggesting that biological, as opposed to photochemical, factors are important in determining the spectrum of induced mutation, these data reinforce the view that it is difficult, and perhaps impossible, to make inferences about causal lesions from such results. In neither set of data do the yields of mutations correspond to the yields of lesions at the various bipyrimidine sites, as determined by Brash and Haseltine (3).

M13 HYBRID PHAGE CONSTRUCTS CONTAINING A SINGLE SPECIFIC LESION

In light of the preceding discussion, we conclude that a full understanding of nucleotide sequence data is possible only with additional information. At the minimum, it is necessary to know the relative error rate per replication cycle of each of the different photochemical lesions and the spectrum of sequence changes they induce. To that end, we have developed methods for the construction of single- and double-stranded vectors that contain a single defined lesion at a single defined site in all molecules. Greatest emphasis was placed on single-stranded constructs, because these provide the clearest evidence for translesion synthesis. In addition, we wished to avoid detection of mutations dependent on changes in gene function, so that error rates and unbiased spectra could be obtained. The method used is quick, flexible, and generally applicable to the study of a variety of lesions. It involves the linearization of single-stranded DNA from M13mp7 hybrid phage with $EcoRI$, the annealing of the cut ends to a "scaffold" synthetic oligonucleotide to create a gapped circular structure, and the ligation of a synthetic oligonucleotide, modified at a specific site, into this gap. Single-stranded DNA of M13mp7 can be restriction cut because the cloning site forms a double-stranded hairpin structure (1). The terminal 20 nucleotides at each end of the scaffold oligonucleotide are complementary to the ends of the linearized vector, while the central region of the scaffold molecule is complementary to an oligonucleotide that is designed for site-specific modification. To study UV lesions, we use a purine-rich 11-mer containing a central bipyrimidine. The length of this oligomer was a compromise between annealing stability and ease of purification and was chosen also because it creates a $lacZ$ frameshift in the vector.

Our aim is to study the properties of cyclobutane and 6,4-PyO dimers in each of the four bipyrimidine sequences. At present we have preliminary data for the T-T cyclobutane lesion. Modification of the T-T site in the 11-mer was carried out by exposure to sun lamp radiation, filtered to remove wavelengths below 320 nm, in the presence of about 10^{-2} M acetophenone and the absence of oxygen. The dimer-containing species was purified from the parent material and other photoproducts by high-pressure liquid chromatography and then characterized by enzymatic photoreactivation with pure *Escherichia coli* photolyase, by short-wavelength photoreversal, and by hydrolysis with single-strand-specific nucleases. Efficient ligation into the annealed, gapped structure was accomplished by using 100× molar excess of oligomer. The DNA was then heat denatured and transfected into a *uvrA6* strain, and the DNA from a sample of progeny phage was sequenced by the dideoxy method. The scaffold 51-mer contained a C-C mismatch opposite the T-T dimer site to allow detection of any progeny phage making use of this information.

Preliminary results showed that the toxicity of the construct is high in uninduced cells (90% or more), but low or nonexistent in SOS-induced cultures (less than 50%). Five out of 54 progeny phage sequenced contained mutations, all single nucleotide substitutions targeted at the 3' site of the bipyrimidine. Two of the substitutions were T-to-C transitions, and three were T-to-A transversions. No untargeted mutations were detected in a region of 20 nucleotides or more on either side of the T-T site. We tentatively conclude that SOS-induced cells possess a high capacity to tolerate a single T-T cyclobutane dimer, that tolerance probably depends on translesion synthesis, and that the dimerized pyrimidines, particularly the 5' nucleotide, retain a considerable capacity for correct base-pairing. The level of replication accuracy seems too high to ascribe it entirely to a polymerase bias for adenine insertion (18).

In addition to the above data, we also have sequence results from 88 progeny phage in which the M13mp7 hairpin structure is present at the 3' end of the 11-mer. These constructs are obtained when *Eco*RI makes only a single cut at the 5' site of its recognition sequence. Nine of the 88 sequenced phage genomes were mutant, all contained a single substitution targeted at the 3' nucleotide of the dimer, and all were T-to-A transversions. Further data are required to determine whether the hairpin influences the spectrum of mutations.

We anticipate that the development of this approach, including analysis of double-stranded and plasmid vectors, different mutant strains and organisms, and in vitro replication studies, will provide some of the information necessary to interpret mutation data from UV-irradiated cells and supply the basis for in vitro reconstruction of mutagenic mechanisms.

ACKNOWLEDGMENTS. This work was supported by Public Health Service grants GM32885, GM21858, and GM27817 from the National Institutes of Health and by U.S. Department of Energy grant DEFG 0285ER60281.

LITERATURE CITED

1. **Been, M. D., and J. J. Champoux.** 1983. Cutting of M13mp7 DNA and excision of cloned single-stranded sequences by restriction endonucleases. *Methods Enzymol.* **101:**90–98.
2. **Blanco, M., G. Herrera, P. Collado, J. Rebollo, and L. M. Botella.** 1982. Influence of RecA protein on induced mutagenesis. *Biochimie* **64:**633–636.
3. **Brash, D. E., and W. A. Haseltine.** 1982. UV-induced mutation hotspots occur at DNA damage hotspots. *Nature* (London) **298:**189–192.
4. **Bridges, B. A., and R. Woodgate.** 1984. Mutagenic repair in *Escherichia coli.* X. The *UmuC* gene product may be required for replication past pyrimidine dimer but not for the coding error in UV-mutagenesis. *Mol. Gen. Genet.* **196:**364–366.
5. **Bridges, B. A., and R. Woodgate.** 1985. Mutagenic repair in *Escherichia coli*: products of the *recA* gene and of the *umuD* and *umuC* genes act at different steps in UV-induced mutagenesis. *Proc. Natl. Acad. Sci. USA* **82:**4193–4197.
6. **Christensen, R. B., J. R. Christensen, I. Koenig, and C. W. Lawrence.** 1985. Untargeted mutagenesis induced by UV in the *lacI* gene of *Escherichia coli. Mol. Gen. Genet.* **201:**30–34.
7. **Elledge, S. J., and G. C. Walker.** 1983. Proteins

required for ultraviolet light and chemical mutagenesis. *J. Mol. Biol.* 164:175–192.
8. **Ennis, D. G., B. Fisher, S. Edmiston, and D. W. Mount.** 1985. Dual role for *Escherichia coli* recA protein in SOS mutagenesis. *Proc. Natl. Acad. Sci. USA* 82:3325–3329.
9. **Kato, T., and Y. Shinoura.** 1977. Isolation and characterization of mutants of *Escherichia coli* deficient in induction of mutations by ultraviolet light. *Mol. Gen. Genet.* 156:121–131.
10. **Lawrence, C. W.** 1981. Are pyrimidine dimers non-instructive lesions? *Mol. Gen. Genet.* 182:511–513.
11. **LeClerc, J. E., and N. L. Istock.** 1982. Specificity of UV mutagenesis in the *lac* promoter of M13*lac* hybrid phage DNA. *Nature* (London) 297:596–598.
12. **LeClerc, J. E., N. L. Istock, B. R. Saran, and R. Allen.** 1984. Sequence analysis of ultraviolet-induced mutations in M13*lac*Z hybrid phage DNA. *J. Mol. Biol.* 180:217–237.
13. **Lu, C., R. H. Scheuerman, and H. Echols.** 1986. Capacity of RecA protein to bind preferentially to UV lesions and inhibit the editing subunit (ϵ) of DNA polymerase III: a possible mechanism for SOS-induced targeted mutagenesis. *Proc. Natl. Acad. Sci. USA* 83:619–623.
14. **Maenhaut-Michel, G., and P. Caillet-Fauquet.** 1984. Effect of *UmuC* mutations on targeted and untargeted ultraviolet mutagenesis in bacteriophage λ. *J. Mol. Biol.* 177:181–187.
15. **Miller, J. H.** 1985. Mutagenic specificity of ultraviolet light. *J. Mol. Biol.* 182:45–68.
16. **Rupp, W. D., and G. Ihler.** 1968. Strand selection during bacterial mating. *Cold Spring Harbor Symp. Quant. Biol.* 33:647–650.
17. **Steinborn, G.** 1978. Uvm mutants of *Escherichia coli* K12 deficient in UV mutagenesis. I. Isolation of *uvm* mutants and their phenotypical characterization in DNA repair and mutagenesis. *Mol. Gen. Genet.* 165:87–93.
18. **Strauss, B., S. Rabkin, D. Sagher, and P. Moore.** 1982. The role of DNA polymerase in base substitution mutagenesis on non-instructional templates. *Biochimie* 64:829–838.
19. **Tessman, I.** 1985. UV-induced mutagenesis of phage S13 can occur in the absence of the RecA and UmuC proteins of *Escherichia coli*. *Proc. Natl. Acad. Sci. USA* 82:6614–6618.
20. **Walker, G. C.** 1984. Mutagenesis and inducible responses to deoxyribonucleic acid damage in *Escherichia coli*. *Microbiol. Rev.* 48:60–93.
21. **Wood, R. D., T. R. Skopek, and F. Hutchinson.** 1984. Changes in DNA base sequence induced by targeted mutagenesis of lambda phage by ultraviolet light. *J. Mol. Biol.* 173:273–291.

Chapter 42

SOS Mutagenesis in *Escherichia coli* Occurs Primarily, Perhaps Exclusively, at Sites of DNA Damage

Eric Eisenstadt

Efficient mutagenesis by UV and other DNA-damaging agents proceeds via an inducible pathway called SOS mutagenesis (7, 31, 35). SOS mutagenesis in *Escherichia coli* is genetically defined by its dependence on three inducible gene products, RecA, UmuD, and UmuC, whose regulation is controlled by the LexA repressor (34).

Key events in SOS mutagenesis are: (i) DNA lesions (e.g., AP sites, UV photoproducts) directly block progression of the DNA replication fork and lead to activation of RecA (27, 34); (ii) activated RecA assists in the proteolytic cleavage of LexA repressor, which leads to derepression of the SOS regulon, including *recA*, *umuD*, and *umuC* (34); and (iii) UmuD is proteolytically cleaved (3a, 32a) to yield an active form of the protein (27a). Beyond this point the steps leading to mutation fixation are ill defined. Ultimately, DNA lesions are processed and DNA replication resumes. Resumption of DNA replication after UV irradiation requires protein synthesis and RecA, but not UmuDC (15). In *recA718* strains, however, resumption of DNA replication also requires UmuDC (36). The relationship between resumption of DNA replication and lesion bypass is not clearly defined.

An attractive model for UV mutagenesis has been developed by Bridges and his colleagues (1–3). They propose that DNA polymerase III holoenzyme, with assistance from RecA, UmuD, and UmuC, bypasses the lesion on the template strand. This bypass replication is the inherently error-prone step that is directly responsible for the induction of mutations by DNA damage. Bridges and Woodgate (2) suggested that UV mutagenesis occurs in two discrete steps. The first event is a highly error-prone event catalyzed by DNA polymerase III holoenzyme when it encounters a UV photoproduct on the template strand. In this step, a base is incorporated opposite the lesion, and a mismatch between the lesion and a newly incorporated base is generated. The second step (DNA replication to extend new strand synthesis beyond the site of the mismatch) requires induced levels of UmuD and UmuC.

The second event may require attenuation or inactivation of the proofreading activity (3'-to-5' exonuclease) provided by the

---

Eric Eisenstadt • Office of Naval Research, Arlington, Virginia 22217.

dnaQ gene product to permit the replication complex to tolerate mismatches involving damaged or missing bases on the template strand (2, 3, 19, 32, 33). Foster (P. L. Foster, submitted for publication) has genetic evidence suggesting that interactions among DnaQ, RecA, and UmuDC can affect the fidelity of DNA replication. Loss of proofreading activity does not, however, appear to be sufficient for replicative bypass of UV photoproducts (3, 39).

Induction of SOS in the absence of DNA damage leads to increased rates of mutagenesis (14, 37). These observations suggested that relaxation of fidelity during DNA replication plays a major role in the induction of mutations at sites of DNA damage and that untargeted or hitchhiking mutagenesis—mutations occurring via replication errors made at, presumably, undamaged sites—is a consequence of a persistent inactivation of proofreading activity (2, 3, 32).

Untargeted mutagenesis contributes only negligibly, if at all, to the mutations induced in *E. coli* via SOS mutagenesis. Four observations argue in support of this idea, as follows. (i) It is difficult to establish the absence of DNA damage in metabolically active cells. For example, overproduction of oxygen radical-scavenging enzymes lowers the spontaneous mutation frequency in *Salmonella typhimurium* (32b) and *E. coli* (J. Greenberg and B. Demple, submitted for publication). Thus, the observed mutator activity of SOS-induced cells could be actually due to the fixation of mutations at spontaneously occurring lesions that would not have caused mutations had SOS not been induced. The recent observation that mutations generated by the mutator activity of SOS-induced cells showed both site and mutagenic specificities supports the idea that "untargeted" mutagenesis might be occurring at cryptic DNA lesions (25, 35). (ii) The frequency of untargeted mutagenesis is usually much smaller (by orders of magnitude) than the mutation frequencies that obtain when DNA is deliberately damaged (37). (iii) Targeted and untargeted mutagenesis apparently occur by different mechanisms (4, 20, 38). Whereas targeted UV mutagenesis in bacteria is dependent on the *umuDC* operon (16, 28), untargeted mutagenesis of λ phage is not dependent on *umuDC* (4, 38). The observation that *recA441*-dependent spontaneous reversion of a *his* ochre allele is blocked in strains carrying the *umuC36* allele (5) is not inconsistent with the results of the λ phage experiments if one accepts the argument that the *recA441* effect is a consequence of SOS processing of cryptic lesions (25, 35). (iv) Finally, as I will discuss in more detail below, the analysis of mutations induced by several SOS-dependent mutagens (8, 10–12, 22–24, 26) revealed that each treatment generated its own characteristic site-specific and mutation-specific spectrum.

A rich source of data on mutational specificity is available from studies on *lacI* nonsense mutations induced in excision repair mutants by several different SOS-dependent mutagens: aflatoxin B_1, benzo[*a*]pyrene, neocarzinostatin, and angelicin (a monofunctional psoralen) (8, 10, 11, 22, 26). Additional data can be drawn from studies of *lacI* mutations induced in repair-proficient cells by 4-nitroquinoline-1-oxide, by UV irradiation, and by induction of SOS in the absence of mutagenic treatment (6, 24, 25). All mutagenic treatments generated distinctly different types and distributions of DNA lesions (9). The data further suggested that the spectrum of base-pair substitutions was determined by the tolerance of the replication apparatus for specific mismatches between recently incorporated bases and lesions on the template strand (26).

FEATURES OF THE *lacI* GENETIC SYSTEM

The *lacI* nonsense system developed by Miller (21) monitors base-substitution mutations, with the exception of A · T-to-G · C transitions, at over 60 different sites in the *lacI* gene. Nonsense mutations arising at each monitorable site have equivalent phenotypes. Thus, there is little selection bias within a collection of nonsense mutants. Sites at which nonsense mutations arise can be viewed as genetic targets for DNA-damaging agents; if the agent both damages a site and induces a monitorable event at that site, a mutation will be scored. If the site is either not damaged or damaged but not mutated to a monitorable event, no mutation will be scored.

The site distribution of *lacI* nonsense mutations can be used to assess the contribution of untargeted or hitchhiking mutations (3, 31, 33) to the collection of mutations that arise via SOS processing of lesions. If mutations are distributed among the monitorable sites in different ways for different mutagens, then this provides prima facie evidence that lesion distribution is an important determinant of the site distribution of mutations. If specific sites are well mutated by some mutagens but not by others, then lesion distribution would appear to be the major and possibly only determinant of the site distribution of mutations.

For example, one extreme hypothesis is that, under SOS conditions, induced mutations are equally likely to arise at any site whether it bears a lesion or not. This hypothesis predicts similar if not identical site distributions of mutations induced by all SOS-dependent mutagens whether they preferentially damage G · C or A · T sites. The other extreme hypothesis is that, again under SOS conditions, mutations arise exclusively at damaged sites. Thus, damage to G · C base pairs would not induce mutations at A · T sites, nor would damage to A · T base pairs induce mutations at G · C sites.

SITE DISTRIBUTION OF *lacI* MUTATIONS INDUCED BY SOS PROCESSING

Each of four different mutagenic treatments—aflatoxin B_1, angelicin, benzo[*a*]pyrene, and induction of SOS via high temperature plus adenine in bacteria carrying the *recA441* allele—induced a distinctly different site distribution of ochre mutations in *lacI* (Fig. 1). Each treatment induced the G · C-to-T · A transversion at high frequencies and as a major proportion of all *lacI* nonsense mutations. The three chemical treatments are known to damage DNA in distinctly different ways (see references in references 8, 13, and 26). The mutations arising in *recA441* cells presumably arose at spontaneously generated "cryptic" lesions, possibly AP sites (25).

A straightforward interpretation of these patterns is that mutations arose primarily via SOS processing at sites of DNA damage. This conclusion is strengthened by considering other evidence derived from studies that utilized the *lacI* system (6, 11, 13), as follows. (i) Neocarzinostatin induced all of the monitorable base changes that give rise to *lacI* nonsense mutations (11). Nonetheless, this compound failed to induce mutations at nearly half of the monitorable sites, a number that was far greater than could have arisen by chance alone ($P < 0.005$). As the "cold spots" included examples of all monitorable base-substitution events, the failure of neocarzinostatin to mutate these sites was not due to an inability to induce particular base substitutions. Nor were these sites refractory to being mutated by other SOS-dependent mutagens (6, 10, 13, 26). The likely explanation for the distribution of neocarzinostatin-induced *lacI* nonsense mutations is that it reflected the sites at which premutational lesions were formed. Indeed, sequences which are hot spots for mutation induction by neocarzinostatin correspond to sites which are hot spots for apyrimidinic site induction by neocarzinostatin (29, 30). (ii) 4-Nitroquinoline-1-oxide was strongly mutagenic, yet, among over 1,000 nonsense mutations, only 5 were detected at A · T sites (6). (iii) G is the primary target for activated aflatoxin B_1, and all but 6 of 189 aflatoxin B_1-induced mutants occurred at

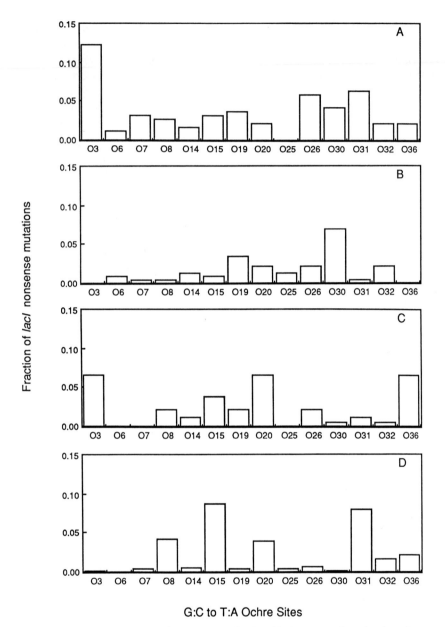

FIGURE 1. The ochre sites in *lacI* arising via G · C-to-T · A transversions that have been induced by SOS mutagenesis. (A) Aflatoxin B_1 (189 mutants) (13); (B) angelicin (233 mutants) (26); (C) benzo[*a*]pyrene (185 mutants) (8, 10); (D) *recA441 sfiA* bacteria at 42°C plus adenine (586 mutants) (25). Bar heights represent the fraction of total *lacI* nonsense mutants that occurred at each of the 14 ochre sites at which G · C-to-T · A events are monitorable.

G · C sites (13). (iv) UV-induced mutations occurred preferentially at sequences containing tandem pyrimidines and consisted almost exclusively of single-base substitutions, even among missense mutations (6, 24). There is good reason to believe that at least one of the major UV photoproducts involving adjacent pyrimidines is the mutagenic lesion (reviewed in reference 24). Thus, even when damage affects two adjacent bases, virtually all of the induced mutations arise opposite only one of the two damaged sites.

If SOS induction causes a persistent inactivation of epsilon that leads to higher-than-normal misincorporation frequencies, then it will be difficult, perhaps impossible, to detect the consequences in vivo. After all, even in the absence of proofreading, procaryotic polymerases misincorporate in vitro at frequencies no higher than 10^{-4} per base (18). If normal mismatch repair processes obtain, even these rare mistakes could be eliminated with high efficiency.

BASE-SUBSTITUTION SPECIFICITY OF SOS PROCESSING

Until now I have focused on the site distribution of mutations. A consideration of the kinds of mutations induced by different pyrimidine-damaging agents suggests that the base-pair substitutions induced by SOS processing are governed by rules that are sensitive to chemical features that are particular to a lesion and its surrounding sequences.

For example, angelicin that has been photodynamically activated by near UV generates adducts to pyrimidines. Angelicin preferentially induced transversion mutations at both A · T and G · C base pairs (26). The induction of transversion mutations by angelicin implies that pyrimidines are preferentially tolerated opposite angelicin-damaged pyrimidines. Had purines been tolerated, such as obtains opposite AP sites (17), then transition mutations at G · C sites would have been a prominent feature of the angelicin-induced *lacI* nonsense spectrum. Toleration of pyrimidines opposite angelicin adducts appears to be a property associated with that particular lesion, as two other pyrimidine-damaging treatments, UV and neocarzinostatin, lead to the preferential toleration of purines opposite their lesions. These results lead to the conclusion that base incorporation via SOS processing is not adequately described by the rule, "when in doubt put in an A." Further examination of the *lacI* data and other in vivo experiments (17, 26) revealed that the specificity of misincorporation varies from site to site in a sequence-dependent fashion (the rules of which are unknown).

It appears, therefore, that bypass replication of a DNA lesion via SOS repair can, in principle, incorporate any of the four common deoxynucleotide monophosphates; the specific deoxynucleotide monophosphate which is incorporated is selected according to rules that are sensitive to the chemical features of a particular lesion and its surrounding bases. Thus, SOS mutagenesis may share features in common with the mechanism by which directly miscoding lesions, such as those produced by base analogs or simple alkylating agents, induce mutations.

SUMMARY

umuDC is a genetic locus in *E. coli* whose expression is required for efficient mutagenesis by UV and other chemical mutagens. When the mutational specificity of several different *umuDC*-dependent mutagens was compared, it was found that (i) mutagens that generated distinctly different patterns of DNA damage induced a distinctly different spatial distribution of mutations within the target gene and (ii) the specificity of base-pair substitutions depended on the chemical nature of the DNA damage (and on surrounding sequences). These observations are interpreted to mean that (i) *umuDC*-dependent mutagenesis is a process that is limited to sites of DNA damage, i.e., is targeted, and (ii) the chemical nature of the damage (and its surrounding sequences) influences which bases will be tolerated by the replication machinery for *umuDC*-dependent bypass of damaged sites.

ACKNOWLEDGMENTS. Research in my laboratory in the Department of Cancer Biology and the Charles A. Dana Laboratory of Toxicology at the Harvard School of Public Health was supported by grants from the National Institutes of Health.

I have greatly enjoyed and benefited from many stimulating discussions about mutagenesis with J. Cairns, P. L. Foster, and J. H. Miller. I am grateful to P. L. Foster for her comments on an earlier draft of the manuscript.

LITERATURE CITED

1. Bridges, B. A., R. P. Mottershead, and S. G. Sedgwick. 1976. Mutagenic DNA repair in *E. coli*. III. Requirement for a function of DNA polymerase III in ultraviolet light mutagenesis. *Mol. Gen. Genet.* 144:53–58.
2. Bridges, B. A., and R. Woodgate. 1984. Mutagenic repair in *Escherichia coli*. X. The *umuC* gene product may be required for replication past pyrimidine dimers but not for the coding error in UV-mutagenesis. *Mol. Gen. Genet.* 196:364–366.
3. Bridges, B. A., R. Woodgate, M. Ruiz-Rubio, F. Sharif, S. G. Sedgwick, and U. Hübscher. 1987. Current understanding of UV-induced base pair substitution mutation in *E. coli* with particular reference to the DNA polymerase III complex. *Mutat. Res.* 181:219–226.
3a. Burkhardt, S. E., R. Woodgate, R. H. Scheuermann, and H. Echols. 1988. The UmuD mutagenesis protein of *Escherichia coli*: overproduction, purification, and cleavage by RecA. *Proc. Natl. Acad. Sci. USA* 85:1811–1815.
4. Caillet-Fauquet, P., G. Maenhaut-Michel, and M. Radman. 1984. SOS mutator effect in *E. coli* mutants deficient in mismatch correction. *EMBO J.* 3:707–712.
5. Ciesla, Z. 1982. Plasmid pKM101-mediated mutagenesis in *Escherichia coli* is inducible. *Mol. Gen. Genet.* 186:298–300.
6. Coulondre, C., and J. H. Miller. 1977. Genetic studies of the *lac* repressor. IV. Mutagenic specificity in the *lacI* gene of *Escherichia coli*. *J. Mol. Biol.* 117:577–606.
7. Defais, M., P. Fauquet, M. Radman, and M. Errera. 1971. Ultraviolet reactivation and ultraviolet mutagenesis of lambda in different genetic systems. *Virology* 43:495–503.
8. Eisenstadt, E. 1985. The mutational consequences of DNA damage induced by benzo[a]pyrene, p. 327–340. *In* R. G. Harvey (ed.), *Polycyclic Hydrocarbons and Cancer*. American Chemical Society, Washington, D.C.
9. Eisenstadt, E. 1987. Analysis of mutagenesis, p. 1016–1033. *In* F. C. Neidhardt, J. L. Ingraham, K. B. Low, B. Magasanik, M. Schaechter, and H. E. Umbarger (ed.), *Escherichia coli and Salmonella typhimurium: Cellular and Molecular Biology*. American Society for Microbiology, Washington, D.C.
10. Eisenstadt, E., A. J. Warren, J. Porter, D. Atkins, and J. H. Miller. 1982. Carcinogenic epoxides of benzo[a]pyrene and cyclopenta[cd]-pyrene induce base substitutions via specific transversions. *Proc. Natl. Acad. Sci. USA* 79:1945–1949.
11. Foster, P. L., and E. Eisenstadt. 1983. Distribution and specificity of mutations induced by neocarzinostatin in the *lacI* gene of *Escherichia coli*. *J. Bacteriol.* 153:379–383.
12. Foster, P. L., E. Eisenstadt, and J. Cairns. 1982. Random components in mutagenesis. *Nature* (London) 299:365–367.
13. Foster, P. L., E. Eisenstadt, and J. H. Miller. 1983. Base substitution mutations induced by metabolically activated aflatoxin B_1. *Proc. Natl. Acad. Sci. USA* 80:2695–2698.
14. George, J., M. Castellazzi, and G. Buttin. 1975. Prophage induction and cell division in *E. coli*. III. Mutations *sfiA* and *sfiB* restore division in *tif* and *lon* strain and permit the expression of mutator properties of *tif*. *Mol. Gen. Genet.* 140:309–332.
15. Khidhir, M. A., S. Casaregola, and I. B. Holland. 1985. Mechanism of transient inhibition of DNA synthesis in ultraviolet-irradiated *E. coli*: inhibition is independent of *recA* whilst recovery requires RecA protein itself and an additional, inducible SOS function. *Mol. Gen. Genet.* 199:133–140.
16. Kitagawa, Y., E. Akaboshi, H. Shinagawa, T. Horii, H. Ogawa, and T. Kato. 1985. Structural analysis of the *umu* operon required for inducible mutagenesis in *Escherichia coli*. *Proc. Natl. Acad. Sci. USA* 82:4336–4340.
17. Kunkel, T. A. 1984. Mutational specificity of depurination. *Proc. Natl. Acad. Sci. USA* 81:1494–1498.
18. Loeb, L. A., and T. A. Kunkel. 1982. Fidelity of DNA synthesis. *Annu. Rev. Biochem.* 51:429–457.
19. Lu, C., R. H. Scheuermann, and H. Echols. 1986. Capacity of RecA protein to bind preferentially to UV lesions and inhibit the editing subunit (ε) of DNA polymerase III; a possible mechanism

for SOS-induced targeted mutagenesis. *Proc. Natl. Acad. Sci. USA* 83:619–623.
20. Maenhaut-Michel, G., and P. Caillet-Fauquet. 1984. Effect of *umuC* mutations on targeted and untargeted ultraviolet mutagenesis in bacteriophage λ. *J. Mol. Biol.* 177:181–187.
21. Miller, J. H. 1978. The *lacI* gene: its role in *lac* operon control and its use as a genetic system, p. 31–88. *In* J. H. Miller and W. S. Reznikoff (ed.), *The Operon*. Cold Spring Harbor Laboratory, Cold Spring Harbor, N.Y.
22. Miller, J. H. 1982. Carcinogens induce targeted mutations. *Cell* 31:5–7.
23. Miller, J. H. 1983. Mutational specificity in bacteria. *Annu. Rev. Genet.* 17:215–238.
24. Miller, J. H. 1985. Mutagenic specificity of ultraviolet light. *J. Mol. Biol.* 182:45–68.
25. Miller, J. H., and K. B. Low. 1984. Specificity of mutagenesis resulting from the induction of SOS system in absence of mutagenic treatment. *Cell* 37:675–682.
26. Miller, S. S., and E. Eisenstadt. 1987. Suppressible base substitution mutations induced by angelicin (isopsoralen) in the *Escherichia coli lacI* gene: implications for the mechanism of SOS mutagenesis. *J. Bacteriol.* 169:2724–2729.
27. Moore, P. D., S. D. Rabkin, and B. S. Strauss. 1982. *In vitro* replication of mutagen-damaged DNA: sites of termination, p. 179–197. *In* J. F. Lemontt and W. M. Generoso (ed.), *Molecular and Cellular Mechanisms of Mutagenesis*. Plenum Publishing Corp., New York.
27a. Nohmi, T., J. R. Battista, L. A. Dodson, and G. C. Walker. 1988. RecA-mediated cleavage activates UmuD for mutagenesis: mechanistic relationship between transcriptional derepression and posttranslational activation. *Proc. Natl. Acad. Sci. USA* 85:1816–1820.
28. Perry, K. L., S. J. Elledge, B. B. Mitchell, L. Marsh, and G. C. Walker. 1985. *umuDC* and *mucAB* operons whose products are required for UV light- and chemical-induced mutagenesis: UmuD, MucA, and LexA proteins share homology. *Proc. Natl. Acad. Sci. USA* 82:4331–4335.
29. Povirk, L. F., and I. H. Goldberg. 1985. Endonuclease resistant apyrimidinic sites formed by neocarzinostatin at cytosine residues in DNA: evidence for a possible role in mutagenesis. *Proc. Natl. Acad. Sci. USA* 82:3182–3186.
30. Povirk, L. F., and I. H. Goldberg. 1986. Base substitution mutations induced in the *cI* gene of lambda phage by neocarzinostatin chromophore: correlation with depyrimidination hotspots at the sequence AGC. *Nucleic Acids Res.* 14:1417–1426.
31. Radman, M. 1974. Phenomenology of an inducible mutagenic DNA repair pathway in *Escherichia coli*: SOS repair hypothesis, p. 128–142. *In* L. Prakash, F. Sherman, M. Miller, C. Lawrence, and H. Tabor (ed.), *Molecular and Environmental Aspects of Mutagenesis*. Charles C Thomas, Publisher, Springfield, Ill.
32. Scheuermann, R., S. Tam, P. M. J. Burgers, C. Lu, and H. Echols. 1983. Identification of the ε subunit of *Escherichia coli* DNA polymerase III holoenzyme as the *dnaQ* gene product: a fidelity subunit for DNA replication. *Proc. Natl. Acad. Sci. USA* 80:7085–7089.
32a. Shinagawa, H., H. Iwasaki, T. Kato, and A. Nakata. 1988. RecA protein-dependent cleavage of UmuD protein upon SOS mutagenesis. *Proc. Natl. Acad. Sci. USA* 85:1806–1810.
32b. Storz, G., M. F. Christman, H. Sties, and B. N. Ames. 1987. Spontaneous mutagenesis and oxidative damage to DNA in *Salmonella typhimurium*. *Proc. Natl. Acad. Sci. USA* 84:8917–8921.
33. Villani, G., S. Bouteux, and M. Radman. 1978. Mechanisms of ultraviolet induced mutagenesis: extent and fidelity of *in vitro* DNA synthesis on irradiated templates. *Proc. Natl. Acad. Sci. USA* 78:3037–3041.
34. Walker, G. C. 1984. Mutagenesis and inducible responses to deoxyribonucleic acid damage in *Escherichia coli*. *Microbiol. Rev.* 48:60–93.
35. Witkin, E. M. 1976. UV mutagenesis and inducible DNA repair in *Escherichia coli*. *Bacteriol. Rev.* 40:869–907.
36. Witkin, E. M., V. Roegner-Maniscalco, J. B. Sweasy, and J. O. McCall. 1987. Recovery from ultraviolet light-induced inhibition of DNA synthesis requires *umuDC* gene products in *recA718* mutant strains but not in *recA*$^+$ strains of *Escherichia coli*. *Proc. Natl. Acad. Sci. USA* 84:6805–6809.
37. Witkin, E. M., and I. E. Wermundsen. 1978. Targeted and untargeted mutagenesis by various inducers of SOS functions in *Escherichia coli*. *Cold Spring Harbor Symp. Quant. Biol.* 48:881–886.
38. Wood, R. D., and F. Hutchinson. 1984. Nontargeted mutagenesis of unirradiated lambda phage in *Escherichia coli* host cells irradiated with ultraviolet light. *J. Mol. Biol.* 173:293–305.
39. Woodgate, R., B. A. Bridges, G. Herrera, and M. Blanco. 1987. Mutagenic DNA repair in *Escherichia coli*. XIII. Proofreading exonuclease of DNA polymerase III holoenzyme is not operational during UV mutagenesis. *Mutat. Res.* 183:31–37.

Chapter 43

Experimental System for the Study of Site-Specific Mutagenesis in Mammalian Cells and Bacteria

M. Moriya, M. Takeshita, K. Peden, and A. P. Grollman

Chemical mutagens form a variety of covalent adducts with DNA (6, 13); unless repaired prior to DNA replication, such lesions lead to nucleotide substitutions, deletions, and chromosomal rearrangements (19). Some fundamental questions can be raised regarding the process by which these mutagenic events occur. For example, which of several DNA adducts formed by a given mutagenic species is responsible for the mutations observed? Does the chemical structure of an adduct correlate with its mutational specificity? What role does nucleotide sequence play in determining the type and frequency of various mutations? To what extent does the presence of DNA adducts affect changes in nucleotide sequence other than at the site of modification?

Conventional mutational analysis does not readily answer the foregoing questions. We chose to approach this problem by studying the mutagenic properties of defined DNA adducts and related lesions introduced at specified positions in the genome. Viral and plasmid vectors containing single adducts have been developed (1, 5, 7, 15, 16); some have been tested for mutagenic properties (1, 9, 11). In this paper, we describe an experimental system which can be used to study site-specific mutagenesis in mammalian cells and bacteria.

EXPERIMENTAL STRATEGY

Our approach is summarized in Fig. 1. Adducts are introduced into DNA in the form of a chemically modified oligodeoxynucleotide, annealed to its complementary strand, and ligated into an appropriately constructed plasmid vector. For these experiments, we developed a simian virus 40 (SV40)-based shuttle plasmid, derived from plasmid pBR322, which replicates in a line of simian kidney (COS) cells that provide the required T antigen (4).

The shuttle vector containing the adduct is used as input DNA to transfect COS cells or to transform strains of bacteria (Fig.

M. Moriya, M. Takeshita, and A. P. Grollman • Department of Pharmacological Sciences, State University of New York at Stony Brook, Stony Brook, New York 11794. K. Peden • Howard Hughes Medical Institute, Department of Molecular Biology and Genetics, Johns Hopkins University School of Medicine, Baltimore, Maryland 21205.

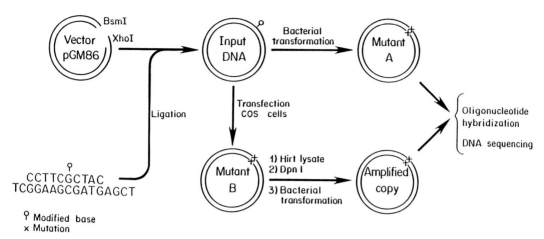

FIGURE 1. General strategy for site-specific mutagenesis.

1). In COS cells, mutations are fixed during replication; the progeny plasmids can be recovered from a cell lysate and subsequently amplified in bacteria. Mutagenesis in bacteria is investigated by direct transformation of competent cells. In both systems, mutations are screened by oligodeoxynucleotide hybridization techniques (18), and the mutational spectrum is determined by DNA sequence analysis.

SYNTHESIS OF OLIGODEOXYNUCLEOTIDES CONTAINING ADDUCTS

Oligodeoxynucleotides are conveniently prepared by automated solid-state methods (10). The adduct under investigation is introduced during assembly of the oligonucleotide or postsynthetically by chemical modification. The utility of automated chemical synthesis for mutagenic studies is limited somewhat since heterogeneity is not readily detected in oligonucleotides greater than 15 bases in length. Furthermore, certain DNA adducts are chemically labile and degrade under the conditions required for chemical deprotection of bases (2). In such cases, it is sometimes feasible to introduce the adduct by postsynthetic modification.

CONSTRUCTION OF PLASMID SHUTTLE VECTOR

The shuttle vector pAG75 (Fig. 2) was constructed by inserting nucleotides 5155 to 119 of the SV40 origin of replication at the EcoRI site of the bacterial plasmid pKP772. Plasmid pKP772 was derived from pBR322 by deletion of nucleotides 1332 to 2520 (K. Peden, unpublished data). The nucleotide sequence of the gene coding for tetracycline resistance was modified by oligonucleotide-directed mutagenesis of bases located at positions 950 and 979, creating the unique restriction sites MluI (position 945) and XhoI (position 978). The 33-base-pair oligonucleotide sequence between the MluI and XhoI sites was replaced by a chemically synthesized oligonucleotide, creating pGM86, which contains additional sites for SnaBI (position 956), NheI (position 964) and BsmI (position 968).

For the mutagenesis experiments described below, the BsmI-XhoI fragment of pGM86 was replaced by a synthetic oligodeoxynucleotide, CCTTCGCTAC (G-10), containing a single deoxyguanosine residue. This sequence, with or without postsynthetic modification, was introduced into pGM86 by standard ligation procedures to create

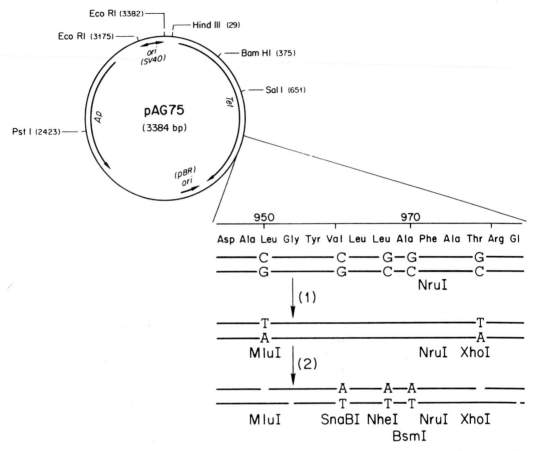

FIGURE 2. Construction of pGM86. (1) Creation of unique *Mlu*I and *Xho*I sites in the gene encoding tetracycline resistance by using oligonucleotide-directed mutagenesis; (2) replacement of the 33-base-pair fragment between *Mlu*I and *Xho*I.

pGM87. All nucleotide changes introduced during construction of these vectors involve conservative codon replacement; thus, pGM86 and pGM87 retain a functional gene coding for tetracycline resistance.

Certain mutations arising from lesions introduced at position 974 (dG) can be detected phenotypically. Deletion of this base inactivates the tetracycline resistance gene. Replacement of dG at this position by dA, dT, or dC leads to substitution of threonine, serine, or proline, respectively, for alanine. Only the proline substitution inactivates the gene encoding tetracycline resistance.

SITE-SELECTIVE MUTAGENESIS BY dG-AAF

Acetoxyacetylaminofluorene (AAAF) was allowed to react with G-10 in citrate buffer for 60 min at 37°C. Analysis of the reaction products revealed a new UV-absorbing peak which could be resolved from unmodified G-10 by high-performance liquid chromatography (HPLC). This modified oligodeoxynucleotide (G-10-AAF) exhibited an absorption maximum at 270 nm with a shoulder at 300 nm. Enzymatic digestion, followed by HPLC, revealed that *N*-(deoxy-

```
        L16:                    R15:
        ACGTATTGCTAGCCTT         TACTCGAGGCTGGAT

          ( 960 )            AAF   ( 980 )
    5' GCTACGTATTGCTAG*CCTTCGCTAC*TCGAGGCTGGAT 3'
    3' CGATGCATAACGAT*CGGAAGCGATGAGCT*CCGACCTA 5'

           C16:    TCGGAAGCGATGAGCT
           G16:    TCGGAAGGGATGAGCT
           A16:    TCGGAAGAGATGAGCT
           T16:    TCGGAAGTGATGAGCT
          Δ16:    TCGGAAG-GATGAGCTC
```

FIGURE 3. Oligonucleotide probes used for the detection of mutants. The inserted sequence is in boldface.

guanosine-8-yl)-acetylaminofluorene (C8-dG-AAF) was the only adduct present (12).

G-10 and G-10-AAF were phosphorylated at the 5' termini by using [γ-^{32}P]ATP and T4 polynucleotide kinase, annealed to the 5'-phosphorylated complementary 16-mer (Fig. 1) and ligated into the BsmI-XhoI site of pGM86. Portions of the ligation mixture, containing 10 ng of vector and a 10-fold molar excess of oligodeoxynucleotide, were used to transform Escherichia coli strains AB1157, AB1886 uvrA, and AB2463 recA. Transformants, recovered from plates containing 80 μg of ampicillin per ml, were analyzed for tetracycline resistance. Mutant plasmids were detected by hybridization techniques with the oligonucleotide probes shown in Fig. 3.

SV40-transformed simian kidney cells (COS-1) were grown at 37°C under 5% CO_2 in Dulbecco modified Eagle medium containing 10% fetal calf serum (12). Twenty-four hours after seeding, cells were transfected with ligation mixture containing 50 ng of DNA in the presence of DEAE dextran. Forty-eight hours following transfection, DNA was extracted, digested with proteinase K, and then treated sequentially with phenol, phenol-chloroform, and chloroform. Following ethanol precipitation, the restriction enzyme DpnI was used to degrade residual input DNA (14). The DNA was then used to transform competent E. coli DH5 cells. Transformants were screened for mutations by oligonucleotide hybridization. Oligonucleotide probes L16 and R15 were used to detect the presence of the inserted sequence.

Approximately 400 transformants, selected from each experiment at random, were analyzed by hybridization techniques for sequence alterations in the region immediately surrounding the AAF-modified base. The presence of a single C8-dG-AAF adduct in a plasmid increased the mutation frequency when introduced into any of several strains of E. coli and COS cells (Table 1). The majority of mutations were observed at the site modified by AAF; 85 to 90% of mutants isolated from E. coli and 80% of mutants recovered from COS cells contained alterations at this position (Table 1).

In E. coli, 70 to 80% of these targeted mutations represented base substitutions; the remainder were single-base deletions. More than 90% of the nucleotide substitutions observed in E. coli were transversions, GC→CG predominating over GC→TA (12). In COS cells, 90% of the targeted mutations were base substitutions and 10% were deletions. Of the base substitutions, 95% were transversions; the two types occurred with similar frequency.

Mutations outside the region covered by the hybridization probes, as well as G→C transversions and deletions of the modified

TABLE 1
Mutations Observed in E. coli and COS cells after Transformation with AAF-Modified DNA

Plasmid	Host strain	Mutation frequency[a]	No. of targeted mutations			No. of other mutations
			Transversions	Transitions	Deletions	
pGM87	AB1157	0.8%	0	0	0	3
pGM87-AAF	AB1157	6.8%	14	2	7	4[b]
pGM87-AAF	AB1886	6.6%	18	0	4	3
pGM87-AAF	AB2463	2.8%	7	0	3	1
pGM87	COS	1.0%	0	0	0	4
pGM87-AAF	COS	8.4%	19	1	2	6[b]

[a] In each experiment 350 to 450 colonies were analyzed.
[b] In one case, accompanied by an additional targeted mutation.

base at position 974, were detected phenotypically by selection for tetracycline sensitivity. No untargeted mutations were observed. In general, bacterial and mammalian cells yield a similar pattern of targeted mutations with transversions predominating over transitions.

SUMMARY

A shuttle plasmid vector system has been developed which should be widely applicable to studies of mutagenic specificity in mammalian cells and bacteria. With the experimental strategy outlined in this paper, DNA adducts or other lesions are introduced at a specific position in this plasmid and may lead to mutations whose nature and position can subsequently be established by sequence analysis. Advantages of this system include the homogeneity of input DNA, lack of bias in the procedure used for detection, and the potential for systematically altering base sequence in the vicinity of the adduct. Moreover, by retaining the integrity of the gene encoding tetracycline resistance, adducts and lesions may be introduced at site 974 which permits rapid phenotypic selection and detection of nontargeted mutations (additions, deletions, and certain base substitutions).

Our experimental system has been applied to a study of AAF mutagenesis (12). The adduct used, C8-dG-AAF, resulted in transversions and single-base deletions in bacteria and mammalian cells, targeted to the site of the modified base. These results contrast with the spectrum of mutations in bacteria described by Koffel-Schwartz et al. (8), in which a fragment of pBR322, modified randomly with AAF, was used to transform E. coli. According to this report, more than 90% of the changes were frameshift mutations. Differences between the two experiments have been discussed elsewhere (12).

Although the purity of the dG-AAF adduct used in this study has been established unambiguously by chemical procedures, it is conceivable that mutagenic changes observed may arise from a different lesion. For example, C8-dG-AAF could be deacetylated by cellular enzymes to form C8-dG-AF, or it could be depurinated to produce the noninformational and presumably mutagenic abasic site (3). These possibilities are currently being explored by incorporating C8-dG-AF adducts and synthetic abasic sites (17) into shuttle plasmids and testing their mutagenic specificities.

ACKNOWLEDGMENTS. Some of the work described in this paper was supported by Public Health Service grants CA17395 and ES04068 from the National Institutes of Health.

LITERATURE CITED

1. **Bhanot, E., and A. Ray.** 1986. The *in vivo* mutagenic frequency and specificity of O^6-methylguanine in φX replicative form DNA. *Proc. Natl. Acad. Sci. USA* **83**:7348–7352.

2. **Borowy-Borowski, H., and R. W. Chambers.** 1987. A study of side reactions occurring during synthesis of oligodeoxynucleotides containing O^6-alkyldeoxyguanosine residues at preselected sites. *Biochemistry* **26**:2465–2471.

3. **Drinkwater, N. R., E. C. Miller, and J. A. Miller.** 1980. Estimation of apurinic/apyrimidinic sites and phosphotriesters in deoxyribonucleic acid treated with electrophilic carcinogens and mutagens. *Biochemistry* **19**:5087–5092.

4. **Gluzman, Y.** 1981. SV40-transformed simian cells support the replication of early SV40 mutants. *Cell* **23**:175–182.

5. **Green, C. L., E. L. Loechler, K. W. Fowler, and J. M. Essigmann.** 1984. Construction and characterization of extrachromosomal probes for mutagenesis by carcinogens: site-specific incorporation of O^6-methylguanine into viral and plasmid genomes. *Proc. Natl. Acad. Sci. USA* **81**:13–17.

6. **Hemminki, K.** 1983. Nucleic acid adducts of chemical carcinogens and mutagens. *Arch. Toxicol.* **52**:249–285.

7. **Johnson, D. L., T. M. Reid, M. S. Lee, C. M. King, and L. J. Romano.** 1986. Preparation and characterization of a viral DNA molecule containing a site-specific 2-aminofluorene adduct: a new probe for mutagenesis by carcinogens. *Biochemistry* **25**:449–456.

8. **Koffel-Schwartz, N., G. Maenhaut-Michel, and R. P. Fuchs.** 1987. Specific strand loss in N-2-acetylaminofluorene-modified DNA. *J. Mol. Biol.* **193**:651–659.

9. **Loechler, E. L., C. L. Green, and J. M. Essigmann.** 1984. *In vivo* mutagenesis by O^6-methyl- guanine bu[...] *Proc. Natl. Ac[...]*

10. **Matteucci, M. [...]** Synthesis of deoxy[...] support. *J. Am. Chem.[...]*

11. **Mitchell, N., and G. St[...]** originating in site-specific [...] *Biol.* **191**:177–180.

12. **Moriya, M., M. Takeshita, F. [...] den, S. Will, and A. P. Grollman.** [...] mutations induced by a single acetylam[...] DNA adduct in mammalian cells and bact[...] *Natl. Acad. Sci. USA* **85**:1586–1589.

13. **Osborne, M. R.** 1984. DNA interactions of r[...] tive intermediates derived from carcinogens, p[...] 485–524. *In* C. E. Searle (ed.), *Chemical Carcinogens*, ACS Monograph 192. American Chemical Society, Washington, D.C.

14. **Peden, K. W. C., S. M. Pipas, S. Pearson-White, and D. Nathans.** 1980. Isolation of mutants of an animal virus in bacteria. *Science* **209**:1392–1396.

15. **Preston, B. D., B. Singer, and L. A. Loeb.** 1987. Comparison of the relative mutagenicities of O-alkylthymines site-specifically incorporated into φX174 DNA. *J. Biol. Chem.* **262**:13821–13827.

16. **Stöhrer, G., J. A. Osband and G. Alvarado-Urbina.** 1983. Site-specific modification of the lactose operator with acetylaminofluorene. *Nucleic Acids Res.* **11**:5093–5102.

17. **Takeshita, M., C.-N. Chang, F. Johnson, S. Will, and A. P. Grollman.** 1987. Oligodeoxynucleotides containing synthetic abasic sites. *J. Biol. Chem.* **262**:10171–10179.

18. **Wallace, R. B., J. Schaffer, R. F. Murphy, J. Bonner, T. Hirose, and K. Itakura.** 1979. Hybridization of synthetic oligodeoxyribonucleotides to φX174 DNA: the effect of a single base pair mismatch. *Nucleic Acids Res.* **6**:3543–3556.

19. **Würgler, F. E.** 1984. Theoretical basis of mutagenesis, p. 113–155. *In* E. H.-Y. Chu and W. M. Generoso (ed.), *Mutation, Cancer and Malformation*. Plenum Publishing Corp., New York.

The DNA sequence analysis of mutations induced by treatment of mammalian cells with chemical or physical agents provides a critical first step toward understanding the mechanisms by which DNA damage results in heritable genetic alterations. Two complementary approaches have been developed for the molecular analysis of mutagenesis in mammalian cells. In the first, mutations are induced in an endogenous gene or in a chromosomally integrated shuttle vector, and after selection for mutant cells, the target gene is molecularly cloned for DNA sequence analysis (2, 17). An alternative approach is based on the use of recombinant DNA shuttle vectors that replicate as plasmids in mammalian cells. These plasmid shuttle vectors contain *cis*-acting sequences derived from DNA tumor viruses that allow replication in mammalian cells, bacterial plasmid sequences required for replication in *Escherichia coli*, and a target gene for analysis of mutagenesis (5, 10, 21). After fixation of mutations by replication of the plasmid in mammalian cells, the plasmid DNA is isolated, and mutant plasmids are selected in *E. coli*. Although plasmid shuttle vectors do not provide the normal chromosomal context of the first approach, they furnish two advantages in the study of mutagenesis. First, the ready isolation of mutants in a form directly suitable for sequence analysis allows the rapid accumulation of the large amount of information required for a detailed analysis of the mutational spectrum for a given agent. Second, the ability to introduce these vectors into a variety of cell lines makes it possible to compare the types of mutations induced as a function of the genetic background of the host cell.

The plasmid shuttle vectors most widely used in studies of mutagenesis have contained replication origins derived from papovaviruses such as simian virus 40 (10, 21). After transfection into mammalian cells, these vectors undergo many rounds of replication, and the amplified plasmids may be isolated after a period of several days. More

Norman R. Drinkwater, Kristin A. Eckert, Caroline A. Ingle, and Donna K. Klinedinst • McArdle Laboratory for Cancer Research, University of Wisconsin, Madison, Wisconsin 53706.

recently, we and others (1, 5, 6) have studied mutagenesis in shuttle vectors that are stably maintained as plasmids in dividing cell populations; the replication origins of these plasmids are derived from Epstein-Barr virus (EBV) or bovine papillomavirus. In this review we describe some of the properties of *oriP-tk* shuttle vectors based on EBV and their use in the analysis of mutations induced in human cells by the alkylating agent *N*-ethyl-*N*-nitrosourea (ENU).

DEVELOPMENT OF *oriP-tk* SHUTTLE VECTORS FOR MUTAGENESIS STUDIES

EBV is a lymphotropic herpesvirus that latently infects and immortalizes human B lymphocytes. In infected lymphoblastoid cell lines, the viral genome is present in multiple copies as a closed circular plasmid of approximately 172,000 base pairs (9). Sugden and co-workers (25, 26) have identified the *cis*- and *trans*-acting EBV sequences required for the replication and stable maintenance of viral or recombinant DNA plasmids. The *cis*-acting element, designated *oriP*, contains two essential sequences: a 65-base-pair region of dyad symmetry that is the putative origin of DNA synthesis and a set of 20 imperfect 30-base-pair direct repeats that function as a transcriptional enhancer *trans*-activated by the viral nuclear antigen, EBNA-1 (22). Expression of the EBNA-1 gene is required for the replication of plasmids containing the *oriP* sequence (26). Studies of the replication of the viral genome or recombinant *oriP* plasmids in human cells have demonstrated that these plasmids replicate during S phase by using the normal cellular replication machinery and that each plasmid undergoes a single replication during each cell cycle (15a). The host range of *oriP* plasmids includes lymphoblastoid, epithelial, or fibroblastic cell lines of human, simian, or canine origin (26).

We have constructed a series of shuttle vectors that contain the *oriP* element (5) (Fig. 1) for studying mutagenesis in EBV-transformed human lymphoblastoid cell lines, which express the required EBNA-1 gene product from the resident viral genome. In addition to the *oriP* sequence, these vectors contain sequences derived from bacterial plasmids that allow replication and expression of resistance to ampicillin or chloramphenicol in *E. coli*, a recombinant hygromycin phosphotransferase gene that is expressed in mammalian cells, and the herpes simplex virus type 1 thymidine kinase gene (HSV *tk*). Because of its relatively large coding sequence (1.1 kilobases) and the ability to select for base-substitution, frameshift, and deletion mutations, the HSV *tk* gene is a sensitive target for mutagenesis studies. In the studies described below, *oriP-tk* plasmids were isolated from mammalian cells, and mutant plasmids were selected after transformation of a 5-fluorodeoxyuridine-resistant derivative of *E. coli* HB101 (FT334 *upp tdk*) and growth of transformed cells on medium containing 5-fluorodeoxyuridine (7). For each mutant plasmid, the approximate location of the induced mutation was determined by deletion mapping and the DNA sequence was obtained by the dideoxynucleotide method.

INTRODUCTION OF *oriP-tk* PLASMIDS INTO HUMAN LYMPHOBLASTOID CELLS

Electroporation is an efficient method for the introduction of *oriP* vectors into human lymphoblastoid cells (24). Electroporation of *oriP-tk* vectors into several normal (721, EAB, 1117, 315) or mutant (GM2783, ataxia telangiectasia; GM2250, xeroderma pigmentosum) human lymphoblastoid cell lines resulted in the expression of hygromycin resistance in 0.2 to 10% of the cells surviving electroporation. When plasmid DNA was isolated after growth of the electroporated cells in hygromycin for a period

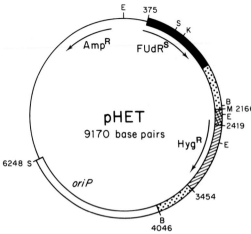

FIGURE 1. Structures of *oriP-tk* shuttle vectors. The construction of the vector shown (pHET) is described in reference 5. The bacterial plasmid portion of the vector (thin line) is a deletion mutant of pBR322 (Δ1420-2490). Resistance of mammalian cells to the antibiotic hygromycin is provided by fusion of the coding sequence of the bacterial *hph* gene (hatched box) to the regions of the HSV *tk* gene that control initiation (cross-hatched box) and termination (stippled box) of transcription. The *oriP* element (open box) from EBV allows replication of the vector as a plasmid in EBV-transformed lymphoblastoid cells. The region of the HSV *tk* gene (*Bgl*II to *Pvu*II) that contains the coding (filled box) and 3' noncoding (stippled box) sequences was inserted at the *Bam*HI site of the bacterial plasmid. Expression of this gene in *E. coli* is controlled by the promoter of the plasmid *tet* gene. The structure of the vector pND112 is identical to that shown except that the nucleotides between 188 and 375 were deleted; this deletion decreases the distance between the HSV *tk* coding sequence and the bacterial *tet* promoter. The vector pND123 was constructed from pND112 by the insertion of the gene conferring resistance to *E. coli* for the antibiotic chloramphenicol at the unique *Bam*HI site in the vector. The nucleotide positions of the junctions between the elements of the vector are indicated. The origin for the map is the *Eco*RI site at the top of the figure. Restriction enzyme cleavage sites are shown for *Eco*RI (E), *Sph*I (S), *Kpn*I (K), *Bst*EII (B), and *Bam*HI (M). The sequences conferring resistance to ampicillin (Amp^r) and hygromycin (Hyg^r), sensitivity to 5-fluorodeoxyuridine (FUdR^s), and the *oriP* element are indicated.

plasmid shuttle vectors demonstrated that transfection of DNA into mammalian cells was mutagenic (10, 21). The mutant frequencies observed when plasmid DNA was isolated from untreated cells ranged from 10^{-4} to 10^{-2} and depended on the construction of the vector, the host cell line, and the method used for transfection. These mutations are presumed to result from damage incurred by the plasmids after entry into the cell and before establishment and replication in the nucleus (16). We have observed a low frequency of mutations in the HSV *tk* gene when *oriP-tk* plasmids were isolated from electroporated populations of 721 cells after selection with hygromycin (5). In seven independent experiments, the mean HSV *tk* mutant frequency was $(1.8 \pm 1.3) \times 10^{-4}$ and was approximately fourfold higher than that for the bacterially derived input DNA. As previously observed for other plasmid shuttle vectors (10, 16, 21), a large majority (approximately 85%) of the mutants isolated from untreated cells contained deletions of all or part of the target gene. Because of their low frequency of occurrence (ca. 3×10^{-5}), we have determined the DNA sequences of only a limited number of background base-substitution mutants. As observed with papovavirus shuttle vectors (16), most of the transfection-induced base substitution mutations occurred at G·C base pairs (70%), with G·C→A·T transition mutations (43%) in greater number than transversions (Table 1).

GENETIC STABILITY OF *oriP-tk* SHUTTLE VECTORS IN HUMAN LYMPHOBLASTOID CELLS

After introduction into human lymphoblastoid cells by electroporation, *oriP-tk* vectors are genetically stable (5). Clones of 721 cells containing *oriP-tk* plasmids were isolated by growth in medium containing 0.35% agarose and propagated in medium containing hygromycin for periods equiva-

of 10 to 14 days, the resistant population of cells contained 10 to 25 copies of shuttle vector DNA per cell (5).

The earliest studies of mutagenesis in

TABLE 1
Types of Mutations Recovered from Untreated and ENU-Treated Human Lymphoblastoid Cells (LCL-721) and from ENU-Treated E. coli

Cell type	No. of mutants	Proportion of mutants (%)					
		G·C→A·T	A·T→G·C	A·T→T·A	A·T→C·G	G·C→T·A	G·C→C·G
Untreated LCL-721	14	43	14	14	0	21	8
ENU-treated LCL-721	46	48	17	20	9	4	2
E. coli	21	68	22	5	0	0	0

lent to 34 to 60 population doublings. Plasmid DNA was then isolated from 11 independent clonal populations of cells and analyzed for the frequency of HSV tk mutations (Table 2). In the six clones grown for 34 to 36 population doublings, the observed mutant frequencies ranged from 2×10^{-5} to 7×10^{-5}. Two of the clones, 112.A and 112.B, had unusually high frequencies of mutant plasmids (ca. 7×10^{-4}). For this set of 11 clones, the median rate of accumulation of HSV tk mutations during replication of the shuttle vectors in mammalian cells was 1.9 mutations per 10^6 plasmids per cell generation. This mutation rate is similar to that reported for some autosomal loci in cultured cells, i.e., approximately 10^{-6} mutations per cell generation (15). Similar results have been obtained by DuBridge et al. for the replication of plasmid vectors containing both the $oriP$ and EBNA-1 DNA sequences in human 293 cells (6). The genetic stability of the $oriP$ vectors is not shared by stably maintained plasmids derived from bovine papillomavirus. Ashman and Davidson (1) have observed deletion mutations at a frequency of approximately 1% in bovine papillomavirus plasmid vectors isolated from clonal populations of transfected C127 cells. The higher rate of mutation in the bovine papillomavirus shuttle plasmids may be related to the presence of a large proportion of the viral genome in the vector.

TABLE 2
Genetic Stability of $oriP$-tk Shuttle Vectors in Human Lymphoblastoid Cells[a]

Clone	Population doublings	Total plasmids assayed (10^5)	HSV tk mutant frequency (10^{-5})
HET.7	34	0.42	4.8
HET.22	34	0.21	<5
HET.19	35	0.16	6.3
HET.8	36	0.58	3.4
123.3	36	1.6	7.0
123.4	36	0.53	<1.9
112.A	37	0.85	60
112.B	37	0.68	90
112.F	39	0.68	11
HET.21	58	0.48	16
HET.23	60	1.07	15

[a] Clones of LCL-721 cells bearing $oriP$-tk shuttle vectors were cultured for the indicated periods of time in the presence of the selective agent hygromycin. Plasmid DNA isolated from the cells was assayed for the frequency of HSV tk mutants after transformation of E. coli.

ANALYSIS OF MUTATIONS INDUCED BY ENU IN HUMAN CELLS

ENU is a potent mutagen for cultured mammalian cells (3) and induces tumors at a wide range of sites when administered to experimental animals (8). Treatment of DNA with ENU results in the ethylation of a variety of nucleophilic oxygen and nitrogen sites, including O^6-guanine, O^2- and O^4-thymine, the phosphodiester backbone, and the N-3 and N-7 positions of purine bases (4). We have used the $oriP$-tk shuttle vectors in an attempt to determine the relative contributions of these DNA lesions to the induction of mutations by ENU in mammalian cells.

Treatment of human lymphoblastoid 721 cells containing *oriP-tk* vectors with ENU resulted in a dose-dependent increase in the frequency of HSV *tk* mutations in the plasmids (5). The dose response for the induction of mutations in the shuttle vector closely paralleled that for the induction of mutations at the cellular hypoxanthine-guanine phosphoribosyltransferase locus. We have determined the DNA sequences for 46 independent HSV *tk* point mutants from plasmids isolated from cells treated with ENU under conditions that resulted in a 10- to 20-fold increase in the frequency of mutants relative to that for plasmids isolated from solvent-treated cells (Table 1). Approximately one-half of the observed DNA sequence changes were G·C→A·T transition mutations, while 17% were A·T→G·C transitions. These two transition mutations also accounted for virtually all of the mutations observed by Richardson et al. (23) in plasmid-encoded *gpt* genes and by us in plasmid HSV *tk* genes after treatment of wild-type *E. coli* with ENU. The G·C→A·T and A·T→G·C transition mutations are the predicted results of mispairing of O^6-alkylguanine and O^4-alkylthymine, respectively, during DNA replication (14, 19).

In contrast to the results with *E. coli*, transversion mutations induced at A·T base pairs were a prominent part (29%) of the mutational spectrum for ENU in human cells. The observed A·T→T·A transversions (20%) might have resulted from insertion of an adenine residue opposite an ethyladenine adduct or apurinic site, or from insertion of a thymine residue opposite an ethylthymine adduct or apyrimidinic site. Apurinic/apyrimidinic sites result from the spontaneous hydrolysis of *N*-3- or *N*-7-ethyladenine or O^2-ethylthymine or through the enzymatic removal of these lesions by DNA glycosylases (13). Although the preferential insertion of purine bases by mammalian polymerases opposite noninformational lesions (20) would initially incline one to favor ethyladenine adducts as the origin of the transversion mutations, comparison of the mutational spectra in human cells for related alkylating agents implicates O^2-ethylthymine adducts. *N*-Methyl-*N*-nitrosourea and ethyl methanesulfonate have been found to induce solely G·C→A·T transition mutations in human cells and do not result in transversion mutations at A·T base pairs (6, 11). While ENU, *N*-methyl-*N*-nitrosourea, and ethyl methanesulfonate treatment of DNA yield similar proportions of 1-, 3-, and 7-alkyladenine adducts in relation to total base modification, these agents differ in their abilities to modify the O^2 position of thymine (4). Modification of this site by ENU is approximately 165-fold greater than that by *N*-methyl-*N*-nitrosourea and 1,000-fold greater than that by ethyl methanesulfonate. The biological importance of the A·T→T·A transversion mutations induced by ENU is emphasized by the observation of this mutation in the activation of cellular proto-oncogenes in ENU-induced tumors in vivo and in germ line mutations induced by treatment of mice with ENU (12, 18).

USE OF *oriP-tk* SHUTTLE VECTORS IN MUTAGENESIS STUDIES

The *oriP* shuttle vectors display several properties that render them useful for molecular studies of mutagenesis in mammalian cells. First, the vectors are stably maintained in cells expressing EBNA-1 and are replicated synchronously with cellular DNA (B. Sugden, personal communication). Thus, it is likely that processes which result in fixation of mutations in cellular genes will operate on the plasmid vector. Second, the low background of transfection-induced mutations and genetic stability of the vector allow analysis of mutagenesis under treatment conditions consistent with those used for mutagenesis of cellular genes. The genetically stable maintenance of the plasmids in mammalian cells also makes them well suited to the study of spontaneous mutagenesis. Third, the *oriP* vectors may be introduced into a variety of EBV-transformed human lymphoblastoid cell lines. Such cell lines are readily available for human genetic backgrounds, such as xeroderma pigmentosum, ataxia telangiectasia, or Bloom's syndrome, with abnormal responses to DNA damage by mutagens or carcinogens or having unusually high spontaneous mutation rates.

ACKNOWLEDGMENTS. We thank Julie Ginsler for her excellent technical assistance.

This work was supported by Public Health Service grants CA 37166, CA 07175, CA 09135, and CA 09020 from the National Cancer Institute.

LITERATURE CITED

1. Ashman, C. R., and R. L. Davidson. 1985. High spontaneous mutation frequency of a BPV shuttle vector. *Somatic Cell. Mol. Genet.* 11:499–504.
2. Ashman, C. R., and R. L. Davidson. 1987. Sequence analysis of spontaneous mutations in a shuttle vector gene integrated into mammalian chromosomal DNA. *Proc. Natl. Acad. Sci. USA* 84:3354–3358.
3. Aust, A. E., N. Drinkwater, K. Debien, V. Maher, and J. McCormick. 1984. Comparison of the frequency of diphtheria toxin and thioguanine resistance induced by a series of carcinogens to analyze their mutational specificities in diploid human fibroblasts. *Mutat. Res.* 125:95–104.
4. Beranek, D. T., C. C. Weis, and D. H. Swenson. 1980. A comprehensive quantitative analysis of methylated and ethylated DNA using high pressure liquid chromatography. *Carcinogenesis* 1:595–606.
5. Drinkwater, N. R., and D. K. Klinedinst. 1986. Chemically induced mutagenesis in a shuttle vector with a low-background mutant frequency. *Proc. Natl. Acad. Sci. USA* 83:3402–3406.
6. DuBridge, R. B., P. Tang, H. C. Hsia, P. M. Leong, J. H. Miller, and M. P. Calos. 1987. Analysis of mutation in human cells by using an Epstein-Barr virus shuttle system. *Mol. Cell. Biol.* 7:379–387.
7. Eckert, K. A., and N. R. Drinkwater. 1987. *recA*-dependent and *recA*-independent N-ethyl-N-nitrosourea mutagenesis at a plasmid-encoded herpes simplex virus thymidine kinase gene in *Escherichia coli*. *Mutat. Res.* 178:1–10.
8. International Agency for Research on Cancer (IARC). 1978. *Evaluation of the Carcinogenic Risk of Chemicals to Humans: Some N-Nitroso Compounds*, vol. 17, p. 191–215.
9. Kintner, C., and B. Sugden. 1981. Conservation and progressive methylation of Epstein-Barr viral DNA sequences in transformed cells. *J. Virol.* 38:305–316.
10. Lebkowski, J. S., R. B. DuBridge, E. A. Antell, K. S. Greisen, and M. P. Calos. 1984. Transfected DNA is mutated in monkey, mouse and human cells. *Mol. Cell. Biol.* 4:1951–1960.
11. Lebkowski, J. S., J. H. Miller, and M. P. Calos. 1986. Determination of DNA sequence changes induced by ethyl methanesulfonate in human cells, using a shuttle vector system. *Mol. Cell. Biol.* 6:1838–1842.
12. Lewis, S. E., F. M. Johnson, L. C. Skow, D. Popp, L. B. Barnett, and R. A. Popp. 1985. A mutation in the β-globin gene detected in the progeny of a female mouse treated with ethylnitrosourea. *Proc. Natl. Acad. Sci. USA* 82:5829–5831.
13. Loeb, L. A., and B. D. Preston. 1986. Mutagenesis by apurinic/apyrimidinic sites. *Annu. Rev. Genet.* 20:201–230.
14. Loechler, E. L., C. L. Green, and J. M. Essigmann. 1984. *In vivo* mutagenesis by O^6-methylguanine built into a unique site in a viral genome. *Proc. Natl. Acad. Sci. USA* 81:6271–6275.
15. McKenna, P. G., and P. E. Ward. 1987. Mutation at the APRT locus in Friend erythroleukaemia cells. 1. Mutation rates and properties of mutants. *Mutat. Res.* 180:267–271.
15a. Mecsas, J., and B. Sugden. 1987. Replication of plasmids derived from bovine papilloma virus type 1 and Epstein-Barr virus in cells in culture. *Annu. Rev. Cell Biol.* 3:87–108.
16. Miller, J. H., J. S. Lebkowski, K. S. Greisen, and M. P. Calos. 1984. Specificity of mutations induced in transfected DNA by mammalian cells. *EMBO J.* 3:3117–3121.
17. Nalbantoglu, J., G. Phear, and M. Meuth. 1987. DNA sequence analysis of spontaneous mutations at the *aprt* locus of hamster cells. *Mol. Cell. Biol.* 7:1445–1449.
18. Perantoni, A. O., J. M. Rice, C. D. Reed, M. Watatani, and M. L. Wenk. 1987. Activated *neu* oncogene sequences in primary tumors of the peripheral nervous system induced in rats by transplacental exposure to ethylnitrosourea. *Proc. Natl. Acad. Sci. USA* 84:6317–6321.
19. Preston, B. D., B. Singer, and L. A. Loeb. 1986.

Mutagenic potential of O^4-methylthymine in vivo determined by an enzymatic approach to site-specific mutagenesis. *Proc. Natl. Acad. Sci. USA* **83:**8501–8505.
20. **Randall, S. K., R. Eritja, B. E. Kaplan, J. Petruska, and M. F. Goodman.** 1987. Nucleotide insertion kinetics opposite abasic lesions in DNA. *J. Biol. Chem.* **262:**6864–6870.
21. **Razzaque, A., S. Chakrabarti, S. Joffee, and M. Seidman.** 1984. Mutagenesis of a shuttle vector plasmid in mammalian cells. *Mol. Cell. Biol.* **4:**435–441.
22. **Reisman, D., and B. Sugden.** 1986. *trans*-Activation of an Epstein-Barr viral (EBV) transcriptional enhancer by the EBV nuclear antigen-1. *Mol. Cell. Biol.* **6:**3838–3846.
23. **Richardson, K. K., F. C. Richardson, R. M. Crosby, J. A. Swenberg, and T. R. Skopek.** 1987. DNA base changes and alkylation following in vivo exposure of *Escherichia coli* to N-methyl-N-nitrosourea or N-ethyl-N-nitrosourea. *Proc. Natl. Acad. Sci. USA* **84:**344–348.
24. **Sugden, B., K. Marsh, and J. Yates.** 1985. A vector that replicates as a plasmid can be efficiently selected in B-lymphoblasts transformed by Epstein-Barr virus. *Mol. Cell. Biol.* **5:**410–413.
25. **Yates, J., N. Warren, D. Reisman, and B. Sugden.** 1984. A *cis*-acting element from the Epstein-Barr viral genome that permits stable replication of recombinant plasmids in latently infected cells. *Proc. Natl. Acad. Sci. USA* **81:**3806–3810.
26. **Yates, J. L., N. Warren, and B. Sugden.** 1985. Stable replication of plasmids derived from Epstein-Barr virus in a variety of mammalian cells. *Nature* (London) **313:**812–815.

Chapter 45

DNA Damage Mediated by Reducing Sugars In Vitro and In Vivo

Annette T. Lee and Anthony Cerami

It has been 75 years since L. C. Maillard first observed nonenzymatic browning of proteins by reducing sugars (12). Since then these studies have been continued by food chemists because of the importance of this reaction in the storage of foodstuffs. It has become apparent only in the last decade that this same reaction is occurring in living tissues. The reaction of sugars with proteins begins with the reversible formation of a Schiff base (Fig. 1), which can then undergo a rearrangement to form an Amadori product. With time, these adducts can undergo further rearrangements and dehydrations to form advanced-glycosylation end products (15). These end products are stable, yellow-brown fluorescent pigments which are capable of cross-linking proteins intermolecularly as well as intramolecularly. Proteins cross-linked in this manner show alterations in both their structure and function, similar to those observed in structural proteins isolated from aged individuals. Advanced-glycosylation end products have been noted in many long-lived proteins in vivo such as lens crystallins and collagen (3, 10, 13, 17).

Annette T. Lee and Anthony Cerami • Laboratory of Medical Biochemistry, Rockefeller University, New York, New York 10021.

The reactivity of amino groups of DNA has also been investigated recently. Glucose 6-phosphate has been shown to react nonenzymatically with the amino groups of nucleotides to give chromophores with absorbance in the 300- to 400-nm range and fluorescence spectra similar to those observed with nonenzymatically glycosylated proteins (5). The incubation of glucose 6-phosphate with single- and double-stranded DNA has shown that double-stranded DNA reacts with the sugar at a much slower rate than does single-stranded DNA (5). However, the inclusion of lysine in the incubation of double-stranded DNA with glucose 6-phosphate led to a more rapid reaction and loss of DNA function.

In fact we observed that preincubation of glucose 6-phosphate and lysine resulted in the formation of a reactive intermediate that could rapidly react with either single- or double-stranded DNA with similar rates (11). The reactive intermediate(s) is formed when glucose 6-phosphate is incubated with either lysine or other polyamines prior to the addition of DNA (Table 1). These reactive intermediates accumulate linearly over several days and will react with DNA, within 1 h after DNA is added, to form acid-stable adducts. The reaction of the reactive inter-

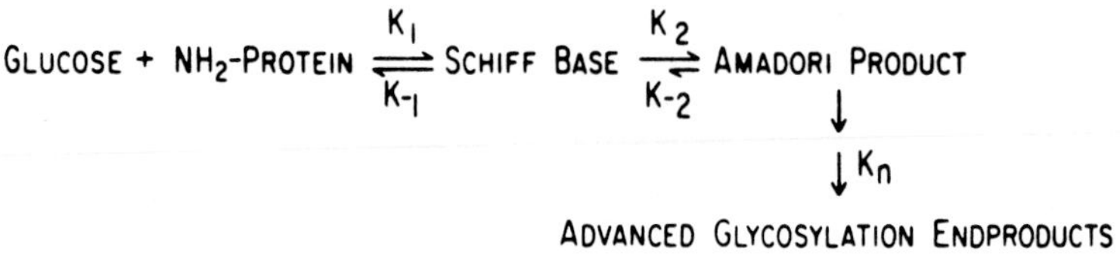

$$\text{GLUCOSE} + \text{NH}_2\text{-PROTEIN} \underset{K_{-1}}{\overset{K_1}{\rightleftharpoons}} \text{SCHIFF BASE} \underset{K_{-2}}{\overset{K_2}{\rightleftharpoons}} \text{AMADORI PRODUCT}$$

$$\downarrow \downarrow K_n$$

$$\text{ADVANCED GLYCOSYLATION ENDPRODUCTS}$$

FIGURE 1. Nonenzymatic reaction of glucose with proteins.

mediates with DNA requires the presence of an amino group on the base since poly(dT) does not react with the reactive intermediates. The structure of the reactive intermediate is not known at this time. Inhibition studies with aminoguanidine, which is known to prevent glucose-mediated protein cross-links (4), and sodium borohydride have indicated that the reactive intermediate does not contain a carbonyl moiety and has gone beyond the Amadori product of lysine and glucose 6-phosphate. Attempts are underway to isolate and characterize the reactive intermediate; however, the small amount of adduct formed (approximately 6 nmol of lysine bound per μmol of DNA bases) makes this task difficult.

The biological consequences of the reaction of DNA with reducing sugars in vitro have been studied (5, 6). When f1 phage DNA was incubated with either glucose or glucose 6-phosphate, a decrease in the ability of the phage DNA to infect *Escherichia coli* was observed. This loss in transfection ability was much more dramatic when phage DNA was incubated with glucose 6-phosphate than with glucose, reflecting the increased reactivity of glucose 6-phosphate with amino groups. The rate of inactivation for both glucose and glucose 6-phosphate was more rapid with f1 phage DNA when the incubations were carried out in the presence of lysine.

In vitro glycosylation studies done with plasmid pBR322, which carries the genes for ampicillin and tetracycline resistance, further confirmed biological inactivation due to the incubation of the DNA with the reducing sugar, glucose 6-phosphate (6). When pBR322 DNA was incubated with increased glucose 6-phosphate concentrations or for increased time periods, a decrease in transformation capacity was observed in an indicator *E. coli* strain in which ampicillin resistance was measured (Fig. 2). The bacteria exhibiting ampicillin resistance were then tested for tetracycline resistance. A surprising observation was that of those colonies

TABLE 1
Requirement for Glucose 6-Phosphate and Lysine in the Preincubation Mixture for Formation of Reactive Intermediates[a]

Mixture component(s)[b] at stage:		[³H]lysine bound (μmol [10⁹]/μmol of DNA bases)
Preincubation	Incubation	
Glc 6-P	*E. coli* DNA + [³H]lysine	3.6
[³H]lysine	Glc 6-P + *E. coli* DNA	2.2
E. coli DNA	Glc 6-P + [³H]lysine	2.8
Glc 6-P + [³H]lysine	*E. coli* DNA	100.0

[a] Preincubation components were incubated at 37°C for 4 days, after which the indicated incubation components were added for 60 min before trichloroacetic acid precipitation (data reproduced from reference 11 with permission from the publisher).
[b] Components were added in the following amounts and concentrations: glucose 6-phosphate (Glc 6-P), 1 M concentration and 1-ml solution; [³H]lysine, 20 μCi (97 Ci/mmol) in 0.1 M HEPES (N-2-hydroxyethylpiperazine-N'-2-ethanesulfonic acid); *E. coli* DNA, 1 ml of 643-μg/ml solution.

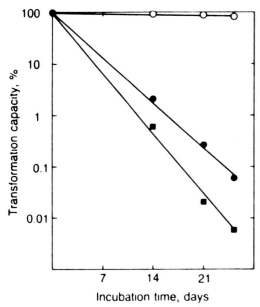

FIGURE 2. Loss of pBR322 transformation capacity with time after incubation with 0 mM glucose 6-phosphate (○), 150 mM glucose 6-phosphate (●), and 200 mM glucose 6-phosphate (■). At the indicated times, competent E. coli cells were transformed with a portion of the incubated solutions. Following overnight growth, ampicillin-resistant colonies were counted (reproduced from reference 6 with permission).

which were Amp^r, a number of them were Tet^s. Plasmid isolation and characterization of these mutants showed that insertions and deletions were mainly responsible for the inactivation of the gene for tetracycline resistance. It appears that the changes in the plasmid were the result of the host's attempt to repair the damaged DNA, since glycosylated DNA transformed into a repair-deficient host (*uvrC*) did not result in plasmid mutations.

To address whether the exposure of DNA to reducing sugars in vivo has deleterious effects, we have developed a model procaryotic system by using *E. coli* mutants in the glycolytic pathway (11a). This model system takes advantage of two *E. coli* mutants, DF40, which lacks phosphoglucose isomerase, and DF2000, which lacks both phosphoglucose isomerase and glucose 6-phosphate dehydrogenase. These strains, as well as the control strain K10, which is not deficient in either of these enzymes, are capable of growing with gluconate as their sole carbon source (7, 8). Under these conditions the mutant organisms do not accumulate significant amounts of glucose 6-phosphate. However, when glucose is added to gluconate minimal medium, strains DF40 and DF2000 accumulate a 20- and 30-fold increase, respectively, in the intracellular levels of glucose 6-phosphate, when compared with the K10 control strain (Table 2). It was hypothesized that exposure of a plasmid to these elevated levels of glucose 6-phosphate in vivo would lead to the forma-

TABLE 2
Glucose 6-phosphate Levels in Cells Grown in Minimal Medium Containing Glucose-Gluconate or Gluconate Alone

E. coli strain	Glucose 6-phosphate [μmol/5 × 10^9 cells (±SD)] in medium containing^a:			
	Glucose/gluconate at ratio:			Gluconate alone^c
	9:1^b	7:3^c	1:1^c	
K10	0.028 (±0.005)	0.0177 (±0.0031)	0.0183 (±0.0029)	0.030 (±0.003)
DF40	0.553 (±0.072)	0.3478 (±0.0230)	0.1636 (±0.0593)	0.005 (±0.002)
DF2000	0.864 (±0.011)	0.6319 (±0.0374)	0.3230 (±0.0796)	0.004 (±0.001)

^{a} Overnight cultures of each strain grown either in gluconate or glucose and gluconate minimal medium were diluted to 10^8 cells per ml. Diluted culture (50 ml) was extracted with perchloric acid and then assayed for glucose 6-phosphate content as described previously (11).
^{b} The results are means of triplicate experiments.
^{c} The results are means of duplicate experiments.

TABLE 3
Relative Mutation Rates of Cells Grown in Gluconate Alone or Glucose and Gluconate

E. coli strain	Relative Lac⁻ mutagenesis per 10^5 transformants (±SD) in minimal medium containing[a]:	
	Glucose/gluconate[b]	Gluconate alone[c]
K10	0.67 (±0.47)	0.5 (±0.5)
DF40	4.84 (±0.65[d])	1.0 (±1.0)
DF2000	8.71 (±1.24[d])	1.5 (±0.5)

[a] Plasmid DNA (50 ng) isolated from cultures grown in gluconate or glucose and gluconate minimal medium was used to transform SB4288 competent cells. Colonies which were Amp^r but had a Lac⁻ phenotype were scored as mutants. Relative mutagenesis was determined by the ratio of mutants found in the mutant strains (DF40 or DF2000)/control (K10 strain).
[b] The results are means of triplicate experiments.
[c] The results are means of duplicate experiments.
[d] The difference in mutation rate between K10 and the mutant strains DF40 and DF2000 is statistically significant ($P < 0.0001$).

tion of sugar-DNA adducts and an increase in DNA mutations.

The plasmid pAM006, which carries the genes for ampicillin resistance, β-galactosidase, and lactose permease production, was transformed into each of these strains. The bacteria were grown under conditions which did or did not prompt the accumulation of glucose 6-phosphate in the DF40 and DF2000 strains. After growth under these conditions for 24 h, plasmid DNA from each strain was isolated and transformed into a Lac⁻ E. coli strain, SB4288. Colonies which were ampicillin resistant but lacked β-galactosidase production were scored as mutants. When compared with the K10 control strain, there was a sevenfold increase in plasmid mutations found in the DF40 strain, which accumulated a 20-fold increase in glucose 6-phosphate. Plasmid DNA isolated from the DF2000 strain, which accumulated a 30-fold increase in glucose 6-phosphate, showed a 13-fold increase in plasmid mutations. However, no increase in the number of plasmid mutations over background was observed when the DF40 and DF2000 strains were grown in minimal medium containing only gluconate (Table 3), implicating the role of the increased intracellular glucose 6-phosphate in DNA damage.

Analysis of the plasmid DNA isolated from the glucose 6-phosphate-induced Amp^r Lac⁻ colonies revealed the plasmids to have an array of large insertions and deletions, as well as smaller mutations. It is interesting that the background mutations found in the K10 strain and those found in the DF40 strain were due mainly to plasmid deletions, while the mutations from the DF2000 strain show a distribution between plasmid size increases and small mutations.

The reason for the insertions and deletions of the plasmid DNA is unknown. It is possible that elevated intracellular glucose 6-phosphate levels could activate a repair and/or recombination mechanism, inducing the SOS response in the mutant bacteria. However, preliminary analysis of the mutant bacteria grown in glucose and gluconate for RecA by Western blot analysis has not shown an increase in this protein. Another possibility is that the DNA is damaged by the formation of DNA adducts either by glucose 6-phosphate alone or by glucose 6-phosphate-lysine intermediates. These cross-links might be recognized by the cell and prompt a repair mechanism that induces the mutations. This latter hypothesis is currently being studied further.

The formation of glucose or glucose-protein adducts on DNA and their accumulation over time could explain some of the DNA dysfunctions associated with aging. Some of these include decreases in DNA replication (14), DNA repair (9), and transcription (1), as well as increases in DNA strand breakage (16). The observation of increased protein-DNA cross-links with increasing age (2) may be explained by the formation of a protein-bound equivalent of the glucose 6-phosphate-lysine intermediate.

Further investigation into the understanding of the formation of glucose and glucose 6-phosphate-lysine adducts should provide insight into the possible mechanism(s) involved in DNA damage observed in vitro and in vivo due to nonenzymatic glycosylation of DNA by reducing sugars. Future studies will no doubt need to focus on DNA mutations in eucaryotic systems and the role that DNA glycosylation might play in causing DNA damage in general and in aging cells.

LITERATURE CITED

1. Berdyshev, G. D., and S. M. Zhelabovskaya. 1972. Composition, template properties and thermostability of liver chromatin from rats of various age at deproteinization by NaCl solutions. *Exp. Gerontol.* 7:321–330.
2. Bojanovic, J. J., A. D. Jevtovic, V. S. Pantic, S. M. Dugandzic, and D. S. Jovanovic. 1970. Thymus nucleoproteins. Thymus histones in young and adult rats. *Gerontologia* 16:304–312.
3. Brownlee, M., H. Vlassara, and A. Cerami. 1984. Nonenzymatic glycosylation and the pathogenesis of diabetic complications. *Ann. Intern. Med.* 101:527–537.
4. Brownlee, M., H. Vlassara, A. Kooney, P. Ulrich, and A. Cerami. 1986. Aminoguanidine prevents diabetes-induced arterial wall protein crosslinking. *Science* 232:1629–1632.
5. Bucala, R., P. Model, and A. Cerami. 1984. Modification of DNA by reducing sugars: a possible mechanism for nucleic acid aging and age-related dysfunction in gene expression. *Proc. Natl. Acad. Sci. USA* 81:105–109.
6. Bucala, R., P. Model, M. Russel, and A. Cerami. 1985. Modification of DNA by glucose 6-phosphate induces DNA rearrangements in an *Escherichia coli* plasmid. *Proc. Natl. Acad. Sci. USA* 82:8439–8442.
7. Fraenkel, D. F. 1968. Selection of *Escherichia coli* mutants lacking glucose 6-phosphate dehydrogenase or gluconate 6-phosphate dehydrogenase. *J. Bacteriol.* 95:1267–1271.
8. Fraenkel, D. F., and S. R. Levisohn. 1967. Glucose and gluconate metabolism in an *Escherichia coli* mutant lacking phosphoglucose isomerase. *J. Bacteriol.* 93:1571–1578.
9. Karran, P., and M. G. Ormerod. 1973. Is the ability to repair damage to DNA related to the proliferative capacity of a cell? The rejoining of x-ray produced strand breaks. *Biochim. Biophys. Acta* 299:54–64.
10. Koenig, R. J., and A. Cerami. 1980. Hemoglobin A_{1c} and diabetes mellitus. *Annu. Rev. Med.* 31:29–34.
11. Lee, A. T., and A. Cerami. 1987. The formation of reactive intermediates of glucose 6-phosphate and lysine capable of rapidly reacting with DNA. *Mutat. Res.* 179:151–158.
11a. Lee, A. T., and A. Cerami. 1987. Elevated glucose 6-phosphate levels are associated with plasmid mutations in vivo. *Proc. Natl. Acad. Sci. USA* 84:8311–8314.
12. Maillard, L. C. 1912. Reaction generale des acides amines sur les sucres: ses consequences biologiques. *C.R. Soc. Biol.* 72:599–601.
13. Monnier, V. M., and A. Cerami. 1981. Nonenzymatic browning *in vivo*: possible process for aging of long-lived proteins. *Science* 211:491–493.
14. Petes, T. D., R. A. Farber, G. M. Tarrant, and R. Holliday. 1974. Altered rate of DNA replication in aging human fibroblast cultures. *Nature* (London) 251:434–436.
15. Pongor, S., P.C. Ulrich, F. Aladar Bencsath, and A. Cerami. 1984. Aging of proteins: isolation and identification of a fluorescent chromophore from the reaction of polypeptides with glucose. *Proc. Natl. Acad. Sci. USA* 81:2684–2688.
16. Price, G. B., S. P. Modak, and T. Makinodan. 1971. Age-associated changes in the DNA of mouse tissue. *Science* 171:917–920.
17. Schnider, S. L., and R. R. Kohn. 1980. Glucosylation of human collagen in aging and diabetes mellitus. *J. Clin. Invest.* 66:1179–1181.

Chapter 46

Mutagenic Activity of Smokeless-Tobacco Extracts

Charles W. Berry

In the United States, smokeless tobacco is produced and used in two general forms: snuff and chewing tobacco. These products contain the leaves of tobacco plants plus a myriad of sweeteners, flavorants, humectants, and scents. There is evidence that smokeless tobacco, principally moist snuff, is gaining popularity with young people. National estimates indicate that approximately 12 million Americans consumed some form of smokeless tobacco during 1985, with use increasing especially among male adolescents and young male adults less than 21 years old (19).

The prevalent appeal of these products assumes a major public health significance because the clinical and epidemiological evidence (3–5, 15) reveals that smokeless tobacco can be associated with leukoplakia, mouth cancer, and other oral health problems such as gingival recession, gingivitis, periodontal disease, discolored teeth, bad breath, and tooth abrasion and can cause nicotine dependence (17) in regular users.

Experimental analyses reveal several potent carcinogens in smokeless tobacco (9, 10). These include polonium-210, a radioactive alpha-emitter and known carcinogen, and representatives of two classes of powerful chemical carcinogens, the polycyclic aromatic hydrocarbons and the nitrosamines. Of the 19 nitrosamines identified in smokeless tobacco, the tobacco-specific nitrosamines N-nitrosonornicotine and 4-(methylnitrosamino)-1-(3-pyridyl)-1-butanone have the highest concentrations, and both are chemically related to nicotine. These nitrosamines often have been detected in tobaccos at levels many times higher than those allowed for U.S. foods and beverages (14).

Outside of the human host, chewing tobacco and extracts from various chewing tobaccos have been tested by oral or topical administration in mice, rats, and hamsters and by subcutaneous and skin application to mice. These bioassays exposing animals to smokeless tobacco have generally failed to demonstrate significantly increased tumor production in the animal model (13, 16). However, some bioassays suggest that snuff may cause oral tumors when tested in animals that are infected with the herpes simplex virus. The evaluation of the carcinogenicity of chewing tobacco and its extracts in animals has been limited because of short application times and low-dose exposures.

Within the past decade, there has been a

Charles W. Berry • Department of Microbiology, Baylor College of Dentistry, Dallas, Texas 75246.

major advance in the methodology for detecting potential carcinogens by alternative procedures through the use of short-term in vitro tests (20). Some of the currently available major short-term tests are mutagenicity assays with bacteria, mammalian cell cultures, and other eucaryotes such as yeasts, *Drosophila melanogaster*, and mice; assays for cell transformation; assays for DNA binding, damage, and repair; and assays for chromosomal abnormalities. The ability of many carcinogens to generate electrophiles that bind to cellular DNA has provided the rationale for utilizing mutagenesis in bacterial or mammalian cell culture as screening tests. One extensively studied and validated assay of this type is the *Salmonella typhimurium* mutagenesis system developed by Ames and colleagues (1), in which highly sensitive tester strains are assayed for either frameshift or base-substitution reversion mutations from histidine auxotrophy to prototrophy. A crucial aspect of this system is inclusion of a microsomal source, the so-called S-9 system, to activate the mutagenic compounds to intermediates that will react with the DNA of the bacteria.

The methodology for this study included screening a total of 10 tobacco extracts for mutagenic potential. Three snuffs, three loose-leaf pouch tobaccos, and three plug tobaccos, along with a tobacco leaf control, were extracted with either distilled water or dimethyl sulfoxide (DMSO). Samples (10 g) of each product were mixed with 35 ml of solvent for 24 h and then centrifuged and filtered. In the standard *Salmonella* mutagenicity spot plate test (12), four strains of *S. typhimurium* were used (TA97, TA98, TA100, and TA1535). Genetic marker quality control tests on each tester strain were performed in each assay. These included tests for the deep rough (*rfa*) mutation, *uvrB* mutation, and R factor. Before use, each tester strain was stored frozen at $-70°C$ in stock culture and then was subsequently grown overnight for each experiment in 10.0-ml tubes of Oxoid Nutrient Broth II in a shaker water bath at 35°C. Tubes containing 2.0 ml of sterile top agar were boiled to melt the agar and then maintained in a 45°C water bath. The molten soft top agar tubes were then seeded with 0.2 ml of 0.5 mM histidine-biotin solution and 0.1 ml of each tester strain, with or without the addition of 0.5 ml of S-9 mix. The S-9 mix was a rat liver microsomal homogenate which was incorporated into the assay for increased efficiency in the detection of mutagens requiring metabolic activation. The seeded molten top agar was then gently mixed and overlaid onto minimal Vogel-Bonner agar plates. After solidification of the top agar overlay, 40 µl of tobacco extract was pipetted into 6-mm filter disks and placed upon the agar surface. Positive and negative controls were run concurrently. Negative controls of either H_2O- or DMSO-saturated 6-mm filter paper disks and positive controls of either 2-methoxy-6-chloro-9-[3-(2-chloroethyl)aminopropylamino] acridine · 2HCl (ICR-191) for strain TA97, 2-aminofluorene for strain TA98, or sodium azide for strains TA100 and TA1535 were included in each experiment. Additionally, the extracts were tested both with and without a microsomal preparation from rat liver. After incubation for 48 h at 37°C, the presence or absence of revertant colonies surrounding each filter paper disk was observed. The results indicated that, in aqueous extract, only one tobacco extract (CH) induced a reversion from histidine dependence to independence by tester strains TA97 and TA1535. However, in DMSO extract, one tobacco extract (CH) caused reversion of histidine dependence by tester strains TA98, TA100, and TA1535.

Since the spot plate test depended upon the ability of a chemical to diffuse through top agar, bacteria in different regions of the plate were exposed to various concentrations of the test compound. The data obtained from this assay were for this reason only semiquantitative. In the plate incorporation test (12), the test chemical, the bacteria, and the S-9 mix were blended into the top agar, thereby increasing the sensitivity of the test. This procedure was used for assays of each tobacco extract, not only those which were positive on the spot plate test. In the plate incorporation experiments the test compound was added directly to a mixture of top agar and bacteria, with and without S-9 mix also included. After mixing, the top agar plus reactants were transferred to Vogel-Bonner plates. Positive and negative controls were run concurrently. After 48 h of incubation, the His^+ revertants appeared as large discrete colonies against a very light background lawn of the His^- auxotrophic mutants. In the absence of a mutagen, each strain produced a characteristic minimal number of spontaneous His^+ revertants. Therefore, a mutagenic effect was noted when the number of His^+ revertants on a test plate was counted and exceeded the number of spontaneous revertants of the control.

The mean number of His^+ revertants was assessed, and then the number of revertants induced by exposure to each tobacco in aqueous and DMSO extract was compared with the number of revertants induced spontaneously in the negative control or induced by the positive control plates. No tobacco product had a dramatic mutagenic effect on any of the *S. typhimurium* tester strains (Table 1); however, a slight effect was observed by one product (a popular moist snuff brand) in aqueous extract on TA1535 cells. This same product was also slightly mutagenic in DMSO extract on TA98 and TA1535 cells (Table 2).

Other investigators have reported the mutagenic potential of only a limited number of smokeless-tobacco extracts. Under acidic conditions of a crude solvent extract, one popular tobacco snuff brand was found to be mutagenic (21). This mutagenic activity was observed only in polar solvent extracts and induced predominantly frameshift mutations. In another report, a possible mechanism of action of a direct-acting mutagen was described (22). Extraction of tobacco snuff with polar and nonpolar solvents, with and without pH adjustment, indicated that the mutagenic potency of the acid-treated extracts was consistent with the content of generated nitroso compounds. The mechanism reported to be responsible for the acid-mediated mutagenicity of snuff extracts was the reduction of the chemical nitrate (found in high concentration in snuff) to nitrite by bacteria of either tobacco or oral origin and the nitrosation of certain constituents in snuff by nitrite under acidic conditions to form the mutagenic nitroso compounds. To emphasize this point, carcinogenic tobacco-specific *N*-nitrosamines which were generated in tobacco snuff during processing were detected previously (8). Also, a mutagenic fraction of tobacco snuff was obtained under conditions which selectively extract flavenoids and then was treated with a strong acid to hydrolyze the sugar conjugates (7).

In another study, aqueous extracts of five smokeless tobacco products were mutagenic without further treatment when assayed by a modification of the *Salmonella*-microsomal assay (6). Most of the mutagenic activity to the base-pair substitution mutant (TA100) failed to extract into organic solvents at neutral pH, and only a slight amount of activity extracted into organic solvents at acidic pH. For mutagenesis to be exhibited at a neutral pH, the aqueous tobacco extracts required metabolic activation and addition of an NADPH-generating system (11).

A study from India using *Nicotiana rustica* tobacco also demonstrated frameshift mutations induced by ethanol extracts of chewing tobacco in TA98 cells and Chinese hamster V79 cells, but not TA100, TA1535, or TA1538 cells (18). The presence of S-9 liver homogenate tended to enhance this mutagenic effect (2).

These and other results demonstrate that smokeless tobaccos contain mutagenic components which may be extracted and potentiated at an acidic pH with polar solvents or at a neutral pH with aqueous sol-

TABLE 1
Mean Number[a] of Histidine-Positive Revertants in Aqueous Extracts of Smokeless Tobacco

Tobacco extract	S9	Mean no. of revertants of strain:			
		TA97	TA98	TA100	TA1535
Snuff					
A	+	659	29	204	23
	−	574	30	209	29
B	+	560	26	214	23
	−	449	17	207	18
C	+	547	32	214	63
	−	486	21	222	53
Pouch					
D	+	551	31	235	18
	−	587	28	209	23
E	+	611	33	224	24
	−	509	22	190	21
F	+	602	31	233	22
	−	571	19	251	25
Plug					
G	+	515	30	241	21
	−	416	29	207	22
H	+	553	33	277	26
	−	468	41	261	19
I	+	621	34	248	25
	−	557	39	233	22
Controls					
TC[b]	+	563	33	242	24
	−	545	28	216	23
(−) Solvent[c]	+	587	38	321	15
	−	528	35	265	19
(+) Mutagen[d]	+	1,580	6,120	3,030	2,340
	−	1,430	196	1,790	2,580

[a] $\bar{X}$ of triplicate analyses.
[b] TC, Pure leaf tobacco.
[c] Water alone.
[d] Positive control mutagens: ICR-191 (TA97), 2-aminofluorene (TA98), and sodium azide (TA100 and TA1535).

vents. Metabolic activation via liver S-9 fraction and incorporation of an NADPH-generating system have also been reported to enhance mutagenesis. It is therefore evident that the N-nitrosamines are responsible for at least some of the mutagenic activity of these extracts. Differences exist in the reported reverse mutations by tobacco extracts induced in S. typhimurium tester strains for base-pair substitution and frameshift mutations. These observations are probably due to the variety of different conditions which were used in the mutagenesis assay systems such that nitrosamine-induced or direct-acting nitroso-induced mutations would be optimized.

In conclusion, these results support the continued use of the Ames test to evaluate the mutagenic potential of tobacco extracts. This study indicated that one brand of moist

TABLE 2
Mean Number[a] of Histidine-Positive Revertants in DMSO Extracts of Smokeless Tobacco

Tobacco extract	S9	Mean no. of revertants in strain:			
		TA97	TA98	TA100	TA1535
Snuff					
A	+	1,110	31	278	16
	−	632	21	257	15
B	+	962	33	370	14
	−	719	27	314	12
C	+	717	53	379	43
	−	598	39	361	35
Pouch					
D	+	1,047	25	341	14
	−	636	30	254	14
E	+	1,060	31	330	15
	−	764	25	254	13
F	+	894	27	408	16
	−	637	32	261	14
Plug					
G	+	821	33	298	11
	−	643	23	252	10
H	+	636	31	298	13
	−	537	21	271	16
I	+	795	23	265	12
	−	596	26	254	9
Controls					
TC[b]	+	674	41	373	20
	−	464	25	292	11
(−) Solvent[c]	+	820	37	280	30
	−	682	21	292	17
(+) Mutagen[d]	+	7,350	10,300	4,260	3,120
	−	3,750	89	3,930	2,400

[a] $\bar{X}$ of triplicate analyses.
[b] TC, Pure leaf tobacco.
[c] DMSO alone.
[d] Positive control mutagens: ICR-191 (TA97), 2-aminofluorene (TA98), and sodium azide (TA100 and TA1535).

snuff was mutagenic in an aqueous extract on TA1535 tester cells and in a DMSO extract on TA98 and TA1535 cells. Although not mandatory in this report for genotoxicity, the presence of S-9 mix slightly enhanced the observed effect. These data demonstrate that differences may exist in the mutagenic potential of a variety of smokeless tobacco products when extracted with water or DMSO.

LITERATURE CITED

1. **Ames, B. N., J. McCann, and E. Yamasaki.** 1975. Methods for detecting carcinogens and mutagens with the Salmonella/mammalian-microsome mutagenicity test. *Mutat. Res.* 31:347–364.
2. **Bhide, S. V., A. S. Shah, J. Nair, and D. Nagarajrow.** 1984. Epidemiological and experimental studies on tobacco related oral cancer in India. *IARC (Int. Agency Res. Cancer) Sci. Publ.* 57:851–857.

3. **Christen, A. G.** 1980. The case against smokeless tobacco: five facts for the health professional to consider. *J. Am. Dent. Assoc.* **101**:464–469.
4. **Christen, A. G., W. R. Armstrong, and R. K. McDaniel.** 1979. Intraoral leukoplakia, abrasion, periodontal breakdown and tooth loss in a snuff dipper. *J. Am. Dent. Assoc.* **98**:584–586.
5. **Greer, R. O., Jr., and T. C. Poulson.** 1983. Oral tissue alterations associated with the use of smokeless tobacco by teenagers. *Oral Surg.* **56**:275–284.
6. **Guttenplan, J. B.** 1987. Mutagenic activity in smokeless tobacco products sold in the USA. *Carcinogenesis* **8**:741–743.
7. **Hardigree, A. A., and J. L. Epler.** 1978. Comparative mutagenesis of plant flavenoids in microbial systems. *Mutat. Res.* **58**:231–239.
8. **Hoffmann, D., and J. D. Adams.** 1981. Carcinogenic tobacco specific N-nitrosamines in snuff and in the saliva of snuff dippers. *Cancer Res.* **41**:4305–4308.
9. **Hoffmann, D., N. H. Harley, I. Fisenne, J. D. Adams, and K. D. Brunnemann.** 1986. Carcinogenic agents in snuff. *JNCI* **76**:435–437.
10. **Hoffmann, D., and S. S. Hecht.** 1985. Nicotine-derived N-nitrosamines and tobacco related cancer: current status and future directions. *Cancer Res.* **45**:935–944.
11. **Lee, S. Y., and J. B. Guttenplan.** 1981. A correlation between mutagenic and carcinogenic potencies in a diverse group of N-nitrosamines: determination of mutagenic activity of weakly mutagenic N-nitrosamines. *Carcinogenesis* **2**:1339–1344.
12. **Maron, P., and B. N. Ames.** 1983. Revised methods for the Salmonella mutagenicity test. *Mutat. Res.* **13**:173–215.
13. **Mody, J. K., and J. K. Ranadive.** 1959. Biological study of tobacco in relation to oral cancer. *Ind. J. Med. Sci.* **13**:1023–1037.
14. **National Institutes of Health.** 1986. *Consensus Development Conference Statement on the Health Implications of Smokeless Tobacco Use, 13–15 January 1986.* National Institutes of Health, Bethesda, Md.
15. **Offenbacher, S., and D. R. Weathers.** 1985. Effects of smokeless tobacco on the periodontal, mucosal and caries status of adolescent males. *J. Oral Pathol.* **14**:169–181.
16. **Peacock, E. E., Jr., and B. W. Brawley.** 1959. An evaluation of snuff and tobacco in the production of mouth cancer. *Plast. Reconstr. Surg.* **23**:628–635.
17. **Russell, M. A. H., M. J. Jarvis, G. Devitt, and C. Feyerabend.** 1981. Nicotine intake by snuff users. *Br. Med. J.* **283**:814–817.
18. **Shah, A. S., A. V. Sarode, and S. V. Bhide.** 1985. Experimental studies on mutagenic and carcinogenic effect of tobacco chewing. *J. Cancer Res. Clin. Oncol.* **109**:203–207.
19. **U.S. Public Health Service.** 1986. *The Health Consequences of Using Smokeless Tobacco. A Report of the Advisory Committee to the Surgeon General.* U.S. Department of Health and Human Services publication no. (NIH) 86-2874. National Institutes of Health, Bethesda, Md.
20. **Weinstein, I. B.** 1981. The scientific basis for carcinogen detection and primary cancer prevention. *Cancer* **47**:1133–1141.
21. **Whong, W.-Z., R. G. Ames, and T. Ong.** 1984. Mutagenicity of tobacco snuff: possible health implication for coal miners. *J. Toxicol. Environ. Health* **14**:491–496.
22. **Whong, W.-Z., J. D. Stewart, Y.-K. Wang, and T. Ong.** 1987. Acid-mediated mutagenicity of tobacco snuff: its possible mechanism. *Mutat. Res.* **177**:241–246.

Chapter 47

Comparisons between the Cellular and Herpesvirus Forms of dUTP Nucleotidohydrolase and Uracil-DNA Glycosylase

Ronald Lirette and Sal Caradonna

Herpesviruses appear to regulate all the necessary elements for nucleotide metabolism and DNA replication. Crucial to the herpesvirus genome, as it appears to be with all thymine-containing DNA, is the exclusion of deoxyuridylate as one of the typical components of DNA. The reason for this appears to be due to the susceptibility of deoxycytidylate residues to deamination (12). This occurrence leads to the creation of deoxyuridylate residues in DNA. If this were left unchecked, a transition mutation would occur (G·C→A·T) during the next round of replication. The suggestion is that uracil-DNA glycosylase evolved to correct this potential mutagenic event (7, 11) by excising the uracil moiety and initiating a repair process involving excision of nucleotides and replacement synthesis by DNA polymerase. Since uracil-DNA glycosylase cannot discriminate between deaminated cytosine residues and incorporated uracil residues, dUMP is disallowed as a normal constituent of DNA. dUTP nucleotidohydrolase (dUTPase) apparently has evolved to keep the levels of dUTP very low in the cell by hydrolyzing dUTP to dUMP and PP_i (20).

Part of our research has focused on demonstrating that herpes simplex virus (HSV) encodes both a uracil-DNA glycosylase and a dUTPase. What we present here is an overview of this work with an emphasis on some interesting observations about these enzymes and some speculation as to why the virus encodes its own version of these enzymatic functions.

dUTPase

A number of reports have appeared, characterizing both the physical and kinetic aspects of the dUTPase enzyme from a variety of sources. Shlomai and Kornberg (18) have purified the enzyme from *Escherichia coli*. This procaryotic enzyme is a zinc-containing tetramer with a protomer molecular

---

Ronald Lirette and Sal Caradonna • Department of Biochemistry, University of Medicine and Dentistry of New Jersey, Piscataway, New Jersey 08854.

weight of 16,000. Progress on the eucaryotic enzyme has established a number of parameters for this function as well (1, 14, 21, 22). We have been able to purify the enzyme from HeLa cells and show that it is a dimer with two equal molecular weight subunits of 22,500 (1). Subunit association occurs in the presence of Mg^{2+} or Mn^{2+} and is apparently only active in the dimeric form and not as a monomer.

The enzyme is highly specific for dUTP. No other naturally occurring nucleotide triphosphate or deoxynucleotide triphosphate serves as a substrate for this enzyme. (This is in contrast to the HSV dUTPase, which can apparently hydrolyze dCTP and dTTP as well as dUTP.) The cellular enzyme can hydrolyze analogs of dUTP, such as 5-fluorodeoxyuridine triphosphate, 6-aza-dUTP, and 2',3'-dideoxy-UTP (2, 10). Monoclonal antibodies were generated against this purified enzyme in an effort at further characterization. This monospecific antibody recognizes the dUTPase enzyme from a variety of human sources, but does not recognize the dUTPase protein from nonhuman sources such as mouse, monkey, or rabbit cells. This appears to indicate a species heterogeneity for at least one antigenic determinant on this protein. Another interesting finding is that the dUTPase derived from human cells is a phosphoprotein (Fig. 1). Phosphoamino acid analysis revealed that phosphorylation occurs at serine residues. Removal of the phosphate group, using calf intestinal alkaline phosphatase, appears to activate hydrolysis of dUTP by this enzyme. In previous work we have shown that dUTPase activity rises and falls with the progression of cells through the S phase of the cell cycle (2). It may be possible that this posttranslational modification is in some way responsible for the fluctuation of dUTPase activity as a function of the DNA synthesis phase of the cell cycle.

Infection of cells with HSV results in a significant elevation in dUTPase activity (14). This was one of the initial observations

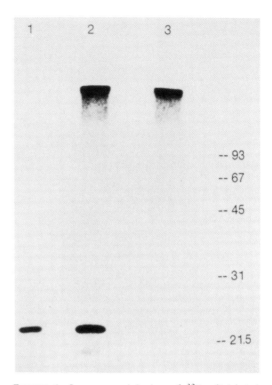

FIGURE 1. Immunoprecipitation of ^{32}P-radiolabeled dUTPase from crude extracts of HeLa S3 cells. HeLa cells were labeled with $^{32}P_i$ for 24 h and extracted according to published procedure (1). Sodium dodecyl sulfate-polyacrylamide gel electrophoresis under reducing conditions was performed on a 12.6% gel. Lane 1, Immunopurified HeLa dUTPase electrophoresed and stained with Coomassie blue. Lane 2, Immunoprecipitation conducted with a monoclonal antibody to HeLa dUTPase (1:1,000 dilution). Lane 3, Immunoprecipitation conducted with nonspecific mouse immunoglobulin G (1:1,000 dilution). The antigen-antibody complex was precipitated with rabbit anti-mouse immunoglobulin G attached to Sepharose beads (Bio-Rad). The gel containing lanes 2 and 3 was dried and exposed to X-ray film for 24 h at −80°C.

leading to the discovery that HSV encodes its own form of dUTPase. This new species of dUTPase has an isoelectric point of 8.0, compared with an isoelectric point of between 6 and 7 for the cellular enzyme (2). This observation has led to the use of phosphocellulose column chromatography for the separation of the cellular form of dUTPase from the HSV form (2). The viral dUTPase is a monomeric protein with a molecu-

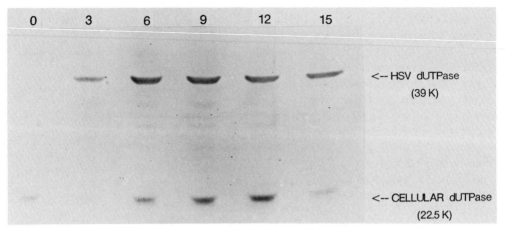

FIGURE 2. Western blot analysis of HeLa cell extracts infected with HSV. HeLa cells were infected with HSV type 1 (strain F) and harvested at hourly time points as indicated at the top of the figure. A 100-μg sample of crude extract was applied to each lane. A 1:500 dilution of each antibody (rabbit anti-HSV dUTPase and mouse anti-cellular dUTPase) was mixed together and allowed to incubate with the blot overnight at 20°C. Second antibody was applied, after washing, at a 1:5,000 dilution for each (goat anti-rabbit and goat anti-mouse immunoglobulin G conjugated to alkaline phosphatase) together and incubated with the blot for 30 min. The blot was then washed and developed. A mouse monoclonal antibody was generated against purified HeLa S3 dUTPase, and a polyclonal antibody was generated in rabbits against a bacterial fusion protein derived from a fusion of the bacterial *trpE* gene and a segment of the HSV gene for dUTPase.

lar weight of about 39,000 (2) and is dependent on Mg^{2+} for its activity. As mentioned above, the viral dUTPase, in addition to hydrolyzing dUTP, will also hydrolyze dCTP and dTTP to the monophosphate level (2). We have not been able to separate these activities by isoelectric focusing, density gradient centrifugation, or ion-exchange column chromatography (2). In addition to this, we have been able to fractionate HSV dUTPase by sodium dodecyl sulfate-polyacrylamide gel electrophoresis and renature the protein (2). Utilizing this renatured protein preparation, we still find hydrolysis of dCTP and dTTP as well as dUTP.

Work by a number of investigators has putatively mapped the gene for the viral dUTPase to a region of the HSV genome having map units of 0.69 to 0.70 (9, 16, 23, 24). Utilizing this information, we have fused a segment of the nucleotide sequence to a bacterial expression vector by using a portion of the *trpE* gene (19). The protein product from this chimeric gene fusion was used to immunize rabbits in an attempt to generate antibody against the HSV dUTPase (Fig. 2). Western blot (immunoblot) analysis, using antiserum from an immunized rabbit, recognized a protein having a molecular weight of 39,000 (Fig. 2). The amount of protein increased as a function of time post-HSV infection from undetectable levels at zero time to peak levels between 6 and 9 h. This antiserum neutralized dUTPase activity from infected cells, but did not neutralize dUTPase activity from uninfected cells.

Utilizing the same Western blot, we also examined the fate of cellular dUTPase as a function of virus infection. We found a 22,500-dalton immunostaining component in addition to the 39,000-dalton band (Fig. 2). In separate experiments we have determined that each antibody does not cross-react with the other species of dUTPase. What was observed in this instance was that cellular dUTPase declined from 0 to 3 h postinfection and then increased again to levels seen at the 0 h time point. We have made an analysis of enzyme activity of the two species as a function of time after virus infection. This was done by separating the two forms of dUTPase by phosphocellulose column chromatography and assaying for activity. What we found in this case was that cellular activity decreased to below measurable limits while the virus activity continually increased with time. It appears that there may be a post-translational modification, possibly virus mediated, which shuts off cellular dUTPase activity. As stated previously, phosphorylation of cellular dUTPase apparently acts to turn off hydrolyzing activity of this enzyme. It is conceivable that the virus may phosphorylate cellular dUTPase to prevent its interference during a virus infection. The question stands as to why the virus would want to turn off cellular dUTPase activity. The answer may lie in the observation that HSV dUTPase has an expanded ability to hydrolyze alternate substrates. The virus may have adapted the hydrolysis of "high-energy" phosphate bonds by the dUTPase to alternative or additional functions, where an energy contribution may be needed. The cellular enzyme may impede this process and in turn would be specifically turned off by the virus. Why the cellular enzyme returns by 6 h after infection is also open to speculation. This may implicate a cellular phenomenon to insure that enough dUTPase exists. Studies with the bacterial enzyme, attempting to eliminate the dUTPase gene entirely from the cell, indicate that dUTPase is an essential function in the cell and cannot be eliminated (21). Some sort of signaling phenomenon within the infected cell (possibly a phosphorylated cellular dUTPase) may activate transcriptional or posttranscriptional elements to augment the synthesis of dUTPase.

URACIL-DNA GLYCOSYLASE

DNA glycosylases belong to the class of repair enzymes which function in the removal of a wide variety of atypical bases from DNA, initiating an excision repair process (11). The function of uracil-DNA glycosylase is in the removal of uracil moieties from DNA created either from the deamination of cytosine or from the infrequent incorporation of dUMP into DNA. This enzyme appears to be ubiquitous in nature and has been well characterized from both procaryotic and eucaryotic systems (6, 8, 11). As stated earlier, the presence of uracil-DNA glycosylase restricts the mutagenic event caused by cytosine deamination. HSV is a DNA-containing virus with a relatively high G+C content (68%) (17). This fact may indicate the necessity for uracil-DNA glycosylase activity during the period of virus production. Previous evidence has indicated that uracil-DNA glycosylase activity is induced in HSV-infected cells (3). This increase could be due to a virus-mediated increase in cellular activity or to the induction of a virus-encoded and distinct species of uracil-DNA glycosylase. We have recently provided evidence indicating that the virus encodes its own species of uracil-DNA glycosylase (4).

Conventional purification strategies have proven unsuccessful at separating a virus form of glycosylase from the cellular form (4). The enzyme, purified from either uninfected cells or cells infected with HSV, behaves identically when subjected to a number of chromatographic procedures. In addition, the enzyme from either source is monomeric, with a molecular weight of 39,000 (4). There does not appear to be any cofactor requirement, and the enzyme is active in the presence of EDTA.

Having failed at the protein level to separate a distinct species of virus glycosylase, we attempted to pursue this question at the DNA level. The initial observation which led us to uncover a virus-encoded glycosylase came from the finding that we could detect uracil-DNA glycosylase activity in in vitro translation reactions which contained either HSV type 1 or type 2 mRNA.

This activity was specific for HSV input message. No activity was detected when uninfected cellular message was used. Since uracil-DNA glycosylase is active in the presence of EDTA, we were able to eliminate nuclease activities of the rabbit reticulocyte lysate as well as the alkaline exonuclease activity encoded by HSV (15). By using a cDNA library constructed from mRNA derived from HSV-infected cells, hybrid arrest, and in vitro translation, we were able to isolate a cDNA which encodes uracil-DNA glycosylase activity (4). Evidence indicates that this cDNA originates from the herpesvirus genome and that this region encodes a family of transcripts which are transcribed from the same DNA strand and partially overlap. We are currently in the process of finding out which one of these transcripts encodes uracil-DNA glycosylase.

The very similar monomeric molecular weights of these two species of glycosylase, along with the inability to separate each from the other by chromatographic procedures, indicate similar physical properties between the two enzymes. This is unlike other HSV enzymes and their cellular counterparts, such as thymidine kinase and DNA polymerase. This similarity may have evolved to allow for the virus glycosylase to interact with other repair enzymes of the cell, which are not coded for by the virus. Recently, evidence has been put forth showing significant amino acid sequence homology between the mouse ribonucleotide reductase and the HSV ribonucleotide reductase (5). Caras et al. (5) indicate that the C-terminal domain and a central portion are common to both types of reductase. Conservation of amino acid sequences between the cellular and virus forms of this crucial enzyme may allow (as was postulated for glycosylase) for viral reductase interaction with cellular components which are not coded for by the virus.

It appears that the induction of the viral glycosylase is coordinate with DNA synthesis, since its activity increases during the time of viral DNA replication. We have shown a similar coordination of cellular uracil-DNA glycosylase activity with the S phase of the cell cycle (2). Since uracil-DNA glycosylase appears to have a surveillance function (removal of deaminated cytosine residues) rather than an editing function (removal of misincorporated dUMP residues), one would expect its activity to be constitutive in the cell. The editing function would be minimized to a very significant extent by the action of dUTPase to prevent dUMP incorporation. This observation seems even more incongruent due to recent evidence that apurinic-apyrimidinic (AP) sites may be mutagenic. Recent information put forth in a review by Loeb and Preston (13) indicates that eucaryotic DNA polymerases can copy over AP sites and that dAMP is the most likely deoxynucleotide to be incorporated into the daughter strand across from the AP site in the template. Incorporation at the AP site is apparently random, since the fidelity created by base pairing is absent. The activity of uracil-DNA glycosylase during semiconservative replication may create AP sites which could be either copied (erroneously) or repaired. This would imply a rather complex and coordinated association between the repair functions and the functions serving in semiconservative replication.

HSV may provide a good model for the study of a number of questions raised here. It may be able to provide information about cellular protein turnover and the signals involved in activation of gene expression. Genetic heterogeneity, within the human population, of the uracil-DNA glycosylase gene may indicate predispositions to certain types of cancer. HSV may provide a significant model to determine how essential uracil-DNA glycosylase is in this regard.

ACKNOWLEDGMENTS. Research in the authors' laboratory is supported by Public Health Service grants CA42605 and DE07710 from the National Institutes of Health.

LITERATURE CITED

1. Caradonna, S., and D. M. Adamkiewicz. 1984. Purification and properties of the deoxyuridine triphosphate nucleotidohydrolase enzyme derived from HeLa S3 cells: comparison to a distinct dUTP nucleotidohydrolase induced in herpes simplex virus infected HeLa S3 cells. *J. Biol. Chem.* 259:5459–5464.
2. Caradonna, S., and Y. C. Cheng. 1980. The role of deoxyuridine triphosphate nucleotidohydrolase, uracil-DNA glycosylase and DNA polymerase alpha in the metabolism of FUdR in human tumor cells. *Mol. Pharmacol.* 18:513–520.
3. Caradonna, S., and Y. C. Cheng. 1981. Induction of uracil-DNA glycosylase and dUTP nucleotidohydrolase activity in herpes simplex virus infected human cells. *J. Biol. Chem.* 256:9834–9837.
4. Caradonna, S., D. Worrad, and R. Lirette. 1987. Isolation of a herpes simplex virus cDNA encoding the DNA repair enzyme uracil DNA glycosylase. *J. Virol.* 61:3040–3047.
5. Caras, I. W., B. B. Levinson, M. Fabry, S. R. Williams, and D. W. Martin. 1985. Cloned mouse ribonucleotide reductase subunit M1 cDNA reveals amino acid sequence homology with *Escherichia coli* and herpesvirus ribonucleotide reductases. *J. Biol. Chem.* 260:7015–7022.
6. Cone, R., J. Duncan, L. Hamilton, and E. C. Friedberg. 1977. Partial purification and characterization of a uracil-DNA N-glycosidase from *Bacillus subtilis*. *Biochemistry* 16:3194–3201.
7. Duncan, B. K., and B. Weiss. 1982. Specific mutator effects of *ung* (uracil-DNA glycosylase) mutations in *Escherichia coli*. *J. Bacteriol.* 151:750–755.
8. Friedberg, E. C. 1985. DNA Repair, p. 141–212. W. H. Freeman & Co., New York.
9. Hall, L. M., K. G. Draper, R. J. Frink, R. H. Costa, and E. K. Wagner. 1982. Herpes simplex virus mRNA species mapping in *Eco* RI fragment I. *J. Virol.* 43:594–607.
10. Ingraham, H. A., and M. Goulian. 1982. Deoxyuridine triphosphatase: a potential site of interaction with pyrimidine nucleotide analogues. *Biochem. Biophys. Res. Commun.* 109:746–752.
11. Lindahl, T. 1982. DNA repair enzymes. *Annu. Rev. Biochem.* 51:61–78.
12. Lindahl, T., and B. Nyberg. 1974. Heat-induced deamination of cytosine residues in deoxyribonucleic acid. *Biochemistry* 13:3405–3410.
13. Loeb, L., and B. D. Preston. 1986. Mutagenesis by apurinic/apyrimidinic sites. *Annu. Rev. Genet.* 20:201–230.
14. Mahagaokar, S., P. N., Rao, and A. Orengo. 1980. Deoxyuridine triphosphate nucleotidohydrolase of HeLa cells. *Int. J. Biochem.* 11:415–421.
15. McGeoch, D. J., A. Dolan, and M. C. Frame. 1986. DNA sequence of the region in the genome of herpes simplex virus type 1 containing the exonuclease gene and neighboring genes. *Nucleic Acids Res.* 14:3435–3448.
16. Preston, V. G., and F. B. Fisher. 1984. Identification of the herpes simplex virus type 1 gene encoding the dUTPase. *Virology* 138:58–68.
17. Roizman, B. 1979. The organization of the herpes simplex virus genomes. *Annu. Rev. Genet.* 13:25–57.
18. Shlomai, J., and A. Kornberg. 1978. Deoxyuridine triphosphatase of *Escherichia coli*. *J. Biol. Chem.* 253:3305–3312.
19. Spindler, K. R., D. S. E. Rosser, and A. J. Berk. 1984. Analysis of adenovirus transforming proteins from early regions 1A and 1B with antisera to inducible fusion antigens produced in *Escherichia coli*. *J. Virol.* 49:132–141.
20. Tye, B. K., P. O. Nyman, I. R. Lehman, S. Hochhauser, and B. Weiss. 1977. Transient accumulation of Okazaki fragments as a result of uracil incorporation into nascent DNA. *Proc. Natl. Acad. Sci. USA* 74:154–157.
21. Weiss, B., and H. H. El-Hajj. 1986. The repair of uracil containing DNA, p. 349–356. *In* M. G. Simic, L. Grossman, and A. C. Upton (ed.) *Mechanisms of DNA Damage and Repair*. Plenum Publishing Corp., New York.
22. Williams, M. V., and Y. C. Cheng. 1979. Human deoxyuridine triphosphate nucleotidohydrolase. *J. Biol. Chem.* 254:2897–2901.
23. Wohlrab, F., and B. Francke. 1980. Deoxyribopyrimidine triphosphatase activity specific for cells infected with herpes simplex virus type 1. *Proc. Natl. Acad. Sci. USA* 77:1872–1876.
24. Wohlrab, F., B. K. Garrett, and B. Francke. 1982. Control of expression of the herpes simplex virus-induced deoxypyrimidine triphosphatase in cells infected with mutants of herpes simplex virus types 1 and 2 and intertypic recombinants. *J. Virol.* 43:935–942.

Chapter 48

Uracil-DNA Glycosylase Activity
Relationship to Proposed Biased Mutation Pressure in the Class *Mollicutes*

Marshall V. Williams and J. Dennis Pollack

The class *Mollicutes* (mycoplasmas), with five genera, contains procaryotic microorganisms which are characterized by the lack of a cell wall (8). The mollicutes are the smallest self-replicating organisms known (20), and they possess the smallest genomes of any free-living organism (20). The genome size of *Mycoplasma* spp. and *Ureaplasma* spp. is approximately 0.5×10^9 daltons, while that of *Acholeplasma*, *Spiroplasma*, and *Anaeroplasma* spp. is approximately 1×10^9 daltons (21). It has been suggested that members of this class must be constrained by their limited genome content (22), and this hypothesis is supported by studies which have demonstrated that some mycoplasmas lack major portions of metabolic pathways involved in the synthesis of cellular macromolecules (19; J. T. Manolukas, M. F. Barile, D. K. F. Chandler, and J. D. Pollack, *J. Gen. Microbiol.*, in press). Thus, these organisms may represent minimal cellular life forms which are capable of self-replication (16).

In addition to their small genome size,

Marshall V. Williams and J. Dennis Pollack • Department of Medical Microbiology and Immunology and Comprehensive Cancer Center, Ohio State University, Columbus, Ohio 43210.

members of the class *Mollicutes* have the lowest genome (DNA) G+C content known (13, 21): *Mycoplasma* spp., 23 to 41%; *Acholeplasma* spp., 24 to 35%; *Anaeroplasma* spp., 29 to 34%; *Ureaplasma* spp., 28%; and *Spiroplasma* spp., 26% (13, 21). The low G+C content observed in some of these organisms could be due to a biased mutation rate towards A+T pairs (17). It has been suggested that biased mutation pressure may be related to the levels of enzymes involved in methylation or deamination and to the activities of enzymes involved with DNA repair (17) and nucleic acid metabolism (25).

There have been very few studies concerning mutagenesis and DNA repair in the mollicutes. Studies with *Mycoplasma orale* and *Mycoplasma hyorhinis* have reported that there is a single species of DNA polymerase in these cells and that it lacks associated endo- and exoDNase activity (1, 15). Thus, DNA synthesis in these organisms may be error prone due to the lack of the 3'→5' exonuclease activity that is associated with efficient proofreading (24). Studies have also demonstrated that the mollicutes are extremely sensitive to UV, X-ray, and gamma-ray irradiations (3, 9, 14). While some mem-

TABLE 1
Effect of EDTA and Uracil on the UNG and DNase Activities in *Mollicutes* Crude Extracts[a]

Organism	Enzyme activity[b]					
	Control (no additives)		+EDTA		+Uracil	
	DNase	UNG	DNase	UNG	DNase	UNG
Acholeplasma laidlawii B-PG9	15.7 ± 3.3	5.2 ± 1.4	5.6 ± 2.3*	8.4 ± 1.8*	11.4 ± 5.4†	1.4 ± 0.9*
A. morum S2	83.0 ± 4.4	58.7 ± 8.1	1.2 ± 0.8*	49.2 ± 16.6†	78.6 ± 9.8†	6.8 ± 3.2*
Mycoplasma sp. strain Let. 1	12.3 ± 2.1	87.4 ± 9.4	0.6 ± 0.8*	83.4 ± 6.6†	14.4 ± 7.2†	7.7 ± 2.4*
Mycoplasma pulmonis KOH	30.3 ± 4.9	17.4 ± 3.1	0.8 ± 1.0*	18.6 ± 6.6†	27.6 ± 10.8†	2.2 ± 2.1*
M. arthritidis 07	23.2 ± 3.1	19.5 ± 2.1	0.2 ± 0.4*	22.8 ± 7.8†	24.6 ± 3.6†	3.5 ± 1.4*
M. pneumoniae FH	9.3 ± 3.4	6.6 ± 2.4	0.1 ± 0.4*	4.6 ± 1.8†	6.0 ± 1.8†	0.9 ± 0.8*
M. gallisepticum S6	0.8 ± 0.3	<0.001	<0.001*	<0.001†	0.9 ± 0.1†	<0.001†
M. capricolum 14	0.6 ± 0.2	<0.001	<0.001*	<0.001†	0.4 ± 0.2†	<0.001†
Escherichia coli UNG[c]	<0.001	23.0 ± 2.1	<0.001†	21.0 ± 2.4†	<0.001†	1.2 ± 0.8*

[a] Reaction mixtures were incubated at 37°C for 60 min and terminated as described in the text. [^{3}H]uracil-labeled calf thymus DNA was used as a substrate. By thin-layer chromatography, 95 to 98% of the radioactive products were isolated and identified as various amounts of free [^{3}H]uracil and [^{3}H]uracil-labeled oligonucleotides.
[b] Enzyme activity is expressed as the nanomoles of [^{3}H]uracil present in the products formed in 60 min per milligram of protein. Values represent the average of three experiments ± the standard deviation. Symbols: *, significant difference ($P < 0.05$), and †, no significant difference ($P > 0.05$), when the indicated value is compared with its respective control (no additives) value. Substitution of [^{3}H]uracil-labeled *Mycoplasma* sp. strain Let. 1 DNA as a substrate resulted in essentially identical results.
[c] A purified *E. coli* UNG was used as a positive control and was prepared as described in the text.

bers of the *Mollicutes* such as *Acholeplasma laidlawii* possess both photoreactivation and dark (excision) repair (4), others, such as *Mycoplasma gallisepticum*, do not contain these repair systems (5). The lack of these DNA repair systems in *M. gallisepticum* has been used to explain the rapid evolution of the class *Mollicutes* (29, 30).

Uracil residues can be formed in DNA either by the direct incorporation of dUTP into DNA by DNA polymerase or by the deamination of dCMP residues in the DNA. There are several consequences of biological significance that occur upon the incorporation or formation of uracil residues in DNA, one of which is an increased mutation frequency (6). The enzyme responsible for preventing dUTP from being incorporated into DNA is deoxyuridine triphosphate nucleotidohydrolase (dUTPase; EC 3.6.1.23), which hydrolyzes dUTP to dUMP and pyrophosphate (23). Conversely, the enzyme uracil-DNA glycosylase (UNG) is responsible for removing uracil residues directly from DNA. The resulting apyrimidinic site is then repaired by a base-excision repair process (2, 11). These enzymes are ubiquitous in nature and have been observed in all procaryotic and eucaryotic organisms that have been examined. This suggests that these enzymes are important in maintaining normal cellular function.

We have reported previously that while organisms in the genus *Acholeplasma* have dUTPase activity, all species in the genus *Mycoplasma* lack this enzyme activity (28). To continue this study concerning deoxyuridine metabolism in the class *Mollicutes*, we examined crude cell extracts from eight mollicutes (two *Acholeplasma* spp. and six *Mycoplasma* spp.) for the presence of dUTPase (26), UNG (2), and nonspecific DNase activities (7). Only six mollicutes possessed UNG activity; *M. gallisepticum* and *Mycoplasma capricolum* did not (Table 1). The UNG activity was not inhibited by EDTA (10 mM), but it was inhibited by uracil (10 mM). The UNG hydrolyzed only [^{3}H]dUMP-containing DNA, and the product was [^{3}H]uracil. Preliminary studies on the UNG activity partially purified from *Mycoplasma* sp. strain Let. 1 demonstrated

TABLE 2
dUTPase Activity in Members of the Class *Mollicutes*

Organism	Enzyme activity[a] (n)
Acholeplasma laidlawii B-PG9	10.3 ± 3.1 (12)
A. morum S2	7.1 ± 0.6 (13)
Mycoplasma sp. (Let. 1)	<0.001[b] (5)
Mycoplasma pulmonis KOH	<0.001 (2)
M. arthriditis 07	<0.001 (3)
M. pneumoniae FH	<0.001 (6)
M. gallisepticum S6	<0.001 (6)
M. capricolum 14	<0.001 (6)

[a] Assays for dUTPase activity were performed as described previously (26). Enzyme activity is expressed as nanomoles of dUMP formed per minute per milligram of protein (mean ± standard deviation). The numbers in parentheses are the number of different batches of cells examined.
[b] The minimum detectable amount was 0.001 nmol of dUMP formed per mg of protein.

that it was similar to the UNG activities from *Escherichia coli* (11) and *Micrococcus luteus* (10). All eight mollicutes possessed nonspecific DNase activity that hydrolyzed both [^{3}H]dUMP-containing DNA and [^{14}C]TMP-containing DNA (Table 1). This nonspecific DNase activity was not inhibited by uracil (10 mM), but was inhibited by greater than 90% in all organisms except *A. laidlawii* B-PG9 by EDTA (10 mM). None of the *Mycoplasma* spp. had dUTPase activity, whereas members of the genus *Acholeplasma* did possess dUTPase activity (Table 2).

The results of this study demonstrate that there is diversity within the class *Mollicutes* with respect to their ability to prevent the incorporation of dUTP into their DNA and to remove uracil residues from their DNA. The *Acholeplasma* spp. appear to be similar to other procaryotic organisms in that they possess both dUTPase and UNG activities. However, none of the *Mycoplasma* spp. that we examined possessed dUTPase activity, and two species, *M. gallisepticum* and *M. capricolum*, also lacked UNG activity. These *Mycoplasma* spp. are the only self-replicating organisms known in nature that lack these enzyme activities, and this unique deficiency could have a number of consequences that are biologically significant.

Studies in *E. coli* have demonstrated that the frequency at which dUTP enters DNA is dependent upon the ratio of dUTP to TTP pools and that under normal conditions, approximately 1 dUTP nucleotide is incorporated per 1,200 nucleotides polymerized (23). Studies concerning the enzymes involved with pyrimidine deoxyribonucleotide metabolism have demonstrated that the mollicutes have the potential to synthesize dUTP (18, 27, 28). This suggests, in the case of the *Mycoplasma* spp., that they could incorporate significant levels of dUTP into their DNA. The incorporation of dUTP in place of TTP into their DNA would not be mutagenic since dUMP exhibits normal base pairing with dAMP. However, since the UNG base excision repair system is error prone, this would lead to a higher mutation rate in those *Mycoplasma* spp. that lack dUTPase activity but possess UNG activity.

Uracil residues can also be formed in DNA due to the spontaneous deamination of dCMP in DNA (12). If the uracil is not removed by UNG, this would result in the formation of a stable dU·dG base pair and, after subsequent rounds of replication, lead to a T·dA transition mutation. This process would result in a higher mutation rate that is biased in favor of A·T base pair formation and ultimately in the formation of an organism whose DNA has a low amount of G·C base pairs. *M. gallisepticum* and *M. capricolum* may be examples of such organisms since they lack both dUTPase and UNG activities, have high biased mutation rates (5, 17), and contain low amounts of G·C base pairs in their DNA (13, 21).

The class *Mollicutes* are thought to be evolved from low-G·C-containing gram-positive bacteria through a degenerative process (29, 30). They also exhibit considerable diversity with respect to possessing dUTPase and UNG activities. It is possible that those organisms, e.g., *A. laidlawii*, which possess both dUTPase and UNG activities are the least evolved from their ancestral gram-positive bacteria, whereas organisms such as *M.*

gallisepticum that lack both enzyme activities may have evolved further and are at a "terminal" stage of their evolution, in that additional mutations in these genome-limited, low-G+C-containing organisms may be lethal. While it is apparent that additional studies are required, our study demonstrates that the class *Mollicutes* can be used as a model system for examining the role of dUTPase and UNG in normal cellular metabolism and emphasizes their possible role in evolutionary processes.

ACKNOWLEDGMENTS. This work was supported in part by Public Health Service grant DE-06866 from the National Institute for Dental Research (M.V.W.) and by The Ohio State University Comprehensive Cancer Center (Core Grant P30CA 1605813).

LITERATURE CITED

1. Boxer, L. M., and D. Korn. 1979. Structural and enzymological characterization of homogeneous deoxyribonucleic acid polymerase from *Mycoplasma orale. Biochemistry* 18:4742–4749.
2. Caradonna, S. J., and Y. C. Cheng. 1980. Uracil-DNA glycosylase. Purification and properties of this enzyme isolated from blast cells of acute myelocytic leukemia patients. *J. Biol. Chem.* 255:2293–2300.
3. Chelton, E. T., S. Jones, and R. T. Walker. 1979. The sensitivity of *Mycoplasma mycoides* var. *capri* cells to gamma-radiation after growth in medium containing the thymine analogue, 5-vinyl uracil. *Biochem. J.* 181:783–785.
4. Das, J., J. Maniloff, and S. B. Bhattacharjee. 1972. Dark and light repair in ultraviolet-irradiated *Acholeplasma laidlawii. Biochim. Biophys. Acta* 159:189–197.
5. Ghosh, A., J. Das, and J. Maniloff. 1977. Lack of repair of ultraviolet light damage in *Mycoplasma gallisepticum. J. Mol. Biol.* 116:337–344.
6. Hochhauser, S. J., and B. Weiss. 1978. *Escherichia coli* mutants deficient in deoxyuridine triphosphatase. *J. Bacteriol.* 134:157–166.
7. Hoffmann, P. J., and Y. C. Cheng. 1978. The deoxyribonuclease induced after infection of KB cells by herpes simplex virus type 1 or type 2. Purification and characterization of the enzyme. *J. Biol. Chem.* 253:3551–3562.
8. Krieg, N. R., and J. G. Holt (ed.). 1984. *Bergey's Manual of Systematic Bacteriology*, vol. 1, p. 740–793. The Williams & Wilkins Co., Baltimore.
9. Labarere, J., and G. Barroso. 1984. Ultraviolet radiation mutagenesis and recombination in *Spiroplasma citri. Isr. J. Med. Sci.* 20:826–829.
10. Leblanc, J. P., B. Martin, J. Cadet, and J. Laval. 1982. Uracil DNA glycosylase. Purification and properties of uracil-DNA glycosylase from *Micrococcus luteus. J. Biol. Chem.* 257:3477–3483.
11. Lindahl, T. 1974. An N-glycosidase from *Escherichia coli* that releases free uracil from DNA containing deaminated cytosine residues. *Proc. Natl. Acad. Sci. USA* 71:3649–3653.
12. Lindahl, T., and B. Nyberg. 1974. Heat induced deamination of cytosine residues in deoxyribonucleic acid. *Biochemistry* 13:3405–3410.
13. Maniloff, J. 1983. Evolution of wall-less prokaryotes. *Annu. Rev. Microbiol.* 37:477–499.
14. Maniloff, J., and H. J. Morowitz. 1972. Cell biology of the mycoplasmas. *Bacteriol. Rev.* 36:263–290.
15. Mills, L. B., E. J. Stanbridge, W. D. Sedwick, and D. Korn. 1977. Purification and partial characterization of the principal deoxyribonucleic acid polymerase from *Mycoplasmatales. J. Bacteriol.* 132:641–649.
16. Morowitz, H. J. 1967. Biological self-replicating systems. *Prog. Theor. Biol.* 1:35–58.
17. Muto, A., and S. Osawa. 1987. The guanine and cytosine content of genomic DNA and bacterial evolution. *Proc. Natl. Acad. Sci. USA* 84:166–169.
18. Neale, G. A. M., A. Mitchell, and L. R. Finch. 1983. Enzymes of pyrimidine deoxyribonucleotide biosynthesis in *Mycoplasma mycoides* subsp. *mycoides. J. Bacteriol.* 156:1001–1005.
19. Pollack, J. D., V. V. Tryon, and K. D. Beaman. 1983. The metabolic pathways of *Acholeplasma* and *Mycoplasma*: an overview. *Yale J. Biol. Med.* 56:709–716.
20. Razin, S. 1978. The mycoplasmas. *Microbiol. Rev.* 42:414–470.
21. Razin, S., M. F. Barile, R. Harasawa, D. Amikam, and G. Glaser. 1983. Characterization of the mycoplasma genome. *Yale J. Biol. Med.* 56:357–366.
22. Rogers, M. J., J. Simmons, R. T. Walker, W. G. Weisburg, C. R. Woese, R. J. Tanner, I. M. Robinson, D. A. Stahl, G. Olsen, R. H. Leach, and J. Maniloff. 1985. Construction of the mycoplasma evolutionary tree from 5S rRNA sequence data. *Proc. Natl. Acad. Sci. USA* 82:1160–1164.
23. Shlomai, J., and A. Kornberg. 1978. Deoxyuridine triphosphatase of *Escherichia coli*. Purification, properties and use as a reagent to reduce uracil incorporation into DNA. *J. Biol. Chem.* 253:3305–3312.
24. Stanbridge, E. J., and M. E. Reff. 1979. The

molecular biology of mycoplasmas, p. 157–185. *In* M. F. Barile and S. Razin (ed.), *Cell Biology: The Mycoplasmas*, vol. 1. Academic Press, Inc., New York.

25. **Tryon, V. V., and J. D. Pollack.** 1985. Distinctions in *Mollicutes* purine metabolism: pyrophosphate-dependent nucleoside kinase and dependence on guanylate salvage. *Int. J. Syst. Bacteriol.* **35**:497–501.

26. **Williams, M. V., and J. D. Pollack.** 1984. Purification and characterization of a dUTPase from *Acholeplasma laidlawii* B-PG9. *J. Bacteriol.* **159**:278–282.

27. **Williams, M. V., and J. D. Pollack.** 1985. Pyrimidine deoxyribonucleotide metabolism in *Acholeplasma laidlawii* B-PG9. *J. Bacteriol.* **161**:1029–1033.

28. **Williams, M. V., and J. D. Pollack.** 1985. Pyrimidine deoxyribonucleotide metabolism in members of the class *Mollicutes*. *Int. J. Syst. Bacteriol.* **35**:227–230.

29. **Woese, C. R., J. Maniloff, and L. B. Zablen.** 1980. Phylogenetic analysis of the mycoplasmas. *Proc. Natl. Acad. Sci. USA* **77**:494–498.

30. **Woese, C. R., E. Stackenbrandt, and W. Ledwig.** 1985. What are mycoplasmas: the relationship of tempo and mode in bacterial evolution. *J. Mol. Evol.* **21**:305–316.

Chapter 49

Expression of *Escherichia coli* DNA Alkylation Repair Genes in Human Cells

G. William Rebeck, Bruce Derfler, Patrick Carroll, and Leona Samson

In *Escherichia coli*, some DNA methylation products lead to mutation by mispairing during replication, while others cause cell death by presenting blocks to DNA replication. One of our goals is to determine whether these particular adducts have the same consequences in eukaryotic organisms, i.e., whether the induction of mutation and cell death by alkylating agents proceeds via the same molecular mechanisms in human cells as in *E. coli*. One of our approaches to determining the molecular mechanisms of alkylation-induced mutation and cell death in human cells has been to generate human cell variants of known genotype by stable transformation and expression of *E. coli* DNA repair enzymes in human tissue culture cells.

DNA METHYLATION DAMAGE AND REPAIR IN *E. COLI*

Methylating agents react with a number of oxygen and nitrogen atoms of DNA, and some of these lesions have been specifically linked to the induction of mutation and cell death in *E. coli*. For instance, O^6-methylguanine (O^6MeG) is a mutagenic lesion, causing G·C-to-A·T transitions by base mispairing during DNA replication (10, 25). O^4-Methylthymine (O^4MeT) is also a mutagenic lesion, causing A·T-to-G·C transitions (31). Neither of these DNA lesions appears to lead to bacterial cell killing (12). N3-Methyladenine (N3MeA), on the other hand, is a lethal lesion (20), inhibiting the progression of DNA polymerase during replication (3). Other lesions, such as N7-methylguanine and methylphosphotriesters (MePT), appear to be innocuous to bacteria, having neither mutagenic nor lethal effects (26, 28).

Four DNA repair enzymes have been identified which specifically repair DNA alkylation damage. A constitutively expressed enzyme, N3MeA-DNA glycosylase I (the *tag* gene product), releases N3MeA lesions (20). We recently found another constitutively expressed enzyme, DNA methyltransferase II, which repairs O^6MeG and O^4MeT lesions. In addition, two inducible DNA alkylation repair enzymes have been identified. DNA methyltransferase (the *ada*

G. William Rebeck, Bruce Derfler, Patrick Carroll, and Leona Samson • Laboratory of Toxicology, Harvard School of Public Health, Boston, Massachusetts 02115.

gene product) repairs O^6MeG, O^4MeT, and MePT lesions (9, 19, 37), and N3MeA-DNA glycosylase II (the *alkA* gene product) repairs N3MeA, O^2-methylthymine, O^2-methylcytosine, and N3-methylguanine (18, 27, 29).

The Ada and AlkA enzymes participate in the adaptive response to alkylation damage. *E. coli*, when chronically exposed to low doses of alkylating agents, adapts to become tremendously resistant to the lethal and mutagenic effects of these agents (33), and the Ada protein plays a central role in the regulation of this response (40). Ada repairs specific DNA damages via the direct transfer of the unwanted methyl groups from DNA to specific cysteine residues within Ada in an irreversible manner (24). After transfer of a methyl group from a phosphotriester lesion to an active site near its amino terminus, Ada acts as a positive regulator for the expression of its own gene and for the expression of a number of other genes, including *alkA* (40). *alkB* shares an operon with *ada* and is therefore coordinately regulated with *ada*. The AlkB protein provides resistance to killing by the methylating agent methyl methanesulfonate (MMS) via an as yet unknown mechanism (22). Thus, Ada is directly responsible for cellular resistance to methylation-induced mutation through the repair of O^6MeG and O^4MeT lesions, but only indirectly responsible for resistance to killing through regulation of the *alkA* and *alkB* genes. Another gene of unknown function, *aidB*, is also induced in an *ada*-dependent fashion (42).

STUDIES OF DNA METHYLATION DAMAGE AND REPAIR IN HUMAN CELLS WITH *E. COLI* REPAIR GENES

The determination of the biological effects of particular alkylated DNA lesions in *E. coli* was made possible by the isolation of mutants specifically altered in their ability to repair those lesions. The isolation of mammalian cell variants known to be altered in a single genetic locus is much more difficult. A number of mammalian cell lines with increased sensitivities to alkylating agents have been isolated (32, 41) or identified (1, 8, 14). HeLa S3 Mer⁻ human cells are extremely sensitive to alkylation-induced killing, mutation, and sister chromatid exchange (SCE) (1, 7). These cells have been shown to lack the human DNA methyltransferase (7), a 25-kilodalton enzyme which repairs O^6MeG lesions (30), which may or may not repair O^4MeT lesions (2, 43), but which does not repair MePT lesions (43). However, it has been suggested that HeLa Mer⁻ cells are deficient in other aspects of alkylation repair, although the deficient repair of specific DNA alkyl lesions has not yet been identified. Interestingly, a cell line derived from HeLa S3 Mer⁻ cells (HeLa A6) was isolated which had become resistant to alkylation-induced cell killing while remaining sensitive to alkylation-induced SCE and mutation (1). Assays of extracts of A6 cells have shown that, even though A6 cells can resist the lethal effects of DNA methylation damage, they have not gained any DNA methyltransferase activity (35). Thus, multiple changes in genotype might account for the alkylation-sensitive phenotype in HeLa Mer⁻ cells.

Human and rodent cells chronically pretreated with low levels of methylating agents acquire increased resistance to the mutagenic, lethal, and SCE-inducing effects of methylating agents (13, 16, 23, 36). Chronic exposure of rodent cells to low doses of methylating agents has also been reported to elevate the levels of DNA methyltransferase (23). However, it seems possible that, like bacteria, mammalian cells would adapt via the induction of several genes, and so it would seem premature to conclude that elevated methyltransferase levels in adapted mammalian cells provide resistance to all the effects of alkylating agents, namely killing, mutation, and SCE induction. Consequently, the questions remain as to which DNA lesions are responsible for mutagenicity in human cells and which lesions are responsible for lethality.

Because of the possibilities of multiple changes in mammalian cell variants and in adapted mammalian cells, we and several other groups have generated variant cell lines altered in the expression of known genes by the transfection and expression of bacterial DNA repair genes in alkylation-sensitive mammalian cells (4, 15, 21, 34). If the bacterial enzymes are stable in mammalian cells and if they can be transported into the nucleus and act upon chromatin, then such transformants would acquire DNA repair capabilities for specific alkylated lesions. By studying the resultant changes in alkylation sensitivity, we should be able to deduce the effects of these unrepaired lesions in mammalian cells. Our prototype experiment was to transfect HeLa S3 Mer$^-$ cells with the E. coli ada-alkB operon in the pSV2 mammalian expression vector (34). The ada gene seemed particularly suitable for use in this type of study because Ada acts through the direct reversal of DNA alkylation damage and does not require interactions with complex enzyme systems such as those required for excision repair. The alkB gene, however, was an unknown quantity.

One transformant, S3-9, had several integrated copies of pSV2 ada-alkB, it transcribed ada-alkB mRNA, and it expressed very high levels of the Ada DNA methyltransferase. Since there is no functional assay for the AlkB protein, we could not determine whether alkB was also expressed. Preliminary experiments, using AlkB antiserum, now indicate that if alkB is being expressed, it is at very low levels. S3-9 cells acquired a high level of resistance to killing, SCE, and mutation induced by the methylating agent N-methyl-N'-nitro-N-nitrosoguanidine (MNNG). It was unclear which resistances were caused by the expression of ada and which might be caused by the possible expression of alkB. Since Ada provides killing resistance in bacteria only through the induction of other gene products, it would be surprising if it was responsible for the S3-9 killing resistance. Ada provides mutation resistance in E. coli via repair of O^6MeG and O^4MeT, and the repair of these adducts may be responsible for the S3-9 mutation resistance. To address the possibility that AlkB was providing killing resistance in S3 ada-alkB transformants, it became necessary to separate the functions of Ada from the possible functions of AlkB. We took two approaches to achieving this separation. First, we attempted to eliminate the role of Ada by depleting levels of the ada gene product in S3-9 cells. Secondly, we transformed S3 cells with only the ada gene or with only the alkB gene.

DEPLETION OF METHYLTRANSFERASE ACTIVITY WITH O^6MeG FREE BASE

To address the question of whether the ada gene provides resistance to alkylation-induced killing in human cells, we depleted DNA methyltransferase levels in S3-9 ada alkB transformants by growth in the O^6MeG free base. This compound had previously been shown to deplete cellular levels of the mammalian DNA methyltransferase, possibly by acting as a weak substrate for the enzyme (11, 17). We have shown that growth of S3-9 cells in 1 mM O^6MeG overnight reduces the bacterial DNA methyltransferase activity by greater than 90%; subsequent growth in 0.1 mM O^6MeG maintains methyltransferase depletion without any apparent toxic effects. Methyltransferase-depleted S3-9 cells became almost as sensitive to MNNG-induced SCE and mutation as the parental cell line, supporting the notion that the Ada protein provides resistance to alkylation-induced mutation and SCE. However, methyltransferase-depleted S3-9 cells remained fully resistant to MNNG-induced killing. Thus, although these cells have a greatly reduced ability to repair O^6MeG DNA lesions, they remain able to resist the lethal effects of MNNG.

TRANSFORMATION OF HUMAN REPAIR-DEFICIENT CELLS WITH THE *ada* GENE

In our second approach to determining whether *alkB* provides resistance to alkylation-induced killing in S3-9 *ada-alkB*-transformed cells, we cotransfected HeLa S3 cells with pSV2neo and with a pSV2 vector containing either the *ada* gene or the *alkB* gene. Our results with the expression of AlkB in HeLa S3 cells are still in the preliminary stages and need not be considered here. Of 80 G418-resistant colonies isolated from the pSV2 *ada* transfection, two clones, S3-A10 and S3-A39, were found to express the Ada methyltransferase. The level of Ada expression in S3-A39 was about three times higher than in S3-A10, but was about 10-fold lower than that in the S3-9 *ada-alkB* transformants. S3-A10 cells had acquired little if any resistance to MNNG-induced killing, but S3-A39 cells were very resistant to MNNG-induced killing and SCE. Thus, it appears that the expression of Ada alone can confer killing and SCE resistance to human cells, and it seems less likely that AlkB provides killing resistance in the *ada-alkB*-transfected cells. The *ada* gene has been transfected into other alkylation-sensitive mammalian cell lines at several other labs, and similar results have been obtained (3, 15, 21).

That *ada* alone provides resistance to MNNG killing in mammalian cells appears to contradict the finding that methyltransferase-depleted S3-9 *ada-alkB* cells did not become sensitive to MNNG killing. Thus, we measured the effect of depleting methyltransferase by 90% in the S3 *ada* transformant upon the sensitivity of these cells to MNNG killing. Growth of S3-A39 cells in O^6MeG free base did not diminish their resistance to MNNG killing. It seemed pos-

TABLE 1
Products of DNA Alkylation with Different Methylating Agents

DNA alkylation product	% Product produced[a] by:	
	MNNG	MMS
N7-Methylguanine	66.4	81
MePT	12.1	0.8
O^6MeG	5.4	0.25
O^4MeT	0.7	
N3MeA	8.4	11.3
N3-Methylguanine	0.6	0.6
Other	6.4	6

[a] Data is from reference 6.

sible that the 10% residual level of methyltransferase in the free-base-treated S3-A39 cells (and S3-9 cells) is still sufficient to provide complete killing resistance. However, this level of methyltransferase is actually less than that found in the S3-A10 *ada* cells, and S3-A10 cells show almost no resistance to MNNG killing. Thus, the level of functionally active Ada methyltransferase does not correlate with the level of resistance to MNNG killing. These results support the argument that O^6MeG is not a lethal lesion in mammalian cells and suggest the possibility that Ada might provide killing resistance through a mechanism independent of its repair functions.

CHALLENGE OF *ada* TRANSFORMANTS WITH MMS AND DMS

Different methylating agents produce different spectra of DNA damages. The spectra of MNNG- and MMS-induced DNA alkylation products are listed in Table 1. It is clear that MMS almost exclusively methylates purine nitrogens, while MNNG produces a significant proportion of lesions at oxygen atoms. MMS produces 20-fold-less O^6MeG lesions than MNNG, but about the same proportion of N3MeA lesions. Dimethyl sulfate (DMS) is a methylating agent similar to MMS in that it produces very few lesions at DNA oxygen atoms (38). HeLa Mer⁻ cells are sensitive to the killing effects of both MMS and DMS. In the *ada*-transformed S3-A39 cells, there is a marked increase in resistance to MMS- and DMS-induced killing. Indeed, at doses of MMS that kill 90% of the parent S3 cells, S3-A39 *ada* cells are completely resistant. Since a very small proportion of the lesions formed by MMS are substrates for the Ada protein, its seems unlikely that the resistance is due to the repair functions of Ada. If, however, O^6MeG or O^4MeT does turn out to be lethal in human cells, one must argue that, at doses causing 90% cell killing, these are the only unrepaired lethal lesions being formed since the Ada methyltransferase can prevent all of this cell death. The very high levels of N3MeA formed by MMS and DMS at these doses might be repaired so efficiently in HeLa Mer⁻ cells that they cause no cell death; on the other hand, contrary to the situation in *E. coli*, N3MeA might be a nonlethal lesion in human cells.

SUMMARY

We have introduced the *E. coli ada* gene into mammalian cells and have observed an increase in resistance to the lethal, mutagenic, and SCE-inducing effects of alkylating agents. The DNA repair functions of the Ada protein involve the transfer of methyl groups from MePT DNA lesions (which are thought to have no toxic effects) and from O^6MeG and O^4MeT DNA lesions (which are thought to be mutagenic but not lethal). Ada is also responsible for providing killing resistance to methylating agents in bacteria, but this is accomplished by the action of Ada as a positive regulator of other genes that are directly responsible for resistance to killing by alkylating agents. Thus, the provision of killing resistance by the expression of *ada* in human cells was surprising.

Others have transfected the *ada* gene, and truncated versions of the *ada* gene, into various alkylation sensitive mammalian cell lines. Brennand and Margison (4) introduced *ada* into Mex⁻ Chinese hamster ovary fibroblasts and found increased resistance to alkylation-induced SCE and killing (although they noted no differences in MMS-induced killing). They also introduced a truncated version of *ada* which encoded a carboxyterminal peptide fragment able to repair O^6MeG and O^4MeT but not MePT lesions and observed the same pattern of resistances (5). Thus, the killing and SCE resistances of transformed cells seem to be provided entirely by the carboxy fragment of Ada. This result was supported by Kataoka et al. (21), who transfected Chinese hamster ovary cells with a truncated version of *ada* which encoded a peptide capable only of MePT repair. These transformants were not resistant to MNNG-induced killing and SCE. Finally, Ishizaki et al. (15) also transfected *ada* into HeLa Mer⁻ cells and reported killing and SCE resistances to MNNG.

Our results are consistent with these studies (except for MMS killing sensitivity), but go further in the use of O^6MeG free base to deplete methyltransferase activity. When levels of Ada DNA methyltransferase in S3-A39 cells have been depleted with free base, cells remain resistant to MNNG killing. Thus, O^6MeG-treated *ada* transformants are analogous to the HeLa A6 cells, which lack DNA methyltransferase repair activity but which are resistant to alkylation-induced killing. We are currently examining whether killing resistance in the *ada* transformants could be due to induction of endogenous DNA alkylation repair enzymes by Ada or by some protein fragment of Ada. Alternatively, it remains a possibility that O^6MeG (or O^4MeT) is actually lethal to mammalian cells; however, this conclusion is not consistent with the methyltransferase depletion experiments and seems unlikely in light of the Ada-conferred resistance to MMS and DMS.

ACKNOWLEDGMENTS. This work was supported by American Cancer Society research grant NP448 and National Institute of Environmental Health Science grant 1-PO1-ES03926. L.S. was supported by an American Cancer Society Scholar Award and then a Faculty Research Award. G.W.R. was supported by a National Science Foundation Graduate Research Fellowship. P.C. was supported by a Dana Foundation Training Program for Scholars in Toxicology.

LITERATURE CITED

1. Baker, R. M., W. C. Van Voorhis, and L. A. Spencer. 1979. HeLa cell variants that differ in sensitivity to monofunctional alkylating agents, with independence of cytotoxic and mutagenic responses. *Proc. Natl. Acad. Sci. USA* 76:5249–5253.
2. Becker, R. A., and R. Montesano. 1985. Repair of O^4 methyldeoxythymidine residues in DNA by mammalian liver extracts. *Carcinogenesis* 6:313–317.
3. Boiteux, S., O. Huisman, and J. Laval. 1984. 3-Methyladenine residues in DNA induce the SOS function *sfiA* in *Escherichia coli*. *EMBO J.* 3:2569–2573.
4. Brennand, J., and G. P. Margison. 1986. Reduction of the toxicity and mutagenicity of alkylating agents in mammalian cells harboring the *Escherichia coli* alkyltransferase gene. *Proc. Natl. Acad. Sci. USA* 83:6292–6296.
5. Brennand, J., and G. P. Margison. 1986. Expression in mammalian cells of a truncated *Escherichia coli* gene coding for O6-alkylguanine alkyltransferase reduces the toxic effects of alkylating agents. *Carcinogenesis* 7:2081–2084.
6. Day, R. S., III, M. A. Babich, D. B. Yarosh, and D. A. Scudiero. 1987. The role of O6methylguanine in human cell killing, sister chromatid exchange induction and mutagenesis: a review. *J. Cell Sci.* 6(Suppl):333–353.
7. Day, R. S., III, C. H. J. Ziolkowski, D. A. Scudiero, S. A. Meyer, A. S. Lubiniecki, A. J. Girardi, S. M. Galloway, and G. D. Bynum. 1980. Defective repair of alkylated DNA by human tumour and SV40-transformed human cell strains. *Nature* (London) 288:724–727.
8. Day, R. S., III, C. H. J. Ziolkowski, D. A. Scudiero, S. A. Meyer, and M. R. Mattern. 1980. Human tumor cell strains defective in the repair of alkylation damage. *Carcinogenesis* 1:21–32.
9. Demple, B., A. Jacobsson, M. Olsson, P. Robins, and T. Lindahl. 1982. Repair of alkylated DNA in *Escherichia coli*. *J. Biol. Chem.* 257:13776–13780.
10. Dodson, L. A., R. S. Foote, S. Mitra, and W. E. Masker. 1982. Mutagenesis of bacteriophage T7 *in vitro* by incorporation of O^6 methylguanine during DNA synthesis. *Proc. Natl. Acad. Sci. USA* 79:7440–7444.
11. Dolan, M. E., C. D. Corsico, and A. E. Pegg. 1985. Exposure of HeLa cells to O6-alkylguanines increases sensitivity to the cytotoxic effects of alkylating agents. *Biochem. Biophys. Res. Commun.* 132:178–185.
12. Evensen, G., and E. Seeberg. 1982. Adaptation to alkylation resistance involves the induction of a DNA glycosylase. *Nature* (London) 296:773–775.
13. Frosina, G., and A. Abbondandolo. 1985. The current evidence for an adaptive response to alkylating agents in mammalian cells, with special reference to experiments with in vitro cell cultures. *Mutat. Res.* 154:85–100.
14. Goth-Goldstein, R. 1980. Inability of Chinese hamster ovary cells to excise O6-alkylguanine. *Cancer Res.* 40:2623–2624.
15. Ishizaki, K., T. Tsujimura, H. Yawata, C. Fujio, Y. Nakabeppu, M. Sekiguchi, and M. Ikenaga. 1986. Transfer of the *E. coli* O6-methylguanine methyltransferase genes into repair deficient human cells and restoration of cellular resistance to N-methyl-N'-nitro-N-nitrosoguanidine. *Mutat. Res.* 166:135–141.
16. Kaina, B. 1982. Enhanced survival and reduced mutation and aberration frequencies induced in V79 Chinese hamster cells pre-exposed to low levels of methylating agents. *Mutat. Res.* 93:195–211.
17. Karran, P. 1985. Possible depletion of a DNA repair enzyme in human lymphoma cells by subversive repair. *Proc. Natl. Acad. Sci. USA* 82:5285–5289.
18. Karran, P., T. Hjelmgren, and T. Lindahl. 1982. Induction of a DNA glycosylase for N-methylated purines is part of the adaptive response to alkylating agents. *Nature* (London) 296:770–773.
19. Karran, P., T. Lindahl, and B. Griffin. 1979. Adaptive response to alkylating agents involves alteration *in situ* of O6-methylguanine residues in DNA. *Nature* (London) 280:76–77.
20. Karran, P., T. Lindahl, I. Ofsteng, G. B. Evensen, and E. Seeberg. 1980. *Escherichia coli* mutants deficient in 3-methyladenine-DNA glycosylase. *J. Mol. Biol.* 140:101–127.
21. Kataoka, H., J. Hall, and P. Karran. 1986. Complementation of sensitivity to alkylating agents in *Escherichia coli* and Chinese hamster ovary cells by expression of a cloned bacterial DNA repair gene. *EMBO J.* 5:3195–3200.

22. Kataoka, H., Y. Yamamoto, and M. Sekiguchi. 1983. A new gene (*alkB*) of *Escherichia coli* that controls sensitivity to methyl methanesulfonate. *J. Bacteriol.* 153:1301–1307.
23. Laval, F., and J. Laval. 1984. Adaptive response in mammalian cells: crossreactivity of different pretreatments on cytotoxicity as contrasted to mutagenicity. *Proc. Natl. Acad. Sci. USA* 81:1062–1066.
24. Lindahl, T., B. Demple, and P. Robins. 1982. Suicide inactivation of the *E. coli* O6-methylguanine-DNA methyltransferase. *EMBO J.* 1:1359–1363.
25. Loechler, E. L., C. L. Green, and J. M. Essigman. 1984. *In vivo* mutagenesis by O6-methylguanine built into a unique site in a viral genome. *Proc. Natl. Acad. Sci. USA* 79:7440–7444.
26. Loveless, A. 1969. Possible relevance of O-6 alkylation deoxyguanosine to the mutagenicity and carcinogenicity of nitrosamines and nitrosamides. *Nature* (London) 233:206–207.
27. McCarthy, T. V., P. Karran, and T. Lindahl. 1984. Inducible repair of O-alkylated DNA pyrimidines in *Escherichia coli*. *EMBO J.* 3:545–550.
28. McCarthy, T. V., and T. Lindahl. 1985. Methyl phosphotriesters in alkylated DNA are repaired by the Ada regulatory protein in *E. coli*. *Nucleic Acids Res.* 13:2683–2698.
29. Nakabeppu, Y., H. Kondo, and M. Sekiguchi. 1984. Cloning and characterization of the *alk*A gene of *Escherichia coli* that encodes 3-methyladenine DNA glycosylase II. *J. Biol. Chem.* 259:13723–13729.
30. Pegg, A. E., M. Roberfroid, C. vonBahr, R. S. Foote, S. Mitra, H. Bresil, A. Likkacher, and R. Montesano. 1982. Removal of O^6 methylguanine from DNA by human liver fractions. *Proc. Natl. Acad. Sci. USA* 79:5162–5165.
31. Preston, B. D., B. Singer, and L. A. Loeb. 1986. Mutagenic potential of O4-methylthymine *in vivo* determined by an enzymatic approach to site-specific mutagenesis. *Proc. Natl. Acad. Sci. USA* 83:8501–8505.
32. Robson, C. N., and I. D. Hickson. 1987. Isolation of alkylating agent-sensitive Chinese hamster ovary cell lines. *Carcinogenesis* 8:601–605.
33. Samson, L., and J. Cairns. 1977. A new pathway for DNA repair in *Escherichia coli*. *Nature* (London) 267:281–283.
34. Samson, L., B. Derfler, and E. A. Waldstein. 1986. Suppression of human DNA alkylation-repair defects by *Escherichia coli* DNA-repair genes. *Proc. Natl. Acad. Sci. USA* 83:5607–5610.
35. Samson, L., and S. Linn. 1987. DNA alkylation repair and the induction of cell death and sister chromatid exchange in human cells. *Carcinogenesis* 8:227–230.
36. Samson, L., and J. Schwartz. 1980. Evidence for an adaptive DNA repair pathway in CHO and human skin fibroblast cell lines. *Nature* (London) 287:861–863.
37. Schendel, P. F., and P. E. Robins. 1978. Repair of O^6-methylguanine in adapted *Escherichia coli*. *Proc. Natl. Acad. Sci. USA* 75:6017–6020.
38. Swenson, D. H., and P. D. Lawley. 1978. Alkylation of deoxyribonucleic acid by carcinogens dimethyl sulphate, ethyl methanesulphonate, N-ethyl-N-nitrosourea and N-methyl-N-nitrosourea. *Biochem. J.* 171:575–587.
39. Teo, I., B. Sedgwick, B. Demple, B. Li, and T. Lindahl. 1984. Induction of resistance to alkylating agents in *E. coli*: the ada^+ gene product serves both as a regulatory protein and as an enzyme for repair of mutagenic damage. *EMBO J.* 3:2151–2157.
40. Teo, I., B. Sedgwick, M. W. Kilpatrick, T. V. McCarthy, and T. Lindahl. 1986. The intracellular signal for induction of resistance to alkylating agents in *E. coli*. *Cell* 45:315–324.
41. Thomson, L. H., K. W. Brookman, A. V. Dillehay, A. V. Carrano, J. A. Mazrimas, C. L. Mooney, and J. L. Minker. 1982. A CHO-cell strain having hypersensitivity to mutagens, a defect in DNA strand break repair, and an extraordinary baseline frequency of sister chromatid exchange. *Mutat. Res.* 95:427–440.
42. Volkert, M. R., D. C. Nguyen, and K. C. Beard. 1986. *Escherichia coli* gene induction by alkylation treatment. *Genetics* 112:11–26.
43. Yarosh, D. B., A. J. Fornance, and R. S. Day III. 1985. Human O^6-alkylguanine-DNA alkyltransferase fails to repair O^4 methylthymine and methyl phosphotriesters in DNA as efficiently as does the alkyltransferase from *Escherichia coli*. *Carcinogenesis* 6:949–953.

VII. Induced Mutagenesis

Induced Mutagenesis: Introduction

In this section the nature of induced mutagenesis is reviewed as it relates to mammalian cells. Since the discovery of the induction of specific genes in procaryotes (e.g., the SOS system and *din* genes), there has been interest in the question of the existence in mammalian cells of genes that are inducible by DNA damage. The activation of a mutagenic pathway in mammalian cells is consistent with much circumstantial evidence. However, because of the lack of detailed genetic analysis, conclusive evidence has been lacking. Nevertheless, this section contains results that are in quite remarkable agreement in their general conclusions. Using both biochemical and genetic approaches, the authors present evidence in support of inducible processes in mammalian cells that play a role in the mutagenic response. It is still uncertain which genes are involved and which specific pathways might be used. However, biochemical and genetic studies are now feasible, and the basic phenomena have been elucidated for further study.

An area of particular interest is the utility of inducible mutagenic and DNA repair processes to a complex multicellular organism. In contrast to unicellular bacteria and yeasts, the cells of higher organisms are not individually essential to the survival of the genotype. Thus, rather than trying to survive at all costs as in the SOS-induced bacterium, it might be better for a damaged cell to die and let the less damaged cells of the same genotype carry on the work of the organism. Indeed, current data can be interpreted to suggest that mammalian cells exhibit an "altruistic" response whereby they secrete paracrine growth factors to help neighboring cells better respond to damage as they, themselves, die.

The response to DNA damage by mammalian cells may have some features in common with the heat shock response or the stress response of mammalian cells in that these processes all disrupt the cell replication cycle, can perturb DNA-protein interactions, and so on. Since the lesions produced by UV are quite specific and can be quantitated, it may be that understanding the detailed pathway of UV-induced gene expression will help illuminate the mechanisms underlying these other responses to more general, probably nonspecific damage.

The chapters in this section each emphasize the questions to be solved by future work. Among the key problems are the following.

(i) How is DNA damage detected by

the cell? Is this question analogous to the problem of the activation of RecA protein in bacteria?

(ii) Is there a final common pathway in the induction of genes by different kinds of cellular damage?

(iii) Is mutagenesis in mammalian cells regulated so that it is an essential step in the differentiation of certain cell types, e.g., in generation of lymphocyte diversity?

(iv) How do the extracellular factors induced by DNA damage function; are these factors of physiological significance?

The Editors

Chapter 50

Endogenous Cellular Contributions to Mammalian Mutagenesis

Peter Herrlich

Bacteria have taught us the surprising lesson that induced mutagenesis requires the active participation of the cells to be mutated (reviewed in references 11, 17, 34, and 35). This cellular contribution consists of both activation of *trans*-acting proteins and de novo synthesis of gene products. This process has been named the SOS response (22), implying that the response made "sense" at least for the population of unicellular bacteria. The benefit to bacteria may be dual: the activated and induced gene products help bacteria bypass mutagen-induced replication blocks to template DNA and thus enhance survival at the expense of accumulating mutations at such sites; in addition, targeted and nontargeted mutations may rescue the rare cell from the effects of a hostile environment by conferring a required new ability. In principle, bypassing replication blocks at the expense of mutation may also be advantageous for proliferating somatic cells of a diploid, multicellular organism. It is not obvious, however, how a multicellular organism would benefit from a potential for increasing the mutation rate of somatic cells.

Attempts to detect an SOS response of mammalian cells analogous to that in *Escherichia coli* have yielded variable results (see reviews in references 11, 26, and 28). In considering the question of induced mutagenesis in mammalian cells, I have set aside possible similarities to the bacterial SOS response, arguing instead from mammalian experimental evidence in favor of invoking endogenous cellular contributions to mutagenesis. I then discuss how these contributions may operate and describe some evidence in support of these views. The relevant genetic response of mammalian cells is named operationally mammalian UV response or mammalian stress response (9).

ARE SITE AND TISSUE SPECIFICITIES IN INDUCED CARCINOGENESIS BROUGHT ABOUT BY INDUCED MUTAGENIC ACTIVITIES?

Cancer is a major consequence of somatic mammalian mutagenesis. Some observations on its generation are most readily explained by assuming active cellular and

---

Peter Herrlich • Kernforschungszentrum Karlsruhe, Institut für Genetik und Toxikologie, and Institut für Genetik, Universität Karlsruhe, D-7500 Karlsruhe, Federal Republic of Germany.

organismic contributions to mutagenesis. A key point is the apparently nonrandom or preferential site-specific induction of point mutations in some cases of chemical carcinogenesis. A normal primary cell needs to acquire at least two (see, e.g., references 14 and 15) and possibly several carcinogenic mutations to become a fully transformed tumor cell. Several additional events, probably also mutations, will confer to the tumor cell the potential to metastasize and progress. These carcinogenic mutations are of diverse nature but affect a limited number of loci, and they can occur with considerable lag periods, e.g., up to 40 years apart (37).

The puzzling features of some induced carcinogenic mutations are that, although induced by agents which a priori should act in a nucleotide- but not site-specific way, the mutations appear to be distributed nonrandomly and their frequencies are influenced by organismic and external conditions commonly circumscribed by the term tumor promotion. The organism does not exert its influence by differential control of mutagen uptake into different organs. For instance, in rats, deoxyguanosine-O^6-alkylation of DNA of brain and liver reached equal levels 1 h after ethylnitrosourea application (20). At the doses usually administered in these carcinogenesis experiments, between 1 and 10 in every 10^6 dG (20) or roughly one specific dG in 1 out of 4×10^5 to 4×10^6 cells is alkylated. O^6-Alkyl-dG is considered the major carcinogenic lesion (31). Although alkylation is statistical, the tumorigenic response can be extremely tissue specific. For example, injection of the short-lived mutagens ethylnitrosourea and methylnitrosourea into rats at different developmental stages will produce, at high efficiency, entirely different specific tumors. Prenatal transplacental exposure will cause predominantly neural tumors (12, 23) and tumors of epithelial and mesenchymal origin (M. Barbacid, unpublished data). When exposed during puberty or in the neonatal period, rats develop mammary adenocarcinomas (reviewed in reference 5). If, however, sexual development is blocked by tamoxifen and ovarectomy, many animals develop mesenchymal kidney tumors 7 to 9 months later (5). Thus, organismic and cellular conditions modify strongly the course of carcinogenesis; i.e., they determine sites where tumorigenic mutations can be observed and tissues in which tumors may develop. The nitrosomethylurea-induced neural tumors contained an activated *neu* oncogene, the kidney tumors activated Ki-*ras*, and the mammary adenocarcinomas only codon-12 mutants of Ha-*ras*-1 (5). The specific G→A transition observed in the second base of codon 12 would occur once in every 10^6 cells, assuming random alkylation and fixation prior to DNA repair. Although the issue is not completely resolved whether *ras* mutations are always very early events in cancer initiation, the carcinogen-specific types and sites of mutations suggest their establishment immediately after mutagen treatment (at least for the pulse or short-lived carcinogens). One or two concurrent or subsequent mutational steps would be required to convert the Ha-*ras*-1 mutated cell to full tumorigenicity. Even with 10 anlagen, a young rat is unlikely to possess 10^6 precursor cells for breast adenoepithelium, a number much too low to account for the observed tumor incidence on the basis of two or more independent mutational events. Equally unexplained is why only one or the other cell lineage would participate in carcinogenesis. What has been said about methylnitrosourea-induced tumors could be complemented by similar data on other chemical and X-ray-induced carcinogenesis (4, 6, 8).

Human oncogenesis adds to these puzzles. Evidence for tissue specificity is abundant, e.g., the B-lineage-specific carcinogenic events that lead to Burkitt's lymphoma or the detection of rare hereditary alleles for almost any human cancer. The combination of c-*myc* oncogene activation and Epstein-Barr virus infection is the critical event driving exclusively B cells to transformation (16, 18). One of the carcinogenic mutations in human oncogenesis apparently can be acquired through the germ line (reviewed in reference 15). Although present in all somatic cells, the inherited alleles predispose only one or a few cell types to tumorigenesis. Lineage-specific mechanisms must exist which operate with remarkable efficiency. For instance, the secondary mutation(s) required to transform an rb⁻ (hereditary retinoblastoma) retinoblast must occur between the 10-stem-cell stage and the completion of monolayer formation, with about 10^6 retinoblasts, in the eye of the child.

The endogenous processes suggested by these observations on in vivo carcinogenesis may overlap with those explored in experimental tumor promotion (19, 38). Tumor promotion is defined by the ability of agents to cause or enhance the transition from a carcinogen-treated normal cell to the hyperplastic stage. Chemical agents not known to interact with DNA directly have been isolated for their ability to promote tumorigenesis in mouse skin, e.g., the phorbol ester tumor promoters. Some of these establish a "memory" in the mouse that decays with a half-life of 10 to 12 weeks (7). In more general terms, tumor promotion shortens the lag periods between carcinogenic mutations and increases the probability of mutation. By fluctuation analysis one can show that a single dose of X rays (13) or ethylnitrosourea (Maher et al., this volume) will confer to cells in culture a proneness to mutation which holds for several rounds of replication (again a memory) such that a fraction of the mutations occur at a time when the mutagenic agent is no longer present. Sexual hormones in the methylnitrosourea-induced mammary carcinoma in rats, described above, apparently influence the chance of contracting the second mutation(s). These states of increased disposition to mutation must be based on the activation of genetically preformed components by the promoting agents. I will discuss briefly what components and mechanisms could be involved.

VARIATIONS OF THE CHANCE TO CONTRACT A RANDOM MUTATIONAL EVENT OR MODULATED MUTATOR FUNCTIONS

Endogenous cellular contributions to mutagenesis could fall into two classes: passive and unavoidable consequences of proliferation and active modulation of mutator functions. The two extreme views are further described as follows.

In the first group, the mutations occur by a strictly random mechanism, and it is the conditions under which they occur which make them appear to be introduced nonrandomly. The phenotypic expression of the mutation may be unmasked in a highly tissue-specific manner, or the introduction of mutations by randomly acting mechanisms may be influenced by tissue-specific or program-dependent access to certain parts of chromatin (as suggested by preliminary data from reference 24) or simply by the fact that critical cells are induced to multiple rounds of division due to the developmental program, mutagenic wound response, or action of a tumor-promoting external agent.

In the second type of cellular contributions to mutagenesis, mutagenesis depends on the action of mutator functions which are subjected to modulation by physiologic and nonphysiologic mediators.

From the above-mentioned points, both views have merits, and a combination of passive and active contributions may in fact be realized. The intracellular pathways leading to passive or to active mutagenesis may even be similar.

MUTATIONS IN trans

The essential feature of modulated mutator mechanisms can be summarized by the term "mutations in trans." The site of excitation from which the modulation is elicited is an entity different from the site where the

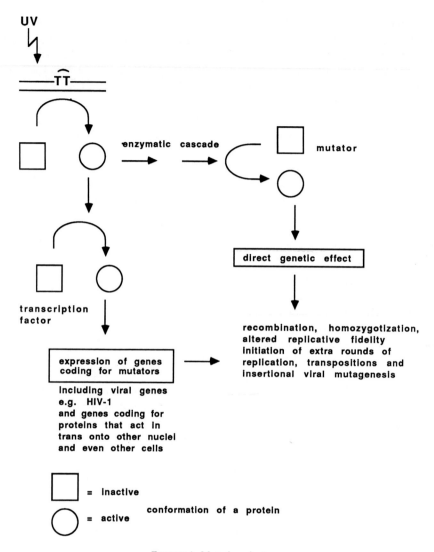

FIGURE 1. Mutations in *trans*.

mutation occurs, and *trans*-acting molecules serve to communicate between the two sites. For instance, a change in DNA structure, e.g., a thymine dimer, could be detected by a protein which, upon contact with the altered DNA, changes conformation and stimulates an enzymatic cascade ending at the activation of a preexisting mutator protein (Fig. 1). Alternatively, the endpoint of a cascade could be a specific transcription factor which then acts in expressing, de novo, a mutator protein. Tumor-promoting agents that do not directly cause DNA damage could activate similar signal cascades starting from another entry. Purposefully, the mutator functions are exemplified by an array of mechanisms (Fig. 1). The mutator functions may operate on any DNA (nontargeted) or, instead, show preferences for DNA sequence or structure.

UV- AND PHORBOL ESTER-INDUCED EXPRESSION OF GENES AS COMPONENTS OF SIGNAL TRANSDUCTION

One type of signal cascade shown in Fig. 1 does indeed exist. Both UV irradiation and phorbol ester treatment of mammalian cells in culture lead to the expression of the same new proteins (30) and RNAs (3). Some of these sequences have been isolated as cDNA clones and shown to code for portions of the human fibroblast collagenase and of metallothionein (1, 3). The collagenase cDNA served to isolate a genomic clone comprising the regulatory and promoter sequences (1). These were systematically mutated, and an 8-nucleotide sequence was identified as sufficient to mediate a phorbol ester response of the collagenase gene or, after insertion, of promoters normally not responsive to phorbol esters (1, 2). The 8-nucleotide sequence is recognized by a transcription factor, AP1, whose binding is a prerequisite of phorbol ester action on the promoter (2). The same factor also appears to receive signals triggered by UV irradiation and signals from the interaction of several growth factors with receptors at the cell surface (B. Stein, J. Heath, and P. Herrlich, unpublished data). The activation of AP1 cannot be blocked by the presence of cycloheximide, suggesting that AP1 acquires new properties by posttranslational modifications. For AP1 these properties include increased binding to the DNA (2; B. Stein, unpublished data) and the ability to stimulate transcriptional initiation.

Subsequently, we and others have detected several other transcription factors which receive signals in response to UV and phorbol ester treatment (reviewed by M. Karin and P. Herrlich, in N. H. Colburn, ed., *Genes and Signal Transduction in Multistage Carcinogenesis*, in press; M. Buescher, unpublished data; Stein, unpublished data). These factors include the serum response factor (32) and a transcription factor which binds the human immunodeficiency virus-1 (HIV-1) promoter (21; Stein, unpublished data; H. Dinter, R. Chiu, M. Imagawa, M. Karin, and K. A. Jones, *EMBO J.*, in press). In addition to UV and phorbol esters, these factors are activated by growth factors. Thus, several transcription factors are at the receiving ends of signal transduction pathways which originate from the binding to the plasma membrane of a phorbol ester tumor promoter, from the absorption of UV, or from the binding to the respective receptor of a growth factor.

By using minimal promoter constructs carrying the AP1 binding site as probes, critical components of the signal transduction pathways to AP1 have recently been identified (A. Schönthal, B. Stein, H. Ponta, and P. Herrlich, unpublished data). These data will be published elsewhere and are only summarized here. Activation of the minimal promoters by all treatments examined absolutely requires interaction with the product of the cellular oncogene c-*fos*, no matter which signal cascade is stimulated. The signal flux initiated by phorbol ester, a growth factor, or UV can be replaced by the elevated expression of either one of a number of oncogenes, including c-*fos*, c-*mos*, v-*src*, c-Ha-*ras*, v-Ki-*ras*, or activated c-Ha-*ras* (EJ). Antisense *fos* constructs block all pathways tested so far, the inductions of minimal promoters by phorbol ester, c-*fos*, c-Ha-*ras*, and c-*mos*. These results suggest that the *fos*-encoded protein serves as the last component in a hierarchy of signal-transmitting oncogene proteins (Schönthal et al., unpublished data). Protein kinase inhibitors block the signal transductions and distinguish at least two types of transductions; a protein kinase C-like enzyme is required for the effects of phorbol ester or of elevated *ras* expression but not for the induction mediated by *mos*. The presumed origins of growth factor and phorbol ester stimulation are at the plasma membrane. Based on comparisons of collagenase and HIV-1 inductions in cells from normal individuals and from a patient with xeroderma pigmentosum and based on absorption spectra of relevant energy at various wavelengths, it is clear that UV is absorbed by DNA and that UV-induced DNA damage is a necessary intermediate in generating the signal to the genes in UV-induced expression. Another intermediate seems to be a protein kinase.

In conclusion, the components of signal cascades, as suggested for the induction of mutator functions, indeed exist. Do they also activate mutators or induce the expression of mutator functions? Mammalian cells seem to possess a physiologically modulated mutator mechanism which is active in certain stages of B-cell development. The mutations appear to be single-base changes. The nature of the physiological targeting of the mutations to the neighborhood of immunoglobulin V genes is not understood. Neither is it known whether the mutator is accessible to nonphysiologic modulation (e.g., by carcinogen or tumor promoters).

A potent mutator mechanism is represented by the activation of retroviral genes such as those of HIV-1. HIV-1 transcription is strongly induced by treatment of cells with phorbol esters or UV (Dinter et al., in press; Stein, unpublished data). Retroviral transcription leads to reintegration of viral genes at other sites. Integration of viral DNA has been shown to be mutagenic both by direct mutagenesis or promoter-enhancer insertion (29, 33) and by the generation of point mutations in adjacent genes (36). Thus, the promoting agents induce indirectly a variety of mutations. Since promoter activity of the new integrate depends on the site (e.g., enhancers nearby), the promoter will often remain active, and several rounds of retroviral transcription and insertional mutagenesis will follow. This is reminiscent of the memory effect which occurs after carcinogen or phorbol ester treatment (7, 13; Maher et al., this volume).

Finally, UV-irradiated cells secrete a factor (EPIF) which causes nonirradiated cells to adapt the whole of the UV response, including the transcription of collagenase (30; M. Kraemer, unpublished data). The cells treated with the factor for an amazingly short time accumulate mutations in several genes (10; Maher et al., this volume). In another cell system presumably the same factor (termed UVIS) (27) enhances targeted mutagenesis (see Dixon et al., this volume). This suggests that the UV response includes the activation of still another mutator function and that this activation is passed on to other cells by an extracellular factor. This mutator introduces point mutations (Maher et al., this volume) and is presumably unrelated to retroviral stimulation. Interestingly, however, the factor also induces the transcription from the HIV-1 promoter (Stein, unpublished data). This raises the possibility that UV-treated human skin cells activate HIV-1 transcription in infected T cells via an extracellular mediator. Mechanisms possibly relevant for tumor promotion may thus also influence latency periods in the development of viral disease.

ACKNOWLEDGMENTS. I am grateful to my co-workers in Karlsruhe, who have made part of this presentation possible, and to W. Sauerbier, with whom I shared many discussions during his sabbatical term in my Karlsruhe laboratory.

LITERATURE CITED

1. Angel, P., I. Baumann, B. Stein, H. Delius, H. J. Rahmsdorf, and P. Herrlich. 1987. 12-O-Tetradecanoyl-phorbol-13-acetate induction of the human collagenase gene is mediated by an inducible enhancer element located in the 5'-flanking region. *Mol. Cell. Biol.* 7:2256–2266.
2. Angel, P., M. Imagawa, R. Chiu, B. Stein, R. J. Imbra, H. J. Rahmsdorf, C. Jonat, P. Herrlich, and M. Karin. 1987. Phorbolester-inducible genes contain a common cis element recognized by a TPA-modulated trans-acting factor. *Cell* 49:729–739.
3. Angel, P., A. Pöting, U. Mallick, H. J. Rahmsdorf, M. Schorpp, and P. Herrlich. 1986. Induc-

tion of metallothionein and other mRNA species by carcinogens and tumor promoters in primary human skin fibroblasts. *Mol. Cell. Biol.* 6:1760–1766.
4. Balmain, A., M. Ramsden, G. T. Bowden, and J. Smith. 1984. Activation of the mouse cellular Harvey-ras gene in chemically induced benign skin papillomas. *Nature* (London) 307:658–660.
5. Barbacid, M. 1987. ras genes. *Annu. Rev. Biochem.* 56:779–827.
6. Bizub, D., A. W. Wood, and A. M. Skalka. 1986. Mutagenesis of the Ha-ras oncogene in mouse skin tumors induced by polycyclic hydrocarbons. *Proc. Natl. Acad. Sci. USA* 83:6048–6052.
7. Fürstenberger, G., B. Sorg, and F. Marks. 1983. Tumor promotion by phorbol esters in skin: evidence for a memory effect. *Science* 220:89–91.
8. Guerrero, I., P. Calzada, A. Mayer, and A. Pellicer. 1984. A molecular approach to leukemogenesis: mouse lymphomas contain an activated c-ras oncogene. *Proc. Natl. Acad. Sci. USA* 81:202–205.
9. Herrlich, P., P. Angel, H. J. Rahmsdorf, U. Mallick, A. Pöting, L. Hieber, C. Lücke-Huhle, and M. Schorpp. 1986. The mammalian genetic stress response. *Adv. Enzyme Regul.* 25:485–504.
10. Herrlich, P., M. Imagawa, V. Maher, K. Sato, J. J. McCormick, P. Angel, M. Karin, I. Baumann, C. Lücke-Huhle, and H. J. Rahmsdorf. 1987. cis- and trans-acting elements responsible for gene induction, p. 95–105. *In* H. zur Hausen and J. R. Schlehofer (ed.), *Accomplishments in Oncology—the Role of DNA Amplification in Carcinogenesis*, vol. 2. J. B. Lippincott Co., Philadelphia.
11. Herrlich, P., U. Mallick, H. Ponta, and H. J. Rahmsdorf. 1984. Genetic changes in mammalian cells reminiscent of an SOS response. *Human Genet.* 67:360–369.
12. Ivanovic, S., and H. Druckrey. 1986. Transplacentare Erzeugung maligner Tumoren des Nervensystems. I. Äthyl-Nitrosoharnstoff (ÄNH) an BDIX Ratten. *Z. Krebsforsch.* 71:320–360.
13. Kennedy, A. R., J. Cairns, and J. B. Little. 1984. Timing of the steps in transformation of C3H 10 T1/2 cells by X-irradiation. *Nature* (London) 307:85–86.
14. Knudson, A. G. 1971. Mutation and cancer: statistical study of retinoblastoma. *Proc. Natl. Acad. Sci. USA* 68:820–823.
15. Knudson, A. G. 1986. Genetics of human cancer. *Annu. Rev. Genet.* 20:231–251.
16. Lenoir, G. M., and G. W. Bornkamm. 1987. Burkitt's lymphoma, a human cancer model for the study of the multistep development of cancer: proposal for a new scenario. *Adv. Viral Oncol.* 7:173–206.
17. Little, J. W., and D. W. Mount. 1982. The SOS regulatory system of *Escherichia coli*. *Cell* 29:11–22.
18. Lombardi, L., E. W. Newcomb, and R. Dalla-Favera. 1987. Pathogenesis of Burkitt lymphoma: expression of an activated c-myc oncogene causes the tumorigenic conversion of EBV-infected human B lymphoblasts. *Cell* 49:161–170.
19. Marks, F., and G. Fürstenberger. 1987. Multistage carcinogenesis—the mouse skin model, p. 18–30. *In* H. zur Hausen and J. R. Schlehofer (ed.), *Accomplishments in Oncology—the Role of DNA Amplification in Carcinogenesis*, vol. 2. J. B. Lippincott Co., Philadelphia.
20. Müller, R., and M. F. Rajewski. 1980. Immunological quantitation by high-affinity antibodies of O^6-ethyldeoxyguanosine in DNA exposed to N-ethyl-N-nitrosourea. *Cancer Res.* 40:887–896.
21. Nabel, G., and D. Baltimore. 1987. An inducible transcription factor activates expression of human immunodeficiency virus in T cells. *Nature* (London) 326:711–714.
22. Radman, M. 1974. Phenomenology of an inducible mutagenic DNA repair pathway in *Escherichia coli*: SOS repair hypothesis, p. 128–242. *In* L. Prakash, F. Sherman, M. W. Miller, C. U. Lawrence, and H. W. Taber (ed.), *Molecular and Environmental Aspects of Mutagenesis*. Charles C Thomas, Publisher, Springfield, Ill.
23. Rajewski, M. F., L. H. Augenlicht, H. Biessmann, R. Goth, D. F. Hülser, O. D. Laerum, and L. Y. Lomakina. 1977. Nervous system-specific carcinogenesis by ethylnitrosourea in the rat: molecular and cellular mechanisms. *Cold Spring Harbor Conf. Cell Proliferation* 4:709–726.
24. Rajewski, M. F., and P. Nehls. 1986. Structural and functional properties of genomic DNA contributing to the non-random formation and repair of carcinogen-DNA adducts. *Pontif. Acad. Sci. Scr. Varia* 70:131–146.
25. Reynolds, S. H., S. J. Stowers, R. M. Patterson, R. R. Maronpot, S. A. Aaronson, and M. W. Anderson. 1987. Activated oncogenes in B6C3F1 mouse liver tumors: implication for risk assessment. *Science* 237:1309–1316.
26. Rossman, T. G., and C. B. Klein. 1985. Mammalian SOS system: a case of misplaced analogies. *Cancer Invest.* 3:175–187.
27. Rotem, N., J. H. Axelrod, and R. Miskin. 1987. Induction of urokinase-type plasminogen activator by UV light in human fetal fibroblasts is mediated through a UV-induced secreted protein. *Mol. Cell. Biol.* 7:622–631.
28. Sarasin, A. 1985. SOS response in mammalian cells. *Cancer Invest.* 3:163–174.
29. Schnieke, A., K. Harbers, and R. Jaenisch. 1983. Embryonic lethal mutation in mice induced by

retrovirus insertion into the α1(I) collagen gene. *Nature* (London) **304**:315–320.
30. **Schorpp, M., U. Mallick, H. J. Rahmsdorf, and P. Herrlich.** 1984. UV-induced extracellular factor from human fibroblasts communicates the UV response to nonirradiated cells. *Cell* **37**:861–868.
31. **Singer, B.** 1979. N-Nitroso-alkylating agents: formation and persistence of alkyl derivatives in mammalian nucleic acids as contributing factors in carcinogenesis. *J. Natl. Cancer Inst.* **62**:1329–1339.
32. **Treisman, R.** 1987. Identification and purification of a polypeptide that binds to the c-fos serum responsive element. *EMBO J.* **6**:2711–2717.
33. **Varmus, H. E., and R. Swanstrom.** 1982. Replication of retroviruses, p. 369–512. *In* R. Weiss, N. Teich, H. E. Varmus, and J. Coffin (ed.), *RNA Tumor Viruses*. Cold Spring Harbor Laboratory, Cold Spring Harbor, N.Y.
34. **Walker, G. C.** 1984. Ultraviolet mutagenesis and inducible DNA repair in *Escherichia coli*. *Microbiol. Rev.* **48**:60–93.
35. **Walker, G. C.** 1985. Inducible DNA repair systems. *Annu. Rev. Biochem.* **54**:425–457.
36. **Westaway, D., G. Payne, and H. E. Varmus.** 1984. Proviral deletions and oncogene base-substitutions in insertionally mutagenized c-myc alleles may contribute to the progression of avian bursal tumors. *Proc. Natl. Acad. Sci. USA* **81**:843–847.
37. **Yamamoto, T., I. Nishimori, E. Tahara, and I. Sekine.** 1986. Malignant tumors in atomic bomb survivors with special reference to the pathology of stomach and lung cancer. *GANN Monogr. Cancer Res.* **32**:143–154.
38. **Yuspa, S. H.** 1987. Tumor promotion, p. 169–180. *In* J. G. Fortner and J. E. Rhoads (ed.), *Accomplishments in Cancer Research, 1986*. J. B. Lippincott Co., Philadelphia.

Chapter 51

Evidence of Inducible Error-Prone Mechanisms in Diploid Human Fibroblasts

Veronica M. Maher, Kenji Sato, Suzanne Kateley-Kohler, Harvey Thomas, Sonya Michaud, J. Justin McCormick, Marcus Kraemer, Hans J. Rahmsdorf, and Peter Herrlich

In bacteria, exposure to agents that damage DNA or interfere with DNA replication gives rise to the expression of a series of coordinately regulated genes whose products are involved in a whole host of responses (31, 33), collectively referred to as the "SOS response," named by Radman (23) after the international distress signal. This inducible response in bacteria controls the expression of genes involved in excision repair, daughter-strand gap repair, *trans*-lesion synthesis, etc., and the phenomena caused by these gene products include increased cell survival, reactivation of damaged viruses, and the production of mutations. In fact, in bacteria the induction of mutations by some agents is completely dependent upon the SOS response (33).

Over the past several years, a number of investigators have provided evidence that damage-inducible genes also exist in mammalian cells (2, 10–13, 19–22, 24, 29). For example, exposure of diploid human fibroblasts to DNA-damaging agents, such as UV radiation, induces the synthesis of a series of proteins over the next 24 to 48 h (29). Some of these DNA damage-induced proteins are secreted into the surrounding medium, and if the conditioned medium containing these secreted proteins is transferred to untreated cells, synthesis of the same series of new proteins is induced immediately and proteins are also secreted by these cells. The secreted protein or proteins responsible for this induction were named the extracellular protein-inducing factor (EPIF) (29). An ap-

---

Veronica M. Maher, Kenji Sato, Suzanne Kateley-Kohler, Harvey Thomas, Sonya Michaud, and J. Justin McCormick • Carcinogenesis Laboratory, Department of Microbiology, and Department of Biochemistry, Michigan State University, East Lansing, Michigan 48824. *Marcus Kraemer, Hans J. Rahmsdorf, and Peter Herrlich* • Kernforschungszentrum Karlsruhe, Institut für Genetik und für Toxikologie, and Institut für Genetik der Universität Karlsruhe, D-7500 Karlsruhe, Federal Republic of Germany.

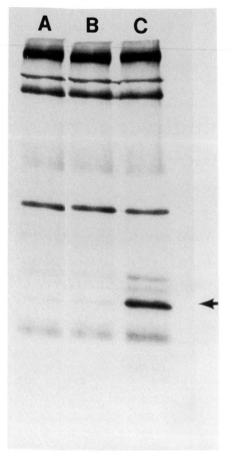

FIGURE 1. Induction of proteins by exposure to EPIF medium. The medium on cultures of human fibroblasts was replaced with regular culture medium (lane A), conditioned medium taken from treated cells after 48 h (lane B), or EPIF medium taken from cells irradiated with 30 J/m^2 48 h previously (lane C). After 7 h, the cells were pulse-labeled with [^{35}S]methionine and the labeled proteins that were secreted into the medium were analyzed by sodium dodecyl sulfate-gel electrophoresis and autoradiography. The protein indicated by the arrow is collagenase.

parently similar conditioned medium is called UV-induced secreted protein (UVIS) by others (24).

Figure 1 shows a sodium dodecyl sulfate-gel electrophoretic analysis of proteins labeled with a pulse of [^{35}S]methionine and secreted by cells into the medium 6 to 8 h after the cells had been given conditioned medium containing EPIF, i.e., medium taken from human cells that had been irradiated with 30 J of UV per m^2 48 h earlier. As compared with ^{35}S-labeled protein in the medium taken from cells that received regular culture medium (Fig. 1, lane A) or control medium, i.e., conditioned medium taken from unirradiated cells after 48 h (lane B), there are at least two additional bands of newly synthesized protein. The darkest band in the right lane represents collagenase. Treatment of human cells with other agents such as interleukin-1 and tumor necrosis factor also induces the synthesis of collagenase, along with other proteins, but the response of human cells to EPIF in conditioned medium taken from cells that were UV irradiated 48 h previously is distinct from their response to these other agents (data not shown).

EPIF was maximally induced in DNA excision-repair-deficient xeroderma pigmentosum cells (group A) at a dose of 2 J/m^2, whereas the same level of EPIF induction in repair-proficient normal cells required 30 J/m^2 (M. Kraemer, H. J. Rahmsdorf, and P. Herrlich, unpublished studies). This difference in the amount of radiation needed to generate EPIF resembles the experimental conditions needed to induce the synthesis of several other UV-induced proteins, including collagenase (referred to in reference 29 as XHF1; see Fig. 2 of that reference). A major difference between the two human cell populations is the inability of the xeroderma pigmentosum cells to remove UV-induced damage from their DNA and the rapid excision repair of such damage by the normal cells. Therefore, the observed difference in dose of UV radiation required to induce these proteins in repair-proficient and repair-deficient human cells strongly suggests that the induction of the proteins is triggered by DNA damage remaining unexcised in the cellular DNA over a period of time following the initial exposure to the damaging agent.

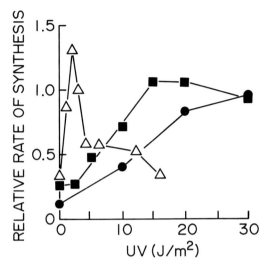

FIGURE 2. Protein synthesis induced as a function of dose of UV radiation. Human fibroblasts were irradiated with the indicated doses of UV radiation, and after 27 h the cells were pulse-labeled with [^{35}S]methionine for 30 min. The proteins were resolved by two-dimensional gel electrophoresis, and autoradiograms were evaluated by microdensitometer. The radioactivity in the individual protein spot corresponding to collagenase was determined by scintillation spectrometry, normalized against a standard protein, and plotted as relative rate of synthesis. Symbols: △, XP12BE cells from complementation group A; ■, XP6DU cells from a xeroderma pigmentosum variant patient; ●, normal human cells. The data have been taken from reference 29, with permission.

FREQUENCY OF MUTANTS INDUCED BY EPIF IN ENDOGENOUS GENES IN HUMAN FIBROBLASTS

In experiments patterned after those conducted with bacteria by many investigators, beginning with Weigle (32), several groups have demonstrated that if damaged viruses are introduced into host mammalian cells that have themselves been treated with DNA-damaging agents, such as UV, 24 to 48 h previously, the survival of the virus is higher than if the host cells were not pretreated (enhanced reactivation) (1, 3, 5-8, 15, 16, 25-27, 30; see also the references in the review in reference 7). An increase in the frequency of mutants (enhanced mutagenesis) was also observed when undamaged or damaged virus was allowed to replicate in pretreated rather than in untreated host cells (1, 3, 5-8, 15, 16, 25-27, 30). In addition to using mammalian cell viruses as probes for mutagenesis (6), Sarkar et al. examined the mutagenic effect of pretreating host cells with the DNA-damaging agent ethyl methanesulfonate (EMS) and introducing undamaged or UV-irradiated plasmids, i.e., shuttle vectors, capable of replicating in mammalian cells as well as in bacteria (28). These investigators found a significant increase in mutations when the plasmid replicated in a monkey kidney cell line that had been exposed to EMS 48 h before being transfected with plasmid, as compared with the frequency seen with nonpretreated host cells (28). The results of the cited studies with viral or plasmid probes are consistent with the hypothesis that pretreatment of host cells induces the production of some gene product, such as a modified DNA polymerase, which allows the cell to replicate damaged DNA more efficiently, resulting in enhanced survival (reactivation), but at the cost of a decrease in fidelity (mutagenesis). In addition, the mechanism can introduce errors into nondamaged DNA.

In the experiments referred to above, the DNA of the host cells was damaged to induce an SOS response, and the target genes for mutation induction were extrachromosomal. Enhanced mutagenesis in the genome of the mammalian host cells themselves by an induced mutator factor might be detectable if the induced response can be separated from the direct attack on DNA by UV radiation or a chemical mutagen. To carry out such a test, diploid human fibroblasts were treated with EPIF medium to induce the series of new proteins, including a putative mutator gene product, and were assayed to determine whether they exhibited a higher frequency of mutations than cells exposed to control medium. For these experiments, we used resistance to 6-thioguanine (TG) as one genetic marker and resistance to diphtheria toxin (DT) as the other. Resistance to TG, a forward mutational assay, results from any kind of alteration (base substitutions, frameshifts, deletions, insertions, etc.) that eliminates the function of the enzyme hypoxanthine(guanine)phosphoribosyltransferase. Resistance to DT, on the other hand, selects for cells that contain only minor alterations in the structure of the elongation factor 2 gene, i.e., a change that prevents elongation factor 2 from interacting with DT, but still permits the factor to function in protein synthesis. Cells that receive major alterations, deletions, etc., in the elongation factor 2 gene cannot survive selection in DT (9).

Foreskin-derived normal fibroblasts were irradiated with 30 J of UV per m^2 or sham irradiated (control). After 48 h, the EPIF medium and the control medium from these cells were harvested, freed of cellular debris by centrifugation, and transferred to two sets of target cells. After an 8-h exposure, these conditioned media were removed, the dishes were rinsed, and the cells from one dish of each set were assayed for survival. Cells (1 × 10^7 to 2 × 10^7) from each set were harvested and assayed for induction of collagenase mRNA as a marker for EPIF. The remainder of the cells (2 × 10^6 to 3 × 10^6) were allowed to undergo expression of DT or TG resistance or both (9, 18).

The results of a series of experiments showed that exposure to EPIF medium caused a significant increase in the frequency of mutations, ranging from 2.1- to 4.0-fold for TG resistance in four experiments and 1.5- to 3.1-fold for DT resistance in two experiments. Exposure to EPIF medium did not affect cell survival. The increase in frequency of TG resistance correlated with the increase in collagenase mRNA. However, we cannot tell whether the same protein factor is responsible for both phenomena until we have achieved its complete purification.

We also tested the effect of EPIF on cells exposed to DNA-damaging agents. Exposure of human fibroblasts to EPIF medium for 8 h before irradiation with a low dose of UV (5 J/m^2) did not result in a higher frequency of mutations than that seen in cells pretreated for 8 h with control medium and then irradiated with the same dose of UV. As expected, the unirradiated target cells exposed to EPIF medium exhibited a two- to fourfold increase above the background, as described above, but the frequency of mutations to TG resistance was the same in both of the populations irradiated with 5 J/m^2, i.e., 70 × 10^6 to 80 × 10^{-6}, a frequency similar to what we routinely observe in normal human fibroblasts irradiated with 5 J of UV per m^2 (18). One explanation for the lack of an additive effect of EPIF is that exposure of the target cells to this dose of UV itself induces an SOS response, and, therefore, the mutant frequency which correlates with this dose reflects the DNA changes resulting from the presence of the DNA photoproducts plus the effect of the error-prone factor.

FREQUENCY OF MUTANTS INDUCED IN AN EXOGENOUS GENE WHEN A PLASMID REPLICATES IN EPIF-TREATED HUMAN CELLS

To determine whether exposure of human cells to EPIF medium would also cause an increase in the frequency of mutations induced in a gene replicating extrachromosomally, we made use of a shuttle vector, pZ189, carrying the *supF* gene as the target for mutagenesis, and the established human cell line 293, which supports efficient replication of the plasmid without resulting in a high background of *supF* mutants (34). The *supF* gene, which codes for the tyrosine suppressor tRNA, is exceptionally responsive to mutations since a change in any of at least 63 of the 85 nucleotides making up the tRNA structure results in a detectable phenotypic change. Therefore, use of the *supF* gene provides a forward mutation assay with almost no silent mutations. The small size of the *supF* gene makes it ideal for sequencing. Therefore, if we found a significant increase in mutant frequency above the background of 10^{-4} after the plasmid replicated in cells treated with EPIF medium, we could analyze these mutants for the types of mutations, using agarose gel electrophoresis, and sequence the *supF* gene in those plasmids exhibiting no major structural changes and, thus, presumed to contain point mutations.

For this study, we introduced untreated plasmids into the human cells by transfection 6 h before treating the host cells with EPIF medium or control medium. The plasmids were allowed to replicate in the host cells for 40 h, and progeny plasmids were then isolated and assayed for mutations as described (34). The results showed a twofold increase in the frequency of the mutations. The background frequency in the progeny plasmids from the cells that received control medium was 0.95×10^{-4}; in those from cells exposed to EPIF medium, the frequency was 1.84×10^{-4}. Analysis of the mutant plasmids for altered gel mobility, which is evidence of gross rearrangements, deletions, or insertions, showed only a slight increase over what we find for control plasmids (34); i.e., 20% of the mutants from the control and 27% of the plasmids that had replicated in EPIF-treated human cells showed such rearrangements. The remaining mutants in each set are presumably point mutations or have alterations of a size less than 150 bp. We have not sequenced the *supF* genes in these mutants because the increase over background was only twofold. Further experiments are in progress.

INCREASE IN THE RATE OF UNTARGETED MUTATIONS INDUCED IN HUMAN CELLS BY DNA-DAMAGING AGENTS

These mutation studies employing EPIF medium suggest that human fibroblasts can be brought to a state in which they have an increased probability of forming mutations in chromosomal or extrachromosomal genes. They further suggest that this state can be induced by exposure to UV or to a UV-induced factor. Another approach to showing that such an error-prone state can be induced in human cells by DNA-damaging agents is by using fluctuation analysis to measure increases in the mutation rate in such cells, in addition to the targeted direct mutagenesis observed. We have carried out such studies using ethylnitrosourea (ENU) as the mutagen.

Diploid human fibroblasts were exposed for 1 h to 2.3 mM ENU, a concentration that we had determined would decrease cell survival to 50% of that of the untreated control and increase the frequency of TG-resistant cells to approximately 450×10^{-6} from a background frequency of 20×10^{-6} to 30×10^{-6} (17). ENU was chosen because it has a short half-life, because we knew from our unpublished studies that human fibroblasts can remove the potentially mutagenic lesions induced by ENU, and because it was used for similar studies in an animal cell line by J. W. I. M. Simons (State University of Leiden, Leiden, The Netherlands) who suggested the experimental approach to us in view of his own results.

After treatment with ENU, a portion of the cells were plated at cloning densities to assay survival as percentage of the untreated control. The rest were given fresh culture medium and allowed to replicate and introduce ENU-induced targeted mutations into their DNA. After 2 days, the cells in the untreated control and those in the treated population were pooled, and 10^6 cells from each group were diluted and plated into five 150-mm-diameter dishes to continue expression of TG resistance before being assayed for mutations in the usual manner (17, 18). The remaining cells were used for fluctuation tests (14) to determine the rate of mutations, by plating a small number of cells into each of a series of wells and allowing them to undergo 15 to 16 population doublings before assaying each subpopulation for the frequency of TG-resistant cells. A total of 24 or more wells were used for the treated and untreated population, and the progeny cells were transferred into 24 large dishes when the wells were full. When the number of cells reached 2×10^6 to 3×10^6, 0.5×10^6 cells from each dish were assayed for mutants. Progeny of ENU-treated cells which exhibited high frequencies ("jackpots") were considered to reflect direct mutagenesis and were excluded. The mutation rate was then estimated from the data in the remaining dishes, using the formula of Capizzi and Jameson (4).

These studies are still in progress, but the results to date indicate a background mutation rate of 0.5×10^{-6} to 1.0×10^{-6} per cell generation and an experimental mutation rate of 10×10^{-6} to 20×10^{-6} per cell generation. We interpret the increased mutation rate observed in the progeny of cells that were damaged by ENU as resulting from a change in the state of the cells so that they are more error-prone than the control cells. This is because we consider it unlikely that the mutations that arose 10 or more population doublings after exposure to ENU represent direct mutagenesis by that agent.

CONCLUSION

From our studies with diploid human fibroblasts given EPIF medium and our studies with the shuttle vector replicating in a human cell line that was treated with EPIF medium, we conclude that exposure to this factor can cause mutations in endogenous genes and in extrachromosomal target genes. The data with the shuttle vector are preliminary, but additional studies are being carried out. From the fluctuation test studies with diploid human fibroblasts exposed to ENU, we conclude that this agent not only induces targeted direct mutations, but also alters the state of the cell so as to increase the frequency of mutations in cells by increasing the rate of making errors. Taken together, the data support the hypothesis that an inducible error-prone mechanism exists in human cells.

ACKNOWLEDGMENTS. We thank Dennis Fry for his advice and assistance on the detection of collagenase mRNA. The excellent technical assistance of Bernard Schroeter is gratefully acknowledged.

This research was supported in part by Public Health Service grant CA21253 from the National Cancer Institute, by a grant from the Elsa U. Pardee Foundation, and by the Deutsche Forschungsgemeinschaft.

LITERATURE CITED

1. **Abrahams, P. J., B. A. Huitema, and A. J. van der Eb.** 1984. Enhanced reactivation and enhanced mutagenesis of herpes simplex virus in normal human and xeroderma pigmentosum cells. *Mol. Cell. Biol.* 4:2341–2346.
2. **Angel, P., A. Poting, U. Mallick, H. J. Rahmsdorf, M. Schorpp, and P. Herrlich.** 1986. Induction of metallothionein and other mRNA species by carcinogens and tumor promoters in primary human skin fibroblasts. *Mol. Cell. Biol.* 6:1760–1766.
3. **Bourre, F., and A. Sarasin.** 1983. Targeted mutagenesis of SV40 DNA by UV light. *Nature* (London) 305:68–70.
4. **Capizzi, R. L., and J. W. Jameson.** 1973. A table for the estimation of the spontaneous mutation rate of cells in culture. *Mutat. Res.* 17:147–148.
5. **Cornelis, J. J., Z. Z. Su, and J. Rommelaere.** 1982. Direct and indirect effects of ultraviolet light on the mutagenesis of parvovirus H-1 in human cells. *EMBO J.* 1:693–699.
6. **DasGupta, U. B., and W. C. Summers.** 1978. Ultraviolet reactivation of herpes simplex virus is mutagenesis and inducible in mammalian cells. *Proc. Natl. Acad. Sci. USA* 75:2378–2381.
7. **Defais, M. J., P. C. Hanawalt, and A. Sarasin.** 1983. Viral probes for DNA repair processes. *Adv. Radiat. Biol.* 10:1–37.
8. **Dinsart, C., J. J. Cornelis, M. Decaesstecker, J. van der Lubbe, A. J. van der Eb, and J. Rommelaere.** 1985. Differential effect of ultraviolet light on the induction of simian virus 40 and a cellular mutator phenotype in transformed mammalian cells. *Mutat. Res.* 151:9–14.

9. Drinkwater, N. R., R. C. Corner, J. J. McCormick, and V. M. Maher. 1982. An *in situ* assay for induced diphtheria toxin-resistant mutants of diploid human fibroblasts. *Mutat. Res.* 106:277–289.
10. Hagedorn, R., H. W. Thielmann, and H. Fischer. 1985. SOS-type functions in mammalian cells. *J. Can. Res. Clin. Oncol.* 109:89–92.
11. Herrlich, P., M. Imagawa, V. Maher, K. Sato, J. J. McCormick, P. Angel, M. Karin, I. Baumann, C. Luecke-Huhle, and H. J. Rahmsdorf. 1987. The molecular basis for the UV response, *cis*- and *trans*-acting elements responsible for gene induction, p. 95–105. *In* H. zur Hausen and J. R. Schlehofer (ed.), *Accomplishments in Oncology: the Role of DNA Amplification in Carcinogenesis.* J. B. Lippincott Co., Philadelphia.
12. Herrlich, P., U. Mallick, H. Ponta, and H. J. Rahmsdorf. 1984. Genetic changes in mammalian cells reminiscent of an SOS response. *Hum. Genet.* 67:360–368.
13. Lieberman, M. W., L. R. Beach, and R. D. Palmiter. 1983. Ultraviolet radiation induced metallothionein-I gene activation is associated with extensive DNA demethylation. *Cell* 35:207–213.
14. Luria, S. E., and M. Delbruck. 1943. Mutations of bacteria from virus sensitivity to virus resistance. *Genetics* 28:491–511.
15. Lytle, C. D., J. G. Goddard, and C. H. Lin. 1980. Repair and mutagenesis of herpes simplex virus in UV-irradiated monkey cells. *Mutat. Res.* 70:139–149.
16. Lytle, C. D., and D. C. Knott. 1982. Enhanced mutagenesis parallels enhanced reactivation of herpes virus in a human cell line. *EMBO J.* 1:701–703.
17. Maher, V. M., J. Domoradzki, R. C. Corner, and J. J. McCormick. 1986. Correlation between O^6-alkylguanine-DNA-alkyltransferase activity and resistance of human cells to the cytotoxic and mutagenic effects of methylating and ethylating agents, p. 411–418. *In* C. Harris (ed.), *Biochemical and Molecular Epidemiology of Cancer.* Alan R. Liss, Inc., New York.
18. Maher, V. M., D. J. Dorney, A. L. Mendrala, B. Konze-Thomas, and J. J. McCormick. 1979. DNA excision repair processes in human cells can eliminate the cytotoxic and mutagenetic consequences of ultraviolet irradiation. *Mutat. Res.* 62:311–323.
19. Mallick, U., H. J. Rahmsdorf, N. Yamamoto, H. Ponta, R.-D. Wegner, and P. Herrlich. 1982. 12-O-tetradecanoylphorbol 13-acetate-inducible proteins are synthesized at an increased rate in Bloom syndrome fibroblasts. *Proc. Natl. Acad. Sci. USA* 79:7886–7890.
20. Mezzina, M., S. Nocentini, and A. Sarasin. 1982. DNA ligase activity in carcinogen-treated human fibroblasts. *Biochimie* 64:743–748.
21. Miskin, R., and R. Ben-Ishai. 1981. Induction of plasminogen activator by UV light in normal and xeroderma pigmentosum fibrobasts. *Proc. Natl. Acad. Sci. USA* 78:6236–6240.
22. Miskin, R., and E. Reich. 1980. Plasminogen activator: induction of synthesis by DNA damage. *Cell* 19:217–224.
23. Radman, M. 1974. Phenomenology of an inducible mutagenic DNA repair pathway in E. coli: SOS repair hypothesis, p. 129–142. *In* L. Prakas, F. Sherman, M. Miller, C. Lawrence, and H. W. Tabor (ed.), *Molecular and Environmental Aspects of Mutagenesis.* Charles C Thomas, Publisher, Springfield, Ill.
24. Rotem, N., J. H. Axelrod, and R. Miskin. 1987. Induction of urokinase-type plasminogen activator by UV light in human fetal fibroblasts is mediated through a UV-induced secreted protein. *Mol. Cell. Biol.* 7:622–631.
25. Sarasin, A. 1985. SOS response in mammalian cells. *Cancer Invest.* 3:163–174.
26. Sarasin, A., and A. Benoit. 1980. Induction of an error-prone mode of DNA repair in UV-irradiated monkey kidney cells. *Mutat. Res.* 70:71–81.
27. Sarasin, A., F. Bourre, and A. Benoit. 1982. Error-prone replication of ultraviolet irradiated simian virus 40 in carcinogen-treated monkey kidney cells. *Biochimie* 64:815–821.
28. Sarkar, S., U. B. Dasgupta, and W. C. Summers. 1984. Error-prone mutagenesis detected in mammalian cells by a shuttle vector containing the *supF* gene of Escherichia coli. *Mol. Cell. Biol.* 4:2227–2230.
29. Schorpp, M., U. Mallick, H. J. Rahmsdorf, and P. Herrlich. 1984. UV-induced extracellular factor from human fibroblasts communicates the UV response to nonirradiated cells. *Cell* 37:861–868.
30. Su, Z. Z., B. Avalosse, J.-M. Vos, J. J. Cornelis, and J. Rommelaere. 1985. Mutagenesis at putative apurinic sites in alkylated single-stranded DNA of parvovirus H-1 propagated in human cells. *Mutat. Res.* 149:1–8.
31. Walker, G. 1984. Mutagenesis and inducible responses to deoxyribonucleic acid damage in *Escherichia coli. Microbiol. Rev.* 48:60–93.
32. Weigle, J. J. 1953. Induction of mutation in a bacterial virus. *Proc. Natl. Acad. Sci. USA* 39:628–636.
33. Witkin, E. M. 1976. Ultraviolet mutagenesis and inducible DNA repair in *Escherichia coli. Bacteriol. Rev.* 40:869–907.
34. Yang, J.-L., V. M. Maher, and J. J. McCormick. 1987. Kinds of mutations formed when a shuttle vector containing adducts of (+)-7,8-dihydroxy-9,10-epoxy-7,8,9,10-tetrahydroxybenzo[a]pyrene replicates in human cells. *Proc. Natl. Acad. Sci. USA* 84:3787–3791.

Chapter 52

Analysis of Induced Mutagenesis in Mammalian Cells, Using a Simian Virus 40-Based Shuttle Vector

Kathleen Dixon, Emmanuel Roilides, Ruth Miskin, and Arthur S. Levine

When bacterial cells are treated with DNA-damaging agents such as UV radiation or mitomycin C, a battery of genes under the control of *recA* and *lexA* are induced (reviewed in references 2, 6, and 17). The induction of this regulatory pathway, referred to as the SOS pathway or the SOS response, leads to enhanced excision repair, enhanced mutagenesis, enhanced recombination, and enhanced survival. It has been suggested that at least some elements of this pathway might be conserved in mammalian cells (12). Evidence for an SOS-like pathway in mammalian cells comes primarily from studies on viral reactivation (reviewed in references 2 and 12). When mammalian cells are pretreated with a DNA-damaging agent before infection with UV-irradiated virus, virus survival is enhanced over that observed in untreated cells. Although this enhancement of virus survival may be accompanied by an enhancement of mutagenesis, this correlation has not been observed in all systems (1, 10, 12, 13, 16). It is not known whether induction of the SOS-like pathway in mammalian cells is mediated by a regulatory system similar to that described for bacterial cells. However, it has been demonstrated that a number of cellular genes are induced following treatment of mammalian cells with carcinogens (11, 14). Furthermore, this treatment results in the secretion of a factor(s) that can mimic DNA-damaging agents in inducing certain aspects of the SOS-like response in other cells (11, 14). Despite this suggestive evidence, the role that gene induction may play in enhanced mammalian mutagenesis has not been established. Moreover, no information has been obtained about the molecular mechanisms involved in either enhanced virus reactivation or enhanced mutagenesis in mammalian cells.

THE pZ189 MUTAGENESIS TEST SYSTEM

To gain a better understanding of enhanced mutagenesis in mammalian cells, we

---

Kathleen Dixon, Emmanuel Roilides, and Arthur S. Levine • Section on Viruses and Cellular Biology, National Institute of Child Health and Human Development, Bethesda, Maryland 20892. *Ruth Miskin* • Biochemistry Department, Weizmann Institute of Science, Rehovot 76100, Israel.

have used the simian virus 40-based pZ189 shuttle vector (15; Fig. 1) to analyze the frequency and sequence characteristics of mutants induced when UV-damaged or undamaged templates replicate in carcinogen-treated monkey cells. The pZ189 vector contains the early region of simian virus 40, the pBR327 replication functions and beta-lactamase gene, and the bacterial *supF* tRNA suppressor gene. When pZ189 is introduced into monkey cells by transfection, it replicates in the nucleus like simian virus 40 DNA. Mutations that occur in the *supF* gene can be identified by subsequent transformation of bacterial cells with the replicated vector DNA. The *supF* region of each mutant vector clone can then be sequenced to determine the characteristics of the mutational changes that occur during replication of the vector in the monkey cells. By comparing the frequency and sequence characteristics of mutants generated when either carcinogen-treated or untreated monkey cells are transfected with UV-irradiated or unirradiated pZ189, we can determine whether pretreatment of the cells causes a change in the pattern and therefore the mechanism of mutagenesis. Furthermore, by using conditioned medium from carcinogen-treated cells to induce a response in untreated cells, we can determine whether the factor(s) secreted by damaged cells can induce an enhancement of mutagenesis in undamaged cells and, if so, what the mechanism of this enhancement might be.

TABLE 1
Effect of Mitomycin C Pretreatment of Cells on the Frequency of Spontaneous and UV-Induced *supF* Mutants[a]

UV fluence to vector (J/m^2)	Mutant frequency (%) after mitomycin C treatment:	
	Mock	1 μg/ml
0[b]	0.032	0.014
200[c]	0.29	0.57
500	1.18	1.82

[a] Results compiled from E. Roilides, P. J. Munson, A. S. Levine, and K. Dixon, submitted for publication.
[b] Average of two experiments.
[c] Average of four experiments.

ENHANCEMENT OF TARGETED MUTAGENESIS IN CARCINOGEN-TREATED CELLS

When monkey cells were treated with mitomycin C 24 h before transfection with undamaged pZ189, there appeared to be no increase in the frequency of *supF* mutants generated during replication of the vector in the treated cells (Table 1). In fact, it appeared that there might be some decrease in the level of the spontaneous background. We postulated previously (3) that this spontaneous background might be due to error-prone repair of vector DNA, damaged by nucleases during the transfection process. If so, the reduction in the spontaneous background may indicate that the input DNA (i) is less damaged in treated cells, (ii) is less likely to be repaired and to replicate, or (iii) is repaired by a more error-free mechanism. Interestingly, when pretreated cells were infected with UV-damaged vector DNA, a twofold increase in mutation frequency occurred (Table 1). These results clearly indicate that mutagenesis (presumably targeted to damaged sites) is enhanced by carcinogen treatment of mammalian cells. Furthermore, the results suggest that the mechanism of enhancement involves a change in the magnitude of the cell's response to damaged templates, leading to additional targeted mutations, rather than an overall reduction in replication fidelity leading to untargeted mutations.

ENHANCEMENT OF MUTAGENESIS IN CELLS TREATED WITH CONDITIONED MEDIUM FROM OTHER CARCINOGEN-TREATED CELLS

In untreated cells incubated in the presence of conditioned medium from UV-

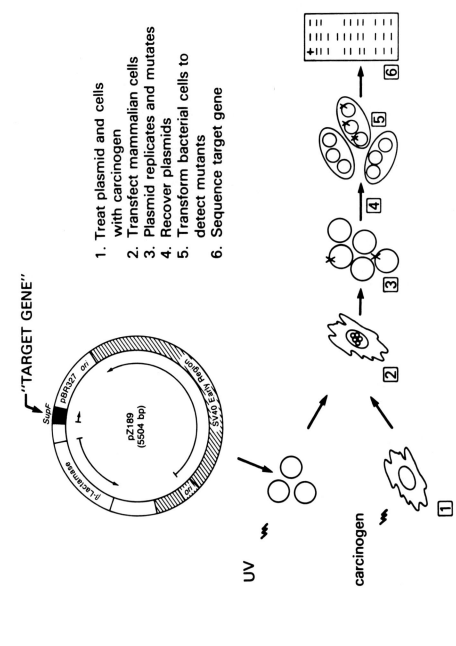

FIGURE 1. The pZ189 vector and schematic representation of the experimental procedure for studying enhanced mutagenesis. The pZ189 vector (15) is treated with UV radiation (254 nm) in vitro and then introduced into TC7 cells (derived from CV-1 African green monkey kidney cells) by DEAE dextran-facilitated transfection as described previously (4). For carcinogen pretreatment, the TC7 cells are incubated in the presence of mitomycin C (usually 1 µg/ml) for 2.5 h at 24 h before transfection. Vector DNA is recovered from the monkey cells 48 h after transfection, purified, and introduced into *Escherichia coli* MBM7070 as described previously (3, 4). Transformed colonies that contain pZ189 *supF* mutants are white or light blue on the appropriate indicator plates, and colonies containing pZ189 *supF*+ are blue. Mutant frequencies are calculated as white plus light blue colonies per total colonies. Independent *supF* mutants are selected and the DNA sequence is determined as described previously (4).

TABLE 2
Enhancement of Mutagenesis with Conditioned Medium from UV-Treated Cells[a]

UV fluence to vector (J/m^2)	Mutant frequency (%) with conditioned medium:		
	Control medium	Conditioned medium without UV	Conditioned medium plus UV
None	0.014	0.009	0.026
200	0.23	0.25	0.43

[a] Conditioned medium, collected 60 h after treatment of monkey cells with 17 J of UV radiation per m^2, was added to fresh monkey cells 8 h before transfection with pZ189.

treated cells, the frequency of *supF* mutants generated during replication of UV-irradiated pZ189 was also increased about twofold (Table 2). This result indicates that the enhancement of mutagenesis which accompanies the induction of an SOS-like response in mammalian cells can be mediated by a secreted factor(s). Presently, we do not know whether this effect of conditioned medium on mutagenesis is related to the UV-induced, secreted factor (UVIS) (11) or the extracellular protein synthesis-inducing factor (EPIF) (14), secreted factors which have been shown to be involved in induction of certain genes by DNA-damaging agents. We have observed a slight elevation of mutagenesis with undamaged vector in cells treated with conditioned medium from irradiated cells, but this finding is of uncertain significance.

ENHANCEMENT OF MUTAGENESIS IN CARCINOGEN-TREATED XERODERMA PIGMENTOSUM CELLS

To determine whether enhanced mutagenesis is mediated by an excision repair process, we determined the capacity of repair-deficient xeroderma pigmentosum (group A) cells to exhibit enhanced mutagenesis in response to carcinogen treatment. Since an enhancement of mutagenesis was observed in pretreated cells (Table 3) and we have not observed an enhancement in excision repair in these cells under similar conditions (M Protić-Sabljić, E. Roilides, A. S. Levine, and K. Dixon, *Somatic Cell Mol. Genet.*, in press), we conclude that the mechanism of mutagenesis enhancement probably does not involve excision repair.

CHARACTERISTICS OF UV-INDUCED MUTATIONS GENERATED IN CARCINOGEN-TREATED CELLS

To gain a better understanding of the mechanism of mutagenesis enhancement in carcinogen-treated cells, we determined the DNA sequence of the *supF* gene in more than 100 independent *supF* mutants generated during replication of pZ189 in mitomycin C-pretreated cells. Although the positions of these mutations within the *supF* gene did not differ dramatically from the

TABLE 3
Effect of Mitomycin C Pretreatment on Frequency of UV-Induced Mutants in Xeroderma Pigmentosum Cells (XP12BeSV40)[a]

UV fluence to vector (J/m^2)	Mutant frequency (%) after treatment:	
	Mock	Mitomycin C[b]
25	0.11	0.20
50[c]	0.19	0.39

[a] Results compiled from Roilides et al., submitted.
[b] Cells were treated with 0.5 µg of mitomycin C per ml for 2.5 h at 24 h before transfection.
[c] Average of two experiments.

TABLE 4
Comparison of UV-Induced Mutations in Mitomycin C-Pretreated Cells with
Spontaneous and UV-Induced Mutations in Untreated Cells

Type of mutation	% of mutations induced by:		
	Spontaneous[a]	Control, UV[b]	Pretreated, UV[c]
Single-base changes	26	56	71
Tandem doubles	4	17	8
Multiples	48	24	9
Small deletions	22	3	11

[a] A total of 23 mutants were sequenced (4).
[b] A total of 96 mutants were sequenced (4).
[c] Results compiled from Roilides et al., submitted.

positions of UV-induced mutations generated in untreated cells (3, 4), there were some differences that deserve comment. First, there appeared to be fewer multiple mutations in the collection from pretreated cells (Table 4). Second, when the base changes that occurred in only single and tandem double mutants were compared, it appears that there were relatively fewer G·C→A·T transitions and relatively more G·C→T·A transversions in the collection from pretreated cells (Roilides et al., submitted). These two observations are discussed below in the context of proposed mechanisms for enhanced mutagenesis.

POSSIBLE MECHANISMS OF ENHANCED MUTAGENESIS

The high level of G·C→A·T transitions in untreated cells has been attributed to the "A insertion rule" (9); the DNA polymerase appears to preferentially insert A opposite noncoding lesions (in this case, UV photoproducts) in the template DNA. Thus G·C→A·T transitions arise from insertion of A opposite C in pyrimidine dimers; insertion of A opposite T in pyrimidine dimers does not result in a mutation. It is possible that the increased mutagenesis observed in pretreated cells results from a relaxation of the A insertion rule such that T is more frequently inserted. This would account for the rise in G·C→T·A transversions. This relaxation might be accompanied by an increase in the frequency with which the DNA polymerase can carry out transdimer synthesis, leading to enhanced mutagenesis.

We see no evidence for a more generalized reduction in the fidelity of DNA synthesis on either damaged or undamaged templates. In untreated cells, there is a relatively high proportion of UV-induced multiple mutations, which appear to arise from a localized loss of fidelity in DNA synthesis in the vicinity of UV photoproducts. In pretreated cells, the proportion of multiple UV-induced mutations is reduced, suggesting that there may actually be an increase in the fidelity of replication in the vicinity of UV photoproducts, despite the fact that more mutations occur at sites of damage. The level of spontaneous mutagenesis may be somewhat reduced in pretreated cells, suggesting that pretreatment may also increase the fidelity of DNA synthesis occurring during repair of nuclease-damaged templates.

COMPARISON WITH BACTERIAL SYSTEMS

In bacteria, little DNA damage-induced mutagenesis is observed in the absence of induction of the SOS pathway (17). When the SOS pathway is induced, UV-induced mutagenesis is greatly enhanced and the mutations appear to be targeted to sites of UV-induced damage in the DNA template (5, 8). In mammalian cells, UV-induced targeted mutagenesis appears to occur in the absence of induction of the SOS-like response. Although it is possible that introduction of UV-damaged viruses and plasmid DNAs into untreated cells partially induces an SOS-like response (1), it appears more likely that mammalian cells constitutively express at least some elements of mutagenic DNA damage processing that are only apparent after induction in bacterial cells (14). The twofold enhancement of targeted mutagenesis we observe in pretreated cells probably reflects the result of maximal induction of the pathway, although a higher level of enhancement has been observed when a simian virus 40 test system has been used (13).

In bacterial systems, induction of the SOS pathway also increases the level of untargeted spontaneous mutations. Although we did not observe an increase in the frequency of spontaneous mutations in pretreated cells in our system, it is possible that other test systems might reveal a change in spontaneous mutagenesis. The spontaneous mutations observed after transfection of pZ189 into mammalian cells appear to occur as a consequence of error-prone repair of nuclease-damaged plasmids. Spontaneous mutagenesis that occurs during normal replication is likely to occur at a very low rate, and a small change in this rate due to induction of an SOS-like response would not be revealed in our system.

CONCLUSIONS

Through the use of the pZ189 shuttle vector system we have analyzed enhanced mutagenesis in carcinogen-treated mammalian cells. We have found that targeted mutagenesis is enhanced in both normal and repair-deficient cells that have been pretreated with mitomycin C. This enhancement of mutagenesis can also be induced by treatment of cells with conditioned medium from other UV-irradiated cells, suggesting that enhanced mutagenesis may be mediated by a secreted cellular factor(s). An analysis of the characteristics of the mutations that occur under these enhancing conditions suggests that there is not a generalized loss of fidelity of DNA replication in these cells, but that the response of the DNA polymerase to damage in the template DNA may be altered, causing more misincorporation of bases at sites of damage and less misincorporation in the vicinity of the damage in pretreated cells. These results give further support to the notion that an SOS-like pathway is induced in mammalian cells in response to DNA-damaging agents.

LITERATURE CITED

1. **Cornelis, J. J., Z. Z. Su, D. C. Ward, and J. Rommelaere.** 1981. Indirect induction of mutagenesis of intact parvovirus H-1 in mammalian cells treated with UV light or with UV-irradiated H-1 or simian virus 40. *Proc. Natl. Acad. Sci. USA* **78**:4480–4484.
2. **Hall, J. D., and D. W. Mount.** 1981. Mechanisms of DNA replication and mutagenesis in ultraviolet-irradiated bacteria and mammalian cells. *Prog. Nucleic Acid Res. Mol. Biol.* **25**:53–126.
3. **Hauser, J., A. S. Levine, and K. Dixon.** 1987. Unique pattern of mutations arising after gene transfer into mammalian cells. *EMBO J.* **6**:63–67.
4. **Hauser, J., M. M. Seidman, K. Sidur, and K. Dixon.** 1986. Sequence specificity of point mutations induced during passage of a UV-irradiated shuttle vector plasmid in monkey cells. *Mol. Cell. Biol.* **6**:277–285.
5. **Kunz, B. A., and B. W. Glickman.** 1984. The role of pyrimidine dimers as premutagenic lesions: a study of targeted vs. untargeted mutagenesis in the lacI gene of *Escherichia coli*. *Genetics* **106**:347–364.
6. **Little, J. W., and D. W. Mount.** 1982. The SOS regulatory system of *Escherichia coli*. *Cell* **29**:11–22.
7. **Miller, J. H.** 1985. Carcinogens induce targeted mutations in *Escherichia coli*. *Cell* **31**:5–7.
8. **Miller, J. H., and K. B. Low.** Specificity of mutagenesis resulting from the induction of the SOS system in the absence of mutagenic treatment. *Cell* **37**:675–682.
9. **Munson, P. J., J. Hauser, A. S. Levine, and K.

Dixon. 1987. Test of models for the sequence specificity of UV-induced mutations in mammalian cells. *Mutat. Res.* **179**:103–114.
10. Radman, M. 1980. Is there SOS induction in mammalian cells? *Photochem. Photobiol.* **32**:823–830.
11. Rotem, N., J. H. Axelrod, and R. Miskin. 1987. Induction of urokinase-type plasminogen activator by UV light in human fetal fibroblasts is mediated through a UV-induced secreted protein. *Mol. Cell. Biol.* **7**:622–631.
12. Sarasin, A. 1985. SOS response in mammalian cells. *Cancer Invest.* **3**:163–174.
13. Sarasin, A., and A. Benoit. 1986. Enhanced mutagenesis of UV-irradiated simian virus 40 occurs in mitomycin C-treated host cells only at a low multiplicity of infection. *Mol. Cell. Biol.* **6**:1102–1107.
14. Schorpp, M., U. Mallick, H. J. Rahmsdorf, and P. Herrlich. 1984. UV-induced extracellular factor from human fibroblasts communicates the UV response to nonirradiated cells. *Cell* **37**:861–868.
15. Seidman, M. M., K. Dixon, A. Razzaque, R. Zagursky, and M. L. Berman. 1985. A shuttle vector plasmid for studying carcinogen-induced point mutations in mammalian cells. *Gene* **38**:233–237.
16. Taylor, W. D., L. E. Bockstahler, J. Montes, M. A. Babich, and C. D. Lytle. 1982. Further evidence that ultraviolet radiation-enhanced reactivation of simian virus 40 in monkey kidney cells is not accompanied by mutagenesis. *Mutat. Res.* **105**:291–298.
17. Walker, G. C. 1984. Mutagenesis and inducible responses to deoxyribonucleic acid damage in *Escherichia coli. Microbiol. Rev.* **48**:60–93.

Chapter 53

Damage-Induced Genes in Mammalian Cells and Their Role in Mutagenesis

Nella A. Greggio, Peter M. Glazer, and William C. Summers

Although genes which are induced in response to DNA damage (*din*, damage inducible) are well known in procaryotes, little is known about this phenomenon in mammalian cells. Even though they are part of a multicellular economy, the individual cells in higher organisms suffer damage and environmental insults and may have developed specific or general responses to such damage. In particular, we are interested in the cellular response to DNA damage. These responses, of course, may overlap, partially or entirely, with responses to other types of cell damage.

The first question to be addressed relates to the existence of *din* genes in mammalian cells. Related questions concern the mechanism by which DNA is sensed and what the "purposes" of the specific responses might be.

Nella A. Greggio • Department of Pediatrics, School of Medicine, University of Padua, Padua, Italy. *Peter M. Glazer and William C. Summers* • Departments of Therapeutic Radiology, Human Genetics, and Molecular Biophysics and Biochemistry, Yale University School of Medicine, New Haven, Connecticut 06510.

ARE THERE *din* GENES IN MAMMALIAN CELLS?

Several lines of evidence support the conclusion that mammalian cells have mechanisms to detect DNA damage and to respond in a variety of ways. The major experimental approaches to the study of this problem include virus reactivation, which is limited to nuclear replicating DNA viruses and in which reactivation appears to be error prone; induction of specific proteins; amplification of specific genes; indirect induction of viruses; and induction of cell mutants by introduction of damaged DNA.

Results from many laboratories are consistent in detecting UV reactivation of nuclear replicating DNA viruses (reviewed in reference 14). This experimental approach involves measuring the survival of UV-damaged viral genomes in normal cells and in cells which have received a rather low, "inducing" dose of UV (or other DNA-damaging agent). The increase in survival of damaged virus grown on UV-treated cells is called UV reactivation (or Weigle reactivation, after Jean Weigle who described this sort of experiment with lambda phage [17]). The concept of such experimental approaches is to separate the inducing lesion (on the damaged genome) from the target (on another undamaged genome in the same cell). The interpretation of such experiments is that the inducing damage results in activation of an inducible repair system which acts in *trans* on the damaged DNA of the infecting virus. That only DNA viruses that replicate in the nucleus are subject to this effect suggests that the activated repair system may be physiologically relevant to the cell since it is compartmentalized with the cell genome.

Another observation which suggests that UV reactivation is a distinct phenomenon in mammalian cells is the finding that viruses repaired by the activated repair system suffer increased mutagenesis, i.e., the UV reactivation seems to be error prone (1). The mechanism of this type of mutagenesis is not yet understood, although it is tempting to suppose that the fidelity of repair replication is somehow altered after activation of the repair system. This could come about by direct modifications of existing repair enzymes, by de novo systhesis of new error-prone activities, or by perturbations in the reaction conditions (e.g., pool sizes, ion concentrations, etc.) in the UV-treated cell.

In experiments which are reciprocal in design and corollary in logic, mutations in cellular DNA have been induced by infection with damaged viral DNA (11). In this case, the induction of the error-prone state is the result of the cell sensing the presence of damaged DNA, now in the viral genome, and activating the repair system to replicate the cell DNA in a mutagenic way.

Variant experiments of this type of *trans*-activation by UV damage were carried out by Hampar and his colleagues (3), who found that UV-irradiated herpesviruses could activate the expression of endogenous retrovirus genomes. Presumably, at least one of the targets for gene activation is the viral long terminal repeat, where the promoter for viral RNA synthesis is located. Very recent results (P. Herrlich, this volume) support this notion since the human immunodeficiency virus long terminal repeat promoter can be activated by UV irradiation.

To investigate the mutagenic mechanisms involved in UV reactivation, we (10) employed an experimental system somewhat simpler that that using intact viruses. In our previous work with herpes simplex virus (HSV), it was not possible to exclude possible viral contributions to the mutagenesis because this large virus encodes its own replicative functions as well as possible repair and recombination activities. With a shuttle vector based on the simian virus 40 origin of replication and with the *supF* (tyrosine suppressor tRNA) gene of *Escherichia coli* as the target for mutagenesis, it was shown that DNA damage to the host cell resulted in increased mutagenesis of the plasmid vector and that the induced mutants were qualitatively different from the "spontaneous" mutations which occurred upon replication of the shuttle vector in control, undamaged cells (10). This system has been used recently by other laboratories as well, to study error-prone mutagenesis (K. Dixon, E. Roilides, R. Miskin, and A. S. Levine, this volume).

Another approach to test whether there are *din* genes in mammalian cells is to search directly for the increased synthesis of specific mRNAs or proteins. A number of reports in the literature suggest that cells respond to DNA damage by activating pathways that lead to the synthesis of new proteins. The exact mechanisms of this induction are still unclear, but recent work by Karin and Herrlich and their colleagues suggests that for at least some genes, the induction is rapid and likely to be the direct result of activation of specific promoters (M. Karin and P. Herrlich, in N. Colburn, ed., *Genes and Signal Transduction in Multistage Carcinogenesis*, in press; Herrlich, this volume).

In a series of parallel studies, both Miskin and her colleagues (6, 9) and Herrlich and his colleagues (13; Karin and Herrlich, in press) have found that cells can respond to damaging agents by secretion into the medium of a factor (or factors) which can activate the expression of other genes. In Miskin's work, she first noted that plasminogen activator (urokinase) was synthesized in response to UV damage. Subsequent analysis showed that the urokinase was not the primary gene activated by the UV damage, but that it was induced as a consequence of secretion of a material she called UVIS (UV-induced substance) (9). Herrlich and co-workers found similar phenomena and have designated the extracellular factor EPIF (extracellular protein-inducing factor) (12). It is not yet known whether UVIS and EPIF are identical or even similar in molecular mechanism. Since neither of these factors has been characterized, it is still uncertain whether they are the products of the primary genes activated by the UV damage. Maher et al. (V. M. Maher, K. Sato, S. Kateley-Kohler, H. Thomas, S. Michaud, J. J. McCormick, M. Kraemer, H. J. Rahmsdorf, and P. Herrlich, this volume) have presented evidence to suggest that EPIF-treated cells exhibit increased mutagenesis, both to cellular and episomal target genes.

Karin and Herrlich and their colleagues have found that some specific genes activated by phorbol ester (e.g., c-*fos*, collagenase) are also activated by UV damage. Since they find that the kinetics of induction is very rapid, it is possible that these genes are direct targets of UV activation. The overlap between phorbol ester and UV damage is interesting but not easily understood. Presumably, the phorbol ester results in long-term, constitutive activation of protein kinase C and modulates (both activates and depresses) many cellular processes (7).

Other activities have been detected after DNA damage. Specific replication of localized regions of the genome has led to gene amplification (4). Activation of an enzyme without new synthesis of protein appears to be the case for NAD^+-ADP-ribosyltransferase (13), which is allosterically activated by nicked DNA in a reaction reminiscent of the activation of RecA protein in bacteria. Alternatively, such activation may result from a posttranslational modification such as a phosphorylation; this class of mechanisms suggests a more proximate step by which UV damage controls the modification reaction.

In an approach based on analogies with procaryotic systems, we reasoned that proteins that are directly and specifically induced by DNA damage might be proteins that interact with DNA for some functional reason. Thus, we used the technique of DNA-protein blotting ("southwestern blot") to search for proteins that were induced by DNA damage and that could be detected as proteins which bind labeled DNA after being fractionated by one-dimensional gel electrophoresis and electroblotted onto nitrocellulose sheets (5).

Figure 1 shows the result of one such analysis. Clearly this method allows the detection of proteins that bind double-stranded DNA and single-stranded DNA only after exposure of the cells to UV damage. Although the various proteins are as yet unidentified, their specificity for DNA conformation (double versus single strand), their induction in a UV dose-dependent fashion, and the time course of induction (maximal induction at 18 to 24 h after UV) all indicate that they may play a physiological role in the response of the cells to the DNA damage (P. M. Glazer, N. A. Greggio, J. E. Meatherall, and W. C. Summers, unpublished results). Further studies on the nature of these proteins and the mechanism of their induction by UV damage are in progress.

WHAT IS THE MECHANISM FOR SENSING DNA DAMAGE?

Ascertaining the mechanism for sensing DNA damage is of key importance in this field, if only because we have so little information at the present time. The rather slow induction kinetics observed for some UV-inducible processes (e.g., UV reactivation

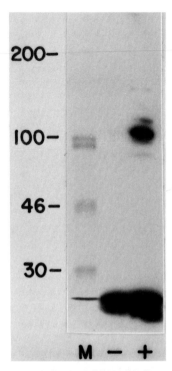

FIGURE 1. UV induction of DNA-binding proteins in HeLa cells. Cells were irradiated with 12 J/m², and after 18 h, extracts of nuclear proteins were prepared from irradiated (+) and mock-irradiated (−) cultures. These extracts were fractionated by gel electrophoresis, transferred to nitrocellulose filters, and blotted with labeled double-stranded DNA to allow autoradiographic detection of DNA-binding proteins (5). Molecular weight markers are indicated (lane M).

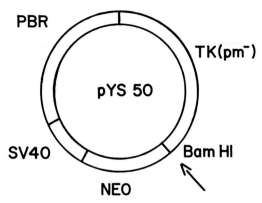

FIGURE 2. Promoter-trap plasmid. The plasmid has the *neo* gene from *E. coli* expressed under the control of the simian virus 40 early promoter and has the HSV-1 TK gene from the *Bgl*II site to the *Eco*RI site (16). A *Bam*HI linker was ligated to the *Bgl*II site to recreate a unique *Bam*HI site in the 5′ untranslated leader region of the normal TK transcript. The promoter sequences from the HSV gene have been removed. Fragments inserted into the *Bam*HI site can be tested for promoter activity by assay or selection for expression of TK under a variety of inducing conditions.

and induction of the DNA-binding proteins as mentioned above) suggests that new protein synthesis may be required as in true enzyme induction. This idea is supported by the experimental finding that inhibition of cellular protein synthesis by cycloheximide abrogates the UV reactivation of HSV (1). The slow kinetics may also be a consequence of the slow accumulation of inducing signal. If this is true, then the signal cannot be the presence of DNA damage, by itself, but rather the metabolic consequences of the damage. The accumulation of replication forks at sites of damage might be expected to be a slow process with a time course of the same magnitude as the cell cycle. Such "jammed" replication forks might result in the formation or accumulation of some product, either a small molecule or a protein, which could signal the presence of DNA damage. An alternative model would suppose that during the repair of the DNA damage some intermediate is formed that can signal the cell to activate specific genes; for example, in the case of recovery from alkylation damage, the presence of alkylated O^6-alkylguanine alkyltransferase (8) or its breakdown products might constitute a signal.

In the case of a few genes (e.g., c-*fos*) the induction is rapid (Karin and Herrlich, in press), so the cell must have a mechanism for detection of DNA damage itself. The lesions themselves might be sites for high-affinity binding of key regulatory proteins. These proteins could be either activated or inactivated by such binding, but in any case would result in perturbation of a cellular regulatory circuit.

We have recently constructed a "promoter-trap" vector for study of the intracellular DNA damage pathway (Fig. 2). This vector will not express the HSV type 1 (HSV-1) thymidine kinase (TK) unless the sequences inserted into the *Bam*HI site can function as a promoter. Under conditions of selection for the *neo* gene this plasmid can be maintained in the cell in multiple copies (N. A. Greggio and W. C. Summers, unpublished results). DNA sequences known to be responsive to activation by UV, such as the c-*fos* promoter, can be inserted at the *Bam*HI site and tested for activation by measurement of the TK activity induced (15). We are currently testing this construction to see whether cell mutants that are incapable of responding to UV damage can be isolated as plasmid-carrying cells that survive in the presence of negative selection for TK expression from the UV-responsive promoter.

WHAT ARE THE PHYSIOLOGICAL ROLES OF *din* GENES?

Many *din* genes in procaryotes appear to be related to the pathways for recovery from DNA damage. Thus, the function of the induced protein is clearly related to the nature of the inducing signal. However, there may be other functions that are less obvious but of important evolutionary significance. Echols (2) has advanced the notion that error-prone repair and replication may be the basis for increasing the rate of evolution in times of global stress. He pointed out that an increase in mutation rate results in more individuals with extreme genotypes and that some such individuals might have increased fitness in new environments. This argument clearly requires the long-term retention of a very basic cellular property, i.e., induction of the error-prone response, over evolutionary time.

Although enhanced individual cell survival may not be the result of DNA damage-induced processes in multicellular organisms, the genes activated by DNA damage might have a different function than those in procaryotes. It may be important for damaged cells to be able to signal to adjacent cells or in some way to enhance their ability to grow. Such an "altruistic" response to DNA damage might allow damaged cells to produce extracellular growth factors that stimulate nearby, undamaged cells presumably of similar cell type and function. Although the individual cell dies, it contributes to the survival of the multicellular system and thereby increases its fitness.

Another question still to be answered is the extent of overlap between genes induced by DNA damage and genes induced by other cellular insults such as heat, membrane damage, oxidation, and so on. If there is significant overlap, it will be of interest to determine whether these inducing agents have different ways of sensing the damage, but rely on a "final common pathway" to activate an entire set of genes for cellular recovery. The overlap between phorbol esters and UV reported by Karin and Herrlich (in press) suggests such a mechanism. A *cis*-acting site on the DNA which appears to be involved in both UV and phorbol ester activation is the binding site for the AP-1 transcription factor (Herrlich, this volume). Thus, in some way both phorbol esters and UV result in increased binding of AP-1 (or AP-1 analogs). Better understanding of AP-1 action and its regulation should clarify this situation.

CONCLUSION

Over the past 10 years there has been increasing experimental support for the notion that DNA damage can induce gene expression in mammalian cells and that one consequence of such induction is a error-prone repair or replication pathway (14). The relative simplicity of the bacterial SOS system is not yet apparent for these phenomena in mammalian cells.

Biochemical approaches to identify specific genes and gene products have now shown that proteins with key cellular roles (e.g., c-*fos* and specific DNA-binding activities) are induced by UV damage. One step in the induction pathway has been identified with the activity of the transcription factor AP-1. Genetic dissection of the induction pathway(s) is now possible by using approaches described in this paper. These combined approaches should illuminate the main features of the mammalian UV response.

The cellular utility of such responses to DNA damage is still unclear. Evolutionary arguments suggest that inducible mutagenesis may have advantages even for multicellular organisms. If there is substantial overlap between cell repair pathways induced by DNA damage and other stresses, the inducibility of mutagenesis may be an evolutionarily useful correlate of the more immediately required response to environmental stresses such as heat and cold in homeotherms and oxidation in obligate aerobic organisms.

ACKNOWLEDGMENTS. This work was supported by Public Health Service grant CA 39238 from the National Institutes of Health. N.A.G. was a Fellow of the Leukemia Society of America.

We are grateful to Wilma P. Summers for crucial advice and help.

LITERATURE CITED

1. **Dasgupta, U. B., and W. C. Summers.** 1978. UV reactivation of herpes simplex virus is mutagenic and inducible in mammalian cells. *Proc. Natl. Acad. Sci. USA* 75:2378–2381.
2. **Echols, H.** 1981. SOS functions, cancer and inducible evolution. *Cell* 23:1–2.
3. **Hampar, B., S. A. Aaronson, J. G. Derge, M. Chakrabarty, S. D. Showalter, and C. Y. Dunn.** 1976. Activation of an endogenous mouse C-type virus by ultraviolet-irradiated herpes simplex virus types 1 and 2. *Proc. Natl. Acad. Sci. USA* 73:646–650.
4. **Lavi, S.** 1981. Carcinogen-mediated amplification of viral DNA sequences in simian virus 40-transformed Chinese hamster embryo cells. *Proc. Natl. Acad. Sci. USA* 78:6144–6148.
5. **Miskimins, W. K., M. P. Roberts, A. McClelland, and F. H. Ruddle.** 1985. Use of a protein-blotting procedure and a specific DNA probe to identify nuclear proteins that recognize the promoter region of the transferrin gene. *Proc. Natl. Acad. Sci. USA* 82:6741–6744.
6. **Miskin, R., and R. Ben-Ishai.** 1981. Induction of plasminogen activator by UV light in normal and xeroderma pigmentosum fibroblasts. *Proc. Natl. Acad. Sci. USA* 78:6236–6240.
7. **Nishizuka, Y.** 1984. Turnover of inositol phospholipids and signal transduction. *Science* 225:1365–1370.
8. **Olsson, M., and T. Lindahl.** 1980. Repair of alkylated DNA in *Escherichia coli*: methyl group transfer from O^6-methylguanine to a protein cysteine residue. *J. Biol. Chem.* 255:10569–10571.
9. **Rotem, N., J. H. Axelrod, and R. Miskin.** 1987. Induction of urokinase-type plasminogen activator by UV light in human fetal fibroblasts is mediated through a UV-induced secreted protein. *Mol. Cell. Biol.* 7:622–631.
10. **Sarkar, S. N., U. B. Dasgupta, and W. C. Summers.** 1984. Error-prone mutagenesis detected in mammalian cells by a shuttle vector containing the *supF* gene of *Escherichia coli*. *Mol. Cell. Biol.* 4:2227–2230.
11. **Schlehofer, J. R., and H. zur Hausen.** 1982. Induction of mutations within the host cell genome by partially inactivated herpes simplex virus type 1. *Virology* 122:471–475.
12. **Schorpp, M., U. Mallick, H. J. Rahmsdorf, and P. Herrlich.** 1984. UV-induced extracellular factor from human fibroblasts communicates the UV response to unirradiated cells. *Cell* 37:861–868.
13. **Smulson, M. E., P. Schein, D. W. Mullins, Jr., and S. Sudhakar.** 1977. A putative role of nicotinamide dinucleotide-promoted nuclear protein modification in the antitumor activity of N-methyl-N-nitrosourea. *Cancer Res.* 37:3006–3012.
14. **Summers, W. C., S. N. Sarkar, and P. M. Glazer.** 1985. Direct and inducible mutagenesis in mammalian cells. *Cancer Surv.* 4:517–528.
15. **Summers, W. C., and W. P. Summers.** 1977. ^{125}I-deoxyuridine used in a rapid, sensitive and specific assay for herpes simplex type 1 thymidine kinase. *J. Virol.* 24:314–318.
16. **Wagner, M. J., J. A. Sharp, and W. C. Summers.** 1981. Nucleotide sequence of the thymidine kinase gene of herpes simplex virus type 1. *Proc. Natl. Acad. Sci. USA* 78:1441–1445.
17. **Weigle, J. J.** 1953. Induction of mutations in a bacterial virus. *Proc. Natl. Acad. Sci. USA* 39:628–636.

Index

Abasic sites
  bypass synthesis, 292
  E. coli DNA pol III, 297
  nucleotide insertion
    D. melanogaster DNA pol α, 291
    D. melanogaster DNA pol α and N-ethylmaleimide, 292
Acetophenone, 401
Acetylaminofluorene, 263, 317, 318, 412
  inhibition of T4 $3' \to 5'$ exonuclease, 265
ada-alkB operon, 446, 447
  expression in HeLa cells, 448
Adaptive response, 446
Adenine phosphoribosyltransferase mutations, 258
aidB, *Escherichia coli*, 446
alkA, *Escherichia coli*, 446
Allosteric interaction
  E. coli DNA pol III, 17, 18, 310
  Hill coefficient, 18
  nucleotide triphosphate binding site, 197
  SSB protein, 159
Altruistic response to DNA damage, 483
Amadori products, 423
Ames test, tobacco extracts, 429
Amino acid sequence
  DNA pol β, 56
  phage φ29 DNA pol and DNA pol α, 126
  τ and γ subunits, 29
3-Aminobenzamide
  effect on sister chromatid exchanges, 260
  inhibition of poly(ADP-ribose) synthesis, 260
2-Aminopurine deoxyribose triphosphate, substrate for T4 DNA pol, 38
Antibodies, *see also* Monoclonal antibodies
  B. subtilis membrane proteins, 141
  DNA pol α, 43
  DNA pol β, 56
  DNA pol δ, 71, 72
  E. coli τ and γ accessory proteins, 28
  herpes simplex virus dUTPase, 436
Antimutator, 35
AP-1 transcription factor, 461, 483
Ap_4A, binding to HeLa multienzyme complex, 43, 44
Aphidicolin
  hypersensitivity mutations in mammalian cells, 167, 168
    UV sensitivity, 168
  inhibition
    DNA pol α and δ, 70, 75, 76, 163
    DNA pol I of yeast, 99

DNA pol of phage φ29, 124
$3' \to 5'$ exonuclease of DNA pol δ, 70
repair synthesis, 163
nonmutagenicity, 164
resistance in mammalian cells, 164, 165
resistance of mutagenesis in yeast, 357
AraC
  effect on DNA pol I, 214
  synthesis in oligonucleotides, 213
A-rule
  apurinic sites, 300, 303, 314
  blunt-end additions, 217
  UV mutagenesis, 282, 300, 314, 401, 407, 476
Ataxia telangectasia, 417
ATPase
  DNA pol III accessory protein τ, 19, 30
  interaction with HeLa multienzyme complex, 45, 46
  phage T7 gene 4 protein, 90
ATPγS, DNA pol III initiation complex, 15
Autonomously replicating sequences, *Saccharomyces cerevisiae*, 100
Auxiliary subunits
  β (*dnaN*), 14, 24
  DNA pol δ, proliferating cell nuclear antigen, 72
  HeLa multienzyme complex, 45
  T4 DNA pol, 237, 242, 245
  τ, γ (*dnaX*, *dnaZ*), 18–20, 23, 27, 31

Bacillus subtilis
  homologies to *oriS* and *oriC*, 151
  membrane proteins, 140
Bacteriophage λ
  groEL, 160
  induction in *ssb* mutants, 154
  λ p*lac* Mu9, cloning *pcbA1*, 309
  nonimmune exclusion, 373
  replication initiation, 367
  replicative killing, 367
Bacteriophage N4, 130
  dbp, *dnp*, *dns*, 132
  DNA polymerase, 132
    $3' \to 5'$ exonuclease, 132
    $5' \to 3'$ exonuclease, 133
  replication, 132, 133, 135
Bacteriophage φ29
  DNA polymerase, $3' \to 5'$ exonuclease, 122
  DNA replication, 112
  terminal protein, 112
Bacteriophage PRD1, sequence homology to phage φ29 DNA pol, 127
Bacteriophage T4

DNA polymerase, 34, 237
  accessory proteins, 237
  araC in template, 214
  etheno-dATP substrate, 237, 238
  homologies, herpes simplex virus, 35, 127
  homologies, vaccinia, 127
  mutations, 34, 37
 gene 41-gene 61 interaction, 91
 mutator phenotypes, 34, 37, 227, 237
Bacteriophage T7, 85
 DNA polymerase, 86, 95
  bypass synthesis, 316
  $3' \rightarrow 5'$ exonuclease, 89
  $3' \rightarrow 5'$ exonuclease, control by oxidation of thioredoxin, 316
  Sequenase, 316
 gene 2.5 protein, SSB-like, 88, 93
 gene 4 protein, helicase-primase, 86, 87, 95
 gene 5 protein, DNA pol, 86
 RNA polymerase, role in DNA replication, 12
 thioredoxin, $E.\ coli$
  complex with T7 gene 5 protein, DNA pol, 86, 89
  $3' \rightarrow 5'$ exonuclease, 316
Base analogs
 fidelity of replication, 196
 5-fluorouracil, 210
 substrates
  DNA-dependent DNA polymerases, 203
  reverse transcriptase, 202
 template for $E.\ coli$ Klenow DNA pol, 199
Base modifications
 acetylaminofluorene, 412
 O-alkylation, 64, 199, 202
 bisulfite and hydrazine reactions with cytosine, 339, 340
 deamination, 284, 437
 effect on DNA pol β, 64
 etheno-dAMP production by chloroacetaldehyde, 237
 methylation, 445
 reducing sugars, 423
 thymine glycol, 212
 UV and acetophenone, 401
Base pairing
 5-bromouracil and misincorporation, 255, 259
  DNA pol I, 221
 etheno-dAMP, 238, 241
 fidelity of replication
  DNA melting temperature, 290
  free energy, 196, 289, 290, 291
 5-fluorouracil in DNA, 212
 5-fluorouracil:A analysis by nuclear magnetic resonance, 211
 misincorporation, 285, 288
  5-bromouracil, 255, 259
  UV mutagenesis, 281

 mispairing
  A:G wobble, 329
  deletion formation, 380
Bent DNA, replication origins, 101, 119
β Subunit ($dnaN$), see Auxiliary subunits, 24
Bleomycin, $pcbA1\ dnaE$ mutants, 306
Blunt-end addition reaction
 $E.\ coli$ Klenow DNA pol, $3' \rightarrow 5'$ exo def, 216
 eucaryotic DNA polymerases, 217
 frameshift mutagenesis, 217
 $T.\ aquaticus$ DNA pol, 217
5-Bromodeoxyuridine, 254
2-(p-n-Butylanilino)-9-(2-deoxy-β-D-ribofuranosyl)adenine 5′-triphosphate, inhibition
 DNA polymerase α, 76
 phage φ29 DNA pol, 124
N^2-(p-n-Butylphenyl)-9-(2-deoxy-β-D-ribofuranosyl)guanine 5′-triphosphate, differential inhibition
 DNA pol α and DNA pol δ, 50, 75
 DNA pol I and DNA pol II of yeast, 99
Bypass replication, see also SOS
 abasic sites, 291, 292, 297
 N^4-aminocytosine, 339
  role of $recA$ protein, 341
 aminofluorene adducts, 319, 321
 DNA pol β, 64
 failure
  acetylaminofluorene adducts, 318
  N^4-amino-5,6-dihydrocytosine 6-sulfonate, 340
 pass-fail option, 281, 296, 307
 photodimers in absence of SSB protein, 298
 T4 DNA pol, 39
 UV mutagenesis, 278, 281, 297, 397

Calf thymus, DNA pol δ, 70
Carcinogens
 mutagenic activity, 237, 263, 458, 473
 tobacco, 428
Cell cycle
 aphidicolin-resistant mutants, 166
 control
  $E.\ coli$, 151
  $E.\ coli$, methylation at $oriC$, 109
  $S.\ cerevisiae$, 98
 regulation of HeLa and herpes simplex virus uracil-DNA glycosylases, 438
c-fos, 461
Chloramphenicol acetyltransferase, mutations, 380
Chloroacetaldehyde, phage T4 DNA pol, 237
cis-acting protein, plasmid R1 replication, 121
Cisplatin, 263
c-mos, 461
c-myc, Burkitt's lymphoma, 458
ColE1 plasmid, 103
Collagenase, 461, 466
Computer modeling
 DNA mispairing and mutagenesis, 330

E. coli Klenow DNA pol interaction with DNA, 222
thymine glycol in DNA, 213
Context effect, mutagenesis, 227
copAB, Escherichia coli, 113, 147, 150
cro, bacteriophage λ, 369
Cryptic lesions, SOS processing, 405
Cycling, 15
  E. coli pol III holoenzyme, 298
  phage T7, 95
  plasmid R1, 119

dam methylase, Escherichia coli
  oriC, 104
  recA730, 335
Damage-induced genes, see also din, mammalian cells, 461, 479
DDR, Saccharomyces cerevisiae, 356
Deamination, see Base modifications
Deletion mutations
  alternating GC sequences, 269
  direct repeats, 231, 387
  DNA secondary structure, 314, 378
  inverted repeats, 387
  plasmid-based assay, 378, 382
  recA and SOS induction, 383
  replicative killing resistance, 371
Deoxynucleotide pools, imbalance and mutagenesis, 254
Dimerization, see also Allosteric interaction, DNA polymerase III, E. coli, 15, 310, 311
Dimethyl sulfate, alkB in HeLa cells and resistance, 448
DIN
  mammalian cells, 479
  S. cerevisiae, 356
dinB, Escherichia coli, 333, 363
dinD, Escherichia coli, 363
Diphtheria toxin resistance, elongation factor 2 mutation, induction by extracellular protein-inducing factor, 468
Direct repeat sequences, deletion mutations and DNA pol I, 231, 379
DNA ligase
  araC in substrate, 215
  HeLa multienzyme complex, 42, 46
  phage N4 replication, 132
DNA methyltransferase, 445–447
DNA polymerase
  bacteriophage N4, $3' \rightarrow 5'$ exonuclease, 132
  bacteriophage φ29, $3' \rightarrow 5'$ exonuclease, 122
  bacteriophage T4
    accessory proteins, 237
    araC in template, 214
    etheno-dATP substrate, 237, 238
    homologies, herpes simplex virus, 35, 127
    homologies, vaccinia, 127
    mutations, 34, 37
  bacteriophage T7
    bypass synthesis, 316
    $3' \rightarrow 5'$ exonuclease, 89
    $3' \rightarrow 5'$ exonuclease, control by oxidation of thioredoxin, 316
    Sequenase, 316
  pol α, eucaryotic cells, 41, 42, 50, 68, 100, 163, 185
    aphidicolin binding, 169
    araC in template, 214
    blunt-end addition reaction, 217
    D. melanogaster $3' \rightarrow 5'$ exonuclease, 187
    fidelity, 165, 185, 286–288
    mutations affecting dCTP binding, 169
    placenta, 72
    purification of multienzyme complex, 42
  pol β, eucaryotic cells, 55, 163, 185
    antibody, 56
    araC in template, 214
    base modification in template, 64
    blunt-end addition reaction, 217
    fidelity, 185
    homology to terminal transferase, 66
    pausing, 64
    purification, 56
    sequence of cDNA clone, 56
  pol Cab, E. coli, 293
  pol δ, eucaryotic cells, 50, 68, 100, 163, 170
    purification and characterization, 69
  pol γ, eucaryotic cells, 103
  pol I, E. coli, see also Klenow fragment of Escherichia coli DNA pol I
    bypass at abasic site, 292
    bypass at chemical adducts, 316
    frameshift mutations, 229, 231
    nucleotide monophosphate binding site, 37
  pol I, S. cerevisiae, 99, 100
    blunt-end addition reaction, 217
    dispensability in mutagenesis, 357
    temperature-sensitive mutants, 99
  pol II, E. coli
    pol Cab, 294
    SSB protein, 155
  pol II, S. cerevisiae, 99, 100
  pol III*, E. coli, 292
  pol III core, E. coli, 14, 27
  pol III holoenzyme, E. coli, 9, 14, 27, 115
    bypass at abasic site, 292
    bypass at photodimers, 297
    dimerization, 15, 310, 311
    SOS mutagenesis, 292
    UV mutagenesis, 277, 279, 306
  pol III holoenzyme, S. typhimurium, 247
dnaA box, 8, 116
dnaA protein, 8
  oriC and P1, 103, 104
  oriR, 114, 115

suppression by *groE* mutants, 160
dnaB protein, 9, 90, 91
  oriR, 115
  phage λ *P*-gene protein, 375
DNA-binding protein, 480
dnaC protein, 9
  oriR, 115
dnaE protein, 14, 23, 27, 248, 249, 277
  mutT, 328
  suppression by *pcbA1*, 305
dnaG protein, *see* Primase
dnaN, *see* Auxiliary subunits, 24
dnaQ protein, *Salmonella typhimurium*, 247
  suppressor (*spq*), 248
DNase
  Mollicutes, 442
  S. cerevisiae, 101
DNase footprinting
  autonomous replication sequence-binding protein, 100
  repA protein, 115
dnaZ protein, *see* Auxiliary subunits
Double-strand breaks
  E. coli, mismatch repair, 336
  S. cerevisiae, 359
    DNA pol I and pol II, 100
    mismatches cut by S1 nuclease, 358
*Dpn*I assay, 183
dUTP nucleotidohydrolase, 434, 441
  herpes simplex virus, map location, 436

EBNA-1, 417
Editing, *see* 3'→5' Exonucleases
Electroporation, 417
Elongation
  abasic sites, 291, 297
  araC in template, 215
  photodimers, 297, 399
Endonuclease P1, 9, 119
EPIF, 462, 465, 475, 481
  mutagenesis, 467
ε Subunit, *see also mutD*, 14, 27, 250, 279
  SOS, 407
Epstein-Barr virus
  Burkitt's lymphoma, 458
  shuttle vectors, 416
Error-prone repair, *see also* SOS, 281, 284, 292, 465
1,N^6-Etheno-dATP
  base pairing, 241
  misincorporation by T4 DNA polymerase, 237, 238, 240
  nucleotide binding sites, 202
Ethyl methanesulfonate, mutagenesis and DNA pol III, 306
Ethyl nitrosourea, 420, 469
Evolution
  homologies, *see* Sequence homology
  inducible, 483

Excision repair
  bacteriophage λ untargeted mutagenesis, 336
  Mollicutes (mycoplasmas), 441
  S. cerevisiae mutants, 355–357
Exconjugants, UV mutation spectrum, 399
Exonuclease I
  SOS induction, 362, 363
  SSB protein, 155
Exonuclease V, *recBC*, 363, 388
Exonuclease VII, SOS, 362
5'→3' Exonucleases, bacteriophage N4, 131, 133
3'→5' Exonucleases, editing
  bacteriophage N4 DNA pol, 132
  bacteriophage φ29 DNA pol, 122
  bacteriophage T4
    DNA pol, 34
    inhibition by *N*-2-acetylaminofluorene adduct, 265
    mutations, 38
  bacteriophage T7 DNA pol, 89
    oxidation of thioredoxin, 316
  E. coli
    blunt-end addition reaction, 217
    DNA pol I, 37, 220
    DNA pol III and bypass, 300
    DNA pol III and mutagenesis, 278, 300
    idling and turnover reaction, 199
    Klenow polymerase active site, 222, 224
  eucaryotic cells
    D. melanogaster DNA pol α, 187
    DNA pol δ, 71, 187
    HeLa multienzyme complex, 42
    S. cerevisiae DNA pol I and pol II, 99
  inhibition
    dGMP, 316
    recA, 300
  replication fidelity, 197, 198, 314, 404
  S. typhimurium, *dnaQ* protein, 247

F miniplasmid, 107, 146
F' plasmid, 400
Fidelity of replication, *see also* Mutator phenotype, 38, 250, 327
  aphidicolin-resistant DNA pol α, 165
  bacteriophage N4 DNA pol, 132
  base analogs, 196, 197
  DNA pol α and β, 185, 197
  DNA pol I, *E. coli*, 220
  kinetic analysis, 287, 288
  kinetic mechanisms, 198
  mutT, *E. coli*, 327, 328
  simian virus 40, 182–185
Filter binding assay, *repA* protein and DNA, 115
5-Fluorouracil
  base pairing, 210
  DNA template, 209
Frameshift mutations, 227
  DNA secondary structure, 314

induction by carcinogens, 263
primer foldback hypothesis, 234

Gap filling, DNA pol β, 61
Gel retardation assay, autonomous replication sequence-binding proteins, 100
Gene amplification, replicative killing-resistant mutant, 371
Gene conversion, mismatch repair in *Saccharomyces cerevisiae*, 356
Gene N, bacteriophage λ, 367
Gene 4 protein
  bacteriophage T7, 87, 95
  dimorphism, 89
Gene 32 protein, bacteriophage T4
  binding to DNA pol, 245
  incorporation of etheno-dATP, 242
Genes O and P, bacteriophage λ, 367, 372, 375
Genetic regulation, DNA polymerase III holoenzyme, 23
Glucose-6-phosphate, reaction with DNA, 423
groE
  E. coli, 156, 159
    bacteriophage λ, 160
    dnaA protein, 160
    replication, 159
Growth rate
  aphidicolin-resistant mutant, 166
  control in *E. coli*, 149, 151
Gyrase
  bacteriophage N4 replication, 130, 131
  oriR function, 115

Heat shock
  groEL protein and SSB protein, 158
  mammalian cells, 483
  RNA polymerase, 150
HeLa cells
  dUTPase, 434
  in vitro DNA replication system, 183
  Mer phenotype, 446
  uracil-DNA glycosylase, 434
Helicase
  bacteriophage N4, 132
  bacteriophage φ29, 124
  bacteriophage T7 gene 4, 86, 90
Herpes simplex virus
  DNA polymerase
    active site, 36
    homology to φ29 DNA pol, 127
    homology to T4 DNA pol, 35
  dUTPase, 434
    map location, 436
  error-prone UV reactivation, 480
  thymidine kinase gene, 417
  uracil-DNA glycosylase, 434, 437
HGPRT

mutations, 256, 284
  induced by ethylnitrosourea, 469
  induced by extracellular protein-inducing factor, 467
H_2O_2, sensitivity of *pcbA1 dnaE* mutant, 306
Hormones, 459
Hot spots
  aminofluorene-induced mutations, 320
  chemical reactivity, 265
  DNA structure, 263, 407
  mutations, 405
HU protein, 8, 11
Human immunodeficiency virus, 459
Hybrid, DNA-RNA, activation of template, 12
Hydrodynamic studies, DNA polymerase III shape factors, 21
Hydroxyurea, *see* Ribonucleotide reductase
Hygromycin resistance, 417

Immunoglobulin locus, 462
INC mutagenesis, *see* Incorporational mutagenesis
incA, *Escherichia coli*, 104
Incorporational mutagenesis, 254
Initiation of replication, 6, 42, 119, 143, 151, 367
Insertional mutagenesis, retrovirus, 462
Insertion-denaturation model,
  N-2-acetylaminofluorene frameshift mutagenesis, 266
Integration
  polyomavirus DNA and transformation, 175
  recombination, 176, 177
Interallelic complementation, *dnaE*(Ts), 310
Interleukin-1, 466
int-kil, *Escherichia coli*, 369, 374
Inverted repeated sequences
  deletion formation, 378
  transposition, 378
In vitro replication
  bacteriophage N4, 135, 137
  bacteriophage T7, 85
  plasmid P1, 107
  plasmid R1, 120
Ionizing radiation, repair of damage, *Saccharomyces cerevisiae*, 358
IS2, induction of transposition, 374

Kinetic mechanisms
  DNA polymerase α, *D. melanogaster*, 286
  DNA polymerase I, *E. coli*, 223
  fidelity, 198, 285, 291
Klenow fragment of *Escherichia coli* DNA pol I
  araC in template, 214
  blunt-end addition reaction, 216, 217
  bypass of base analogs, 199
  3'→5' exonuclease active site, 222, 224
  frameshift mutagenesis on T4 DNA, 228–231
  pseudo-template-directed addition, 217

site-directed mutations in enzyme, 222
structure, 222
substrate site and primer terminus, 222
X-ray crystallography, 222

LacZ fusion protein, τ protein, 27
lacZα, mutation, 183
λ p*lac* Mu9, cloning *pcbA1*, 309
Lethal sectoring
DNA pol I, 251
ε subunit of DNA pol III, 250
lexA, *Escherichia coli*
homology to *mucA*, 349
UV mutagenesis, 277, 332
Lipid A disaccharide synthase, see *lpxB*
lpxB, *Escherichia coli*, *dnaE*, 23
LTR, see Retrovirus
Lymphoblastoid cell lines, 417
Lysogeny, phage P1 and methylation, 109

Macromolecular synthesis operon, 23
Mammalian cells, see DNA polymerases α, β, γ, and δ
Mammalian stress response, 457, 483
Maxicells, *sup-411* (*groEL*), 155
Membranes
DNA binding, 141
DNA replication, 140
Mer phenotype, 446
Metallothionein, 461
Methyl methanesulfonate
alkB and resistance, 446, 448
sensitivity of *pcbA1 dnaE* mutant, 306
N3-Methyladenine DNA transferase, 445
Methylation
in vitro replication of plasmid P1, 107–109
mismatch correction, 334
oriC, 103, 104, 108
DNA secondary structure, 110
pause sites, 64
removal in *E. coli* and mammalian cells, 445
N3-Methyldeoxythymidine, DNA polymerase β, 64
N-Methyl-N'-nitro-N-nitrosoguanidine, sensitivity
alkB in HeLa cells, 448
DNA methyltransferase, 447
Mex phenotype, 449
Mismatch correction, 280, 314
double-strand breaks, 336
methylation, 334
mutH, *mutL*, *mutS*, and *mutU*, 334
mutT, 328, 329
S. cerevisiae, 356, 357
Mispairing, see Base pairing
Mitomycin, 473, 475
Mollicutes
base composition and mutagenesis, 440, 442
DNase, 442
dUTPase, 441

photoreactivation and excision repair, 441
uracil-DNA glycosylase, 440
Monoclonal antibodies
DNA polymerase δ, 72
HeLa dUTPase, 435
HeLa multienzyme complex, 47
τ subunit of *E. coli* DNA pol III, 20
mucAB, *Salmonella typhimurium*, 349
Mutation in *trans*, 459
Mutation spectrum
acetylaminofluorene adducts in mammalian cells, 413
aminofluorene adducts, 319, 320
apurinic sites, 302
deletion in chloramphenicol acetyltransferase gene, 384
DNA damage, 403, 404
DNA structure, 263
N-ethyl-N-nitrosourea and Epstein-Barr virus shuttle vector, 419, 420
glucose-6-phosphate, 426
mammalian SOS protocol, 480
mutT, 328
recA441 induced, 333
replicative killing-resistant mutants, 370
spontaneous mutants
DNA pol α, 185
shuttle vectors, 418, 480
uracil-containing template, 319, 320
UV mutagenesis
shuttle vector, 475, 476
umuC, 397, 398
Mutator phenotype
bacteriophage T4, 34, 37, 231
deletion formation (*dli*), 384
dnaE486 and *polC74*, 281
mammalian cells, 459
mutT, 325
pms mutants, *S. cerevisiae*, 356
SOS effects, 332
mutD, *Escherichia coli*, 14, 250, 251
mutS, *Escherichia coli*, 185
mutT, *Escherichia coli*, 325
action on single-strand DNA, 326
cloning, 326
Mycoplasmas, see *Mollicutes*

Na-K ATPase locus, mutations, 256
Nitrosamines, 428
Nuclear magnetic resonance spectroscopy
A:G base pairing and wobble, 329
5-fluorouracil:A base pair stability, 211
Nucleotide binding site
allosteric model, 197
DNA polymerases, 197, 202, 222

Okazaki fragments, 15, 85, 89, 131, 139
Oncogene
　mutagenic activation, 458
　overexpression, 461
oop, bacteriophage λ, 369
orf$_{23}$, *dnaE*, 23
ori, *see* Origin of replication
Origin of replication
　B. subtilis, 141
　bacteriophage P1, 107, 108
　　sequence, 106
　bacteriophage T7, 86
　DNA secondary structure and methylation, 110
　oriC, 6, 103, 146, 151
　　sequence, 7
　oriP, Epstein-Barr virus, 417
　oriR, 113, 114
　oriS, 146, 151
　oriV, 151
　S. cerevisiae, 100
　simian virus 40, 42, 182
Ouabain resistance, *see* Na-K ATPase locus
Outer membrane, 23

P1 plasmid replication, 104
Pause sites
　DNA pol β, 59, 61
　methylation, 64
　substrate concentration, 202
pcbA1, *Escherichia coli*, 305
　mapping, 305
　suppression of *dnaE* mutation, 305
Phorbol esters, *see* Tumor promoters
Phosphoamino acids, HeLa dUTPase, 435
Phosphonoacetic acid, bacteriophage ϕ29
　binding site homology to α DNA pol, 125
　inhibition of DNA pol, 124
Photodimers
　bypass by DNA pol III, 297
　UV mutagenesis, 398
Photoreactivation, *Mollicutes*, 441
Photoreversal, UV mutagenesis, 279
Placenta, DNA pol α and δ, 68
Plasmid shuffling, construction of conditional lethal mutations, 99
pms, *Saccharomyces cerevisiae*, 356
Poly(ADP-ribose), inhibition of synthesis by 3-aminobenzamide, 260
Polycyclic aromatic hydrocarbons, 428
Polyomavirus, 173
Prepriming complex, 10
Primase
　bacteriophage N4, 132
　bacteriophage T4 gene 61/41 proteins, 91
　bacteriophage T7 gene 4 protein, 86
　dnaG protein
　　homology to T7 gene 4 protein, 90
　　interaction with *dnaB* protein, 91

HeLa multienzyme complex, 45
oriR, 115
Primer, bacteriophage T7 RNA, 93
Primer foldback hypothesis, frameshift mutations, 234
Processivity, 18, 27, 31
　bacteriophage N4 DNA pol, 132
　bacteriophage ϕ29 DNA pol, 123
　bacteriophage T7 DNA pol, 89, 92
　DNA polymerase β, 59
　DNA polymerase III holoenzyme, *E. coli*, 298
Proliferating cell nuclear antigen, auxiliary factor for DNA pol δ, 72
Promoter trap plasmid, 482
Proofreading, *see* 3′→5′ Exonucleases
Protein kinase C, 461
Protein n, SSB protein, 155
Proteolysis
　τ and γ auxiliary subunits, 30
　umuD protein cleavage, 349
Pseudo-template-directed end addition, frameshift mutations, 217
Psoralen, sensitivity of *pcbA1 dnaE* mutant, 306
Pulsed-field gel electrophoresis, *Saccharomyces cerevisiae*, DNA strand breaks, 358, 359

R1 plasmid, 113
　regulation of replication, 118
RAD6, *Saccharomyces cerevisiae*, 358
ras oncogenes, 458, 461
Rate of replication, 15
　bacteriophage T7, 92
　E. coli DNA pol I, 223
recA
　AAF −1 frameshift mutagenesis, 265
　cleavage of *umuD* protein, 281, 352
　deletion mutagenesis, 384
　inhibition of DNA pol III 3′→5′ exonuclease, 300
　proofreading, 282
　SSB protein, 154, 155
　UV mutagenesis, 277, 278, 332, 349
recF pathway, 363
Recombination
　deletion formation, 388
　integration, 176, 177
　interviral, polyomavirus, 175
　studied by restriction site polymorphisms, 177
REP mutagenesis, *see* Replicational mutagenesis
repA, *Escherichia coli*, 104, 113, 114
　autoregulation, 105
　DNA binding, 115
Repair synthesis
　aphidicolin inhibition, 163, 164
　fidelity in aphidicolin-resistant mutant, 166
repE, *Escherichia coli*, 147
　transcription, 150
Replication blockage by misinsertion, 250, 298, 299
Replicational mutagenesis, 255, 256, 258

Replicative intermediates, bacteriophage N4, 133, 134
Replicative killing, 367, 374
 SOS, 375
Repressor, *Bacillus subtilis*, initiation of replication, 144
Restriction site polymorphisms (RFLP), polyomavirus recombination, 177
Reticulocyte, DNA pol δ, 70
Retinoblastoma, hereditary, 459
Retrovirus
 activation by UV, 480
 integration and mutagenesis, 462
REV2, *Saccharomyces cerevisiae*, 357
 cloning, 358
Reverse transcriptase
 fidelity, 197
 use in synthesis of specific oligonucleotides, 200, 202
rho-dependent transcription termination, R1 replication, 114
Ribonucleotide reductase
 bacteriophage N4 replication, 131
 nucleotide pools, 257
 S. cerevisiae, hydroxyurea inhibition and mutagenesis, 357
Rifampin, *see also* RNA polymerase, 11
 resistant mutations, 335
*rII*B, bacteriophage T4, frameshift mutagenesis, 228
RK mutatest system, 369, 370
RNA polymerase
 bacteriophage N4, 130, 133
 bacteriophage T7, 12, 86
 heat shock, sigma-32, 150
 independence of *oriR*, 115
 requirement for in vitro replication, 11
RNase H, 9
 HeLa multienzyme complex, 42, 45
Rolling circle model, bacteriophage T7 replication, 92
Runaway plasmid, 148

S phase, *Saccharomyces cerevisiae*, 98
S1 mitochondrial DNA, *Z. maize*, homology to φ29 DNA pol, 125
S1 nuclease, *see* Single-strand DNA
Saccharomyces cerevisiae
 DNA strand break repair, 358, 359
 inducible mutagenesis, 355
 mismatch correction and UV mutagenesis, 355
 replication, 98
 REV2 mutant, 357
Saturation mutagenesis, error-prone reverse transcriptase, 200, 201
Sequenase, *see also* Bacteriophage T7 DNA polymerase, 316
Sequence homology
 bacteriophage φ29 DNA pol
  α-like DNA pol, 125–127

DNA pol α, 125–127
 pGKL1 killer plasmid, 125
 bacteriophage T4 DNA pol and herpes simplex virus DNA pol, 35
 bacteriophage T7 gene 4 protein, *dnaG* and *dnaB* proteins, 90
 DNA pol β and terminal transferase, 66
lex protein
 mucA protein, 349
 umuD protein, 350
τ protein
 ATPase, 19
 cowpea mosaic virus, 20
 gene *B*, phage Mu, 19
 histone H1, 20
Serum response factor, 461
Shuttle vectors
 Epstein-Barr virus, 416
 simian virus 40, 410, 411, 447, 468, 473, 480
Simian virus 40
 fidelity of replication, 182
 in vitro replication, 42, 50, 182
 shuttle vectors, 410, 411, 447, 468, 473
Single-strand break repair, *Saccharomyces cerevisiae*, 359
 DNA pol I and DNA pol II, 100
Single-strand DNA
 DNA pol III, 17
 SOS induction, 362
 S1-sensitive sites in mispaired DNA, 358
 stimulation of τ ATPase, 30
Site-specific mutations
 chemical synthesis, 208, 410, 411
 engineered by error-prone reverse transcriptase, 200, 202
 SOS mutagenesis, 403
 umuD, 350, 352
Snuff, *see* Tobacco
SOS induction, 250, 284, 296, 332, 333, 401, 403, 404, 457, 465, 467, 472, 473, 479, 480
 deletion formation, 383
 dnaE, 24
 exonuclease I, 362
 mammalian cells, 482
 replicative killing, 375
 S. cerevisiae, 355
Southwestern blot, DNA-binding proteins
 membranes, 140
 UV induced, 481
Spontaneous mutations
 base substitutions, 284
 DNA sequence and deletions, 269
 ε subunit of DNA pol III, 251
 in vitro DNA synthesis, 316
 shuttle vectors
  Epstein-Barr virus, 418
  simian virus 40, 480
 simian virus 40, 182

spq, see *dnaE* protein and *dnaQ* protein
SSB, 9, 154
  bacteriophage N4, 130
  bacteriophage T7, 86, 93
  bypass synthesis, 298
  DNA pol III, exo I, and protein n, 165
  gene structure, 158
  groEL411
    allosteric effects, 159
    supression of *ssb-1*, 155
  oriR, 115
  recA, 155
  ssb-1 and *ssb-113*, 154
Streisinger model
  deletion mutagenesis, 387
  frameshift mutagenesis, 266, 314
Stress response, see Mammalian stress response
Sugar, adducts with DNA and protein, 423
Superhelix density, RNA polymerase and in vitro replication, 11
supF, mutagenesis
  induction by extracellular protein-inducing factor, 468
  induction by UV, 476
  induction by UV-induced, secreted factor, 475
  SOS in mammalian cells, 480
Suppression
  dnaE by *mutT*, 328
  dnaE by *pcbA1*, 305
  heat shock, 158
  recBC by *sbc*, 363
  ssb-1 by *groEL411*, 155

T antigen, simian virus 40 and replication, 42, 50
Terminal protein, phage φ29, 122
tet, mutations, 263
TG resistance, see HGPRT
θ Subunit, see also DNA pol III core, 14, 27
α-Thio-dNTP
  DNA pol I, 200
  synthesis of specific oligonucleotides, 200
Thioredoxin, *Escherichia coli*, bacteriophage T7 DNA pol, 86, 89
  oxidation and 3'→5' exonuclease activity, 316
Thymidine kinase
  mutations, 258, 420
  promoter trap plasmid, 482
Thymine glycol
  DNA structure, 213
  synthesis and structure, 212
Tobacco, 428, 429
Topoisomerase I, 18
  HeLa multienzyme complex, 45, 46
Topology, DNA
  polyomavirus integration, 173
  replication fork, 17

TPA, see Tumor promoters
Transcription, see also RNA polymerase, 11, 12
  autoregulation of *repA*, 105
  factors, 460, 461
Transformation, polyomavirus replication, 173
Transposition
  excision of Tn*10*, 379, 388
  inverted repeated sequences, 378
  tex mutant, 386
Tumor necrosis factor, 466
Tumor promoters, 459, 461, 481

umuCD
  acetylaminofluorene-induced −1 frameshift mutations, 265, 266
  cleavage, 350, 351
  DNA pol Cab, 294
  homology to *lexA*, 350
  UV mutagenesis, 277, 281, 332, 349, 350, 397, 398
ung-dut, 318
  HeLa and herpes simplex virus, 434
  in vitro assay for mutagenesis, 317
  Mollicutes, 441
  uracil in DNA template, 317
Untargeted mutagenesis, 333, 334, 404, 460, 469, 477, 480
Uracil-DNA glycosylase, 434, 437, 440
UV mutagenesis
  DNA pol III, 227, 281, 306
  recA, 278, 332, 336
  S. cerevisiae, 355
  SOS, 403
  umuCD, 350, 351, 397
UV sensitivity, aphidicolin-sensitive mutants, 168
UV-induced, secreted factor, 462, 465, 481
  mutagenesis, 475

Vinyl chloride, see Chloroacetaldehyde
v-*src*, 461

Weigle reactivation, 479
  error prone in virus reactivation, 480
  ssb mutants, 154

X rays, see Ionizing radiation
Xeroderma pigmentosum, 417, 466, 475
xonA, *Escherichia coli*, 363
X-ray crystallography, structure of Klenow DNA pol, 222, 225

Yeast killer plasmid pGKL1, homology to phage φ29 DNA pol, 125

Z-DNA, mutagenesis, 267, 269
Zinc finger proteins, bacteriophage T7 gene 4 protein, 90